Craftsman's Illustrated

Dictionary of Construction Terms

by James T. Frane

Craftsman Book Company
6058 Corte del Cedro / P.O. Box 6500 / Carlsbad, CA 92018

To Tré, for her continuing confidence and encouragement

The Library of Congress has cataloged this book as follows:

Library of Congress Cataloging-in-Publication Data

Frane, James T.
Encyclopedia of construction terms / by James T. Frane

p. cm.
ISBN 1-57218-008-0
1. Building--Dictionaries. I. Title.
TH9.F6517 1994 94-38747
690'.03--dc20 CIP

Second printing 1997

Acknowledgments

The author and the publisher gratefully acknowledge the following companies for providing pictures of their products used to illustrate this encyclopedia:

American Welding Society
550 N.W. LeJeune Road
P.O. Box 351040
Miami, FL 33135

Babcock & Wilcox
20 S. Van Buren Avenue
P.O. Box 351
Barberton, OH 44203-0351

Berger Instruments
4 River Street
Boston, MA 02126

CMI Corporation
I-40 and Morgan Road
P.O. Box 1985
Oklahoma City, OK 73101

Coast Crane & Equipment Company
1835 N.E. Columbia Boulevard
P.O. Box 11209
Portland, OR 97211

Cooper Industries, Crouse-Hinds Division
P.O. Box 4999
Syracuse, NY 13221-4999

Delta International Machinery Corp.
246 Alpha Drive
Pittsburgh, PA 15238

Foster Wheeler Energy Corporation
Perryville Corporate Park
Clinton, NJ 08809-4000

L-Tec Welding and Cutting Systems
P.O. Box 100545
Florence, SC 29501-0545

Mueller Company
500 West Eldorado Street
P.O. Box 671
Decatur, IL 62525

O-Z/Gedney, a Unit of General Signal
Main Street
Terryville, CT 08788

Stanley Tools, Division of the Stanley Works
New Britain, CT 06050

Thermadyne Welding Products of Canada Ltd.
2220 Wyecroft Road
Oakville, Ontario, Canada L6L 5V6

Trencor, Inc.
1400 E. Highway 26
Grapevine, TX 76051

W. J. Savage Company
P.O. Box 157
Knoxville, TN 37901

Illustrations not provided by the manufacturers are the work of the author or come from various Craftsman publications. These publications are listed in the order form in the back of this book.

A: Aaron's rod to azimuth

Aaron's rod – a straight or rounded molding with a scroll or leaf design in the form of a single serpent entwined about the rod; used as a building ornament.

abaciscus – a small, thick square or rectangular plate (small abacus) forming the top of a column.

abacus – a thick square or rectangular plate of any size forming the top of a column.

abandon – to leave unused in place, without intending to return; as in the case of pipe that has been capped off and left in place rather than being removed when it is no longer needed.

abate – 1) to remove or shape material to form a relief design; 2) to hammer metal; 3) to end or suppress; 4) to reduce or lessen.

abatement – 1) the decreasing of the strength of a timber as it is shaped to the proper size for use; 2) the wastage of wood when lumber is planed to size; 3) a diminishing, as in strength; 4) the shaping of metal by hammering.

abatjour – 1) a skylight or other means of admitting light into a building and deflecting the light downward; 2) a movable screen; 3) a movable slat.

abeyance – a lapse in succession during which title to a piece of real property is not clearly established.

ablate – removal of a material by melting and/or burning away.

ablative surface – a surface designed to melt or burn away at a controlled rate, as in the heat shield of a space capsule.

Abney level – a hand-held level used in surveying to determine elevations and slope angles.

abode – residence.

abrade – to wear away or erode, as to wear down with sandpaper or emery cloth.

abrasion – wearing away.

abrasion soldering – an intentional mechanical abrasion of the base metal during soldering.

abrasive – any material used for grinding, sanding, polishing or the wearing away of another material. Aluminum oxide, garnet and silicon-based compounds are commonly used as abrasives for sanding and smoothing wood. *See Box.*

abrasive flapper – strips of material impregnated with an abrasive that are attached to a hub with a shank. When the shank is inserted into the chuck of a power drill, the spinning strips can be used to sand irregular surfaces.

abrasive stones – grinding stones used to sharpen metal blades. The blades are either rubbed along the stone or the stone is spun and the metal held against it, wearing down the metal and creating a sharp edge.

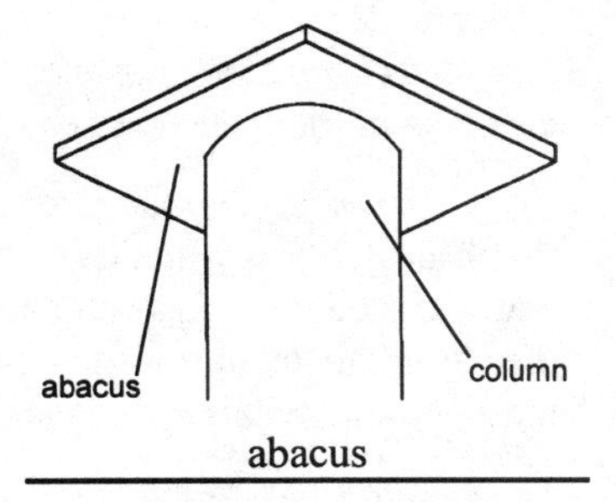

abacus

Abrasives

Abrasive materials are used for sanding, polishing, or wearing away the surface of another material by friction. Most abrasives in common use today are manufactured, rather than the naturally-occurring products. By manufacturing them, control can be exercised over their properties, such as their uniformity, hardness, particle size and cutting characteristics. For instance, diamonds can now be manufactured from carbon, usually in the form of graphite. The process requires high pressure and temperatures, but the shape of the crystals produced can be closely controlled. As a result, diamond abrasives are now available in uniform sizes that are far cheaper to use than diamonds mined and cut from a natural state.

The following are the most commonly used abrasives or abrasive materials:

Boron carbide powder is a fine powder that is compressed into a solid and used on wool polishing bonnets or bonded to paper or cloth. It is used for grinding and lapping objects into a final shape.

Boron nitride is an abrasive powder used for grinding hard tool steels and high alloy steels. The abrasive is so hard that it cuts with very little effort, thus keeping heat generation to a minimum.

Chemically purified aluminum oxides are used in manufacturing abrasives for precise and fine grinding. Other oxides, such as ferric oxide and silicon dioxide can be added to produce tougher abrasive compounds for the abrading of high tensile strength metals, such as steel.

Crushed steel is made from high-grade crucible steel that is heated to a white heat and then quenched in cold water to produce hard steel fragments. The fragments are then crushed to the desired particle size for grinding masonry and metals.

Crystallized aluminum oxide, called corundum, is a mineral containing about 93 to 97 percent aluminum oxide. It is used far less frequently than manufactured aluminum oxide.

Emery is a natural-occurring aluminum oxide. It is a powder that is often bonded to cloth and is still commonly used for polishing metal and glass.

Garnet, a red semi-precious stone, is used for making sandpaper, as is *quartz*. Garnet costs more than quartz, but lasts longer.

Abrasive Minerals

Natural minerals

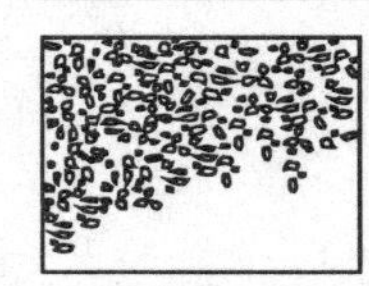

Flint:
This mineral is quartz, white in color. Its appearance is similar to white sand.

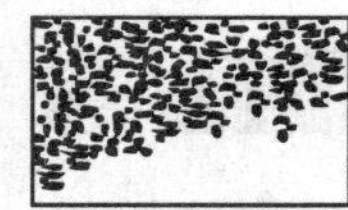

Emery:
Dull black in color. It is hard and blocky in structure.

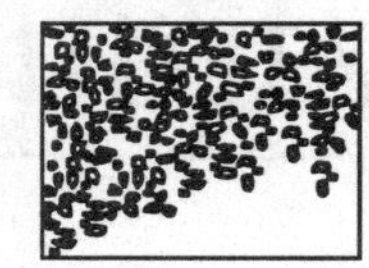

Garnet:
A reddish brown colored mineral of medium hardness with good cutting edges.

Synthetic minerals

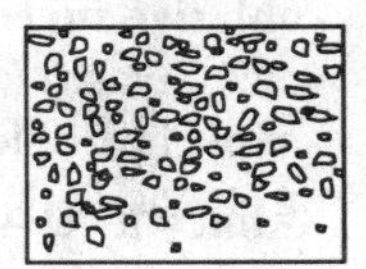

Aluminum oxide:
An off-white to gray-brown colored mineral that is extremely tough, durable and resistant to wear. It is capable of penetrating almost any surface.

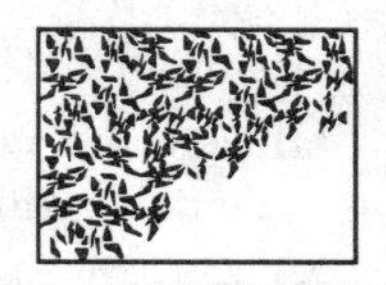

Silicon carbide:
It is shiny black in color and due to its brittle qualities fractures into sharp, sliver-like wedges.

Natural diamonds, crushed and graded, are bonded with resin, metal powder, or glass-based compounds and used for grinding carbide cutting tools, glass, stone and ceramic materials. They are gradually being replaced by manufactured diamonds.

Pumice is a volcanic material used mainly in granules or powder form in mild abrasives and polishes.

Rouge and *crocus* are fine powders manufactured from iron oxides and used for polishing metal, stone or gemstones.

Silicon carbide is a hard abrasive made from coke (coal held at a high temperature in the absence of oxygen) and sand which are fused together with the addition of sawdust. The sawdust burns away, leaving escape passages in the material for gases, such as carbon monoxide, created during the fusing process. Silicon carbide is used for grinding materials with low tensile strength, such as copper alloys, stone, concrete and glass.

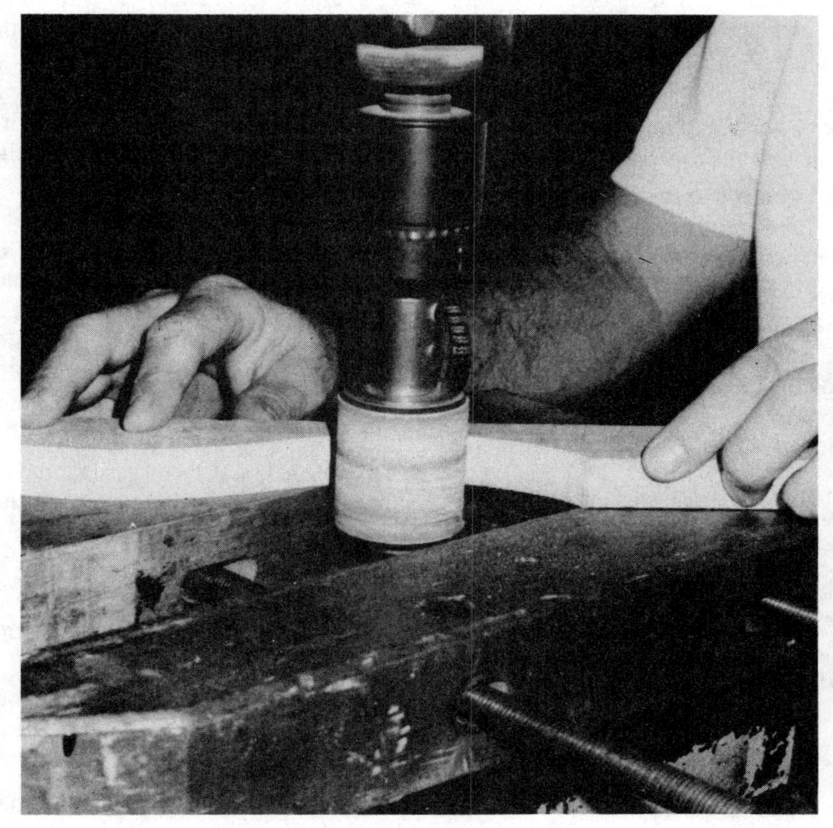

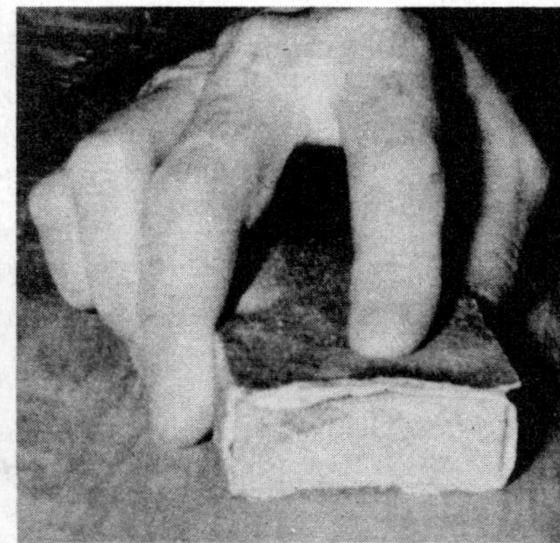

Left: Drum sander
Top: Block sander

Abrasive Tools

As the technology of abrasives continues to progress, more products and improved methods for cutting and shaping materials are being introduced. In some cases, where it is practical and economical, metal machining is being replaced by abrasive shaping. Abrasive tools include grindstones made of pure sandstone, whetstone, or emery wheels. Natural oilstones, the majority of which are quarried in the state of Arkansas, are used for putting a fine edge on cutting steel.

Abrasive waterjets use water pressure at up to 60 psig to carry solid abrasive material through a wear-resistant nozzle that directs the stream onto the material to be cut. They are used for hard-to-cut materials such as glass, ceramics, graphite or titanium. Abrasive waterjets cut at a high rate of speed without generating dust or stress in the material being cut.

The most commonly used abrasive tool is an abrasive paper. It is generally called sandpaper, though it may also be referred to by the type of abrasive used, such as garnet paper or emery paper. Abrasive paper is made from crushed abrasives that are glued to a backing material such as heavy paper or cloth and used for smoothing and polishing. Such coated papers are graded by grit numbers, from coarse to fine, the grade being determined by the size of the grit as measured by a mesh system. They are also graded as to the openness of the coating. *Open coat* has fewer abrasive grains on the backing and *closed coat* has more grains. Open-coat papers are used with soft woods that tend to clog up the paper. Abrasive papers can be hand held or attached to a block of wood for manual use. They can also be used with sanding machines such as belt sanders, disk sanders or drum sanders for various woodworking applications.

acanthus

abrasive wheel – a non-metallic disc that is impregnated with an abrasive, such as Carborundum, and used in a power saw to cut masonry and metal.

abreuvoir – a mortar joint between stones in a stone structure.

ABS (acrylonitrile-butadiene-styrene) – plastic pipe and fittings used for plumbing drains and vents.

absolute – 1) an exact amount; 2) a value that does not change when other factors change, such as a constant in a mathematical equation; 3) something that is pure, free from imperfection; 4) unconditional, without restriction, such as the absolute power of a dictator.

absolute humidity – the mass of water in a given volume of air.

absolute scale – a temperature scale using absolute zero as the zero point on the scale.

absolute zero – the lowest temperature on the absolute, or Kelvin, scale, equal to -459.7 degrees F.

absorb – to fill or soak up, as when a sponge soaks up water.

absorption – 1) the process of drawing a fluid or gas into a porous material, such as a sponge soaking up water; 2) masonry absorption is the weight of water that can be absorbed by a brick immersed in water.

absorption field – a system of perforated or open joint underground piping laid in a bed of coarse rock through which the effluent from a septic tank is piped for distribution and absorption into the soil.

abstract of title – a history of the ownership record and other facts related to the title of a piece of real property.

abut – to join or rest against another member; to butt up against.

abutment – the supporting structure at the end of a bridge, arch or vault.

abutment piece – the sole or bottom plate of a wall.

acanthus – an architectural ornament in the shape of the acanthus leaf.

acceleration loss – energy used in an air-moving system to accelerate the air to the required velocity.

accelerator – 1) a chemical substance added to a mixture to speed up a reaction. For example, adding small amounts of calcium chloride to concrete batches in cold weather will accelerate the curing time of the concrete; 2) the device that adjusts the speed on a machine.

accelerometer – an instrument that measures acceleration. It is attached to the surface of the object that will experience the acceleration. As the object moves, it causes an electric current to flow which measures the amount of acceleration that is taking place and indicates that amount on the calibrated accelerometer.

acceptable – 1) satisfactory performance; 2) meeting the conditions under which a project is judged to be complete.

access – 1) ability to reach something; 2) a passageway or means of reaching an area of a building or a piece of equipment that is concealed.

access court – an open area onto which the exits from two or more buildings lead.

accessible – within reach.

accession – 1) increase by something added; 2) acquisition of additional property by growth or increase in the existing property. Accession can be a natural process, such as the change in the course of a river which adds land area to the adjacent property, or it can be accomplished through the purchase of an adjacent lot or portion of a lot.

accessory – 1) any device that supplements the usefulness of a system or machine; 2) a natural feature, such as a rock formation, used as a reference point in surveying.

access panel – a removable panel that permits entry to an area that is normally sealed, such as a pipe chase.

accident – 1) an unplanned event; 2) a relatively small, but distinct surface irregularity in land.

accolade – a molding or decoration with the approximate shape of an ogee cut in the flat surface of an arch or lintel. An ogee is an S-shaped curve on a surface formed by a convex and concave surface curve joined together.

accordion door – a type of door that is pleated its full height into many vertical folds and supported by rollers inserted in a track at the top. As it is closed, the folds of the door resemble the bellows of an accordion.

account – 1) a financial fund; 2) a record of financial transactions; 3) credit extended to a customer; 4) a customer; 5) a commercial relationship involving credit.

accountant – a person who specializes in working with financial records or accounts.

accounts payable – payments you owe to someone else.

accounts receivable – payments owed to you.

accouple – to join together.

accouplement – 1) a timber brace, or a tie of timber between two structural members; 2) pairs of columns placed close together.

accretion – an increase in real property by a natural cause, such as the deposit of soil along a riverbank or receding water in a lake.

accrual accounting – recording income as accounts receivable when earned, and recording debts as accounts payable when they are incurred. Also called accrual basis accounting.

accrual basis – see accrual accounting.

accrue – 1) increase; 2) to accumulate or be added periodically; 3) to come about as a natural growth; 4) to become enforceable.

accumulator – a storage chamber for gas pressure. The accumulator contains pressurized gas which can be fed into the item to which it is connected, such as a hydraulic or pneumatic system. The energy produced by the release of the gas serves as a motive force in the system to do the work, such as moving the fluid or actuating a valve.

accuracy – correctness; agreement with a standard.

accurate – without error; conforming to a standard.

acetate – a salt or ester of acetic acid used in the manufacture of such products as synthetic fibers and plastics.

acetic acid – a colorless acid which is the essential part of vinegar; it is used in the production of synthetic materials and solvents.

acetone – a volatile solvent often used for cleaning.

acetylene – a hydrocarbon gas used in welding and flame-cutting operations; it generates high heat when used with gaseous oxygen under pressure.

acid – a sour tasting compound that can chemically neutralize bases. For example, muriatic acid is commonly added to swimming pools to inhibit algae growth and to keep the water pH balanced.

acid brick – a chemical-resistant brick made from hard-burned shale, often used as flooring in areas where chemical spills are likely to occur.

acid cleaning – washing concrete with a 5 to 10 percent solution of muriatic acid to clean it. This can be done after the cement has cured for at least two weeks.

acid-core solder – solder with an acid core designed solely for soldering metal together in non-electrical applications, such as joining pieces of flashing together.

acoustic – 1) pertaining to hearing or sound; 2) relating to music not electronically amplified.

acoustical board – any rigid material that attenuates or absorbs sound.

acoustical ceiling coating – a rough decorative coating, sometimes called popcorn because of its appearance. The coating is sprayed on, and when applied to acoustical board, aids in the reduction of reflected sound. Acoustical ceiling coating applied over drywall has less sound damping properties, but requires less drywall preparation than other textures, offering both time, labor and cost savings for the builder.

acoustical cloud – acoustically reflective panels used in concert halls to control the sonic properties of the hall.

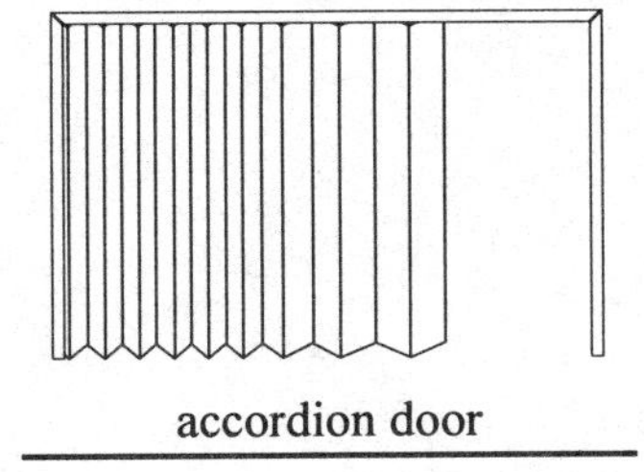

accordion door

acoustical material – any material that absorbs and decreases the volume of sound, rather than reflecting it. Acoustical materials are used in building whenever the reduction of sound transmission is desired, such as in the walls of a laundry room adjacent to a living area.

acoustical plaster – plaster that is mixed with a sound-absorbing material.

acoustic impedance – an indicator of the ability of a material to transmit sound.

acoustic ohm – the unit used to measure a material's ability to resist the transmission of sound.

acoustic reactance – resistance to the passage of sound through a medium, either solid, liquid or gas. It is caused by the internal and elastic properties of the medium, which allows it to absorb the sound.

acoustic resistance – resistance to the passage of sound through a substance or medium caused by internal friction of the medium.

acoustics – science of sound and its behavior.

acoustic tile – wall and ceiling tile of a sound-absorbent material.

acquest – to acquire property by purchase.

acquiescence – accepting or complying without objection, thus implying the waiver of the right to legal action.

acre – a measure of land area equal to 43,560 square feet.

acroter – a small pedestal for a statue; the statue on such a pedestal.

acroterium – see acroter.

acrylic plastic – noncrystalline thermoplastic that is clear, weather-resistant, and shatter-resistant.

acrylic resin – a thermoplastic resin.

action – 1) the process of doing; 2) a mechanism; 3) a legal proceeding.

actionable – affording grounds for a law suit.

activate – 1) to put in motion; 2) to make something more reactive; 3) to speed up a reaction, such as the curing of concrete or the solidifying of epoxy.

activator – a substance that acts as a catalyst, speeding up a chemical reaction without being affected by the reaction.

active – 1) disposed to action; energetic; 2) engaged in present operation or movement; 3) requiring a power input to operate, such as a fan-driven air circulation system.

act of God – an unforeseeable and sudden natural force, such as a tornado, lightning strike or flash flood. Insurance companies often exclude "acts of God" from insurance protection.

actuator – a device for moving or controlling something indirectly, such as an air actuator used to remotely open or close a valve.

adapt – to make fit.

adapter – a device, mechanism or fitting used to mate two parts of different design, shape or size to each other. For example, a socket adapter permits the use of a ⅜-inch drive ratchet with a ½-inch drive socket. The fitting has a female ⅜-inch square on one side into which the ratchet drive fits, and a male ½-inch square on the other side onto which the drive socket slides.

addendum – 1) a supplement or addition to something; 2) the distance from the tip of a gear tooth to the pitch line or diameter, measured radially on a circular gear.

addition – a part added, as a new room built onto and attached to an existing structure.

additive – a substance added to another to enhance or change its performance. An example would be the detergents added to lubricating oil that keep contaminants in suspension so that they can be removed with the oil when the oil is drained.

address – 1) a physical location; 2) a code or path used to locate stored information in a computer.

adherence – 1) holding fast or sticking; 2) to follow faithfully, as in a set of plans or rules.

adherend – one of the surfaces held to another by an adhesive.

adhesion – 1) a firm attachment; 2) the molecular attraction exerted between two surfaces in contact, as occurs between the rails and the moving wheels of a train, holding the train to the track.

adhesive – a glue or mastic compound used to fasten materials together. *See Box.*

adhesive bond – an firm attachment between materials or between an adhesive and the material to which it is applied.

adhesive spreader – a notched board or trowel which spreads adhesive in ridges along a surface; the spaces between the ridges allow the adhesive to spread when the surfaces are pressed together.

ad hoc – for this (Latin); for the particular purpose at hand, without consideration for a wider application.

ad infinitum – without end (Latin).

adit – 1) a doorway or passageway; 2) a horizontal passage into a mine.

adjacent – next to.

adjoin – to contact or lie close to.

adjudge – 1) to formally declare; 2) to judicially decide or award.

adjudicate – 1) to act as judge; 2) to settle judicially.

adjustable hook spanner wrench – a wrench with a hinged, hook-shaped section that can be fitted into the rectangular slots on the edges of a variety of nut sizes. With the hook securely held in the slot, the wrench head fits around the nut, and can be used to exert torque on the nut.

adjustable pliers – pliers that have a slot or tongue and groove joint that allows the jaws to widen and grip different-sized objects. The jaw opening is adjusted by opening the pliers fully and selecting the desired groove or hinge pin position.

adjustable wrench – a wrench with a pair of jaws, one of which is movable by means of a screw adjuster. The wrench can be made to fit different-sized fasteners.

administration – the management of all the activities that allow a system or organization to function.

admiralty metal – an alloy that is at least 70 percent copper and 1 percent tin, with the remainder being zinc and often small amounts of other elements. The metal is resistant to deterioration in the presence of salt air or saltwater and is used commonly where those conditions exist. Includes admiralty bronze and admiralty brass.

admix or admixture – 1) a material, other than the basic constituents of grout or mortar, that is added to improve the mix; 2) an element or compound that is added to a mixture; 3) the mixture to which something has been added.

adobe – a type of heavy clay used in making bricks.

adobe brick – an unfired, sun-dried brick made from adobe clay and straw.

adorn – to beautify or decorate.

adsorb – to take up and collect molecules of one substance to the surface layer of a solid or liquid with which the molecules come in contact.

adsorbate – a substance that is adsorbed.

ad valorem – according to value (Latin).

advection – a horizontal movement of air.

adverse possession – possession of real property by continuous occupancy without the consent of the owner.

adze or adz – a tool resembling an axe, with a thin arched blade set at a right angle to the handle, used for cutting, shaping or smoothing wood.

adze eye – a short channel extending along the eye of a metal hammer head that provides additional bracing around the handle as it fits into the eye.

aerated concrete – concrete with air bubbles incorporated into the mix, making it relatively lightweight.

aerator – 1) a mechanism used to introduce air into water, soil, sewage or other substances; 2) a device installed on the end of a sink spout to mix air with the water, thus reducing splattering.

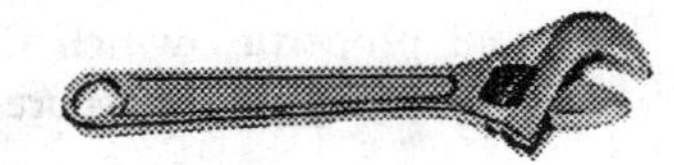

adjustable wrench

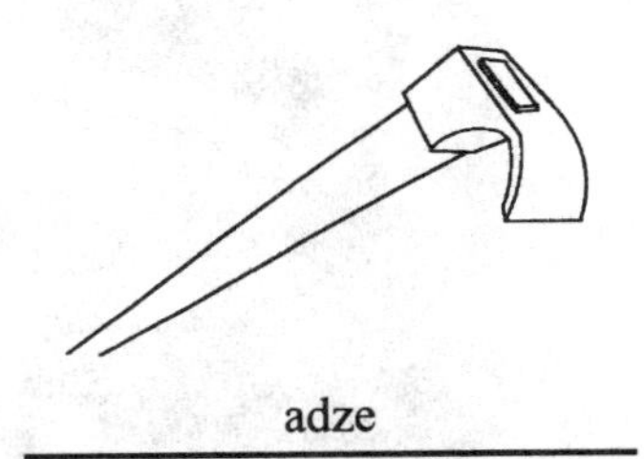

adze

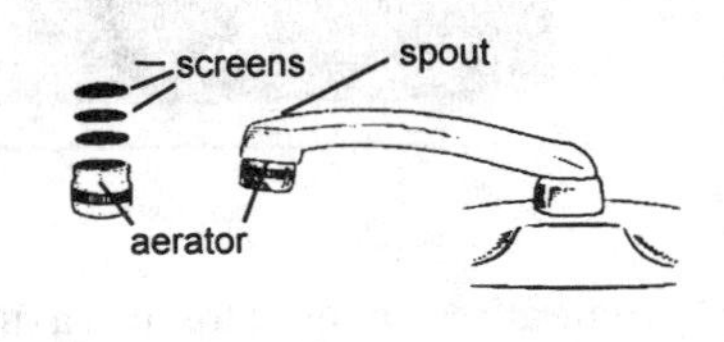

aerator

Adhesives

Adhesives are used to hold materials together by establishing a bond between the two material surfaces to be attached. The materials may be of different types or textures, and often require some surface preparation, such as cleaning or sanding, in order to permit the bond. Adhesives distribute the stresses between materials uniformly, permit rapid and economical assembly of materials or parts, often provide some vibration damping and insulation, and may reduce galvanic action between materials by eliminating metal to metal contact.

There are various types of adhesives, each with different properties which fit particular applications. Some may require long cure times, have temperature limitations, may be toxic in a fire or deteriorate with time or under conditions of stress, but still have other redeeming characteristics which keep them in use. It is advisable to know the properties of an adhesive before using it on a particular project.

Left: Neoprene panel adhesive
Top: Joint adhesive

Thermoplastic adhesives are plastic-based adhesives with long-chain polymers that soften when heated. They have good holding properties and can be used with a variety of materials, such as cloth, rubber, plastics, wood, paper products, even steel and concrete, depending of the specific manufacturer's recommendations. This type of adhesive is manufactured in the form of glue sticks for use with hot glue guns.

Thermosetting adhesives are plastic-based adhesives with cross-linked polymers that won't soften once they've solidified. Both thermoplastic and thermosetting adhesives may be cured, or solidified, by a variety of actions, including heat, chemical reaction or solvent evaporation, depending on the characteristics desired in the particular adhesive used. Thermosetting adhesives are commonly used with wood, plastics, glass, metals, stone and concrete.

Elastomers are thermoplastic adhesives that remain flexible or elastic. They act as sealants as well as adhesives, and are commonly used in bonding carpet, rubber, wood, glass, ceramics and metals.

Anaerobic adhesives are thermoplastic adhesives that set up only in the absence of air. There are two types of these polyacrylate adhesives, one known as machinery, which has only shear strength, and the other, which is structural, and has both shear and tensile strength. They are used in situations where high shock resistance is needed. Anaerobic adhesives are commonly used with glass, plastics and metals, and particularly as a bond with threaded fasteners.

Pressure-sensitive adhesives are permanently tacky materials which form an immediate bond when two surfaces coated with these adhesives are brought into contact under pressure. Contact cement is an example of this type of adhesive. Pressure-sensitive adhesives are used in construction for such applications as laminating countertops.

aerial photograph – a photograph taken from the air, usually from a plane.

aerial survey – a survey made from a high point, usually an airplane, or with the use of aerial photography.

aerodynamics – the study of the effects of air and other gases on bodies where relative motion exists between them.

aerosol – the suspension of small particles of solid or liquid in gas; the mixture is held in a pressurized container and dispensed through a special nozzle that atomizes it into a fine spray.

aesthetic – pleasing to the eye.

affidavit – a written statement sworn to under oath before a recognized authority.

affix – to attach physically.

A-frame – a building with a steep sloped roof that reaches almost to the ground. The profile of the structure resembles the letter A.

aftergrowth – a second growth of timber in an area that has been harvested or where the trees have otherwise been removed or destroyed.

agar – a gelatin substance taken from seaweed that is used, among other things, as paper sizing and an emulsifier for adhesives.

agata – a type of stained glass with a mottled rose and white color.

aged Mediterranean texture – texturing stucco with areas of flattened blotches troweled on a furrowed surface.

agent – one who is authorized to act for another.

agglomerate – 1) clustered together; 2) volcanic fragments fused into a rock.

agglutinate – to glue together.

aggregate – 1) inert filler material made up of sand, stone, or gravel which is used to strengthen cement and form concrete. *See Box;* 2) the whole.

aggregate base – aggregate of course material, $1/4$ inch and larger.

aging list – a summary of accounts receivable and the amount of time they have been owed to you. The summary allows a view of several accounts and their payment histories to determine which accounts are past due.

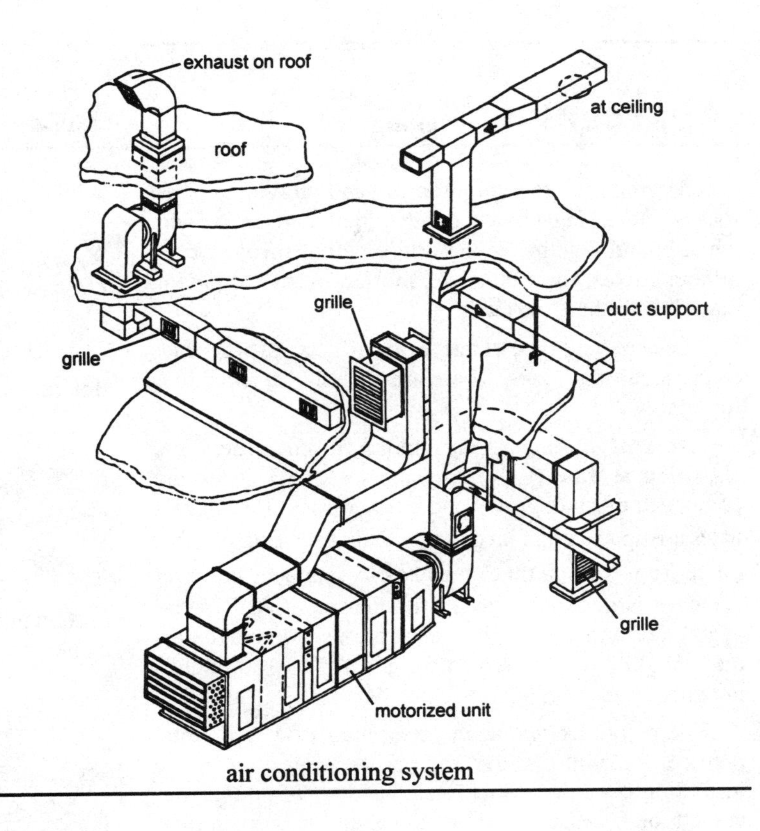

air conditioning system

agitator – a mechanical device for mixing.

agreement – a statement of common purpose between two or more parties.

aiguille – a device for boring holes in stone or masonry.

air acetylene welding (ACW) – a welding process, with or without filler metal, using a heat source produced by the combustion of acetylene with air.

air brick – a hollow brick, open at both ends, used to permit air to pass through a wall. Air bricks are placed at various locations as air vents.

airbrush – a device using compressed air and a fine nozzle to apply paint. The application can be precisely controlled to cover small areas.

air carbon arc cutting (AAC) – a metal cutting process that melts the metal with an electric arc from a carbon electrode and removes it with a jet of air.

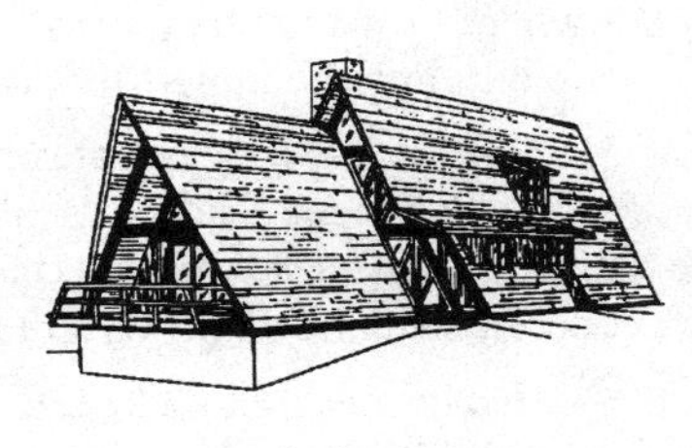
A-frame

Aggregates

Aggregates are composed of sand, gravel, crushed rock or stone, slag, cinders or other inert materials which, when bound together in a mixture with cement or bituminous mixes, form concrete, mortar, plaster or paving materials for hard surfaces.

The size and type of aggregate affects the properties of the resulting mixes. There are several classifications for aggregates:

Natural aggregates are graded for coarseness and classified as fine (F.A.) and coarse (C.A.). They are composed of sand, gravel and crushed rock. Fine aggregates are up to 1/4 inch in diameter, and coarse aggregates range from 3/8 inch up to 6 inches in diameter, however 3 inches is generally the maximum size used. Natural aggregates and blast furnace slag are used to make normal-weight concrete, which weighs 135 to 160 pounds per cubic foot.

Very lightweight aggregates such as vermiculite, pumice, diatomite, scoria and perlite are used to make insulating concrete, which weighs 15 to 90 pounds per cubic foot. It is used where deadweight loads are limited or where its insulating properties can be beneficial, such as in building the upper floors of buildings.

Lightweight is aggregate of low bulk specific gravity such as expanded shale, slate, slag, clay, vermiculite or pumice. It is used for making plaster or lightweight concrete, which weighs 85 to 115 pounds per cubic foot.

Heavyweight is aggregate of high specific gravity such as barite, limonite, hematite and magnetite. It is used to make heavyweight concrete, which weighs from 180 to 380 pounds per cubic foot. Heavyweight concrete is used for radiation shielding or as ballast in ship building.

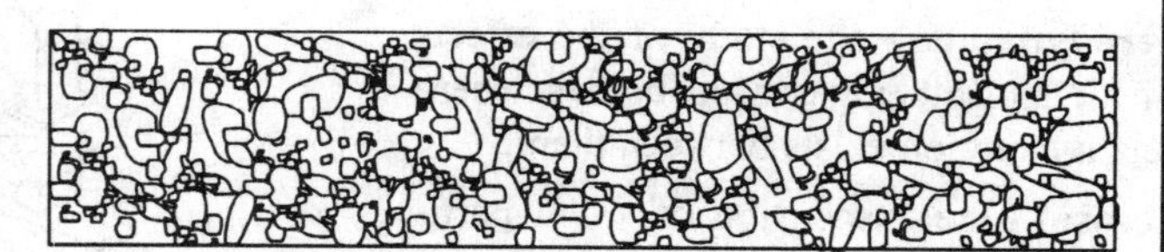

Before separation into various sizes

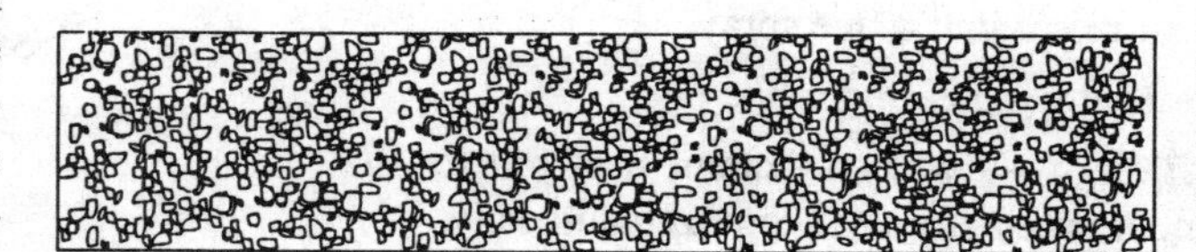

Portion passing 3/8" mesh screen but retained on No. 4 screen

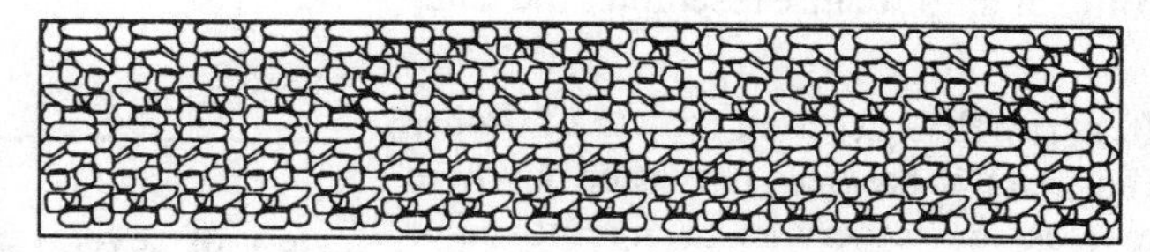

Portion passing 3/4" mesh screen but retained on 3/8" mesh

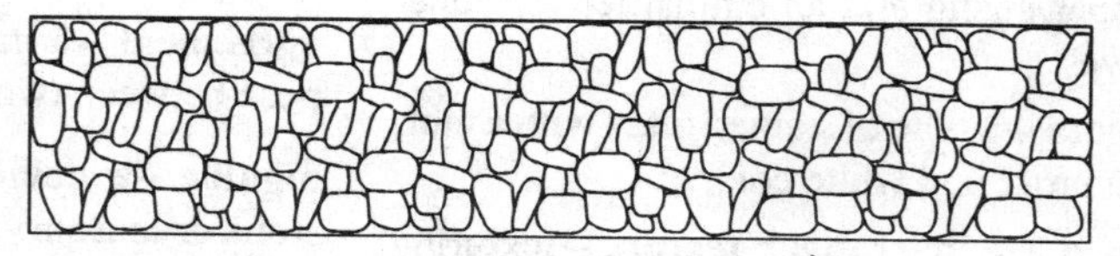

Portion passing 1 1/2" mesh screen but retained on 3/4" mesh

air change – replacement of a volume of air in a confined space with fresh air, as in a room.

air cleaner – a filtering device for removing particles from the air, particularly air at the intake of an engine or other mechanical device.

air conditioner – a machine or system for circulating air and controlling its temperature. It may also control humidity and remove airborne particles or impurities.

air conditioning – the process of controlling the movement, temperature and humidity of air in a confined space.

air cooled – a reduction in temperature brought about by the transfer of heat to air as it circulates.

air-cooled cooling system condenser – a finned, tube-type condenser. Refrigerant is circulated through the tubes and gives up heat to air passing over the outside of the finned tubes, thus cooling the refrigerant.

Macadam is a course aggregate of uniform size, usually small compacted stones or gravel which is mixed with a grout or mortar to make a wear surface for walkways or driveways. Uniformly sized crushed stone, gravel or slag is mixed with asphalt to make road pavement.

Well-graded is aggregate that is graded in size from the maximum down to fine filler with the object of creating an asphalt mix with a controlled void content and high stability for use in various types of paving, such as highway base courses or overlaying existing pavement.

Aggregate testing: The properties of the aggregate used to make concrete should be studied and tested before use. Concrete is a mixture of portland cement, water and aggregates, with aggregates comprising approximately 60 to 80 percent of the total volume. The cement and water mixture coats the aggregates and binds them together as it sets up. The greater the proportion of aggregate to cement-water mixture, the more economical the mix and the less shrinkage that will occur as the cement sets and cures. Shrinkage reduces the water-tightness and strength of the concrete. However, too much aggregate and not enough cement will create a poor bond and weak concrete.

Weak aggregate will also produce weak concrete. Large aggregate should be examined for its potential to fracture under pressure. ASTM C131 is a test for the abrasion resistance of aggregates, which is a measure of its soundness for use in concrete.

Aggregates used to make concrete should be reasonably clean and free of silt. Otherwise more cement will have to be added to get a good mix. Usually, visual inspection can be used to determine the presence of silt, but a silt test should be performed for fine aggregate. Use a one quart jar, such as a canning jar, to perform the test. Fill the jar to a depth of 2 inches with a sample of the fine aggregate to be tested. Add water until the jar is three quarters full, cover and shake it well. Let the mixture sit until the silt settles out of the water. The silt will form a layer on top of the aggregate. Measure the depth of the silt layer. If the layer is less than ⅛ inch thick, the aggregate is acceptable. If it's more than ⅛ inch thick, the aggregate should be cleaned, using screens, vibrators or tumblers with a water wash or air flow, before it is used.

Aggregates should also be tested for fine organic matter, such as decaying material or other soil constituents, which will slow down or even prevent concrete from setting up properly. The presence of organic matter can be determined using a colorimetric test. Put 4 ounces of aggregate in a container. Add 7½ ounces of a 3 percent solution of sodium hydroxide (NaOH). Shake the mixture and let it stand and settle for 24 hours. If the resulting color is darker than a straw color, there's too much organic matter in the aggregate to use it for concrete. It must be cleaned before used.

There are a number of other tests that can be used to determine the various properties of aggregate before using them to make concrete. Among them are ASTM C127, which is a test for moisture content in fine grade natural aggregate, and CSA A23.2.6, set by the Concrete Standards Association, which determines the specific gravity of aggregate. Consult a manual on concrete for more details.

air cushion tank – see expansion tank.

air-dried lumber – lumber dried by exposure to natural air currents without the aid of any heating devices.

air-driven nailer – a machine that uses compressed air to drive nails. Also called a pneumatic nailer.

air duct – a large thin-walled conduit, usually sheet metal or fiberglass, used for directing the passage of air through buildings.

air ejector – a device used to remove noncondensible gases from steam condensers in order to maintain a vacuum and improve the operating efficiency of the condenser.

air-entrained concrete – concrete that has an additive included which traps small air bubbles within the mixture. The air bubbles improve the workability of the concrete and increase its ability to resist damage due to temperature changes.

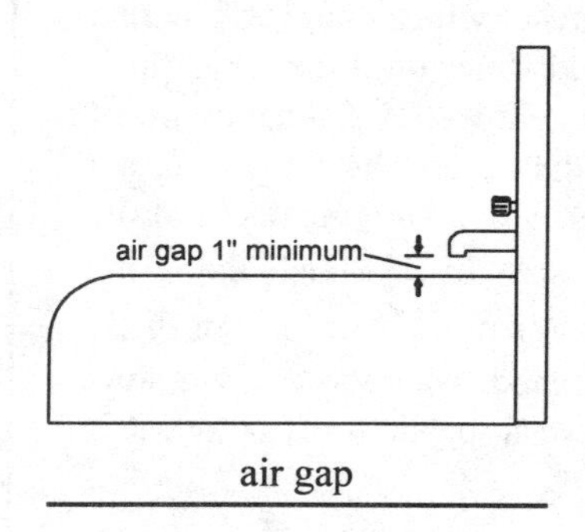

air gap

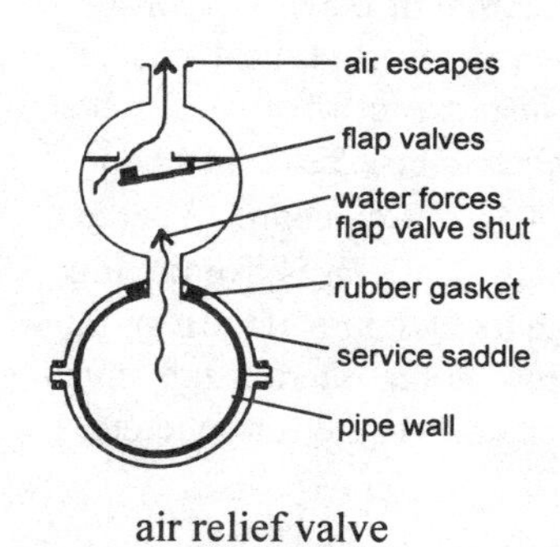

air relief valve

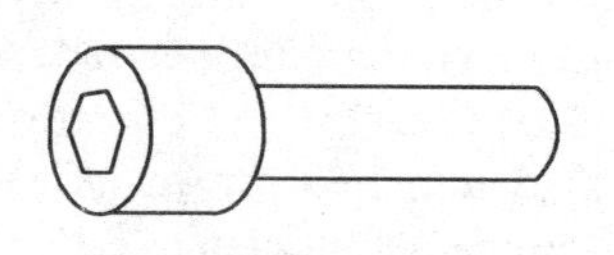
Allen screw

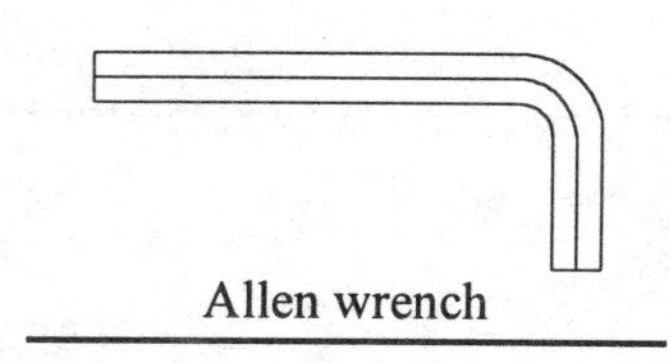
Allen wrench

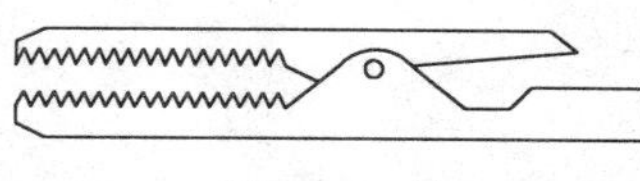
alligator clip

air entrapment – air bubbles formed on a new paint surface causing a defect in the finish.

air feed – a disbursing system for powdered material which is to be thermally deposited on a surface. The powder is carried by an air stream and sprayed on the surface to be coated. Under the appropriate heat conditions the material bonds to that surface.

airfoil – a device shaped to provide a lifting force when moved through an air current.

airfoil fan – a fan that has blades with an airfoil cross section.

air gap – the unobstructed vertical distance between the flood level in a plumbing fixture and lowest water supply inlet; the distance between the rim of the tub and the faucet, for example.

air gun – a nozzle, powered by compressed air, used in the spray application of paint or other commodities.

air hammer – a hammer driven by compressed air; a pneumatic nailer.

air handling unit – a unit, utilizing a fan system, designed to move air and sometimes also change its temperature. The units are used in air distribution systems to circulate air throughout buildings by means of a series of ducts.

airless sprayer – a spray gun for applying paint or other liquid finishes that uses pressure from a mechanical pump to pressurize the liquid, or spins it from a wheel to spray.

air lock – 1) a blockage in the flow of liquid, particularly on the suction side of a pump, caused by a gas bubble trapped in the line; 2) a gas-tight chamber with an entrance and an exit that is used as a passageway into another space. Also called a vapor lock.

air pocket – a void where a poured commodity, such as concrete or insulating foam, has failed to flow evenly and fill the space as it should.

air relief valve – a valve to vent air out of a water system.

air space – a space or cavity between structural members, such as between walls or wall surfaces, that is created as part of the architectural design.

air tight – a vessel or structure constructed to contain gas pressure.

air vent – a valve in a water system that automatically vents entrapped air from the system.

aisle – a passageway between structures or between sections within a building or room, such as between sections of seats in a church or theater.

alabaster – fine-grained gypsum often used for carved ornaments.

alarm check valve – a check valve located in the riser of a wet-pipe fire sprinkler system; the valve sounds an alarm when water starts to flow through it.

alcove – a small recessed portion of a room.

alette – a buttress or a wing of a building.

alidade – 1) a telescopic instrument that is used to take bearings; 2) the upper portion of a transit.

alienation – a transfer of property.

align – to bring two or more objects into line with each other.

aligned section – see section, aligned.

aliphatic glue – a yellow wood glue that is strong, but not waterproof.

alkyd – a thermoplastic or synthetic resin-based paint.

alkyd binder – oil-modified phthalate resin used in paint; it speeds up the drying process by reacting to oxygen in the air.

Allen screw – a screw or bolt with a hexagonal socket in the head.

Allen wrench – a hexagonal cross section rod, often with one part bent at a right angle to provide leverage, designed to fit a hex socket or Allen screw.

alley – a narrow street through the middle of a block that provides access to the rear of buildings.

alligator clip – a spring-loaded clip with alligator-like jaws used to make temporary connections in an electrical circuit, or to hold small objects.

alligatoring – a coarse pattern of cracks in paint resembling alligator skin.

allodium – land that is owned outright with no rental or other fees due a landlord.

allotment – 1) apportionment 2) divide into shares.

allowable span – see span, allowable.

allowable stress – see stress, allowable.

alloy – a combination of metals and/or other elements designed to improve certain properties of the materials to be produced from the alloy.

all-purpose joint compound – see joint compound, all-purpose.

alluvial fan – a fan-shaped deposit of soil and/or rock left at the point where water flows through a narrow gorge that opens onto a plain, or at the junction where a small stream joins a larger one.

alluvion – a gradual increase in land along the shore of a moving body of water caused by alluvium deposits from upstream.

alluvium – sand, clay, gravel or other material deposited along the shore by a moving body of water.

all-weather wood foundation – a building foundation fabricated from pressure-treated wood that can be placed in direct contact with a gravel base (for drainage).

allyl resin – a thermosetting resin; one that can be heated only once and then cooled into a temperature-resistant solid. Used in laminate adhesives.

alnico – a permanent-magnet alloy of iron, aluminum, nickel and cobalt.

alteration – a change or modification to an existing body or structure.

alternate power source – a generator used to provide back-up electrical power in the event the normal power supply fails.

alternating current (AC) – an electrical current that reverses its direction at regularly recurring intervals.

alternator – an electric generator that produces alternating electrical current.

aluminum – a silver-white lightweight, rustproof metal, non-magnetic but a good thermal and electrical conductor, found abundantly in the earth's crust; its atomic symbol is Al, atomic number 13, and its atomic weight is 26.98.

aluminum nails – lightweight, corrosion-resistant aluminum nails made in a variety of sizes and shapes; they are used in situations where other metals might cause corrosion problems, such as in fastening aluminum flashing in place.

aluminum oxide – an abrasive used for grinding and polishing.

aluminum panel nails – aluminum nails with a large diameter head and an elastomer gasket or washer under the head; they are designed for use with aluminum, such as in fastening an aluminum skylight rim to the skylight curbing.

aluminum sheathed cable (ALS cable) – individually insulated electrical conductors housed in a close-fitting aluminum tubing.

amalgam – 1) an alloy containing mercury. The hardness of the alloy at a given temperature is determined by the amount of mercury it contains. With sufficient quantity of mercury, the amalgam may be liquid at room temperature; 2) a mixture of different elements.

amalgamate – 1) to mix or merge into a single body; 2) to chemically recondition old paint or lacquer.

ambient – surrounding on all sides.

ambient temperature – the temperature of the air in a location.

American bond – a brick-laying pattern that has five layers of stretchers and one course of alternating headers. Also called common bond.

American National Standards Institute (ANSI) – the sponsor/publisher of many standards used in industry.

American Plywood Association (APA) – an association of plywood manufacturers which, among other things, sets the grading standards for plywood.

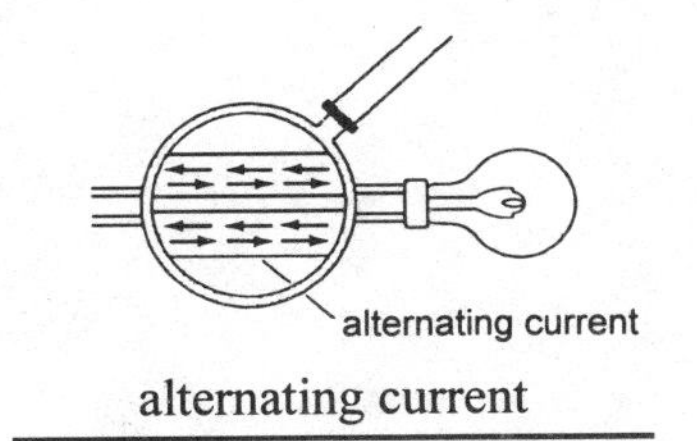

alternating current

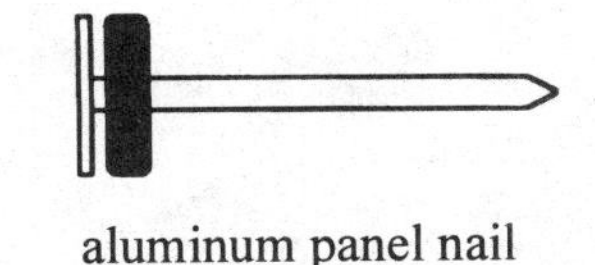

aluminum panel nail

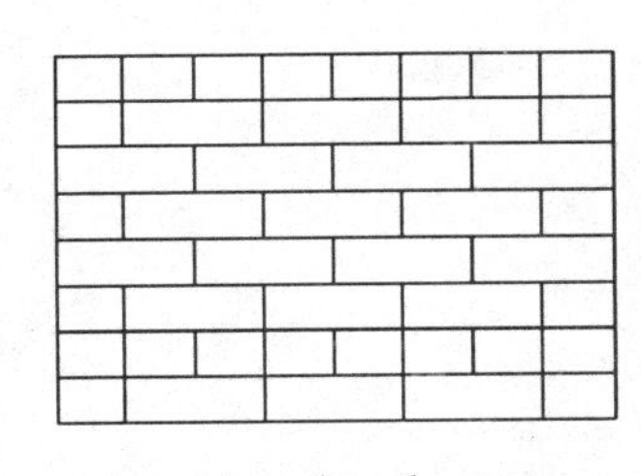

American bond pattern

American Standard Code for Information Interchange (ASCII) – the code used in computers to represent letters, symbols and use instructions.

American Wire Gauge (AWG) – a series of numbers used to identify the diameters of electrical wire manufactured in the U.S. Gauge designations are from 46 to 4/0 (0000), with 46 being the smallest and 4/0 the largest. Many of the wires available in the larger sizes, particularly 4 through 4/0, are stranded for flexibility. Wires larger than 4/0 are designated by cross-sectional area in circular mils. The size following 4/0 would be 250 MCM, then 300 MCM, with the larger sizes designated by larger numbers as the sizes increase because the numbers now equal the cross sectional area of the wire.

amide – a curing agent for epoxy resin.

ammeter – an electrical meter that measures current flow in an electrical circuit. The ammeter is connected in series with the power going into the circuit so that the electricity flows through the meter. This current causes an indicating needle to deflect an amount proportional to the current. The needle's position on a graduated scale shows the amount of electrical current flowing through the circuit in amperes, or amps.

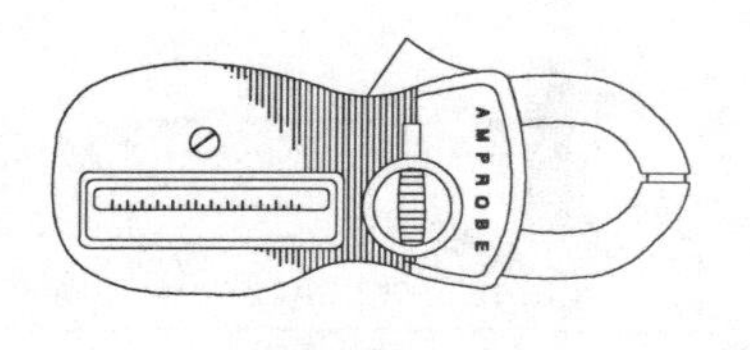

ammeter, clamp-on

ammeter, clamp-on type – an ammeter which reads current flow through a conductor by measuring the strength of the magnetic field around the conductor. The current flow through a conductor creates the magnetic field. The conductor is placed so that it is between the jaws of the meter, and the magnetic field produced by the current is translated by the meter into amperage that can be read on the meter's dial. This type of meter is convenient to use because it does not have to be connected to the circuit in order to measure the current flow.

ammonia – a strong-smelling chemical compound (NH3) used in cleaning solutions, the manufacture of chemicals, and, in gaseous form, as a refrigerant in refrigeration systems. In refrigeration systems, ammonia is used much the same as Freon. However, Freon is preferred by manufacturers because it is more stable and less corrosive than ammonia.

amorphous – having no defined form.

amortization – spreading a large cost over a period of years.

amortize – to gradually eliminate a financial obligation by setting up a system of periodic payments over a specified time.

ampacity – the current-carrying capacity of an electrical conductor.

amperage – a measured amount of electrical current expressed in amps.

ampere (amp) – a unit of electrical current, analogous to the amount of flow in a hydraulic circuit measured in gallons per minute.

ampere-hour – the rate of electric current flow in a one-hour time span.

ampere-turn – the amount of magnetomotive force that is produced in one turn of a coil by one amp passing through that turn.

amphitheater – a stadium, room, or other structure with seats arranged around a central arena where shows, lectures, or other activities are performed for an audience.

amplidyne – a machine consisting of a direct-current generator that uses a small amount of DC power applied to its field windings to control the generator output.

amplifier – a device that amplifies.

amplify – to increase in intensity or power or amplitude.

amplitude – the maximum value of an alternating current.

anaerobic – living without air or free oxygen. The bacteria that break down the waste in a septic tank are anaerobic.

analysis – an investigation into the nature of something; analyses may be performed by a variety of means, including chemical and mathematical.

anchor – a sturdy attachment point. An anchor in a piping system is a six-way restraint against pipe movement.

anchorage – the point at which something is anchored.

anchor block – a block of wood included in a masonry structure as a nailing surface for the attachment of wood structural members to the masonry wall.

anchor bolts – bolts used to fasten a wood sill to concrete or masonry; the bolts are often bent on one end to form a "J" or "L" so that they resist being pulled out of the concrete. They are set in wet concrete or in grout placed in a hole bored into the concrete or masonry.

anchor, post – a preformed metal piece used to anchor the bottom of a vertical post slightly above the foundation so that it will not be rotted by contact with surface moisture. The post anchor is made of galvanized sheet metal with vertical sections bent up to fit around the bottom of the post to be anchored. Fasteners hold the post securely in place.

anchor strap – a metal strap embedded in a concrete pour. Holes in the strap allow the framing to be fastened securely to the concrete.

ancon – 1) a curved or rounded support between a cornice on a building and a wall; 2) a projection from a block of masonry; 3) a bracket.

andiron – metal supports which hold the logs in a fireplace at a slightly elevated position, permitting the air to flow freely around them.

anemometer – a device for measuring air flow velocity.

anemometer, hot-wire – a device that measures force or speed by measuring the resistance change of a heated wire; the resistance changes with temperature, and the temperature changes with air flow velocity.

anemometer, rotating vane – a lightweight propeller-type vane that is rotated by air flow; its rotational speed is proportional to the air velocity.

anemometer, swinging vane – a lightweight spring-loaded metal vane held in a housing that is deflected by air flow, moving a pointer on a scale which indicates the amount of flow.

angle – two straight lines that diverge from the same point.

angle bay window – multiple windows arranged as a projection from the wall in the shape of a polygon.

angle brace – a support brace spanning the angle formed by two other members.

angle brick – a brick that is shaped with a taper.

angledozer – earth moving equipment with a blade mounted at an angle to the direction of travel, used to scrape soil off an excavation site or move dirt to a new location. The earth is moved at an angle to the vehicle so as not to pile up in front of the blade, delaying forward movement. It may also be used for smoothing and leveling a building site.

angle dozing – moving earth with the dozer blade set at an angle that allows the earth to be moved forward and off to one side.

angle iron – a metal support used to reinforce joints forming a 90 degree angel.

angle of incidence – the angle formed between the line along which something (such as a laser beam) touches a surface, and a line that is perpendicular to that surface.

angle of repose – the slope of an angle at which a given soil type will tend to stay at rest, rather than slip due to the force of gravity.

angle paddle – a plastering tool used to finish a corner after the plaster has been floated.

angle plane – a multibladed scraping tool used on plaster to remove high spots before the finish coat is applied.

angles, adjacent – two angles sharing a common side and having a common intersection point.

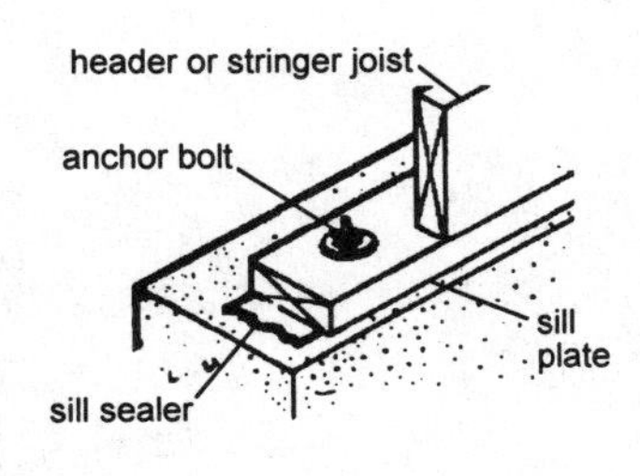

anchor bolt

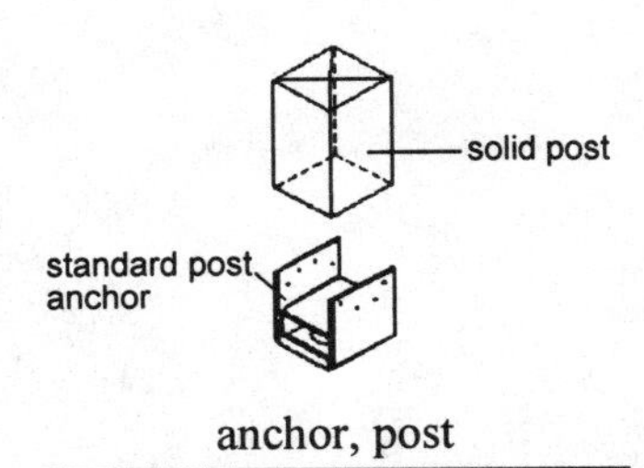

anchor, post

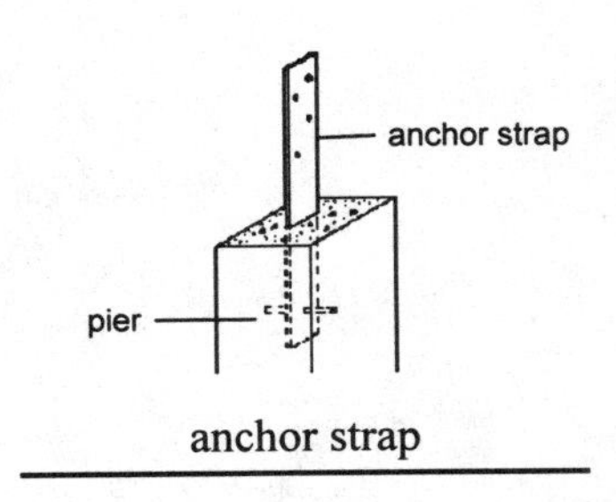

anchor strap

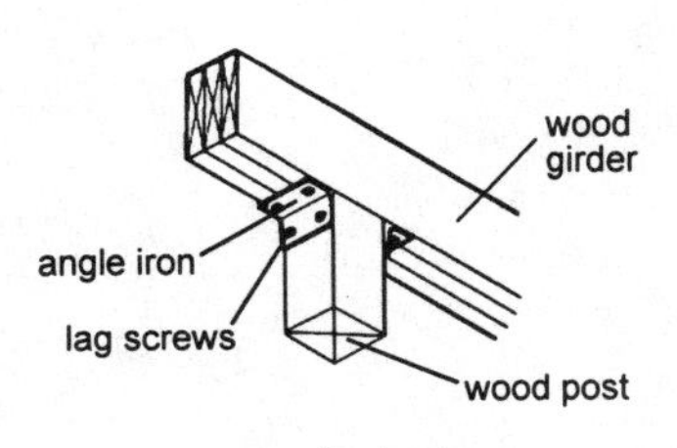

angle iron

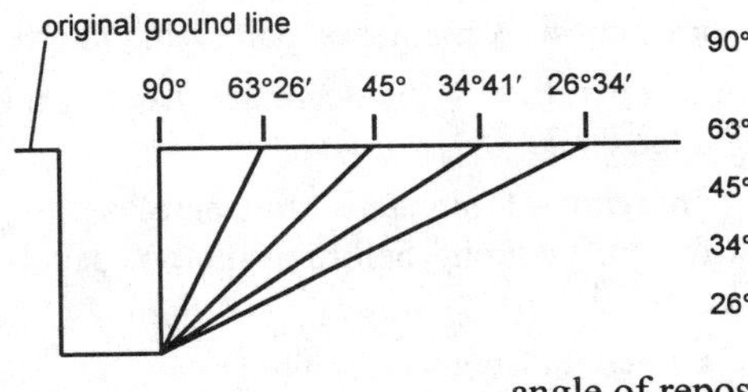

angle of repose

angle valve – a globe valve with the inlet and outlet at right angles to one another.

angular – containing or composed of angles.

anhydrous – having no water content.

aniline dye – a blue dye made from nitrobenzine, a derivative of coal tar, used to dye fabrics.

anion – an atom or ion with a negative charge.

anneal – a process in which metal is heated and allowed to cool slowly, removing stress and softening the material. The process improves the machineability of metals. *See Box.*

annual – occurring every year.

annual percentage rate – the interest percent per year.

annual ring – growth rings in a tree; they can be seen in a horizontal cross section of a tree when it has been cut down.

annular – in the shape of a ring, such as a ring around a shaft.

annular bit – see bit, annular.

annular nail – see nail, annular.

annulus – the space in a tank or other container above the surface of the contents of the tank.

annunciator – any form of signal or indicator used to show when an action or condition occurs.

anode – the positive electrical terminal.

anta – a rectangular pillar, often placed at the end of a masonry wall.

antechamber – an outer room, often used as a waiting area, leading to another room. Also called an anteroom.

antefix – upright ornaments along the eves or cornices to conceal tiling ridge ends.

antemion – ornamental palm leaf pattern.

antependium – the decorated front of an alter.

anterior – 1) situated before or toward the front; 2) coming before in time or development.

anteroom – see antechamber.

ante-solarium – a balcony facing the sun.

anthracite – a shiny black, compact, hard natural coal which contains little volatile matter. It is difficult to ignite, but when ignited, burns with a smokeless blue flame. It is used for heating, powering the boilers in steam power plants, and as a source of carbon. It is also blended with bituminous coal to make coke or coke substitutes, and to sinter iron ore. See also meta-anthracite and semianthracite.

anthropometry – the study of the size and proportions of the human body; this information is applied to the proper sizing and proportioning of objects meant to be used by people, such as chairs or control panels.

anticatalyst – an inhibitor to a chemical reaction; the opposite of a catalyst, which encourages a chemical reaction.

anticlinal – 1) inclining in opposite directions; 2) a geological formation in which stratified rock slopes downward from each side of a ridge or arch.

antifouling – a material or substance applied to objects used underwater, such as ship bottoms or diving apparatus, to prevent marine growth on their surfaces.

antifreeze – a liquid that has a lower freezing temperature (and often a higher boiling temperature) than water; it is usually mixed with water and added to a machine cooling system, such as an automotive cooling system, to improve its performance in extreme temperatures.

anti-kickback assembly – a device to prevent the kickback of wood being cut by a power saw. It consists of teeth on pawls that are hinged so that the wood may be freely pushed in one direction, but cannot move in the opposite direction; this device is found on radial arm and table saws.

antimagnetic – material that does not magnetize to a significant degree.

antimony – a metal often added to tin or lead as an alloying agent to increase their hardness; its atomic symbol is Sb, atomic number is 51, and atomic weight is 121.75. Antimony is alloyed with lead, in a ratio of 6 to 7 percent, to make the plates in

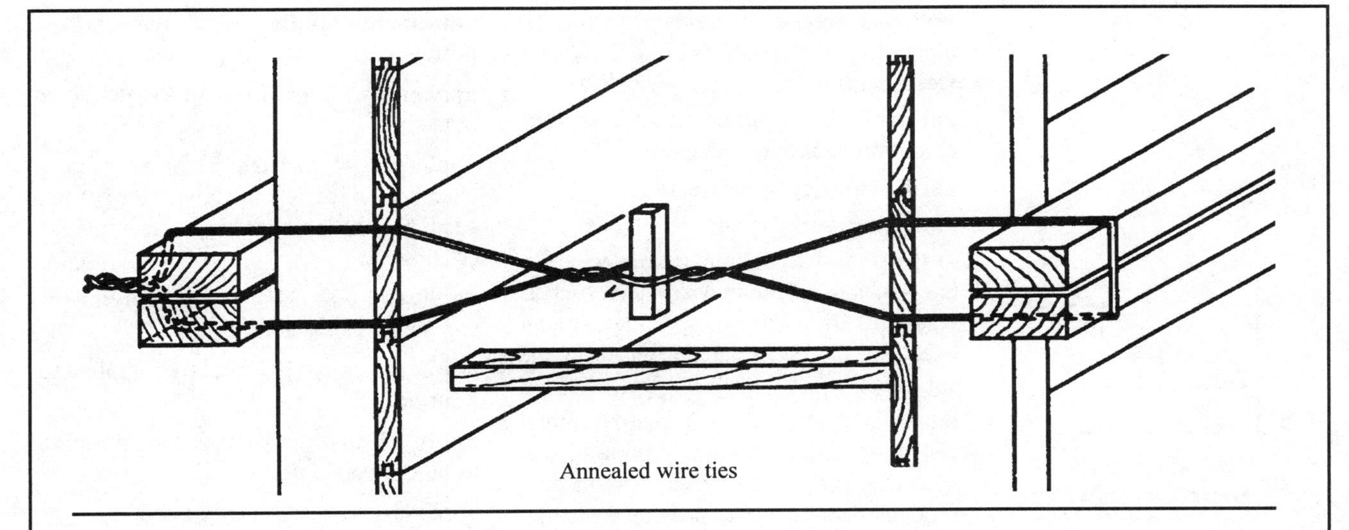
Annealed wire ties

Annealing

Annealing is a process applied to metals and metal alloys to soften and alter their properties. The metals are heated and then allowed to cool slowly. Different heating temperatures, holding times at a specific temperatures, and cooling methods are used to attain the desired results with a particular metal or alloy. Annealing can be used to reduce stresses locked in solidified metal, to make a metal softer and more pliable and improve its machineability, to change its physical properties, such as improving its conductivity or toughness, to remove gasses trapped in the metal, or to change its crystalline structure for a specific application. The various annealing processes are classified and generally defined by the industry as follows:

Full annealing involves heating a metal to a specific temperature, plus or minus a stated tolerance level, holding it at that temperature for a measured amount of time, and then cooling it slowly until the temperature is under the low end of the stated tolerance level. Metals being treated in this manner are sometimes cooled in the furnace in which they were brought up to temperature. This process reduces stress and minimizes the tendency for the metal to crack.

Process annealing involves heating iron base alloys to between 1000 and 1300 degrees F, which is just below their critical temperature (the temperature at which carbides are dissolved) and then cooling them. This process improves the formability of the alloys and is often used in the manufacture of wire and sheet metal.

Normalizing involves heating iron base alloys to about 100 degrees F above their critical temperature, dissolving their carbides. The dissolved carbides become a homogeneous part of the metal's structure, altering its properties. The alloy is then cooled in still air to below the critical temperature.

Patenting is a process that consists of heating iron base alloys above the critical temperature, the temperature at which carbides are dissolved, and then cooling them down below the critical temperature to about 800 to 1050 degrees F, depending on the required end properties and the carbon content of the alloy. This treatment is used to improve the ease with which wire is drawn and reduce tearing in the process.

Spheroidizing is an annealing process designed to produce spherical carbides, and may consist of prolonged heating below the critical temperature and then slowly cooling the metal; or heating to temperatures that alternate and vary above and below the critical temperature; or heating to about 1380 to 1480 degrees F and holding for one or more hours followed by slow furnace cooling. This process produces a ductile iron that is less subject to brittle fracture.

Tempering or drawing is a process for reheating hardened steel to below the critical range and then cooling. This makes the steel tougher and more resilient and is used in making springs and a variety of tools.

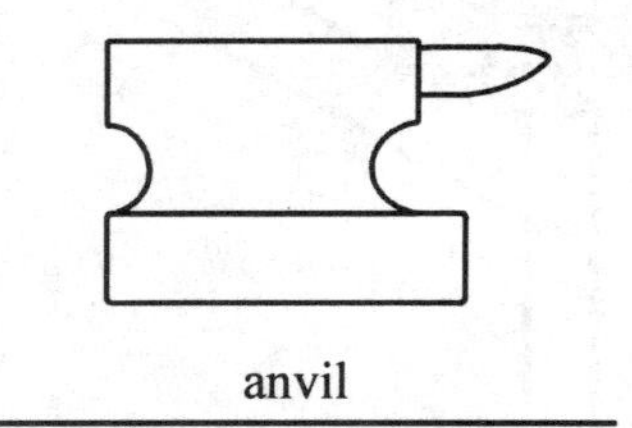
anvil

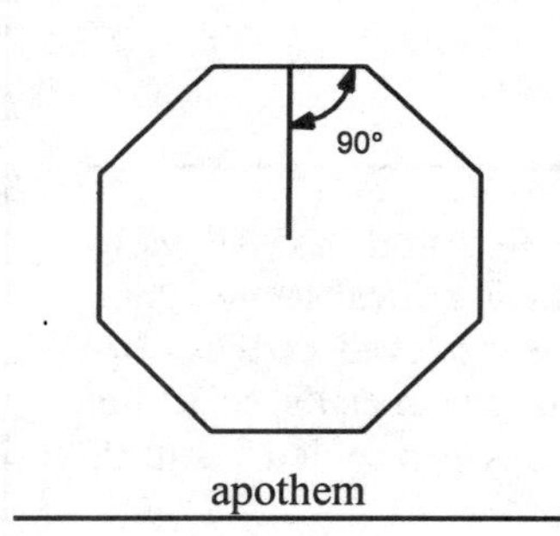

apothem

lead-acid batteries. In a one percent combination with lead, it is used to sheath electrical cables.

antioxidant – a compound that inhibits oxidation of another material.

antique oil – see Danish oil.

antiquing – see glazing.

anti-siphon – a device to prevent the back-flow of a fluid into a system, such as water trickling from a sprinkler system back into the water supply that feeds the sprinklers.

anvil – a steel block on which hot metal can be shaped by pounding with a hammer. One end of the anvil tapers to a near point and is rounded to provide a shaping surface; the top of the anvil is flat.

anzac siding – a type of tapered bevel siding.

apartment house – a structure containing multiple living units which are rented to tenants. Also called an apartment building.

APA Sturd-I-Wall – a siding material that can be applied directly to building studs as exterior siding. APA Sturd-I-Wall is the trade name for APA 303 plywood panel building siding.

aperture – 1) an opening, such as in a wall; 2) the opening in a photographic lens that allows light to enter.

apex – the uppermost part of a structure.

apex stone – a triangular stone at the top of a gable or dome.

apophyge – a small curvature at the bottom or top of a Corinthian or Ionic column where the column expands to join the base.

apothem – a perpendicular line from the side of a regular polygon to the center of the polygon.

apparatus – an assembly designed to perform a specific task.

apparent power – the calculated and theoretical power (watts) in an alternating current circuit. The apparent power is often greater than the actual power in the circuit because there are losses in the circuit due to the effects of reactance in the circuit. The reactance opposes the flow of AC current, using up some of the available power.

appearance lumber – see lumber, appearance.

appendage – an addition to a structure or object.

appliance – a mechanical or electromechanical device designed to perform work such as heating, cooking, cleaning, or cooling.

applicator – an object for applying or spreading a substance.

applique – an ornamental trim applied to a surface.

apply – 1) to lay or spread on a surface; 2) to put into operation.

appraisal – an estimate of value.

appreciation – 1) increase in value, as in property values increasing over time; 2) an expression of admiration or gratitude.

apprentice – one who is learning a trade through practical experience by working under the guidance or supervision of an experienced craftsman.

approximate – nearly correct or exact.

apron – 1) a section of pavement outside the entrance to a building, such as at a loading dock; 2) a horizontal trim piece under a window sill; 3) a covering worn to protect clothing and/or to hold parts and small tools while working; 4) a protection for an earth surface; 5) the section of a theater stage that is in front of the line of the curtain; 6) a surface onto which the water from a dam overflow falls.

apron piece – the horizontal beam that supports the upper end of a stair carriage or stringer, also called the pitching piece.

apron wall – the section of a panel wall that is between the window sill and the base of the wall.

apse – 1) a semi-circular projection from a building; 2) the curved or angular and vaulted part of church behind the alter.

aqueduct – a man-made channel for carrying water.

aqueous – having water as part of its make-up.

arabesque – ornamentation using the shapes of plants, flowers, and vases.

arbor – 1) a lattice with climbing plants, such as grape vines; 2) a shaft for holding a rotating cutting tool.

arboretum – a land area used for growing trees and plants.

arbor press – a press that is used to force an arbor, mandrel, or shaft into a hole that has been bored in a material. The arbor can then be fastened into a lathe chuck so that the work can be turned on the lathe.

arc – 1) a portion of the circumference of a circle; 2) the flow of electrical current through a gas, such as air, from one pole to another.

arc gouging – the process of using an electric arc to shape the edge of metal.

arch – a curved structure that spans an opening and is supported at both ends.

arch, back – see back arch.

arch brick – a wedge-shaped brick used in the building of an arch.

arch, discharging – see receiving arch.

arch, Gothic – see major arch.

architect – a person trained in building design and building strength analysis who uses that knowledge to design structures.

architectonics – the science of designing and building structures.

architect's scale – a type of ruler that shows smaller units representing feet, such as ½ inch or ¼ inch equals 1 foot. The units are used so that all building measurements will be of equal proportions to the actual measurements, but on a smaller size scale for the drawings. This gives the architect an accurate reduced-scale representation of the building.

architectural drawings – floor plans with the foundation details, the wall, floor, ceiling and roof construction details, door, window and partition locations, and sketches of the proposed exterior face of the building.

architrave – molding around a window or a door.

archivolt – decorative molding around an arch.

arch, jack – see jack arch.

arch, major – see major arch.

arch, minor – see minor arch.

arch, parabolic – see major arch.

arch, receiving – see receiving arch

arch rib – a raised band or ridge that projects along the line of an arch.

arch, segmented – see minor arch.

arch, semi-circular – see major arch.

arch stone – wedge-shaped stone used in the building of an arch.

arch, trimmer – see trimmer arch.

arch, Tudor – see major arch.

archway – 1) an opening under an arch; 2) an arched entry or other opening.

arc light – a light that uses an electric arc as the light source. Also called an arc lamp.

arc strike – a surface defect or blemish on metal caused by the initial strike of an electric welding arc.

arc welding – a process that utilizes the heat created as an electric arc jumps from an electrode to metal to weld the metal.

area – 1) the surface measure of a defined shape expressed in square units, such as square feet; 2) a defined space; 3) an unoccupied plot of land; 4) a building site; 5) land around a building.

area drain – a receptacle or drain for collecting water runoff from a particular area.

area wall – a retaining wall that surrounds a basement window located below grade.

areaway – an open space, below grade, which allows light or access to a basement door or window.

arena – 1) an area in which entertaining events are conducted; 2) a structure that houses such an area; 3) the performance area in an amphitheater.

arena theater – a theater that has seats on three or four sides of the stage.

argon – an inert gas with the atomic symbol Ar, atomic number 18, and the atomic weight of 39.948.

armature – a cylinder constructed by winding insulated wire on a core of conductive material, such as copper. There are usually multiple layers of windings held in

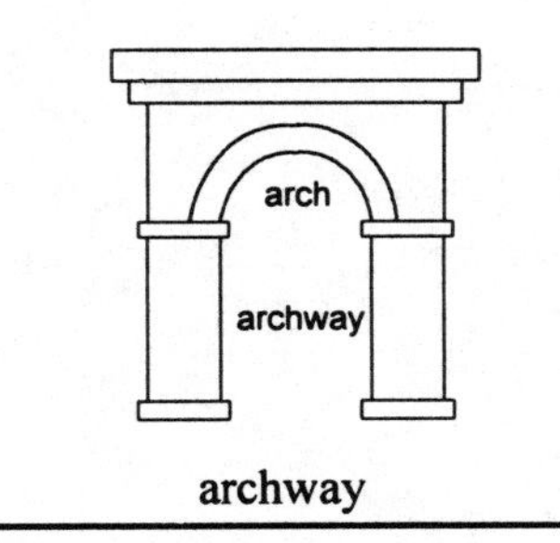

archway

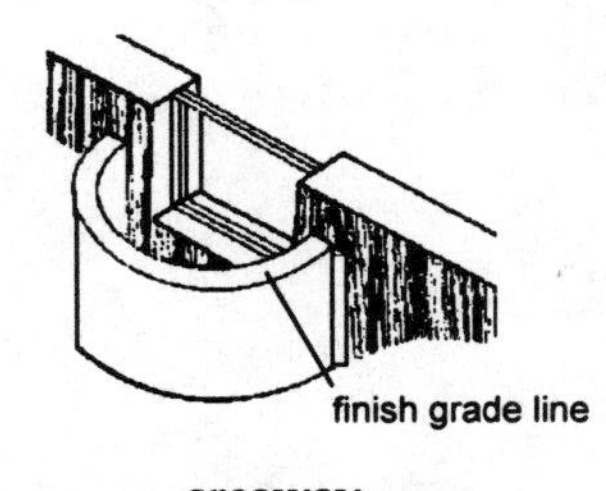

areaway

slots in the conductive material. The armature is part of an electric motor and generator. A generator converts the mechanical energy used to drive the generator into electrical energy. A motor converts electrical energy that is applied to it into mechanical energy as it rotates. Both machines work by electrical conductors passing through lines of magnetic force. These conductors are carried on the surface of the armature. The armature may be fixed or rotating, and the conductors pass through the lines of magnetic force from the field. The field is stationary if the armature rotates, as it does in DC machines. If the armature is stationary, the field rotates, as in AC machines.

armor – a protective covering.

armored cable – insulated electrical wiring inside a spiral-wound, flexible outer covering.

arnott valve – a one-way or check valve operated by gravity, and located at the top of a room or space, that permits the escape of air from that space.

arrester – 1) a wire-mesh screen covering an outlet that prevents sparks or burning material from escaping a heat or spark-producing source such as a chimney, incinerator or engine exhaust pipe; 2) a wire mesh screen that protects a fuel tank vent from flame propagation.

arris – the edge that is formed where two surfaces intersect or join.

arris fillet – a strip of wood running along the roof of a building at the eave line; the strip raises the end of the first course of shingles, tiles, or other roof covering so that water will run cleanly off the roof edge.

artesian well – a well in which natural underground water is forced to the surface by subterranean pressure.

artisan – a craftsman.

asbestine – a trade name for magnesium silicate, an agent used in paint as a binder to retain pigment.

asbestos – a natural mineral fiber used for fireproofing.

asbestos cement – a combination of portland cement and asbestos fibers used to make a fire-resistant cement.

asbestos cement conduit – an electrical conduit made from asbestos cement.

asbestos shingles – shingles made of asbestos fibers and portland cement or asbestos and other compounds.

as-brazed – a joint that has had no other operations except brazing performed on it.

as-built drawings – drawings that show the actual installed dimensions and locations of commodities, piping systems, etc.

ASCII – see American Standard Code for Information Interchange.

ash – 1) residue from burning; 2) a tree with tough wood resembling oak in appearance.

ash dump – see ash pit.

ashlar – precisely-cut building stone that can be used with thin mortar joints.

ashlar brick – a rough-surfaced brick made to resemble stone.

ash pit – a pit at the bottom of a fireplace into which ashes are swept and collected for removal. Also called an ash dump.

askew – not lined up with respect to some reference.

aslope – in a slanting position or direction.

aspect ratio – the ratio between the width of an object, such as a vehicle tire, and its height. For example, if a tire has an aspect ratio of 45, the distance from the outside diameter of the tread to the inside diameter of the tire is 45 percent of the width of the tread.

aspen – a poplar tree.

asphalt – a black, thick hydrocarbon (bituminous) substance found in the earth naturally, and manmade as a by-product of oil refining. It is commonly used for waterproofing and paving. *See Box.*

asphalt felt – a material made of pressed fibers that has been saturated with asphalt for waterproofing.

Asphalt

Natural asphalt has been used in construction for hundreds of years. It is a bituminous substance, composed primarily of hydrocarbons, which bubbles to the earth's surface forming pools of sticky, brownish-black liquid. It is also called mineral pitch or tar. Asphalt taken from natural beds is often called Lake Asphalt.

Most asphalt used today is produced as a by-product of petroleum refining. It remains after the more volatile parts of crude oil are distilled out. There are a variety of types and grades of asphalt products, each having different characteristics and uses. Sol-type asphalts are semi-liquid and very ductile. They change in liquidity with temperature changes, and harden with age. Gel-type asphalts are solid or hardened, and not easily worked. Semi-solid or intermediate types have characteristics in between the liquid and hard types. The hardness or consistency of asphalts is measured by a penetration test. The test, conducted under established conditions, determines the distance a standard needle will penetrate a given sample. A rating less than 10 indicates a solid, and a rating of 10 to 300 indicates an intermediate sample. Anything over 300 is liquid, and beyond the range of measurement by the penetration test. Asphalt is used primarily as a paving material for roadways, a waterproofing agent in roofing products or wall coatings, and as a protective coating or lining for piping.

Asphalt cement is made of asphalt specifically refined for paving. It has a semi-solid consistency, and is heated and mixed with aggregate to make a paving compound. The mix usually consists of 14½ percent asphalt, 44 percent coarse aggregate, 33 percent fine aggregate, 5 percent mineral dust, and 3½ percent air by volume. The smaller portions of aggregate fill spaces between the larger pieces, providing structural support and adding strength to the mixture. The asphalt binds the mix together. The asphalt mixture is usually "cut back" with an additive to liquify the material for easier blending and application. Rapid-curing asphalt has gasoline or Naphtha added to it. Both are very volatile and evaporate quickly, resulting in a fast-curing, solid asphalt cement. Medium-curing asphalt has kerosene added to it. Since kerosene is less volatile, the evaporation process is slower. Slow-curing asphalt is mixed with an oil that has low volatility, and even slower evaporation. Emulsified asphaltic cement is mixed with water and detergent, which keeps the mixture in a liquid state longer. With exposure to air, the water slowly evaporates, leaving a stable dense mix. After the asphaltic cement mixtures are applied, the material is tightly compacted to further strengthen the finished pavement.

Asphaltic mastic is a mixture of asphalt and fine mineral material that may be applied hot or cold, depending on the proportions. It can be poured onto a surface and compacted by troweling smooth; mixed with aggregate as a bond for roofing; or, it can be hotmopped on with layers of felt to make asphalt roofing.

Asphalt paint is a liquid asphaltic product sometimes mixed with additives such as lampblack, aluminum flakes or mineral pigments. It is used as a protective coating or waterproofing agent.

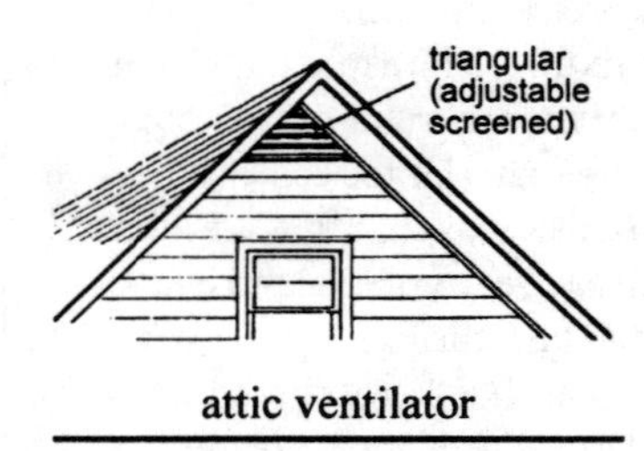

attic ventilator

asphaltic plastic cement – an asphalt-based roof cement that remains pliable at low temperatures. Also called flashing cement.

asphalt lute – a rake with blunt teeth on one side and a straight edge opposite the teeth. It is used for spreading and smoothing asphalt.

asphalt mastic – an asphalt-based adhesive used for roofing installation.

asphalt paint – a waterproofing paint with an asphalt base.

asphalt shingles – shingles made from felt saturated with asphalt. The exterior surface is coated with mineral granules to protect the felt from sunlight and weather.

asphaltum – a very thin asphalt compound commonly used as a coating to waterproof materials.

aspirate – to remove a fluid or other material by suction.

aspirator – a device consisting of a suction tube and some means of creating suction to remove fluids or materials from a vessel or container.

assay – an analysis to determine relative quantities of metals in an ore sample.

assemble – to put pieces of a unit together.

assembly – an item made of two or more parts fitted together.

assembly drawing – a drawing that shows how parts of a machine fit together. It is usually an exploded view with lines drawn between the parts indicating how each fits with the next.

assembly line – stations in series through which a production item moves, with parts being added at each station, until the item is completely assembled.

assembly section – see section, assembly.

assess – 1) to formally estimate a value; 2) to levy a tax or other fee on something.

asset – property that is owned.

assize – a cylinder-shaped block of stone.

ASTM Standards Index – a list of all of the standards set by the American Society of Testing and Materials by subject and by number.

astragal – 1) a molding attached to one of a pair of swinging doors, against which the other door strikes. When closed, the molding covers the space between the doors and provides a weather seal; 2) a molding resembling a string of beads.

as-welded – the condition of a weld without any further treatment such as postweld heat or grinding.

atmometer – an instrument used to measure the rate at which water evaporates.

atmospheric pressure – the pressure that the earth's atmosphere imposes, which is 760 mm or 29.92 inches of mercury at sea level under standard conditions.

atom – the smallest complete part of an element.

atomize – 1) to change a liquid into small droplets; 2) the separation of molten metal, like that at the end of a weld rod, into small droplets or particles.

atrium – a glass enclosed room or courtyard, usually in the center of a building, which is open to the sky. It is often used for indoor light and plants.

attached column – a column with three quarters of its diameter projecting from an adjacent wall.

attached garage – a garage that is part of the same building as the dwelling.

attenuate – 1) to decrease in intensity or volume, as in reducing the volume of a sound level; 2) to taper.

attenuator – an electrical device to vary the strength of a signal.

attic – the space in a building between the rafters and the ceiling of the top story.

attic fan – a fan, mounted either on the roof or side wall of a building, which ventilates the attic space. The air exhausted through the fan is replaced by air drawn in through openings, or attic ventilators.

attic ventilators – screened openings that permit the flow of air into an attic space.

audible – within the range of normal human hearing, which is generally accepted as being from 20Hz to 20kHz.

audio – referring to sound or hearing.

audio frequency – a frequency within the range of human hearing.

auditorium – a large building or hall used for public or group lectures or presentations.

auger – 1) a tool used to bore holes into wood. It may be manual or power driven; 2) a spiral-shaped drill bit used for boring holes in wood.

austenite – the crystalline structure of steel, formed under temperatures of between 1670 and 2535 degrees Fahrenheit, containing elements, such as carbon, in solid solution.

autecology – ecology that deals with organisms.

auto – a prefix meaning self.

autogenous healing – a natural process through which masonry cracks eventually fill themselves over a long period of time.

autogenous weld – a weld that is made without adding filler metal.

automate – to make a machine self-controlling.

automatic – self-moving and/or self-regulating.

automatic irrigation control station – an adjustable timing device to turn solenoid-actuated irrigation system valves on and off.

automatic level – a type of transit, which, once set roughly level, automatically maintains itself level using a suspended compensator.

automatic sprinkler system – a fire protection system which is actuated automatically by intense heat.

automatic vent damper – a balanced vent damper that opens with a small air flow and closes when there is no air flow.

automatic welding – a process that has a machine-controlled welding wire feed and weld head movement.

autotransformer – a transformer that has a combined primary and secondary coil. It has very low losses, and is used to step up an electrical voltage from a generator to a higher voltage for efficient transmission over power lines. Autotransformers are also used to create a three-wire electrical system from a two-wire electrical supply.

auxiliary – supplementary; backup; standby.

auxiliary hoist – a second crane hoist to assist the main hoist.

avoirdupois weight – a weight measurement based on 1 pound equaling 16 ounces and an ounce equaling 16 drams.

awl – a sharp pointed steel tool with a handle which is used for piercing holes, scribing lines, or making marks.

awning – a covering, usually made of canvas or metal, projecting out above a window or doorway to provide shade and/or weather protection.

awning window – a type of window in which the sash is hinged at the top and swings open at the bottom.

axe or ax – a single or double bladed metal chopping tool with a long handle. Used primarily for chopping down trees and cutting wood.

axes – the plural of axis.

axial – along the axis of an object.

axial-flow fan – a fan in which the movement of air is parallel to the fan shaft.

axial force – a force along the axis of a body or object.

axis – 1) a straight line about which a body or geometric figure rotates or appears to rotate; 2) a line with respect to which a body is symmetrical.

axle – a shaft used to support one or more wheels.

axle, fixed – a non-rotating axle.

axle, rotating – an axle which is fixed at the wheel and which rotates on bearings.

azimuth – a clockwise horizontal angle measured from the north or south meridian to a fixed point or object.

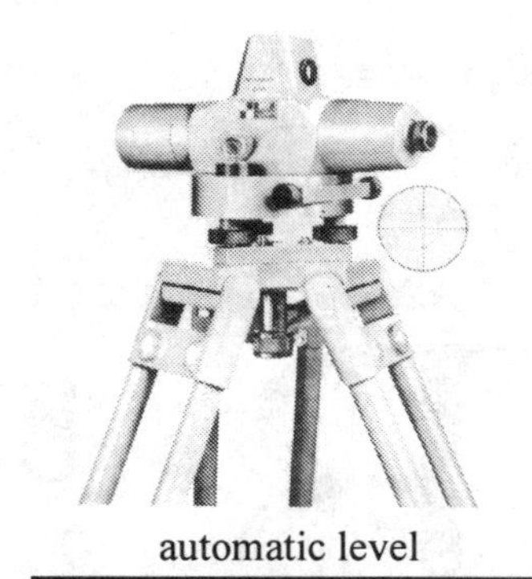

automatic level

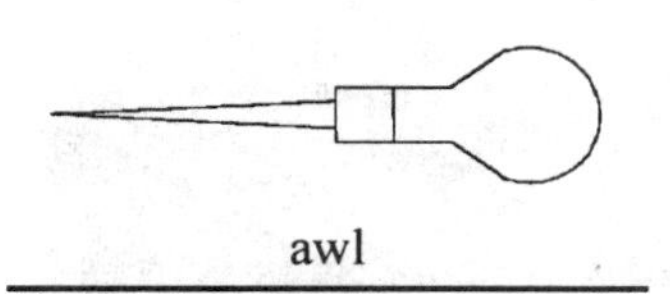

awl

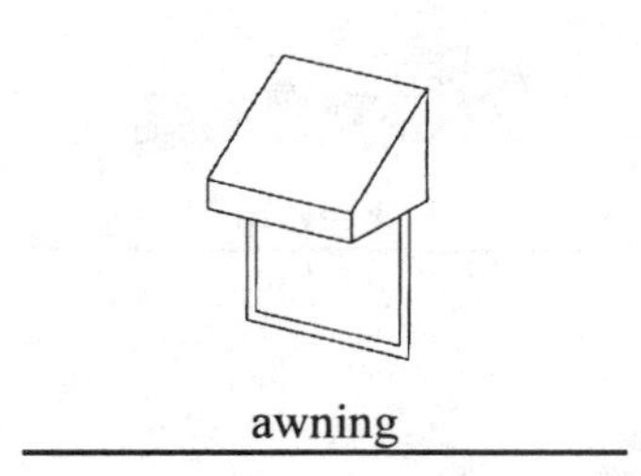

awning

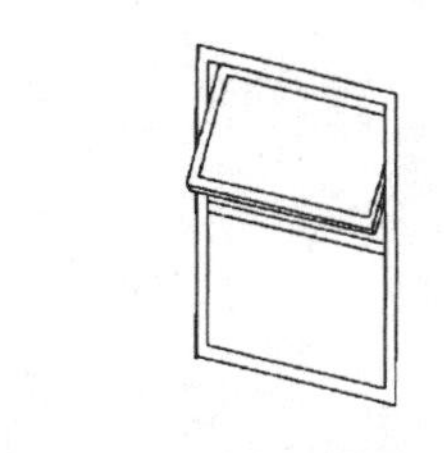

awning window

B: babbitt metal to bypass door

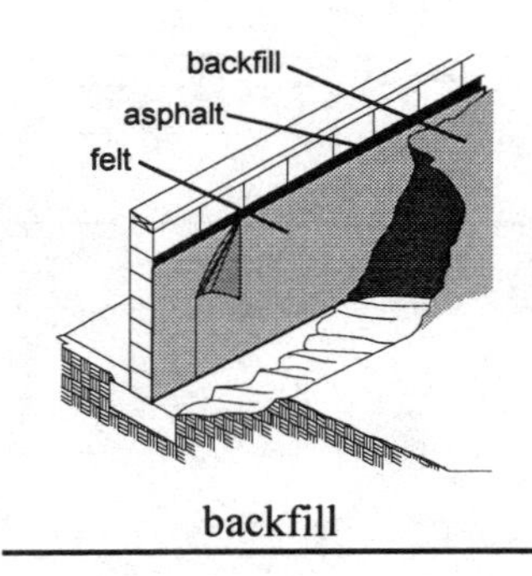

backfill

babbitt metal – a material made of various alloys of tin with antimony and copper which is used in the manufacture of bearings. A film of lubricating oil between the bearings and the surface of the moving and stationary parts prevents friction and metal-to-metal contact.

baby Roman brick – see brick, baby Roman.

back – the side of a gypsum wallboard panel that is designed to be in contact with the framing.

back arch – an arch-shaped structural support for an interior or veneer wall. The exterior support of the wall is provided by a lintel.

back band – a molding used around the outer edge of a rectangular window or door casing.

back blocking – a piece of gypsum wallboard laminated to the backs of two gypsum panels to reinforce a joint between framing members. Adhesive or joint compound may be used to laminate the piece of wallboard to the panels.

back clips – clips attached to the back of gypsum wallboard that fit into slots in the framing and hold the wallboard panels in place. They are used in installations where wallboard panels are intended to be demountable.

back cut – a cut on the back of a panel or board so the piece will lie flat.

backerboard – a gypsum wallboard designed to be the first layer in a multiple-layer wall system, or the base layer to which acoustical ceiling tile is applied. Base ply wall panels are installed with the long edges parallel to the wall studs, unless the wall height exceeds 97 inches or decorated wall panels are to be installed. The face ply, or second layer, and base ply panels need to be staggered so they don't fall in the same place and weaken the joint. Base ceiling panels are applied with the long edges perpendicular to the framing members or ceiling joints. The face ply is then installed with the long edges of the panels parallel to the ceiling joists. They are fastened through the base ply panels into the joists. Backerboard is also called backing board.

backfill – 1) material used to fill any excavation, such as a trench or the area around a foundation after the foundation has been installed; 2) rough masonry used as support behind a finished masonry wall.

backflow – reverse flow; water flowing from a distribution point back into a supply system.

backflow preventer – a plumbing device that prevents water from flowing back into the water system and possibly contaminating the supply. Backflow preventers work by using an air gap or a check valve. An air gap is a vertical distance between the outlet of the supply and the overflow rim of the fixture into which the supply flows, such as the distance between the faucet and the sink bowl. Water on the high pressure side can flow across an air gap, but it is physically impossible for the water to return to the supply. A check valve is a one-way valve which permits flow in only one direction. It does this by means of a hinged flapper or guided poppet that opens with the flow in one direction, but closes if the flow reverses. Sometimes a spring is included to aid valve closure. Most plumbing codes require backflow preventers be installed between the water main and branch connections to individual service.

back gouging – removing material from the back side of a weld.

backhand welding – welding with the torch, stinger or weld rod pointing toward the weld as it is being laid. The torch faces the weld and is moved backward as the weld bead moves forward.

back hearth – the floor of a fireplace combustion chamber.

backhoe – a powered digging machine with a bucket suspended on a movable boom.

backing – 1) material used behind a facing to add strength, stiffness or structural support; 2) material on the back side of the root which supports the weld metal during welding.

backing, hip rafter – chamfering or beveling the top edge of a hip rafter to the slopes of the adjoining sections of roof. The top of the hip rafter is cut at an angle along its length, sloping downward from a centerline on its top edge. These angled slopes match the slopes of the adjoining sections of roof.

backing ring – metal ring placed inside a pipe butt joint before welding to provide backing for the molten weld metal until it solidifies.

backing strip – weld backing material that comes in strips.

backing weld – a weld root pass that serves as a backing for a groove weld.

backlash – the play between parts of a machine, particularly the gears.

backlighting – lighting behind or to the side of an object.

back pressure – pressure in a piping system that resists the flow of fluid.

back rake – the angle at the top cutting edge of a lathe tool bit measured from the horizontal axis of the tool.

backsaw – a fine-tooth hand saw with a stiffening rib along the back edge of the blade. This type of saw is often used with a miter box to cut angles.

backset – the distance from the face of a lock, at the edge of the door, to the centerline of the knob or lock cylinder.

back siphonage – negative pressure in a potable water system that permits fluids to flow into the system.

back surface – the side of a panel that will be in contact with the framing members of a structure.

back-to-back – two or more similar items that are placed in mirror image to one another, such as plumbing fixtures in bathrooms that share a common wall. This arrangement saves installation and material costs by permitting the sharing of both the water supply and the drain, waste and vent piping. It also saves space within the building by eliminating the need to duplicate piping runs to two different locations.

backup – 1) a stationary object against which an operator of earth-moving equipment can push; 2) a standby or auxiliary unit; 3) a structure behind a facade; 4) a second storage of computer data on a disk or other storage media.

back vent – see individual vent type.

backward inclined fan – a fan with blades inclined opposite to the air flow.

back weld – a weld applied to the back side of a joint. Also called a backing weld.

baffle – an obstruction placed to control the flow of a gas, liquid, light, or sound, or to control the splashing of a liquid in a tank or other container.

balance – 1) an amount remaining; 2) to make equal.

balance sheet – a financial statement that lists assets, liabilities and net worth.

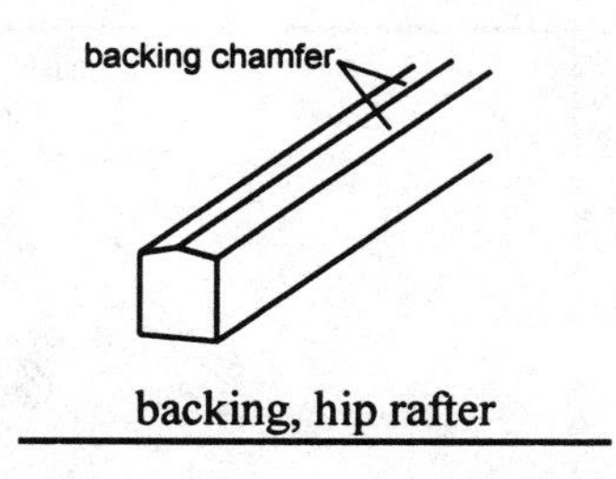

backing, hip rafter

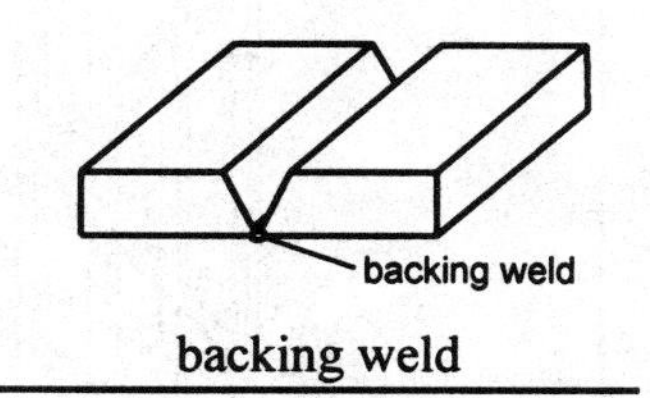

backing weld

backsaw

Balloon framing

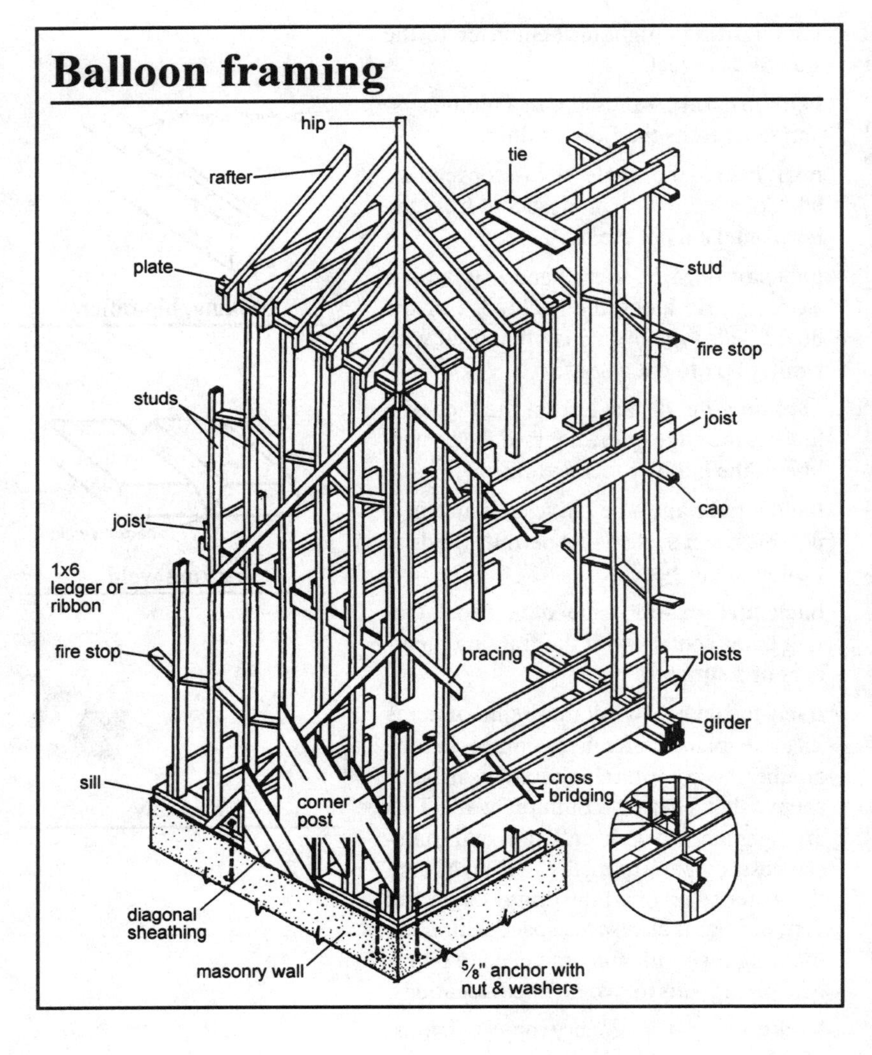

balancing – bringing into a state of equilibrium.

balancing by dampers – using air dampers, which are movable vanes that control air flow, to regulate the flow within each portion of an HVAC system.

balancing by static pressure – balancing air flow within an HVAC system by properly sizing the ducts.

balancing ell – an elbow plumbing fitting used to control water flow through a heat distribution unit.

balancing subgrade – the final grading of a subgrade surface prior to the fine trimming.

balancing valve – a valve used in a hot water heating system to control water flow and balance heat distribution. The hot or chilled water usually comes from a common source and is distributed to many different locations. The initial design of the system considers the rate of usage and distance from the source to each of these locations by properly sizing the different piping runs branching off the main. Balancing valves are used to fine tune the flow distribution to account for the variables that may have been unknown at the time of design, such as added flow restrictions resulting from more bends in the pipe line. Balancing valves generally have an indicator that shows the degree to which they are open in order to permit the proper water flow. This allows them to be closed for maintenance and then reset to the same opening, simplifying the resetting of the system after down time.

balcony – a platform projecting from the wall of a building, usually enclosed with a safety rail.

ball-and-socket joint – a joint in which a partial sphere on the end of a shaft is free to move within a closely fitting spherical socket. This allows free movement, within the limits of the socket, in any angular direction. This type of joint is used in a joystick.

ballast – 1) broken stone or gravel used beneath railroad ties to provide drainage and stability; 2) a transformer that controls the current supply to a fluorescent tube; 3) weight added to boats and ships for stability, and to submersibles for descent.

ballast transformer – a transformer used in fluorescent lights to supply voltage to the light tubes. It provides a high starting voltage, and then limits the current through the tube.

ball bearing – a type of bearing used with rotating parts that has an inner and an outer race separated by steel spheres that are free to turn. These spheres or balls give this type of low-friction bearing its name.

ball check valve – a non-return valve that uses a ball to seal against a seat and stop the flow in one direction.

Ball valve

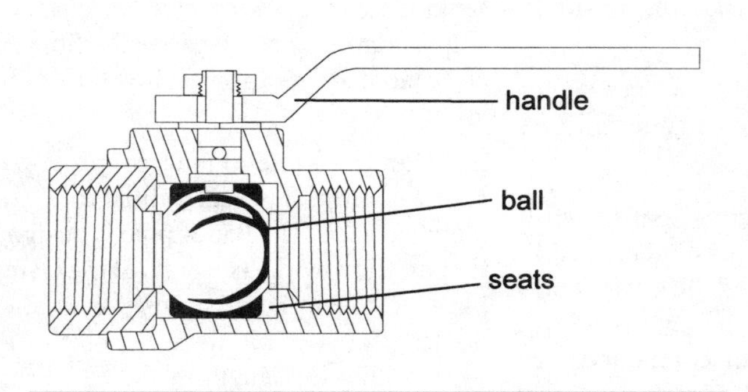

A type of valve used to regulate or stop the flow in a fluid system. It consists of a ball with a hole bored through its diameter, mounted on a spindle within the body of the valve. When the valve is closed, the hole through the ball is at right angles to the valve body, so no flow can take place. A quarter turn of the spindle opens the valve, allowing the fluid to pass through the hole. When the hole through the ball is fully in line with the fluid, the flow through the valve is almost entirely unrestricted. The degree of opening regulates or stops the flow of fluid through the valve. Some ball valves are designed to be maintained in a partially-open position so that they can be used as flow control valves. This design generally has a stem that attaches to the top and the bottom of the ball so that the ball is supported top and bottom for structural stability, particularly in the partially-open position.

Ball valves are commonly used at the inlets and outlets of heat exchangers in HVAC cooling systems. They can be used to regulate the flow through the heat exchangers and balance the cooling capacity in order to meet the total requirements of a system. Because each piece of equipment or room on the system may have different cooling needs, the ability to regulate the cooling flow ensures adequate cooling for all areas. A ball valve used for flow control is generally supplied with a position indicator to show the percent that the valve is open, and with detents to hold the valve in the desired position. Once the system has been balanced, the positions of the control valves are noted. When a valve is replaced or rebuilt at a later date, or the system shut down, the valves can easily be reset to the proper flow positions for restart.

ball cock – a type of valve used to regulate liquid levels in a tank by means of a ball-shaped float which rises with the liquid and closes the valve by lifting a lever.

ball joint – a flexible plumbing joint with a spherical end resting in a socket.

ball mill – a rotating drum containing metal balls that is used as a paint-mixing device.

balloon framing – a framing method in which one-piece studs extend from the foundation to the roof to form the walls of both stories in a two-story structure.

balloon payment – a final payment larger than the periodic payments, paid at the end of a loan.

ball peen hammer – a hammer with one flat end and one rounded or hemispherical end on the hammer head, used primarily for metal work.

ball valve – a valve used to regulate or stop the flow in a fluid system. *See Box.*

balsa – a very lightweight but strong wood from a tropical American tree.

baluster – a small spindle or vertical member that supports a rail. Balusters form the main support for the handrails along a stairway or around a balcony. Also called a banister.

balustrade – a row of balusters supporting a handrail along a stairway.

bamboo – the hollow shafts of large tropical grasses used for furniture and as a building material.

band clamp – a flexible band (which may be made of a variety of materials) with a cinching device used to secure objects.

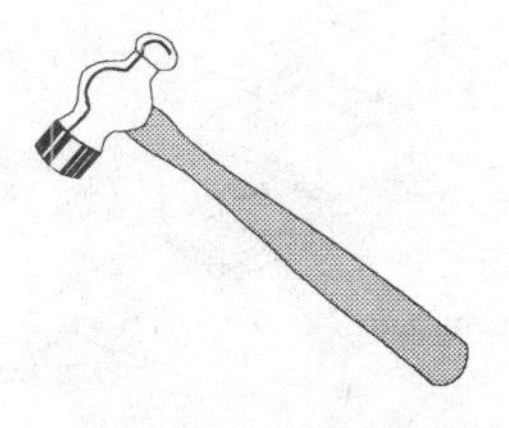

ball peen hammer

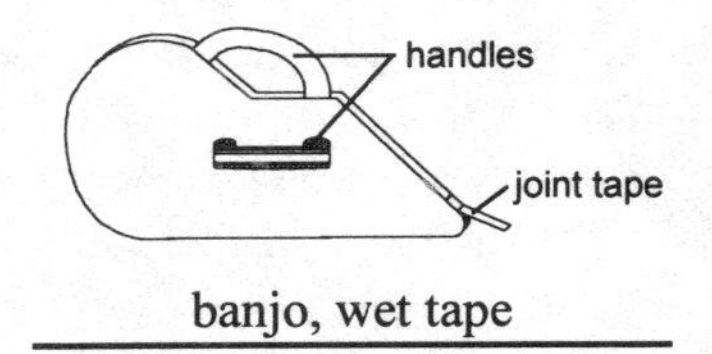

banjo, wet tape

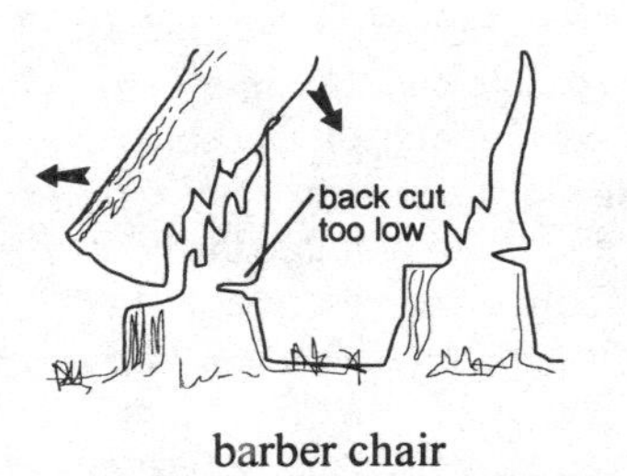

barber chair

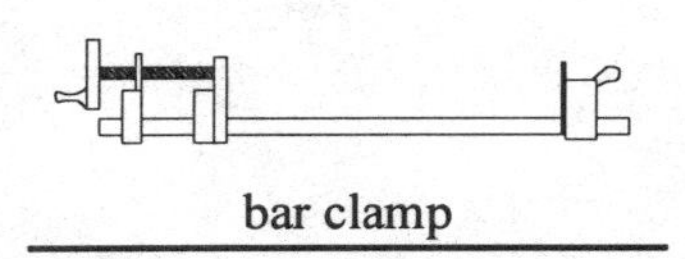
bar clamp

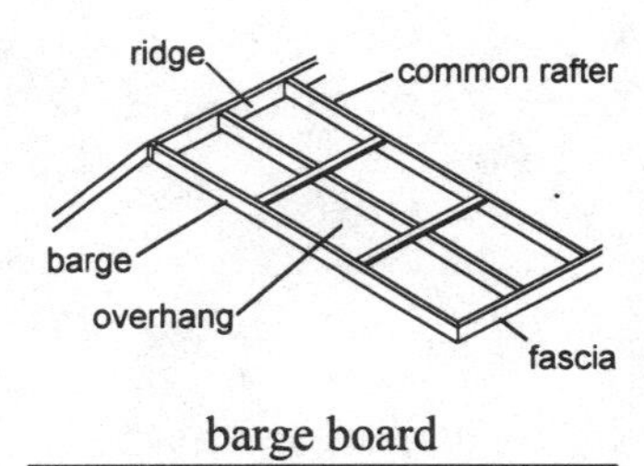

barge board

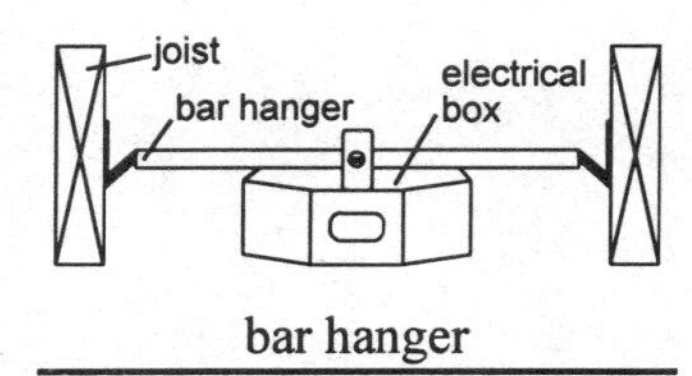

bar hanger

band saw – a power saw with a continuous, flexible blade that revolves around two wheels. Both straight and curved cuts can be made with the band saw. Work is held against the moving blade and maneuvered as required to make the desired contour of the cut.

bands of color – a paint defect in which the painted surface is covered unevenly.

banister – see baluster.

banjo – an automatic taping tool for applying joint tape and joint compound together. Both the tape and the compound are held in the tool. As the tape is withdrawn, it is coated with compound. A blade at the tape exit point cuts the tape when the end of the joint is reached. There are two types of banjos, wet and dry. A wet banjo holds the roll of tape and the joint compound in the same compartment. The dry banjo holds them in separate compartments, so that any unused tape can be used at a later date. Both types of tools offer a faster and easier method of applying joint tape and compound than manual application.

bank – 1) a sloped area adjacent to a stream, roadway or flat area; 2) the grouping of objects in rows or tiers.

banker – a mason's bench used for laying out and shaping arches.

bank plug – a 2 x 4 wood wedge marker that is set (driven into the ground so that about 2 feet of the wedge is above ground) prior to a final earth grading operation. A string is attached and stretched between the bank plugs to indicate to the grading machine operators the final grading level that is required.

bankrupt – financially insolvent.

bank statement – a bank report, usually sent out monthly, that summarizes money transactions and the status of accounts.

bar bending table – a work table with holders designed to make the task of bending rebar easier.

barber chair – a split in a tree trunk that can occur during tree felling when the felling cut is not properly made.

bar chair – see chair.

bar chart – a graph that uses horizontal or vertical bars to indicate relative values, such as might be used to schedule activities. The duration of each of the activities can be shown by the length of the representative bar. Looking at a bar chart can help a contractor to visualize the time and craft labor needed for each activity on a job.

bar clamp – a metal bar or pipe with screw clamps and an adjustable stop mounted on it. The bar or pipe is designed to span wide work. The parts of the clamping mechanism may be positioned along the length of the bar to fit the size of the work within the length of the bar. Screw threads on the clamp permit fine adjustment, and the application of clamping pressure.

bar doors – swinging doors, usually louvered, that are not the full height of the doorway in which they are installed.

bare conductor – an electrical conductor that is not insulated.

bare electrode – a welding electrode without a flux coating.

barge board – a fascia that follows the roof rake from the horizontal fascia to the roof peak, forming (in appearance) the outermost exposed rafter.

barge course – a course of masonry units that are laid on edge along the top of a wall to form a coping for the wall.

barge rafter – see fly rafter.

bar graph – a representation of change using bars on a graph to show the rate of change as compared to another quantity, such as time. The length of the bars represents quantity or value. Tools like these enable people to quickly visualize changing patterns over time or to see how closely goals correlate with actual activity.

bar hanger – a hanger designed to support an electrical box. The hanger spans between structural members, such as joists, and suspends the box between the members.

barites or barytes – barium sulfate, a material opaque to X-rays used as a paint pigment.

bark – the outer protective covering on a tree.

barrel bolt – a cylindrical sliding bolt used as a door lock.

barricade – a barrier to keep the public out of construction areas or any area where they might be in danger.

barrier – a device that slows or stops a transfer or movement from one place to another.

bartizan – a small turret overhanging a tower or a wall.

barytes – see barites

basalt – a dark-colored lava rock.

bascule – a hinged cantilever which is counterbalanced by a weight.

base – the bottom or first layer of a structure or object.

baseboard – a trim board placed against the wall at the junction of a wall and the floor.

base cabinet – a floor-mounted cabinet.

base coat – the first coat when multiple coats of paint or other coverings are applied.

base course – the first course of masonry units.

base leg – the horizontal run length of a rafter.

base level – the lowest possible level of land erosion by water.

baseline – 1) a reference line from which measurements can be taken; 2) an established east-west reference line used with a north-south meridian line to plot townships on U.S. government survey maps.

base material – 1) material used in paving to compensate for ground swell, provide drainage, and to support the pavers. Base material might be gravel, concrete or asphalt. 2) the material on which a brazing, soldering, cutting or welding operation is performed.

basement – the lowermost portion of a building, located partially or completely below ground level.

base metal – 1) metal upon which welding or brazing is to be performed; 2) a non-precious metal; 3) a structurally inferior metal.

base molding – a trim piece along the edge of a baseboard.

base ply – 1) the first layer of gypsum wallboard in a multi-ply construction; 2) the first layer in a multi-layer system, such as in built-up roofing.

base shoe – 1) a molding strip attached to the lower portion of a baseboard; 2) a strip to hold down the edge of carpet.

basin – 1) a shallow sink for washing; 2) a depression in the surface of the land.

basin wrench – a wrench with an extended handle that allows the jaws to reach and tighten the nuts which fasten the faucets to an already installed sink.

basket grip – a cylindrical gripping mechanism that slips over the end of a wire to hold it firmly, and connects to a fish tape so that the wire may be pulled by the fish tape.

basket grips

basket pattern – a masonry pattern in which alternating sets of brick are laid vertical/horizontal/vertical to give the appearance of a basket weave.

basket strainer – a basket with holes and/or slots that fits into a sink drain and catches food particles before they enter the drain. See also strainer.

bas-relief – a three-dimensional sculpture slightly projecting from the surface of the surrounding area.

bastard file – a coarse-toothed flat file used when a lot of stock is to be removed. Because it does not produce a fine finish, it is often followed with a finer-toothed file.

bat – approximately half a brick or less. A bat is cut or broken from a whole brick and used to fill in a section where a whole brick will not fit.

batch – 1) a portion of material mixed at one time, such as cement; 2) a production run of something; 3) a grouping.

batch plant – a plant for producing concrete in large quantities.

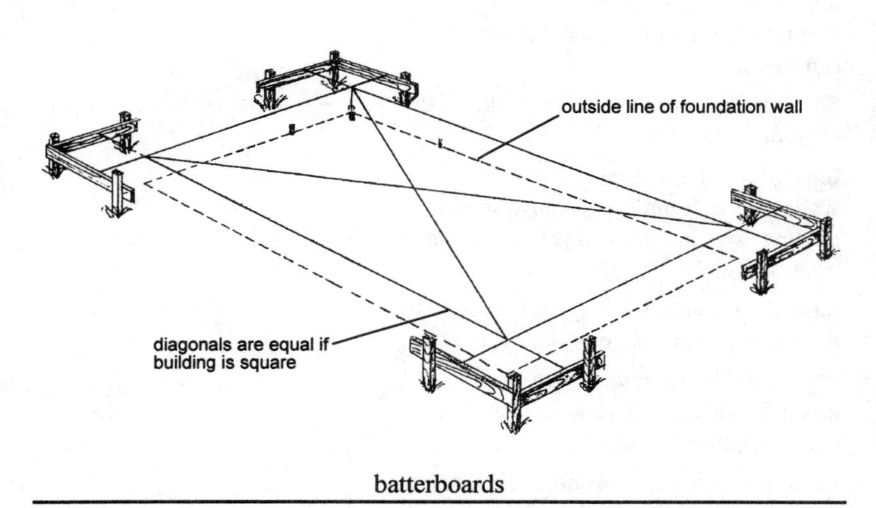

batterboards

bath – a liquid in which an item can be immersed for cleaning, coating or chemical processing.

bathroom accessories – towel holders, soap dishes and other items manufactured for use in a bathroom.

bathtub – a plumbing fixture large enough for bathing the human body.

bathtub, recessed – a tub with both ends and one side built into a wall.

batt – a precut piece of insulation designed to fit between framing members.

batten – a wood strip used to cover the joint cracks between boards, as in siding or fencing.

batten plate – a steel structural plate attached to another member or members to add stiffness or strength.

batter – masonry that slopes back in courses. A batter is the reverse of a corbel.

batterboard – 1) horizontal board nailed to the corner posts at an excavation to indicate the desired level; 2) the fastening point for strings that are stretched to indicate the outline of a foundation wall.

batter brace – an angled support near the end of a truss.

battery – 1) a chemical device that stores direct current electricity; 2) a grouping of objects.

battery of fixtures – a group of plumbing fixtures discharging into a common waste line.

bay – 1) an area within a structure that is marked or walled off; 2) a projection outward from a wall which affords more interior space or frames a bay window.

bay window – a window, divided into three walled sections, that projects out from a wall in an arc. It is designed to give a room a larger appearance by extending the window area out from the wall. Bay windows are supported from the wall or floor structure of the building, and usually do not extend as far up as the roof line. Because of this, they need their own roof projection from the wall. Ready-made bay window units are available. They come in preassembled components that are easily installed into the wall opening.

bay window, angle – multiple windows arranged as a projection from the wall.

baywood – a type of mahogany from Mexico.

bead – 1) a preformed strip used to reinforce the corners or ends of wallboard panels, or the raised metal section at the corner of this strip. 2) the surface of a weld; 3) an edge or border; 4) a trim molding with a convex, partially circular, cross section; 5) a ridge formed in sheet metal; 6) a corner reinforcement made of stiff woven wire used in stucco application; 7) a line of adhesive or caulking compound.

bead and reel – a convex trim molding shaped like a series of small disks on edge alternating between beads that are round or oval.

beading – bead molding.

beading plane – a wood plane designed to cut a bead or semicircular pattern.

beam – a structural member installed horizontally to support loads over an opening.

beam anchor – a metal fixture used at the end of a beam to secure the beam to the structure.

beam and slab construction – a reinforced concrete slab that is supported by structural beams.

beam compass – a bar, with a point at one end and a pen or pencil at the other, used for drawing large diameter circles.

bearing – 1) a support for a moving part of a machine; 2) the length of the rafter in the seat cut that rests, or bears, on the rafter plate; 3) the area of a load-bearing surface; 4) an angle of less than 90 degrees measured by a surveyor from either the north or south meridian.

bearing area – the area of a surface on which a force acts.

bearing cap – a structural member between the top of a column and the bottom joint of another structural member. It is of larger dimension than the top of the column and serves to distribute the load.

bearing life – the minimum life expectancy of bearings as measured by observing a control group operating under a prescribed set of conditions. Each bearing produced is slightly different due to manufacturing tolerances, surface smoothness, and overall roundness, so their life expectancies vary slightly as well. With the information provided by the control group, a designer can be reasonably certain of the minimum expected bearing life under studied conditions, and can adjust the design accordingly. In this way, overall machine performance can be reasonably predicted. Bearing life is represented in hours.

bearing, lifetime-lubricated – a bearing, lubricated with a high-stability grease, and designed with seals to eliminate the need for periodic lubrication.

bearing partition – a partition that acts as a vertical structural support.

bearing plane – the load carrying surface.

bearing plate – a metal plate that is placed under a load-bearing structural member to spread the load over a larger area.

bearing seat – the end of a joist or beam where the load is transferred to the member on which it rests.

bearing wall – walls that support the vertical load of a structure. The bearing walls are essential for the stability of the structure. They must be designed to carry the combination of live and dead loads that are expected for the intended building service. Bearing walls may be limited to the exterior walls, but often also include one or more interior walls as well. Openings that are cut into bearing walls for doors or windows must be reinforced and an alternate load path developed through the use of headers which carry the loads across the tops of the openings.

beaverboard – a type of wood-fiber sheet used as building material. Also called Beaver Board, a trademark name.

bed – 1) an area especially prepared to act as a base, such as a layer of mortar in which brick or stone is embedded; 2) soil prepared for planting.

bed coat – the first coat of joint compound into which the joint tape for gypsum drywall joints is embedded.

bedding – 1) fill material on which buried pipeline rests in a trench; 2) mortar which is placed under masonry units to seat them in position.

bed joint – the layer of mortar in which a masonry unit is set.

bed molding – a molding used for an inside angle of a building.

bedrock – solid rock under the soil layer.

bed stone – a large stone that is used to support a beam.

bell – 1) a resonant object which vibrates to make a sound when struck; 2) the enlarged end of a pipe or fitting designed to receive the straight end of a length of pipe.

bell and spigot piping – gravity-rated vent and drain piping, or pressure piping, with a plain or straight section at one end and an enlarged or bell-shaped section at the other. The enlarged end is designed to receive the straight end of the length of pipe as they are being laid. Also called hub and spigot piping. *See Box.*

bell caisson – see caisson, bell.

bell mouthing – a hole bored in a piece of work that is flared at the opening so that the mouth of the hole is larger than the inside diameter of the hole.

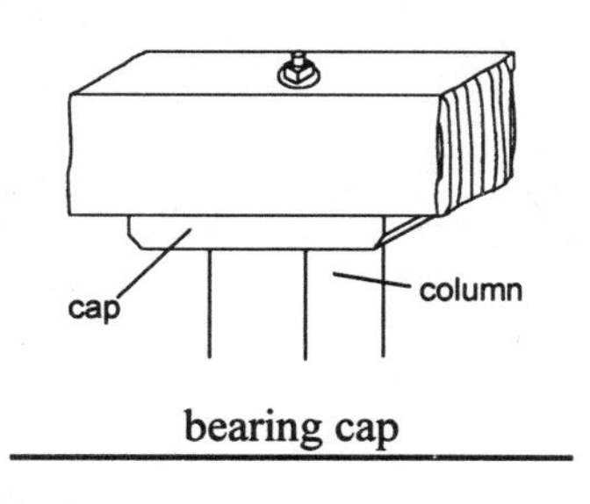

bearing cap

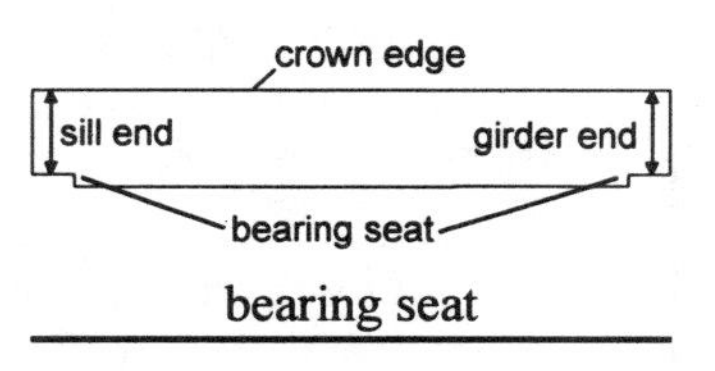

bearing seat

Bell and spigot piping

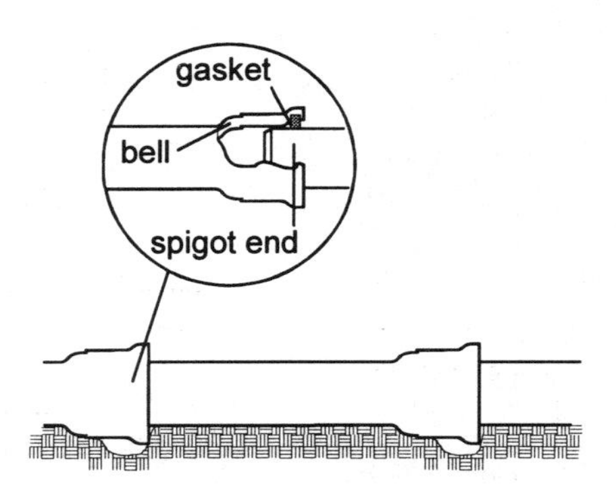

Bell and spigot piping

Gravity-rated vent and drain piping, or pressure piping, with a plain or straight section at one end and an enlarged or bell-shaped section at the other. Bell and spigot piping is generally used for underground sewer and potable water distribution. When the pipe is laid, the straight end is set into the enlarged end of the adjoining pipe. The bell end is fitted with a rubber gasket for sealing, and the spigot end is beveled so that it will slip easily into the gasket. As the spigot is inserted past its beveled section, the gasket is compressed in the bell, making a tight seal. Though the seal is tight, it is subject to being pulled apart by earth movement if not properly restrained. Concrete thrust blocks, or other types of mechanical restraints, must be installed at elbows and other changes of direction to secure underground joints. When the pipe trenches are dug, an excavation deeper than the rest of the trench is made at any point where there is a change of direction. These excavations become the forms in which the concrete thrust blocks are poured. The size of the thrust block depends on design guides and standards based on the amount of pressure the joint must withstand. The thrust blocks prevent the pipe from moving and spread the thrust load over a larger area.

Bell and spigot piping is commonly made of PVC (polyvinyl chloride), vitrified clay, asbestos cement and cast iron or cement-lined ductile iron. The joints on cast or ductile iron piping used for sanitary sewer drains may be caulked with oakum, a rope fiber made from hemp, and sealed with molten lead. Because this is an expensive and time-consuming process, it is seldom used in modern construction.

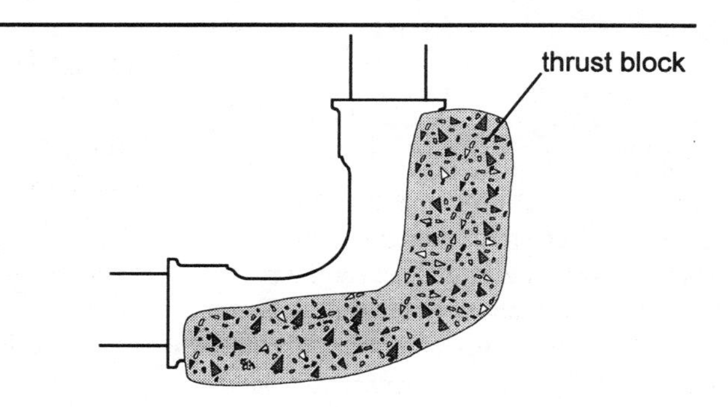

Thrust block supporting bell and spigot pipe joint

bellows – 1) a device for moving air by drawing it through a one-way valve into a flexible chamber and then forcing the air out of a nozzle by collapsing the chamber; 2) a flexible section of an expansion joint.

bell wire – small size electrical wire used for low voltage applications. One of its primary uses is to connect the door bell to its power source and switch, which is how it came by its name.

belt course – a decorative horizontal masonry band, also called a string course or sill course.

belt sander – a power sander having a continuous revolving abrasive belt. At the point of contact with the work, the belt is supported by a flat metal shoe that holds it against the work to be sanded. This type of sander can rapidly produce a smooth, flat surface, but because of its cutting speed, it must be used with great care.

belvedere – a building, such as a summerhouse or gallery, designed to give the occupants an attractive view.

ben – 1) the second room of a two-room dwelling; 2) an inner room or parlor.

bench – 1) a work surface; 2) a long seating surface; 3) a shelf of rock with steep areas above and below; 4) an elevation in a mine.

bench dog – a peg set into a hole in a work surface for the purpose of holding the work in place.

bench grinder – a motorized grinding wheel (or wheels) mounted either on the motor shaft or a separate shaft supported on bearings and driven by a motor through a belt. The grinder is attached to a stable surface, such as a bench, to raise it to a convenient working height. It is used to shape and sharpen metal, usually steel and steel alloys. Softer metals will not hold a sharp edge and tend to clog the grinding surface of the wheels.

benching – flattening the area around an excavation so that the soil at the cut line has less tendency to slip into the excavation.

bench lathe – a small wood or metal lathe that mounts on a workbench.

bench mark – a permanent marker used as a reference point from which all measurements are taken. In surveying, a bench mark may be placed in the ground so that all future surveys can be made from that starting point. In a similar manner, a bench mark is established when building a structure so that all measurements are made from a common point.

bench saw – see table saw.

bench stone – a sharpening stone.

bench stop – any device used in conjunction with a work surface to hold work in place.

bench vise – a vise attached to a work surface.

bend – a formed curve in a section of pipe, rebar, conduit, or similar material.

bender – a tool or device for making bends, such as a conduit bender.

bending moment – a force working against a structural member in such a way as to make, or tend to make, the member to curve.

bending schedule – a table showing the required sizes and radii of bends determined to be needed in pipe, conduit or rebar for a specific project. Each of the bends listed in the table is keyed to the overall drawing in some manner so that it is clear which bend (or bent section) is to be installed in each location. The bending schedule helps the installer plan his work and keep track of the correct sequence of delivery and installation of the bend sections.

beneficiary – a person who receives the benefit.

bent – a structural support system that extends across a building centerline, such as a framework used to strengthen the width of a bridge.

bent cap – the uppermost structural member in a bent.

bent-nose pliers – needle-nose pliers with the ends of the jaws bent at right angles to the tool axis. They are designed to reach into tight places.

bentwood – curved wood that has been permanently bent along the grain.

benzene – a flammable compound that is colorless, volatile, and toxic, used in the manufacturing of chemicals and dyes, and as a solvent for resins and fats.

berm – a dike or earth embankment used to provide containment, isolation, and/or protection. Berms are often built up along roadways.

beryllium – a hard, lightweight, gray metallic element; atomic number 4; atomic weight 9.0122; atomic symbol Be.

bevel – a corner with two angles and a flat plane between; a chamfer; an edge cut at an angle.

bevel angle – the angle to which the bevel is cut. This angle is measured from a plane perpendicular to the axis of the material.

beveled corner box – an electrical box with the top and bottom back edges beveled so that the box fits into the wall openings of older buildings.

beveled edge – the tapered factory edge of a gypsum wallboard panel.

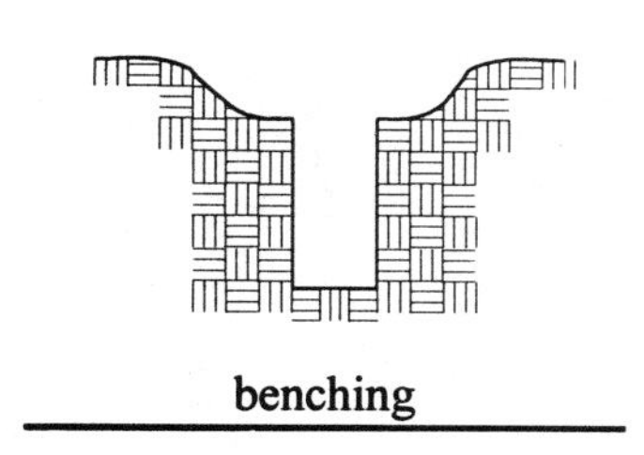
benching

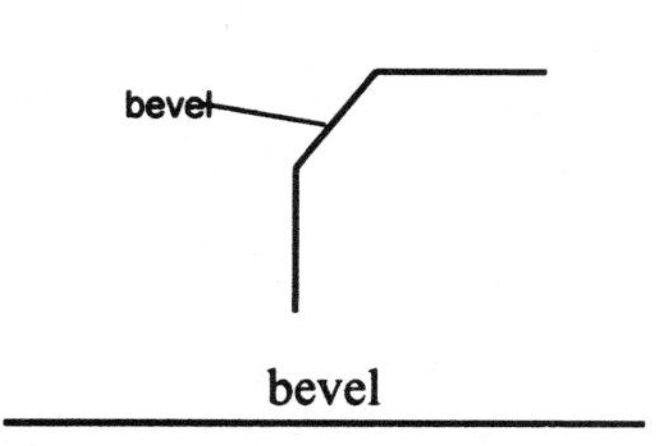

bevel

bent-nose pliers

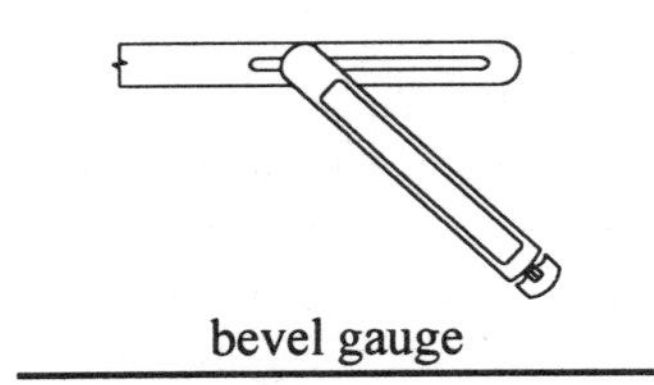
bevel gauge

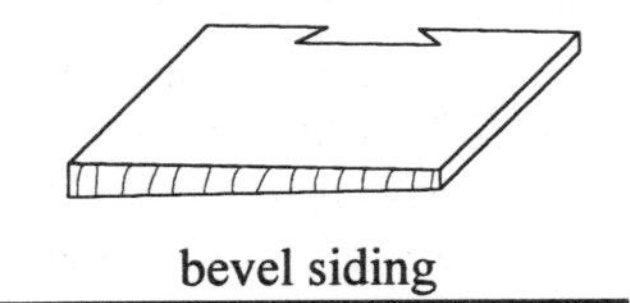
bevel siding

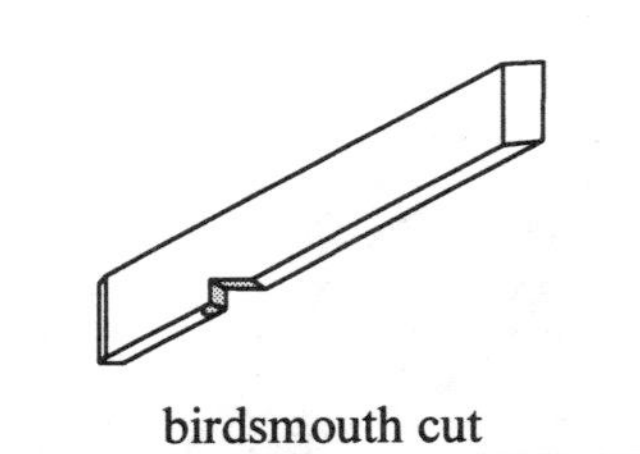
birdsmouth cut

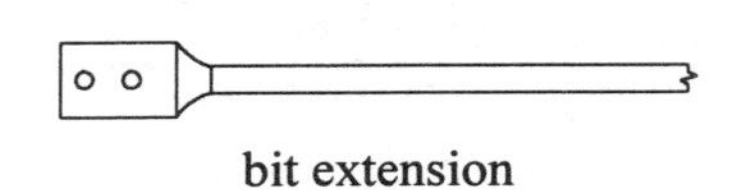
bit extension

bevel gauge – a measuring tool with a sliding blade that can be adjusted and locked into place along the handle. The blade is adjusted to the dimensions of an angle and then locked in so that the exact measurement of the angle can be marked and transferred to another piece. Also called a bevel square or sliding T-bevel.

bevel gear – a gear with the teeth cut on a bevel with relation to the axis of the gear. This allows it to mesh with another bevel gear, providing a quieter transmission of power than with spur gears.

bevel joint – the point where two pieces of material join at an angle to the axis of one or both items.

bevel siding – a type of horizontal siding that is tapered in cross section, with each board overlapping the edge of the board below.

bevel square – see bevel gauge.

bezel – the diagonal face at the cutting end of a tool such as a drill bit or chisel.

biangular – an object with two angles.

bibb or bib – a faucet with the outlet at a downward angle, generally used for hose connections on the outside of a building. Also called a bibcock.

biconvex – convex on both sides.

bid – an offer to do a job at a stated price.

bid abstract – a document that shows the unit prices from jobs that government agencies have awarded.

bid bond – a guarantee by a bonding company that a bidder will enter into a contract if awarded the job and will supply performance and payment bonds.

bidet – a bathroom fixture similar in appearance to a toilet, but with water jets for bathing the lower torso.

bid opening – the opening and reading of bids that have been submitted for a job.

bid time – the amount of time between the solicitation of a bid and when the bid is due to be submitted.

bifold door – see door, bifold.

bill of material – a list, usually on a drawing, of the parts of the item depicted by the drawing.

bimetallic strip – a device composed of two different metals that have different rates of expansion. *See Box.*

bimetallic thermometer – see thermometer, dry bulb.

bind – 1) to confine or restrain; 2) seizure between moving parts, preventing further movement; 3) to sew the edge of carpeting to prevent raveling.

binder – 1) an additive, usually starch, used to increase cohesion in the gypsum core of wallboard or to increase the bond between the paper and the core of the board; 2) the carrier liquid in paint that contains the pigment in solution or suspension; 3) a compound used to hold pieces of material together; 4) any device for holding multiple pieces in place.

binder bar – a metal strip laid over the edge of carpeting at a doorway or entry to reduce the hazard of tripping over a loose carpet edge.

bird screen – a screen which covers an opening, such as a vent, in a building and prevents the entry of birds.

bird's-eye – a type of wood grain in which the grain forms small circles or ellipses.

birdsmouth – the triangular cutout in a rafter made so that it will fit flat against the top plate of the wall.

bisect – to divide into two equal parts.

bit – 1) the portion of a screw gun or power screwdriver that fits into the head of the screw; 2) a drilling, boring or cutting tool; 3) the heating tip of a soldering iron.

bit, annular – a drill bit that is in the form of a hollow cylinder and used to cut plugs of wood.

bitch pot – an asphaltic emulsion container.

bit extension – a steel rod with a coupling to hold a drill bit on one end. The other end fits into the chuck of a drill. This extension effectively lengthens the drill bit for hard to reach places and deep holes.

bit gauge – a device that clamps on a drill bit to limit the depth of the hole being drilled.

bitumen – a natural asphalt (hydrocarbon) compound.

Bimetallic strip

Two metals that have different rates of expansion bonded together in a strip. The strip is coiled into a spiral and as the metal with the greater rate of expansion expands or contracts, the spiral becomes looser or tighter. The differential expansion and contraction causes the end of the bimetallic strip to move. Because of this, it can be used to turn a switch, such as a thermostat, on and off. One way to accomplish this is to connect the end of the strip to a closed vial partially filled with mercury. Wires projecting part way into the vial make an electrical connection by means of the mercury. If the strip moves and the vial is tipped so that the mercury no longer aids in the connection between the wires, the circuit is broken and the power is shut off to the device being controlled. The bimetallic strip can also be connected to a pointer and used to indicate temperature changes.

bituminous – containing asphalt.

bituminous coal – a soft coal containing volatile hydrocarbons.

bituminous paving – paving using asphalt and aggregate.

black cherry – a hard red-brown wood used for making furniture.

blacktop – asphalt paving.

black walnut – a hard brown wood used for making furniture.

blade – a cutting, scraping or shaping tool.

blade and combination – a combination square with a center head attached which can slide along the blade of the square.

blade load – the material being moved by a dozer blade.

blanch – 1) to bleach; 2) to acid-clean a surface.

blanket – insulation batts. Also called blanket insulation.

blast – 1) a sudden rise in pressure caused by an instantaneous release of energy; 2) air forced into a furnace under pressure.

blast doors – doors designed to withstand or mitigate the pressure of a blast.

blast furnace – an iron ore smelting furnace.

blast gates – sliding dampers in an heating, ventilating and air conditioning (HVAC) system that can be adjusted to regulate air flow.

blasting – 1) the use of high explosives to break up or move rock and earth so that it can be reworked by conventional means; 2) using a stream of abrasive material suspended in a fluid to clean and/or abrade a surface.

blasting cap – a device containing a very sensitive explosive charge in a measured amount. This charge is easily ignited and powerful enough to set off the main explosive charge.

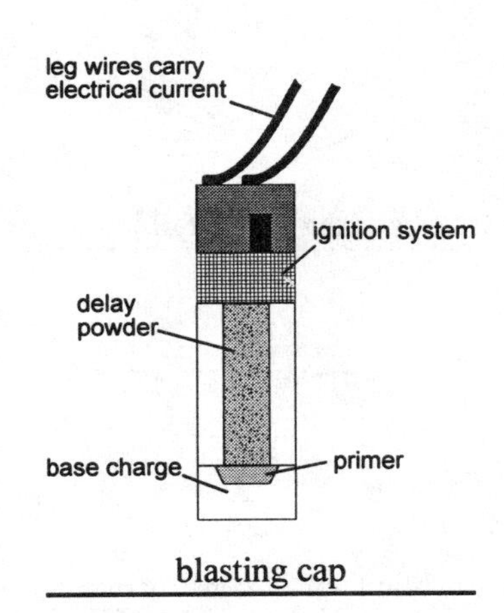

blasting cap

bleach – 1) to lighten or whiten using a chemical agent; 2) the bleaching agent.

bleachers – stepped seating benches which allow an area of less than 3 square feet per person.

bleed – to seep out, as in a stain or pigment seeping to the surface.

bleeding – 1) a defect in gypsum wallboard installation in which discoloration seeps or bleeds through to the surface.

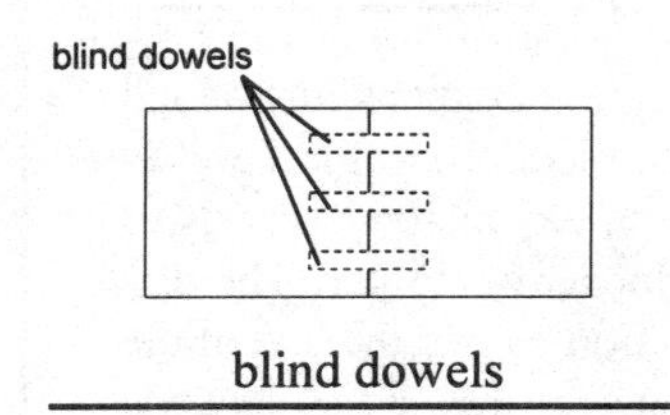

blind dowels

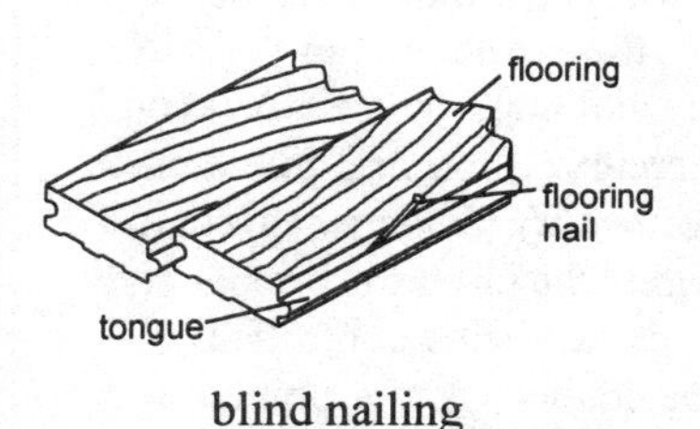

blind nailing

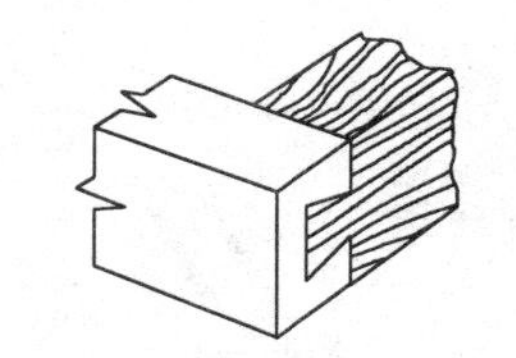

blind single dovetail joint

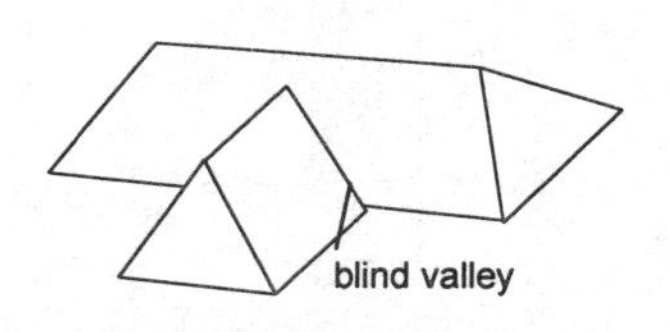

blind valley construction

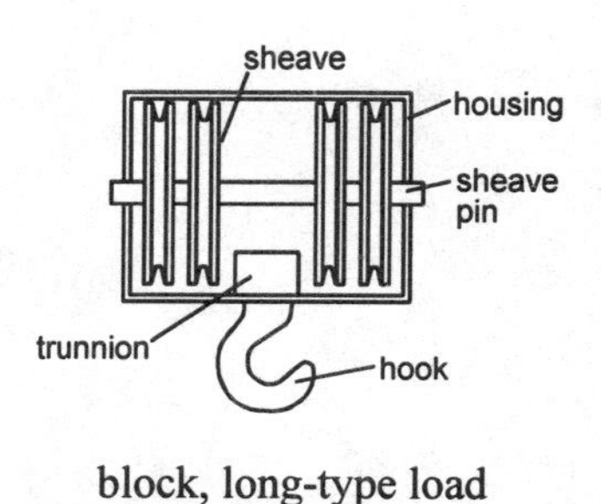

block, long-type load

This usually occurs at a joint; 2) a paint defect in which color from the underlying surface comes through and discolors the paint covering it; 3) removing air from a liquid system, such as a hydraulic system, by opening a valve at a high point in the system.

blemish – a defect that mars appearance.

blend – to mix together.

blind corner cabinet – see cabinet, blind corner.

blind dowel – a dowel that extends into holes drilled in adjoining pieces, with neither hole completely through its respective piece. The dowel joins the two pieces together, but cannot be seen once the joint is completed.

blind header – a concealed brick header.

blind hole – a hole that is not drilled completely through the material.

blind joint – a joint that is not visible.

blind nailing – nailing so that the nail heads are not visible on the finished surface.

blind single dovetail joint – a dovetail joint in which the dovetail passes only part way through the other member.

blind stop – a rectangular molding, part of a window frame, that acts as a stop for a storm window and/or a screen.

blind valley – a valley added on to an already framed and sheathed roof slope.

blister – 1) a portion of the gypsum drywall facing paper (or joint tape) that comes unbonded from the surface of the core (or the joint); 2) a paint defect in which a bubble forms under the paint surface, lifting the paint away from the material underneath.

block – 1) a building unit; 2) a housing containing one or more pulleys; 3) a piece of wood used as a spacer; 4) a portion of a subdivision or city divided by streets.

block and tackle – two or more blocks with pulleys connected by rope that are used to gain a mechanical advantage for lifting a heavy load. One fixed pulley serves to change direction, but does not give a mechanical advantage for lifting. A movable pulley, one that moves with the load, gives a two to one mechanical advantage, because as the rope is moved, the load moves only half as much as the rope. The force exerted is twice the force expended. In a block and tackle, the mechanical advantage is equal to the number of ropes that support the load, so heavy loads can be moved with more ropes, but very little force. The load will move much slower (in proportion to the mechanical advantage gained) than the ropes moves.

block bond – see stack bond.

block brazing – a brazing method that derives the necessary heat from heated blocks adjacent to the parts being brazed.

block bridging – bridging between joists using wood of the same size as the joist material and installed so that its vertical orientation is the same as the joists. See also bridging, solid.

blocking – horizontal stock of the same size lumber as the studs; the blocking is installed between studs at the approximate midpoint of the studs.

block, load – the assembly of a crane hook, with the swivel, sheaves, bearings, pins, and frame suspended from the hoist ropes.

block, long-type load – a load block in which a crane hook is mounted on a trunion connected to a housing, and the sheaves are mounted on the sheave pin.

block plane – a wood plane for cutting across the grain.

block sequence – a welding process in which layers of weld metal are deposited in a section built up along the length of a weld area, leaving gaps between sections. These gaps are later filled in with weld metal deposited to finish the weld. See also block welding.

block, short-type load – load block in which a crane hook and sheaves are mounted on the swivel.

block, upper – the sheave, bearing, pin and frame located on a crane trolley cross member that support the load block and load.

Block welding

Block welding involves the application of weld metal in several layers along sections spaced out the length of a weld. For example, a ten-foot-long butt weld between two three-inch-thick steel plates might be block welded using the following method:

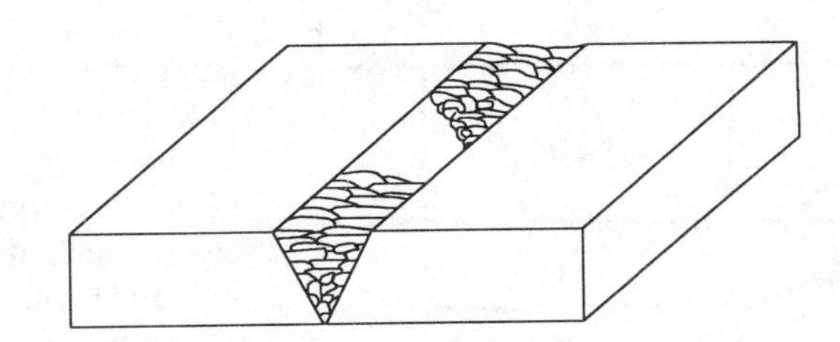

- A bevel is cut along the length of the weld on each of the plates so that a groove is formed when the plates are butted together. This groove provides access for the weld rod through the entire thickness of the plates.
- The bottom of the groove is tack welded in several places to hold the plates together during the welding process.
- Successive layers of weld bead are applied along a short portion, perhaps a 12-inch-long section, of the length of the groove. The layers may partially or completely fill the groove from the bottom to the top of the base material.
- A space is left, and then successive layers of weld bead are applied to the next 12-inch section, and repeated along the length of the plates.
- Once these short sections have been block welded, the remaining gaps between the block welds may be filled in with weld metal.

block welding – the application of weld material in layers along sections of the weld. *See Box.*

bloom – 1) multi-colored appearance of a painted surface that has not completely dried. It is a temporary condition caused by excessive humidity; 2) an efflorescence on a masonry wall.

blowdown – 1) the pressure drop in a pressurized vessel from the opening of a safety relief valve to the resetting (closing) of the valve; 2) the partial venting/draining of the water side of a boiler under pressure to remove or decrease unwanted contaminants.

blown oil – a vegetable oil that has been saturated with air to increase the viscosity of the oil.

blowpipe – a brazing/soldering torch with a small flame that can be precisely directed for use in jewelry making and similar work having fine joints.

blowtorch – a torch that uses pressurized fuel and atmospheric air to generate a flame used for soldering or similar applications.

blue lead – a combination of lead sulfide and carbon used as a rust preventative.

blueprint – a term used to denote a construction drawing. It is so named because the original common mode for reproducing such drawings resulted in a blue drawing with white lines.

blue stain – a discoloration of sapwood caused by fungi.

blue top – an earth grading hub or stake with a blue-painted top used to mark fine grading or trimming. See also guinea.

blunt – thick-edged; dull.

blushing – a defect in clear finishes caused by excess humidity which whitens or clouds the finish.

board – a piece of wood that has been sawn or otherwise milled to a final regular shape. A board is nominally less than 2 inches thick and 1 inch or more wide.

Bogie

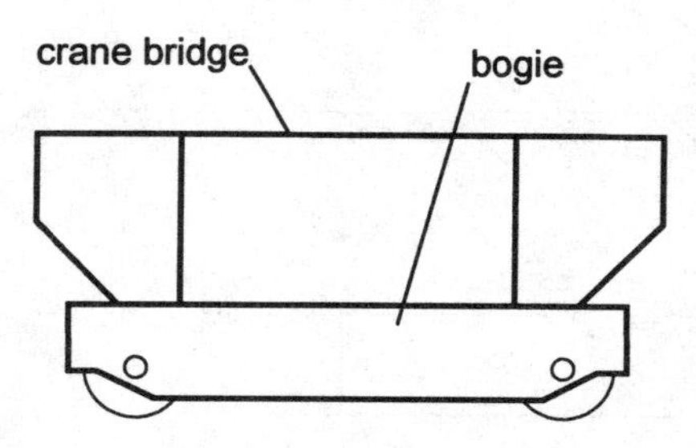

A two-axle truck, with two wheels per axle and a short wheel base or short distance between wheel axles, used to support the ends of the girders of a bridge crane. It has steel wheels that ride on the rails and permit the movement of the bridge along the rails. When more than four wheels are required on a crane, a bogie is used at each end of each girder to reduce the load on individual wheels. There are various types of bogies for specialized needs. An equalizing bogie is flexibly connected with a pin to one bridge girder and used to support the girder ends of a bridge crane. The pin connection allows the truck to rotate about and equalize the load on the wheels. A fixed bogie is rigidly connected to one girder or connecting member, with a flexible end tie between girders to equalize wheel loads. It is used to support the girder ends of a bridge crane, but only where the rails are in very good condition, making it unlikely that the wheels of the bogie would be unevenly loaded.

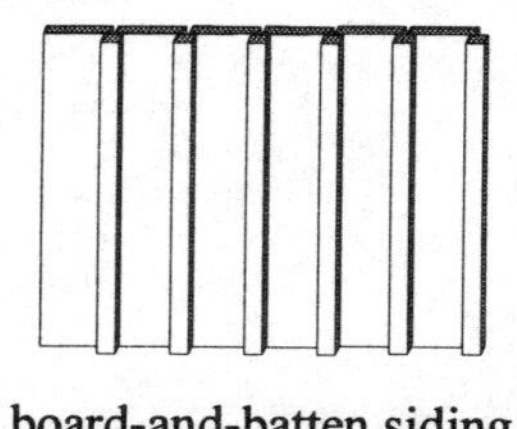
board-and-batten siding

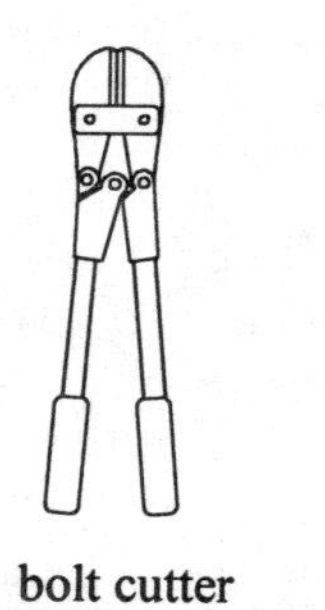
bolt cutter

board-and-batten siding – a type of vertically installed building siding using wide boards with narrow boards covering the joints between the wide boards.

board foot – a lumber measurement unit that is 12 inches wide by 12 inches long by 1 inch thick, or a combination of dimensions that equal the same volume. The number of board feet can be calculated by multiplying the thickness in inches by the width in inches, dividing the total by 12 inches, and then multiplying that by the length in feet.

board measure – lumber measurement using board feet as the measurement units.

board-on-board siding – a building exterior siding technique where gaps slightly narrower than the boards are left between boards; these gaps are then covered with boards the same size as the first boards which are nailed on so as to span the gaps.

boardwalk – a wide walkway made of boards and often elevated on piers. A boardwalk is usually located at a beach or along a sandy shoreline.

boaster – a wide chisel used for cutting brick or stone. See also bolster.

bodied linseed oil – thickened linseed oil, used in the manufacture of paints or as a treatment for wood surfaces, especially wood furniture.

body – 1) the viscosity of a liquid; 2) the longest blade of a framing square.

bogie – a two-axle truck used to support the ends of the girders of a bridge crane. *See Box.*

boiled linseed oil – linseed oil with additives to make it harden faster.

boiler – a sealed tank connected to a heat source used to convert water into steam for heating or power purposes. There are a variety of heating methods, sizes, and applications of boilers. *See Box.*

bolection – a heavy raised molding that is applied to a door surface to create the appearance of a carved surface.

bolster – 1) a short horizontal beam on the top of a column to support other beams; 2) a wide chisel used for cutting brick or stone, also called a boaster, especially when used for stone; 3) see high chair.

bolt – 1) a fastener that is threaded with machine screw threads so that a nut can be threaded onto it. It has a head or socket at one end that can be gripped by a wrench; 2) a short section of tree trunk; 3) a short log used for making veneer; 4) a sliding lock.

bolt cutter – a hinged cutting device with cutting jaws operated by long handles which provide leverage; used to cut metal rod, cable, bolts, etc.

Boiler

A boiler is a sealed tank connected to a heat source used to convert water into steam for heating or power purposes. Boilers of the 19th and early 20th centuries were fire-tube type boilers. The water was contained in a vessel with the tubes passing through the water. Heat from the firebox was passed through the tubes, eventually turning the water to steam in the upper part of the boiler. Make-up water was added at the bottom to replace that converted to steam, so that there was a continuous source of power. In the early days, the heat in the firebox was generated by burning wood. Coal was introduced later as an efficient and economical source of fuel.

Even though boilers provided an efficient supply of heat and power, water and steam under pressure also created a potentially dangerous situation. Failure of the vessel wall within the boiler would suddenly release the pressure. The hot water, with the pressure relieved, instantly turned to steam, expanding to many times its former volume. In effect, it created an explosion, causing damage to the area and serious injury or death to people in the vicinity. Working around boilers was risky business.

The invention of relief devices, called safety valves, made steam safer for common use. A safety valve is a spring-loaded valve that is adjusted so that it will lift off its seat at a predetermined pressure. The opened valve releases the pressure building up within the boiler. The predetermined pressure setting is far below the potential rupture point of the boiler shell in order to protect it from an uncontrolled or explosive release of pressure. Fire-tube boilers have been used to power railroad locomotives and steamships, provide motive power for machines in factories, power the prime movers used to rotate generators for making electricity, and for heating systems in buildings.

The water-tube boiler was introduced later in this century. As the name implies, it contains water in tubes surrounded by heat from the combustion in the firebox. If a water tube bursts, generally no serious damage occurs. The tube can be plugged and the boiler restarted and used until it is convenient to repair the damaged tube. Boiler start-up is faster with the water-tube boiler, as the amount of water surface exposed to the heat is much greater than in the fire tube design. However, if the source of combustion is lost, the fire-tube boiler has much more reserve heat and can continue to generate steam longer than the water tube type can.

Both types of boilers are used today. Coal, natural gas, oil, geothermal sources and heat resulting from the combustion of another machine, such as a diesel or gas turbine engine, are used as fuel. Steam power is used commonly today as a motive force for generating electricity, as well as to drive many of the world's large ocean-going ships.

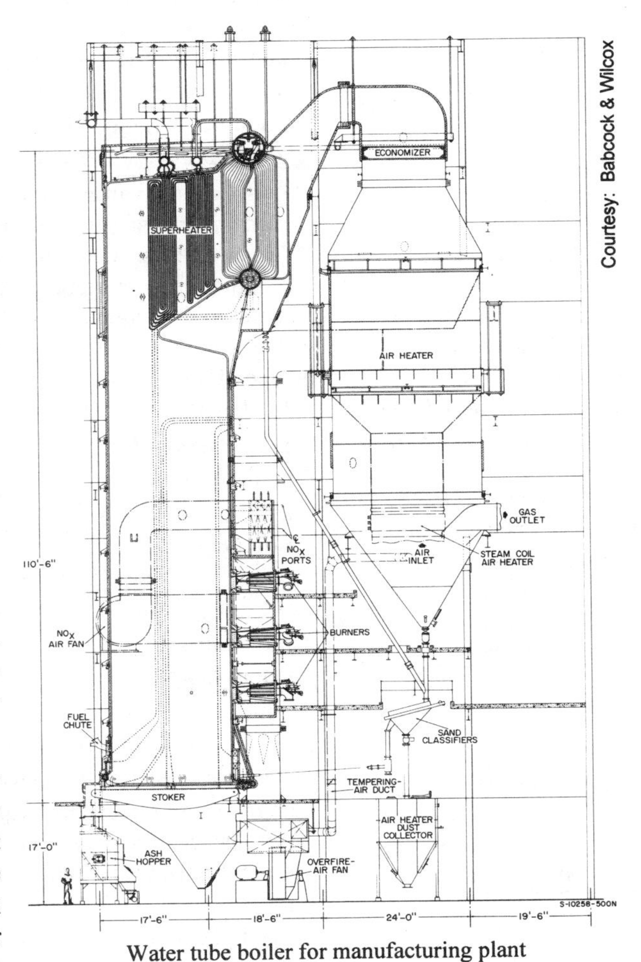

Water tube boiler for manufacturing plant

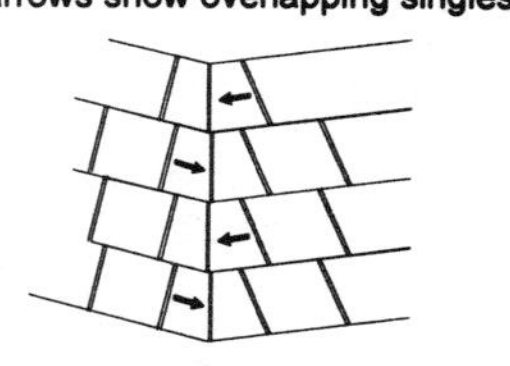

Boston hip

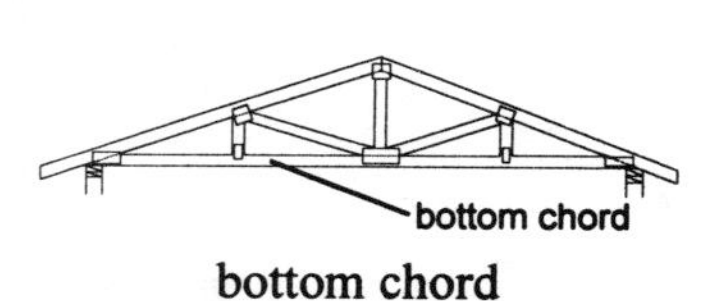

bottom chord

bolt extractor – a thin, tapered bolt-like rod with a very steep left-hand thread used to remove bolts that have broken off in a material or object. The bolt extractor is inserted into a hole drilled in a broken-off bolt and turned counterclockwise to extract the broken-off bolt.

bona fide – in good faith (Latin); made in good faith without fraud; genuine.

bond – 1) an arrangement of masonry blocks, bricks, etc.; 2) an adhesion between objects; 3) tying masonry wythes together by overlapping the units or by metal ties; 4) joining metallic conductors to provide a path for electric current; 5) a type of insurance against default or fraud; 6) a length of cord for tying items together.

bond beam – horizontally reinforced courses of masonry units mortared together or reinforced with concrete.

bond coat – a primer coat to improve the bond between the final coat and the material being coated.

bond course – a row of masonry units in a masonry structure.

bonder – see header.

bonding jumper – an electrical conductor connecting two metal parts together to complete an electrical circuit between the parts, such as connecting both parts to the same electrical ground.

bonding wire – a wire that grounds electrical boxes back to the service entrance. Also called bond wire.

bond line – the plane along which materials are bonded together.

bond stone – a stone that extends through the thickness of a stone wall, tying the wall together.

bones – a slang term for rocks which have come to the surface of an aggregate base.

boom – 1) a projecting beam used for lifting or guiding; 2) the main lifting structure of a backhoe mechanism.

boot – a lath set near a hub stake or marker that is used to sight grades when obstructions block the view between hub stakes.

boot truck – an asphalt tank truck that is used to spray a tack coat of asphalt on old asphalt surfaces before new asphalt is placed. The tack coat consists of emulsified asphalt and ensures a bond between the old and new asphalt surfaces.

Bordeaux mixture – a copper sulfate, lime and water compound that is used as a fungicide.

border – an outside edge, or trim.

bore – to drill or cut a hole.

bored deadlock – see deadlock, bored.

boring bar – a bar that holds a boring bit for boring holes in metal. The bar is held by an arbor, or shaft, that is gripped in the chuck of a lathe or other motor-driven rotary machine. The boring bar is at a right angle to the arbor, and as the arbor is rotated, the bit in the boring bar cuts a circle in the material. The size of the circle is adjustable by sliding the boring bar in the arbor to change the distance of the cutting tool from the arbor.

boring jig – a device used to ensure the proper alignment of a hole that is being drilled. The jig maintains the desired position between the drill and the work.

boring log – a record of information about the soil from bore samples taken from an excavation site.

borrow site – an area from which soil is removed for use at another location. Also called a borrow pit.

boss – 1) a projecting rock formation; 2) a protruding ornamental fixture; 3) a stone that is placed and will later be carved; 4) a projecting pipe connector that is welded to a large diameter run of pipe for connecting a smaller pipe, an instrument, etc.

Boston hip – a roofing method with shingles butted tightly against one another at the hip and alternating shingle courses lapping over the edges of the opposing shingles.

bottom chord – the lowest structural member of a truss, a framed structure designed to act as a beam.

bottom plate – the board on which the bottoms of wall framing studs rest; also called a sole plate.

boulder – a rock of large dimensions.

bow – a warp in a panel or structural member.

bow compass – a device for drawing circles. It consists of two legs that are joined together with a hinge joint. The end of one of the legs is pointed and the end of the other leg holds a pen or pencil. The point of the one leg is held in one place and the other leg is rotated about the point to scribe a circle.

bow dividers – a device, similar to a compass, used for measuring and stepping off dimensions. It consists of two legs that are joined together with a hinge joint. The ends of both legs are pointed. The distance between the points can be adjusted and set at the desired dimension for measuring. Once set at a particular measurement, the divider can be used to transfer that dimension or to step off multiples of the dimension on another surface. The distance between the points will remain fixed until changed.

bow-tie paint pattern – a paint defect in which two well-covered circular areas are joined by a section that is poorly covered.

bow window, or bow-bay window – see bay window.

box – 1) a container; 2) an enclosure for electrical equipment and connections. See also junction box.

box beam – a hollow structural member built up of separate structural members joined along their lengths, with structural intermediate bracing between the members.

box column – a hollow column built up of individual structural members used in a vertical position to support a downward load.

box corner joint – two members with alternating straight fingers cut in their edges which interlock to form a joint. The two members are usually glued in place, producing a very strong joint.

box cornice – a closed-in cornice at the edge of a roof.

boxing – 1) pouring paint back and forth between containers to mix the paint; 2) a term for cornice trim work; 3) continuing a fillet weld around a corner of the pieces being welded.

box nail – a flat-head nail with a thin shank diameter used on thin material where a common nail, which has a larger shank diameter, might split the wood.

box wrench – a wrench designed to fit a specific nut size; the portion of the wrench which grips the nut is a closed circle and is placed over the top of the nut or bolt head. Also called a box-end wrench.

brace – 1) a hand-powered crank device used with a drill bit or auger to bore holes, or with a screw driver to turn screws. Some designs are equipped with a ratchet that allows a series of partial turns to be used, making work in tight places easier. There are also models with a shoulder pad to permit the application of more force behind the brace. Also called a brace and bit; 2) a stiffener, often fastened at an angle to the structure it is stiffening or strengthening. A brace can be made of a strong metal or it can be a structural member that is added to strengthen other structural members.

brace and bit – a hand-powered crank-type device used with a drill bit or auger to bore holes.

brace table – a table, located on the back of the tongue of a framing square, that gives lengths of 45-degree braces for various situations.

bracket – 1) a relatively small piece of material, often metal, used to support an object; 2) projections supporting a shelf; 3) an angle reinforcement.

brad – a finishing nail with a small diameter head and shank.

brad awl – a pointed tool used to make holes into which brads are inserted.

brad-point drill bit – a twist drill bit with a sharp point that can be forced into the work to start the drill at a precise position. The point prevents the drill bit from wandering.

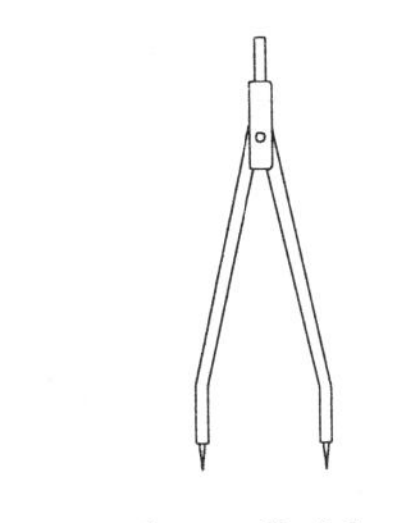

bow dividers

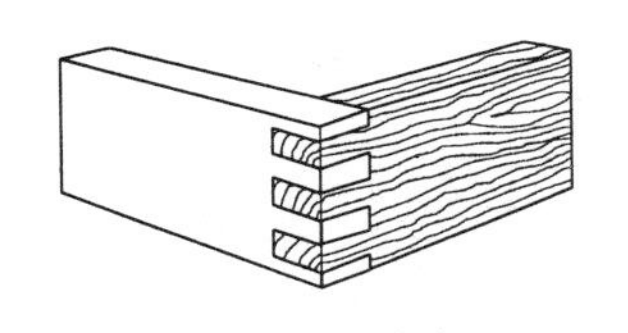

box corner joint

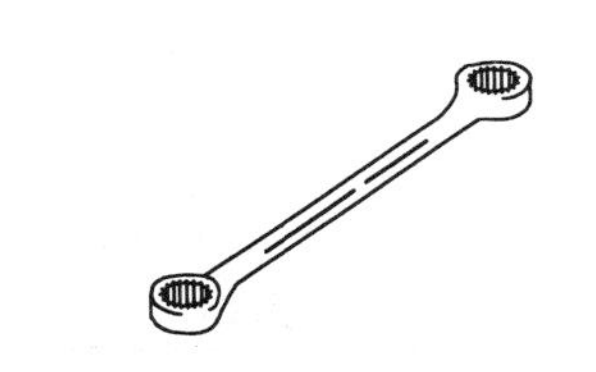

box wrench

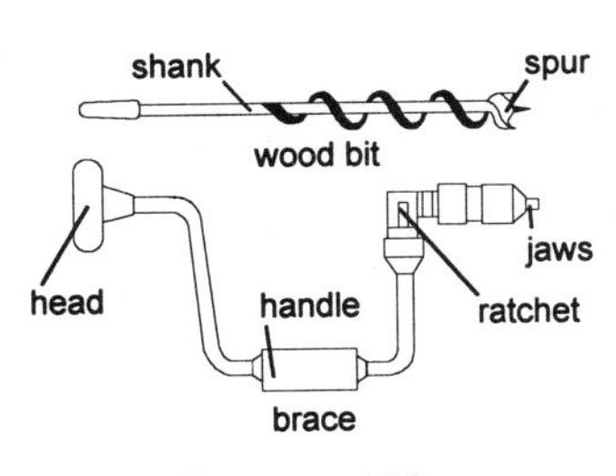

brace and bit

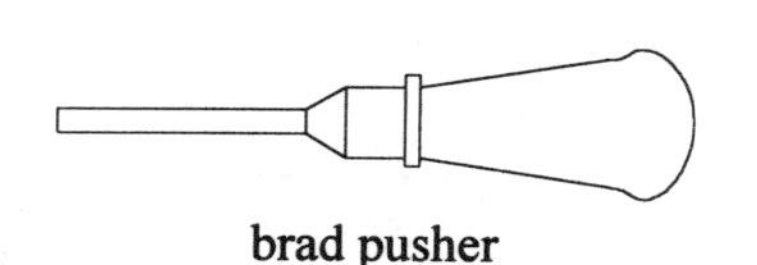
brad pusher

brad pusher – a tool with a magnetic shank with a recess in one end of the shank and a handle on the other. This tool is used to push a brad into a piece of wood.

brake – 1) a device to slow or stop motion, such as on the wheels of a vehicle; 2) a machine for bending sheet metal.

brake, disk – a type of brake in which friction is created by pads pressing against both sides of a rotating disk fixed to the wheel that is to be stopped or slowed.

brake, drum – a type of brake in which friction is created by shoes pressing outward against the inside of a rotating drum fixed to the wheel that is to be stopped.

brake, eddy current – a means to control motor speed by placing a supplemental load on the motor using the interaction of magnetic fields produced by a variable direct current in motor stator coils. Eddy currents are often used on power saws to rapidly slow the rotating blade when the power switch is turned off.

brake, holding – a lockable friction brake for a hoist.

brake horsepower – a measure of an engine's power equal to 33,000 foot-pounds per minute.

brake horsepower curve – a representation of the horsepower needed to operate a machine, such as a fan, under different conditions, different speeds, air flow rates, etc.

brake, parking – a lockable friction brake to hold a vehicle stationary.

brake, service – a momentary friction brake with force applied only as long as the operator desires it. An example is the foot brake on a vehicle.

branch – a part of a system stemming off the main run of the system, as in a plumbing branch.

branch circuit – the electrical wiring from a fuse or a circuit breaker to individual loads, such as receptacles or lights. An electrical system may be made up of several branches stemming off from the main power source.

branch circuit, appliance – a special 115 volt, 20 amp electrical circuit which permits the use of large appliances drawing relatively heavy current without tripping the circuit breaker. Most household circuits are 15 amps.

branch circuit, general purpose – 115 volt electrical wiring to lights and receptacles.

branch circuit, individual – an electrical circuit to only one appliance.

branch drain – a drain from a plumbing fixture to the main drain line.

braze – a type of high-temperature soldering which uses a bronze filler rod to make strong, durable metal joints. Brazing differs from soldering by using a filler metal with a higher (840 degrees F) melting point.

brick – a rectangular building block of nonmetallic inorganic materials that has been hardened by either heat or a chemical process. Common brick is made of burned clay, shale or clinker. See also listings under brick types.

brick, baby Roman – a brick measuring 2 x 4 x 8 inches.

brick bond – 1) the mortar joint between bricks; 2) the style or pattern in which bricks are laid out. See listings under individual bond types.

bridge crane – see crane, bridge.

bridging – braces, made from wood or metal, used to stiffen and distribute the load between joists, studs, or trusses.

bronze – a copper base alloy containing tin or other elements. Tin bronzes contain 5, 8, and 10 percent tin, and are called Alloys A, C, and D, respectively. Tin bronzes may also contain up to 4 percent phosphor to improve their casting qualities and elastic properties. Tin and phosphor bronzes are used for electrical terminals and springs. They are strong, resist corrosion and have good machinability characteristics. See also cupronickel.

bronze, aluminum – an alloy of copper and aluminum. Bronze alloys that contain 5 to 8 percent aluminum have high strength and corrosion resistance. Those containing 10 percent or greater amounts of aluminum have very high strength and remain plastic when hot. Aluminum bronzes are used for piping and valve materials in sea water, and for other parts that may be exposed to salt or brackish water.

bronze, silicon – an alloy of copper and silicon that is corrosion resistant, workable when hot, has high strength and is readily weldable. It is used for valve bodies and valve seats when both strength and corrosion resistance are required.

bronze, tin – see bronze.

builder's hardware – all hardware used in the construction of a building, such as joist hangers, brackets, structural straps, hinges, door locks, and so on.

builder's level – a device consisting of a telescope and bubble level mounted on a tripod base. The instrument is leveled with the aid of the bubble level, and the telescope is used to establish work lines and elevation points and measure horizontal angles using a horizontal vernier scale.

building – an enclosed structure for habitation, storage, manufacture, etc.

building code – the requirements to be met during construction as set forth by a recognized authority, usually the city in which the building will take place. The Uniform Building Code is a nationally recognized code and is widely used throughout the country. Laws in individual municipalities or other jurisdictions may vary slightly, or add additional requirements or modifications to the Uniform Building Code.

building drain – the main drainage system within a building.

building line – the outside edge of the rafter plate; the perimeter of the outside of the building.

building paper – a heavy reinforced waterproof paper used for installation around windows and doors, under siding, etc.

building permit – an authorization from the local building inspection agency to construct or modify a building from an approved set of plans.

building restrictions – limitations and regulations based on local building codes.

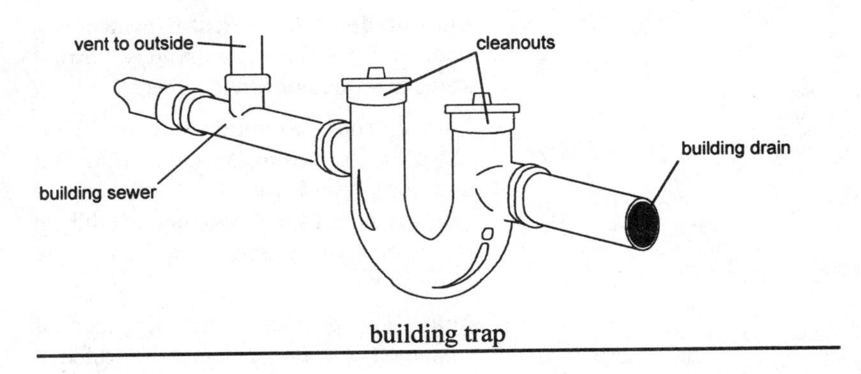

building trap

building sewer – see private sewer.

building storm drain – a drain for rainwater and other runoff.

building storm sewer – the line that connects the building storm drain to a disposal point, such as a public storm sewer.

building trades – the various skilled and semi-skilled workers involved in building and construction, such as carpenters, plumbers, and electricians.

building trap – a water seal trap, such as a P-trap, in a building drainage system line from the building to the main sewer line. This trap services the cumulative drain lines from the building. At one time, a building trap had to be placed in each building drain line. Today most codes do not require one in each line.

buildup – 1) increase in paint thickness from multiple paint coats; 2) any increase.

built-up – constructed of many parts.

built-up beam – a horizontal structural member made by fastening two or more pieces of lumber together for added strength. A built-up beam may be a composite of many lengths of material laminated together. The individual pieces of wood can be oriented so that their grain shapes and crowns are opposite. Grains often crown in one direction, so by alternating them you can eliminate the weak spots, making a glue-laminated wood beam a very stable structural member.

built-up column – a vertical structural member made by fastening two or more pieces of lumber together for added strength.

built-up door stile – vertical member of a door frame made of relatively narrow strips of wood bonded together.

built-up roof – a multiple layer roof, usually using succeeding layers of roofing felt and hot-mopped asphalt, covered with a final protective gravel coating. A built-up roof is most often used on roofs with little or no slope.

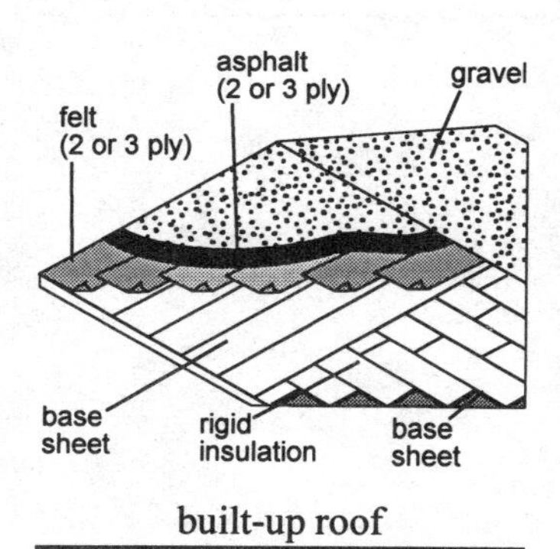

built-up roof

bulb – 1) a spherical or partially spherical shape; 2) a glass container in which a filament is heated by an electrical current in a vacuum causing it to glow and give off light.

bulkhead – 1) a structural wall; 2) an outside door that is either horizontal or at an angle over the entrance to a stairway into a cellar; 3) a structure located on a roof which encloses a stairwell access to the roof.

bulk unit weight – the weight of concrete aggregate that will fill a 1 cubic foot container.

bulldozer – a tractor which moves on treads, with an adjustable front-mounted scraper blade. The blade moves soil by pushing it in front of the vehicle and is used for scraping off, smoothing and/or leveling a building site.

bull float – a flat device with a long handle used to smooth wet concrete.

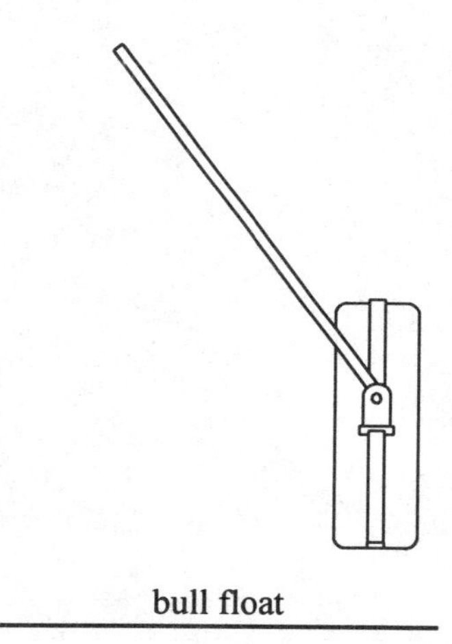
bull float

bull header – a masonry unit used to tie rows of masonry together.

bullnose – having a rounded edge.

bullnose plane – a tool used to shave and smooth wood surfaces with its cutting blade placed close to the front end so that it can be used in tight places, such as corners.

bullnose tile – tile that has one rounded edge. Bull nose tiles are used as the edge row in a tile installation.

bull pin – a tapered steel pin used for aligning holes in structural members.

bumper – a device designed to absorb shock between vehicles, or between a vehicle and a fixed object. A bumper has some means of moving to dissipate the force imparted to it. It may be made of a resilient material, or it may have hydraulic or pneumatic cylinders that permit the bumper to move in response to a force.

bumping – removing dents from sheet metal by striking it with a hammer; shaping sheet metal with a hammer.

bumping hammer – a hammer designed to be used to form sheet metal, or remove dents from sheet metal.

bundle – 1) two pieces of gypsum wallboard packaged face to face with a piece of paper tape along each short edge to hold them together; 2) a quantity of shingles or shakes held together by wire or paper for ease of handling; 3) any grouping of items that are bound together.

bunker – 1) a bin which is a container for bulk material; 2) a reinforced structure that is partially below ground.

burl – a section of a tree with a twisted and/or swirled grain.

burlap – a coarse jute or hemp fabric.

burned brick pavers – brick made from clay or shale and used for outdoor paving.

burner – a device for metering and mixing a fuel with air so that the fuel burns efficiently. Burners are found in many heating devices such as boilers, stoves and heaters.

burnish – to smooth and polish a surface by friction.

burnishing tool – a smooth metal tool for smoothing and polishing surfaces using friction.

burr – 1) rough projection(s) remaining on a cut edge; 2) a portable power mill for cutting metal.

bus – an electrical conductor for carrying large currents. Also called bus bar.

bus duct – electrical conductors mounted in a sheet metal enclosure for routing high current capacity through an area.

bush hammer – an impact device used to spall, chip and roughen a cured concrete surface in preparation for the next concrete pour. Each pour needs a rough surface in order to bond to the cured concrete.

bushing – a hollow cylinder used to increase an outside diameter, or decrease an inside diameter, and/or effect a better fit between cylindrical parts where one part fits inside the other.

busway – a metal enclosure containing factory-installed electrical conductors that may or may not be insulated, used for service and feeder equipment.

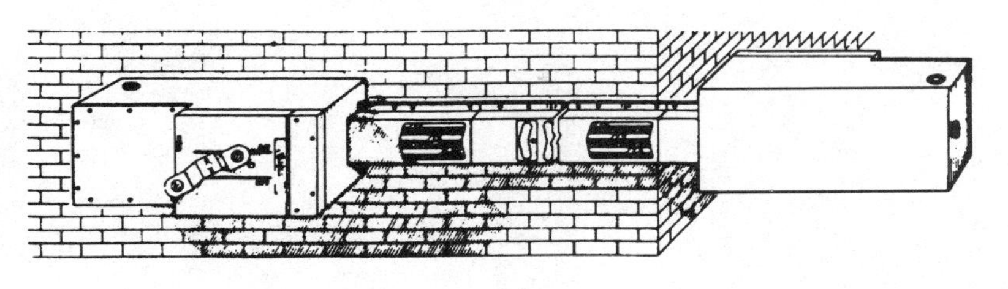
a busway system

busway plug – device to permit a power takeoff from a plug-in busway.

busway, plug-in – an industrial plant busway where high current feeder circuits with low voltage drop are needed to supply numerous power takeoff points.

butane – a flammable gas.

butcher block – an assembly of strips or pieces of hardwood laminated together.

butt – 1) the thick end of a shingle; 2) the large end of a handle or other object; 3) a large barrel.

butterfly bolt – see toggle bolt.

butterfly valve – a valve consisting of a square, rectangular, or round disc mounted on a shaft inside a body of the same shape; a 90 degree rotation of the shaft will move the valve from fully open to fully closed.

buttering – 1) applying mortar to masonry; 2) applying any bonding compound to the surfaces to be joined; 3) applying a layer of weld metal alloy to metal surfaces before the joint is welded so that the alloy in the weld zone is not diluted by the base metal being welded.

butt hinge – a hinge consisting of two rectangular metal plates which are joined together by a pivot pin.

butt joint – 1) the joint formed when two pieces of wood, gypsum wallboard or other material are butted together without an overlap; 2) the joint between two pieces of metal that have been welded together without an overlap.

buttress – an external reinforcement to a wall.

butts – butt hinges.

butt weld – a weld between two pieces of metal that are joined together without overlapping one another.

butyl rubber – a sealant made from polybutene in combination with resins, oils and solvents. It can be designed to accommodate movement of up to ten percent of the joint width for use in the construction industry.

buzz saw – a circular saw.

BX cable – a trade name for a type of residential electrical wiring in which the wire bundle, consisting of individually insulated conductors, is covered by a flexible spiral-wound metal armor. Also called metal clad (MC) cable.

bypass – a piping line or electrical wire that permits a fluid or current to flow around an object rather than through it.

bypass door – see door, bypass.

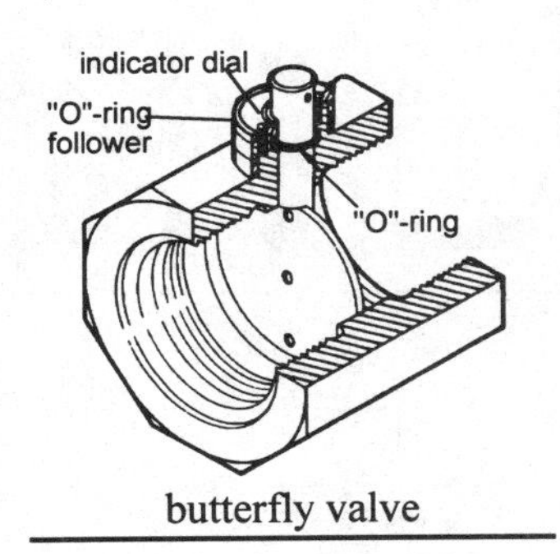

butterfly valve

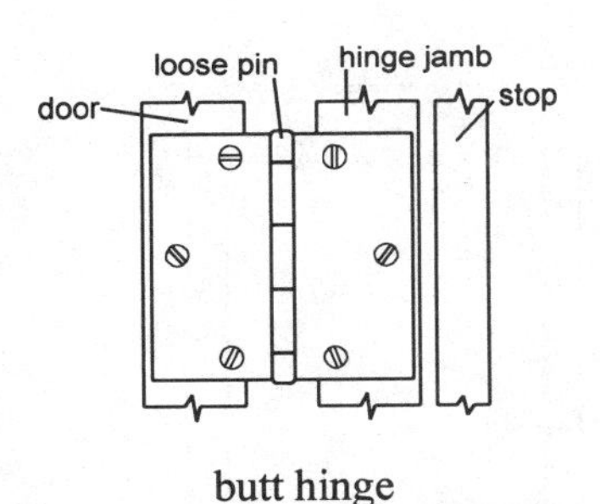

butt hinge

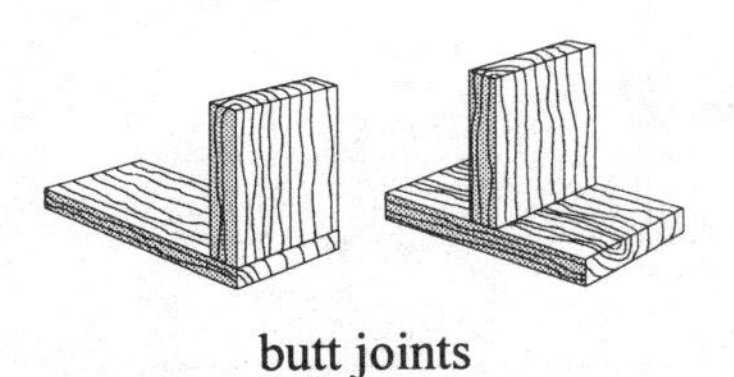
butt joints

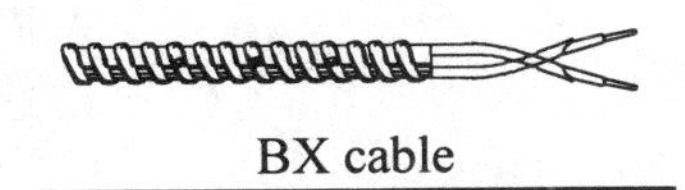
BX cable

C: cab to cypress, red

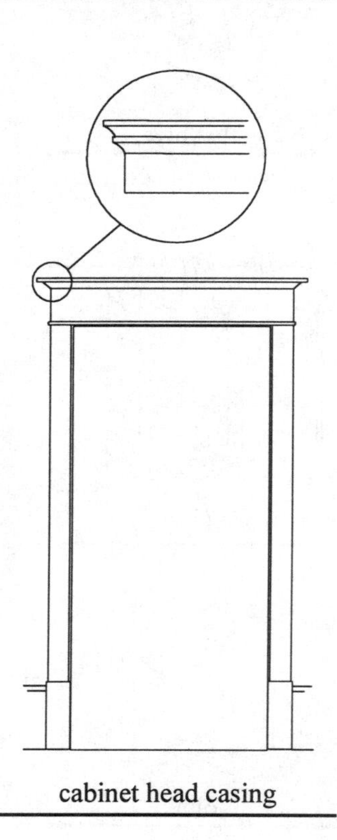
cabinet head casing

cab – an enclosure or housing for the operator of a piece of machinery. The cab is part of the machine and moves with it.

cabin – a small, often simple or rustic dwelling.

cabinet – 1) an enclosure, usually for storage; 2) a shower stall or other specific use enclosure.

cabinet, base – a storage unit or enclosure that is installed on the floor and often serves as a base for a countertop. Standard base cabinets are 34½ inches high, not including the countertop, and 24 inches deep, not including the thickness of the cabinet door.

cabinet, blind corner – a cabinet that fits into a corner with cabinets abutting it on both of the intersecting walls. Access to a blind cabinet is very limited. Often it is designed with a face that is at a 45 degree angle so that the shelve space is more functional. Another means of improving the usefulness of a blind corner cabinet is by installing a lazy Susan section inside so that materials can be stored on revolving shelves.

cabinet combination saw blade – a circular combination saw blade used in a power saw for making smooth cuts when both ripping and cross-cutting wood.

cabinet head casing – an elaborate trim molding for a door or window consisting of two or more pieces of molding joined together along their lengths. This gives the effect of a single complex molding.

cabinet, wall – a storage unit or enclosure that is installed off the floor and against a wall, usually in a kitchen or laundry room. Standard wall cabinets are 12 inches deep, but vary in width and height. Kitchen wall cabinet units are installed about 18 inches above base cabinets and may extend to the ceiling or stop six or more inches short of the ceiling. A soffit may be built above the cabinets to fill in the empty space or to conceal indirect lighting. Often the space above the cabinets is left open and used as a shelf to display decorative items. Laundry room wall cabinets are mounted 18 inches above washer/dryer units or the laundry tub. The cabinets are mounted with screws through the cabinet backs into the wall studs and their weight is cantilevered from the wall.

cable – 1) two or more electrical wires within a common insulator cover; 2) heavy rope, often of stranded metal.

cable box – an electrical box which protects wire terminations and connections and serves as a mounting device for outlets and switches. *See Box.*

cable box connectors – fittings through which electrical cables can be passed into cable boxes. The fittings provide a smooth passage past the edge of the box and are sometimes designed to grip the cable.

cable cutter – a device similar to a bolt cutter, but with jaws having curved cutting edges. The cutter has long handles for leverage and is designed for cutting metal rope-type cable.

cable plan – a drawing that shows cable routing, electrical terminations and the number of wires for each box. The drawing is made by the architect, building designer or electrical system designer for the electrician to use in wiring the building.

cable, shielded – one or more insulated electrical wires surrounded by a woven metal sheath. The sheath provides electromagnetic shielding.

cable tap box – a device that permits the connection of an electrical cable to an electrical busway, the enclosure for the electrical bus bars or rods. The cable top box mounts on the busway and provides a secure entryway for an electrical cable to enter the busway.

cable tray – a tray designed to support electrical cables where many cables are run side by side. The cable tray is supported by the structure and becomes a permanent part of the structure where it is used. There are different types of cable trays to fit different requirements. One has a solid bottom and side rails; another has a ladder-type bottom and side rails.

cable tray support – structures to take the weight of, and otherwise restrain, cable trays.

cabriole – a style of furniture leg that curves down to an enlarged and rounded foot, which is often shaped like an animal foot or claw. This type of furniture dates back to the early 18th century.

CADD – see computer aided design and drafting.

Cable box

An electrical box in which wire terminations or connections can be made and to which devices, such as outlets or switches, can be mounted. The boxes are available in a variety of sizes, styles and materials for different applications. They are made of metal, usually galvanized steel, plastic or fiberglass. Non-metallic boxes may have a tab in each cable entrance point that is sprung to one side when the cable is pushed into the box. Once through the opening, the tab then grips the cable, preventing it from sliding back out. Some boxes have screw clamps that grip the cable where it enters the box, and others have knockouts designed to accept a cable bulkhead connector. The bulkhead connector is held in the knockout hole with a nut. When installed in new construction, cable boxes are usually mounted directly to the studs. Some boxes come with nails already mounted for easy installation. In existing construction, the boxes may be supported by the wall covering. Boxes may be ganged, or joined together, to make larger boxes in situations where a number of switches are to be mounted at the same location. The *National Electrical Code* specifies the number and size of wires that can enter a box of given dimensions.

cage – 1) a framework of rebar wired together; 2) an assembly of vertical metal bars surrounding an area or object, such as a ladder, for the safety of the worker on the ladder; 3) an enclosure of slats or bars.

caging – framing around pipes or other protrusions in a wall surface.

caisson – 1) a recessed ceiling panel; 2) a watertight box used to surround work, such as a building or bridge foundation, which is being constructed below water level.

caisson, bell – a caisson pile with a 60-degree flare at the bottom which provides more surface bearing area.

caisson, bored – a large diameter caisson pile set into a deep hole; a 10-foot diameter hole, 150 feet deep is practical for cohesive soils. Used to support the foundation of a structure built on soil that requires reinforcement in order to support the weight of the structure.

metal

plastic

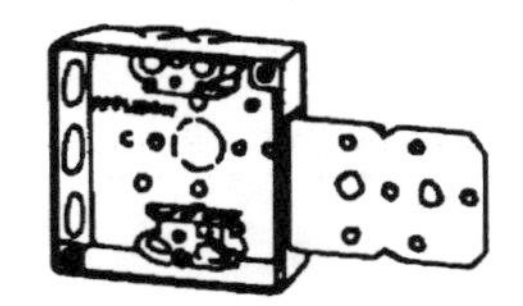
internal within box

cable connectors

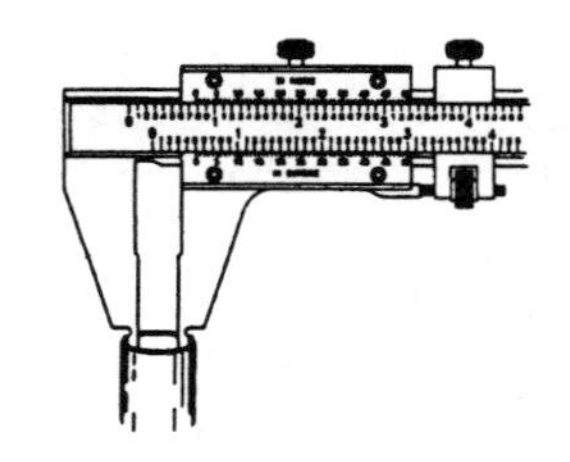

caliper, inside

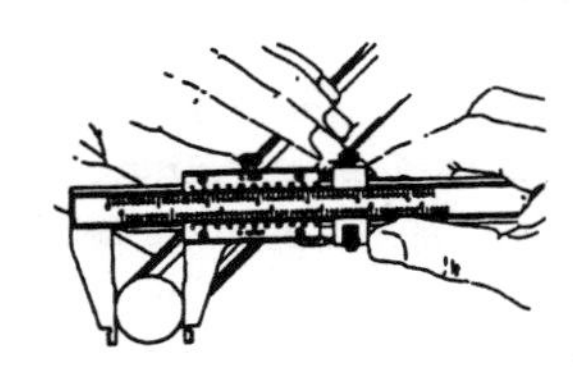

caliper, outside

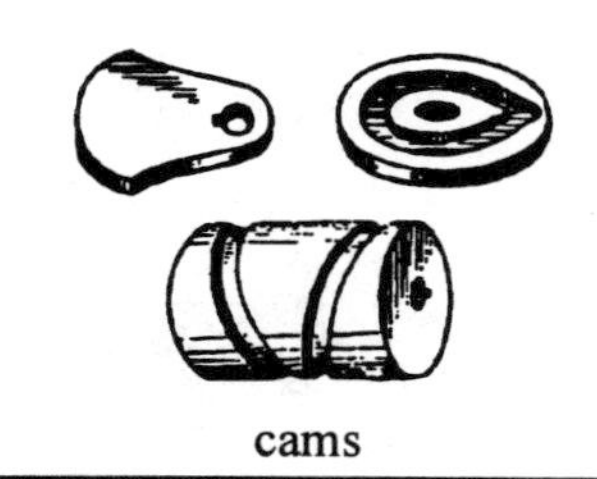

cams

caisson, jacked-in – a hollow cylinder pile forced into the soil with jacks while it is rotated back and forth to reduce the friction with the soil. Once the cylinder is in place, the soil in the center is excavated, and then it is filled with concrete for added structural strength.

caisson pile – a large concrete or reinforced concrete piling sunk into the ground or into the bed of a body of water to support a structure such as a building or bridge. Caissons are made of cast-in-place concrete poured into a cylindrical hole in the ground or poured into a hollow cylinder that has been driven into the ground with a pile driver. Drilled-in caissons have a steel shoe welded to the bottom of the cylinder to reinforce the end.

caisson, slurry – a caisson pile made by digging a rectangular-shaped vertical pit and filling it with concrete while a cured concrete slurry holds the walls of the excavation in place.

caking – the solidification of paint pigments in the bottom of a can of paint.

calamander – a hard wood originating in India and Ceylon used for cabinets.

calcimine – a type of white or colored wash used for coating surfaces, such as interior walls.

calcination – the process of breaking down chemical compounds by heat.

calcine – to heat gypsum (or other material) to a point where the water is released from the crystallized gypsum.

calcium – a white metallic element that forms part of limestone, chalk and gypsum; atomic number 20; atomic weight 40.08; atomic symbol Ca.

calcium chloride – an admixture used with concrete or mortar to decrease the time it takes to set up.

calendared paper – paper with a hard, smooth finish.

calibrate – to adjust to match a standard.

California – a gable roof end used to close off the end of an overshooting roof ridge.

California ranch style – a one-story house with a low pitched roof patterned after the early Mexican ranchero style.

caliper – 1) a tool with two legs joined at one end, one of which is adjustable and may be moved in relation to the other, used to measure a dimension, such as an inside or outside diameter; 2) thickness or depth.

caliper, inside – a caliper with the ends of the legs curved outward for measuring inside dimensions.

caliper, outside – a caliper with the ends of the legs curved inward for measuring outside dimensions.

caliper, vernier – see vernier caliper.

calorie – a measure of the heat required to raise 1 gram of water, 1 degree C, to 15.5 degrees at atmospheric pressure.

cam – a shaft or disk with a shape other than round. It can impart a rocking motion, or motion away from the centerline of the shaft on which it is mounted, to a part in contact with it. A cam is used in a machine to change rotating motion to reciprocal motion. For example, in the valve train of an automotive engine, a series of cams mounted on a camshaft operate the engine valves by bearing directly or indirectly on the ends of the valve stems. As the camshaft is rotated, each cam forces a valve stem away from the camshaft centerline until the cam is at the maximum point. Further rotation of the cam beyond this point allows the end of the valve stem to move back toward the camshaft centerline, assisted by a spring. Cams are also used to actuate switches. The rotating of the cam can be sequenced to coincide with some other action. A cam mounted on a shaft can be geared to the rotation of a motorized valve operator and adjusted to open a switch to cut off electrical power to that operator when the valve has reached its fully open or closed position. Such switches are called limit switches.

camber – 1) an orientation that forms an angle with respect to true vertical or plumb; 2) a deviation from trueness to compensate for a load on a structure, such as a horizontal structural member with a slight upward bow (camber) when installed. When the designed load is put on this member, the load pushes the bow down, making the member straight. If the structural member did not have a camber, when the load was placed on it, it would sag.

cambium – the soft wood layer directly under the bark of a tree.

came – a thin lead bar with grooves that hold pieces of stained or other decorative glass. It is used in the making of stained or leaded glass windows.

campanile – a bell tower with openings around the top so that the sound of the bells is not restricted.

camshaft – a shaft with one or more cams along its length. Motion can be imparted to an object that is in contact with a cam on the camshaft as the shaft is rotated. For example, a camshaft in an internal combustion engine is used to impart linear motion to the valves to open them.

candlepower – a measure of light intensity based on a burning wax candle and the approximate corresponding amount of light given off by a light bulb. *See Box.*

canopy – 1) a covering, usually of fabric, extended from the wall of a building to shelter an entrance or window. An awning; 2) a covering suspended over a bed.

cant – 1) a tilted surface; 2) a slant; 3) a sudden overturning movement.

cant brick – a brick that has a taper on one side.

cant hook – an assembly consisting of a hook on a pole or rod used for turning logs.

cantilever – a structural member or beam supported on only one end that projects out from a wall or other structure.

cant strip – a strip of wood placed under shingles or boards to hold them at an angle to the surface to which they are attached.

cap – 1) the top part of a column, pilaster, etc.; 2) the uppermost part; 3) a closure; 4) the top of a wall.

capacitance – a measure of the amount of electrical charge that can be stored by a capacitor.

Candlepower

A measure of light intensity based on a burning wax candle and the approximate corresponding amount of light given off by a light bulb. A moderate size 25-watt light bulb has a light intensity equivalent to 25 candles. The light from an unshaded bulb or other light source spreads out equally in all directions where the light energy is not impeded. The area on which the light energy shines increases as the square of the distance of that area from the source. This means that the light will shine on an area of 25 square feet if that area is 5 feet from

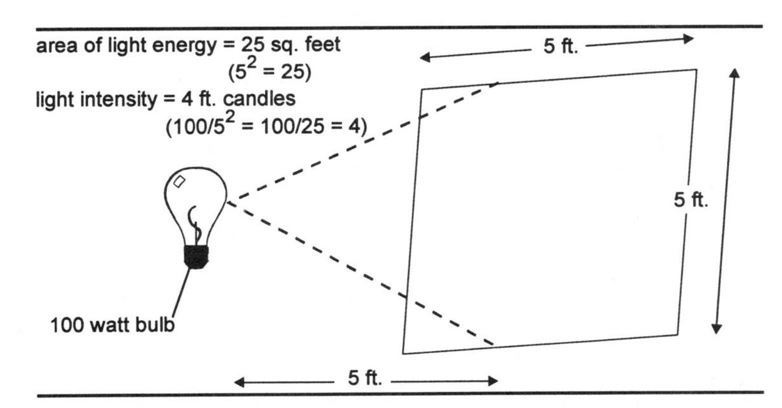

the source. As the area increases, the light intensity decreases at the same rate. This light intensity is sometimes measured in foot candles. The number of foot candles of light shining on an area, at right angles to the area, is a function of the intensity of the light source divided by the number of feet squared that the area is from the source. For example, if an area on which light shines is 5 feet from a 100 candlepower light, the light intensity shining on that area is equal to 4 foot candles ($100/5 \times 5 = 100/25 = 4$).

capacitive reactance – the opposition of a capacitor to changes in voltage, measured in ohms.

capacitor – an electrical device that can store an electrical charge. It consists of several layers of electrical conductors in the form of flat sheets separated by layers of dielectrics, or insulators. Also called a condenser.

cap, acorn – a metal cap, with a rounded top resembling an acorn, that is fitted to the top of a metal fence post for decoration and weather protection.

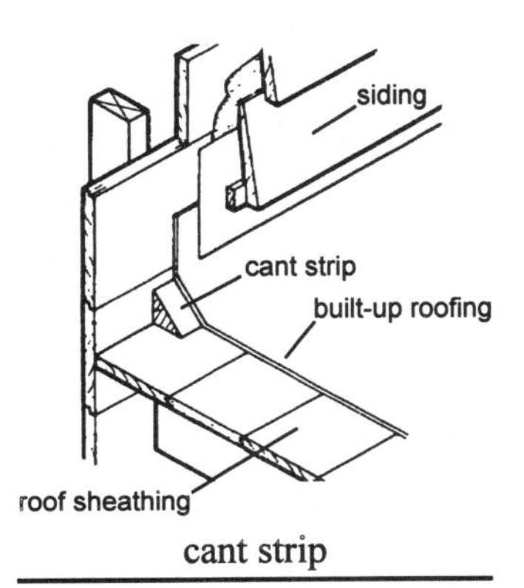

cant strip

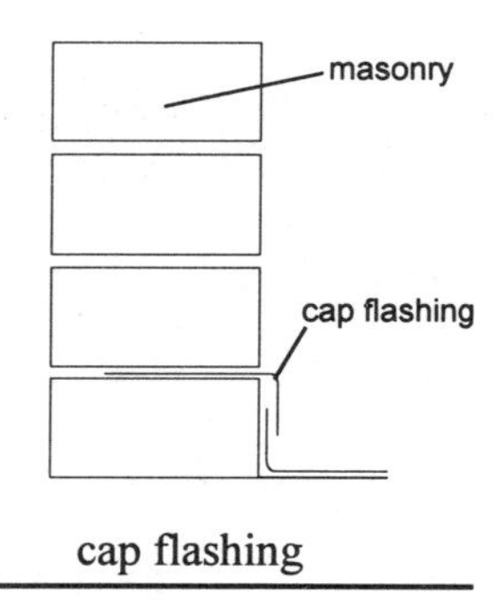

cap flashing

cap block – a flat, thin masonry unit used to cap off a wall. Also called a paving block.

cape chisel – a metal chisel with a cutting edge that is narrower than the body of the chisel, used for cutting slots.

cap flashing – flashing set in mortar and bent down to overlap another piece of flashing.

capillary action – the movement of a liquid into small spaces by molecular attraction.

capital – 1) the top portion of a column or pillar; 2) net worth; 3) accumulated possessions which will bring in income.

capital expenditure – the total amount of money or assets invested in a business venture.

capping brick – brick made as a finishing layer for the top of a wall or other brick structure.

cap sheet – the top layer, often covered with a protective mineral coating, of a built-up roofing system.

capstan – a windlass, or hoist, consisting of a cable or rope for lifting which is wound around a drum. A crank or motor is attached to the drum to turn it, and raise the load. A capstan has a vertical axis of rotation, and is used to wind or unwind a cable or chain with a load attached.

capture velocity – the air velocity needed to draw contaminants into the filter of a ventilation system.

capturing hood – a hood with enough air suction to draw in fumes from outside the area of the hood. The fumes are then exhausted out of the area by the ventilation system. Capturing hoods are used in industry and in laboratories to control fumes for safety and/or health reasons. Fumes captured in the ventilation system can be processed and neutralized to remove harmful vapors before they are released into the atmosphere.

carbarn – a large garage for buses, streetcars, etc.

carbide tipped – a steel tool with a tungsten carbide cutting edge, usually made by brazing a piece of tungsten carbide metal to the end or edge of the steel tool. The ends of drill bits or the teeth of a saw blade are examples of tools that are carbide tipped for a long-lasting edge.

carbide tipped saw blade – a circular saw blade for a power saw with carbide tips that are slightly wider than the thickness of the blade. This added thickness allows the teeth to cut freely without the blade binding in the work. Also called a tipped tooth blade.

carboloy – see tungsten carbide.

carbon – an element that combines with other elements, and exists in pure form as graphite or a diamond; atomic number 6; atomic weight 12.011; atomic symbol C.

carbon arc cutting (CAC) – process in which metal is cut by the heat of an electric arc between the base metal and a carbon electrode.

carbon arc welding (CAW) – a welding process using an electric arc from a carbon electrode to the work as the heat source, with or without the addition of filler metal.

carbon arc welding, shielded (CAW-S) – an electric arc welding process using a carbon electrode shielded by a blanket of flux on the work or the combustion of a solid material or both.

carbon arc welding, twin (CAW-T) – an electric arc welding process using an arc between two carbon electrodes as the heat source with no shielding, and with or without pressure and/or filler metal.

carbon electrode – carbon in the form of a rod used as an electrode.

Carborundum – a trade name for a tough abrasive made of alumina, silicon carbide, and other materials.

Carborundum cloth – a fabric embedded with Carborundum-type abrasives, used for smoothing and finishing.

Carborundum finish limestone – see limestone, Carborundum finish.

Carborundum stone – a manufactured abrasive stone using carbon-silicon abrasives as part of its formulation.

carboy – a large glass bottle which is often cushioned in a special container.

carburize – the process of adding carbon, for hardness, to the surface of steel by heating the steel in contact with carbon. Also called case hardening.

carcinogen – a substance known to cause cancer.

carnauba – a type of hard wax from the leaves of the wax palm plant. It is one of the primary ingredients in paste wax, used to coat surfaces such as wood floors.

carpenter – a person who builds with wood.

carpenter's pencil – a flat pencil with a wide durable lead, used by carpenters for marking measurements on the job.

carpenter's square – see framing square.

carpentry – the art or trade of building with wood.

carpet – a floor covering made of natural or synthetic woven fibers.

carpet flange – a decorative edge or trim piece around an electrical box that is set flush with the floor.

carpet pan – an electrical junction box extension in an underfloor raceway designed to bring the junction box cover up to the level of the carpet or other flooring.

carport – a shelter for cars that has a roof, but not necessarily side enclosures.

carriage bolt – a bolt with a smooth, partially spherical, domed head and a short section of square shank under the head designed to prevent the bolt from turning.

carriage, built-up – stair stringer made of triangular blocks fastened to a continuous piece of lumber.

carriage, sawed-out – stair stringer cut with steps so that the stair treads can rest on them.

carriage, stair – stringer to which stair treads are attached.

cartouche – a convex and rounded decorative trim.

cartridge fuse – a fuse with terminals on each end.

cartridge fuse, ferrule type – a cartridge fuse in which the end terminals are the end caps of the fuse.

cartridge fuse, knife blade type – a cartridge fuse in which the end terminals have flat, rectangular-shaped blades extending from the end caps.

cartridge fuse, replaceable link type – a cartridge fuse in which the internal fuse element may be replaced.

carve – to shape by cutting.

carving – forms cut into, or from, a material for ornamentation.

caryatid – a column shaped in the form of a female figure.

cased opening – a finished interior opening in a building wall without a door.

case hardening – the process of carburizing, combining or impregnating a ductile material, such as a metal, with carbon. The heated material is exposed to the carbon, then hardened by quenching, or slow cooling, reheated and quenched again. This process is used on tools which require a hard surface, but need to retain the resilience of a soft metal so they won't crack or break during use. It is also used where it is easier or more economical to shape a softer material and carburize it to add the required wear resistance to the tool than to use a harder metal. The center of a lathe tailstock as well as the cutting tools used for metal lathes or milling machines are often carburized for wear resistance. Another application would be strengthening the edge on a tool which must remain sharp to be effective, but be resistant to impact blows, such as a chisel. There are different methods of case hardening. *Pack carburizing* uses solid carbon, or another carburizing agent such as barium carbonate and/or coke. The metal is heated to a predetermined temperature, usually between 1550 and 1700 degrees F, in contact with the carburizing material. The metal must be carefully packed in the carbon source as it will only be impregnated with the carbon where it is in direct contact with the carbon. The metal is quenched in water or oil. This method has high labor costs associated with the placing and unpacking of the material being treated. *Gas carburizing* uses such gases as carbon monoxide, methane, ethane or propane as

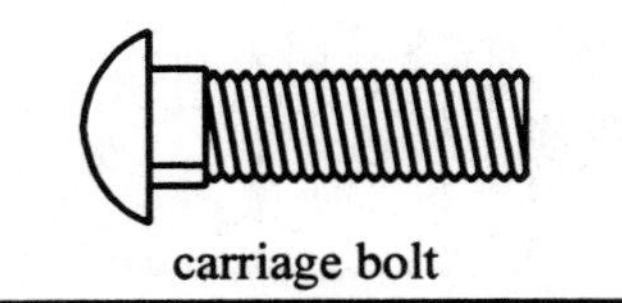

carriage bolt

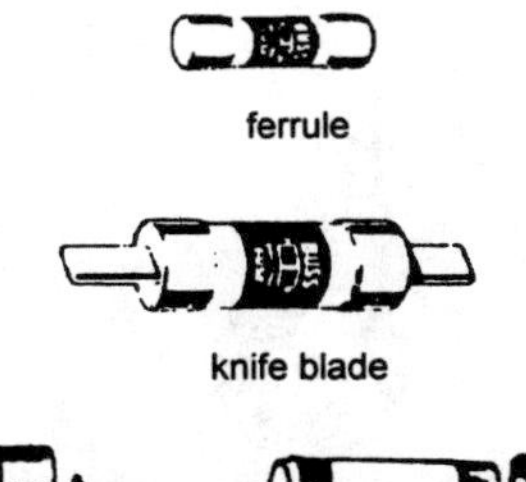

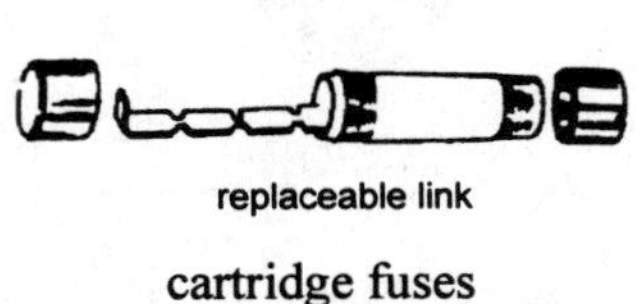

cartridge fuses

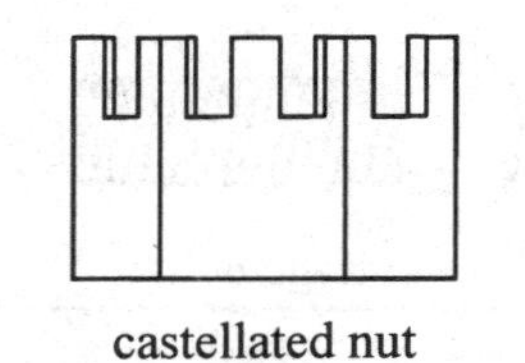
castellated nut

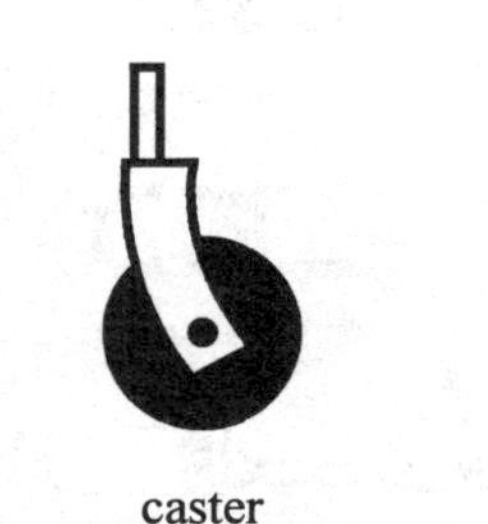
caster

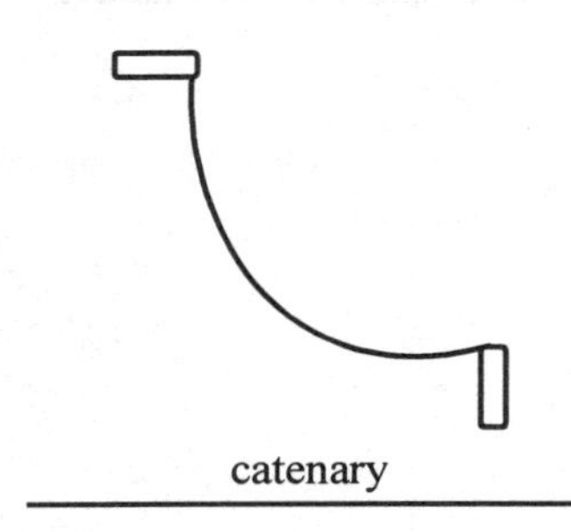
catenary

the carbon source. In this method the carbon can be closely controlled, resulting in uniform carburization of the metal. The quenching process is automatic, and the labor costs are low. *Liquid carburizing* uses calcium cyanamid and sodium or potassium cyanide as the carbon source. As with gas carburizing, the quenching process is automatic, and the labor costs are low. See also carburize.

casement – frames hinged on the side to allow them to swing open.

casement windows – windows that have side-hinged sashes that swing open. The windows may be pushed opened or they may be operated by a cranking mechanism. The crank operated windows have a crank shaft attached to a gear which turns a quadrant gear. The quadrant gear (1/4 of a gear wheel) forces a lever to open or close the window. The window can be opened to any position, and the gearing will hold in that position. Some of these gearing mechanisms are operated by motors with remote controllers which allow the windows to be opened and closed from a distance.

cash accounting – the recording of transactions dealing with money paid out or received.

cash flow – measure of the liquidity of a business.

cash flow statement – a financial statement that shows the receipt and dispersal of cash during a given time period.

casing – molding used to trim door and window openings.

casing bead – a metal plaster stop fastened to framing around doors and windows to provide a smooth edge and eliminate the need for trimming the plaster.

casino – a building designed for entertainment, often gambling.

cast – to make an object or tool by pouring molten metal into a mold and allowing the metal to solidify.

castellated – 1) having a notched top edge or surface; an edge with merlons and crenels (square notches); 2) built in the form of a castle.

castellated nut – a nut whose top surface has notches (merlons and crenels). The nut can be locked in place by means of a cotter pin inserted through a hole in the shaft on which the nut is threaded. The cotter pin fits into the crenels, or notches, on the top surface of the nut, preventing the nut from turning.

caster – a swiveled wheel used on the bottom of furniture legs to permit easy movement of the furniture.

cast iron – an iron alloy used for vent and drain fittings and steam and water valves. The metal is cast in a mold to the desired shape.

cast iron soil pipe – a pipe made of cast iron used for gravity-rated vent and drain applications.

cast iron valve – a valve made of cast iron used in applications such as water systems.

catalyst – an additive used to start a chemical reaction.

catalytic coating – a paint that hardens by chemical reaction.

catch – a latch for a door or gate that automatically secures when the door or gate closes.

catch basin – a basin for collecting water. The basin is usually the collection point for a piping system that routes the water to another location.

catch point – see hinge point.

catenary – the curve or sag in a cable, rope or line due to the force of gravity, when the line is suspended from two points that are at different heights.

cathode ray tube (CRT) – the monitor screen for a computer.

cation – an atom or ion with a positive charge.

cat's eyes – see fish eyes.

cat's paw – a rounded cross section pry bar with notched and flattened ends for pulling nails. One end is bent at right angles to the bar.

cattle gate – a grate made of parallel bars covering a trench that extends between gate posts in a fence. Cattle are afraid to walk over the grate because they can see into the trench below.

caulk – 1) a sealant used to fill small gaps in surfaces or between joints. The compound is available in tubes or in cartridges that fit a caulking gun; 2) to apply sealant.

caulking gun – a tool that holds a cartridge of caulk and has either an air- or mechanically-driven trigger plunger that forces the caulk from the cartridge.

caulking iron, inside and outside – a tool for caulking lead and oakum joint leaks in cast iron drainage piping.

caveat – a caution or warning against certain acts or practices.

cave-in – the collapse of an excavation.

cavetto – a concave quarter round molding.

cavil – a heavy mason's hammer with a head that is pointed on one end and flat on the other. It is used for rough finishing stone.

cavity – a hole or recess.

cavity, ceiling – the lighting zone from the light fixture level to the ceiling. It is one of the three light zones used in calculating a design for an overall room lighting system. The value of the light reflection of a zone is used in a formula to identify the number and type of light fixtures required to light the room.

cavity, floor – the lighting zone from the floor to the work level. It is one of the three light zones used in calculating a design for an overall room lighting system. See cavity, ceiling.

cavity, room – the lighting zone from the light fixture level to the work level. It is one of the three light zones used in calculating a design for an overall room lighting system. See cavity, ceiling.

cavity wall – a double masonry wall with an air space between the two sides. The air space between the walls may be filled with insulation material to improve temperature control. Both walls are separately reinforced for seismic resistance.

C/B ratio – the ratio of the weight of water that is absorbed by a masonry unit in cold water to the weight absorbed in boiling water. Also called the saturation coefficient.

C-clamp – a metal clamp resembling the letter C with an adjustable threaded screw that bridges the opening in the C and clamps onto a piece of work.

cedar – one of several varieties of rot-resistant conifer tree.

cedar, incense – a red-colored cedar wood with streaks of off-white-to-yellow color, having uniform texture, a spicy odor, good thermal insulating properties and good decay resistance. The wood is also repellant to insects, including moths. One of its most common uses is as a moth-resistant lining for closets and chests designed for clothes storage.

cedar shake – a thick roofing shingle (called a shake) split from cedar. Shakes are 18 or 24 inches long and are either of constant thickness or may be tapered from a sharp edge at one end to a blunt end at the other. The blunt end may be from ½ to 1¼ inches thick, depending on whether it is a medium or heavy shake. If properly installed, cedar can withstand many years of exposure to the weather without deterioration. The only change will be in its color. As they age, shakes turn from red-tan to a gray-brown. A shake roof must have good drainage. If it does not shed water properly, it may leak. For this reason shakes are not suitable for installation on a roof with a pitch of less than 4 in 12. Shake shingles should be installed with no more than 10 inches of the thicker end of the shake exposed to the weather, although some straight-split shakes can be exposed up to 16 inches without a problem. The remainder of the shake is protected by the overlapping shakes as the rows progress up to the peak of the roof. Shakes are installed over roofing felt and may be laid on spaced or solid roof sheathing. Shakes add to the structural stiffness of the roof.

cedar shingle – a shingle sawn from cedar. See cedar shake.

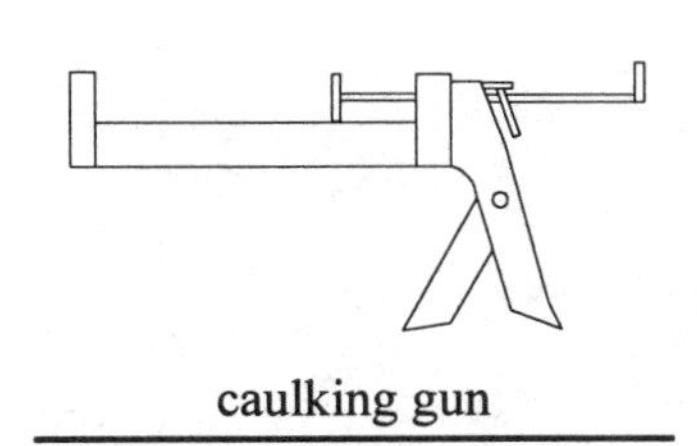
caulking gun

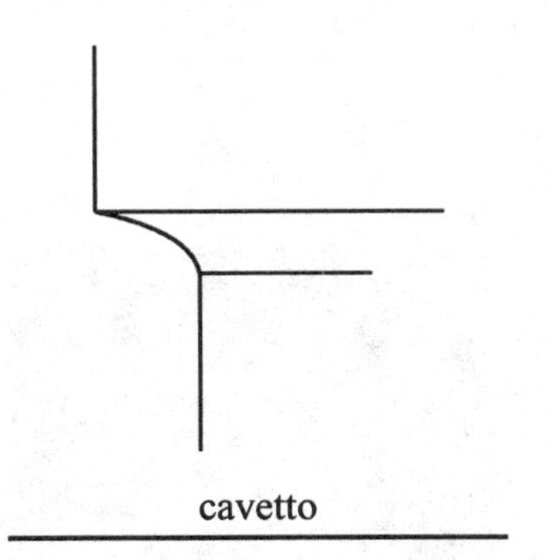
cavetto

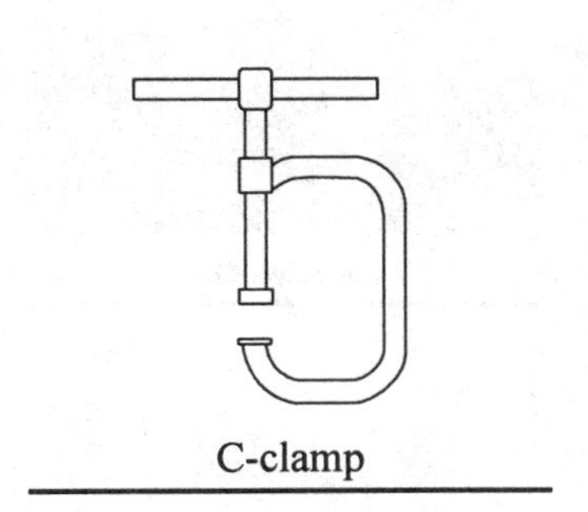
C-clamp

cedar, wormy

cedar, western red – a light, even-grained wood that is a good thermal insulator and useful for moist applications such as building siding and roofing shingles and shakes. Red cedar contains an oil that makes it resistant to rot and insect attack.

cedar, wormy – cedar wood with random channels and grooves eaten through it.

ceiling – 1) the inside top surface in a structure. It is often horizontal, but can also be angled, as in the case of a vaulted or open-beam ceiling; 2) an upper limit.

ceiling exposure limit – the maximum limit, set by OSHA, of a contaminant that workers can be exposed to in the workplace.

ceiling fan – 1) a ceiling-mounted exhaust fan that is ducted to the outside; 2) a ceiling-mounted fan used for air circulation.

ceiling fixture – an electrical fixture mounted on the ceiling.

ceiling fixture box – an electrical box mounted in the ceiling for attachment of a lighting fixture.

ceiling joist – a horizontal structural member spanning the top plates and to which the ceiling covering is attached. When the roof structure is built with the rafters at a pitch or angle, the downward weight of the roof tends to push the tops of the walls outward. The ceiling joists tie the tops of the walls together so that this does not happen. They redirect the load so that instead of pushing out, the load goes down through the walls to the foundation. In a multi-story building, ceiling joists also serve as the floor joists for the next floor up. They provide a space through which utilities, such as plumbing and electrical wiring, can be run between floors. Ceiling joists are sized according to the length of the unsupported distance they must span and the load they must carry. Formulas may be used to calculate the required size based on the allowable stress for the material used. Tables based on these calculations are also available.

ceiling panel – lightweight, and usually sound-absorbing, pieces of material used to cover a ceiling area.

ceiling panel suspension grid – a suspended framework of metal in which ceiling panels are placed to form a ceiling surface.

ceiling tile – see ceiling panel.

cella – an enclosed area in a temple containing a statue of a deity.

cellar – the lowest floor of a building, located completely or partially underground.

cellarage – the volume or space in a cellar.

cellaret or cellarette – a wine cabinet

cellular concrete – a low-density, lightweight concrete containing entrained air.

cellular trim – vinyl trim.

cellulose acetate – a paint and varnish binder made from an acetic or sulfuric acid reaction with cellulose fibers.

cellulose nitrate – a paint or varnish binder made from a nitric acid reaction with cellulose fibers. Also called pyroxylin or nitrocellulose.

Celsius – the centigrade temperature scale, based on the freezing and boiling points of pure water. See centigrade.

cement – see specific cement type, such as natural cement and portland cement.

cement, ASTM – see ASTM cement type listings in the appendix.

cement base paint – a coating containing portland cement, calcium chloride, aluminum stearate, hydrated lime, and zinc sulfide or titanium dioxide used for covering masonry surfaces.

cement board – a ½-inch thick sheet material of concrete and fiberglass used as underlayment in showers, on countertops, and other places where a moisture barrier is desired. The sheets are installed in much the same manner as gypsum drywall. The 3 x 5 foot sheets are cut to size and fastened to the structural backing, such as the cabinet base or wall studs, using drywall screws. Holes may be cut in the cement board using a hole saw, or by scoring the outline of the hole with a utility knife on one side of the board and then breaking the hole out with a hammer. Joints between panels are sealed with drywall joint tape and thinset mortar or whatever adhesive is to be used to set the tile. The surface covering is then applied according manufacturer's instructions.

cement brick – a rectangular building block made of cement and sand that can only be used where there are no acids or alkaline extremes that could attack the brick. See also brick.

cement-coated nails or sinkers – nails with a coating of plastic cement which makes them easier to drive and increases their holding power.

cement joint pipe – drainage pipe laid with joints sealed by cement mortar.

cement mason – a mason who builds structures, driveways, sidewalks and other items using concrete.

cement mixer – see concrete mixer.

cement mortar – a mixture of portland cement, sand and water.

cement tile – a tile made with portland cement, white sand and color pigments that is cast in polished metal molds to yield a matte tile surface that can be used for both interior and exterior walls and floors.

cement trowel – a flat metal plate with a handle used to finish cement work.

cement, white portland – see white portland cement.

center – 1) a tool with a tapered shank and a conical tip that is placed in the tailstock of a lathe for holding the work to be turned; 2) a temporary form that is used to support masonry arches or lintels during construction.

center bit – a drill bit with a small, sharp point extending from the cutting end which is used to precisely locate the center of the hole to be drilled. The sharp point focuses the force of the drill, boring a small diameter hole. This prevents the bit from walking or drifting off center and also eliminates the need for a center punch. It is also called a brad point bit.

center drill – a drill with a short shank and a short cutting section tapered on the end, that is designed to drill a shallow, flared hole in the end of a piece of metal which is to be turned in a lathe. The lathe tailstock center fits in the center-drilled hole in the metal.

center gauge – a template with an external V on one end and an internal V on the other and two smaller V's on the side. Each of the V's is precision ground and lapped. The gauge is used to check the shapes of threading tool bits, for setting tool bits in the tool post of the lathe, and for checking the included angles of machine centers. Also called a thread gauge.

center head – a V-shaped device that can be mounted on a ruler (such as the blade of a combination square) and placed against a circular object so that the object is in the center of the V, permitting a measurement of the radius of the circular object.

center hung sash – a window sash that pivots horizontally about suspension points in the center.

centerline – a line on a drawing through the center of an object.

center of gravity – the point within any object around which its mass is evenly distributed.

center punch – a hardened steel punch with a conical tip used to mark an indentation in metal, such as that used to keep a drill bit from wandering at the start of drilling.

center-to-center – a measurement from the centerline of one object to the centerline of another.

centigrade scale – a temperature scale with 100 divisions between the freezing point of water at 0 degrees and the boiling point of water at 100 degrees. Temperatures using the centigrade scale are indicated by the letter C, as in 30 degrees C.

centimeter – one one-hundredth (0.001) of a meter in the metric measurement system, and equal to 0.3937 inch.

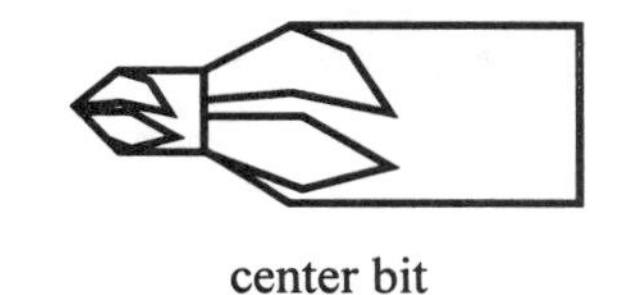

center bit

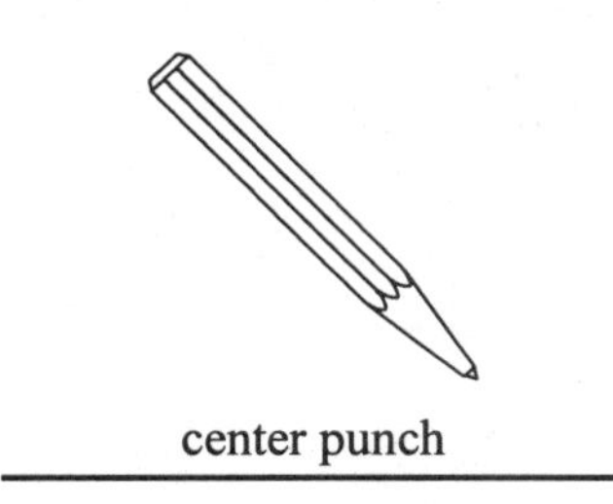

center punch

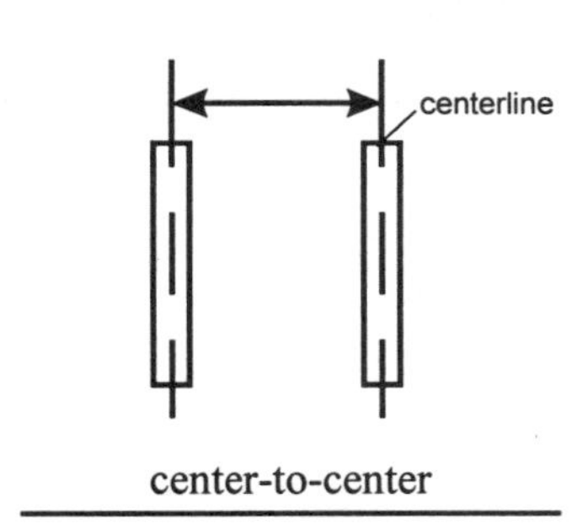

center-to-center

Ceramic tile

Ceramic tile countertop

Flat tiles made from clay and other silicon materials, such as sand or quartz. The materials are formed into the desired size and shape and fired in a kiln to fuse them into a solid and cohesive structure. Ceramic tiles are available in a wide variety of colors, sizes, shapes and finishes to suit different needs and decors. They can be installed and used for floor and wall surfaces, countertops, bath and shower surrounds, or in decorative applications, such as wall mosaics. They are commonly used in homes, businesses, schools, hospitals and industries because they are impervious to water, resistant to stains, easy to clean and disinfect, and they wear well even in heavy traffic areas. Ceramic tile is generally installed in evenly-spaced rows, cemented to a firm backing such as plywood, with the joints filled with grout. Smaller tiles, under 4 inches in diameter, can be purchased in sheets consisting of 16 or more tiles attached to a flexible mesh backing. These are easy to install and save time and labor costs.

central processing unit (cpu) – the section of a computer that contains the memory and processing circuitry for the operations the computer performs.

centrifugal – moving outward from the axis of rotation.

centrifugal compressor – a type of compressor using centrifugal force to compress gas. The gas is fed into a wheel with radial vanes. The wheel is sealed inside a cylinder and spun. As the wheel rotates, the gas is thrown outward, away from the center of the wheel. This outward spinning motion compresses the gas.

centrifugal fan – a fan that moves air outward in a direction perpendicular to the direction of air entering the fan.

centrifugal force – a force that tends to move a mass away from a center of rotation.

centrifuge – 1) a machine that employs centrifugal force and the difference in densities of materials to separate those materials; 2) a machine, used for testing purposes, that spins and imparts a centrifugal force to an object attached to it.

centripetal – moving inward toward the axis of rotation.

centroid – the center mass of an object.

ceramic – made from clay and/or like materials.

ceramic color glaze – an opaque glaze for ceramic tile, made of a silica or similar material, and fused to the masonry at temperatures above 1500 degrees F. ASTM Standard C126 controls the quality of the finished ceramic products and ensures that they are tested and meet specified requirements.

ceramic tile – flat tiles made from clay and other silicon materials, such as sand or quartz. *See Box.*

cesspool – see septic tank.

cetane number – a measure of diesel fuel ignition quality using a comparison process. Cetane and alpha-methyl-napthalene are mixed. They are both hydrocarbons and burn, but cetane has very good ignition quality and alpha-methyl-napthalene has very poor ignition quality. The proportions of this mixture are adjusted so that it has the same ignition quality as the diesel fuel being tested. The percentage of cetane in the comparison mixture is the cetane number given to the diesel fuel.

chain – interlocking metal links which can be assembled in various lengths suitable for uses such as holding, lifting or transmitting power.

chain guard – a safety covering over a power-transmitting chain.

chain hoist – a hoist that uses gearing to gain a mechanical advantage for lifting a load. The gearing is coupled to a drum on which a chain with a hook is attached. The load is hooked onto the chain and the gearing turns the drum, winding the chain onto the drum and lifting the load.

chain intermittent weld – a series of intermittent welds along a joint that alternate from one side of the joint to the other along the length of the joint.

chain link, barbed edge – twisted and cut wire connection at the edge of a chain link fence.

chain link fence – a type of see-through fence that is formed of interlocking heavy wire lengths stretched between metal poles.

chain link, smooth edge – a looped wire connection at the edge of a chain link fence.

chain operator – a closed loop chain which operates a manual valve in a fluid system. The chain is looped around teeth on the outside of the wheel used to turn the valve. The chain loop is long enough so that it can be reached by a person working several feet below the valve. An operator, by pulling on one side of the chain loop or the other in a pulley-like motion, can open, close, or adjust the valve.

chain saw – a power driven hand-held saw that makes rough cuts in wood and lumber. The cutting edge is a chain with sharp teeth that rotates around a flat metal projecting arm.

chain vise – a pipe vise with a lower jaw and a chain to pass around the upper circumference of a length of pipe. The chain serves the same function as the upper vise jaw. It is held in slots and can be tightened to hold the pipe.

chain wrench – a pipe wrench that consists of a length of chain permanently attached to a handle at one end while the other end can be secured at various positions on the handle. The chain can be looped around the pipe to grip and secure it.

chair – a metal device, often bent from rebar, used to hold rebar in position when concrete is poured. It keeps the rebar elevated in the center, rather than allowing it to rest on the bottom of the pour.

chair rail – a horizontal molding around the interior walls of a room located at the approximate height of a chair back.

chalk – soft limestone.

chalk box – a metal or plastic box that contains powdered chalk and string on a reel. As the string is withdrawn from the box, it is coated with the chalk. The string is stretched tightly between two points on a surface, pulled away from the surface, and then allowed to snap back. This leaves a straight marking line of chalk dust between the two points.

chalking – the formation of loose powder on the surface of paint as the paint ages and deteriorates.

chalking paint – a paint formulated to oxidize and form a powder on the surface that can be washed off. Dirt is washed off with the oxidized layer, leaving a clean surface.

chalk line – the straight line formed by snapping a string covered with chalk powder against a surface.

chamber – 1) a private room; 2) a compartment; 3) the space between gates in a canal lock.

chamfer – a beveled edge; an edge cut at an angle, particularly at 45 degrees.

chain link, barbed edge

chain link, smooth edge

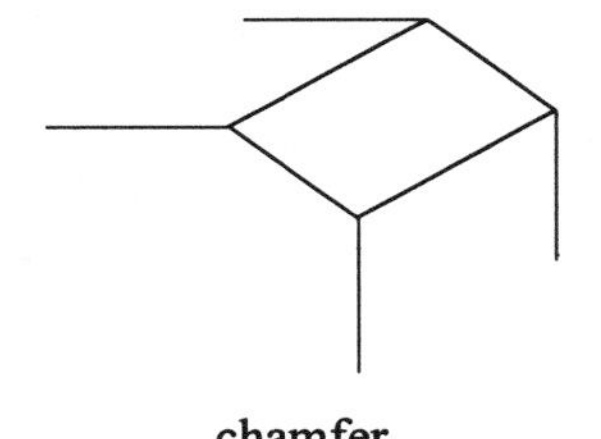

chamfer

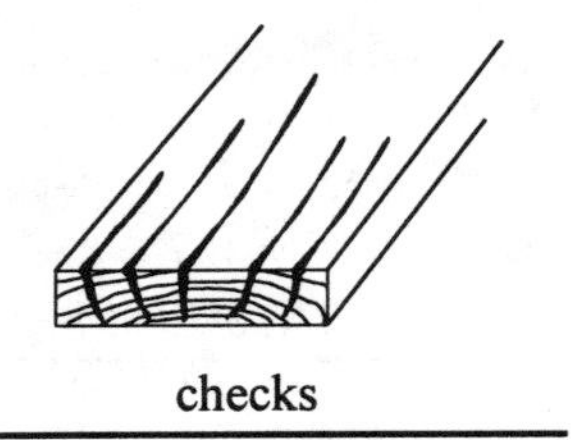

checks

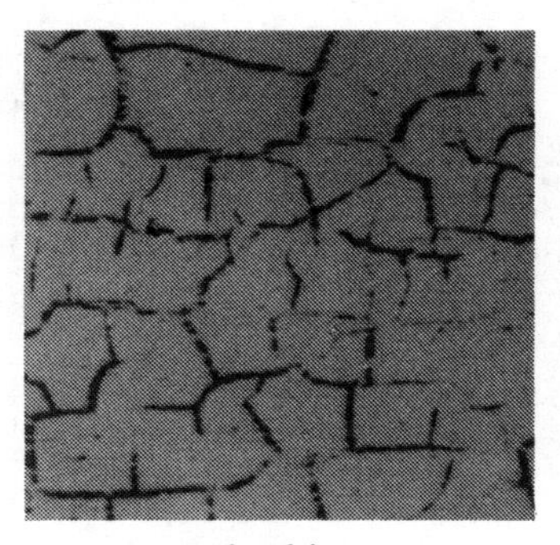

checking

change order – an authorization to make a change in formal plans, specifications, or a contract which is signed by all the parties involved.

channel – 1) a metal structural shape with a cross section resembling the letter C. Used as a supporting member in a variety of installations such as window and door headers, piping support, and cable tray support; 2) a groove or flute; 3) a waterway; 4) a selectable radio frequency.

channel block – a concrete masonry unit with a channel or groove along its length. Rebar can be mortared into the channel to strengthen the installation. Also called bond beam block.

channel cable tray – a tray with a solid, one-piece bottom and side rails used to support electrical cables where many cables are run side by side.

channel iron – a metal structural shape with a cross section resembling the letter C. Used as a supporting member in a variety of installations such as window and door headers, piping support, and cable tray support.

charcoal – a form of carbon material made by burning organic substances without air. It makes a slow- and even-burning fuel for use in applications where steady, long-term heat is desired.

charge – 1) to fill; 2) to add electrical energy to a storage medium; 3) to purchase on credit with a promise to pay later; 4) to place materials in a furnace; 5) to load a mixer.

chase – 1) a groove; 2) a trench; 3) a space through which piping or electrical service is run; 4) to work the surface of metal into decorative patterns.

chase wall – a wall around a space through which piping or electrical wiring is run. This interior space is a chase.

chasm – 1) a large fissure in the earth; 2) a structural breach.

chassis – a structural framework, such as that of a vehicle, piece of electronic gear or cabinet.

chat – limestone gravel.

chat-sawn finish limestone – see limestone, rough shot-sawn finish.

chattel – tangible property that is portable.

check – a separation in a piece of lumber that runs along the grain of the wood and across the annual rings.

checking – a defect in a paint surface consisting of a pattern of squares made up of small, regular cracks.

check rail – the top horizontal sash member of the bottom window and the bottom horizontal sash member of the top window in a double hung window that mate together tightly when the window is closed so no air passes between the sashes. Also called meeting rails.

check valve – a valve that permits flow in only one direction.

cheek – the side of an object.

cheek cut – a chamfer cut along the wide portion of a board end that allows the board to fit tightly against another framing member at an angle, such as at the end of a hip rafter.

cheek wall – a low wall along a stairway.

chemical – a substance that is used in or produced by a chemical process.

chemical resistance – the ability of a material to withstand change from contact with chemicals.

chime – the paint can lip to which the can lid seals.

chimney – the vent from a source of combustion, such as a furnace or fireplace, which goes through the roof to conduct smoke and combustion products out of the building.

chimney cap – concrete cap around the flue liner and on top of the brickwork of a chimney.

chimney cricket – a slightly peaked ridge or projection built adjacent to a chimney to redirect water flow away from the chimney and prevent puddling on the roof.

chimney effect – an upward flow through a channel, so-called because of the way heated air rises through a chimney.

china clay – an abrasion-resistant fine clay paint pigment.

China wood oil – see tung oil.

chink – a crack or narrow opening.

chipping hammer – a hammer with a steel head that is pointed on one end and is tapered to a horizontal edge on the other. It is used to remove slag from a weld or to remove scale or paint from a metal surface.

chip seal – a layer of finely crushed rock spread over an asphalt oil base and then rolled smooth.

chisel – a steel cutting tool shaped into a flat bar with one end edge sharpened to a wedge, and a handle on the other. There are chisels made for cutting wood, metal, stone, and masonry.

chisel, cold – a chisel designed for cutting metal. This type of chisel has a striking surface and handle that are integral with the shank of the chisel. A hammer is used to strike a cold chisel. About a third of the length of the chisel is wedge shaped and tapers down to the hardened cutting edge.

chisel tooth saw blade – a cutting blade for wood. The teeth and the spaces between the teeth are designed to both cross cut and rip cut. The resulting cut is not very smooth, and is best for rough carpentry applications such as cutting framing members.

chisel, wood – a chisel designed for cutting wood. This chisel can be used with the force of a hand for shaving, or can be struck by a mallet for heavier cutting.

chlorinated polyvinyl chloride (CPVC) – a plastic material used for pipe and fittings that has a higher thermal operating capability and a higher pressure rating at any given temperature than PVC (polyvinyl chloride).

chock – 1) a wedge; 2) a heavy metal fitting used as a guide for a rope or cable.

choke – 1) a flow restriction; 2) an electrical coil that adds impedance, or resistance, to the flow of current in an AC circuit.

Chop saw

An electric circular saw mounted on a short arm attached to a portable platform. The arm has a spring-loaded hinge which allows the saw to be lowered for cutting through work on the platform. The arm automatically raises when pressure is released from the saw. The saw blade is usually 10 or 12 inches in diameter. On sophisticated models, the blade can be moved horizontally as well as vertically, adding considerable versatility to its use. Unlike a radial-arm saw, a chop saw can easily be taken to the job site and moved from place to place where it is needed. It offers the precision of a table or radial-arm saw and the portability of a circular saw. Chop saws are also known as power miter boxes.

Chop saw

choker – a shoulder that is to remain higher than the subgrade when grading is done.

chop saw – an electric circular saw mounted on a platform and hinged to allow the front of the saw to be raised and lowered for cutting through the work placed on the platform. *See Box.*

chord – 1) a line between points on a curve; 2) the main structural member of a roof truss.

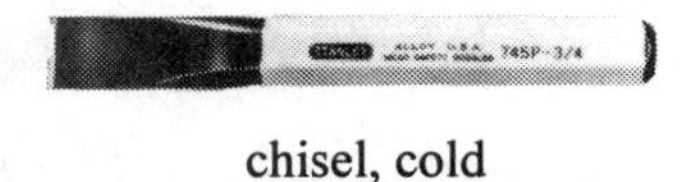

chisel, cold

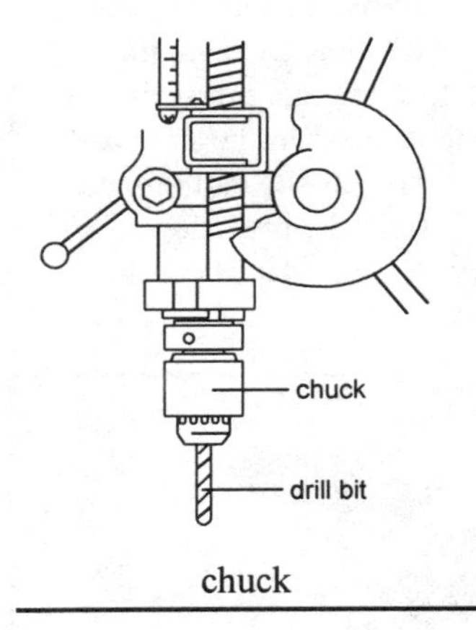

chuck

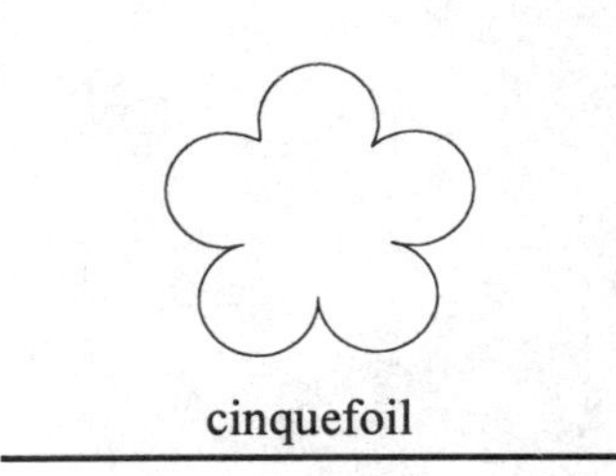
cinquefoil

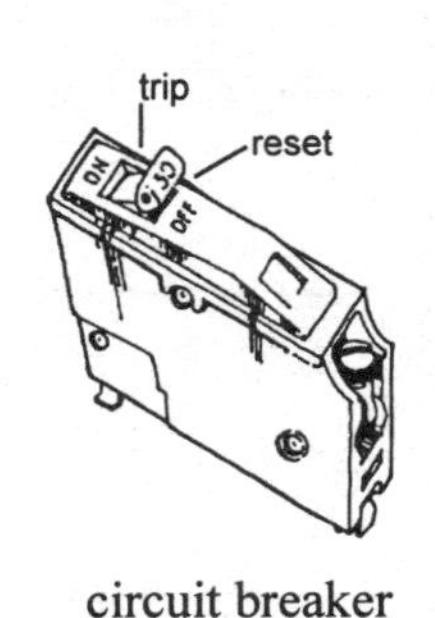

circuit breaker

chroma – color purity or intensity.

chuck – a device on the end of a drill or lathe that is adjustable and locks to hold an arbor or drill bit tightly in its jaws.

chuck, combination – see combination chuck.

chuck, independent – see independent chuck.

chuck, universal – see universal chuck.

chute – a metal trough, set at an angle to the ground, used to guide the flow or movement of a commodity from one area to another, such as to guide the flow of concrete from the truck into the forms.

cinder – residue from the high temperature burning of coal, coke or other hydrocarbon.

cinder block – a masonry building unit made from cement with cinders used as aggregate.

cinquefoil – a flat decoration consisting of five lobes that radiate from one center.

cipolin – impure marble having white and green areas and layers.

circle – 1) a symmetrical plane curved figure in which all points are an equal distance from a center point; 2) the circular plate on which the telescope of a transit rides.

circle cutter – a tool with a center shaft in which a drill bit is mounted, and an adjustable cutter blade mounted on an adjustable arm perpendicular to the center shaft. The center shaft is mounted in a drill motor, and as the center shaft rotates, the cutter scribes a circle on the work and cuts its way through the work. The cutting blade is generally carbide tipped and cuts through a variety of materials. The tool can be used to cut a wide range of diameters, however, the speed of the cutting must be closely controlled. The rotating cutter blade is an off-center mass and is inherently out of balance, making it difficult to guide.

circuit – 1) electrical conductors arranged to permit the flow of electrical current; 2) a complete flow path in a fluid system.

circuit breaker – an automatic electrical switch that interrupts an electrical circuit when the current exceeds safe limits. The circuit breaker can be reset when the problem that caused it to interrupt the current has been corrected.

circuitry – the composite of all the electrical circuits in any device.

circuit tester – a small lamp with test leads that can be inserted into an electrical receptacle or touched to wires to test if the circuit is live. The lamp will glow if live current flows through it from the circuit being tested.

circuit vent – a vertical vent from a plumbing drainage system. It is run from the last two fixture traps on a horizontal drain line of a building drainage system to the main vent stack of the system.

circuit voltage – the amount of voltage in a specific electrical circuit.

circular measure – measure in portions of a circle, such as radians, degrees, minutes, or seconds.

circular mils (cm) – the area, in thousandths of an inch, of an electrical conductor.

circular saw – a portable hand-held power saw having a circular, toothed blade, a handle with trigger switch, and a platform to rest on the work to be cut. The platform is adjustable for angularity and depth of cut.

circular stairs – see spiral staircase.

circulate – to move around in a pattern.

circulation – the pattern of movement of a fluid or gas in a system, such as the flow of air in a building.

circumference – the outer perimeter of a circle.

circumscribe – to draw a line around a figure or object.

cistern – a storage reservoir for water, often located underground.

civil action – a non-criminal court case.

civil engineering – that branch of engineering which specializes in the design of structures such as roadways, buildings, and bridges.

clad – to be covered with another material that is bonded to the base material, such as a rubber cladding on a metal pipe, or a corrosion-resistant metal cladding on a non-resistant metal. Cladding may lower the cost or improve the overall performance of the material.

claim – 1) a formal request or demand; 2) the right to the possession of something.

clamp – mechanical device used to hold two or more items together, usually temporarily.

clamp, adjustable – a clamp with a screw thread, or other type of adjustment, that allows the jaws to fit and firmly grasp materials of various sizes. The screw thread can be turned to move the jaws of the clamp against the work to be held.

clamp and step block – a device used to hold metal in place on the work table. It consists of a stepped block that is used to support one end of a horizontal lever, pivoted off-center along its length, on a bolt that is held in a slot in the work table. The other end of the lever is laid on the work, with the stepped block end of the lever on the proper step so that the lever is horizontal. The long leg of the lever is placed on the stepped block so that there is a mechanical advantage clamping the work in place. Tightening the nut on the bolt exerts force on the step and on the work and holds the work in place. The step block permits adjustment to fit over and hold different thicknesses of material, as thicker material can be placed under the section of the block having the largest cutout.

clamp, C – see C-clamp.

clamp, corner – see corner clamp.

clamping time – the length of time two members have to be held, with pressure applied to the joint between them, by a clamp. The time is determined by the curing time of the adhesive used to join the two members.

clamshell – a powered soil-excavating bucket that can be closed upon a load of soil and then opened to dump the load.

clapboards – wood siding board that is applied horizontally. The boards are tapered in thickness, with the thick lower edge of one board designed to overlap the thinner top edge of the board below.

clasp nail – a heavy U-shaped staple.

Class M buildings – farm buildings used for machinery storage, crop storage, livestock, produce processing or farm shops, as classified by the National Frame Builders Association.

clastic – a rock composed of pieces of older rock.

claw – a two-tined curved end on a bar or hammer head used to grip nails and remove them from a material or surface.

claw hammer – a hammer for driving nails that has a claw for pulling nails opposite the striking portion of the head.

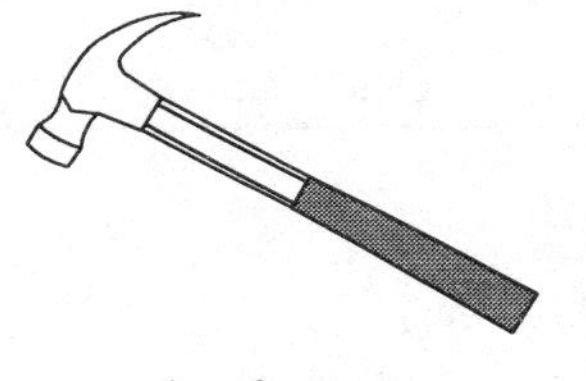
claw hammer

clay – a natural soil or mineral aggregate high in hydrous aluminum silicate which dries hard, but can be worked and molded when moist.

clay masonry – brick made from natural clay: Grade NW is used for walls; Grade SW for floors; structural load-bearing clay tile is Grade LB or LBX.

clay pipe – gravity rated drain pipe and fittings made from clay. Gravity rated means that they are not designed to have more pressure put on them than is exerted by fluid flowing under the natural force of gravity.

clay stone – igneous rock that has decomposed.

clay tile – tile made from clay. Clay tile can be made into many forms, such as curved to form part of a cylinder or flat for floor tile.

cleanout – a wye plumbing fitting with a plug in one branch of the "Y" which provides access to the pipe in case of clogs in the line.

cleanout plug – a plug that closes the access to the cleanout fitting when it is not in use.

clearance – the minimum distance needed between a moving part or machine and another object.

clear and grub – to clear a site of vegetation and obstructions prior to grading or construction.

clear lumber – lumber that does not have knots or defects in it.

clear span – the span of a structural member between supports. Also referred to as allowable span.

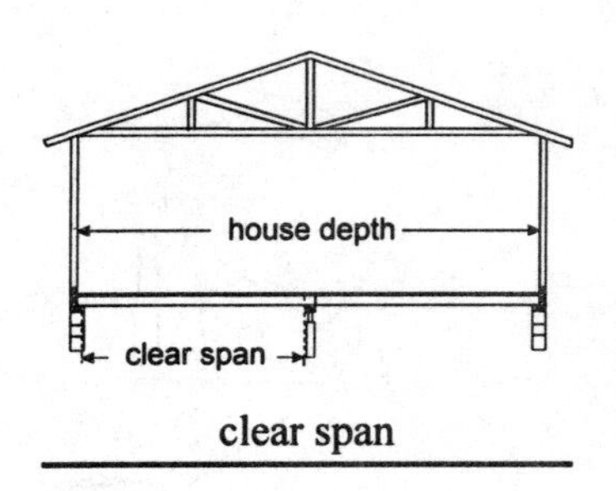

clear span

cleat – 1) a piece of wood or metal fastened to a structural member, such as a stair stringer, to support or provide an attachment point for another member, such as a stair tread; 2) a wedge fastened to a surface; 3) a device with horizontal projections around which a rope can be wound to tie it off; 4) a length of material fastened to a surface for traction; 5) a stiffener used to strengthen a structural member.

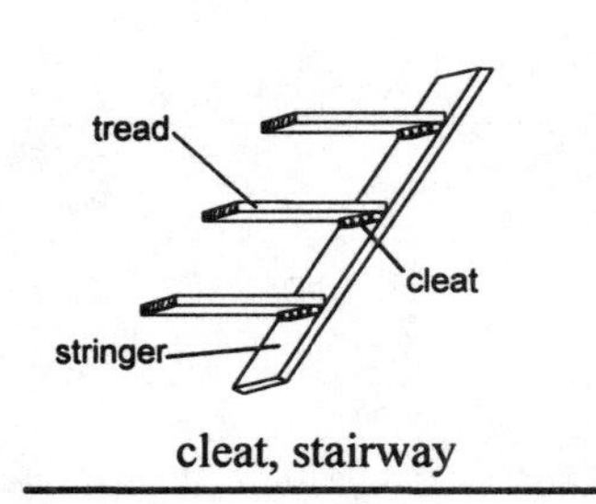

cleat, stairway

cleating method, joist installation – a method of installing cleats on existing joists and beams to support new joists that are added.

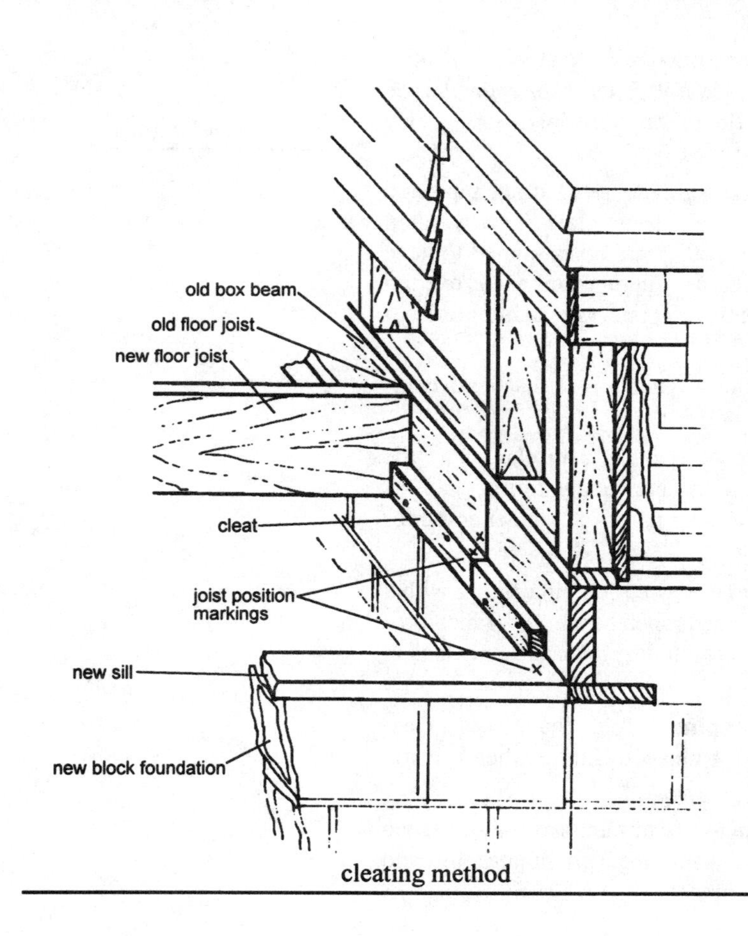

cleating method

clerestory – 1) a vertical wall between roofs of different heights often used to provide light and/or ventilation; 2) a raised section of roof with windows.

clevis – a U-shaped fixture, with a pin or bolt spanning the legs of the U, used to attach a rod to a plate, beam, etc.

client – the party who hires others to do work for him.

clinch or clench – to secure; such as bending the protruding end of a nail down and hammering it flat to prevent the nail from working loose.

clinker brick – a hard burned brick with its shape distorted by the vitrification.

clip – 1) a metal fastener used to secure one object to another; 2) a small section of brick trimmed to a specific length.

clipped header – a brick that has been cut down in size, often in half, used to establish the brick pattern. Also called false header.

clockwise – moving in the same direction as the hands of a clock.

close – 1) to shut; 2) the originating and ending point at which the perimeter lines of a survey come together, or "close;" 3) nearby.

close-coupled closet tank – a water closet tank that is entirely supported by the bowl.

closed circuit – an electrical circuit that is complete and continuous.

closed circuit television – a system of television cameras and monitors that are interconnected so that remote locations can be monitored from one point, usually for security reasons.

closed-cut valley – a roof valley where the courses of shingles from one slope overlap the valley and the courses from the opposite slope are cut along the valley centerline.

closed joint – a joint between sections of concrete that permits differential movement between the sections, but is sealed to keep moisture out.

closed loop – a system that forms a continuous loop, such as a refrigeration system, where the refrigerant is circulated, compressed, and expanded within the system.

closed-loop contour lines – a contour line that forms a closed loop, indicating a depression or a mound.

closed-string stairway – a stairway with finish stringers covering the ends of the treads and risers.

closed valley – a roof valley where the shingle courses meet to cover the valley. The rows of shingles are interwoven and overlap in the valley so that water runs off.

close grain – wood with small, tightly spaced grain lines.

closer – the final brick or other masonry unit in a course. Also called closure.

closet – a small room or enclosed space designed for storage.

closet auger – a device used to clear a plugged toilet. It consists of a coiled spring cable with a crank handle on one end. The auger is rotated and maneuvered down into the drain to clear the passage. Also called a closet snake.

closet bend – a plumbing drainage fitting, that makes a 90 degree turn, to which the outlet of a water closet is connected.

closet flange – a flange used to connect the closet bend to the water closet. The closet flange is set slightly above the surface of the subfloor, so that the bottom of the water closet will be flush with the finished floor. The flange is attached to a closet bend by cementing or soldering, or by a lead and oakum joint, depending on the type of drainage system material that is used. A wax gasket is fitted between the water closet and the closet flange. The bolts are tightened to hold the water closet firmly in place and to compress the gasket to make a tight seal. Closet flanges can be made of metal or plastic. Also called closet floor flange.

closing a survey – finding the area of a piece of land using the measurements obtained during the survey.

closing costs – service charges for processing a transfer of ownership on property.

closure – 1) a cap on the end of a length of pipe; 2) any object or structure used to close off an opening; 3) short masonry units that are placed at the ends of courses to maintain the bond pattern. See also closer; 4) material used to conceal a joint.

closure strip – a resilient strip used to fill voids between panels or other objects.

clothesline units – retractable clotheslines in spring-rewind housings.

clouding – an uneven luster or color in a painted surface.

cluster – a grouping of similar objects.

clutch – 1) a means of engaging and disengaging a power drive using friction plates; 2) a device for gripping an object.

coach – a vehicle, such as a motor home or rail car.

coagulate – a liquid thickening or clotting into a semi-solid mass.

coalesce – to unite or grow together.

coal-fired furnace – a heating furnace that is designed to burn coal as a fuel.

coal tar – a bituminous substance applied to a surface for waterproofing or corrosion resistance.

coarse – rough; not smooth.

coarse aggregate – see aggregate, coarse.

coarse grain – a wood grain in which the rings are relatively far apart.

coarse thread – a machine screw with a relatively few number of threads per inch as compared to a fine-thread screw with many.

coat – a covering, such as a coat of paint or primer.

coating in – applying a uniform coat of paint to a surface.

coaxial – two or more objects mounted, fabricated, or installed so that their axes are the same. For example, coaxial cable, often used to connect an antenna to a TV, has an inner conductor surrounded by an insulator, which is surrounded by a braided wire or foil conductor, which is covered by another layer of insulation.

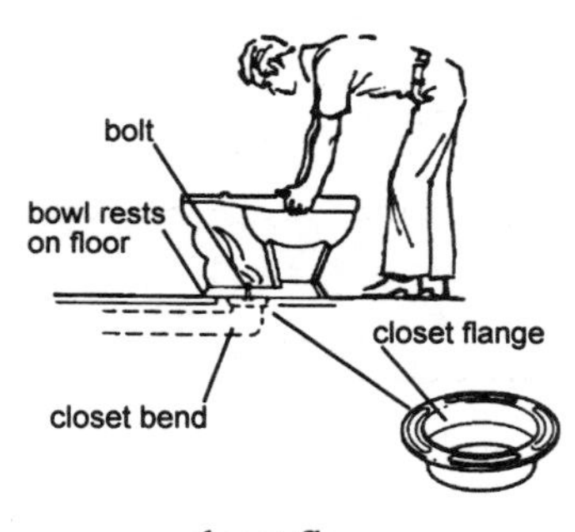

closet flange

The central axis of each of the parts is the same, so that one part is concentric around another, from the innermost insulation outward.

cobalt – a metallic element; atomic number 27; atomic weight 58.933; atomic symbol Co.

cobble – a paving stone used most often today in decorative applications, such as for masonry walls or walkways.

cobweb finish paint – a texture paint whose surface resembles cobwebs.

cock – 1) to turn on one side or to turn up; 2) to set a spring-loaded device into position to be actuated; 3) a valve.

cockle – a wrinkle or depression in a wallboard surface occurring during the manufacturing of the product.

collar beam

code – 1) a standard, or a legal document, that sets forth minimum requirements which must be met for a design, fabrication, or construction to be considered acceptable; 2) the marking on the back of gypsum wallboard, plywood, or other material that identifies type, manufacturer, thickness or quality.

codicil – a supplement to a will.

coefficient of energy – the actual airflow compared to the theoretical air flow in an HVAC system under set conditions.

coextrusion welding – a welding process using heat and the pressure of extruding the materials being welded through a die.

coffer – 1) a sunken panel in an overhead surface; 2) a box-like structure.

cofferdam – a watertight structure placed around a construction area to enable dewatering.

cog – a projection or tooth on a gear.

cog railway – a railway for steep inclines, using a gearwheel and cogged rail for traction.

cogwheel – a gearwheel.

cohesion – a bonding together.

coign – see quoin.

coil – 1) to wind uniformly in a circle; 2) a helix or spiral; 3) a spiral-wound electrical conductor.

coil spring – spring material that is wound in a spiral.

coke – the carbon left from maintaining coal at a high temperature in the absence of air.

cold-applied – a substance applied without the addition of heat.

cold checking – checking, a pattern of small, regular squares appearing in a paint surface, caused by excessively cold temperatures.

cold chisel – see chisel, cold.

cold cracking – cracks in paint caused by excessively cold temperatures.

cold drawn – metal that is drawn through dies to form it without first adding heat to the metal.

cold forming – a method of shaping metal without first heating it.

cold joint – 1) a solder joint in which the solder has not fused to the base metal because the base metal was either dirty or not hot enough; 2) a mortar or concrete joint where mortar or concrete is placed against mortar or concrete that has already set.

cold rolled – metal plate that is rolled without first adding heat to it.

cold solder joint – see cold joint.

cold steam cleaning – cleaning with a mixture of steam and water under pressure.

cold water paint – glue or casein dissolved in water and used as a sealer coating for plaster or masonry.

cold weld – a welding process using pressure at room temperature to fuse materials together.

collar – a reinforcement around the periphery of an object.

collar beam – a board connecting or bracing together pairs of rafters above the plate line.

collar joint – a vertical joint between masonry wythes; a joint between back-to-back bricks.

collar tie – see collar beam.

collateral – 1) objects that are situated parallel to each other; 2) accompanying as subordinate to another; 3) an addition to; 4) security pledged by a borrower to protect the interests of the lender.

collector, runway – an electrical contact that moves with a crane bridge, and is used to conduct electrical power from an electrical busway to the bridge, and then to the motor(s).

collector, solar – a heat exchanger that uses heat absorbed from the sun to warm fluids, such as potable water for a building hot water system. General domestic water is heated with copper tubing (often with fins to increase the surface area) which is wound back and forth in a flat glass box mounted to face and absorb the heat of the sun. The tubing is usually painted black and the glass box insulated on the bottom and sides to hold in the heat. Water is circulated, with the aide of a small energy-efficient pump, through the tubing and heated as it passes through the glass box. The heated water is then collected in an insulated tank, similar to a water heater, and stored for household use. Black plastic collectors are usually used for heating the water for swimming pools and hot tubs. The purpose is to use solar energy for heat and conserve other fuel sources.

collector, trolley – an electrical contact that moves with a crane trolley, and is used to conduct electrical power from an electrical busway to the trolley, and then to the motor(s).

collet – 1) a split holder for drill bits, shafts, etc, that is tightened by compression, drawing the split sections closer together to grip the bit tightly; 2) an encircling band.

colloid – a suspension of fine solid particles in a liquid.

colonial texture – a plaster or stucco texture of semi-circular swirls created with a stiff-bristled brush.

colonnade – a row of columns.

color – one of the wavelengths of reflected light in the visible light spectrum.

color-coding – the use of various colors to identify items of different value or application for ease of sorting.

color-fast – non-fading color.

column – a vertical structural pillar or post.

column anchor – a metal support and fastener used to firmly attach the base of a column to a footing or foundation.

column clamp – a device used to encircle column forms and hold them together against the force of the wet concrete poured into the forms. Also called a column form clamp.

column form – a temporary structure used to hold wet concrete in the form of a column until the concrete has set.

column reinforcement – reinforcing steel added to a concrete column to prevent lateral expansion of the column under load.

combination – a combined wye plumbing drainage fitting and ⅛ bend.

combination chuck – a chuck, which is the device on the end of a lathe to hold an arbor or drill bit tightly, with three jaws that move simultaneously to lock the bit in place.

combination doors and windows – coverings for window or door openings that have glass or screen inserts. The glass inserts provide insulation and light in the winter and the screen inserts provide ventilation in the summer.

combination faucet – a sink faucet with two valves, one for hot water and one for cold water, discharging into a single spout.

combination pressure and temperature relief valve – a valve on hot water heaters which opens when either the pressure or the temperature rises above a set level.

combination saw blade – a wood cutting blade for a power or hand saw that is suitable for both ripping and crosscut work.

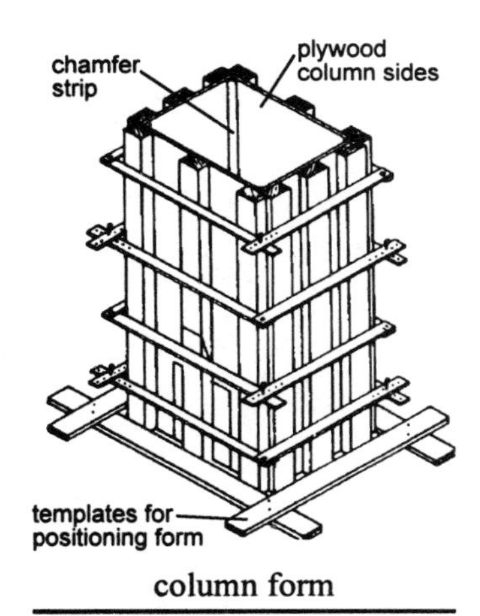

column form

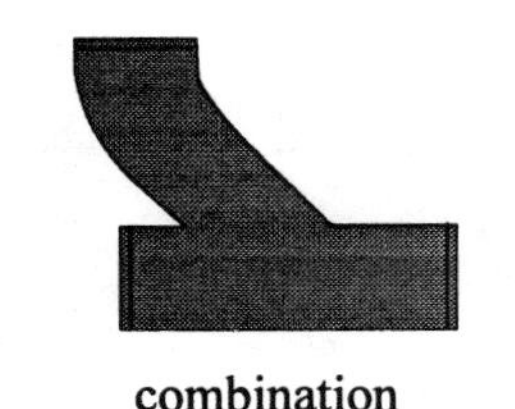
combination

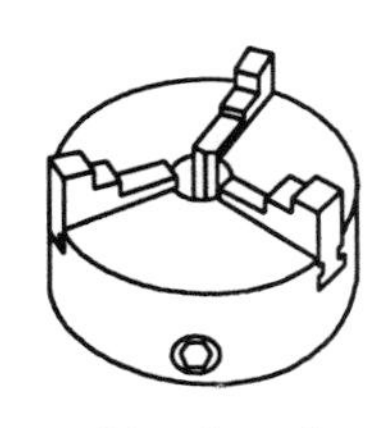
combination chuck

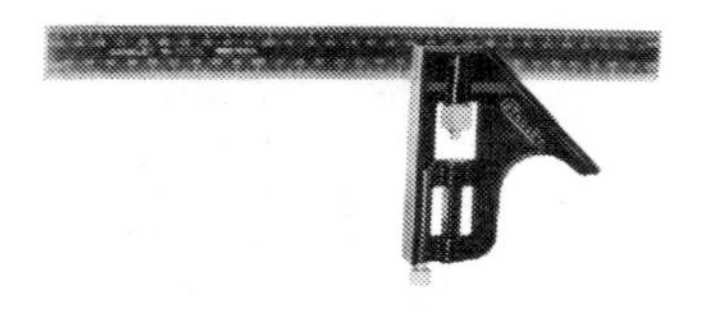
combination square

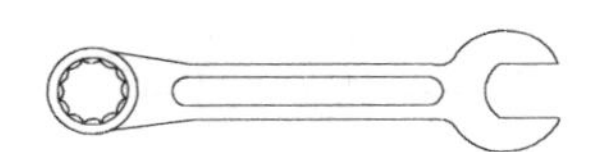
combination wrench

full headers

Flemish headers

common bond

combination square – an adjustable square which has a level protractor as part of the sliding head. Useful in carpentry and cabinetmaking to check level, plumb, squareness, and angles. It can also be used for drawing a line parallel to an edge by sliding the head to the proper dimension, locking it into place and then sliding the square along the edge while holding a pencil at the end of the tongue.

combination switch and receptacle – an electrical switch and receptacle in the same unit.

combination wrench – a wrench with a box end and an open end.

combined building sewer – a sewer disposal system designed to carry both sanitary waste and water drainage from a building.

comb swirl – a textured wall finish created by using the teeth of a hair comb in random semi-circular swirls. This technique can be applied to plaster, stucco, or concrete surfaces.

combustion air – the air required for combustion in a device that burns fuel, such as a furnace.

combustion analyzer – an instrument that measures the contents of flue gas from a boiler, furnace, or other device that burns fuel in order to test the fuel-air mixture and tell whether the device is operating correctly.

comfort ventilation – ventilation which provides a comfortable air temperature and humidity level for the occupants of an area.

commercial property – property developed for businesses.

commodity – a definable and quantifiable group of economic goods, such as piping, electrical cable, or food stuffs.

common bond – an arrangement of brickwork in which the majority of the courses are stretchers with either full headers or Flemish headers every sixth course.

common brick – a standard clay brick. See also brick.

common nail – a nail with a smooth shank and a flat head used to join wood members in general carpentry. They come in a variety of lengths.

common rafter – a rafter that extends from the wall top plate to the roof ridge.

common vent – a vertical vent line in a building drain system that is connected to two or more fixture branches on the same level.

common wall – a wall that is common to two living spaces.

commutator – 1) a device to reverse the direction of electrical current; 2) a cylindrical conductor, in contact with stationary conductive brushes, mounted on the shafts of DC motors and generators.

compaction – the process of using water and/or weights to tamp down earth or other fill.

compactor – a soil compacting machine. The machine has a motor which vibrates a weighted flat plate that rests on the ground. The combination of the weight and the vibration works to compact the soil.

compass – 1) a drafting tool shaped like the letter A, with a hinge at the apex of the A. The legs adjust for the size of the circle to be drawn. One leg has a sharp pointed end and the other holds a pen or pencil. A circle may be drawn by placing the sharp leg at a fixed point and rotating the penciled leg around it. See also bow compass; 2) a magnetized floating needle suspended on a point bearing that rotates freely to indicate magnetic north.

compass brick – factory made curved brick.

competitive bid – a bid solicited so the job can be awarded to the contractor with the best bid.

complete fusion – fusion between materials over the entire welded area.

complete joint penetration – weld penetration, with complete fusion, to the root of the weld.

completion-of-contract accounting – an accounting method in which money received is not counted as income until the contract has been 100 percent completed.

component – a part of a whole.

composite – built up of different parts, pieces, or materials.

composite electrode – a filler metal welding electrode made of more than one material.

composite joint – a joint made by combining welding and another means of fastening.

composite wall – a masonry wall made up of wythes of two or more different types of masonry units.

composition – the aggregate parts that together make a whole.

composition shingle – a shingle made from asphalt and structural fibers bonded together and covered with a protective granular coating.

compound circuit – an electrical circuit that has both parallel and series sections.

compression – a pushing force that puts inward pressure on an object or material.

compression coupler – a pipe repair clamp that spans and joins two sections of pipe after a damaged piece has been removed. The surrounding screw clamp is tightened around each end of the coupler and exerts compressive force as it is tightened. This force tightens the rubber around the outside diameter of the pipe effecting a tight seal. There are different types of compression couplers, but each works in a similar fashion. The compression coupler can be used as a temporary or permanent repair to a leak or crimped section of pipe. However, care must be taken to ensure that the pipe is well supported on both sides of the coupler because compression couplers are not designed to take much stress.

compression fitting – a tube fitting that uses a nut with wrench flats, a compression ring (a sleeve that is thicker in the center and tapered toward the ends), and a mated fitting with male threads and an internal taper. As the nut is tightened on the male threads of the mated fitting, the compression ring is squeezed between the nut and the fitting and pressed against the outside diameter of the tubing making a tight seal. Compression fittings are used in plumbing applications for smaller diameter tubing.

compression valve – a valve that is opened or closed by raising or lowering a horizontal disc by means of a threaded stem. An elastomer washer on the end of the stem seals against the valve seat, closing off water flow. This type of valve is most commonly used for water faucets.

compressive fracture test – a test of the strength of concrete in which a concrete cylinder is cast under prescribed conditions and a specified level of compressive load is applied to the cylinder after the concrete has cured.

compressive strength – the ability of a material to sustain a compressive, or heavy weight, load.

compressor – a mechanical device for decreasing the volume of a gas by increasing the pressure upon the gas. The pressurized gas is usually stored in tanks or other containers for later use, such as for powering air-driven tools.

compressor, centrifugal – a compressor that relies on speed and centrifugal force to compress a gas.

compressor, reciprocating – a compressor that uses a piston moving in a cylinder to compress gas.

compressor, roots type – a positive displacement compressor that uses two spiral intermeshed rotors within a tight-fitting housing to compress the gas.

computer aided design and drafting (CADD) – a sophisticated graphics program for use on computers. The drawings can be done in two or three dimensions. The three dimensional designs (used on larger computers) can be animated so that the viewer can look at the design from any angle, or even move inside the drawing and look around. Animation can be applied to show various parts being moved, as they would during installation or maintenance procedures. Program animation

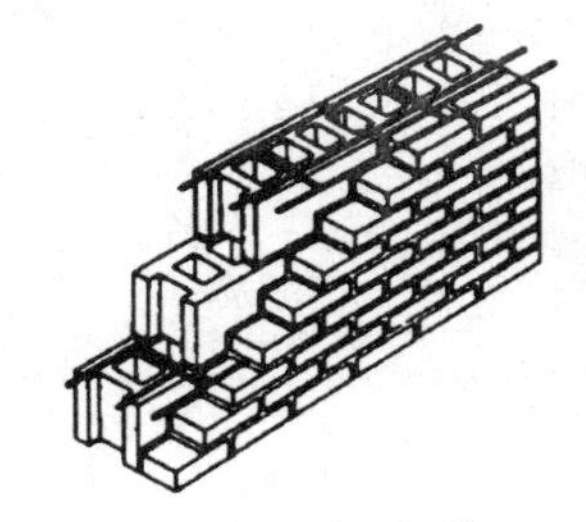

composite wall

compression coupler

can be used for demonstrations, simulations, situation re-enactments and training and education. Items, systems and structures of any size and complexity can be designed using CADD programs. They are used increasingly for designing and planning construction sequences and logistics for large and small building projects, as well as landscaping and maintenance designs. Drawings can be made to scale, with dimensions shown by the computer, and measurements can be taken electronically. Because personal computers have become more powerful and faster, there are many CADD programs available for home and small business use.

concave – a surface that is curved or dished inward.

concealed hinges – hinges that are not visible when the door which they support is closed.

concealed nail method of installing roll roofing – an installation method in which each succeeding course of roll roofing overlaps and covers the nails in the preceding course. Roofing cement is used to hold down the edges of the courses and bond them to the preceding course.

concealed wiring – electrical wiring that is run inside the structural elements of a building, such as inside the walls.

concentric – having a common center, as in a circle within a circle.

conceptual design – a design in sufficient detail to define the basics of the object or system.

concrete – a mixture of cement, sand, and aggregate which, when mixed with the proper proportion of water, solidifies through chemical reaction into a structural material with good compressive strength.

concrete, aerated – see air-entrained concrete.

concrete block – building blocks that have been cast from concrete, which is made of portland cement and aggregates. These blocks come in a wide variety of shapes.

concrete block wall – a wall made up of concrete blocks held together with mortar.

concrete brick – solid, brick-sized masonry blocks of concrete made of portland cement and various aggregates such as sand or cinders.

concrete finishing – screeding, floating and troweling concrete before it has set so that the final cured surface of the concrete will be hard and have the desired finish texture.

concrete forms – molds with adequate structural strength to shape and hold the poured concrete until it has set.

concrete, insulating – concrete with expanded mica aggregate (6 to 1, aggregate to cement) which provides an R-factor (resistance to temperature transmission) of about 1.1 per inch of thickness.

concrete machine – a machine that pours, spreads and smooths a concrete pavement onto a graded area.

concrete masonry – masonry structures built using concrete masonry units.

concrete masonry unit (CMU) – individual concrete bricks or blocks available in many shapes and sizes.

concrete mix – the proportions of cement, aggregate, and water required for the amount of concrete needed.

concrete mixer – a machine-rotated drum with internal paddles into which the components of concrete are placed for mixing.

concrete nail – a thick-shanked nail with a flat head that can be driven into set concrete.

concrete paving – roadway paving using concrete as the road surface.

concrete pipe – pipe made of concrete with steel reinforcement.

concrete, precast architectural – a precast concrete that is used for finished parts of a building. See box under precast architectural concrete.

concrete pumper – a pumping unit used to move concrete from the mixer or truck to a remote and/or elevated location. The concrete is moved from the truck to the intake hopper of the pump. Hoses are run from the outlet of the pumper to the forms to be filled, with booms supporting and directing the hoses. The concrete pumper is powered by an internal combustion engine and mounted on a trailer or truck so it can be easily moved into place.

concrete, reinforced – see reinforced concrete.

concrete saw – a power saw, with a Carborundum or diamond blade, used for cutting hardened concrete. This type of saw is often water-cooled.

concrete slab – a flat, horizontal concrete pour.

concrete tamper – see jitterbug.

concrete trowel – see cement trowel.

concrete vibrator, electric – a motor-powered vibrator with a long, flexible wand that can be inserted into fresh concrete to consolidate it and eliminate all the voids in the forms.

concrete wall, poured – a concrete wall that has been poured in place into forms and allowed to set up at its permanent location.

concrete workability – the character and consistency of a concrete mix that determines how well it can be placed and finished.

concurrent heating – applying a second source of heat to a material during a welding or thermal cutting operation.

condemn – 1) to judge unfit for use; 2) to secure by right of eminent domain.

condensate – water that has been condensed from steam, or other fluid condensed from a vapor. In most systems where steam is generated, whether as a driving force for an engine or as heating medium, the steam is condensed after use and then recovered as water to be used again.

condensation – changing a substance from a vapor to a liquid state by removing the heat. The condensate shows up on surfaces as a film or drops of water.

condense – 1) to change to a liquid or solid from a gas; 2) to reduce in volume.

condenser – 1) a device for cooling a gas sufficiently to change it into a liquid state; 2) an electrical capacitor.

condition – 1) a situation; 2) a requirement that is part of an agreement; 3) to prepare; 4) a state of fitness or readiness for use.

conditioner – a device that brings an object or commodity to a desired state.

condominium – individual ownership of a living unit in a multi-unit structure, with shared or communal ownership of the common areas and grounds.

conduct – to carry or direct a flow.

conductance – the capacity of a material to pass or carry an electrical current.

conduction – 1) the transfer of heat between objects of different temperature; 2) the flow of electricity or fluid.

conductor – 1) a material that allows electricity to pass through it; 2) a water pipe entering a building; 3) a material that conducts heat, sound or other energy forms.

conduit – 1) a passageway or flow path; 2) a tube made of metal, plastic, or other material used to protect electrical wiring from damage or moisture. *See Box.*

conduit bender – a handle with a curved shoe on the end. The shoe is similar to a length of pipe that has been split lengthwise and bent into a 90 degree curve. A pawl on the end of the shoe grips a piece of electrical conduit. The shoe is held against the conduit and force is exerted on the handle to bend the conduit around the curve of the shoe. The shape of the curved shoe prevents crimping in the conduit as it is bent.

conduit box – an electrical box at which electrical conduit is terminated. The conduit box serves as both the termination point for the conduit and as a container in which electrical wires can be terminated, redirected, or joined. Also called a junction box.

conduit fitting – a fitting used to join and/or change the direction of conduit, or to join conduit to a termination point, such as a panel or box.

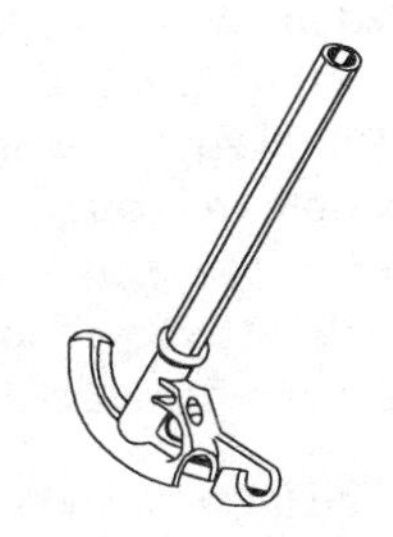

conduit bender

conduit boxes

Conduit

A tube made of metal, plastic, or other material used to protect long runs of electrical wiring from damage or moisture. Conduit should be used any time the cost is acceptable, but always in situations where the wiring may be exposed to hazards or when a wiring insertion path is needed. Wiring may be inserted in conduit and later removed or replaced. There are several different types of conduit, each with its own specific use or purpose.

Electrical metallic tubing (EMT) conduit is a thin-walled metal conduit used to protect electrical wiring in commercial and industrial applications, as well as in residential applications where the wiring is exposed and accessible and therefore in need of protection. It may be used for concealed and exposed installations, as long as it won't be exposed to severe damage, such as from an impact or other hazard that could deform the conduit.

EMT is relatively lightweight and inexpensive, and metal conduit acts as a ground path. It is suitable for burial in concrete and in fire-resistant construction. If it is to be installed in a corrosive environment, such as a tannery, pulp mill or fertilizer plant, it must be made of corrosion-resistant material. Corrosion-resistant EMT is made of copper, silicon bronze or plastic-coated steel. When the conduit is to be located in wet areas, such as a laundry, the entire conduit system must be constructed in a manner that will prevent the entrance of water. The materials used for these locations must be approved by local codes, and the conduit system must be mounted with at least a ¼-inch space between the conduit parts and the surface on which it is mounted.

EMT can be joined using several types of threadless fittings. One type uses a setscrew to clamp the tubing to the connect, and another uses compression rings and nuts on the fitting to squeeze the ring to the outside diameter of the tubing. The fittings are classified in two categories: concrete-tight fittings or rain and concrete-tight fittings.

Intermediate metal conduit (IMC) has a thicker wall and is heavier than EMT, but thinner than the rigid metal type. It can be used in any location in which rigid metal conduit is used, provided the thinner wall can withstand whatever physical abuse may be anticipated at the installation location. Since it is thinner than rigid conduit, it is somewhat easier to bend, giving it an advantage for use in certain locations.

Rigid metal conduit is made of galvanized steel, plastic coated steel, aluminum, or silicon bronze piping. All except galvanized steel are known as corrosion-resistant conduit and will withstand harsher environments than galvanized steel. Silicon bronze can be exposed to severe weather and relatively harsh, corrosive environments where aluminum would not be suitable, such as piers, docks, bridges, and underwater, as well as refineries and chemical plants. Rigid aluminum conduit has the advantage of being lightweight and easy to work with, reducing labor installation costs. It cannot be used where it might come into contact with chloride compounds, as they severely corrode aluminum. Polyvinyl chloride coated steel conduit is used in locations such as paper plants, meat packing plants, industrial organic and inorganic product plants, canneries, food industries and fertilizer plants.

Rigid steel conduit may be painted with enamel (referred to as "black"), galvanized ("white"), or Sheradized ("green"). Sheradizing is a trade name for a corrosion-resistant coating.

Rigid metal conduit has all burrs and sharp edges removed to prevent damage to the wire insulation. It is manufactured in 10-foot lengths with factory-threaded ends which can be joined using threaded connectors. A variety of fittings are available for use with the conduit, as well as special lubricants that make pulling wire easier.

Rigid nonmetallic conduit is made of polyvinyl chloride (PVC) or fiberglass. It can be used in walls, floors and ceilings, in wet locations such as dairies and laundries, and in areas where there may be a corrosive atmosphere. It cannot be used in hazardous locations where it may be exposed to physical damage, such as around heavy mobile equipment like forklifts, or in areas where ultraviolet light may cause the pipe to deteriorate. It also cannot be used where the temperatures may exceed the limitations of the plastic or where there is the possibility of fire or explosion.

Rigid nonmetallic conduit is available in schedule 40 and schedule 80 wall thicknesses. Manufacturers' catalogs, handbooks and several standards list the actual wall thicknesses for the schedule designations. Wall thicknesses increase with conduit size. Schedule 80 is called "extra-heavy wall" by Underwriter's Laboratories. In addition to these two thickness designations, there is the thinner Type A, which can be used for underground applications, or may be encased in concrete. Schedule 40 and 80 conduit are suitable for above ground installation.

Lengths of PVC conduit can be cut with a fine-toothed saw and joined using special cement that fuses the surface of the pipe and fitting together. PVC can also be bent if properly heated. Fiberglass conduit can be cut with a fine-toothed saw and the lengths joined by fusing them in a socket-type fitting using the same type of epoxy or vinyl ester used in the manufacture of the pipe.

Flexible metal conduit can be used in most circumstances where you could use rigid conduit. However, it has the advantage of being flexible and therefore easier to install, especially in tight spaces or where tight bends make using rigid conduit impractical or even impossible. Flexible metal conduit is made of galvanized steel or aluminum that has been spiral wound with an interlocking seam. It is not water or gas tight, so it should not be used in a situation where the environment may be detrimental to the wiring or in wet locations, unless the wiring inside is designed for wet exposure. Although metal-clad flexible conduit can protect wiring from ordinary physical damage, it should not be used in locations where it will be subject to heavy traffic or constant use. It can be used in indoor locations, but not underground, embedded in concrete or in hazardous locations. The National Electrical Code contains specific limitations and exceptions for its use. There is a watertight flexible conduit, known as "Greenfield," which is available for use in wet locations.

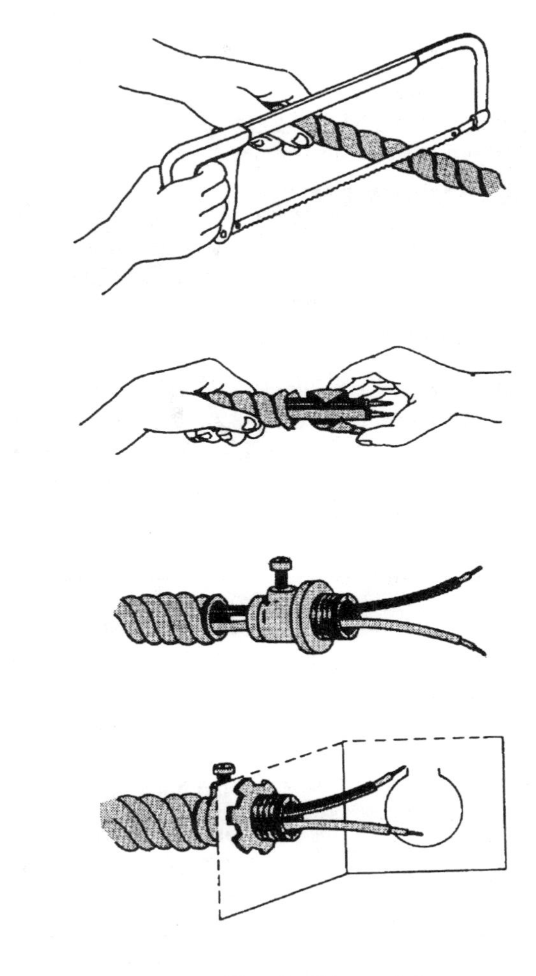

Installing flexible conduit

Surface raceway is a type of conduit that mounts on the surface of walls or ceilings. It is used when it is impractical to conceal the conduit inside the walls or ceiling, such as in room or office remodeling. Surface raceway and fittings are made of metal or plastic and come in a variety of colors to match or complement a room. It can also be painted to match the surface to which it is attached.

There is also a surface raceway that is made of plastic in the form of a baseboard. It serves both as a baseboard and a means of concealing the wiring. Receptacles are mounted in the baseboard, and the wiring is held by a strip within the baseboard raceway. The baseboard is made of paintable material and held in place with clips which attach to a backing plate mounted on the base of the wall.

Surface raceway can only be used indoors in dry locations because it does not seal the wiring against contact from the environment. It is only used to conceal the wiring and physically isolate it. The *National Electrical Code* has specific requirements for surface raceway. Two of these requirements are that it be flame retardant and crush resistant.

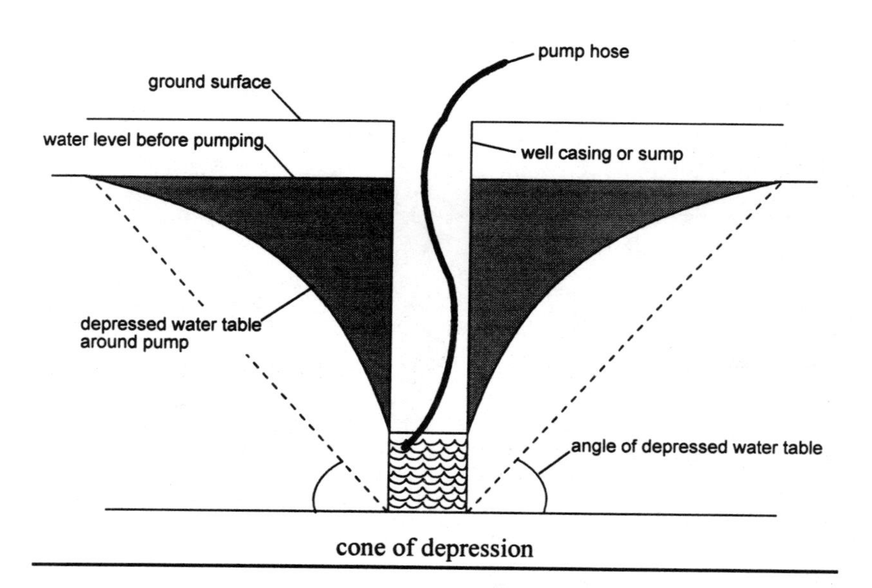

cone of depression

conduit strap – a support strap that fits around conduit and can be fastened to a structure to support the conduit.

cone – a three-dimensional geometric figure with a circular base and sides that taper to a point at the top.

cone of depression – the angle to which a water table is depressed during dewatering.

conform – 1) to meet a requirement; 2) to match a contour.

conformable strata – strata in the earth in layers that are similar in contour and thickness.

conglomerate – gravel or pebbles cemented together to form a type of rock. Also called pudding stone.

congruent – matching, as superimposed figures that match at all points.

conic – cone-shaped.

conic projection – a map made in the form of a projection of part of the earth's surface onto a cone. The cone is unrolled so that the map can be viewed flat. The flat map shows the actual shapes and relationships of the features of the earth's surface.

conic section – a section through a cone, forming an ellipse, circle, hyperbola or parabola.

conifer – an evergreen tree with needle-type leaves and seed cones.

connect – 1) to join together; 2) to bridge between.

connecting rod – a link between rotating and reciprocating parts of a machine. The connecting rod has a pinned connection at each end so it is free to rotate about the pin. A connecting rod connects the piston in an automobile engine to the crankshaft. The piston moves in a reciprocating motion within a cylinder, and the connecting rod imparts this motion to the crankshaft causing it to rotate.

connector – 1) a device to connect an electrical wire to another wire or to a piece of equipment; 2) a device to join two objects together.

connector plate – plates, usually made of steel, used to join structural members, such as a truss to a wall.

connector, pressure – an electrical connector that maintains contact by mechanical means. Also called a solderless connector or a wire nut.

consequential damage – damage incurred to other items as a result of the failure of one item.

conserve – to preserve from loss; save.

consistency – the thickness, cohesiveness and/or smoothness of a mix.

construction – all the various operations involved in building a structure, including electrical and plumbing systems, interior and exterior finishes, flooring, trim and possibly landscaping, completed and the structure ready for occupancy.

construction adhesive – a strong adhesive used to hold structural and/or non-structural members together. In its most convenient form, construction adhesive is available in cartridges to fit a caulking gun.

construction classification – a method of categorizing structures in terms of their fire resistance.

construction documents – the drawings and specifications for building a structure and/or other tasks to be completed as part of a construction effort.

construction joint – a temporary stopping place between successive concrete pours designed to form a good bond with the next section to be placed.

construction management contract – a contract in which the prime contractor is responsible for obtaining and managing subcontractors who perform the various tasks needed to complete the job.

construction specifications – see specifications.

consumable insert – a filler metal laid in the weld joint before welding. The filler metal melts and becomes an integral part of the weld joint during welding.

contact cement – a type of adhesive that is applied to both objects to be joined and then allowed to set up. When the desired state of dryness is reached, the objects are touched together and the coated surfaces bond on contact. One of the most common uses for contact cement is in bonding laminated plastic sheet to counters and other surfaces.

contain – to hold within boundaries.

container – a vessel used to hold a commodity.

contaminate – to infect something with unsuitable, undesirable and/or harmful material.

continuity – an electrical circuit that is complete, having no open circuits.

continuity tester – a meter or other indicator used to determine if there are open or broken connections or grounds in an electrical circuit.

continuous – without a break or termination.

continuous load – an electrical load of three or more hours in duration.

continuous power system – an electrical system that is independent of an alternate power source and supplies power without an interruption of more than one cycle. Used where an interruption of power would be extremely costly or hazardous.

continuous ridge vent – a vent into an attic located along the ridge of a roof.

continuous vent – the vertical section of vent piping in a building drain system that extends upward from the drain to which it is connected.

continuous waste – a drain that connects a fixture with multiple compartments or multiple fixtures to one common trap.

continuous weld – a weld along the complete length of the joint.

contour – an outline of a shape.

contour gauge – a hand tool consisting of a series of stiff metal wires held in a frame that allows them to slide back and forth. The wires can be pressed against an irregularly shaped object and clamped into place so that their arrangement conforms to the shape of the object. This allows the transfer of the shape to another member or location, such as transferring the contours of a door jamb and trim to be cut out in vinyl flooring.

contour interval – the amount of rise between contour lines on a contour map which shows the difference in elevation, such as 1 foot of rise between lines.

contour line – a line on a topographic map which passes through points at the same elevation. The different contour lines on a map show the changes in elevation of the land.

contour map – a map with contour lines that indicate changes in elevation. Also called a topographic map.

contour plan – a plan view of a site showing the finished grade contours indicated by contour lines.

contract – 1) an agreement between two or more parties to perform a specific task; 2) to shrink with temperature change.

contractor – a person who has agreed to perform a defined scope of work in accordance with the terms of a contract. A contractor may be the prime contractor, one who has contracted directly with an owner for work and is responsible for the entire job, or a subcontractor, one who contracts with a prime or general contractor for part of the work on a job requiring multiple skills.

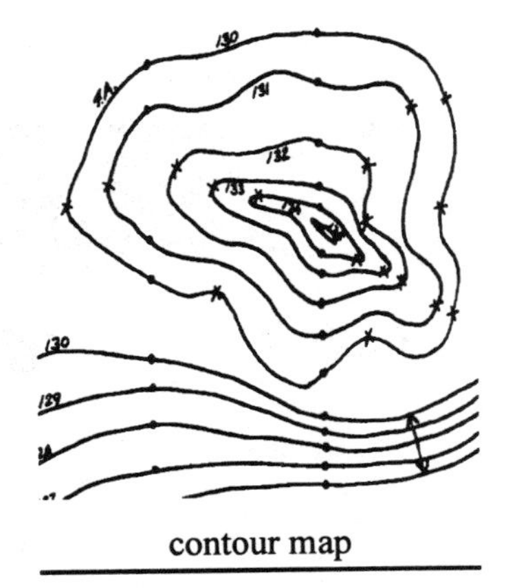

contour map

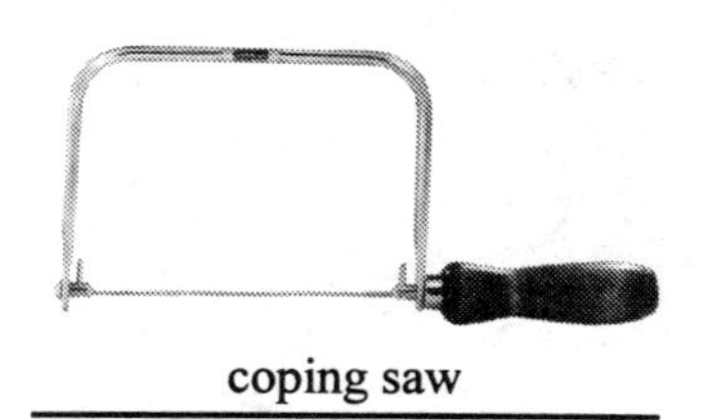
coping saw

control fittings – devices in a fluid system that start or stop flow and control the rate of flow and/or the pressure.

control joint – 1) a joint between sections of a structure designed to permit differential movement between the structures; 2) a space left in a large expanse of wallboard to allow for expansion and contraction of the surface. The space is filled with a preformed section of metal that can be finished and will allow movement to prevent cracking and buckling of the wallboard; 3) a groove in a masonry wall designed to control cracking.

controller – a device that governs the power that is delivered to an electric motor.

convection – thermal (heat) transfer through a gas or a liquid.

convector – an enclosed, free-standing heating device with openings at the top and bottom for air passage. The air passing through the openings is heated by hot water or other fluids passing through interior tubes over which the air passes. The warmed air flows out of the convector to heat the room.

convector cover – a cover that conceals a convector to make it more aesthetically acceptable.

convenience outlet – a 120-volt electrical receptacle into which appliances and other devices that use electricity may be plugged. Also called an electrical outlet.

conversion – changing from one standard or form (such as a unit of measurement) to another.

conversion chart – a chart showing equivalent units of measurement from one standard to another, such as from feet to meters.

converter – 1) a device that changes something from one form to another; for example, a device that converts analog to digital; 2) a metal refining furnace that utilizes air blasts as part of the refining process.

convex – a surface that curves outward.

conveyor – a mechanical device using a belt, or a variation of a belt, for moving a commodity from one place to another. Also called a conveyor belt.

convolute – a coiled or twisted form.

cooktop – a smooth ceramic stove top with heating elements under the ceramic.

coolant – 1) a medium used to extract heat; 2) a fluid used for cooling.

cooler – a refrigeration or heat extraction unit.

cooler nail – a gypsum drywall nail.

cooling tower – a device for removing the heat from water used in steam condensers or water used to cool down equipment. There are two types of cooling towers, the forced draft and the natural draft type. Cooling towers are used at stationary steam power plants, refineries, and in other industries requiring the cooling of large amounts of water.

cool time – the time that elapses between multiple weld passes during which heat is lost from the weld.

coop – a sheltered enclosure for small animals.

cooperative (co-op) – 1) a multi-unit dwelling owned jointly by tenant/shareholders, each of whom owns enough stock to possess at least one unit in the building; 2) a group working toward a common purpose.

cope – to cut or trim something to fit a specific shape.

coped joint – a joint where both pieces have been specifically cut to fit together.

copestone – the finishing piece of stonework along the top of a structure.

coping – 1) the cap on a masonry wall or other structure which prevents water from penetrating through the top; 2) cutting a member to a shape that fits tightly against another member.

coping saw – a U-shaped saw with a thin, fine-toothed blade stretched across the opening between the ends of the U. A handle, parallel with the blade, extends from the U at the end of the blade. The saw is used for elaborate scrolling cuts.

coplanar – two or more objects in the same plane.

copper – a reddish metallic element having good thermal and electrical conductivity; atomic number 29; atomic weight 63.54; atomic symbol Cu.

copper tubing or piping – tubing and piping made from copper alloy. They are available as thin (Type M), medium (Type L), and heavy wall (Type K), in flexible (annealed tubing) and rigid (drawn tempered pipe), with outside diameters ranging from ¼ inch to 12 inches. Connections can be made using flare fittings, compression fittings, and solder joints. Threaded joints are generally used for pipe. Standards for copper tubing used in fluid pressure applications are set in ASTM Specification B88. The basic parameters, wall thicknesses required, and minimum pressure capabilities for each type of application is given in the material specifications. When selecting the proper type of tubing for a job, the thickness value must be calculated using the formula and allowable material stress values given in the code. Once the calculated wall thickness that is applicable to the design has been determined, the closest higher thickness that is available for that size tube should be selected from ASTM B88. The outside diameter of the tubing is slightly greater than the nominal size in each size. For example, ½-inch tube has an actual OD of 0.625 inches, and 2-inch tube has an actual OD of 2.125 inches. See also piping, copper.

corbel – a projection from the face of a wall or column, usually projecting more with each succeeding course of masonry units.

corbel arch – corbels that meet across a span.

cord – 1) thin rope or string; 2) small electrical cable; 3) a measure of a quantity of firewood that is 4 feet deep by 4 feet high by 8 feet long.

cordage – rope and cables that constitute the rigging of a nautical vessel.

cordless screwdriver – a screwdriver which uses a built-in rechargeable battery as its power source. The battery allows it to be operated for an extended period without an external connection to a power supply. There are also other cordless tools available, such as drills and saws.

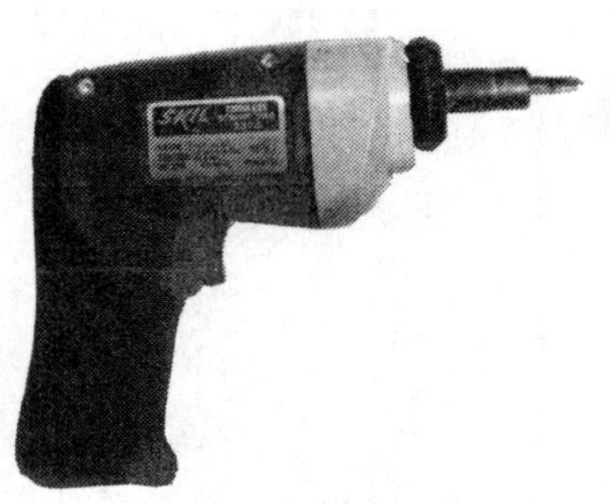

cordless screwdriver

core – 1) the center of an object; 2) a sample drawn or drilled out from the earth or an object, such as a tree trunk or rock; 3) the gypsum in a wallboard panel between the facing and backing papers.

coreboard – the one-inch thick gypsum board designed to fit between wallboard panels in a self-supporting gypsum wall. Coreboard is held in a metal channel at the top and bottom, and gypsum facing panels are fastened to the coreboard using adhesive and screws.

cored brick – masonry units with longitudinal holes running the length of the unit.

core drill – a hollow drill used to remove a section of material.

cored solder – a solder, in wire form, with a center core of flux.

core stock – the center ply in a built-up panel of wood material.

Corian – a trade name for a synthetic marble made of methyl methylacrylate, a type of plastic.

cork – the bark of a type of oak tree.

corkwood – 1) a specific type of small deciduous tree or shrub, with the Latin name leitneria floridana, that has a light, porous wood; 2) any tree or shrub, such as balsa, that has a light, porous wood.

corner – the junction of two planes or surfaces.

corner bead – a formed piece of metal designed to protect an external corner. Also called corner reinforcement. *See Box.*

corner block – a glue block, usually made of wood, that is fitted tightly into an inside corner of a cabinet and fastened with glue and/or screws to reinforce the joint.

corner board – exterior trim used with siding at the corners of a house.

Corner bead

Inside corner bead

sharp fold

A metal strip, bent at a right angle, that wraps around or fits into a corner to protect it and add strength to the corner joint. Corner bead is installed in one continuous piece for a smooth corner finish. It comes with solid metal flanges or expanded metal (mesh) flanges. A clinching tool, drywall nails or screws can be used to fasten corner bead with solid flanges. The mesh flange type is installed with staples. When the bead is secured in place, the first coat of joint compound can be applied with a taping knife wide enough to spread the compound about 2 inches beyond the edge of the flange. The corner bead provides a screeding surface for the taping knife and holds the two edges of the wallboard together at the corner to make a smooth joint. The second and third coats of compound are feathered about 2 inches beyond the edges of the previous coats, providing a smooth finished wall corner.

Outside corner bead

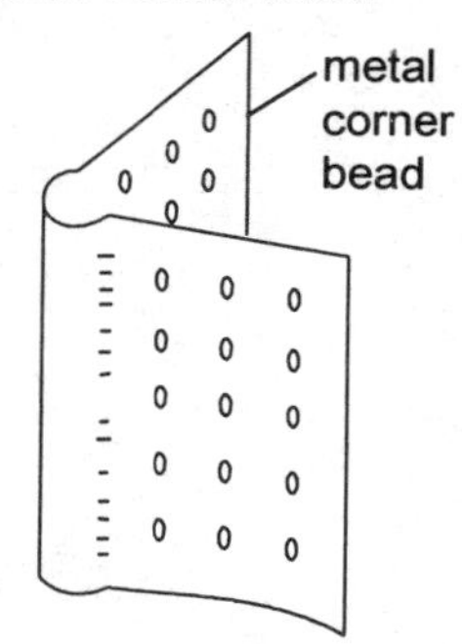

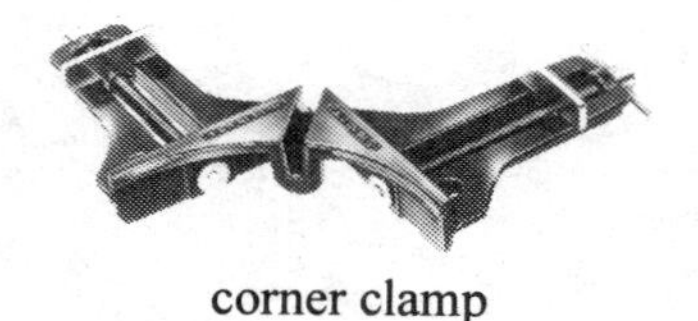

corner clamp

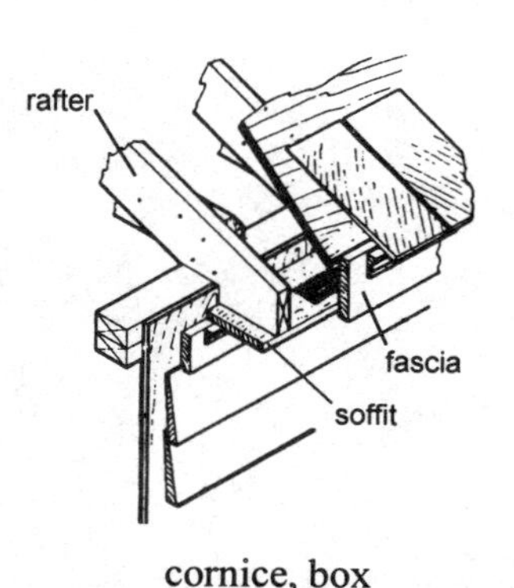

cornice, box

corner brace – 1) a diagonal brace across studs that adds support to wood frame structures; 2) a metal strap bent at right angles and used to reinforce corner joints.

corner chisel – a chisel with two straight cutting edges coming together at point forming a right angle. It can be used to cut both sides of a recessed right angle simultaneously in wood.

corner clamp – a clamp designed to hold members in place that are joined at an angle.

cornerite – a metal mesh lath that is used to add strength at corners in plastering.

corner post – 1) a structural member, built up of 2-by lumber, which supports the wall corner; 2) a fence post supporting a change of direction in the fence.

corner reinforcement – wire or expanded metal formed to use on inside or outside corners that require reinforcement when plastering walls. It is completely concealed by the plaster when it is applied.

cornerstone – a masonry block set at the corner of a building, usually at the base. It often contains information about the structure and may conceal a vault holding items to be saved for the future, like a time capsule.

cornerstone box – a receptacle, of durable metal, for storing items behind a cornerstone for some future generation to find.

corner tool – a tool designed to apply drywall joint compound or plaster simultaneously to adjacent sides of an inside or outside corner.

cornice – 1) a decorative molding placed at the top of a wall; 2) the junction formed by the eaves and walls of a building.

cornice, box – a cornice in which the rafter ends are covered by fascia and soffit.

cornice, closed – a cornice with no rafter projection beyond the walls. Also called a close cornice or flush cornice.

cornice, open – a cornice that is made without a fascia board fastened to the ends of the rafters.

cornice return – the part of the overhang trim members that continue around the end of a gable roof.

corona – 1) the vertical projection of a cornice designed to divert rainwater away from a structure; 2) a chandelier with concentric hoops.

corona discharge – a discharge at the surface of one or more conductors, often with a glowing light. It is caused by a buildup of excess electrons on a conductive surface at high voltage, and is usually accompanied by the ionizing of the air immediately surrounding the conductor.

corp – see corporation stop.

corporation – a group of persons, formed together under law, who share powers, responsibilities and liabilities as a group rather than as separate individuals.

corporation stop – a plug-type valve tapped into a main line, such as a water main, to control the flow to a branch line. Also called a corp or corporation cock.

corrosion – the chemical deterioration or degradation of a metal, such as the rusting of steel.

corrosive – the quality of a substance that can cause the chemical deterioration of a material.

corrugated – a surface that is composed of a series of alternating rounded ridges and valleys.

corrugated fastener – a small, rectangular corrugated fastener with one sharp edge which is placed across a wood joint and driven into the wood to strengthen the joint. Also called a corrugated staple.

corrugated iron – sheet iron that is corrugated and usually galvanized for corrosion protection.

corrugated metal pipe – galvanized steel pipe made from rolled and welded corrugated sheet steel.

corrugated staple – see corrugated fastener.

corrugated steel – sheet steel that is corrugated and usually galvanized for corrosion protection.

corrugation – the series of rounded ridges and valleys that form a corrugated surface.

cosigner – one who jointly makes an obligation.

cost – the price of an item or task.

cost plus fixed fee contract – a contract in which the owner pays for all of the costs plus a fee to the contractor who performs the work.

cost plus percentage contract – a contract in which the owner pays for all of the costs plus a percentage of those costs to the contractor.

cotter – a pin or fitting that slides into a hole or slot to secure an object in place.

cotter pin – a looped wire pin used to secure machinery parts. The pin is inserted through a hole, such as in a shaft, and then the ends are spread or bent to keep the pin in place.

coulisse – a piece of wood with a groove used to guide a sliding panel, such as a sliding door in a piece of furniture.

counter – 1) the top surface of a cabinet or other structure that is used as a worktop, eating area, or other purpose; 2) to offset something.

counteract – to work against a force.

counterbalance – a weight used to offset the effects of another weight, or to balance a wheel or other rotating object.

counterbore – to increase the diameter of a hole for a portion of its length. For example, a hole through a 2-inch thick material may have a 1-inch diameter for ½ inch, and a ¾-inch diameter for the remaining 1½ inches of its length. It was counterbored to a diameter of 1 inch to a depth of ½ inch.

counterclockwise – moving in the opposite direction that the hands of a clock move.

counter electromotive force – the voltage produced by a magnetic field that is in opposition to the magnetic field around a coil. When a coil is cut by a magnetic field, a voltage is induced in the coil. The current that flows through the coil builds a magnetic field around the coil. This field cuts across the turns of that coil and induces a second voltage which builds a magnetic field that opposes the source current. This induced voltage is called counter voltage or counter electromotive force.

counterflashing – flashing used on chimneys and other masonry to prevent water leakage. One edge of the flashing is embedded in the masonry and the other edge overlaps the roof flashing.

counterpoise – equilibrium; balance.

counterrotating – rotating in opposite directions.

countershaft – see jackshaft.

countersink – 1) the cone-shaped depression at the outer end of a hole, made to allow a flathead screw or bolt head to set flush into the surface of the material; 2) to drive the head of a finishing nail or brad below the surface of the wood in order to conceal it.

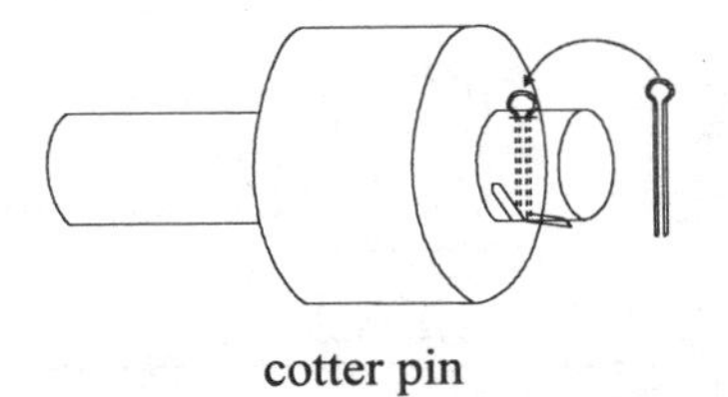

cotter pin

countersink bit

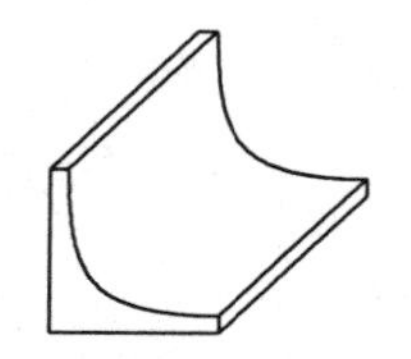
cove molding

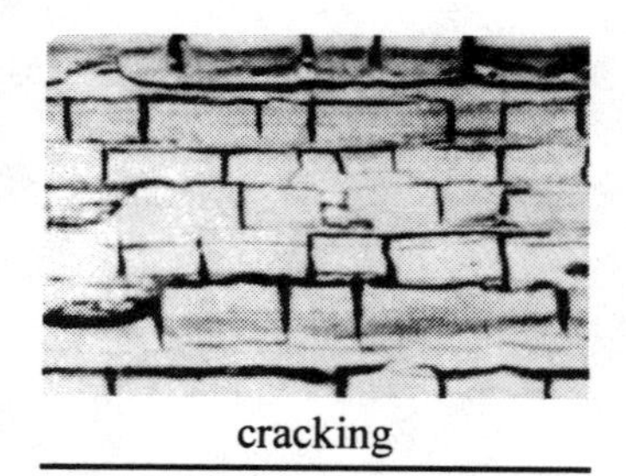
cracking

countersink bit – a conical-shaped cutting edge on the end of a shank that is designed to be rotated in a chuck and used to cut countersinks.

counterthrust – an opposing thrust.

countertop – the top surface, usually about 3 feet high, of a floor-mounted cabinet. The covering material for the surface is usually chosen with the intended use in mind to ensure a durable, appealing and serviceable work area. A wear-resistant material such as ceramic tile or laminated plastic sheeting is common, as well as water-resistant synthetic marbles or non-laminated plastics.

countertorque – a method of braking an electric motor by reversing the power to the motor in order to develop torque opposite to the direction of rotation.

counterweight – a balancing weight used to offset another weight.

coupling – 1) a pipe or tube sleeve fitting with two female ends into which the slightly smaller ends of pipe or tube are inserted to make a joint; 2) a device designed to hold two other objects together to make a joint, such as a motor shaft and a drive shaft.

coupling, bandage type – a wrap-around pipe repair clamp used to seal longitudinal cracks. It consists of an elastomer seal that encircles the pipe and a longitudinally split metal sleeve that fits around the elastomer to provide strength. The two halves of the sleeve are tightened around the elastomer seal with band clamps.

coupon – a sample from a piece of a material used for destructive testing or as an archive sample. ASTM specifications covering destructive testing of a material, such as tensile testing, specify the dimensions of the sample to be used as a coupon. For some specific critical applications, an archive sample of material may be cut and retained for the life of the project. In the event of a material failure, such an archive sample may be invaluable in helping identify changes that took place in the material while in service and assist in determining the cause of the failure.

course – 1) a continuous row of masonry units installed with mortar; 2) a route of movement.

court – 1) an outside area enclosed by walls (courtyard); 2) a spacious interior, usually with large glass areas; 3) a level game area.

cove – 1) a corner with an inside radius; an interior corner; 2) a recessed place.

cove lighting – lighting installed near the junction of the ceiling and walls which is shielded to create indirect light.

cove molding – molding, with a concave face, designed to fit into an interior corner.

covenant – an agreement, or a clause in an agreement, between two or more parties to do or not do something specific. An example would be the Covenants, Conditions and Restrictions governing a housing development and enforced by either the city or a homeowners association. All parties buying into the development must sign a statement that they will comply with this agreement.

coverage – the number of square feet of surface area that can be covered by a specified material.

covered electrode – a consumable metal arc welding rod that is coated with a flux and/or other materials designed to enhance the weld quality.

covering – the outer skin of a metal frame building.

cover plate – a removable metal cover.

cowl – a hood on a vent pipe that prevents water or snow from entering the pipe.

crack – a fissure or fracture in a material.

crack a valve – opening a valve slightly to allow a small flow through it.

cracking – 1) a paint defect caused by the incorrect mixing of materials; 2) a heat refining process.

crackle lacquer paint – paint that will develop an alligator skin appearance when sprayed over a gloss base lacquer. This technique is used most commonly on automobile dashboards and on the outside surfaces of safes, especially small, portable safes.

craft – a skill which involves some manual ability such as carpentry.

craftsman – a person skilled at a craft, such as carpentry.

cramp – an iron or steel rod, with the ends bent at right angles to its length, used to hold blocks of stone together.

crane – a motorized hoisting machine that can lift, lower or move a load horizontally using load cables and a boom. They are used in a variety of jobs in construction such as moving earth during excavation or lifting precast concrete walls in tilt-up construction. There are several types of cranes designed for specialized lifting and/or moving jobs. *See Box.*

crane bridge – the structural member of a bridge crane that spans the distance between the girders or rails.

crane cab – a housing for the crane operator attached to the bridge or trolley of a crane.

crane girders – the structural members on which a bridge crane rides and is supported. Also called crane rails.

crane jib – an extension at the top of a crane tower that gives the crane additional lifting or moving capabilities. There are different types of jibs. A saddle jib is a horizontal extension, at a right angle to the tower, with a hook attached to a trolley. The trolley can be moved along the length of the jib to alter the reach of the hook. A saddle jib can handle loads both at a distance and close to the tower. A luffing jib is an extension that pivots at the top of the tower. It can be held at a variety of angles, from near vertical to near horizontal. The hook is suspended from the end of the jib. A greater lifting height is available from the luffing jib than the saddle jib, and since the height of the luffing jib can be changed, obstacles can be avoided as well. A fixed luff jib also pivots at the tip of the tower, but it is fixed at a horizontal angle. The hook may be mounted on a trolley which can be moved along the length of the jib, or be suspended at the end of the jib.

crane, manually operated – 1) a crane operated by hand using an endless closed-loop chain that works like a pulley; 2) a crane that is not power driven and must be moved to different locations by external means.

crane rail – the structural members on which the ends of a bridge or gantry crane rides.

crane, remote operated – a radio controlled crane.

craning – using a backhoe bucket, with a chain or cable wrapped around the bucket teeth, as a lifting device.

crank – a lever and wheel mechanism to convert between reciprocating and rotary motion. One end of a lever is connected to a wheel near the outside diameter, and goes around with the wheel as it turns. The other end of the lever is a crank which is connected by some means to a piston within a cylinder or to some other device on a guide or rail. When the end of the lever moves with the wheel, the motion of the crank end of the lever is translated into a sliding, or reciprocating, motion.

crankcase – an enclosure for the crankshaft in a machine.

crankpin – the pin at the outer periphery of a crank to which a connecting rod is attached.

crankshaft – the rotating shaft to which connecting rods are attached to convert between rotary and reciprocating motion, used in machinery and piston engines.

crank shaper – a metal shaper in which the reciprocating arm is driven by a crank pin on the main driving gear. See also shaper.

crater – a hole-like defect in a weld or a surface.

crawling – an uneven paint surface caused by shrinkage during drying.

crawl space – the space within the foundation perimeter under a building's flooring that allows access to plumbing pipes and other systems.

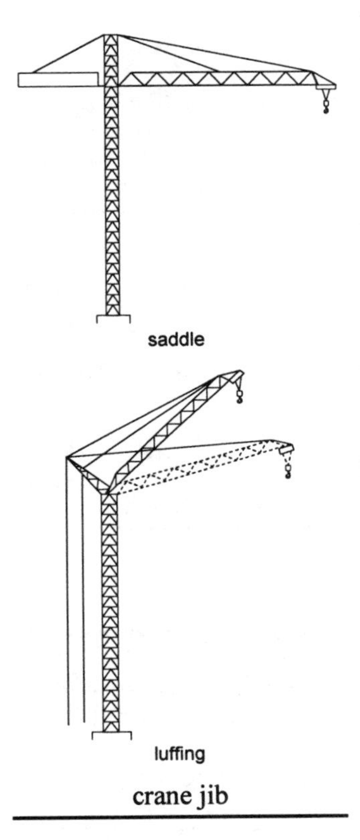

crane jib

Crane

Courtesy: Coast Crane and Equipment

All-terrain crane

The crane is one of the most widely-used pieces of motorized construction equipment. There are several types with a variety of uses, but the basic operating components are the same. They are: the mounting, which may be crawler tracks, tires, a platform or another medium; the deck and cab which sit on the mounting and house the operator and controls for the crane; and the boom, which is the essential lifting and turning device for the crane. The deck and cab on most cranes rotate 360 degrees, as does the boom which is attached to the deck. The boom can be raised or lowered from the deck by cables or by hydraulic power or a combination of both. The purpose of the crane is to lift and move earth, supplies or materials from one location to another.

Vehicle-mounted cranes are permanently mounted on wheeled or tracked vehicles, making them portable. They can be readily moved to and around jobsite locations. They are used primarily in construction activities where a permanent crane is not needed, or to supplement an existing crane. Vehicle-mounted cranes are categorized as truck-mounted, rough-terrain or all-terrain type vehicles. The selection of one of these types depends on the job requirements and the jobsite conditions. Hydraulic power is used to a large extent in mobile crane designs to provide great power in a small space. Diesel power is used for transport as well as for powering the hydraulic pumps. Many vehicles are equipped with outriggers which extend out from the vehicle to rest on the ground and provide a stable base during crane operation.

Vehicle-mounted cranes vary in lifting capacity from a few tons to several hundred tons. *Truck-mounted cranes* are designed to move quickly between jobsites and are capable of highway speeds. Their crane capacities range to about 300 tons. *Rough-terrain cranes* are used in off-road situations. They have a relatively short wheel base and high vehicle clearance. The crane booms have telescoping extension capabilities to reach loads and capacities ranging to about 40 tons. *All-terrain cranes* combine highway speed capabilities with off-road capabilities to increase their versatility. The vehicle axles can move independently on their suspensions to maintain wheel contact over rough surfaces. Automatic crane bed leveling makes set-up faster, and suspension locking features accommodate different lifting requirements. Crane booms telescope so they can be extended as needed to reach and move loads. Crane capacities on this type of vehicle range to over 100 tons.

A *lift crane* may be mounted on a vehicle, a platform or floor. This is the most common type of crane, consisting of hoist cables and a hook and block, and is used to move an object to a new location. It may or may not be equipped with hydraulics to aid in lifting. The same crane may be fitted with other components such as draglines, clamshells and shovels and used for excavation work. Hydraulic lift cranes can be fitted with several boom variations, including a telescoping boom. Lift cranes are also called crane booms or hooks.

Power cranes, with hydraulically operated backhoe and shovel attachments, are also commonly used in excavation work. These have shorter booms than the lift crane and are commonly vehicle-mounted.

Most *tower cranes* are stationary, rather than mobile. They are used when there is a long-term need for a crane and the rental cost will be less than a mobile crane, when site space is limited, when the required lift is very high or long, or when the required lifting and moving is best done from within the structure being built. The stationary tower crane occupies little space and can be set at many different heights. However, there is a maximum free-standing height. If the crane must be raised above this height, it can be supported by tying it to the building being constructed. The lifting capacity of the tower crane decreases with the distance of the hook from the tower, and this limitation must be considered. There are different styles of tower cranes, with different tower configurations. A *mono-tower crane* has a single tower structure. An *inner-and-outer tower crane* has the outer tower as the main supporting structure, on which the jib and inner tower turn. A *telescopic tower crane* has two or more nested sections, which permits changing the crane height without dismantling the crane.

Tower cranes may also be rail mounted, the climbing type, crawler mounted, or truck mounted. *Rail-mounted tower cranes* provide a maximum area of coverage, while taking up little space. They are stable enough to move with a load, but they are expensive, when track installation costs are considered. The *climbing tower crane* is often used with the construction of very tall buildings, but the structure must be strong enough to support the crane. At the beginning of construction, the crane is mounted at grade level, but as the structure height increases, the crane mounting point is lifted to higher elevations within the structure. The *crawler-mounted tower crane* must be located on a level and stable surface when lifting a load. It can travel over smooth ground at a slow speed with the tower partially erected. A *truck-mounted tower crane* must be level and on stable ground and have outriggers extended when lifting. It requires dismantling for moving, and cannot carry loads while moving.

A *bridge crane* is a permanent installation. It is mounted between two overhead tracks or rails with the main structural member forming a bridge between them. The rails are supported by the building structure. The crane bridge can move along the length of the rails, while the trolley containing the hoisting mechanism can move back and forth along the bridge, permitting the lifting hook or hooks to reach virtually any location beneath the bridge and rails. Larger bridge cranes have a trolley mounted cab in which the operator rides. The operator has direct visual contact with the load, as he travels with it. Smaller bridge cranes have a control box on a cable that reaches to the floor of the building. They are controlled by an operator on the floor who walks along with the load as it travels and controls the movement and lifting.

Courtesy: Coast Crane and Equipment

Rough-terrain crane

A bridge crane that moves on overhead rails as described above is often called an *overhead crane*. An *underhung crane* operates like an overhead crane except the bridge hangs from a carriage suspended under the crane rails. Another type of bridge crane is a *gantry crane*. It is mounted on rails that are supported on a series of braced legs, like a trestle. A *semi-gantry crane* has one rail supported on legs and the other rail supported from the building structure.

Bridge cranes are used in buildings where there is a constant or periodic need for lifting and moving loads within a building. One example of this type of use is in a manufacturing facility where large pieces of material, such as airplane parts, must be moved along an assembly line.

A *polar crane* is similar to a bridge crane in many ways. It is a permanent part of the structure in which it is installed. The difference is that it rides on rails that form a circle, rather than extending along a straight line. The crane bridge is located across the diameter of the circle, and the trolly and hook can travel back and forth along the length of the bridge. The bridge rotates about its center, just as a planet rotates about its poles (hence its name, polar crane). The hook can reach any point within the circle of the rails. This type of crane is used primarily in the domed interiors of nuclear power plants. It is used for reactor head removal and replacement, and for refueling operations.

crawl space plenum – a heating system which utilizes the crawl space (or a large portion of it) to accumulate and distribute the heated air by means of fans.

crazing – a network of hairline cracks in the surface of paint, brick, or any other type of surface. Also called stress cracking.

credit – to supply goods or services in trust that the purchaser will pay for them within a specified time allowed by the seller.

credit report – an evaluation of the credit worthiness of an individual or business. It covers the current financial condition and credit history, based on assets and income and past ability to pay credit debts on schedule. Usually made to determine whether a loan or credit extension would be a risk for the lender.

creep – 1) to go very slowly; 2) a slow change over time in the shape of an object or material caused by stress in the form of weight, pressure, and/or prolonged exposure to heat.

creeper – a low wheeled platform on which a person can lie down and slide under a machine, such as an automobile, to work on it.

creeping – a paint defect in which paint runs together in small droplets.

creep speed – a slow and constant crane movement speed.

crenels – evenly spaced square notches along the top edge of a wall. Also called crenellations.

creosote – a chemical preservative with a coal-tar base used to treat wood items such as utility poles and railroad ties to protect them from deterioration and decay caused by insects and rot.

Crescent wrench – the trade name for an adjustable, open-end wrench. The jaws adjust so that one wrench can fit a large number of different size fastener heads.

crest – the top of a ridge.

crib – 1) a small dwelling; 2) a small room or space; 3) a structural framework; 4) storage bin for grain; 5) a barrier projecting into a river which reduces the rate of flow and provides storage for floating logs.

cribbing – 1) closely spaced supporting timbers in an underground shaft, such as a mine; 2) digging a narrow trench with a backhoe, starting near the machine and working away from it. The spoil is deposited in the newly dug trench near the machine and then removed as the trenching progresses. By moving the spoil along the trench and out near the machine, only the trench is affected, and the area on either side of the trench remains undisturbed; 3) wood placed to support a load and keep it from shifting during transit.

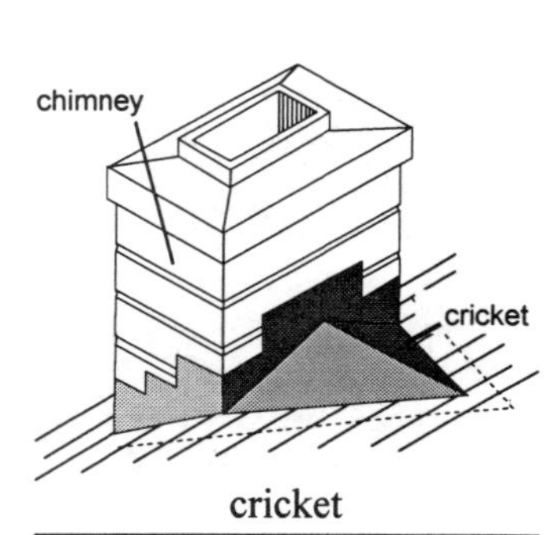

cricket

cricket – a peaked structure, designed to divert water at the juncture of two roof slopes or at a roof slope and a vertical structure, such as a chimney.

crimp – 1) to make evenly spaced alternating folds; 2) to make into corrugations; 3) to fold sheet metal edges for seams; 4) to mechanically clinch pieces of sheet metal together.

crimping tool – 1) plier-like tool used in sheet metal work to make corrugations in the end of a duct or other metal piece in order to reduce its size; 2) a tool designed to clinch corner bead or metal studs in place without separate fasteners; 3) a tool designed to crimp or squeeze an electrical connector tightly around a conductor.

cripple – a short stud between a header and a top plate or a sill and a sole plate in wall framing.

critical load – a radioactive crane load that, if dropped, could lead to the release of unacceptable levels of radiation or impair the safe shutdown of a nuclear power plant.

Critical Path Method (CPM) – a management tool used for work scheduling in which all the major tasks that must be completed to finish a job are laid out in a diagram. This diagram shows the proper sequence of work, with parallel operations indicated, and the time required for each task. This provides a visualization of the work to be done and indicates which operations are critical to the completion of the others.

critical point – the physical condition at which a substance exists in two forms, such as a liquid and a gas.

crocket – an architectural decoration that consists of a plant-like knob on an upward curving stalk. Crockets are used to decorate vertical or inclined surfaces, such as along the eave line of a sloping roof.

crook – a longitudinal warp in a board that creates a curved edge.

cross – a plumbing fitting used to join four lines together.

cross beam – a structural beam spanning across a space.

cross-bedded – irregular earth laminations or strata made up of layers of different types of compressed soil and/or rock. These layers can be readily seen when viewed from the side, after a cut has been made through such an area.

cross bond – a brickwork pattern in which the joints of alternating courses line up with the centers of the courses on either side.

cross brace – see cross bridging.

cross break – a separation across the grain in a piece of lumber.

cross bridging – small wood pieces placed at angles so that they extend from the bottom of one floor joist to the top of the adjacent joist to add stability to the structural members. They literally bridge the floor joists.

cross connection – a connecting line between two piping systems.

crosscut – to cut across the wood grain.

crosscut saw – a saw designed to cut across the grain of the wood.

cross filing – filing with alternating strokes at 90 degree angles.

cross-garnet – a tee-shaped hinge.

cross-grained – irregular wood grain running in a pattern other than along the length of a board.

cross hairs – fine wires in the focus piece of an optical instrument, such as a transit, which provide visual reference lines.

crosshatch – the use of a series of parallel lines to indicate materials in a cross-section drawing. There are conventions for the type of crosshatching used to represent various materials such as concrete, wood or metals.

cross-level – an elevation taken at right angles to the main line of surveying levels.

cross-line – a buried line that crosses the line of an excavation to be made.

cross peen hammer – a metal working hammer that has a head with one end shaped like a cylinder with a flat face and the other end shaped in a wedge. The wedge has a rounded edge. The hammer is used for breaking and chipping hard surfaces.

cross section – a drawing showing a view of the internal construction of an object or structure.

cross sectional area – the surface area of the portion cut from a cross section, or a cut through an object. An example would be the area of the end of a wire that has been cut straight across.

cross shaft – the shaft on a bridge crane that connects the drive motor to a wheel at each end of the bridge. Also called squaring shaft or drive shaft.

cross spraying – paint spraying in directions at right angles to each other. Spray painting in this manner lays an even coat of paint.

cross tees – the members that support the ceiling tiles in a suspended or drop ceiling.

cross tie – a structural member running from side to side across a structure.

cross ventilation – the circulation of air through an area achieved by having windows or vents on opposite sides of the area.

crosswalk – a pedestrian lane crossing a vehicular roadway.

crotch – the angle or juncture formed by the branching out of members from a common base, such as the angles formed by branches growing out from a tree trunk.

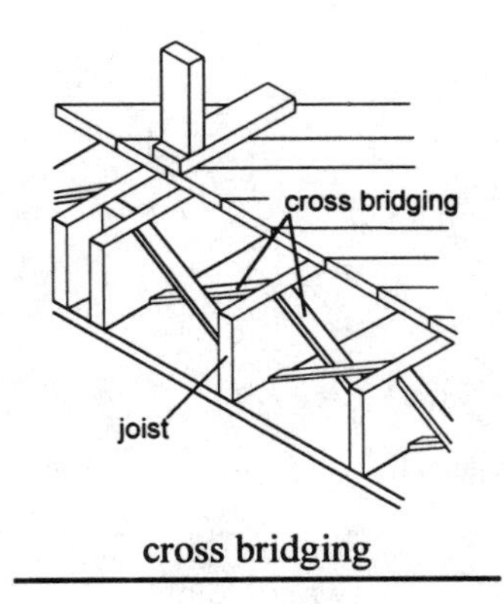

cross bridging

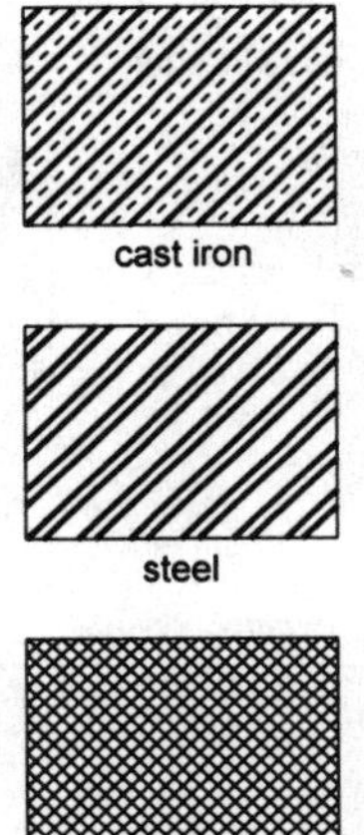

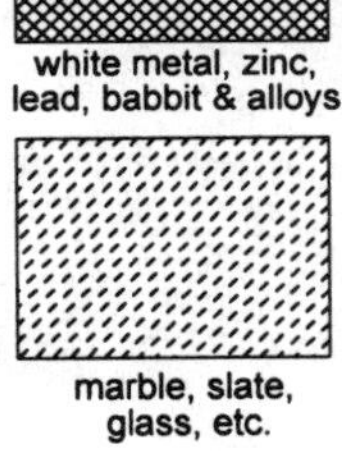

crosshatch material symbols

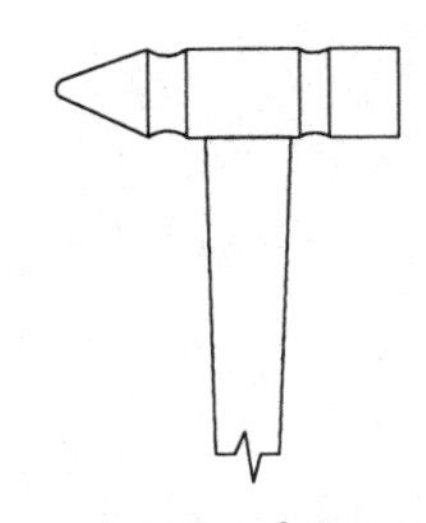

cross peen hammer

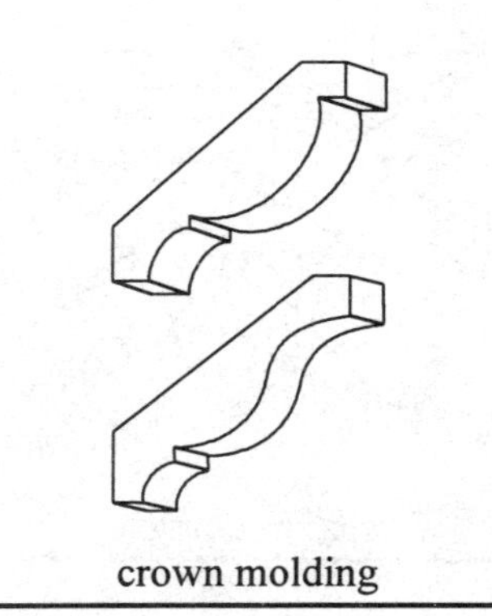
crown molding

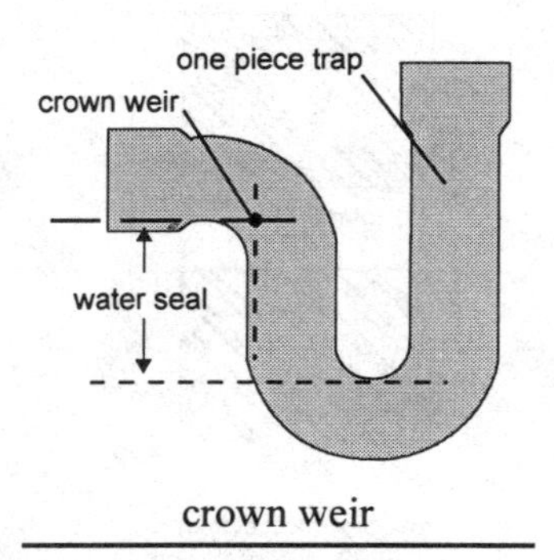

crown weir

crotchet – a small hook or item resembling a hook.

crowbar – a heavy metal pry bar tapered to a wedge on one end, with the other end bent into a hook with a claw foot for pulling nails.

crowd – the moveable arm of a backhoe that connects to the bucket.

crown – 1) the buildup of joint compound over a taped joint in gypsum wallboard; 2) the high center of a roadway; 3) the outward bow of a board.

crowning – installing boards crown up during framing.

crown molding – molding with a curved face and chamfers on the edges used to cover large angles, such as the intersection of a wall and ceiling.

crown plug – a screw-in plug for a floor-mounted electrical box.

crown saw – a hole saw, consisting of cutting teeth in the end of a cylinder. The cylinder is mounted on an arbor and is turned in a drill to cut a hole.

crown weir – an internal projection in the outlet side of a P-trap that assists in maintaining the water seal in the trap.

crows foot – a survey stake marked with the cuts and fills to be performed during earth grading operations.

CRT – see cathode ray tube.

crucible – a container made of heat-resistant material for heating, melting or transferring molten material.

cruciform – arranged or shaped like a cross.

crud – a layer of dirt, grease or foreign matter.

crude – not finished; rough.

crumbing shoe – a shoe fitted to a wheel trencher that forces dirt back into the trencher buckets to keep it from falling into the trench.

crust – the hardened or solidified covering of the earth's surface.

cryogen – a substance which produces very low temperatures, such as liquid nitrogen, liquid oxygen or liquid hydrogen. A liquified gas is very cold and takes up much less volume than when in gas form. Nitrogen becomes liquid at -320 degrees F and oxygen at -297 degrees F.

cryogenic liquid – a liquified gas, such as liquid nitrogen.

cryogenics – the science or technology that deals with the production and effects of very low temperatures.

cryostat – an insulated vessel used to maintain a constant low temperature.

crypt – an underground vault or chamber.

cubature – the cubic content of a container.

cube – 1) a six-sided figure in which all sides are equal and all angles are right angles; 2) raised to the third power.

cubicle – a small, partitioned area.

cubic measurement – 1) the measure of volume in cubic units such as cubic inches, cubic feet or cubic yards; 2) three-dimensional measurement; 3) in the shape of a cube.

cul-de-sac – a street with only one outlet and ending in a circular turnaround area.

culling – sorting and grouping masonry units or other items by size or shape.

culls – items or building materials that are rejected because they do not meet standards.

culvert – a large drainage pipe used to direct the flow of drainage water under a road or other passageway.

cuneiform – in the shape of a wedge.

cup hook – a small open hook, usually of brass, with a threaded shank and often a small escutcheon above the threads. The hook can be screwed into wood, as in a cupboard, up to the escutcheon and cups hung from the hook.

cupola – a small decorative structure shaped like a small house or dome extending upward from the ridge of a roof.

cupped taper – a defect in the tapered edge of a wallboard panel in which the taper is dished, or has a concave indentation, so that the edge of the taper is actually thicker than the center of the taper. Special effort must be taken when finishing a joint with a cupped taper to ensure a well-concealed joint. Extra compound applied to the cupped area and a greater degree of feathering should blend it to the wallboard surface.

cupronickel – an alloy of copper and nickel. It is corrosion resistant, very malleable without being annealed, and has high temperature properties exceeding those of other copper alloys. It is also called nickel silver because of its white color and tarnish-free characteristics.

curb – 1) a rim or raised border, such as along a street, or around a floor-mounted tank; 2) a control over a process or action.

curb cock – see curb stop.

curb edger – a trowel for rounding the edge of freshly poured concrete.

curb machine – a machine that forms and pours a concrete curb as it moves along a road.

curb-mounted skylight – see skylight, curb-mounted.

curb roof – a roof that has two or more slopes on each side of the ridge or high point, the lower slopes having a steeper angle.

curb shoe – a metal shoe that is bolted to the bottom of a scraper blade. The curb shoe is shaped to match the bottom of a curb and is used to trim the slope at the edge of a street in preparation for putting in curbs.

curb stake – a survey stake that locates and identifies the curb grade to be cut.

curb stone – a stone used as a curb.

curb stop – a shutoff valve used typically at building water meters, between the meter and the building.

cure – 1) to change properties by aging; 2) to allow a chemical reaction to complete its process.

curing time – the time required from initial application for a material to attain its design properties.

current – 1) flow in a fluid; 2) the quantity of electrical flow; 3) present time.

current assets – assets that are liquid or can be made liquid within one year.

current liabilities – liabilities that are due to be paid within one year.

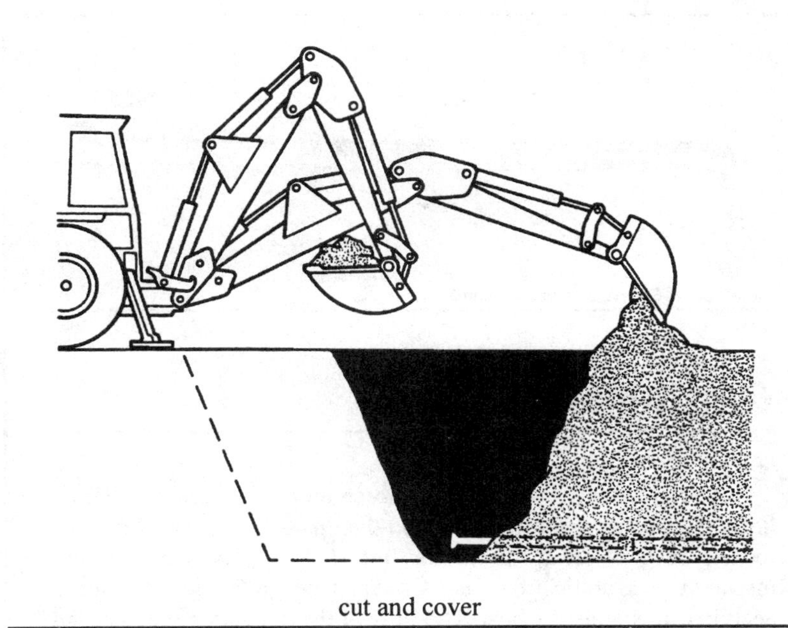

cut and cover

current ratio – the ratio of current assets to current liabilities.

curtain wall – a nonbearing exterior wall that is suspended in front of the structural framing, with its load transferred to the framing.

curvature – a bend or curve.

curve – a bent line without angles.

curved claw hammer – a hammer with a split wedged claw bent into a curve to provide leverage for pulling nails.

curvilinear – described by or consisting of curved lines.

cushion material – materials, such as fine aggregates or layers of roofing felt, used in paving to level and smooth the finished grade.

cusp – a decoration consisting of two curved lines meeting at a point.

custom-built – built to order, usually one of a kind.

customer – client; the person or group for whom work is being performed.

cut – a surveying term indicating a grading level in relation to a reference point. In a cut, earth has to be removed to match the reference point.

cut and cover – a method of trenching and pipe laying in which the spoil from one section of trench is used to bury the

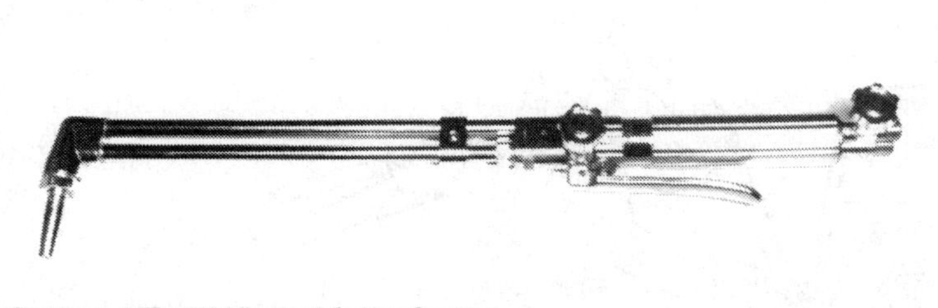

Courtesy: L-Tec Welding and Cutting Systems

Cutting torch

An oxyacetylene or electric arc torch for cutting steel. The flame heats the steel to a molten state and then uses an increased flow of oxygen to blow away the molten metal. Heavy duty gas cutting torches use a bottle of gaseous oxygen and a bottle of gaseous acetylene to supply the heat. The flow of the two gases is controlled by a regulator at the outlet of each tank. They are mixed in a nozzle which is used to direct the flame, control the flow of excess oxygen, and provide a positive shutoff for the gasses. Once the metal has been heated to the require temperature, a lever on the nozzle increases the oxygen flow for cutting. Lighter duty cutters are more portable and use acetylene or propane, along with oxygen from a small bottle or generated by chemical means. An electric arc cutting torch uses the heat of the arc from a nonconsumable electrode, such as a carbon electrode, along with a flow of oxygen to cut the metal. All types of cutting torches mandate the use of very dark glass to shield the eyes of the operator. Cutting lines are usually demarcated by dimples punched along the length of the proposed cut.

pipe and trench in the adjoining section. In this manner, only the first section of spoil has to be removed from the site. The rest of the pipe and trench is buried as the work progresses down the line. Also called dig and set.

cutback asphalt – asphalt to which "cutterstock" petroleum distillates have been added to keep the asphalt in a liquid state when working at lower temperatures.

cut end – the edge of a gypsum wallboard panel where the gypsum core is exposed.

cut-in box – an electrical box, with clamps on the sides, that can be easily installed during remodeling. The box does not have to be attached to a stud, but can be slipped into a hole cut in drywall and fastened in place by tightening screws which secure the clamps to the drywall. The boxes may secure in different manners. One type has sheet metal strips that are folded by the screw so that they stick out as ears from the side of the box. Another type of box has pieces of heavier gauge sheet metal at right angles to the box walls. Turning the screws draws these plates toward the front of the box. In both cases, the screws force the metal protrusions tightly against the back side of the wall covering. Tabs on the front of the box are forced back against the front surface of the wall covering. This action sandwiches the wall covering tightly between the metal protrusions on the front and sides of the box, anchoring the box in place.

cut-in brace – see let-in.

cut joints – masonry joints that have been cut flush with a trowel.

cut nail – a nail, cut from sheet steel, with a square or rectangular cross section shank and a blunted end. Used for heavy timber or masonry.

cutoff – 1) a section that has been severed from the main body of an object; 2) the point in the stroke of a reciprocating steam engine piston where the steam supply is stopped, or cut off.

cut of the varnish – a measure of the oil, in gallons, that is mixed with 100 pounds of resin.

cutout – a notch, or the removal of a piece of material.

cut sheet – a surveying sheet that shows information such as the cuts, elevations and distances of an area of land for use during earth-moving operations.

cut stake – an information stake placed in the ground to indicate cut locations for earth moving.

cut stone – stone masonry units that have been cut and shaped.

cutter bar – 1) a stationary blade against which moving blades pass to make a shearing action; 2) a device that holds a cutting tool in a machine, such as a lathe.

cutterstock – petroleum distillates such as gasoline, kerosene and diesel fuel which can be added to asphalt to keep it liquid at lower temperatures.

cutting off tool – see parting tool.

cutting torch – an oxyacetylene or electric arc torch for cutting steel. *See Box.*

cyaniding – a case-hardening process for steel. The steel is heated while in contact with molten cyanide salt, and then quenched.

cycle – 1) a process or action that repeats periodically; 2) a full wave of an alternating electrical current, first flowing with one polarity and then with the other.

cycloid – 1) circular; 2) a section of a circle along an axis, forming half of a sine wave.

cylinder – 1) a solid geometric figure with parallel sides connecting two circular ends; 2) a pressurized container for the transportation and storage of gases; 3) the circular piston chamber in an engine.

cylinder head – the closure at the end of a cylinder in a reciprocating machine.

cylinder lock – a type of door lock with tumblers enclosed in a cylinder. The tumblers are aligned with a key so that the lock may be turned.

cylindrical – in the shape of a cylinder.

cylindroid – a solid figure in the shape of a cylinder.

cyma – a style of molding having a convex and a concave curve. The double curvature profile may be either *cyma recta*, with the concave portion projecting beyond the convex portion, or *cyma reversa*, with the convex portion projecting beyond the concave portion.

cymatium – the top portion of a classical cornice.

cypress – a weather and rot resistant wood from one of several types of evergreen trees.

cypress, red – a fine-textured wood with a beautiful grain pattern and high weather and rot resistance.

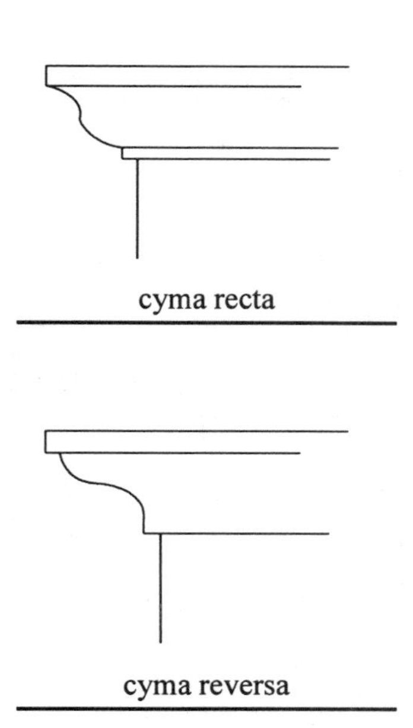

D: dado to dynamometer

dado – a rectangular groove cut into a piece of wood designed to accept the end or edge of another piece, such as a groove cut across a board for inserting and providing support for the end of a shelf.

dado blade – a saw blade, for use with a table or radial arm saw, designed to cut a rectangular groove or dado in a piece of wood; it is adjustable for the width of dado it will cut by either stacking blades, or by dialing in eccentricity so that the blade wobbles back and forth as it rotates.

dado joint – a joint where one member is fit into a groove or dado cut in the other. *See Box.*

dam – 1) an obstruction to the flow of water; 2) any obstruction to the flow of a fluid.

damages – compensation awarded or paid for harm done.

damascening metal – the spot finishing of a metal surface using an abrasive rod turned in a drill motor. This action leaves a series of polished circles on the metal surface.

damp – 1) wet with a small amount of moisture; 2) to retard motion; 3) to retard flow; 4) to deaden; 5) a noxious gas, particularly in a mine.

damp course – a layer of moisture-proof material installed to prevent water being drawn up into a wall or other structure.

damper – 1) adjustable air-flow control device used in a duct; 2) a device used in an electromechanical instrument to limit or prevent rapid oscillations of an indicator.

dampproofing – 1) making moisture resistant; 2) a waterproofing treatment for mortar or masonry to prevent water penetration. Available as a coating or an admixture to concrete or mortar.

Danish oil – a blend of drying oil and plastic in a vehicle, such as mineral spirits, used as a wood finish to preserve the wood and to impart an oiled surface with a low sheen. It is sometimes used to stain the wood. Also called antique oil.

darby – a straight wooden board 3 to 8 feet long and about 4 inches wide, used to smooth, compact, and level wet concrete.

database – a type of computer software that permits the categorization, sorting and manipulation of information. The user can assign categories, enter information, and then use that information in a variety of ways. The information can be sorted in various combinations of categories and hierarchies and then listed in any desired format. The formats can be set up to print, leaving out or including any information desired in the printout. For example, a database may be built up using client information such as names, addresses, phone numbers, types of work, and dates

Dado joint

A dado joint is made by cutting a groove into one piece of wood and fitting the adjoining piece into that groove. It is one of the strongest joints used in cabinetmaking. There are two basic dado joints. One is a straight dado joint in which the groove is cut from edge to edge of the wood piece. The other type has the groove stopping just short of the edge of the material so that the joint does not show from the face of the piece. The adjoining piece must be notched where the dado groove stops, to give the joint a finished look. In making the dado joint, the groove is cut to a width either equal to, or very slightly larger than, the thickness of the member to be inserted. It should be a very tight fit. The depth of the groove depends on the thickness of the board in which the cut is made. The groove depth should not be more than half the thickness of the board, otherwise the board may be weakened by the cut. Glue is spread along the groove, and then the adjoining piece is inserted. This type of joint is used in a variety of carpentry applications, from furniture making to shelving.

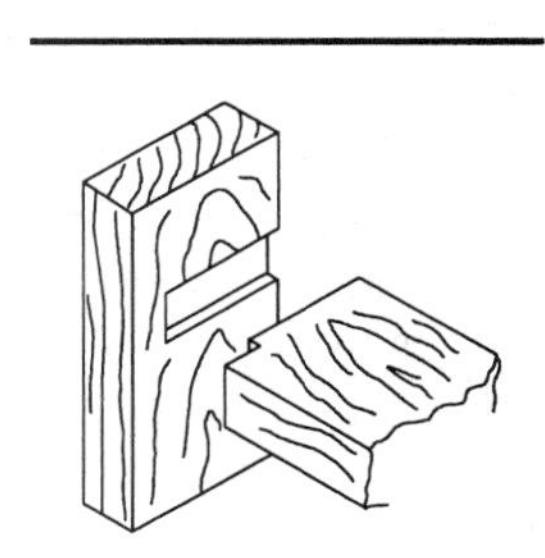

that work was performed. Advertising flyers, letters or bills may be printed out for everyone or just a few people on this database. A mail merge feature will take selected information, like the names and addresses of these people, and print envelopes or mailing labels for them. There are any number of business uses for this type of software.

datum plane – an imaginary level surface shown on a drawing with an elevation of zero.

datum point – a point that has been established as a reference point.

daub – 1) a spot of adhesive; 2) to cover, coat or apply a rough texture with an adhesive material such as plaster.

dead air space – an insulating space containing trapped air that is isolated from circulation. The space provides thermal insulation.

dead blow hammer – a hammer with a head that is hollowed and refilled with lead pellets. The lead reduces the rebound when the head strikes an object.

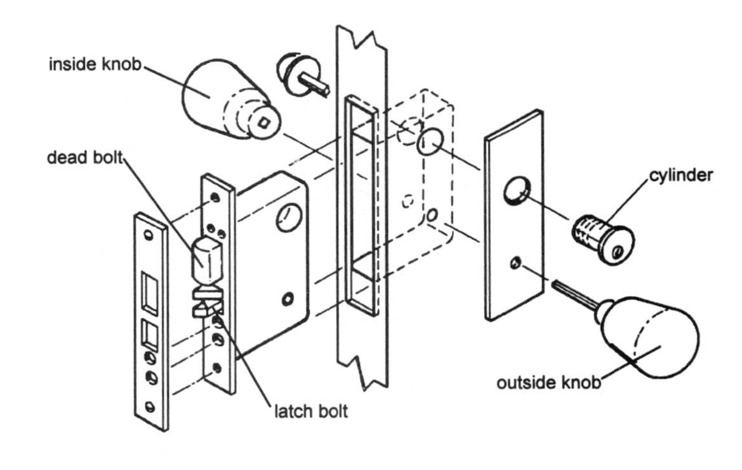

dead bolt, mortise type

dead bolt – a type of lock in which a bolt slides into a receptacle in the door jamb at the turn of a key or turn piece. Also called a deadlock.

deaden – to lessen vibration or reaction by adding damping and/or insulation.

dead end – 1) the termination of a piping system with a closure, such as a cap; 2) a street that is closed off at one end or that terminates without connecting to another street.

Deadman

A deadman is an anchor for a cable, guy line or other member which is fastened to it. A deadman is often used as a means of hidden support for a retaining wall. A beam, concrete block or other mass is buried in the soil behind the wall, and a cable or other load-carrying member is attached to both the wall and the mass. The cable is used to restrain the wall, and keep it from tipping under the pressure of the load it is holding. The weight of the deadman and the resistance of the soil transfers the load back to a more stable area. A deadman "X-type" anchor may also be used to support retaining walls. The anchor is made up of ½-inch rebar in an L-shape, with an X-shaped anchor connected to the long end of the L. The short end of the L is embedded in the wall and the long end is buried in the earth behind the wall. The X section, which should be at least 18 inches wide, extends 8 to 10 feet into the earth and anchors the wall. Several such connections may be needed to support the wall, depending on the soil conditions, the type of wall structure, and the length and height of the wall.

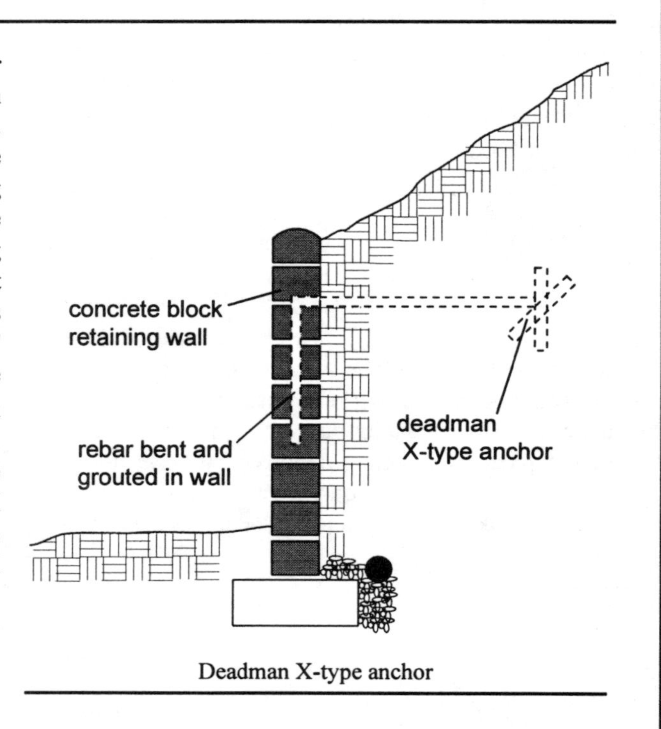

Deadman X-type anchor

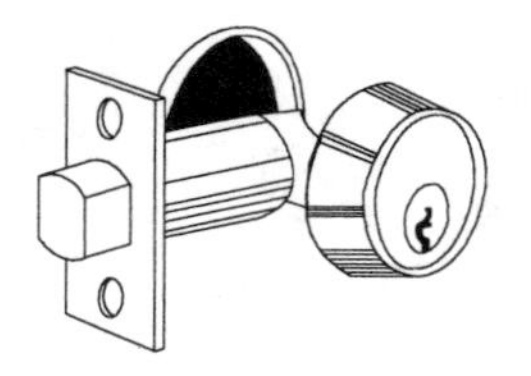

deadlock, bored

deadening board – a sound absorbing fiberboard.

dead flat – a painted surface with no gloss.

deadline – the time when something is due to be completed.

dead load – 1) the load on a building structure caused by the weight of the building structure itself. The dead load on floor joists is the load created by the flooring. In contrast, the live load is the load caused by people, furniture, and anything else that is not part of the building. When planning a building, a designer must take into consideration both the dead loads and the anticipated live loads on the load bearing structural members; 2) the fixed load on the structural members of a crane due to the weight of the crane components.

dead load deflection – the vertical deviation of a crane bridge caused by its own weight and the weight of the permanent attachments to the crane bridge.

deadlock – a type of lock in which a bolt slides into a receptacle in the door jamb at the turn of a key or turn piece. See also dead bolt.

deadlock, bored – a deadlock with an outside key-operated cylinder and an inside knob; the cylinder fits into a hole bored into the door.

deadlock, mortise – a type of dead bolt which is set into a mortise or cavity cut into the door.

deadman – a mass, such as a log or concrete block, buried in the ground to provide an anchor for a guy line or cable. *See Box.*

debouch – land that emerges from a narrow valley onto a plain.

debris – 1) remains of materials that are generally useless; 2) loose rock fragments.

debug – to locate and remove defects, as from a computer system or program.

decagon – a figure with 10 sides and angles.

decalcomania – the process of transferring pictures or designs from specially-treated paper to glass or other objects; applying decals.

decarburization – the loss of carbon from the surface of steel.

decay – decomposition; rot.

decibel (dB) – a standard measure of sound pressure level.

deciduous – a tree that loses its leaves each fall.

decimal – fractions in tenths.

decimal equivalent – a fraction shown as a decimal.

decimal feet – a length measured in feet and decimal portions, or tenths, of a foot; 3.5 feet in decimal feet is the equivalent of 3 feet 6 inches or 3½ feet.

deck – 1) an exterior platform structure, usually made of wood and elevated off the ground; 2) a covering, as over rafters, which provides a surface to which shingles can be attached; 3) the floor of a ship.

deck paint – a wear-resistant paint formulated for use on heavy-wear areas or exterior floors.

declination – 1) bending downward; 2) turning aside or swerving.

declivity – a downward slope, such as a hill.

decompression – a releasing from pressure.

decontaminate – to remove contamination from; to purify.

decorate – 1) to add designs to a surface; 2) to furnish with ornamentation; 3) to add furnishings to a room.

decorated gypsum drywall – drywall panels with a factory-applied finish, such as decorative vinyl or paper.

decoration – the finish used on the surfaces of a building.

decorative metal fencing – ornamental fencing made from metal bar, pipe, and/or other metal shapes. These shapes are welded together into sections of fencing that are installed as connecting units.

decoupling – the separation of elements or components in a structure so that they do not interact with each other.

decrease – to reduce; diminish.

deed – a formal document to transfer ownership.

deep-cut trim plate – a convex electrical cover plate designed to fit over an electrical box that protrudes slightly from the wall.

de facto – 1) actually; in reality; 2) exercising power as if legally constituted.

default – 1) a failure to complete or perform an action; 2) a pre-programmed standard used in the absence of other instructions.

defect – 1) an imperfection that impairs utility; 2) a condition that does not meet specifications or standards.

deflate – 1) to release air or gas from; 2) to reduce in size.

deflection – to deviate from a straight course; the movement of an object when a force is applied to it, such as the movement of a beam when a load is applied to it.

deforestation – the removal of trees from an area.

deform – to spoil or distort the shape; disfigure.

deformation – an alteration in shape or form caused by stress, such as when a load is applied and the shape bends.

defrost – to thaw from a frozen state.

degauss – to demagnetize, or remove the magnetic energy from an object.

degenerate – 1) to deteriorate; 2) to decline in quality or capability.

degradation – a deterioration or lessening in strength.

degree – 1) a unit of measurement used for temperature; 2) a 360th part of the circumference of a circle, the unit of measurement from which all angles are determined.

decorative metal fencing

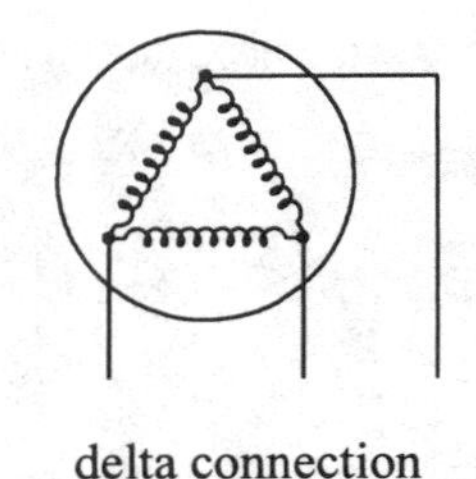
delta connection

degree-day – a unit of measure used to estimate HVAC needs and operating costs based on maintaining a given temperature between the mean and 65 degrees over a 24-hour period.

degree of difficulty production factor – a factor that takes into consideration difficult working conditions when evaluating the time it takes to complete a task, as opposed to the normal time under average working conditions that it takes to complete the same task. A relative degree of difficulty is arrived at to compensate for adverse conditions. Once this is applied and the time to complete the task has been equalized, the rate of production on tasks can be compared on an equal basis.

dehumidifier – a device to remove moisture from the air.

dehumidify – to remove moisture from the air.

dehydrate – to remove moisture from an object.

delamination – 1) a separation of layers, that were or were intended to be fastened together, due to failure of an adhesive bond; 2) a paint defect where paint layers separate.

delay – to slow or defer action.

delayed action fuse – see time-lag fuse.

delayed toggle switch – an electrical switch that leaves power on in a circuit for a predetermined length of time after the switch has been set to the off position.

delineate – to outline or to represent by drawn or painted lines.

deliverable – the product that is due a client with whom one has a contract.

delta connection – an electrical circuit in which the three windings of a generator or transformer are connected end to end.

demagnetize – to remove residual magnetism.

demand factor – the ratio of the maximum electrical demand of a system to the total load connected. This ratio represents the percentage of possible use needed for designing the system.

demolition – the destruction of a structure for removal.

demountable – 1) temporary; 2) movable.

demountable partition – a temporary or movable partition.

dense – 1) compacted; 2) a large mass per unit of volume.

density – the mass of a substance per unit of volume of that substance.

dent – 1) a depression in a surface; 2) a projection, such as a gear tooth.

dentil – 1) square blocks repeated in a pattern; 2) the decorative projections under a cornice.

deoxidize – to remove oxygen from a substance.

departure – the distance east or west of a survey point.

depend – 1) to hang from; 2) to exist due to a relationship of necessity.

depreciation – 1) a reduction in value; 2) discounting the value of property over its estimated life for income tax purposes. It is assumed that the structure, not the land itself, will lose value over time due to deterioration. When the estimated life of the property has elapsed, the structure will be fully depreciated, and further value cannot be deducted from the income of that property.

depression – 1) applying force to something in order to lower it; 2) a dent or hollow; 3) a downward angle from the horizontal in surveying; 4) a low land area.

depressurize – to release pressure from a vessel.

depth – 1) a vertical dimension from the top to the base of an object; 2) a horizontal dimension from front to back of an object.

depth gauge – a measuring device, consisting of a graduated ruler which slides through a crosspiece, used to measure holes and grooves. The projections of the crosspiece rest on the surface and the ruler slides into the hole in the object whose depth is to be measured.

depth micrometer – a device for measuring the depth of a hole. It consists of a precision-threaded spindle with marked graduations held in a flat frame designed to span a hole whose depth is to be measured. The frame rests on the material on either side of the hole, with the spindle set at zero. The spindle is then lowered into the hole until it reaches the bottom. The reading on the spindle indicates the depth of the hole.

depth of cut – the depth to which a drill, saw blade, or other cutting device is set to cut into the work.

depth of fusion – the depth to which a weld penetrates.

derrick – 1) a crane with a boom or jib; 2) an A-frame structure over a well.

desalt – to remove salt from a substance.

descale – to remove scale from metal.

desiccant – a moisture-absorbing compound or material used in refrigeration systems or in containers and packaging with contents that must remain moisture free.

desiccate – to remove moisture; to dehydrate.

design – 1) to plan and provide details for fabrication or construction; 2) a decoration.

design and construct – a contract calling for the design of the job as well as the building of the approved design.

design drawings – drawings, used in all types of building and manufacturing, that detail information regarding the construction or fabrication of a structure or product.

design load – the maximum load which a structure or structural member is designed to support.

design symbols – representations of different materials using symbols.

desk – a table surface designed for writing, drawing, etc., usually having storage spaces as part of its structure.

detach – to separate.

detached garage – a garage structure separate from the dwelling which it serves.

detail – 1) small part; 2) a specific feature; 3) to plan and draw specific features.

detail drawing – a drawing of a portion of a structure or object that shows a high level of detail.

detailer – one who makes detail drawings of structures to be built or detail drawings of objects within structures.

details – portions of a drawing that are enlarged to increase clarity.

detect – to find or identify.

detonate – to cause to explode.

detritus – 1) debris; 2) rock fragments.

developed length – piping length measured along the centerline of the pipe and fittings.

developer – one who plans and makes improvements to real property, usually on a large scale, such as a shopping complex or a housing tract.

development – a large building complex, such as a housing tract or shopping center.

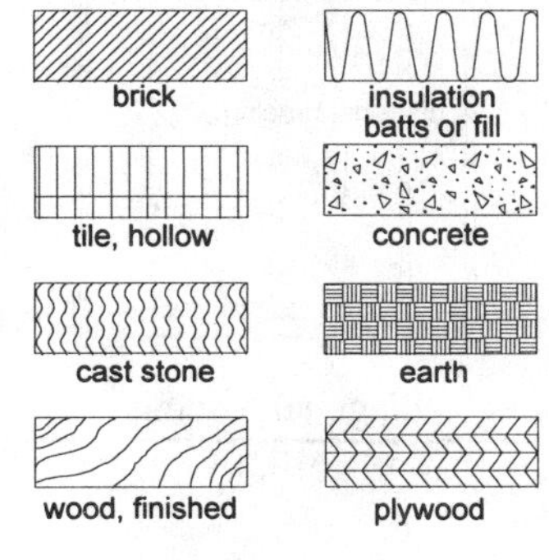

design symbols

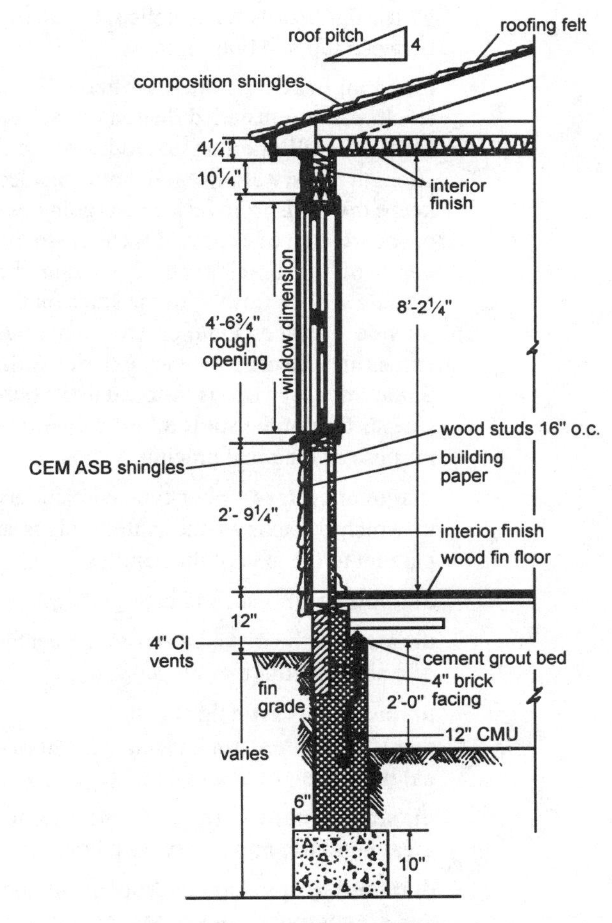

detail drawing of wall section

diagonal bracing

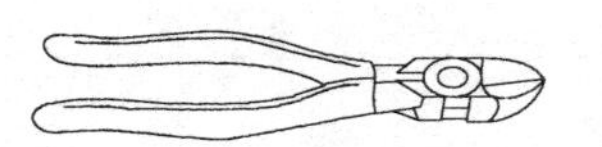

diagonal cutter

device – 1) a mechanism or piece of equipment with a specific purpose; 2) a part of an electrical system that is not intended to carry electrical current, such as a junction box.

device plate – a cover plate for an electrical item, such as for a switch box.

dew – moisture condensed on surfaces.

dewatering – pumping or draining water from a construction area to enable work to proceed.

dew point – the temperature at which moisture in the air condenses on a surface.

diagonal – 1) a line of measurement that runs from the center of one angle to the center of the opposite angle in a rectangular figure; 2) inclined obliquely from a reference line; 3) an inclined straight line or member extending across a surface.

diagonal board fence – a fence style in which flat boards are installed diagonally between top and bottom rails.

diagonal bracing – boards, often of 1 x 4 stock, that are installed diagonally across studs in wall framing to add structural strength to the wall. Diagonal bracing also keeps the walls from racking, or going out of square. Part of each stud is cut out to the depth of the bracing board so that the bracing will be flush with the studs on the outside surface. One or two diagonal braces may be used for each exterior wall. Sometimes plywood is fastened to the outside surface of the studs to serve the same purpose as diagonal bracing.

diagonal cutters – plier-type wire cutters in which the axis of the cutting edges is parallel to the axis of the handles.

dial indicator – see indicating gauge.

diameter – the distance across the center of a circular figure or object.

diameter, breast height (dbh) – diameter of a tree trunk taken at the nominal breast height of an adult.

diamond – pure carbon, formed under great pressure, into a very hard crystal.

diamond drops – small droplets of sap that seep through, and appear on, the surface of painted wood.

diamond point chisel – a metal cutting tool with a V-shaped cutting edge used to cut V-shaped grooves and make sharp corners in holes.

diaphragm – 1) a membrane ; 2) a partitioning surface; 3) structural sheathing material applied over framing, such as between the webs of a bridge crane girder, which acts as a load-bearing unit; 4) a stiffener plate.

diaphragm action – a stressed skin on a structure that provides stiffness and resists movement that might distort the squareness of the structure.

diaphragm tank – an expansion tank with a partition that separates the gas and water in the tank to prevent gas from being absorbed into the water system.

diaphragm valve – a shutoff or metering valve for fluid systems. *See Box.*

diatomaceous earth – a fine white or gray silicon material resembling chalk which is composed of the fossilized remains of small marine life. It is used for filters, thermal insulation for steam piping, as a paint extender, and for making scouring powders, polishes and dynamite. Also called Tripoli, infusorial earth or diatomaceous silica.

diazo – a method of photocopying that uses a coating of a diazo compound which decomposes when exposed to light.

die – 1) a tool used for shaping or forming in cutting, stamping or impressing operations; 2) a mold into which molten metal or other material is forced.

die cast – molten metal cast in a die.

dielectric – a material that is an electrical insulator; a non-conductor.

diesel engine – an internal combustion piston engine which uses a high compression ratio with the resultant heat of compression used to ignite the fuel.

diesel hammer – a pile driver which uses diesel combustion to raise the hammer.

differential – 1) the difference between values; 2) a gear set that transmits power to two shafts and allows the shafts to turn at different speeds.

Diaphragm valve

A shutoff or a fluid control valve with a thin partition or diaphragm designed to prevent pipeline fluids or gases from contacting the operating parts of the valve. The valve has a threaded stem which forces a plug, called a compressor, against the diaphragm. The diaphragm is forced against a raised section, called a weir, in the body of the valve to effect a seal. Since the weir is in the path of the fluid flow, the flow is stopped when the diaphragm is sealed against the weir. Diaphragms and valve bodies are available in a wide variety of materials, which are selected for fluid compatibility. Sometimes metal diaphragm valve bodies are lined with an elastomer material to prevent exposure of the valve body base material to the fluid in the system. Diaphragm valves may be manually, hydraulically, pneumatically or electrically actuated to open or close. The diaphragm valve may be used as a shutoff valve, fully open or fully closed, or it may be used in an intermediate position to throttle and control the flow of fluids.

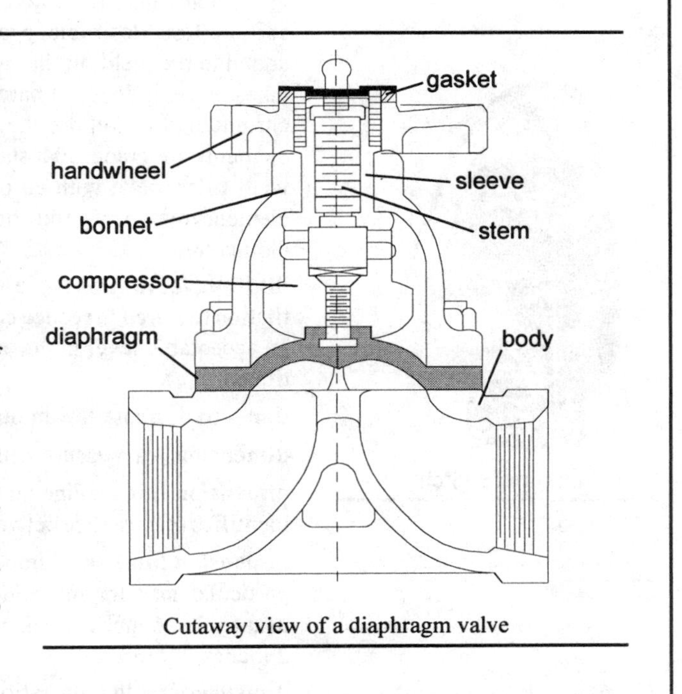

Cutaway view of a diaphragm valve

differential-acting hammer – a pile driver which uses steam or air to raise and drive the hammer.

differential leveling – determining the difference in elevation between two points using a level and a rod.

differential pressure – the difference in pressure between two parts of a system, or between two areas.

diffraction – the redistribution of energy at an obstruction or at an abrupt change in the surface over which it is traveling. For example, a sound wave that travels across a flat plane surface will be at least partially redirected when it reaches the edge of the surface.

diffuse – to spread out.

diffuser – an air outlet in an HVAC system which directs the air in a wide, fan-shaped pattern.

diffusion – dispersing; spreading out.

diffusion welding (DFW) – a welding process that uses heat and pressure to join parts without deforming them. The parts being joined are not melted or moved relative to each other.

dig and set – see cut and cover.

dike – an embankment to control or confine water.

dilapidate – to cause to deteriorate; destroy.

dilate – to spread wide, expand or increase in size.

dilute – to weaken the strength of a fluid by mixing it with another.

dilution – changing the composition of weld filler metal by additions from other weld metal or by the base metal melting and mixing with the weld filler metal. For example, when grade 304 stainless steel is welded to carbon steel, dilution of the stainless steel weld metal can take place at the interface with the carbon steel as the heat of the welding melts not only the filler metal, but the carbon steel base as well.

dimmer switch

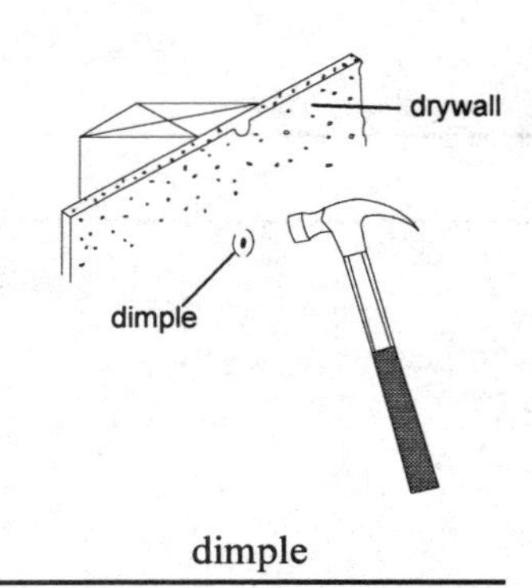

dimple

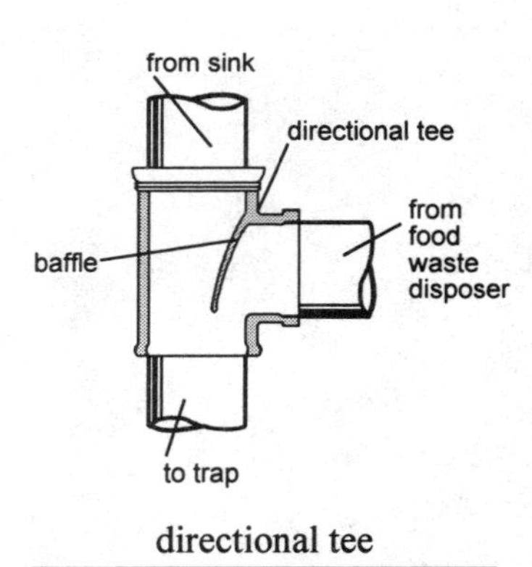

directional tee

The solution to this problem is to use a stainless steel weld filler metal that is rich in elements that, if diluted, would not result in less desirable properties being added to the weld. In that way, if dilution takes place, it does not harm the weld. As chrome and nickel are the main alloying elements of grade 304 stainless steel, a weld filler metal with an excess of these elements may be a good choice for a weld application.

dilution ventilation – the amount of ventilation required to reduce contaminants to an acceptable level in the area being ventilated.

dim – to decrease the amount of light.

dimension – a measurement value.

dimension line – a line on a drawing that identifies the distance between two points.

dimension lumber – lumber precut to a particular size for the building industry, usually 2 to 5 inches thick, with a width of 2 inches or more.

dimension ratio – the ratio of the average outside diameter of plastic pipe to its minimum wall thickness, rounded to the nearest 0.5. Dimension ratios are the common method of defining and identifying the various hub and spigot types of plastic pipe. Standard dimension ratios (SDR) have been established to provide an internationally recognized series of designations for plastic pipe.

dimmer switch – an electrical control to reduce the intensity of light from a fixture.

dimple – 1) the depression left in the surface of a gypsum wallboard panel when a nail is set by a hammer; 2) any depression in a surface.

diode – an electrical component that permits current to pass in only one direction.

dip brazing – a brazing process using heat from molten chemical or metal bath.

dipole – having two electrical poles, such as an electrical switch that switches on two electrical lines simultaneously.

dipping – immersing in a liquid, such as paint.

dip soldering – a soldering process using heat from a molten metal bath of solder.

direct current (DC) – electrical current that flows in one direction, from the negative to the positive terminal of the source.

direct expansion cooling system evaporator – a finned-tube heat exchanger which cools air directly. A refrigerant is expanded into the evaporator coil. The expansion cools the refrigerant. Air passing over the outside of the coil is cooled by heat transfer through the coil.

directional tee – a plumbing tee fitting with an internal baffle to direct flow in one direction.

direct labor – wages earned by employees working on a specific job.

direct lighting – light shining directly on a surface with no intervening reflection between the surface and the light source.

direct nailing – to nail perpendicular to the work surface.

directory board – a listing of building occupants and their locations, usually found in commercial buildings.

direct overhead – overhead costs (the costs of doing business) that are related to a specific job.

direct tap – a device that clamps around a main service line, such as a water main, so that a branch line can be drilled and tapped off. A combination drill and tap is used to cut into the line, and then a corporation stop is installed in the tapped hole to provide fluid flow control. In this way, a connection can be made to a main line without taking the main line out of service.

discharge – 1) the output of a process or machine, such as the flow from the outlet of a pump; 2) to complete an obligation; 3) to dismiss or relieve of an obligation.

discharging arch – see relieving arch.

discontinuity – a lack of uniformity throughout a weld or casting.

disk sander – a sander that uses an abrasive rotating disk.

disperse – 1) to distribute; 2) to scatter.

disrepair – a state of neglect in which deterioration has taken place.

dissipate – to spread out or spread thin to the point of disappearing.

dissolve – to make a solution by combining a solid or liquid substance with a liquid which absorbs it.

distill – to evaporate and then condense a liquid in order to purify it.

distribution – the delivery of a commodity to the proper locations.

distribution panel – an electrical panel from which electrical circuits emanate.

district – a specified area.

ditch – a trench.

ditch stake – a survey stake to mark and identify the location where a ditch is to be cut.

diverge – 1) to change direction; 2) to spread out from a source.

diversion – a drainage channel which directs the flow of water runoff.

divert – to redirect.

divide – to split into two or more parts.

divider – a partition.

dividers – a tool similar to a compass, but with two pointed legs, used for measuring, comparing and transferring dimensions.

divider strips – strips of metal, plastic or other suitable material that are set in a flooring material, such as terrazzo or concrete, to control cracking and to serve as screed surfaces.

dividing head – a device used with a milling machine to divide the work into equally-spaced increments.

dobie – a small concrete block placed to support rebar so it is not forced to the bottom during a concrete pour.

dock – 1) a platform over water from which boats or ships can be loaded or unloaded; 2) a platform at the end or back of a commercial building used as a loading area for goods.

dodecagon – a polygon with 12 angles and sides.

dodecahedron – a solid figure with 12 adjoining plane surfaces.

dog anchor – a heavy metal staple used to form a temporary connection between timbers.

dog-eared board fence – a board fence in which the top corners of the vertical fence boards are cut off at 45-degree angles.

dogleg – a change in direction.

dog's tooth – brick corners projecting from a wall.

dolerite – coarse-grained basalt.

dolly – 1) a platform with wheels used to move objects placed on it; 2) a small locomotive for use in a limited area.

Dolly Varden siding – a type of bevel siding with a rabbet, or two-sided groove along the edge, that overlaps the top of the board below.

dolphin – a vertical structure made of pilings arranged to be in contact with, and provide support to, each other.

dome – a hemispherical projection from a roof, or comprising the roof.

domicile – 1) a legal and permanent residence; 2) a dwelling place.

door – a swinging or sliding covering, usually made of wood or metal, used to close off a cabinet or a passageway into a building or room.

doorbell – an electrical device, located outside the entrance to a building, used by visitors to signal their arrival. When the device is activated by a button or switch, a bell or buzzer sounds inside.

door bevel – a slight bevel cut on the lock-side, vertical edge of a swinging door that permits the door to fit tightly in the jamb but still clear the edge of the jamb as the door swings.

door, bifold – a door made of panels that are hinged vertically in the middle so that they fold back upon themselves.

door buck – a metal or wood jamb for a finish door frame in a masonry wall.

door, bypass – sliding doors that are installed so that one passes in front of the other; frequently used to close off a closet.

door casing – molding used to trim door and window openings. Also called door trim.

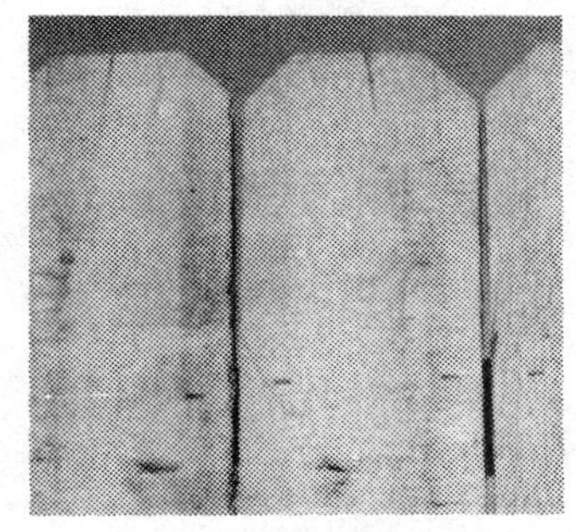
dog-eared board fence

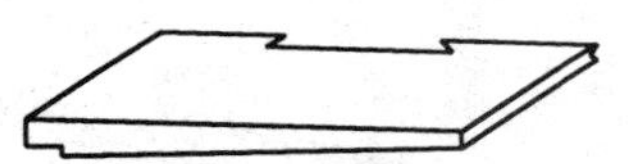
Dolly Varden siding

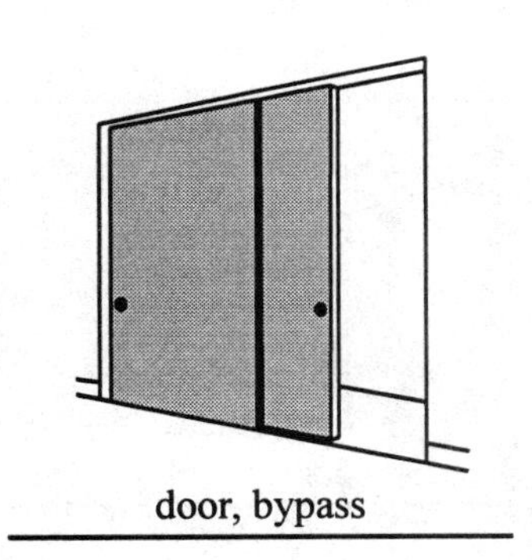
door, bypass

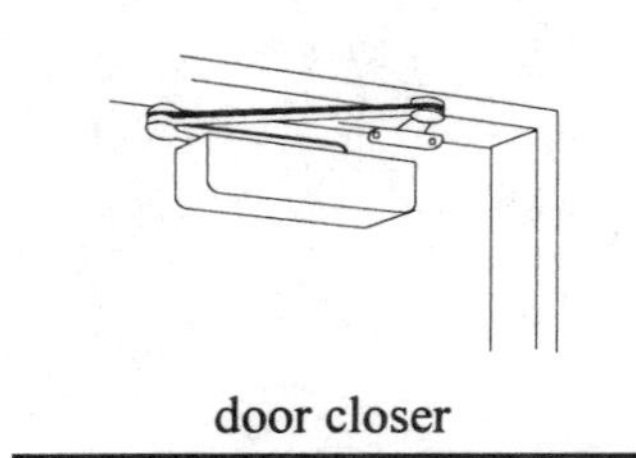

door closer

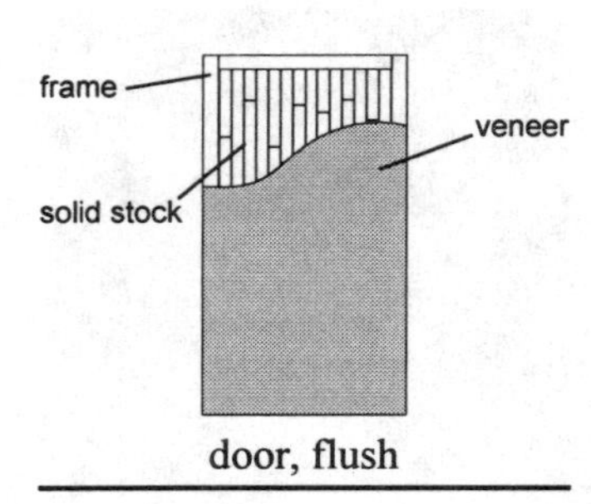

door, flush

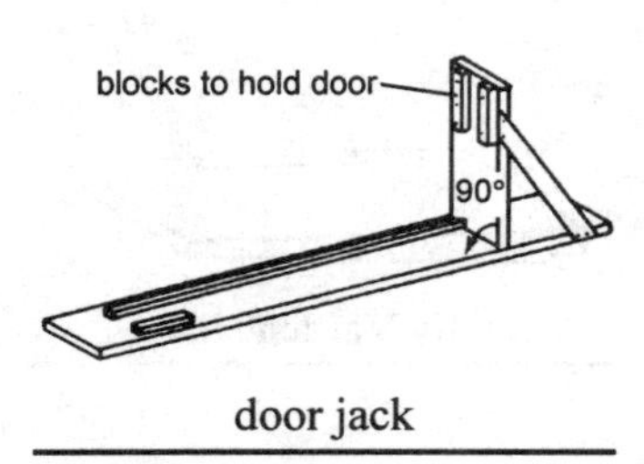

door jack

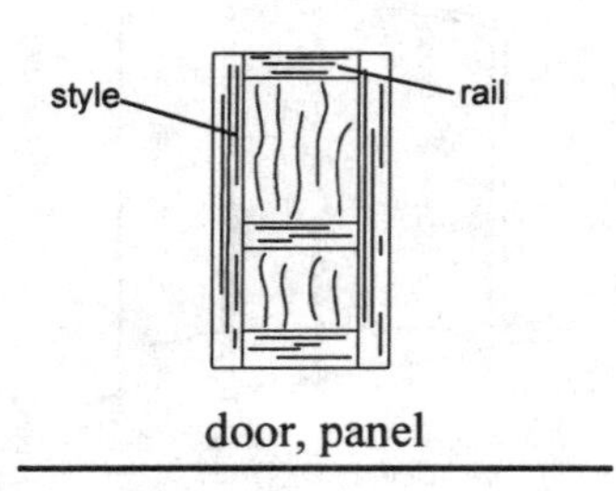

door, panel

door, sliding glass

door chain – a chain link locking device for a door. One end of the chain is fastened to the door jamb or an adjacent wall and the other end slides into a holder attached to the door.

door check – a device to retard the closing of a door.

door chimes – a doorbell sounder that makes the sound of chimes.

door closer – a device that automatically pulls a door closed at a speed usually adjustable by a sweep speed control. There are many types of door closures, such as single acting, double acting, overhead, and floor.

door, exterior – a door designed to be exposed to the elements on at least one side. It is used to close a doorway through the outside wall of a building.

door, fire – see fire door.

door, flush – a door that has a thin plywood facing over a framework and a core of wood or wood products, such as particleboard; or a thin plywood facing over a hollow core framework. The latter is known as a hollow core door.

door frame – the structural framework surrounding the door opening on which the door is hung, and against which the door closes.

door hardware – the knobs, latches and hinges attached to a door which allow the door to be opened and closed.

door head – the head jamb, or topmost horizontal door frame member.

door, hollow core – see door, flush.

door, interior – a door not designed to be exposed to exterior elements. An interior door is used to close a doorway inside a building.

door jack – a device, made of boards, that holds a door erect while it is worked on prior to installation.

door jamb – the surrounding framework of a door opening. Also called a door post.

door knob – a handle, which may or may not operate a door latching mechanism, used to pull or push a door open or close it.

door lock – a latching device used to secure a door in the closed position.

doornail – a large nail with a large, decorative head that a door knocker strikes against. Doornails can be used to strengthen doors or to be simply ornamental.

door, panel – a door with recessed or raised panels installed in the spaces between exposed stiles and rails.

door post – see door jamb.

door, prehung – a door that has been factory-installed in its frame so that it can be set into a structure as a single unit.

door pull – a fixed handle for opening or closing a door. This type of handle, which may be in the shape of a knob, does not operate a latching mechanism.

door, rolling – a door made from interlocking hinged metal panels that can roll up on an overhead drum.

door schedule – a listing of the types, sizes and locations of doors to be installed in a building. It aids in the ordering of the doors and serves as a guide to the installation of the doors.

door shoe – a weather seal at the bottom of a door.

door sill – the bottom portion of an exterior door casing that extends the full depth and width of the casing, from the finish floor on the inside to the outside wall.

door, sliding glass – a metal or wood frame door with glass panels that slide on rollers in a track. Usually there are two such doors, one of which is fixed, with the other sliding past the fixed panel to open.

doorstop – 1) the section of the door jamb against which the door rests when it is closed. The stop prevents the door from moving past the closed position; 2) a wedge used to hold a door open.

door threshold – see threshold.

door trim – see door casing.

doorway – a passageway into a building or room.

dooryard – the exterior yard just outside a door.

dope – 1) a thick paste; 2) an absorbent material; 3) a type of paint used on the fabric of aircraft wings.

dope, pipe – a compound used to lubricate and seal pipe threads prior to making a threaded pipe connection.

Doric – one of the five classic architectural orders, the others being Ionic, Corinthian, Tuscan and Composite. Doric is the oldest and simplest of these architectural forms, originating in the Dorian region of ancient Greece.

dormer – a projection built out from the slope of a roof which provides additional interior space, light and ventilation.

dormer horizontal cornice – horizontal exterior trim installed on a dormer.

dormer rafter – the roof rafters framing a dormer roof.

dormer rake cornice – exterior trim on the roof slope edge of a dormer.

dormer window – a window mounted into the end wall of a dormer.

dormitory – a residence hall designed as a sleeping facility for a large number of people, such as students living on a school campus.

dosimeter – a radiation-measuring device, such as the film badges worn by people working around radioactive materials. The amount of radiation exposure is determined by the extent to which the film has been exposed.

double acting – moving in more than one direction, or performing work in more than one direction. For example, a double-acting-piston pump moves fluid in each direction within the cylinder.

double-acting hinge – a door hinge that permits a door to swing outward and inward.

double-alternate saw tooth set – a fine-tooth hacksaw blade with every other pair of teeth set alternately to the right and left.

double brick – a masonry brick measuring 4 x 5⅓ x 8 inches.

double coursing – the application of siding shingles in double courses or layers.

double-cut file – a file with two sets of teeth cut in rows diagonally crossing each other.

double glazing – a double pane window which provides additional insulation due to the air space between the panes.

double-hung window – a window with an upper and a lower sash, both of which slide vertically, so that the window can be opened at the top or bottom.

double insulated – electrical power tools in which the parts carrying current are insulated from the user by internal insulation plus a non-conducting outer casing.

double nailing – a fastening method for gypsum wallboard in which pairs of nails are driven close together.

double plating – a second structural member or plate installed on top of the first plate to strengthen walls.

double-pole, double-throw switch (DPDT) – a switch designed to simultaneously connect or interrupt two electrical conductors.

double-pole reversing switch – see four-way switch.

double-pole, single-throw switch (DPST) – an electrical switch that connects or disconnects two pairs of terminals.

double-stud wall – a sound isolating wall consisting of two rows of offset studs on a common plate that is wider than the studs. The wall coverings on each side are not mounted on common studs. Because the wall coverings do not share common framing, there is no direct sound transmission from one wall covering to the other. Insulation within the wall cavity provides additional sound damping. Also called a double wall or offset-stud framing. See also offset studs.

double-wall siding – siding in which sheathing is installed and then covered over by exterior siding.

double-welded joint – a joint that is welded from both sides.

double Y-branch – a Y-shaped plumbing fitting with two branches, each at an angle to the main run.

Douglas fir – a yellow to reddish-brown wood, primarily from fir trees in the northwest U.S. It is mostly heartwood, with

dormer, shed-roof type

double-hung window

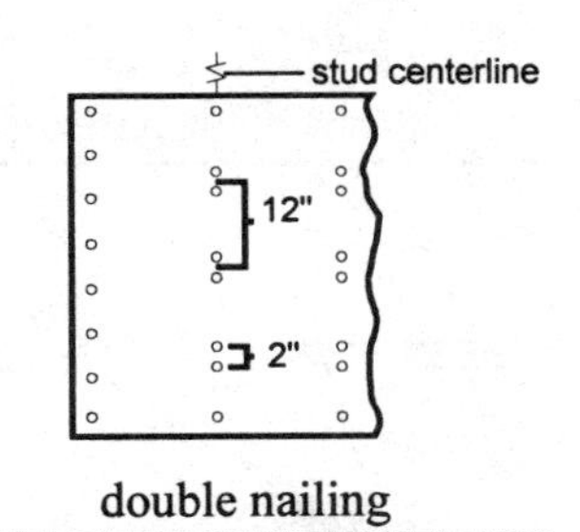

double nailing

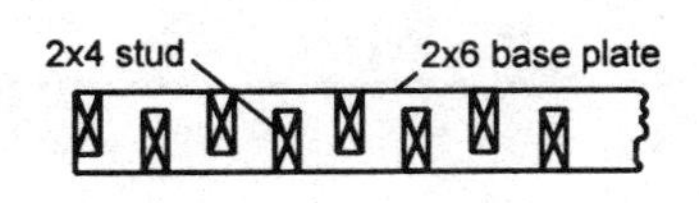

double-stud wall

decay-resistant qualities equal to white oak and long leaf pine. Fir has good bending strength, is stiff, hard, and has good nail-holding ability.

dovetail – a woodworking joint with the interlocking mortises and tenons fanned at the ends like a dove's tail.

dovetail jig – a device for holding the ends of two boards that are to be joined with a dovetail joint so that when cut with the proper router bit, the mortises and tenons will be in the proper locations.

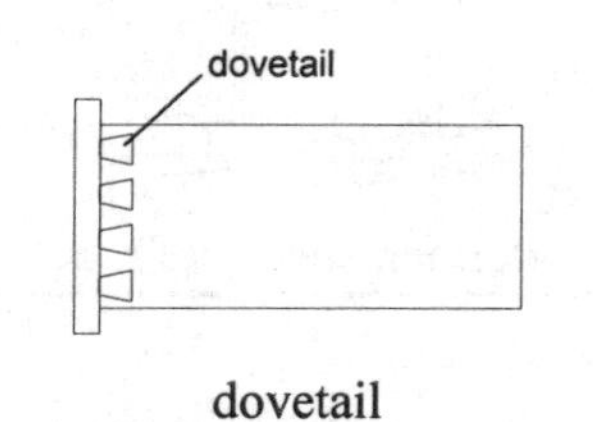

dovetail

dovetail saw – a small, fine-toothed backsaw with a stiff blade and reinforced spine, designed for cutting dovetails by hand.

dowel – a cylindrical rod, often of wood, used to join two members together by inserting it into matching holes drilled into adjacent members.

dowel center – a plug with a pointed center on one end used to position dowel holes. The plug is set into an existing dowel hole and the adjoining piece is butted up to it. The point on the dowel plug marks the position where the adjoining piece must be drilled in order for the two dowel holes to mate properly.

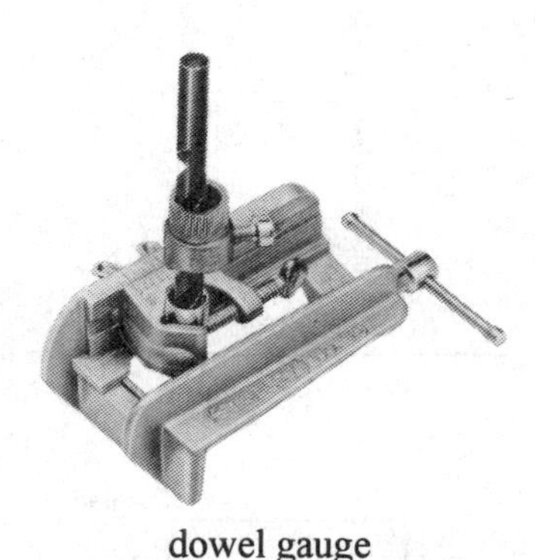

dowel gauge

dowel gauge – a tool for locating and guiding drill bits to the correct position for drilling dowel holes. Also called a dowel jig.

dowel jig – see dowel gauge.

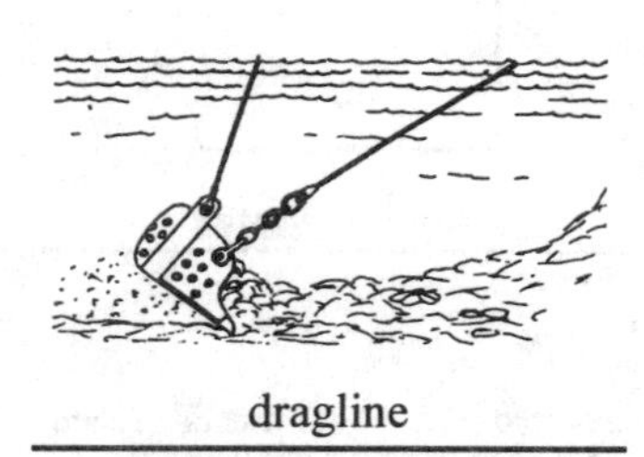

dragline

dowel joint – a joint that is held together partially or completely by dowels.

downdraft – a downward flow of air.

downdrain – a drain designed to carry water runoff from roads and the shoulders of roads.

downhand welding – see flat position.

down payment – an initial payment, made at the time of purchase of a commodity or property, which is a portion of the total price of the item.

downspout – the pipe or duct used to carry water from a gutter to the ground.

downstream – the flow of a water beyond a specific location.

downtime – lost time on a job or in a process while waiting for repairs to be made.

dozer – see bulldozer

dozing – moving material or grading with a bulldozer.

draft – 1) a flow of air; 2) to draw with precision.

drafter – a person who draws in precise detail using drawing tools.

draft gauge – see manometer.

drafting – mechanical drawing; drawing in precise detail using drawing tools.

drafting machine – a mechanism that controls a straightedge over the surface of a drawing so that the straightedge is always at a constant angle.

drafty – having an uncontrolled flow of air through a passage or crack.

drag – 1) an application problem in which paint does not flow smoothly from the brush or roller onto the surface; 2) to move an object by pulling.

dragline – 1) a cable used to pull; 2) a powered excavating bucket used to remove large quantities of soil.

draglink – a connecting link between two cranks that are on different shafts which allows the two cranks to turn in unison.

drain – an outlet or pipe used to carry waste fluids and matter from a collection area.

drainage course – a layer of porous material placed to encourage water drain-off by percolation.

drainage system – the gravity-rated, or non-pressure rated, piping system and fixtures that carry waste from a building or location to a sewer. The drainage system includes the system vents.

drainboard – the slightly-angled area next to a sink that drains into the sink.

drain cock – a valve installed at a low point in a system to allow the system to drain.

drain pan – a pan with a routed drain line, placed under a piece of equipment to catch and contain any fluid that may leak out, and direct it to a specific location.

drain tile – lengths of clay drain pipe used for gravity-rated, or non-pressure, drainage. The pipe sections (tiles) are installed in a trench below grade.

drapery – window covering; curtain.

drapery pocket – trim used to conceal drapery hardware installed at or just below the ceiling line.

draw – 1) to pull; 2) to remove by suction; 3) to attract; 4) to obtain or collect; 5) to create pictures of figures, shapes or items on a surface; 6) to write.

drawer – an open-topped container that slides on a track or guides in and out of a storage cabinet.

drawer pull – a handle on the front of a drawer.

draw filing – filing to a smooth finished surface by alternately drawing and pushing the file across the work.

drawing – 1) the technique of using lines to represent an object; 2) to spread or lengthen metal by hammering or pulling it through dies; 3) to pull toward.

drawing sizes – letter designations that indicate the dimensions of the material on which a drawing is made or reproduced. For example, *A* size equals 8½ x 11 inches and *B* size equals 11 x 17 inches.

drawknife – a blade with a handle on each end, used for trimming, shaping and smoothing wood; the knife blade is laid nearly horizontally on the work and drawn toward the user with both hands to shave the wood. Also called a drawshave.

drawplate – a metal plate with tapered holes through which wire or small-diameter tubing can be formed by drawing the unshaped metal through the holes.

drawtube – a close-fitting tube that slides within another tube.

dredge – an earth-moving machine for removing soil from the bottom of a body of water.

dress – to smooth and finish a surface.

dressed and matched – wood pieces fabricated with a tongue milled on one edge and a groove milled into the other edge. The tongue of one board fits into the groove on the next. Also called tongue and groove.

Dresser coupling – a tradename for a coupling sleeve used to repair a section of pipe. *See Box.*

driers – additives that shorten the drying time of paint or other coatings.

drift – to move slowly away from a designated position.

drift pin – see drift punch.

drift point – a crane motion control point where the electric brake is released and the electric motor is not energized, so that the crane is held in position by its inertia.

drift punch – a tapered metal pin or punch used for aligning holes, such as bolt holes, between two structural mem-

Dresser coupling

Dresser coupling is a tradename for a threaded metal coupling sleeve which can be used to repair a section of pipe. The sleeve has nuts at each end and elastomer gaskets that fit around the pipe and seal against the outside diameter of the pipe. The nuts are threaded onto the ends of the coupling sleeve and tightened to seal the gaskets against the pipe. Different styles of couplings are available for different applications. For example, one style has ends of different diameters to couple iron pipe sizes to steel pipe sizes, which have different outside diameters for a given nominal pipe size.

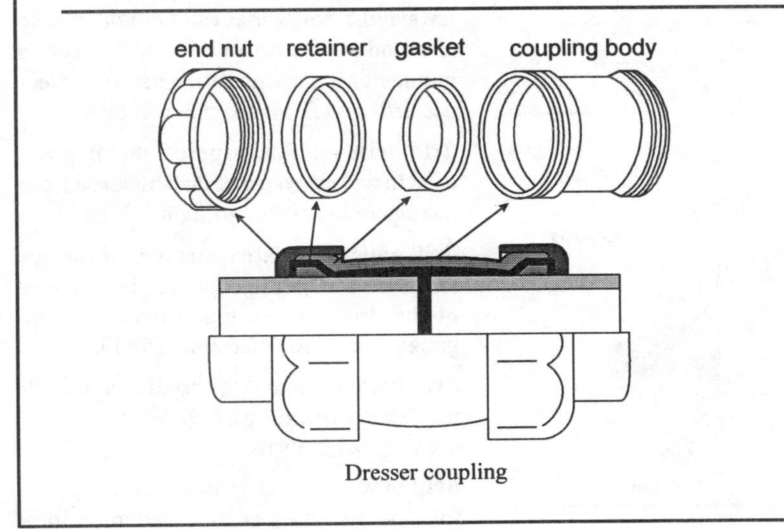

Dresser coupling

bers. Once the holes are aligned, bolts or fasteners are inserted to join the members together.

drill and counterbore bit – a drill bit with an enlarged portion on the shank to make a counterbore with the drilled hole in one operation.

drill bit – a tool used to bore holes. There are many types and configurations of drill bits, many designed for a particular purpose and/or material. May be power operated or manual. Also called a drill.

drill chuck – a mechanical device with jaws and a collar that can be tightened to grip and hold a drill bit. A drill chuck is mounted on a manually-operated or electric drill which rotates the drill bit.

drill drift – a flat, tapered metal piece used to wedge a tapered drill bit loose from the tapered socket holding it.

drill gauge – a plate with holes of various sizes, the size being equal to the diameter of the drill bit that made the hole. The gauge is used to determine drill bit size.

drill holder – a tool for holding a drill bit in a lathe, permitting a hole to be drilled in the work as it turns.

drill motor – an electric motor or air motor, one driven by compressed air, with a chuck or other device used to hold and turn a drill bit.

drill press – a machine used to bore holes at precise angles and precise locations. It consists of a stand, a means of holding a drill bit, and a motor to turn the bit.

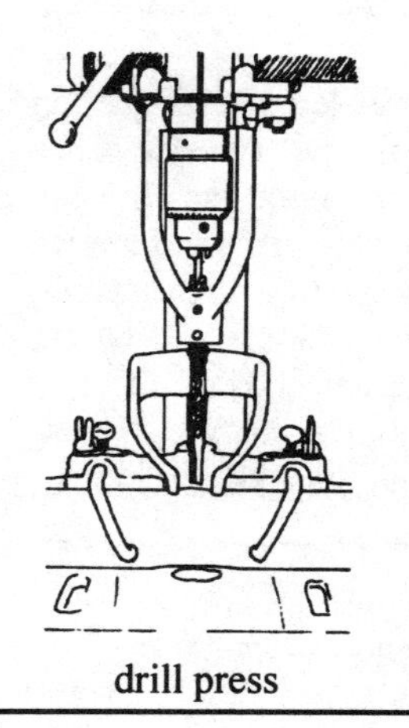

drill press

drill saw – a drill bit with cutting teeth along its shank so that it can be used in a drill motor to drill and then enlarge a hole.

drip – 1) a periodic leak of water, one drop at a time; 2) a groove on the underside, and parallel to the face of, a projecting section of stone. Water will drip from the edge of the groove rather than running along the bottom of the stone toward the wall from which it projects; 3) a groove made on the underside of a threshold or other building part which causes water to drip from the groove rather than allowing it to run off at a wall or other surface.

drip cap – a projection placed at the tops of windows and doors to redirect water.

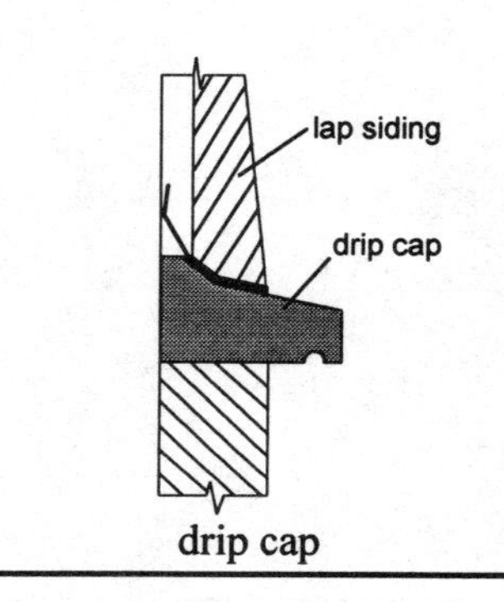

drip cap

drip edge – metal edging placed around the edge of a roof prior to installing the roofing material. It is designed so that water runs off over the drip edge and falls from a slight projection at the bottom edge of the roof rather than running back under, or along, the eaves.

drip molding – a molding around the base of building walls to redirect water so that it does not run from the walls down into the foundation.

drip pan – a pan placed under machinery to catch and contain leaking fluids.

drip screed – a formed metal piece, with a lip that projects downward, used at horizontal exterior corners to divert water away from the underside of a soffit or opening in a wall.

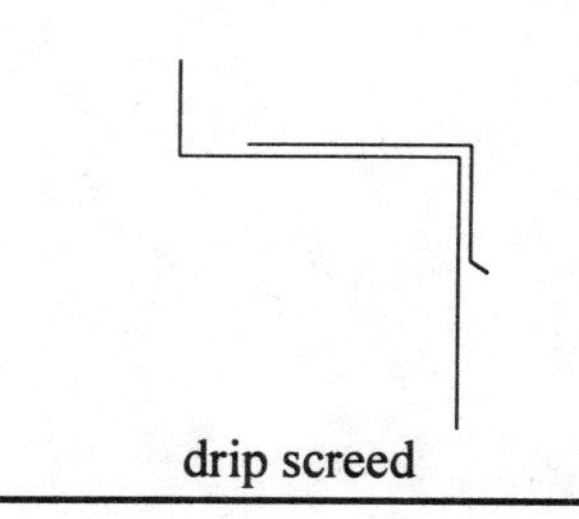

drip screed

drive – 1) any mechanism that transfers power from a motor or other power source to the mechanism being driven; 2) the motor and gearing used to drive a crane.

drive fit – a very tight fit requiring force to mate the parts. Also called an interference fit.

drive screw – a nail with steeply-pitched screw threads along the length of its shank.

drive shaft – a shaft used to transmit power. See also cross shaft.

driveway – a paved area for vehicles to move to and from a parking area or garage.

driving wheel – a wheel that imparts motion and force.

drop – 1) in an HVAC system, the vertical distance in feet that air has fallen below an outlet at the point of the air's terminal velocity, or the point where the air velocity coming from an outlet has decreased to 50 feet per minute; 2) a horizontal section of a vertical scaffold.

drop chain – a small link chain with a weight on one end for feeding an electrical wire through a vertical passage.

drop cloth – a protective covering to keep items clean during operations such as painting.

drop forge – a machine for forging metal using the gravity fall of a weight. The metal is heated to a malleable temperature and the weight is used to hammer the metal as needed to provide the desired size and shape.

drop hammer – a heavy steel casting raised and dropped from a controlled height to drive pilings into the earth.

drop leaf – a hinged table extension that can be folded out of the way when not in use.

drop light – an electric light, suspended at the end of a long cord, that can be moved to a work area to provide direct lighting.

drop-off – a steep decline or cliff.

drop outlet – a gutter system fitting that provides an outlet for the water collected in the gutter, usually feeding into a downspout.

dropped ceiling – a non-structural suspended ceiling installed below the level of the overhead structure.

dropping a hip rafter – cutting a deep seat cut so that the hip rafter sits lower in relation to the surrounding roof framing. This prevents the top edges of the hip rafter from projecting above the plane of the adjoining roof.

drop siding – siding boards installed horizontally. The boards have a tapered cross-section with the top edge being the narrow edge. The top edge fits into a rabbet or groove in the bottom edge of the board above.

drum – 1) a cylindrical shaped container; 2) a power-driven rotating cylinder which can be used for hoisting by wrapping a hoisting rope, attached by means of a pulley or pulleys, to a load.

drum sander – a drum, made of rubber or other material, with an abrasive around the outside surface. The drum is mounted on an arbor which may be held in the chuck of a motor and turned. The work to be sanded is usually held against the drum.

drum trap – a cylindrical plumbing fixture with an inlet at its base and an outlet at its top, which retains water as a seal. It is used where plumbing fixtures are close to the floor and there is no room to install a P-trap.

dry battery – an electrical storage battery that is sealed and does not contain free liquid.

dry bulb thermometer – see thermometer, dry bulb.

dry dock – a structure into which a boat or ship is floated or pulled for maintenance work. Once the ship is in the structure the water is pumped out, providing access to the hull of the ship for cleaning and repairs.

dryer receptacle – a 240 volt, 30 ampere electrical outlet for a clothes dryer.

dry gas – natural gas from which impurities have been removed. Also called sweet gas.

drying oil – any oil that will dry to a solid when exposed to air. These oils are derived from a variety of plants and are good preservatives and protective coatings for wood.

dry masonry – masonry units that are set without mortar.

dry mix – a pre-packaged mix of dry concrete containing cement, sand and gravel.

dry-pipe sprinkler system – a fire protection system which uses air pressure to keep water out of the pipes until the system is activated.

dry-press brick – brick made under high pressure using clay with a 5 to 7 percent moisture content.

dry rot – deterioration of wood by fungi. Dry rot is most likely to occur where wood is exposed to a constant source of moisture, such as fences on which sprinklers spray or in studs behind a leaky shower wall.

dry-set ceramic tile grout – a mixture of portland cement and sand with additives that have water retention abilities, used to grout walls and floors subject to ordinary use.

dry-set ceramic tile mortar – a mixture of portland cement and sand with additives that have water retention abilities, used to bond ceramic tile to a base.

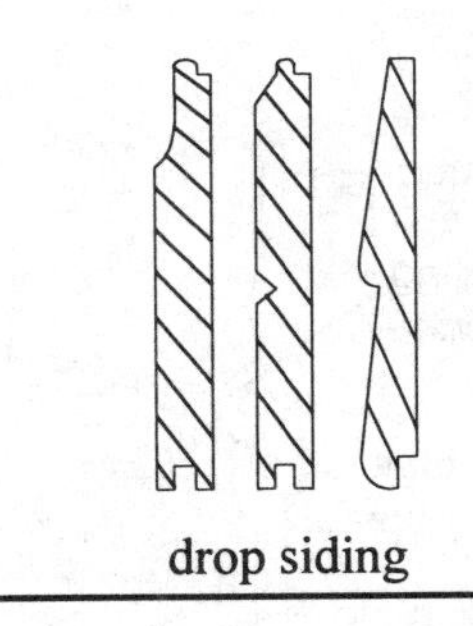

drop siding

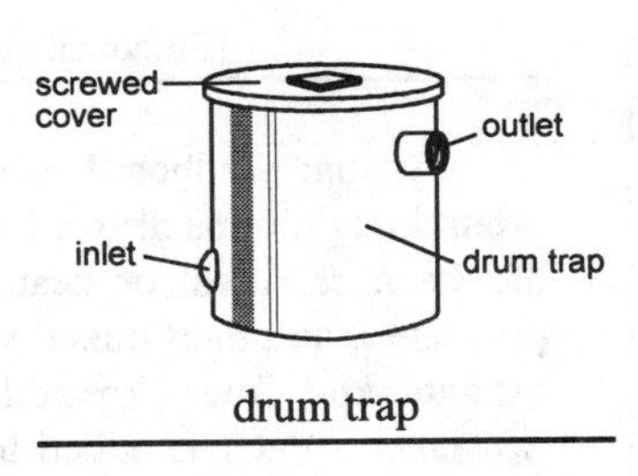

drum trap

Drywall

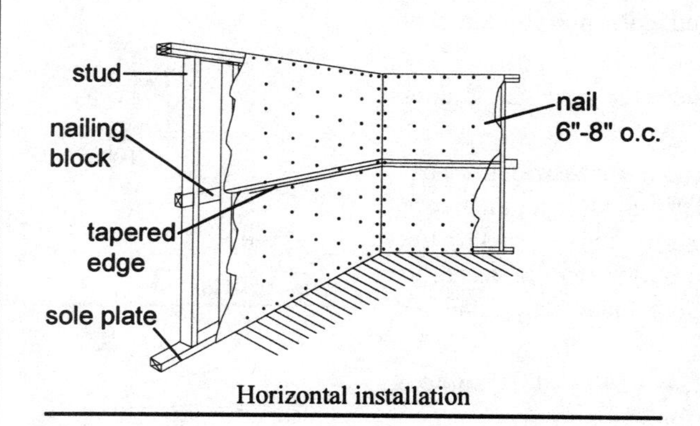

Horizontal installation

Gypsum wallboard is the material that is most often thought of as drywall. It is made of gypsum that has been calcined, or heated to drive off moisture, powdered, and then mixed with additives such as aggregate and fibers to add strength and moisture resistance. Water is added to the gypsum mixture so that it can be molded and shaped into the forms that are needed. As the gypsum mixture dries, it forms interlocking crystals that turn it into a rock-hard board. Before it hardens completely, it is cut to the lengths needed. The drying process is then completed to remove any remaining excess moisture. Drying is important because drywall shrinks as it dries, and too much shrinking after installation will spoil the job. Gypsum drywall has several advantages over plaster. It is inexpensive, easy to install and it offers noise insulation and fire protection. Gypsum drywall panels are normally 4 feet wide and 8, 10, 12, 14 or 16 feet in length. There are six standard thicknesses available, ranging from ¼ inch to 1 inch, to meet the requirements of different conditions. There are also several different types of drywall for a variety of applications.

Standard drywall is intended for most wall covering applications in dry interior areas. It has a smooth paper covering on both sides that will take a variety of finishes.

Water resistant drywall, also called WR panels or greenboard, is designed for high-moisture areas, such as kitchens and baths. Asphalt compounds are added to the gypsum compound to make it moisture resistant. The paper covering is multilayered and chemically treated to resist moisture. The facing paper is a light green, giving the panels their name.

Gypsum drywall is naturally fire resistant, but in Type X panels, the fire resistance is increased by adding glass fibers to the core. It is used where fire-rated walls

drywall knives

dry-stone wall – a stone wall laid without mortar.

dry taping – the application of joint tape for gypsum wallboard using adhesives other than joint compound.

drywall – wall coverings that are applied dry, or without mortar. The term is most often used in reference to gypsum wallboard, although other wall covering materials, such as wood panels, also fall into this category. *See Box.*

drywall blisters – a defect in gypsum wallboard in which the facing paper comes unbonded from the gypsum core.

drywall hammer – a hand tool with a flat striking surface on one end of the head and a hatchet-like surface on the other. The striking surface is designed especially for use with gypsum drywall nails, and the hatchet blade, which is dull, is intended for prying rather than cutting.

drywall knife – a flat-bladed tool that comes in a variety of widths. Also called a putty knife or taping knife.

drywall saw – a saw with coarse teeth and a blunt end used for cutting drywall.

drywall screw gun – see screw gun.

drywall T-square – a large T-square used for laying out cuts on drywall.

duckbill pliers – pliers with the ends of the jaws tapered to a horizontal flat wedge.

duckboard – board laid over muddy ground areas to provide a stable walking surface.

are required, such as between an attached garage and living quarters. The panels come in ½- and ⅝-inch thicknesses. The ½-inch thickness has a fire rating of 45 minutes and the ⅝-inch thickness has a 60-minute rating. Type X panels are also called fire-resistant drywall.

Foil-backed gypsum drywall is made by laminating aluminum foil to the back surface of drywall panels. The foil backing creates a waterproof membrane, making the panels an effective vapor barrier as well as increasing their insulating value. However, they should not be used as a base for tile or in high-moisture areas, as the foil backing will trap and hold the moisture in the panel.

Exterior sheathing panels can be used for indirect exposure to the weather on vertical surfaces. They have a water-resistant core and are covered with water-repellent paper. They can be applied directly to framing members and covered with an exterior finish, such as aluminum siding, shingles, stucco or masonry. Exterior ceiling panels can be used on exterior ceiling-type surfaces, such as soffits and carports. They are also water-resistant as long as they do not have direct exposure to the weather.

Gypsum lath is drywall that is installed as a backing for interior plaster walls and ceilings. Veneer-base drywall, also called blueboard, is used as the base for veneer plaster. It provides a smooth, durable and wear-resistant finish. Decorated drywall is standard drywall covered with decorative paper or vinyl. It is available in a variety of colors, patterns and finishes.

Other specialized types of gypsum panels include backing board, form board, coreboard and partition blocks. Gypsum drywall is also known as plasterboard, wallboard or gypboard.

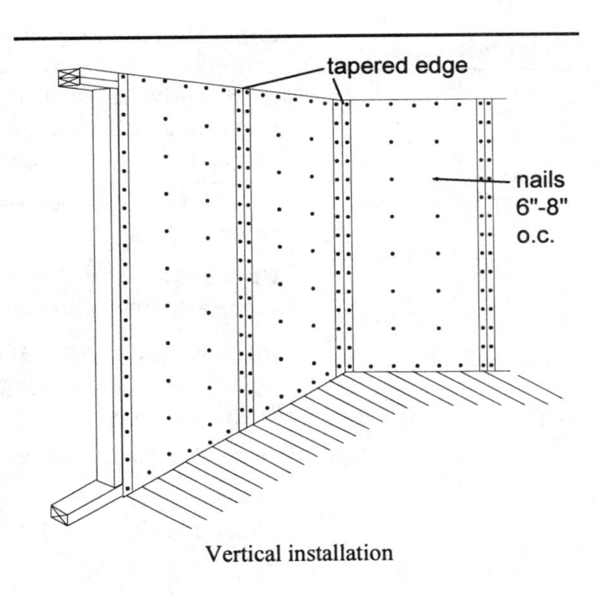

Vertical installation

duct – thin-wall sheet metal or lightweight tubular conduit used for conveying air at low pressure; may be round, square, or rectangular.

duct fan – a fan mounted in a duct to move air through the duct.

ductile – workable, as a metal that can be shaped without breaking.

ductility – a material property that allows the material to be deformed or worked into different shapes.

ductwork – an installed duct system.

ductwork, preformed – ductwork fittings that have been factory-shaped for a specific applications, such as elbows and tees.

due process – the regular working process of the law.

dull – 1) blunt; 2) not shiny.

dull finish – a painted surface with almost no gloss.

dull rub – a finish that has been rubbed to a dull sheen.

dumping – applying paint very quickly.

dumpy level – see level, dumpy.

dunnage – material used to space and protect objects during shipment.

duplex – 1) consisting of two parts; 2) a dwelling with two living units joined by a common wall.

duplex nail – a nail with two heads on the shank, one about ½ inch below the other. The nail is driven into wood until the first head is seated, leaving the second head protruding above the surface so the nail

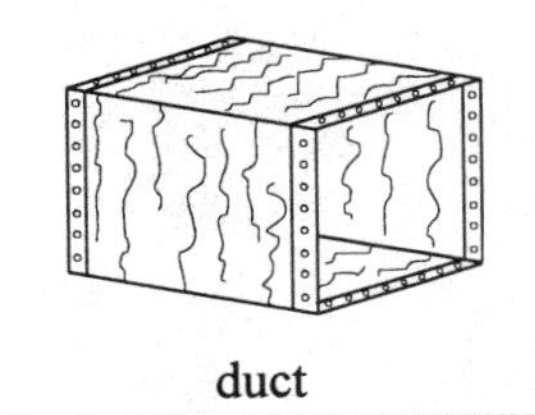

duct

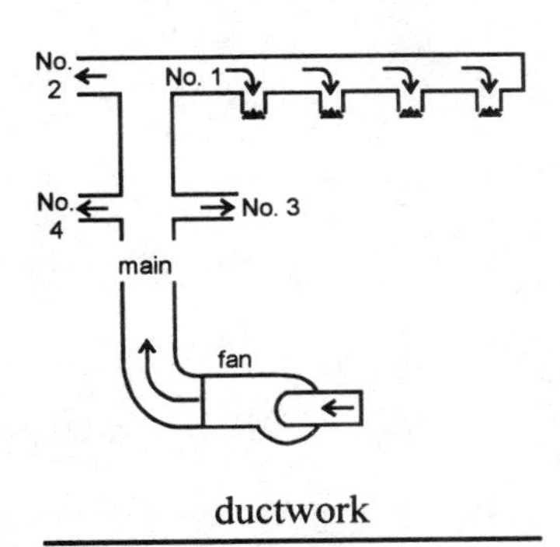

ductwork

can be easily removed with a claw hammer or pry bar. Often used for concrete forms, or similar application where the installation is temporary. Size 16d, used for concrete forms, is the most common size of duplex nails, but other sizes are available.

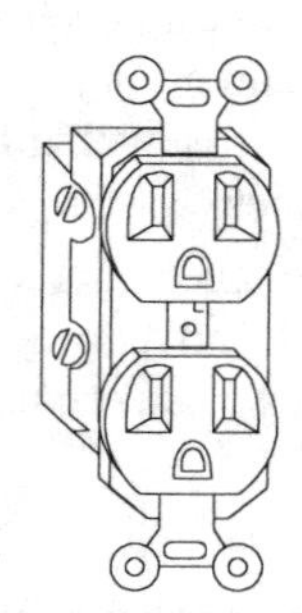

duplex receptacle

duplex receptacle – an electrical outlet designed to allow insertion of two plugs.

duplicate – 1) to copy; 2) redundant.

duplicate switching – a method of wiring multiple lamp fixtures so that half the lamps can be turned on or off independent of the other half. This allows for a reduction in the level of lighting when desired.

durability – resistance to wear or deterioration.

Duralumin – a trademark name for a strong aluminum alloy.

Durham system – a trade name for a rigid, threaded joint piping using recessed drainage fittings.

dust – small solid particles.

dust cover – a cloth or plastic covering which keeps items, such as furniture, clean while stored or not in use.

dust-free – 1) an area where there are no solid floating particles in the air; 2) a painted surface that has dried sufficiently so that dust will not stick to it.

dust mask – a mask worn over the mouth and nose to filter out dust particles from the air being breathed.

Dutch arch – a brick arch that is flat at the top, made from bricks that are rectangular rather than wedge-shaped.

Dutch door – a door that is divided horizontally so that the upper half may swing open while the lower half remains closed.

Dutch hip roof – a roof style in which the upper gable ends of the roof are built with a short hip section.

Dutchman cut – a type of tree felling cut used when a tree must fall in the opposite direction from the tree's natural lean. The tree is notched with an undercut at 90 degrees to the direction of fall. The face cut made for the undercut is then extended. A back cut on the side of the natural lean is made and a wedge is inserted. Then the back cut is extended to fell the tree.

dwarf partition – a partition that does not go all the way to the ceiling. Also called a dwarf wall.

dwelling – a structure in which people live.

DWV tubing – copper or plastic tubing used for drain, waste or vent applications in a plumbing system.

dynamic – moving; changing.

dynamic braking – a means of braking an electric motor. The motor is used as a generator and the electrical energy generated is dissipated in resistors.

dynamo – a dynamoelectric machine; a direct-current electric generator.

dynamometer – a device used to measure mechanical force.

E: earned income to eye screw

earned income – income that is earned on work that has been completed, as opposed to unearned income, which is income on work that is not completed.

earnest money – a deposit made with an offer to buy. It is a good faith commitment towards the purchase.

earth – ground; soil.

earth pigments – mineral pigments, or colors from the earth, that are naturally occurring. Earth pigments can be used to color concrete.

earthquake load – see seismic load.

earthwork – moving and shaping earth based on detailed site drawings. The changes to be made are determined by surveying and staking out the desired contours; excavation, trenching, backfill and compaction are all part of earthwork.

eased edge – the tapered and rounded factory edge of a gypsum wallboard panel. The tapers on the edges of adjoining panels form a shallow depression along the joint. The depression is filled by the application of joint tape and joint compound to seal the joint. Once this is accomplished, the depression along the joint is flush with the surface of the wallboard.

easement – 1) a designated right of way, such as the access guaranteed to utility companies for repair of utilities that are located on, or cross over, private land. An easement cannot be built upon or otherwise obstructed; 2) a curved portion of a staircase handrail used to prevent abrupt direction changes.

eave – the edge of a roof that projects beyond the exterior walls of a structure.

eave course – the first course of shingles along the eave line.

eave flashing – roofing material that is laid along the eave line and extended up the roof under the shingles. *See Box.*

eave height – the distance from the ground to the eave measured at the eave line.

eave line – the overhanging edge of the eave of a roof.

ebony – a dense, dark and hard wood from the tropical Ebonaceae family, used for both its beauty and wear resistance.

eccentric – 1) rotating off-center, such as a cam or inclined plane on a rotating shaft; 2) out of round, oval, elliptical.

ecology – the science dealing with the relationship of living organisms to their environment.

economy brick – a brick measuring 4 x 4 x 8 inches.

eddy – a water or air current moving contrary to the main current. An example would be rotary turbulence in a stream of air. It is a condition of fluid flow affecting the construction of such systems as HVAC or plumbing.

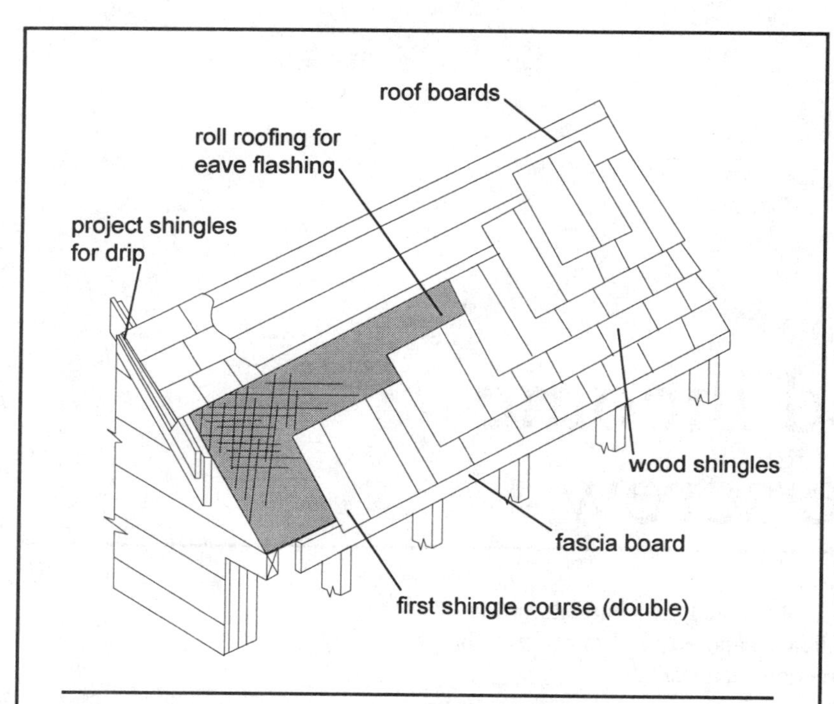

Eave flashing

Eave flashing consists of an undercourse of 36-inch, 50-pound or heavier, roll roofing installed along the eave line. It forms a continuous waterproof membrane extending up from the edge of the roof. Its purpose is to prevent rainwater from backing up under the shingles at the eave line. In cold climates, it helps keep ice from building up and forming ice dams that trap water under the shingles. The trapped water and melting ice may eventually leak through the roof. In areas with very heavy snowfall, two courses of eave flashing may be necessary, as ice dams could reach higher than 36 inches.

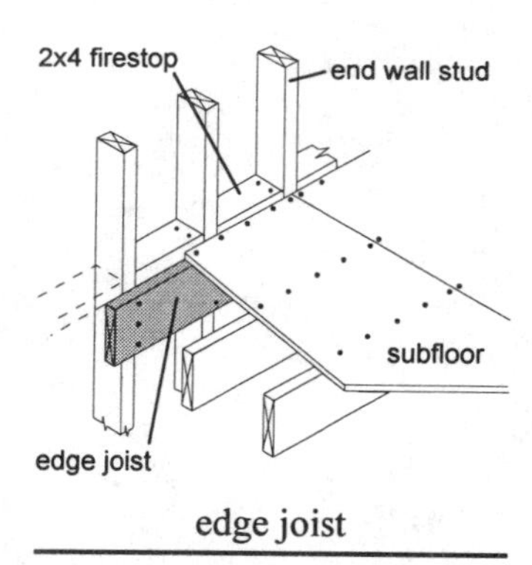

edge joist

eddy current braking – a means of braking an electric motor by inducing a back current to work against motor rotation.

eddy current testing – a nondestructive examination that correlates the electromagnetic properties of an object with its physical properties.

edge – 1) the narrow surface of an object perpendicular to its face; 2) the long edge of a gypsum wallboard panel; 3) the sharp cutting portion of a knife or other blade.

edge clamp – a C-clamp with three adjustable clamping screws, one at each end of the C, and the other one through the middle of the clamp, dividing the C. The clamp is attached to a surface, such as a countertop, and the middle clamping screw is used to apply pressure to edge trim on the surface.

edge cutter – see stripper.

edge grain – wood that has been sawn approximately at right angles to the growth rings of the tree.

edge joint – a weld between the edges of parallel pieces of material.

edge joist – the joist at the inside edge of a building wall.

edge molding – a molding made by shaping the edge of a board with a molding cutter mounted on a power saw.

edger – a hand tool used to round and compact the edges of wet concrete, creating a strong uniform edge.

edge venting – regularly spaced vents around the eave line of a roof which provide ventilation to the attic space under the roof.

edging – 1) painting along an edge; 2) finishing concrete edges with an edging tool; 3) trim along an edge.

edging strips – nine-inch wide strips of roll roofing installed, using the concealed nail application method, along all roof edges and rakes prior to the first course of roofing.

edifice – a large structure or building.

Edison base – a porcelain base with a light bulb or fuse socket and terminals to which electrical wiring can be attached.

effective opening – the smallest diameter or cross-section of piping at a point of discharge.

effective throat – the distance through a weld, from the root of the weld to the face of the weld, after subtracting the amount of weld reinforcement.

effective weld length – the length of the weld that has a configuration and a cross section that is equal to, or greater than, that required by the design. Welds are designed by size and configuration to withstand specific conditions. The effective weld length would meet those conditions.

Elbow

A preformed fitting for pipe, tubing, conduit or duct. Elbows provide a convenient and consistent means of turning corners rather than bending the pipe or conduit itself. The fittings usually form a 45- or 90-degree angle at the bend, though elbows of lesser angles are available in some materials. Elbow fittings are generally joined to the system in the same manner that other connections are being made. For example, if pipe is being joined with flanged connections, then the elbows will also be joined with flanged connections. The fittings are manufactured in various materials for different systems. Elbows are also called ells.

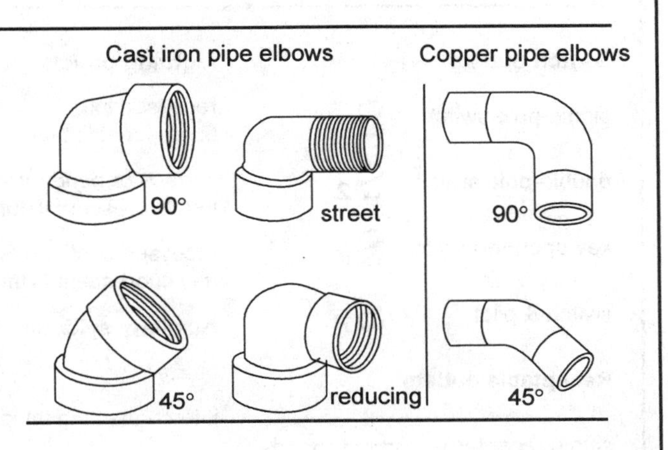

efflorescence – soluble salts that collect on the surface of masonry or concrete, leaving unsightly stains. It is a cosmetic defect and can be removed by cleaning with a dry brush and rinsing in clear water. A heavy collection of salts may require brushing with a solution of muriatic acid diluted in water.

effluent – liquid waste from a septic system.

eggshell – a paint luster that has a low gloss sheen similar to that of the shell of an egg.

elastic – 1) a rubberized material with the ability to return to its original shape after being distorted; 2) capable of recovering shape and size after deformation.

elastomer – a natural or synthetic elastic used for such items as seat washers in faucets, seals in hose connections, protective linings and many different types of gaskets. Elastomer lining material is used to protect valve and pipe bodies from contact with harsh chemicals carried in fluid systems. Elastomer gaskets ensure a tight seal between parts without requiring perfectly machined and fitted metal surfaces.

elastomeric – a term used to describe the elastic properties of a material.

elbow – a curved, preformed fitting designed for use with pipe, tubing, conduit, or duct. Also called an ell. *See Box.*

elbow curvature – the ratio of the bend radius of the elbow to its diameter.

electrical charge – the amount of electrical energy that can be stored in an electrical conductor or semi-conductor, capacitor, or battery.

electrical construction drawings – building drawings that show the installation details for the electrical equipment.

electrical diagrams – drawings showing the wiring and electrical connections for a building.

electrical distribution system – the electrical system from the point of generating the electrical power to its final use point. The electrical power is conducted by busways from the power plant to distribution power lines and then to electrical substations for distribution to local use points. It is transmitted at high voltage to the substations where the voltage is dropped at local transformers before being routed into dwellings and commercial buildings. Electrical power is transmitted at high voltage so that the unavoidable line losses due to electrical resistance are overcome during the long distance that it travels.

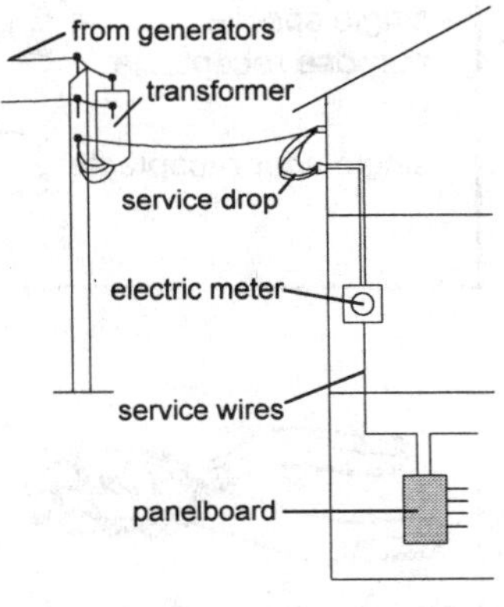

electrical distribution system

Electrical symbols

Switch outlets

single-pole switch S

double-pole switch S2

key operated switch SK

switch & pilot SP

Receptacle outlets

single receptacle G

duplex receptacle G

single special purpose receptacle G

single floor receptacle G

Lighting outlets

recessed ind. fluorescent fixture OR

surface or pendant ind. fluorescent fixture

recessed continuous-row fluorescent fixture OR

Auxiliary systems

buzzer

interconnecting telephone

radio or TV receptacle R TV

Circuiting

wiring concealed in ceiling or wall

wiring concealed in floor

wiring exposed

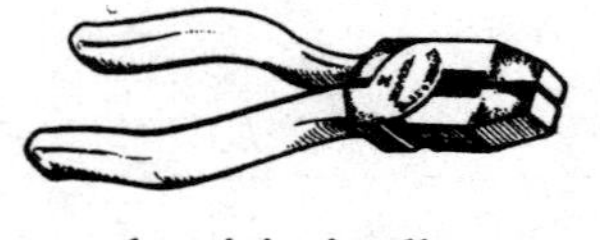

electrician's pliers

electrical drawings – building drawings that show the wiring layout, the electrical equipment types and placement, and the locations of all electrical fixtures. Also called an electrical layout.

electrical induction – the generation of electrical current by passing a conductor through the magnetic field existing between two poles of a magnet.

electrical metallic tubing (EMT) – see conduit.

electrical power – current flow at a voltage, stated in watts (watts = amps x volts).

electrical raceway – a metal or non-metallic trough or conduit through which electrical cables are routed for support.

electrical schematic wiring diagram – a single line wiring diagram indicating wire sizes and the quantities of wires needed.

electrical single line diagrams – drawings that show the electrical equipment arrangement and connection into a circuit.

electrical specifications – written requirements for a building's electrical system.

electrical symbols – icons and/or abbreviations used in wiring drawings to indicate various types of electrical components.

electric arc spraying (EASP) – a process of spraying molten metal as a coating using an electric arc for heat. The arc passes between two consumable electrodes and a gas is used to atomize and spray the molten metal.

electric arc welding – see arc welding.

electric drill – a rotary tool for boring holes, sanding, or similar uses, powered by electricity.

electric heating – the conversion of electrical power into heat by means of a high-temperature wire that is resistant to electrical flow. The amount of resistance and the amount of electrical current flowing into the wire determines the rate at which the electricity is converted to heat. For example, the more electric current that is fed into a heating element, the greater the resistance, and the more heat produced.

electrician – a person trained, skilled and licensed in electrical wiring installations. License requirements and other regulations regarding the electrical profession vary in different locations.

electrician's pliers – pliers with built-in side cutters and insulated handles. Also called lineman's pliers.

electricity – energy in the form of moving charged particles.

electric screwdriver – a hand-held electrically-operated screwdriver; most models are cordless and operate on rechargeable batteries. They come in a variety of models and speeds. Some are shaped like a manual screwdriver and others are shaped like pistols. The more expensive models have variable speeds and greater torque than the less expensive ones, and some come with replaceable battery packs.

electric shock – the flow of an electrical current through a person's body.

electrode – an electrical terminal or a welding rod used in arc welding and cutting.

electrode holder – an insulated clamp that holds a welding electrode and supplies power to the electrode from attached wiring.

electrode lead – the power supply cable to an electrode holder.

electrogas welding (EGW) – an electric arc welding process using a consumable solid or flux-cored electrode, with or without an external shielding gas.

electrolysis – the erosion of a metal by a chemical reaction with another metal in the presence of an electrically-conductive fluid. For example, zinc in contact with steel in saltwater will be eroded away.

electrolyte – an electrically conductive fluid or other substance. Saline is an example of an electrolyte.

electromagnet – a device consisting of a magnetic material wrapped with an electrically-conductive coil through which a current is passed to induce magnetism in the material.

electromechanical – a device that has both electrical and mechanical components, such as an electric motor or an electric fan.

electromotive force – voltage; the force that makes an electrical current flow.

electron – an atomic particle, in orbit around a nucleus, that has a negative electrical charge.

electron beam gun – a mechanism that produces and accelerates electrons for use in welding and cutting operations.

electronic stud finder – a magnetic or ultrasonic device for locating studs behind a finish wall. It senses the presence of a nail or screw head, or the increase in wall density at a stud, and makes a beeping sound and/or activates a light.

electroslag welding (ESW) – a welding process that uses an electric arc to melt the slag initially, and then an electric current to keep the slag molten, which provides heat for welding and also provides an inert shield to protect the weld from oxygen in the air.

electrostatic filter – a filter that has the opposite electrical charge of the particles that it needs to attract. As with a magnet, opposite electrical polarities attract. The particles are drawn to the filter and trapped, held by their electrostatic attraction to the filter.

electrostatic painting – a type of painting in which an electrically-charged powder is sprayed on a surface charged with the opposite electrical charge. The coating is then baked on.

element – 1) a basic substance, like oxygen, hydrogen or carbon, that cannot be made into a simpler substance through chemical reaction; 2) an electrical device; 3) part of a whole.

elevation – 1) the height above sea level; 2) the height above an established reference point, such as a grade reference point on an elevation drawing used in construction.

elevation view – the drawing of a structure showing a side, front, or back view, or a particular portion of the structure.

elevator – a platform or enclosure for moving people and/or commodities from one level of a structure to another.

ell – 1) a fitting that allows a bend in a pipe, conduit, duct, or tube. See also elbow; 2) a wing of a building at right angles to the main building.

ellipse – a closed curve plane figure in which the sum of distances from any point on the curve to two fixed points remains the same.

ellipsoid – a solid figure in which any section through it is a circle or an ellipse.

elongate – to make longer.

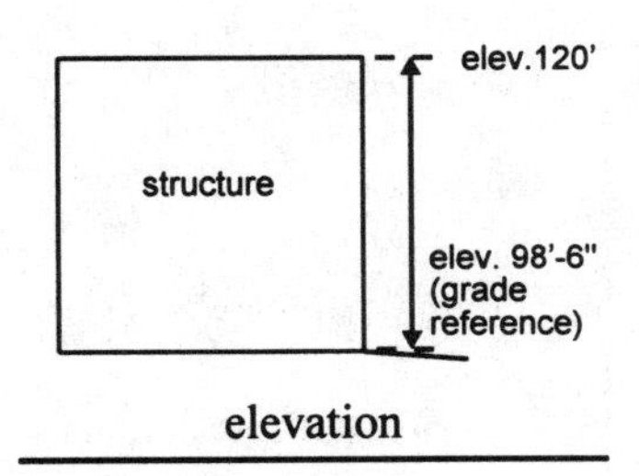

elevation

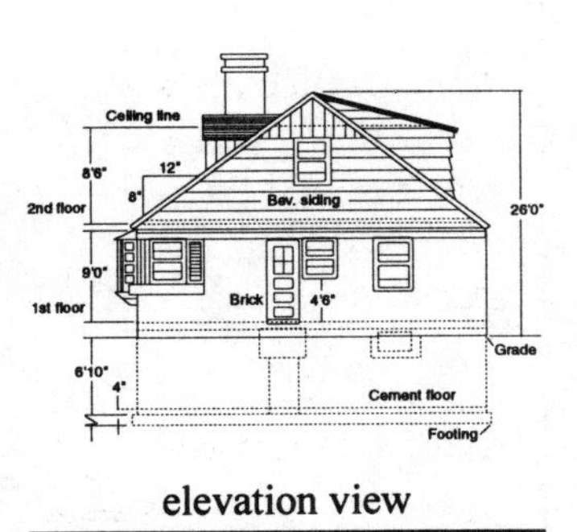

elevation view

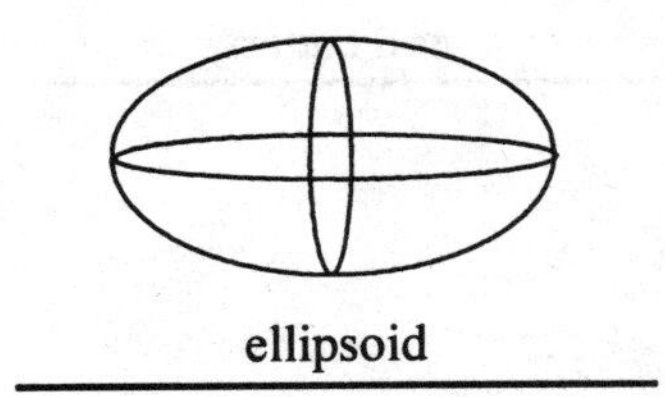

ellipsoid

eluvium – a deposit of decomposed rock or soil.

embankment – soil piled into a mound or ridge used to retain water or as a support along roadways.

embed – 1) to enclose closely. When joint tape is applied over wet joint compound on gypsum wallboard and covered with another layer of joint compound, it is *embedded*; 2) to become an integral part of, as a metal plate set in concrete becomes part of the structure when the concrete is poured. The plate can then be used as an attachment point on the concrete structure.

embedment – the distance that a pier or column is sunk below grade.

emboss – to make a raised design on a surface.

embrasure – the widening or flaring of a window or door opening toward the inside of a structure.

emergency – a situation out of the ordinary that requires special measures to ensure the safety of life.

emergency exit – an exit that is not normally used, but that provides an alternative route out of a structure in special situations.

emergency light – a light, powered by a continuously charged battery, which comes on during the loss of normal electrical power.

emery – an abrasive material of Carborundum and iron oxide.

emery paper – a type of paper having a very hard abrasive coating.

eminent domain – the right of a governing body to take private property for public use, such as for widening or extending a highway. The governing body must pay just compensation to the owner of the property. The property is appraised and the owners are compensated in cash or given property whose value is equal to or greater than that taken.

emission – a discharge, such as carbon monoxide exhaust from an internal combustion engine.

emulsion – 1) one liquid dispersed throughout another; 2) a mixture of water and a liquid bitumen substance.

enamel – a paint that is a mixture of quartz, silica, lead, feldspar, and mineral oxides, used as a decorative and/or protective coating.

enameled nails – nails that have been coated with enamel for protection and color. These nails are used in applications, such as paneling installation, where it is desired that the color of the nails blend with the material being installed.

enamel paint – paints with a varnish, polyurethane, alkyd resin or acrylic base which form a hard, glossy surface when dry.

enclose – to close in or surround.

enclosing hood – a hood that envelopes the source of fumes or contaminants that need to be controlled.

enclosure – a housing or fence surrounding a structure or thing to be protected.

enclosure wall – an exterior wall that is not a bearing wall.

encompass – to encircle or enclose.

encroachment – something that is gradually taken over or possessed by means of trespass. For example, if a fence is built on the neighboring property, and ownership of the fence is not contested, over time the property that the fence is on will be assumed to belong to the owner of the fence.

encumbrance – a burden or claim against property, such as a lien. Also incumbrance.

end – 1) termination point; 2) the short edge of a gypsum wallboard panel that shows the gypsum core.

end-bearing pile or piling – a piling that is loaded on the end so that it acts as a column.

end cutters – a plier-type hinged device for cutting nails or wire by pressing the hinged handles together and pinching the material to be cut. The cutting edge is at the end of the tool and at right angles to the handles. Also called end cutting pliers; end nippers.

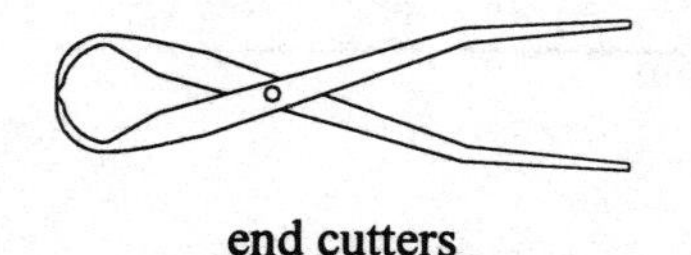

end cutters

end grain – the grain pattern exposed on the end edge when a board is cut across the grain.

end-hook approach – the closest distance that a crane hook can come to the walls of the building that houses the bridge crane. The distance is measured parallel to the rails on which the bridge moves. This distance must be taken into consideration when designing the facility to make maximum use of the crane's reach and movement patterns. The end-hook approach is the major limiting factor in the use of an overhead crane in a manufacturing facility.

end molding – to shape the end of a board into molding with the use of a power saw and a molding cutter.

end nippers – see end cutters.

end-of-line receptacle – the last outlet receptacle in an electrical circuit.

endothermic – a reaction that absorbs heat. Lime is manufactured from limestone by heating it. The limestone absorbs the heat, and in so doing, isolates the lime. This is an endothermic reaction.

end post – the primary anchor post at the end of a fence.

end tie – a structural support member of a crane bridge that braces the bridge members and prevents movement between them. This maintains the squareness of the bridge.

end truck – the wheeled part of the crane bridge that allows the bridge to move along the crane rails.

endwall column – a vertical structural member in the endwall of a building.

ene – suffix used on paint product labels to denote aromatic hydrocarbons such as xylene.

energy-efficient window – a window with two or more parallel panes of glass which retards the heat loss through the window.

energy loss – 1) the heat loss from a structure; 2) the energy required to generate heat to replace that which is lost.

engineered brick – a brick with the dimensions of $3\frac{1}{5}$ x 4 x 8 inches.

engineered masonry – a masonry design based on structural analysis.

engineering – the application of science and mathematics toward understanding how the properties of matter and the sources of energy in nature can be used to benefit man.

engineering controls – the regulators, valves, switches or levers used by an operator to regulate, manipulate or run a system.

engineer's scale – a scale of dimensions that can be made equal to a foot or other unit of measure. The divisions on the scale are in 10's and multiples of 10, such as 100's and 1000's. The engineer's scale is often used for site plans.

English bond – a brickwork pattern using alternated courses of headers and stretchers.

English stipple – a type of texture created by dabbing a wet plaster or joint compound surface with the bristles of a stiff brush to make indentations.

engrave – to carve in a hard surface.

entasis – a slight and gradually curved tapering of a column toward its center to add visual appeal.

entrain – to trap and retain, such as gas bubbles in a solid or a liquid, or air in concrete.

entrance – the passageway into a building. Also called an entryway.

entrance cap – the watertight cap at the top of an electrical mast where the electrical wires from the service drop are run to the electrical meter in a building. The wires droop from the pole to the entrance cap so that the entrance cap is not the low point in the downhill run from the pole. Water will run to the low point on the line and then drip to the ground. The wires enter the entrance cap at an upward angle through a snug-fitting insulator. The entrance angle further limits the possibility of water getting through the entrance cap. Entrance caps are also called rain caps, mast heads or weatherheads.

entryway – see entrance.

envelope – to enclose or surround.

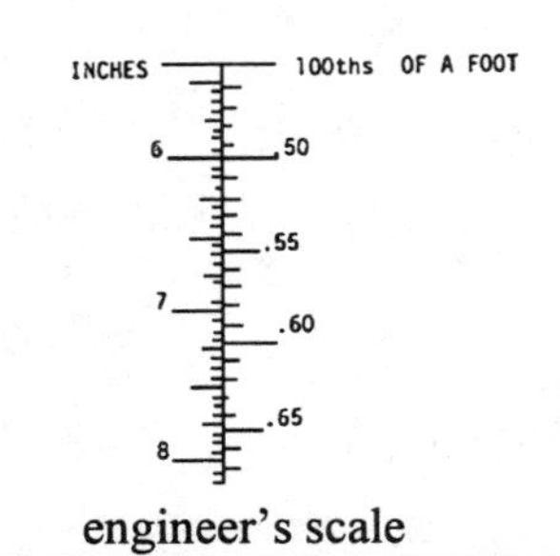

engineer's scale

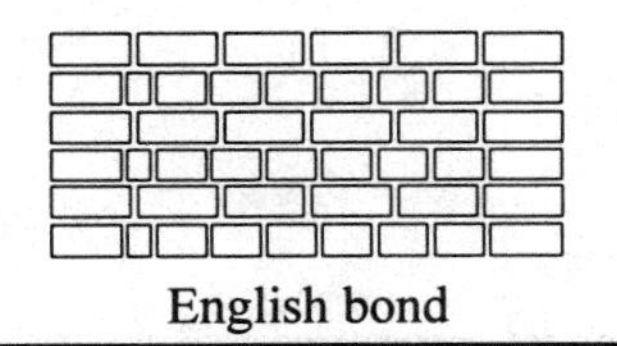
English bond

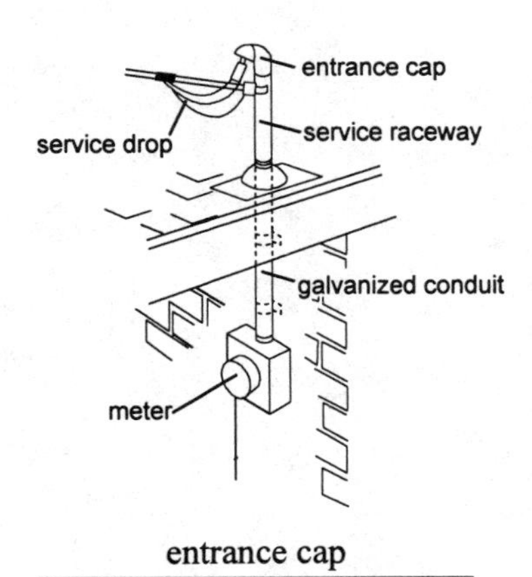

entrance cap

epistyle

epistyle

enveloping vapor barrier – a vapor barrier that is wrapped around an entire wall.

environment – the conditions or surroundings that affect the growth and survival of various life forms within a given area.

environmental impact statement – a report on the results of studies done to analyze the impact, or changes that may occur in the environment, if a proposed construction project is built.

epicyclic train – a cluster of gears with axes that revolve around a central axis, such as the differential in the drive train of an automobile.

epistyle – 1) a cross beam between columns or between a column and another part of the structure; 2) an architrave.

epoxy – a plastic, in liquid or semi-liquid form, to which a catalyst is added to create a chemical reaction that causes it to dry to a hard finish.

epoxy adhesive – a fast-setting adhesive-containing epoxy.

epoxy binder – a binder, consisting of a resin with a polyamide hardener, which causes a chemical reaction when mixed with epoxy paints resulting in the mixture hardening as it dries.

epoxy catalyst – the chemical added to an epoxy that starts the curing and hardening reaction.

epoxy cement – a cement that uses an epoxy to add durability and reduce curing time.

epoxy paint – a two-part paint with an epoxy vehicle that hardens to a gloss by chemical reaction.

equalizer – a device that provides for equal distribution of force, such as sheave or bar used in rigging to compensate for unequal hoist rope length.

equilibrate – to balance or keep in equilibrium, such as keeping a load in balance at the end of a crane hook.

equilibrium – a state of balance where opposing influences or forces are equalized or remain in check. An example is a plumb bob suspended from a cord without movement; any change in its position disrupts its equilibrium.

equipment – tools and other implements for accomplishing a specific task.

equipment ground – a grounding wire, separate from the system ground, to a piece of equipment or to a receptacle. The equipment ground will direct electricity from a faulty motor or other piece of electrical equipment to the grounding wire, thus preventing a person touching the defective equipment from inadvertently becoming the ground and being electrocuted.

equity – 1) the amount of the total value of a tangible item, after all debts on that item are considered; 2) fairness and justness.

equivalent – being the same as or having the same properties as another.

equivalent thickness – the actual thickness of the solid material in a hollow concrete masonry unit.

erasing shield – a thin metal plate with sections cut out in various shapes to permit erasing small, precise sections of a drawing without affecting the other areas.

erect – to construct or raise.

erection – the act or process of constructing a building or structure.

erection drawings – structural drawings detailing the installation and connection of structural members.

erode – to wear away.

erosion – the washing or wearing away of a surface, such as soil along a riverbank.

erosion control – a system of drainage, water diversion or other means, taken to prevent or diminish erosion. Soil erosion control often involves rotating crops and planting wind breaks.

erratic – unstable; random; wandering; deviating from the norm.

escheat – the reversion of land ownership to the government when there are no legal heirs.

escrow – legal documents held in trust by a neutral third party to be delivered upon the fulfillment of a condition.

Estimate

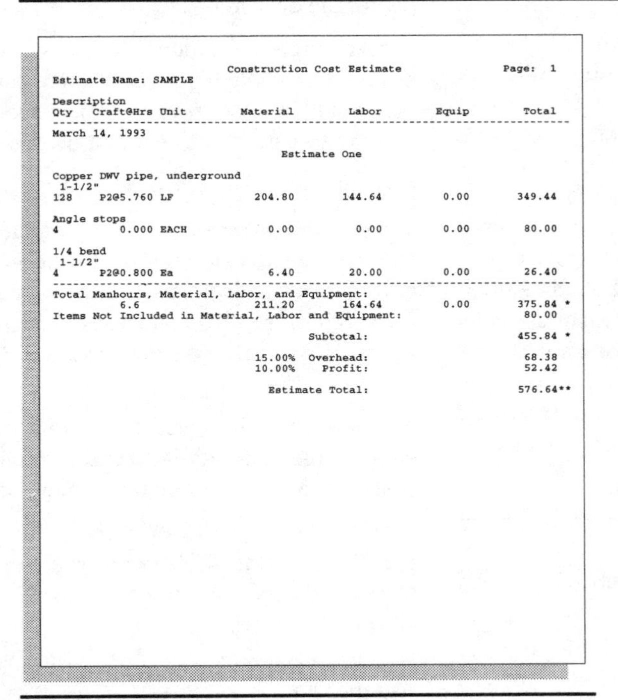

```
                    Construction Cost Estimate                    Page:  1
Estimate Name: SAMPLE

Description
Qty   Craft@Hrs Unit        Material        Labor        Equip        Total
---------------------------------------------------------------------------
March 14, 1993

                              Estimate One

Copper DWV pipe, underground
 1-1/2"
128     P2@5.760 LF           204.80       144.64         0.00       349.44

Angle stops
4          0.000 EACH           0.00         0.00         0.00        80.00

1/4 bend
 1-1/2"
4       P2@0.800 Ea             6.40        20.00         0.00        26.40
---------------------------------------------------------------------------
Total Manhours, Material, Labor, and Equipment:
           6.6                211.20       164.64         0.00       375.84 *
Items Not Included in Material, Labor and Equipment:                  80.00

                                    Subtotal:                        455.84 *

                      15.00%  Overhead:                               68.38
                      10.00%    Profit:                               52.42

                               Estimate Total:                       576.64**
```

To calculate all costs, including labor and materials, that will be involved in completing a construction project. The estimate should come as close to the actual cost as possible. Labor rates and material prices are available in manuals and computer programs to assist the estimator. The estimate values arrived at are usually used to determine the price for bidding a job. If the estimate is too low, the contractor may not make much profit, or worse, may lose money on the job. If the estimate is too high, the contractor will not get the job. The most accurate estimates are made when the contract requirements are fully understood, careful measurements are taken, and a close inspection of the site is made. All circumstances associated with a job and all potential problems or hazards must be taken into consideration. The estimator must have enough experience to recognize possible pitfalls, especially in dealing with remodeling where there are almost always tasks beyond the obvious that must be dealt with. Estimates may be prepared by the contractor who will do the work or by a trained estimator working for the contractor.

escutcheon – a circular trim piece which fits around a pipe, such as the pipe to a shower head or faucet, and covers the hole where the pipe passes through the wall or floor.

escutcheon pin – a decorative nail used in locations where it will remain visible, as on the surface of a strap hinge on a door.

estate – 1) possessions or property; 2) the assets and liabilities left by a person after death.

estimate – to determine, in advance of material purchase or construction, the cost of completing a task. *See Box.*

etch – to eat away part of a surface by chemical means. The portion not to be etched is protected from the chemicals. Etching is used to create decorative patterns on metal or glass.

etched nails – nails that are chemically treated to increase their holding power in wood by creating a rough shank surface.

ethylene propylene diene monomer (EPDM) – a synthetic elastomer material made from ethylene, propylene and diene monomer synthetics. It is sometimes used as gasket and valve seat material.

European style cabinets – frameless cabinets of particle board or plywood that are covered with plastic laminate. Also called Eurostyle cabinets.

eutectic – an alloy formed by that proportion of elements, in combination, that yields the lowest melting point of any of the proportions. Solder is a eutectic alloy.

evacuate – 1) to empty; 2) to withdraw; 3) to create a vacuum.

evaporate – to vaporize or expel moisture from.

evaporative condenser – a device in which air is cooled by being passed through a water spray that flows over the outside of finned tubes containing refrig-

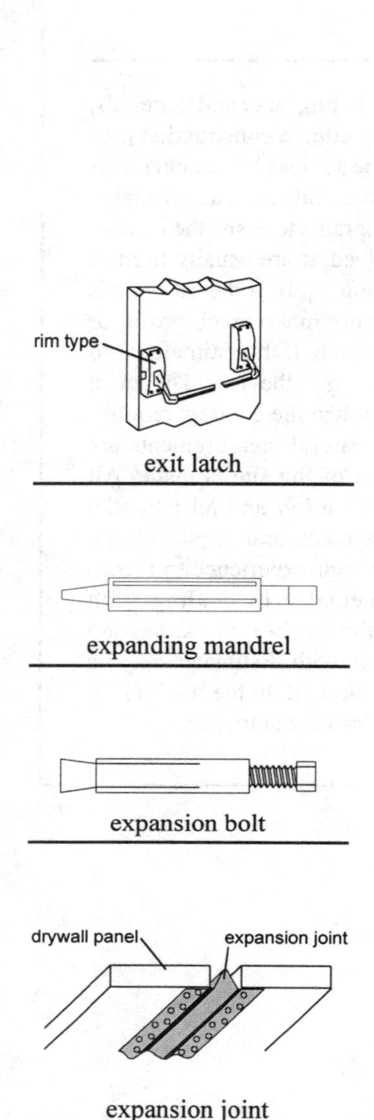

erant. The air causes the water to evaporate and cool the tubes, and thereby cool the refrigerant in the tubes.

evaporator – 1) a device for evaporating a liquid for purification, such as the evaporators used on board ships to distill fresh water from sea water. The fresh water generated is used for drinking water and for boiler feedwater replenishment; 2) a device in a refrigeration system in which the liquid refrigerant is evaporated into a gas, and by heat exchange, absorbs heat from the surrounding fluid. A fan is used to force the air across the evaporator coils to increase the rate of heat transfer and better distribute the cooled air. Fins on the coils of the evaporator increase the surface area exposed to the air, adding to the efficiency of the heat transfer.

evase – a gradually tapering section of an exhaust stack.

evergreen – a tree or shrub that retains green leaves year-round.

evict – to remove a tenant by legal process.

evidence – data presented as proof.

excavate – 1) to dig; 2) to unearth.

excavation – the removal of soil.

exclusive listing – to list property for sale with just one broker.

exfiltration test – a test for leaks in a buried water line. Part of the line is isolated and a manhole filled with water. The water level is then watched to determine if there is any leakage from that section of the system.

exfoliation – a scaling or peeling off of the surface of stone or masonry caused by chemical action or physical weathering. Surface cracks may occur over time due to temperature changes which cause the material to expand and contract. Moisture freezing in the cracks can further expand the cracks, resulting in surface fractures which flake or peel off.

exhaust fan – a fan for drawing air out of a room or an enclosure.

existing grade – the grade at an earth moving job before any work is done.

exit – a passageway out of a structure.

exit latch (door) – a latch operated by a crossbar on the inside of the door. The crossbar is pushed to retract the latch and permit the door to open.

exothermic – a reaction which gives off heat. An example would be when a catalyst is added to resin in fiberglass work. The mixture gives off heat as it hardens.

expand – to increase in size.

expandable paver – a paving machine that can be expanded by using extension tubes and extendable screeds to increase the width of the area that can be paved in one pass. The machine can expand to lay pavement more than 20 feet wide.

expanded metal – sheet metal with a pattern of holes through it so that the holes constitute a large portion of the material.

expanded shale and clay – shale and clay that are heated to a point where they soften and produce internal gases which expand the material.

expanded slag aggregate – blast furnace slag that is treated with water and used as aggregate.

expanding mandrel – a lathe mandrel that has a spring steel sleeve with longitudinal slots. The small end of the mandrel is started in the hole in the work and is pressed into place using an arbor press. A tapered pin is moved further inside the sleeve to expand the sleeve tightly against the inside diameter of the hole.

expansion bit – see expansive bit.

expansion bolt – a type of masonry anchoring device consisting of a bolt and expandable sleeve. The unit is inserted into a hole drilled in the masonry and then tightened, expanding the sleeve into the hole with a wedging action for a tight fit.

expansion joint – 1) strips of rubber or flexible material used to separate units of concrete. Because they can be compressed, they give the concrete room to expand with temperature changes. This prevents cracking; 2) a flexible piece of metal used between gypsum wallboard panels to prevent buckling when temperature changes cause the panels to expand.

expansion tank – a closed tank in a hot water system that provides space for the expansion of water in the system as it is heated. The tank contains both water and air. As the water expands, the air in the tank is compressed.

expansion valve – a valve in a refrigeration system used to control the flow of liquid refrigerant into the evaporator. Most expansion valves are thermostatically controlled. As the liquid refrigerant passes through the expansion valve, it cools to the temperature in the evaporator. A sealed bulb containing refrigerant is clamped to the refrigerant line leaving the evaporator. It is also connected by a tube to a chamber on the expansion valve which is sealed by a diaphragm. The diaphragm is linked to the valve stem, which is forced toward the closed position by a spring. The refrigerant in the bulb, tube and upper chamber on the expansion valve work against the spring to open the expansion valve when the temperature of the line to which the bulb is attached rises sufficiently. When this mechanism opens the expansion valve, more refrigerant is admitted into the evaporator to cool it. The expansion valve operates in response to the cooling load demand on the evaporator.

expansive bit – a drill bit with a cutter head mounted at right angles to the shank of the bit that can be moved in relation to the shank to change the diameter of the hole to be bored. Also called an expansion bit.

expedite – to speed up a process or action, such as following up on and ensuring the prompt delivery of materials to a job site.

expenditure – money paid out or used up.

expense – 1) a financial burden; 2) the cost of doing business.

explosive – a material that creates a near-instantaneous pressure rise.

explosive bolts – fasteners that contain an explosive that can be detonated by remote control to sever a bolted connection.

explosively-driven fastener – a fastener, such as a concrete nail, which is driven into the material by the force of an explosive cartridge in the nail gun.

explosive valve – a valve containing an explosive actuator that can be remotely triggered to change the valve position from open to closed or closed to open.

exposed aggregate – a horizontal concrete surface in which decorative aggregate has been embedded after the concrete is poured but before it has completely set. The top half of the aggregate is washed clean and remains exposed in the finished concrete.

exposed-nail method of roll roofing installation – a method of installing roll roofing in which the roofing nails remain exposed on the bottom edge. The top edge is covered by the overlap from the next roofing course.

exposed wiring – 1) a type of electrical building wiring in which single insulated wires are spaced apart and mounted on insulators; 2) electrical wiring run on a surface or behind removable panels.

exposure – open to the elements; not hidden from view.

extend – to lengthen or stretch.

extended plenum ducting – a large central duct from which smaller branches are routed; used in HVAC systems.

extension – 1) a projection; 2) a longer handle.

extension bit – a metal rod with a means of gripping a drill bit. The extension is inserted in the drill motor chuck to lengthen the effective reach of the drill bit.

extension cord – a long portable electrical cord used to move a source of electricity closer to a work area.

extension line – a line used on a drawing to locate a dimension away from the actual points of the dimension. This method is used when a drawing would be too crowded or cluttered if the dimension were to be shown within the two points.

extension rule – see folding rule.

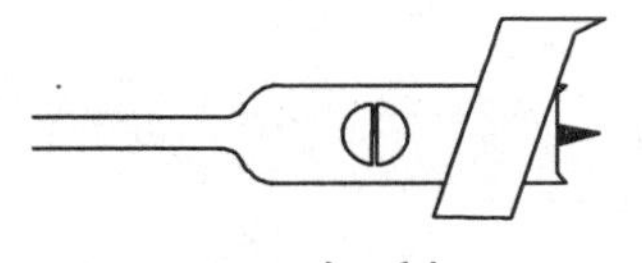

expansive bit

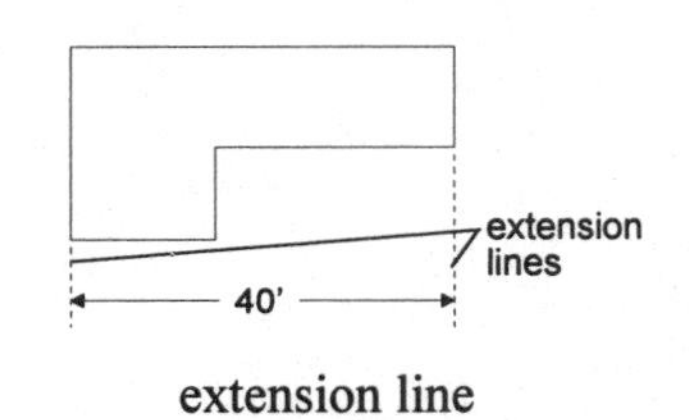

extension line

exterior ceiling panels – gypsum drywall panels designed for outdoor use in areas with limited weather exposure, such as for soffits, carport ceilings or canopies. The panels have a water-resistant core wrapped with water-repellant beige facing paper. The paper covering is suitable for use without additional finishing materials or decoration. Joints are finished with battens or joint tape and water-resistant joint compound. Exterior gypsum panels are 4-feet wide and come in 8- and 12-foot lengths and ½- and ⅝-inch thicknesses. The ⅝-inch thick panels have a fire-resistant rating.

exterior door – a door designed to be exposed to the elements on at least one side. An exterior door is used to close a doorway through the outside wall of a building.

exterior sheathing – gypsum drywall panels that are designed to be installed under finished siding materials on the vertical surfaces of exterior building walls.

exterior wall – a wall in which one side forms the outside exposed surface of a building.

external-combustion engine – a type of engine powered by a combustion process that takes place outside the engine, such as a steam engine that is driven by boiler generated steam.

external corner – a corner which protrudes into a room. Also called an outside corner.

extreme fiber in bending – a measure of the resistance of lumber to an applied load.

extrude – to force a material through a small opening, such as a die, to form or shape it. Wire is made by extruding metals through dies.

extruded metal – metal that has been shaped by forcing it through a die. The metal shapes used in manufacturing sliding glass door sections and tracking are made of extruded metal.

extrusion – forming a shape by forcing it through a die.

eye – an opening in the head of a tool which allows an object to be passed through or fitted into the tool, such as the opening in the head of a hammer for the handle.

eyebolt – a threaded bolt with a loop or eye formed at the end by bending a portion of the bolt into a circle.

eyelet – a reinforced hole in material.

eye protection – safety glasses, face shields, or other means used to prevent contaminants from entering a worker's eyes.

eye screw – a wood screw shank that is bent into a circle on one end.

F: fabric to fusion welding

fabric – cloth woven from fibers, either natural or synthetic.

fabricate – 1) to make from raw materials; 2) to assemble from parts, such as the roof trusses used in roof framing. The trusses are made from structural members cut to size and joined together using metal plates with integral fasteners. They are usually assembled offsite and brought to the building site for framing installation.

facade – 1) the face of a building; 2) a non-structural covering on the face of a building, such as brick or stone.

face – 1) an exposed surface; 2) the side of a gypsum wallboard panel facing away from the framing; 3) the exposed surface of a masonry unit; 4) the exterior surface of a structure.

face brick – bricks that form the outside surface of a wall. Facing bricks are made from clays that produce the desired colors and are chosen for their appearance.

faced masonry – a masonry structure with backing and facing masonry of different materials, such as brick on a concrete wall. The backing and facing are bonded together.

faced wall – a structural wall with a veneer covering.

face harden – a process used to harden the surface of a material. For example, carbon steel is face hardened by heating the steel to approximately 1200 degrees Fahrenheit and immersing it in powdered carbon. Some of the carbon is absorbed into the molecular structure of the surface of the steel and hardens the surface, or face, of the steel.

face nail – nailing perpendicular to the face or top surface of a board.

face of weld – the exposed weld surface on the side of the material on which the weld was applied.

face plate – a metal plate to which the work to be turned on a lathe is attached. The face plate attaches to the lathe headstock, the part of the lathe that turns the work.

face ply – the outer layer of wallboard where there are two or more layers. Also called face layer.

face reinforcement – weld material that is added on the face of the weld.

face shield – a covering, with transparent eye panel, worn to protect the worker's entire face while using a grinder, sander or any other equipment which may throw small particles up into the face or eyes. Face shields designed for welding are opaque with a dark glass eye panel to protect the worker's vision from the bright light of the welding torch.

face velocity – the velocity of air measured at the face of an inlet or outlet in an HVAC system.

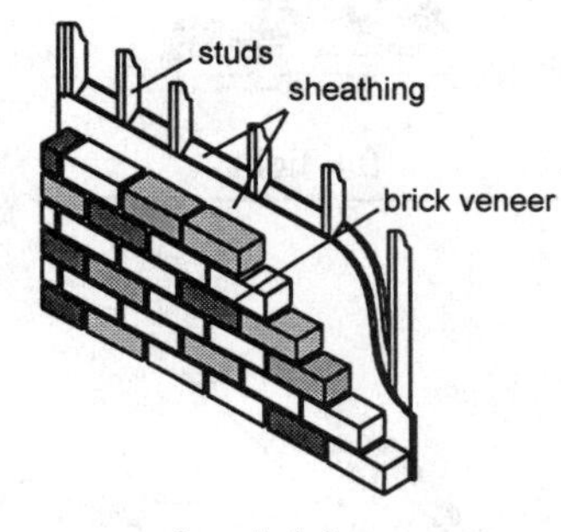

face brick

fan light

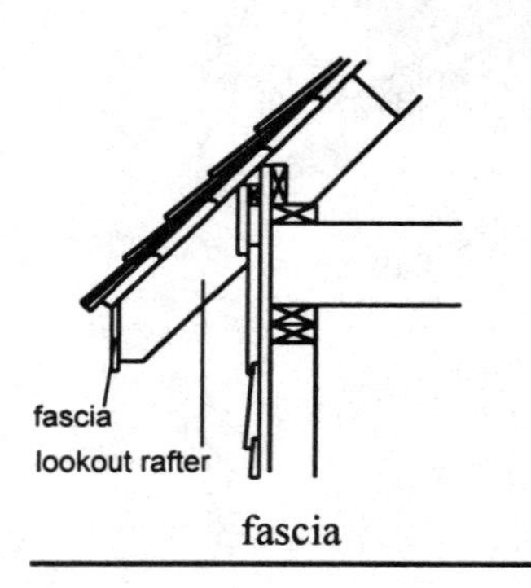

fascia

face veneer – the outer layer of veneer on plywood.

facility – 1) a building or structure designed to perform a specific function, such as a hospital or gymnasium; 2) ease of performance or aptitude for a task.

facing – 1) a finished wall surface; 2) smoothing and finishing.

facing brick – a brick that may have special color or texture which is made specifically for the outside or facing of a wall or surface.

factoring – purchasing the accounts receivable of a business as an investment, or loaning money to a business using their accounts receivable as collateral for the loan.

factor of safety – the ratio of the ultimate strength of an object to its maximum permissible design load. For example, if a pressure tank will fail at 2000 psig (pounds per square inch gauge) and it is used at a maximum pressure of 500 psig, it is said to have a 4 to 1 factor of safety.

factory edge – 1) the long edge of a wallboard panel that comes from the factory covered with paper; 2) the edge of a fabricated item that has been prepared in a factory.

fade – 1) to lose the intensity or brightness of color due to wear or aging; 2) to reduce in strength or intensity.

faggot – 1) a bundle of branches for fuel; 2) a bundle of metal pieces to be welded together.

Fahrenheit – a temperature measurement scale in which pure water freezes at 32 degrees and boils at 212 degrees at sea level. Base zero is the freezing point of a solution of equal parts (by weight) of snow and salt.

faience – a type of glazed pottery or tile, usually with colorful designs.

false header – see clipped header.

false set – a loss in the consistency of concrete shortly after it is mixed.

falsework – temporary structural supports that are used during construction and then removed.

fan – 1) a powered device which uses rotating blades or vanes for moving air in a room or building; 2) a device that moves air or other gases within a machine, such as the forced-draft fans used to force air into the firebox of a boiler.

fan, axial – a fan in which the air flow is parallel to the fan shaft, such as a small portable house fan.

fan, centrifugal – a fan consisting of a rotating, cylinder-shaped cage with longitudinal slots cut around the circumference of its outer wall. Sections of metal are inserted into the slots to form blades. Air enters the center of the cylinder and as the cylinder rotates, the blades move the air out through the slots so that it exits perpendicular to the cage axis. Centrifugal fans are often used in forced air furnaces and air conditioning systems. They are among the quietest type of fan, and so are preferred for use in homes.

fan inlet vanes – blades positioned at a fan inlet to direct and control air flow.

fanlight – a semi-circular window with spacers radiating out from a single point in the shape of a fan, usually placed over a door or window.

farad – a unit of electrical capacitance, the ability of a device to store an electrical charge.

fascia – 1) a flat trim board fastened to the outer ends of roof rafters; 2) horizontal bands that project out from a wall, with each band projecting out a little more than the one below, forming a tiered molding around a door or window.

fasten – attach.

fastener – any mechanical means of holding two materials together, such as a nail, screw, or clip.

fastener depression – a defect in gypsum drywall installation in which improperly-filled depressions remain over the nail or screw heads after the wallboard has been finished.

fat edge – a paint defect resulting from the build-up of paint on the edge of a surface. This occurs when too much paint or an uneven amount of paint is applied to a vertical surface. The excess paint sags and builds up on the edge. This can be avoided by applying thin, even coats of paint.

fatigue factor – a factor relating to the decrease in the actual productivity of a worker as the number of work hours increase.

fat mortar – a sticky mortar that contains a high percentage of cement.

faucet – a soft seat globe valve for controlling water flow to the outlet of a sink or similar appliance.

fault – 1) failure in part of an electrical circuit; 2) a break and dislocation in the continuity of the earth's structure.

faying surfaces – surfaces in close contact with one another, such as the mating surfaces of a tongue and groove joint or the mating surfaces a hinge plate and hinge mortise.

feather – a spline joining two grooved board edges. *See Box.*

feathered edge – the point where one material blends into another, such as the edge of a layer of joint compound that smoothly tapers to the surface of gypsum wallboard and conceals the joint.

feather edge – the thin edge of a tapered piece.

feathering – 1) to blend or taper an edge into the surrounding surface, such as tapering new asphalt over existing pavement; 2) opening and closing valves in a hydraulic control system to make the machine movements, which are controlled by the hydraulic system, smooth.

feathers and wedges – a two-part device for splitting stone. Thin metal pieces, bent into a V-shape, called feathers, are placed in holes bored along a line in a section of stone. A metal wedge is inserted into each of the feathers. The wedges are pounded tight, each in turn, traversing the length of the row and back. Eventually the stone splits along the line of holes from the tensile force the wedges exert on the stone. Since the holes are in a straight line, the force is directed for a clean split.

Feather

A thin wood strip, or spline, used to supplement the strength of a wood butt joint. The spline fits into grooves cut in each of the two board edges to be joined. It is sufficiently narrow to permit the board edges to butt together. Power cutters allow precise grooves to be cut for the feather. The feather should be a tight fit, so that when the glue is added, a strong tight joint results. Without the use of such a supplement, the strength of the joint is limited by the strength of the adhesive joining the pieces and the contact of the surfaces. The feather spline is sometimes called a biscuit because of its flat wafer-like appearance.

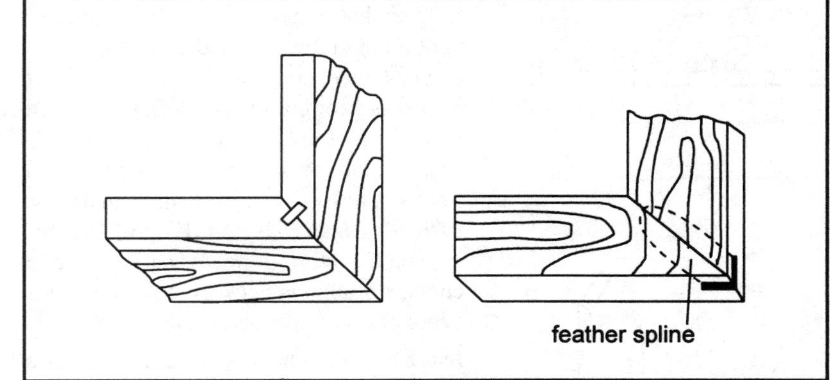

feeder – the electrical conductor from the electrical service entrance, where electricity enters a structure, to the last fuse or circuit breaker for a branch circuit.

feeder busway – a busway for electrical feeder circuits between the service entrance and the main switchboard or load centers.

feed-in box – an electrical junction box for making power connections to lighting busways and trolley busways, which are the enclosures for the electrical busbars that carry electrical power.

feed stop – a clamp on the height adjustment rack of a drill press which limits the depth that the drill can penetrate the work, allowing the hole to be drilled to a preset depth.

feeler gauge – see thickness gauge.

faucet

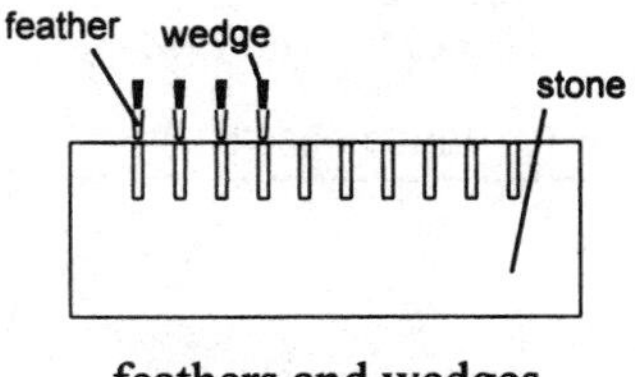

feathers and wedges

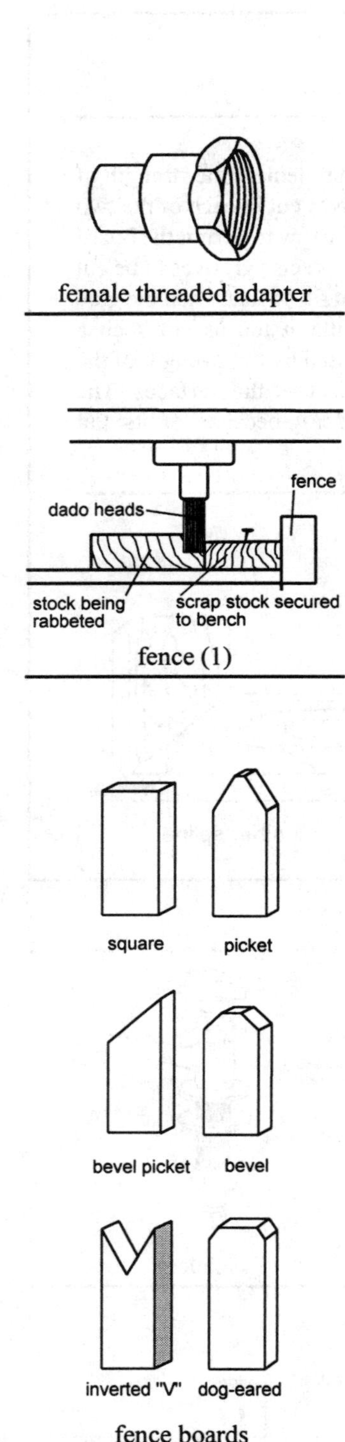

female threaded adapter

fence (1)

fence boards

feet per minute (fpm) – a measure of the velocity of air, or other medium, used in calculations for HVAC systems, piping systems, pump capacity calculations and other systems in which flow measurements are needed.

felt – 1) a compressed, rather than woven, fabric; 2) a heavy building material made of felt impregnated with an asphalt compound used for water resistance under such items as roof shingles or flashing.

female threads – threads on the inside diameter of a fitting, nut, or other object.

fence – 1) a guide on a saw, jointer, or other power tool against which the work rests so that its position can be controlled in relation to the saw blade, cutting blades or bit. Fences are adjustable by different means depending on the design, and can be locked into place for precision cuts; 2) an enclosure or decorative barrier around a section of land. They are made of a wide variety of materials and designs depending on functional needs and aesthetic considerations. Fences are often used as a demarcation of property lines.

fence boards – boards used in making a wood fence. Various board designs are available.

fender washer – a washer with a large outside diameter in relation to the bolt, which allows the washer to distribute the bolt force over a wide area.

ferrite number – a series of numbers assigned by the American Welding Society to represent the ferrite or iron content in austenitic stainless steel weld metal, an alloy consisting of iron, chromium, nickel and sometimes small amounts of other elements (AWS A4.2).

ferrous – material that contains iron.

ferrous sulfate – an iron and sulfur compound used as a paint pigment.

ferrule – 1) a short metal sleeve around a spike used for fastening gutter sections to a building. The ferrule adds structural support, bracing the outer edge of the gutter the proper distance from the inner gutter wall; 2) a short metal reinforcing sleeve that fits over the end of tubing as part of the connector; 3) a strengthening band around the end of a tool handle at the juncture of the tool head and the handle. There is a stress concentration at this juncture when the tool is in use and the band provides needed reinforcement. The band is usually made of metal, often steel because of its strength.

festoon – an architectural decoration resembling a string of flowers, or other plant material, and ribbons. It is a horizontal decoration that appears to be draped between two points, sagging slightly in the middle.

fettle – 1) to remove sand from a sand casting; 2) the removal of casting mold marks from a casting; 3) repairing an open hearth furnace.

fettling – heat-resistant refractory material used to line some types of metal refining furnaces.

fiber – a filament of material.

fiberboard – sheet material, made by compressing wood or other plant fibers with a binder, used in many construction applications such as floor underlayment, wall sheathing under finish walls, and cabinet walls which are to be covered with a veneer.

fiber conduit – wood pulp fiber tube that has been impregnated with a bituminous sealant and preservative. It is used as a nonmetallic electrical conduit, although it has generally been replaced by PVC.

fiberglass – small diameter glass strands that are used loose, compressed into a mat or woven into cloth for insulation, reinforcement and the manufacture of a variety of products. *See Box.*

fictile – moldable; a material property that permits shaping, such as wet clay or plaster.

fidelity – 1) being faithful or trustworthy; 2) accuracy in details.

fidelity bond – a performance bond for employees who handle money or material intended to protect a contractor against embezzlement.

field – 1) the area of a surface, such as gypsum wallboard or another type of panel; 2) an open area; 3) a job site; 4) the volume around a magnet that contains

Fiberglass

Fiberglass is made of spun filaments of glass woven into yarn or roving (twisted strands) and textile materials such as cloth or mats. The roving and textiles can also be hardened with resins to make strong, lightweight construction materials. There are hundreds of manufacturing uses for fiberglass materials.

Common construction uses include the small diameter glass strands used to make glass wool or batting for insulation. Tiny glass fibers are also added to materials such as epoxy resins for reinforcement.

Fiberglass cloth saturated with a plastic (epoxy or vinyl resin) compound can be pressed into molds to make products such as bathtubs and shower enclosures. In manufacturing, as much of the plastic compound as possible is pressed out of the mixture. The strength of the product comes from the glass fibers. The greater the percentage of glass fiber per unit area, the stronger the fiberglass product. The plastic compound, called a matrix, serves simply as a bond to hold the glass fibers together. Fiberglass tubs and shower surrounds are built into many new living units. They are lightweight and easy to install. A plastic gel-coating installed during manufacturing gives them a shiny enamel-like surface.

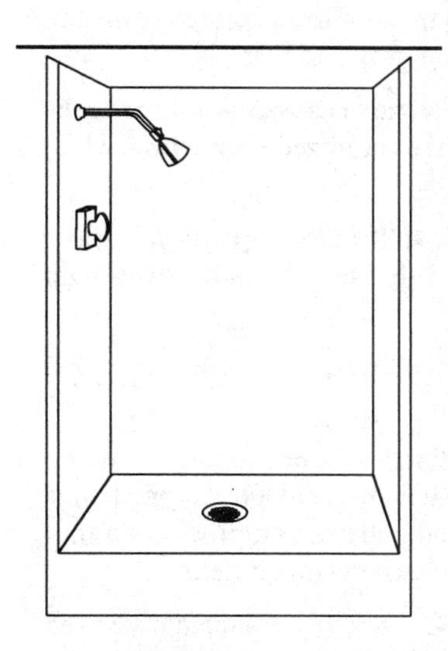

Fiberglass shower stall

Fiberglass pipe is centrifugally cast in a mold or manufactured by laying saturated fiberglass around a mandrel. The inside diameter of the pipe is determined by the size of the mandrel. Fiberglass pipe, pipe fittings and valves are commonly used where chemical resistance is required. The matrix, or plastic that bonds the glass fibers together, can be formulated to resist specific chemicals and/or other environmental conditions.

Corrosion-resistant gratings are also made from fiberglass. They are strong, lightweight, impervious to weather and many chemicals, and they require very little maintenance.

magnetic lines of force; 5) the volume around an electrical device that contains electrical lines of force.

field house – 1) a structure for storage or other uses at an athletic area; 2) a structure in which athletic events are conducted.

fieldstone – a natural stone used in the construction of decorative walls or walkways.

fieldstone, blue frost rubble – a light blue-gray rock with some tan and black veins.

fieldstone, California driftstone – a driftwood- textured stone that varies from medium brown to black.

fieldstone, California travertine – stone chunks marbled with browns, tans, and off-white colors.

fieldstone, drift (pink thins) – a pink-gray lava rock with rust-orange highlights.

fieldstone, ebony – a black rock with white veins and green and gold overtones.

fieldstone, glacier quarry fossil – blue-gray stone with rust- to gold-colored fossils. The stone has been smoothed by glacier action.

fieldstone, Indian Creek – a sandstone with earthtone colored strata.

fieldstone, mountain orchid – an orchid stone with multi-colored overtones of green, yellow, red, and white.

fieldstone, river rounds – smooth granite from river beds.

fieldstone, Santa Fe lava – rusty black lava.

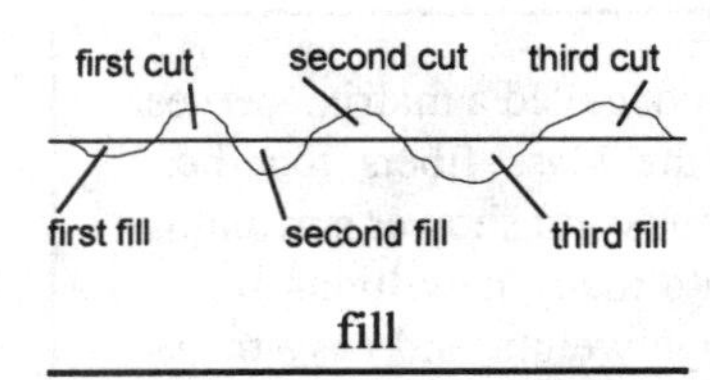

fill

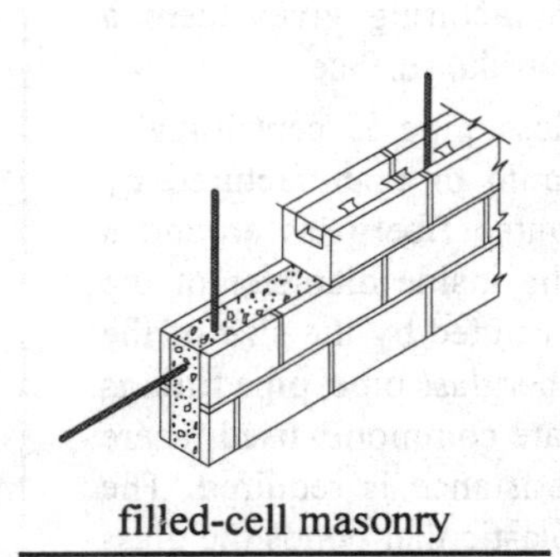
filled-cell masonry

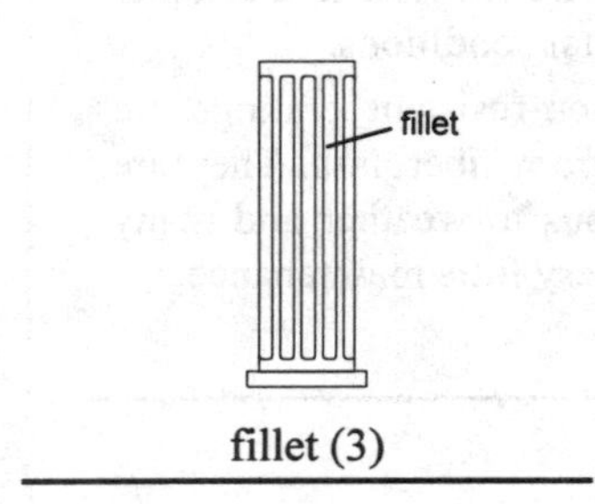

fillet (3)

fillister-head screw

fieldstone, silver green – a green stone with flint and light orchid overtones.

fieldstone, snow mountain crystal – a chunky stone that is pale quartz with veins of silver and green mica through it.

fieldstone, victor log – hard quartzite that varies in color from honey-cream and pink to light gray and rust.

fieldstone, victor red rock – a brownish-red to orange fossilized rock with a white interior.

fieldstone, victor silver gray – a hard quartzite with light to dark silver with sparkles.

field weld – a weld performed at a job site.

field winding – the electrically conductive current path, laid or wrapped in a symmetrical pattern, that produces a motor or generator magnetic field.

filament – a fine fiber. Filaments are used to make cloth, provide the glowing element in a light bulb, as reinforcement for epoxy resin, cement and other matrices, to provide additional fire resistance in Type X gypsum drywall panels, and a number of other applications.

file – 1) an abrasive tool made of a hardened metal bar or rod, with single or double rows of teeth, used to smooth or shape wood, metal or other material. Files are available in many different degrees of coarseness, shapes, and sizes for a variety of applications. Hand-held files are used for fine work. Rotary files, designed to be turned by a drill motor, remove material quickly but are not suited for smoothing flat surfaces; 2) an identified storage location for documents. Document files are placed under directory or subject codes to make them readily identifiable and retrievable at a later date. Computer files are stored on magnetic or optical media. They can be retrieved for viewing on a monitor or printed out on paper.

file card – 1) a brush with short stiff bristles used for cleaning the grooves of a file; 2) a marker card used as a locator in a file of information.

filigree – delicate ornamentation often using fine gold, silver or copper wire on gold or silver.

fill – 1) soil used to take up space around a structure or to fill up an excavation; 2) surveying term indicating a grading level in relation to a reference point. In a fill, earth has to be added to match the reference point.

filled-cell masonry – a vertical wall of hollow masonry units, one unit thick, with all voids filled with grout.

filler – 1) substance used to fill holes and voids in a material, such as joint compound used between sheets of drywall; 2) an additive which changes or improves the properties of the material it is added to, such as fiberglass strands added to the core of some types of gypsum wallboard to increase its fire-resistant qualities; 3) a surface coating to add solidity, such as adding a layer of fiberglass cloth and resin over Styrofoam to add strength.

filler metal – metal that is added during welding, brazing, or soldering to fill the gap in the joint and bond the materials being joined.

filler wall – a nonbearing wall between columns, and supported at each story. A filler wall is used as a partition to separate spaces.

fillet – 1) a rounded surface at the joint between two planes forming an inside angle; 2) a narrow flat molding used in conjunction with other moldings; 3) a narrow raised strip between flutes on a column.

fillet gauge – see radius gauge.

fillet weld – a triangular-shaped weld at the intersection of two pieces of metal joined at angles to each other.

fillister – a groove, or rabbet, cut in a material.

fillister-head screw – a fastener with a threaded base and a slotted, cylindrical, domed head that is slightly larger than its base.

fillister plane – a plane for cutting grooves or rabbets.

fill stake – a survey information stake which indicates where earth fills are to be made.

film – 1) a coating; 2) a sheet or strip of photosensitive material on which images can be made with exposure under the proper conditions; 3) a thin sheet of material, such as clear plastic wrap.

film build – increasing overall coating thickness by the application of successive layers.

film integrity – the continuous and unbroken condition of a coating, such as paint or varnish, or a film, such as a thin plastic sheet.

film thickness – the thickness of a coating.

filter – 1) a device which traps unwanted particles as air or fluid is passed through it, such as a painters mask or an oil filter on an internal combustion engine; 2) an optical device that prevents the passage of certain wave lengths, such as a lens filter on a camera; 3) an electrical device that prevents the passage of certain frequencies, such as a capacitor in series with a loudspeaker that blocks low frequencies.

filter bed – a layer of filter material through which a fluid is passed, such as water that is purified of certain chemical compounds by passing it through a layer of activated charcoal.

filter block – a hollow vitrified clay masonry unit designed as flooring material in a sewage treatment plant.

filter fabric – 1) woven or nonwoven synthetic fabrics used in excavation work to stabilize soil. The fabric is laid on the soil and fastened down in a few places to help prevent erosion and minimize slippage; 2) woven or nonwoven synthetic fabrics used as a sieve to separate fine materials from coarser ones and to filter particles from water or fluids.

filter paper – porous paper used for filtering.

filtrate – filtered fluid, such as oil after it has been passed through an oil filter.

final set – the complete curing of concrete or mortar.

fine – 1) high quality; 2) small particles or pieces.

fine aggregate – see aggregate, fine.

fine grading – grading earth to within specification tolerances.

fine grain – small grain; having an invisible or barely visible grain, such as clear maple wood, or a metal that has been processed to limit grain size.

fine nail – a small finishing nail used for trim or molding or in other installations where it is desirable to have as small a nail hole as possible.

fineness modulus – an index of the average coarseness of aggregate in concrete.

fines – a cement mixture of very fine particles used for finishing masonry surfaces. It is applied with a burlap cloth, or brushed on and then rubbed with a burlap cloth. It yields a uniform, thin coating on the masonry surface.

fine tex – a type of sprayed-on texture in which medium-size droplets of texturing compound are flattened with a trowel to make a finished pattern on a wall or ceiling.

fine-tune – to adjust precisely.

fingering – an airless spray-painting defect in which thick bands of paint have been applied in areas, rather than a uniform coating thickness. Also called tails.

finger joint – a method of joining wood pieces in which narrow extensions are cut out in the ends of two pieces of wood in such a manner that they fit together to make an interlocking joint. *See Box.*

finial – an ornamental shape, such as a spire, at the top of a design, or an ornamental top on a piece of furniture.

finish – the final surface treatment.

finish carpentry – trim installation, such as the moldings and trim around doors and windows, cabinetry, and fascia installation. The last carpentry details done which bring the job to a finished and completed state.

finish coat – the final coat of paint, varnish, plaster, stucco or other surface covering.

finished grade – see finish grade.

Finger joint

A wood joint in which tenons, loosely resembling fingers, are cut in the ends of two pieces of wood so that they interlock to form a joint. The extensions are fitted together to form a level finished surface. Finger joints allow otherwise unusable wood scraps to be salvaged by joining one or more small pieces together into a single larger piece. The large surface area of this type of joint provides sufficient gluing area to make the completed joint at least as strong as the remainder of the wood.

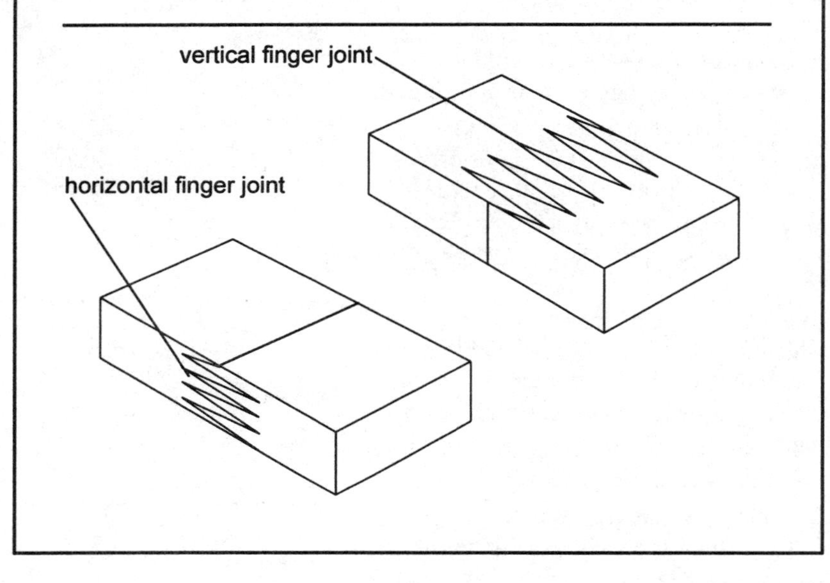

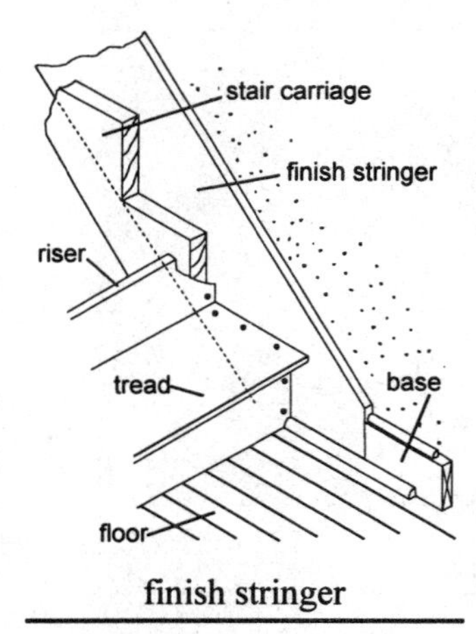

finish stringer

finish flooring – the final floor covering installed over the subfloor.

finish grade – the final surface of a site after earth moving and contouring have been completed.

finish hardware – trim and functional hardware, such as knobs, cabinet latches and drawer pulls, needed to make a job complete.

finishing nail – a slender nail with a semi-spherical head only slightly larger than the shank of the nail, used for trim molding and other installations where it is desirable to keep nail holes as small as possible.

finish plumbing – the complete installation of plumbing fixtures so that they are functional and ready to use.

finish stringer – a trim stringer used to cover the outside of a stair carriage.

finned-tube radiation unit – a hot water heat distribution device that passes hot water through finned tubes to heat the surrounding air. A finned tube is a hollow metal pipe with thin, sheet metal fins bonded perpendicular to the outside of the tube. Heat from hot water passing through the tube is conducted into the fins and dispersed into the room. The fins increase the effective heating surface area of the tube.

fir – an evergreen tree with structurally useful wood.

fire alarm – an alarm system designed to warn building inhabitants of the presence of smoke, flames, or a rapid rise in heat.

fire alarm annunciator panel – a panel which indicates the location of a fire.

fire alarm control panel – a panel that contains the devices to control and test the fire alarm system.

fire block – a solid closure of a concealed space to cut off air and prevent the spread of fire, such as pieces of framing that are placed between wall studs, perpendicular to the studs. They are designed to block the spread of fire by limiting the open airspace within the framing. Also called a fire stop.

fire box – the combustion area of a fireplace, furnace, or boiler.

fire break – plowed or cleared land that has been cleaned of combustible material, such as dry brush, to stop the spread of a fire.

fire brick – heat-resistant refractory clay brick used in fireplaces, boiler fireboxes, or other similar areas. Fire bricks have the capacity to retain their physical properties even when exposed to high temperatures.

fire clay – a heat-resistant clay that is used to make fire brick.

fire control – a system to contain and control fires which involves training in the procedures, materials, and fire suppression apparatus developed for that purpose.

firecracker weld – a shielded metal arc weld using a length of covered electrode. The electrode is laid in the weld groove, and the welding arc consumes it, making it part of the completed weld.

firedamp – combustible gas that forms in underground mines.

fire department connection – a water source to which the fire department can connect a fire hose.

fire division wall – a wall, extending continuously from the bottom to the top of a structure, that is designed and rated to retard the passage of fire in the structure.

fire door – a door designed to resist the passage of fire. Today, fire doors are rated by the amount of time they can resist the penetration of fire. The time ranges from one-half to three hours. Formerly, fire doors were given alphabetical ratings which corresponded to the current range of time ratings; for example, an A-rated door was fire resistant for three hours. Fire doors are used to close openings in fire walls, so that the door area is no more vulnerable to fire than the wall.

fire door, corrugated – a fire-resistant door constructed of two layers of corrugated galvanized steel with 2½-inch corrugations over an asbestos sheet core. The door has a steel frame.

fire door, metal-clad – a fire-resistant door fabricated from 30 gauge, or thicker, sheet steel panels. It has interlocking seams and a core of dry wood. It is the most common type of fire door currently in use. Also called tin clad.

fire escape – a means of egress that is an alternate from the normal method, and is isolated from the building proper to provide protection for people using it.

fire extinguisher – a device containing a fire suppressing material under pressure. When the device is activated, the fire suppressant can be directed at the fire to extinguish it by means of oxygen deprivation and/or cooling. Fire extinguishers are available in four different types: A, B, C, and D, each rated for different types of fires. Class A is for fires in ordinary materials. Class B is for fluid fires. Class C is for electrical fires. Class D is for materials that require an extinguishing compound that absorbs heat and does not react with the fuel.

fire hydrant – a large valve with a high flow rate and hose connections. The valve stem passes through the body of the hydrant to a globe valve-type circular disc which, when opened, permits the flow of water up into the hydrant body to the hose connections. Some hydrants are wet hydrants. They are always filled with water. The hose outlets in these hydrants are individually controlled by separate valves, the stems of which pass horizontally through the hydrant.

fire lines – the wet standpipe fire protection system for a building. A wet standpipe is a vertical pipe, extending to the upper floors of a building, that is always full of water and instantly available to distribute water during a fire emergency.

fireplace – a structure of fireproof material used to contain an open fire within a building.

fireplace base assembly – the foundation and hearth support for a fireplace.

fireplace, multi-face – a fireplace having two or more sides open to a room or rooms.

fireplace, Rumford – a shallow, single-face fireplace with a high opening and flared firebox sides. Its wide sides, shallow depth and large surface opening makes it highly efficient at radiating heat.

fireplace, single-face – a fireplace with a single opening into a room.

fire proofing – to cover with a fire-resistant and heat-resistant covering, such as covering steel structural members with masonry to protect them from the heat of a fire.

fire rating – a rating that indicates a material's or system's ability to resist fire when tested by a recognized laboratory against applicable ASTM standards. For example, one ⅝-inch thick Type X gypsum drywall sheet installed on each side of a wood stud gives the wall a 1-hour fire

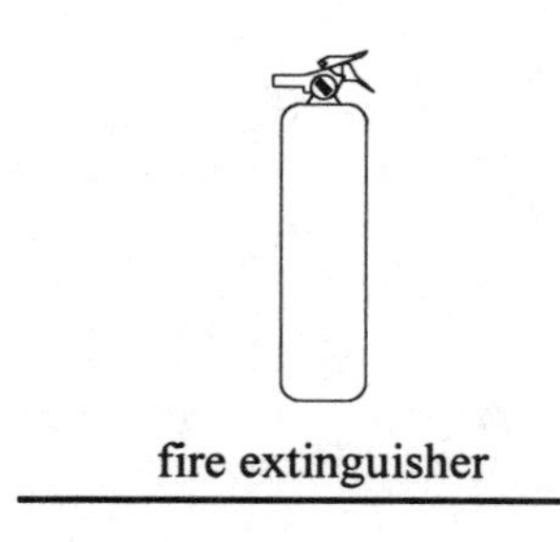
fire extinguisher

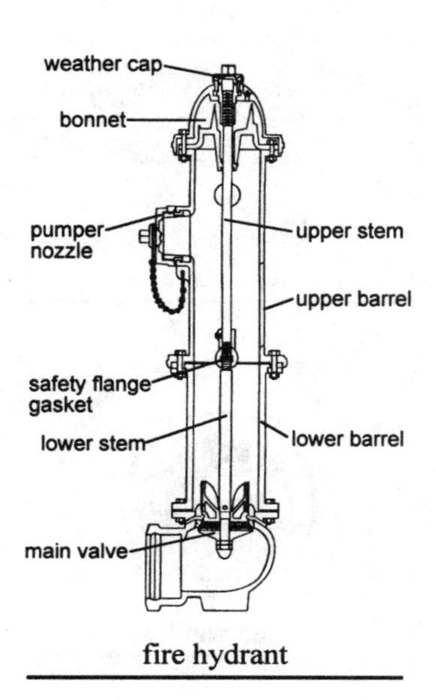

fire hydrant

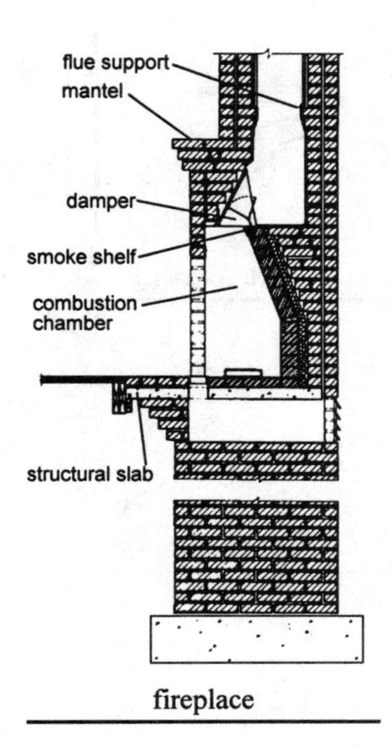

fireplace

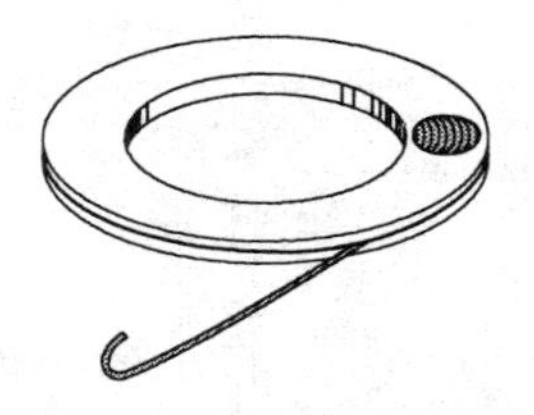
fish tape

fixed window

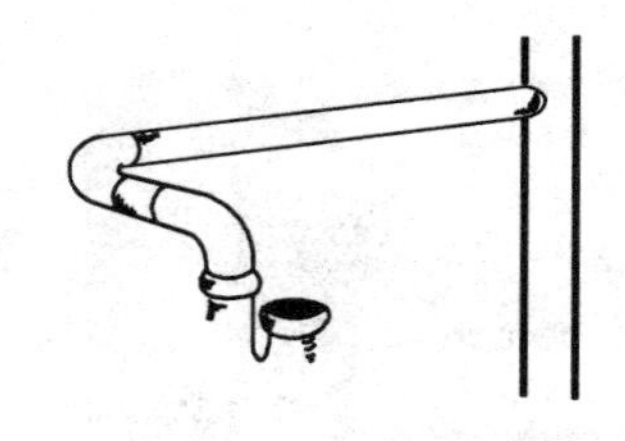
fixture branch

rating. A double thickness of $\frac{5}{8}$-inch Type X gypsum drywall sheets on each side of a wood stud provides a 2-hour fire rating.

fire resistance – the ability of a material to resist combustion or burning in the presence of a fire.

fire-resistant panels – Type X gypsum wallboard which has fiberglass, and other heat-resistant, noncombustible materials, added to the core to increase the natural fire resistance of the gypsum.

fire retardant – chemical or chemicals formulated to resist the spread of fire and/or increase the fire resistance of a material.

fire-retardant paint – paint that contains nitrogen compounds that blister and form insulating gas bubbles when it is exposed to heat. The trapped gas slows the heat transfer, protecting the surface covered by the paint.

fire-retardant treatment – coating or saturating a material with a fire retardant.

firestone – a heat-resistant sandstone or similar material that retains its physical properties in the presence of high temperatures.

fire stop – wood blocking between joists or studs which closes off the passage of air and limits the spread of fire in the structure. Also called a fire block.

fire taping – taping the joints in gypsum wallboard to eliminate a path for fire, without applying the finish coat of joint compound. This method is used where blocking a fire path is important, but appearance is not.

fire wall – a wall rated to withstand the effects of a fire for a period of time, and prevent the further spread of the fire beyond the wall for that length of time.

first mortgage – the mortgage that has the highest priority as a lien over all others except for those imposed by law, such as a tax lien.

fiscal – having to do with financial matters.

fish eyes – 1) a paint defect in which the paint surface has spots with rainbow pattern rings around them; 2) a weld defect where a small hole or piece of foreign matter in the weld is surrounded by a bright circular area. Also called cat's eyes.

fish mouth – an opening in the seam of a roof covering.

fish oil – oil from fish used as a paint drying agent.

fish plate – a wood or metal plate which spans or overlaps a butt joint between two wood members to strengthen the joint.

fish tape – heavy, flexible flat metal wire that can be pushed through conduit or through holes in framing and other hard-to-access locations. Electrical wire is attached to the fish tape and pulled, or fished, through. Also called fish wire.

fission – splitting or breaking apart.

fissure – a crack with a small separation.

fix – 1) to fasten in place; 2) to repair.

fixative – 1) a substance used to increase durability; 2) a substance sprayed on a drawing as a preservative; 3) a compound that retards evaporation.

fixed – not movable.

fixed assets – business assets that can be depreciated, such as office machinery or furnishings. Also called long term assets.

fixed light – a window that does not open. Also called a fixed sash window.

fixture – 1) an electrical lighting device that is directly connected to an electrical current; 2) a plumbing device, such as a sink; 3) a device to hold parts in a specific relation to each other. For example, a jig is designed to hold two parts in a precise position so that they can be welded together in a manner that results in a consistent and dependable product.

fixture branch – the pipe from the trap of a plumbing fixture to the vent pipe, carrying non-body wastes to the building drain. Also called fixture drain, waste pipe or wet vent.

fixture drain – see fixture branch.

fixture unit – a design factor for plumbing systems used to indicate the load produced by each fixture on the overall system.

flag – see flagstone.

flagging – a paving made of flagstone.

flagstone – irregularly-shaped flat stones or pieces of concrete used for paving walkways and patios. Also called flag.

flakeboard – a sheet material made of wood flakes and a binder. It is used for roof and wall sheathing under a finished surface material, or under a veneer for cabinets.

flame – burning gas.

flame-cut – refers to metal that has been cut with a cutting torch.

flame-proof – not burnable; non-combustible.

flame spread – the movement of flame across a combustible surface.

flame spread rating – a measure of how fast and how far flames will spread over surfaces under test conditions as determined in accordance with ASTM Standard E84. The test is used to establish fire-resistant values in building materials so that designers and builders know which materials are best for various applications.

flammable – material that can be easily ignited.

flange – 1) a rim on the end of a length of pipe or duct that provides a connection point to another length of pipe or piece of equipment. There are generally bolt holes arranged in a circle around the outer edge of the rim. Bolts or studs with nuts are used to hold two flanges together, with some type of gasket between them. There are standard pressure/temperature ratings for pipe flanges. The pressure and temperature at which a flange can be used depends on the its thickness, diameter and the material from which the flange is made; 2) a rim that encircles a busway where it passes through a wall, floor or into an electrical equipment enclosure. 3) the side sections of an I-beam.

flank – the side of an object.

flanking paths – sound transmission paths around a structure that is designed to stop or reduce sound transmission. If, for example, a wall was designed and built to resist sound transmission, but a gap was left along one side of the wall, this gap becomes a flanking path. The gap allows sound to pass around the wall. Without the gap, the sound would be resisted.

flap wheel – a sanding wheel used to sand irregular surfaces. The wheel, with short flaps covered with sanding abrasive, attaches to a shank held in the chuck of a drill motor. As the motor turns the wheel, the flaps spin, sanding hard-to-reach areas, such as contoured surfaces.

flare – 1) a cone-shaped enlargement at the end of a tube; 2) a fluctuating flame.

flare bevel-groove weld – a weld between a curved piece and a flat piece.

flare fitting – a flared tube end that can be joined with a male cone-shaped tubing end and sealed with a coupling or nut.

flare header – an exposed header that is painted a darker color than the rest of the wall so that it stands out.

flare V-groove weld – a weld between two curved pieces.

flaring tool – a tool used to flare the end of tubing. The tool has a female die that holds the end of the tubing to be flared, and a clamp that is secured to the die. The clamp has a male cone-shaped mandrel on threaded stem which screws down through the clamp. The edge of the female die is chamfered so that when the cone-shaped mandrel is forced into the end of the tubing, the tubing end is forced out in a flare to match the chamfer in the die.

flash – 1) a sudden bright light; 2) a finished paint surface with uneven color.

flashing – waterproof sheets, often of corrosion-resistant metal or plastic, installed with exterior finishing material to prevent water leakage in places where it is likely to occur, such as at the intersection of a wall and roof or in the valley of a roof.

flashing cap molding – a drip cap fabricated from metal flashing that is installed over a door or window.

flashing cement – see asphaltic plastic cement.

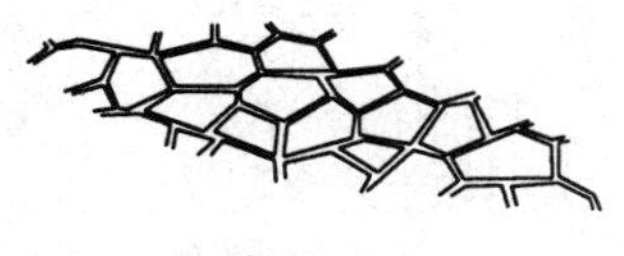

flagstone

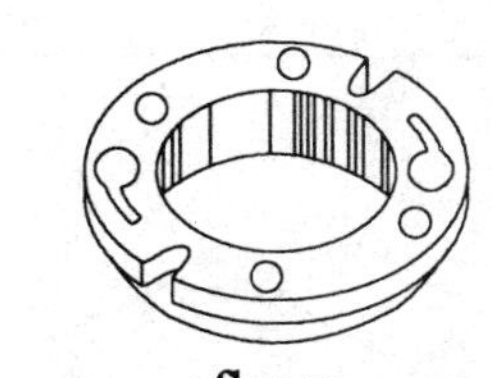

flange

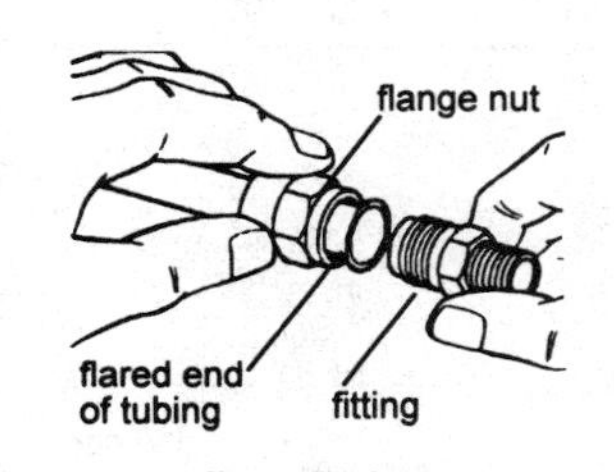

flare fitting

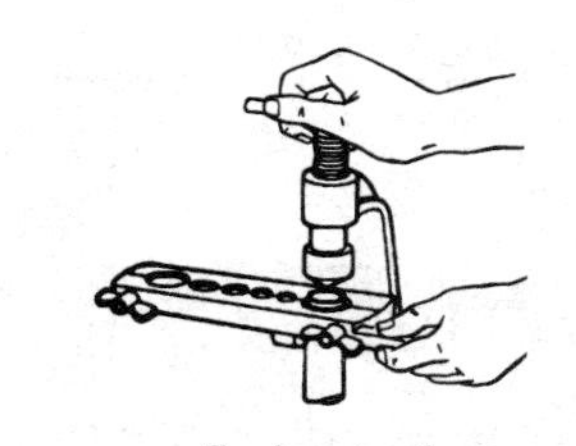

flaring tool

flashing

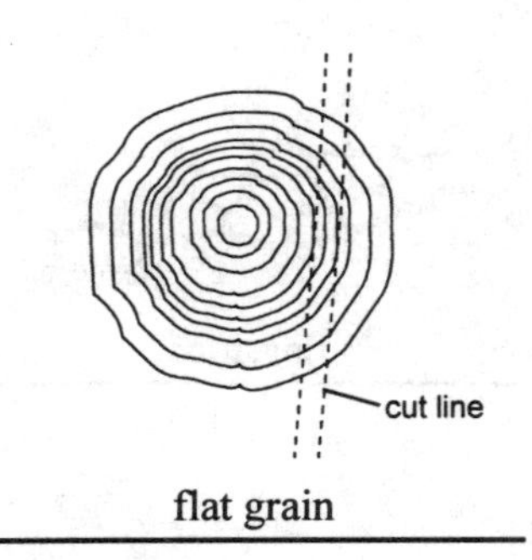

flat grain

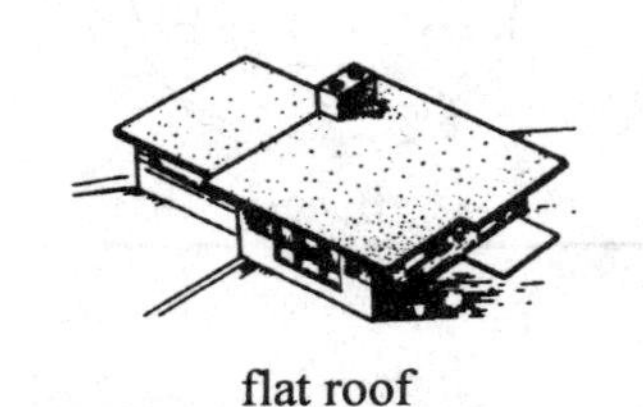
flat roof

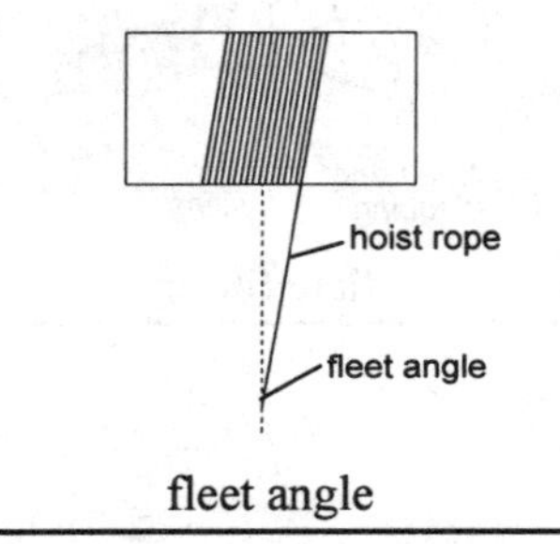

fleet angle

Flemish bond

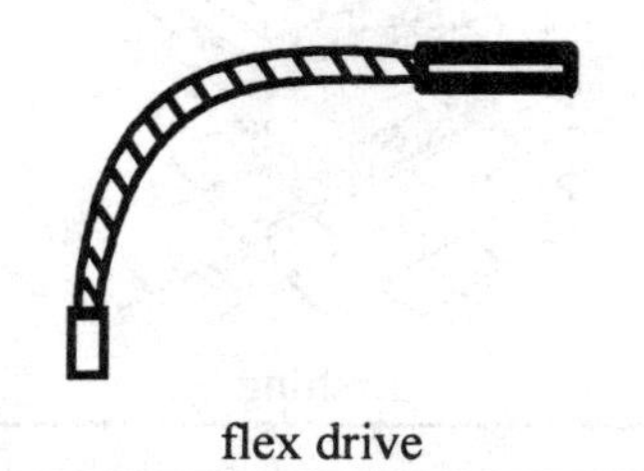
flex drive

flashover – an electrical discharge over the surface of an insulating material caused by a voltage too high to be contained by the insulator.

flash point – the temperature at which vapors given off by a fuel may be ignited.

flash set – accelerated setting of concrete in the presence of too much heat.

flash welding – a welding process in which the parts are joined over the entire faying surface at once by resistance heating with an electric current and pressure applied to the joint.

flat – 1) not glossy; 2) smooth and true, referring to a surface; 3) an apartment.

flat arch – see jack arch.

flat file – a flat bar-shaped wood or metal file with a rectangular cross section.

flat grain – wood that has been sawn at a tangent to the growth rings of the tree.

flat-nose pliers – pliers with jaws that taper in a horizontal wedge to a squared off blunt end.

flat paint – a paint that dries to a finish with little or no gloss. Flat paint has a high ratio of pigment to vehicle.

flat position – a weld position in which the work is laid flat, or nearly so, and the weld made on the upper, horizontal surface.

flat roof – a roof with little or no pitch.

flat slab floor – concrete slabs that are reinforced in two directions or more, and usually isolated from the foundation by rigid insulation.

flatting agent – a paint pigment that is added to reduce the gloss on the finished surface.

flaw – an imperfection or defect.

fleche – a narrow spire on a roof ridge.

fleet angle – the angle that a hoist rope deviates from vertical.

Flemish bond – a brick-laying style in which the short and long sides of the bricks, known as headers and stretchers respectively, are set alternately in a course, and each course is staggered in relation to the one above and below it.

flex – to bend or otherwise deflect without permanently deforming.

flex drive – a flexible shaft for turning sockets at an angle to the driving force.

flexible – free to move within limits.

flexible base – compacted gravel or other aggregate that is capable of changing shape and level to some extent. A flexible base is usually placed under a slab, driveway or walkway in areas where the soil expands and contracts, to prevent the slab from being cracked or otherwise affected by soil movement.

flexible hacksaw blade – a saw with a blade that is hardened only on the teeth. This keeps the teeth sharp for a long period of time, while allowing the blade to remain flexible and resistant to breaking.

flexible metal conduit – bendable conduit made from spiral-wound metal. Also called Greenfield conduit.

flexible metal conduit, liquid tight – flexible metal conduit covered with a waterproof plastic covering.

flight of stairs – a set of steps leading from one landing to the next.

flint – a naturally-occurring hard, silica-based stone that is sometimes used as an abrasive.

flitch – a piece of wood veneer or bundle of pieces of wood veneer.

flitch plate – a metal plate bolted between two pieces of timber to form a girder.

float – 1) a flat plate, usually made of wood, used to smooth the surface of wet concrete or plaster before it is troweled; 2) a sponge-rubber faced trowel for applying grout to tile. The sponge forces the grout into the joints between the tiles without scratching the tile surface; 3) to be buoyed on top of a fluid; 4) a hollow shape used to rise with a fluid and actuate some mechanism, such as a shutoff valve; 5) slack time in a work schedule created when tasks take less time than was allotted for them. See float time.

float glass – a type of plate glass formed on top of molten tin that is flat, smooth and thick. It is used for large windows because of its strength.

floating – 1) applying a second coat of plaster in a three-coat system; 2) smoothing a new concrete surface with a flat plate to solidify and strengthen the surface.

floating angle – a type of gypsum wallboard application which omits some corner fasteners, allowing limited structural movement of the wallboard, which reduces the risks of cracking.

floating dock – 1) a buoyant structure which can be partially submerged while a boat or ship is moved into position, and then raised completely out of the water to provide access to the hull of the boat or ship for cleaning or repairs; 2) a boat dock on floats.

floating edge – a type of wallboard installation in which the long edge of the gypsum wallboard is not supported by a framing member, permitting some shifting and settling to occur without cracking the wallboard joint.

floating joint – a gypsum wallboard panel joint that is located between framing members.

float, power – a motor-powered device for compacting wet concrete.

float time – slack time in a work schedule created when a task is completed ahead of schedule. Excess time, or float time, can be calculated into a job schedule to serve as a buffer for unexpected problems that may delay the completion of a project.

float valve – a valve controlled by a float on a lever that actuates the valve. The float rises with the water level in a tank or other container, causing the valve to shut off the water when the level reaches a predetermined height.

floodgate – a device used to regulate water flow past a barrier or through a waterway.

flood lamp – see floodlight.

flood level rim – the rim of a plumbing fixture from which overflow will spill, such as the top edge of a sink.

floodlight – a non-directional light, so-called because it spreads light over a wide area, often used in exterior applications.

floodlight mount – a weatherproof box with socket and cover for installing a floodlight on the exterior of a building.

flood plain – a low-level flat area of land along a waterway which is subject to flooding.

floor – 1) the horizontal surface of a structure that is designed to be walked on and to accommodate portable and permanent fixtures; 2) the lowermost limit.

floor drain – a drainage fitting designed to fit flush with a floor. The fitting is usually made of gravity-rated pipe, which is piping that is not designed to contain pressure other than that exerted on the fluid flow by the force of gravity.

floor edger – a disk sander designed for use in tight places that a large floor sander cannot reach.

floor framing – the structural members of a floor that rest on the foundation or other structural members that support a vertical load.

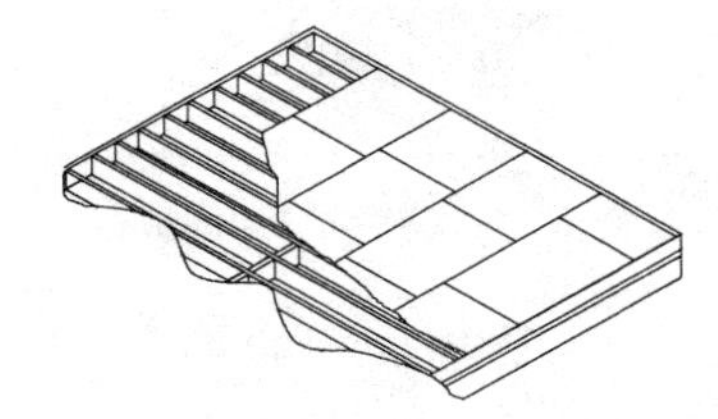

floor framing

flooring – material intended for use as a finish floor surface.

flooring, parquet – wood flooring in patterns.

flooring, plank – random-width wood flooring boards.

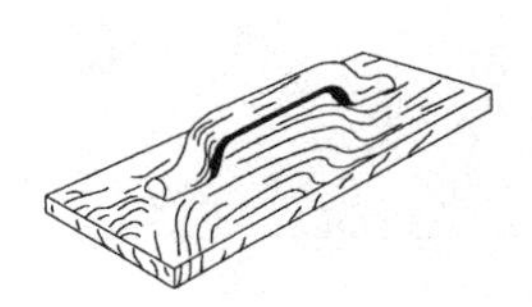

floating (2)

flooring saw – a short hand saw for starting and making a cut in installed flooring. It has teeth for about $\frac{1}{3}$ of the length of the blade along the top, with the bottom of the blade, fully toothed, curving down away from the front tip. The shape, as well as the teeth on top and bottom, allow the saw to cut into a solid surface.

flooring, strip – wood flooring in narrow strips.

flooring, vinyl – a floor covering made primarily of polyvinyl chloride (PVC).

floor jack – a screw jack used to provide temporary support under floor framing members for aligning and/or repairs.

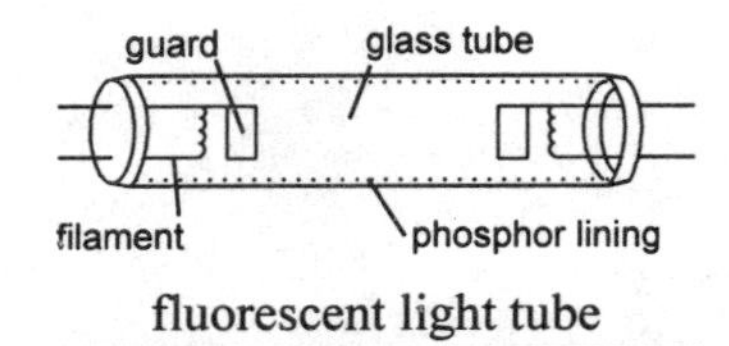

fluorescent light tube

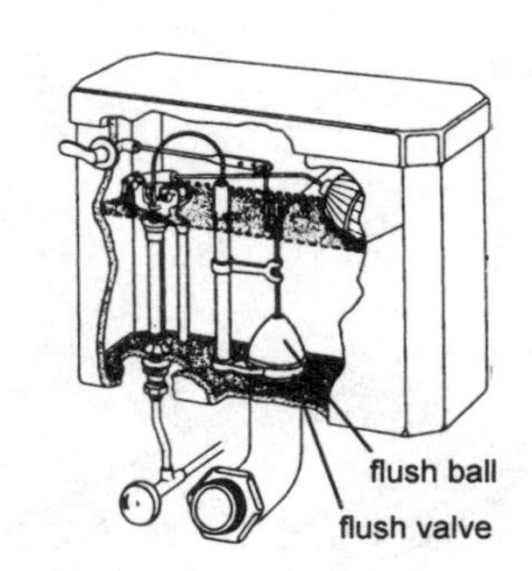

flush ball valve

floor plan – a drawing of a building, looking vertically down at the floor from above. The floor plan shows at least the outline of the wall locations and lengths to scale.

floor sander – a power sander used for finishing and smoothing hardwood floors. It can be a powered drum sander or oscillating disk sander on a moveable carriage with a long handle that can be operated by a person standing up.

floor sink – an enameled receptacle at floor level connected to a trap and drain for receiving floor drainage.

floor varnish – a varnish that is wear-resistant for use on wood floors.

floppy disk – the flexible, magnetic data storage disks used in a computer.

flow – movement, especially of a fluid.

flow brazing (FLB) – a brazing process in which molten brazing material is poured over a joint for a sufficient period of time for brazing temperature to be reached and bonding to take place.

flow chart – a graphical representation of the steps in a process or job from the start to the finish.

flow control valve – a check valve in a hot water heating system used to prevent the reverse flow of water when the system pump is turned off.

flowing sand – granular soil that has become saturated with water and becomes fluid and unstable.

flow meter – a gauge or instrument for measuring the amount of fluid flow in terms of quantity per unit time, such as cubic feet per minute or gallons per minute.

flow welding – a welding process in which molten filler metal is poured into a joint for a sufficient time for the welding temperature to be reached, the joint filled and bonding to take place.

fluctuate – to vary in intensity or movement.

flue – the chimney passageway for smoke and combustion produced in a fireplace or furnace.

flue lining – fire clay or terra cotta pipe used for the heat-resistant inner lining of chimneys.

fluid – a substance that flows, such as a liquid or gas. One characteristic of a fluid is its ability to change shape at a steady rate when force is applied to it.

fluid drive – a power transmission coupling using two impellers (rotors with blades like a turbine wheel), in a housing containing hydraulic fluid to form a coupling between the impellers. The driving impeller transmits power to the driven impeller through the hydraulic fluid.

fluidics – the technology of applying fluids in systems to perform functions. These fluids can be used to control mechanisms using hydraulic pressure and flow. For example, a pump may be used to exert pressure on a fluid, such as oil, in a closed system, and valves can direct the fluid to pistons in cylinders to perform work.

fluid mechanics – the science of the behavior of fluids.

fluid ounce – a measure of fluid volume equal to $\frac{1}{32}$ quart.

flume – a water-conducting channel.

fluorescence – the property of a material that enables it to emit radiation when it receives the proper stimulation. This radiation may be in the form of visible light, such as in a fluorescent lamp, or in the form of a radiation emission that is not visible to the eye.

fluorescent light – a highly-efficient light in which a tube with an interior phosphor coating is made to glow by ultraviolet light generated during the ionization of argon gas and mercury.

fluorocarbon – various substances in which fluorine is combined with carbon. Fluorocarbons are used in lubricants, cleaners, refrigerants, and fire extinguishers.

flush – 1) even with the surrounding surface; 2) to splash or flood with a fluid to cleanse.

flush ball valve – the outlet valve in the water tank of a toilet, consisting of a guided stopper in the tank outlet and a means to connect the stopper to the flushing handle. The toilet is flushed when the handle lifts the stopper out of the water tank outlet, allowing the water in the tank to drain through to the toilet bowl. The stopper is pulled back into the water tank outlet when the water level in the tank is lowered and the handle is released. Once the stopper is back in the outlet, the outlet is sealed and no more water flows out of the tank until the handle is actuated again.

flush device box – electrical box with tapped holes for fastening switches, outlets or other electrical devices flush with the surface.

flush door – a door that has a thin plywood facing over a framework and core of wood or wood products, such as particle board, or a thin plywood facing over a framework without a solid core. The latter is known as a hollowcore door.

flush ell – see water closet ell.

flush joint – a mortar joint that is troweled off flush with the face of the brick. Also called a rough-cut joint.

flush-mount lavatory – see lavatory, flush-mount.

flushometer – a type of toilet valve that meters a predetermined amount of water flow, and then automatically shuts off.

flush plug – a screw-in plug for a floor-mounted electrical box.

flush weld – a weld that is flush with respect to the surrounding base metal.

fluting – a decorative molding style consisting of a series of adjacent rounded convex channels along the length of the molding.

flux – 1) chemicals used to isolate heated metal from oxygen in the air, to prevent or dissolve or aid in removal of oxidation during welding, brazing or soldering, and to permit better metal flow in the joint; 2) rate of particle, fluid, or energy flow; 3) a compound used to combine with impurities in molten metals during refining; 4) a measure of magnetic energy.

flux-cored arc welding – an arc welding process in which filler metal and flux are fed through the center of the electrode from which the arc travels to the base metal.

flux-cored electrode – a hollow consumable welding electrode with a flux core.

fly cutter – a cutting tool or bit on an adjustable arm attached to a shaft or arbor. The arbor is turned by a drill motor and the bit revolves around the arbor and is used to cut circular sections out of material. The adjustable arm can be used to increase or decrease the distance of the bit from the arbor, thus changing the diameter of the hole that can be cut.

flying levels – in surveying, making an approximation of elevations.

fly rafter – a rafter at a gable end that is fastened to the roof sheathing above and the lookout blocks which sit behind the rafter. It is also called barge rafter or barge board.

flywheel – a wheel attached to a rotating shaft which adds weight and reduces fluctuation or pulsations in rotation.

foam – 1) lightweight substance containing large quantities of entrapped gases; 2) small bubbles formed from, or on the surface of, a fluid.

foamed-in-place insulation – an expanding synthetic material with good thermal insulation properties that is mixed and applied, or applied from a pressurized container. *See Box.*

foam rubber – a lightweight rubber with large quantities of entrapped air in the form of small cavities or bubbles within the material.

foam sheathing – a rigid plastic foam insulating board used in or on the walls of a building to increase resistance to the transfer of heat or cold through the walls.

foil – very thin sheet metal.

foil-backed panel – a gypsum drywall panel with a foil vapor barrier on the back. The reflective foil adds to the insulating value of the wallboard.

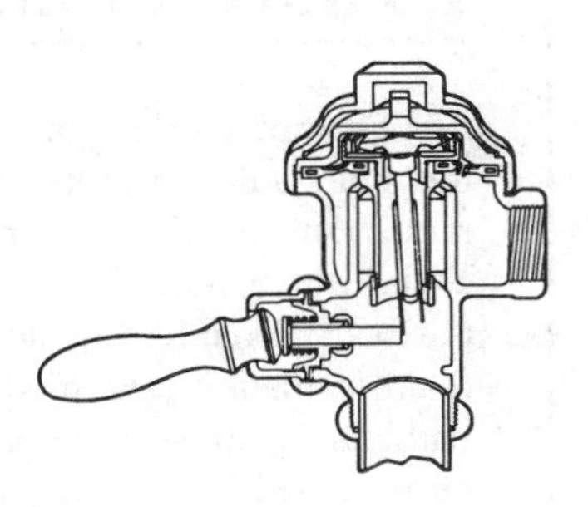
flushometer valve

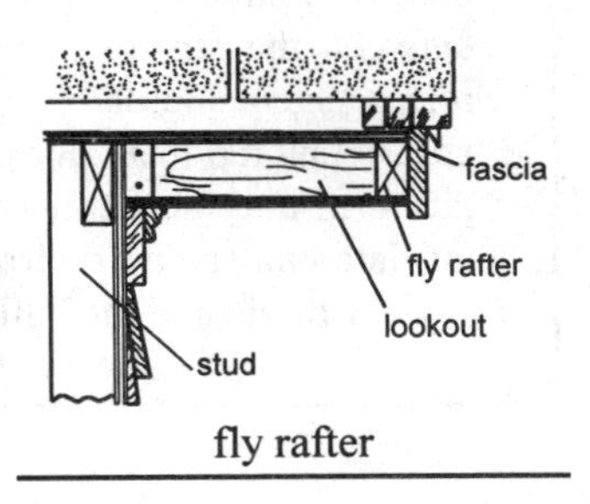

fly rafter

Foamed-in-place insulation

Foamed-in-place insulation is a synthetic material with good thermal insulation properties. It comes in a two-part liquid that is mixed and applied, or applied from a pressurized container. A chemical reaction between the two parts causes rapid expansion of the foam until it fills up the area provided. It then hardens into a rigid foam insulation. The mixture is totally free-form unless confined within a container, form or mold. If a structure, such as a tank, is to be insulated with foam, a form must be constructed around the exterior of the structure to contain the foam and provide the desired final shape and cured foam thickness. The liquid foam mixture can then be poured into the form where it will expand and fill the area between the structure and form. Before the foam is installed, a release agent should be applied to the inside surface of the form to ensure that the foam does not adhere to it. When the foam hardens, the form may be removed. Containerized foam in cans is used to insulate small areas, such as the gap between the outside of a window or door frame and the building frame.

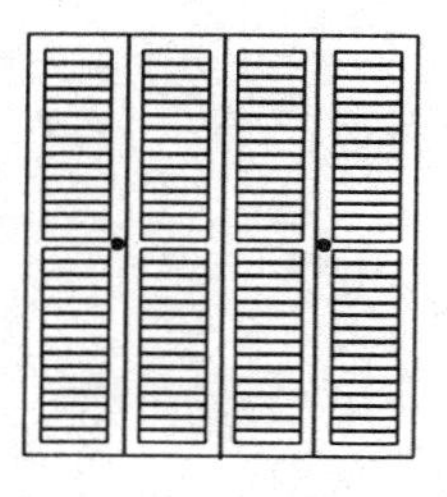
folding door

forced-draft cooling tower

folding door – multiple door sections that are hinged to each other and fold back as they are slid open. They are generally mounted in ceiling tracks which hold the weight of the doors and prevent sagging.

folding partition – a room divider for a large room area, such as a conference room, which folds back to the wall, usually sliding on an overhead track.

folding rule – a long ruler, hinged at intervals into sections, that can be folded up for convenience and extended for measuring.

folding stairs – a retractable stairway for an attic that can be pulled down from the attic opening for use. The stairs fold up into the opening and out of the way when not in use.

foot – a unit of length measurement equal to 12 inches.

foot candle – a measurement of light intensity equivalent to the amount of light produced by a lighted candle that is one foot from the measuring point.

footing – the base or bottom of a foundation pier, wall, or column, that is usually wider than the upper portion of the foundation. The added width at the bottom spreads the load over a wider area.

foot-pound – a unit of work, which is equal to the energy required to move a weight of one pound over a distance of one foot.

footprint – the floor or ground area taken up by an object.

foots – pigments that settle out on the bottom of a paint can.

footscraper – a metal bar mounted horizontally just above floor level outside the entrance to a building for scraping dirt and mud from shoes.

force – energy that is exerted to perform work.

force cup – a device with a long pole handle and a rubber cup used for unclogging plumbing drains. The rubber cup is fitted over the drain and pushed in toward the drain by exerting pressure on the handle. This compresses the air in the cup and exerts force on the obstruction to push it on through and clear the drain. Also called a plumber's friend or plumber's helper.

forced air heating system – a heating system that circulates warm air from a heat source through the ducting by means of a blower fan.

forced-draft cooling tower – a cooling tower that uses a fan to force air through the tower to cool water. The air, which passes through a water spray, cools the water by evaporation. The cooled water is used to cool equipment such as steam condensers or bearings on rotating machinery. The water is continually recycled through the cooling tower to be cooled and used again.

force fit – a fit between parts that are so close to being the same size that force is required to join them. Examples would be the fit between an axle and a bearing or a valve seat and a cylinder head.

foreclosure – the process by which a property owner forfeits his right to redeem the mortgage held on the property. The property is usually sold at auction to cover the mortgage lien and legal costs.

forecourt – a courtyard at the entrance to a building.

forehand welding – welding with the torch or electrode pointing in the direction the weld is being laid.

foreman – a person responsible for supervising the performance of a group of workers.

forfeit – to lose or give up a right.

forge – to shape or form by heating and hammering while hot, such as metal hammered into shapes.

forge welding (FOW) – a welding process that uses the heat of air in a forge, and steady or impact pressure, to cause a permanent deforming of the surfaces to be joined.

form – see formwork

form anchor – any metal device used to attach forms to existing concrete.

formation – a significant geological grouping, such as a large cluster of rocks.

form hanger – a device used to support a concrete form from an existing part of a structure.

Formica – the trade name for a plastic laminate in sheet form.

form jack – a device, with an alterable height adjustment, used to support the bottom of a form.

form lining – any material used to line concrete forms and alter the texture imparted to the concrete as it sets up.

form of a curve – the sharpness of the curve.

form oil – an oil applied to the surfaces of concrete forms which prevents the poured concrete from sticking to the forms after it has set.

form stripping – removing formwork after concrete has set.

form tie – rod, wire, or other tensile members used to hold forms together and resist the pressure of the wet concrete.

formwork – temporary structures used to hold wet concrete in the shape and position desired until it sets up.

Forstner bit – a drill bit designed to cut round holes in wood. It consists of a pair of radial scraper-type blades and a circumferential guiding blade that is the diameter of the hole to be drilled. As the bit is turned, the radial scraper blades remove the wood between the center starter point and the circumferential guiding blade. This blade cuts the exact diameter of the hole being drilled.

forward-curved blade fan – a centrifugal fan with blades curved in the direction of rotation.

found – 1) to melt and pour into a mold; 2) to establish or locate.

foundation – 1) the base or portion of a structure that is in contact with the ground, usually extending below grade; 2) a support on which something stands.

foundation bolt – a bolt set in wet concrete and used to bolt the mudsill to the concrete after the concrete has set.

foundation plan – a drawing showing the placement and dimensions of a building foundation.

foundation vent – openings through the foundation to permit ventilation under a structure. Such vents are screened to prevent the entry of small creatures.

foundation walls – the walls from the footing to the first floor.

foundry – a factory for working and/or casting molten metal.

fountain – a spray or the flowing of a fluid, usually water, over an ornamental or functional structure.

fourpenny – a nail, approximately 1½ inches long. Also called 4d.

four-way switch – an electrical switch used in conjunction with two three-way switches where three points for controlling an electrical circuit are needed.

fraction – a portion of a whole.

fragile – easily broken; delicate.

fragment – a part that has been broken off.

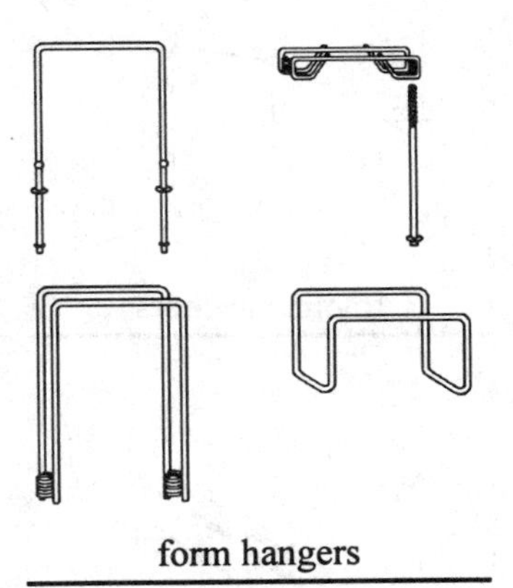
form hangers

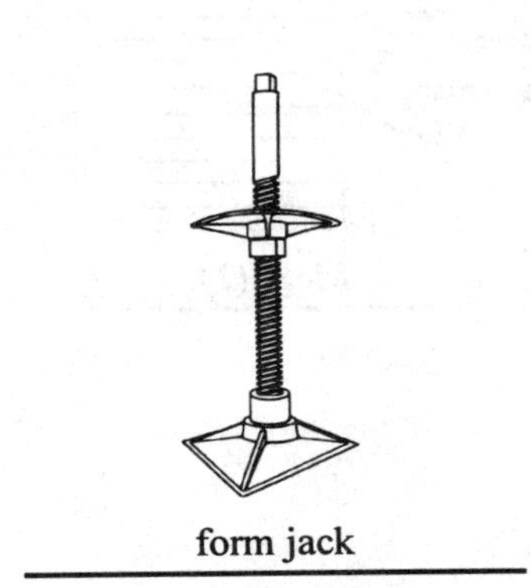
form jack

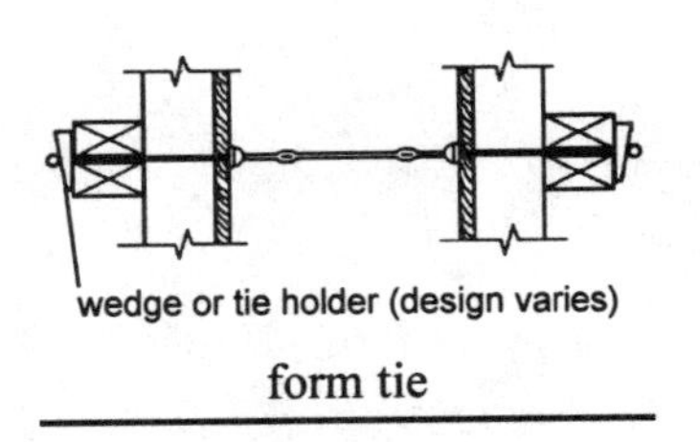

form tie

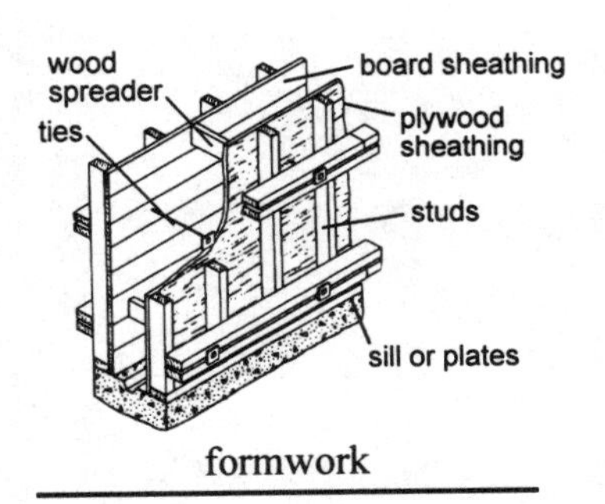

formwork

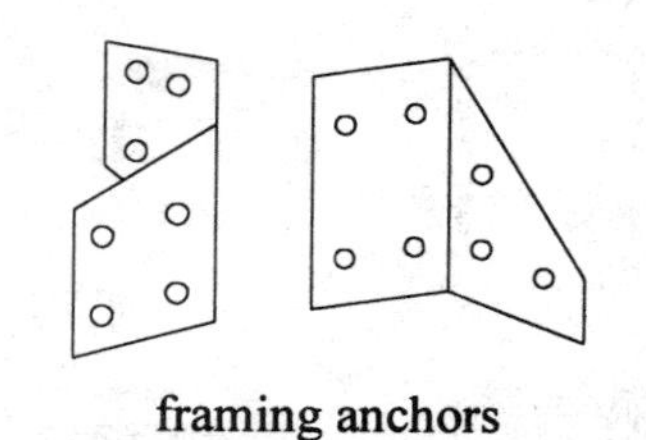
framing anchors

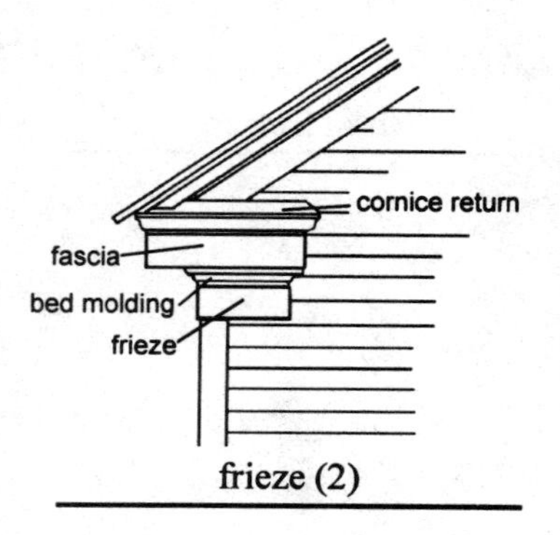

frieze (2)

frame – 1) the basic structural members of a building; 2) a surround for an object, such as a picture, door or window.

framework – the skeleton or structural support.

framing – the structural elements of a building.

framing anchors – metal fasteners used to connect framing members and reinforce the joints. Also called framing clips.

framing, balloon – see balloon framing.

framing in – completing the framework of a structure.

framing square – a flat, metal L-shaped measuring tool used to lay out angles and calculate lengths. The legs of the L, which often have measuring tables and marks for rafter cuts embossed on them, come together at a right angle. Also called carpenter's square.

framing square gauges – attachments or markers that can be clamped to a framing square to help make accurate, repeat measurements.

free area – the total area of inlet/outlet openings for air passage in an HVAC system.

free enterprise – trading in a competitive system.

freestone – a fine-grained homogeneous stone that does not have a tendency to split along a particular plane.

freewheel – the condition that exists when a motor driven device disengages from the drive when its speed exceeds the driving speed. An example is an automobile that coasts downhill faster than the speed generated by the engine.

freezeback – the formation of an ice dam at the eave line of building which forces water back under the shingles.

French arch – a masonry arch whose bricks slope at an outward angle symmetrically from the center.

French curve – a drafting template with multiple changing curvatures.

French door – single or double doors in which the majority of the door area is made up of one or more panes of glass surrounded by a frame. One or both doors swing open on hinges. French doors are traditionally made of wood. Some have metal clad exteriors for increased weather resistance and durability. Tightly sealed double pane french doors are available, some with adjustable blinds between the panes of glass.

Freon – a trade name for a number of fluorinated hydrocarbon compounds used as refrigerants and cleaning compounds.

frequency – electrical cycles per second, called Hertz (Hz).

frequency meter – a device to measure alternating current frequency in cycles per second in an electrical circuit.

frequency modulation – the varying of a radio carrier wave frequency.

fresco – a painting made on wet plaster.

fret – 1) a design having an interlocking angular and repetitive pattern; 2) one of a series of crossbars on the fingerboard of some stringed instruments.

fretwork – see fret.

friable – a material that crumbles easily.

friction – resistance to motion between two joined surfaces or adjoining commodities, such as between a fluid and the piping that it must flow through. Friction can be reduced by smoothing a surface, reducing the viscosity of a fluid by increasing its temperature, or by the application of a lubricant.

friction catch – a mechanism which holds a cabinet door closed using friction. Also called friction latch.

friction pile – a pile or piling that relies on friction with the surrounding soil for its support.

friction tape – a cloth-based, adhesive-backed electrical insulating tape that is impregnated with a water-resistant compound.

frieze – 1) a raised sculpture or ornamental decoration on a building; 2) the horizontal portion of a cornice set vertically against a wall.

frieze ventilator – a screened vent through a frieze board.

frit – a glass material in powdered form that is applied to ceramic tile and then fired in a kiln to fuse it to the tile and create a glazed finish.

frog – a groove in the bed surface of a brick designed to key it to the adjacent bricks with mortar.

front – the face of an object; the forward-facing side of a structure.

frontage – the portion of a piece of property bounded by a public thoroughfare.

front clearance – the clearance at the front of a lathe cutting tool created by the angle at which the front of the tool is ground.

front view – a drawing of the front of an object.

frost heave – the upward movement of an object or structure caused by the freezing and consequent expansion of water in the soil under the object or structure.

frostline – the depth to which frost penetrates the soil.

frow – a tool for splitting wood, such as shingles. The tool has a cutting blade at a right angle to the handle. Also called a froe.

frustum – 1) a flat-topped, cone-shaped figure, with its top parallel to the base; 2) a shortened column.

fuel cell – a device for chemical conversion of fuel to electrical energy.

fulcrum – the pivot point for a lever. A lever is braced against a fulcrum which multiplies the force exerted on the lever to perform work.

fuller – a metal working hammer with a half-round (hemispherical) head.

fuller's earth – an absorbent clay.

full fillet weld – a fillet weld between two pieces in which the weld is the same size as the thinner of the two pieces being welded together.

full section – see section, full.

fulminate – to detonate with a loud sound.

fume – a vapor.

functional – working as intended.

fundament – the geographic characteristics of an area.

funded debt ratio – the ratio of working capital from borrowed funds to the working capital from equity.

fungicide – a chemical treatment to halt or prevent the growth of fungi.

fungus – a type of parasitic plant that does not contain chlorophyll for making its own food. Molds, mildew, yeasts and mushrooms are fungi.

funicular – using a rope or cable.

funicular railway – a railway on a steep slope using cars connected by cables on which ascending and descending cars counterbalance each other. The cars travel up and down the slope on a rail track.

funnel – a cone-shaped device with a spout opening extending from the small end of the cone. The funnel is used to direct the flow of fluid through a small opening.

furane resin – a resin from pine trees that is used in paint.

furnace – a device in which heat is generated by burning a fuel. The heat may be used to process something such as refining an ore, or for heating air for comfort.

furnace soldering – a soldering process which uses the heat of a furnace to raise the temperature of the parts to be soldered sufficiently high for soldering.

furring – wood or other material that is fastened to a surface prior to attaching gypsum wallboard or other wall covering panels. The furring is used to provide a gap between the panels and the building surface. It can also be used to provide a level or plumb attachment plane for the panels. Furring is installed with fasteners, adhesives or a combination of the two.

furring clip – metal clips that are fastened at intervals to a wall which will be covered with masonry veneer. The clips are embedded in the mortar joints of the wall to secure the veneer to the wall behind it.

furrow – a long, narrow trench or depression in the earth.

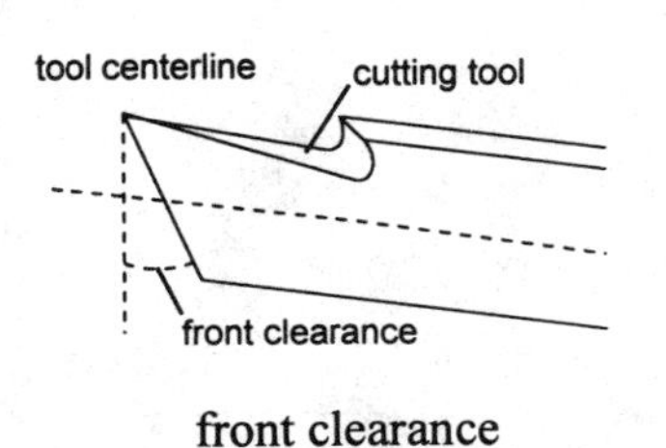

front clearance

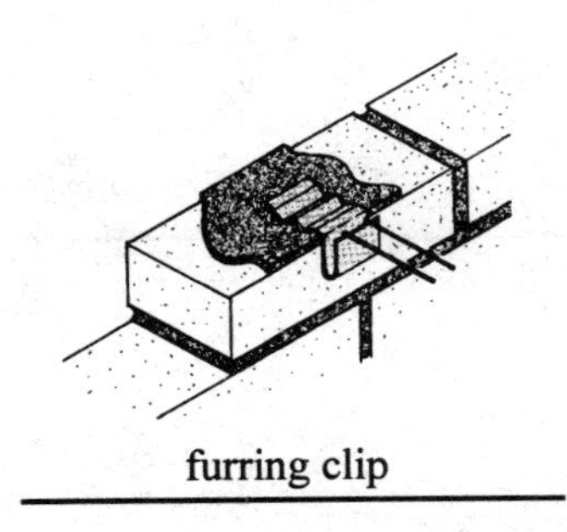
furring clip

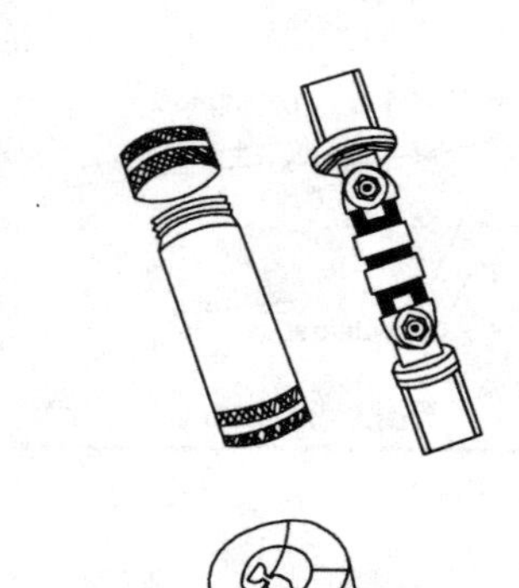

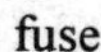

fuse

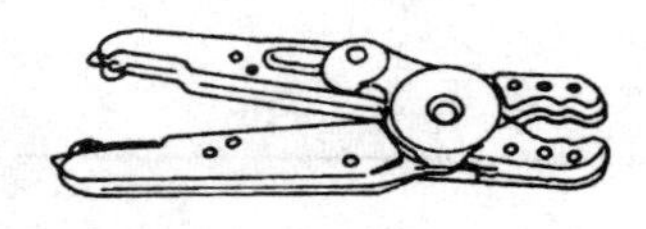

fuse puller

fuse – 1) to melt together; 2) a safety device for an electrical circuit that melts and interrupts the circuit if the current exceeds a predetermined level for a specified amount of time.

fuse box – a distribution box for electrical circuits which contains fuses for the protection of the circuits.

fuse panel – see fuse box.

fuse puller – a non-conducting tool resembling pliers used to pull electrical fuses from fuse holders.

fuse-type relief valve – an outmoded relief valve for hot water heaters in which a fusible metal will melt and open the valve if the water is overheated, releasing the pressure in the system.

fusible metal – a metal with a low melting point.

fusil – melted and cast in a mold.

fusion – 1) merging together; 2) the melting of the base metal, or the base and filler metals.

fusion welding – joining together by melting the materials being joined at the interface between them.

G: gable area to gyroscope

gable area – the surface wall area between the roof slopes of the gable end of a building and the top wall plate.

gable dormer – a dormer with a gable roof.

gable end – the portion of the end wall of a building between the top wall plate (eave to eave) and the point or ridge of the roof.

gable end brace – a brace from the wall top plate at the gable end of a building to the roof ridge board at the gable end. This brace adds stability to the ridge board.

gable end stud – a stud that extends upward from the horizontal gable end wall top plate to the gable end rafter.

gable roof – a roof design in which all rafters are cut to the same length and joined in the center to form a peak, with the two sides of the roof sloping down from that peak.

gable ventilator – a vent at the gable end of a building.

gag – 1) a clamp that physically prevents the safety valve on a boiler from opening to relieve pressure when the boiler is hydrostatically pressure tested. This test is done on a periodic basis to insure that the boiler is safe for operation. The gag allows the valve to be submitted to higher water pressure than the safety valve setting would normally permit; 2) a metal block used with a press or vise to bend or straighten a piece of metal. The block is sized and shaped so that it imparts the desired contours to the metal.

gain – 1) the recess cut into a door edge or jamb which allows the hinge plate to be set flush with the surrounding surface; 2) a mortise or notch; 3) an increase.

galilee – a ground floor porch located at the entrance of a church.

gallery – 1) a narrow covered walkway; 2) a narrow room; 3) a porch or veranda; 4) an interior passageway; 5) an interior platform along a building wall; 6) a raised area in a theater.

gallon – a unit of liquid capacity equal to 4 quarts or 231 cubic inches.

gallons per minute (GPM) – a measure of fluid flow used to measure volumes of intermittently flowing fluids, such as pump discharges.

galvanic action – a flow of electrons from one metal to another which results in the corrosion or the eating away of the metal that is lower on the galvanic scale. Galvanic action requires the presence of two different metals and an electrolyte, a substance that will conduct electricity. An example of galvanic action is found on board ocean-going ships where it is used productively to protect steel condensers from corrosion. Zinc pieces are mounted in the water boxes of the condensers. The saltwater is an electrolyte. Since zinc is

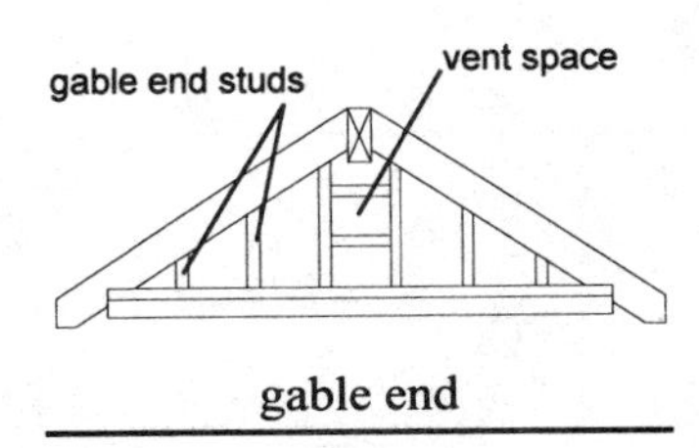

gable end

gable roof

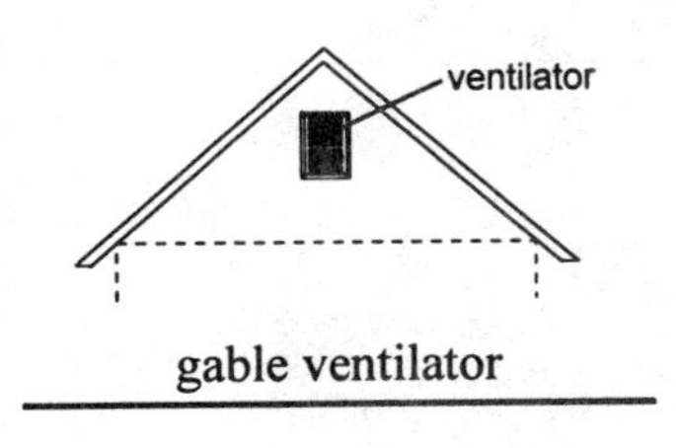

gable ventilator

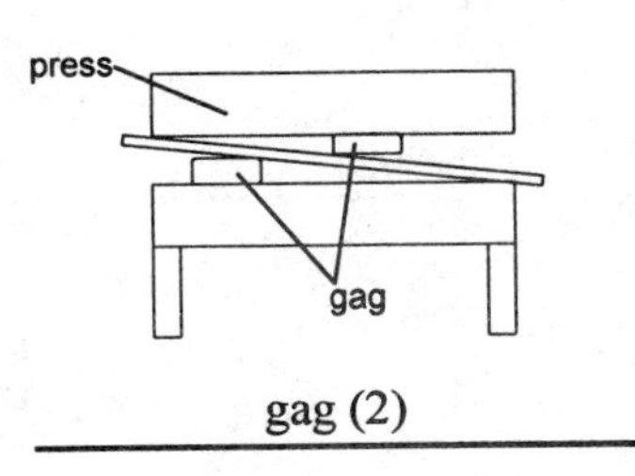

gag (2)

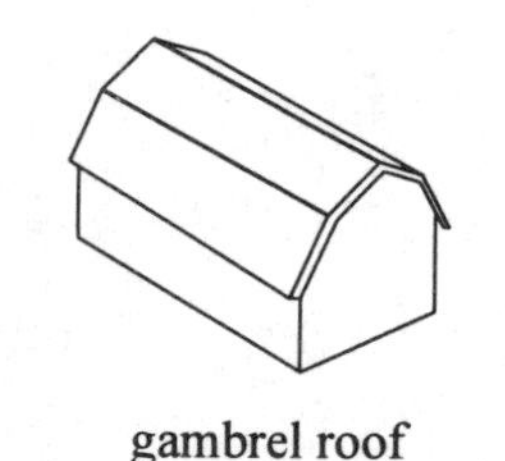

gambrel roof

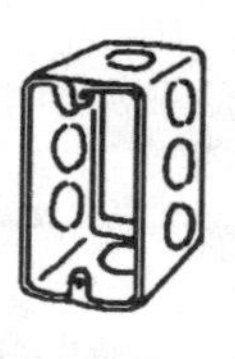

gang box

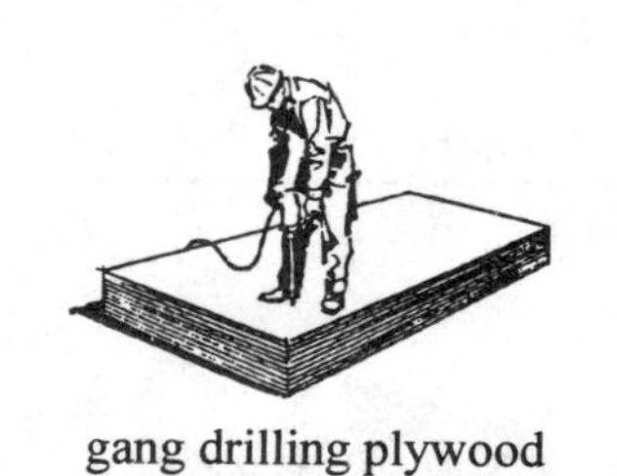

gang drilling plywood

stretcher header

garden bond

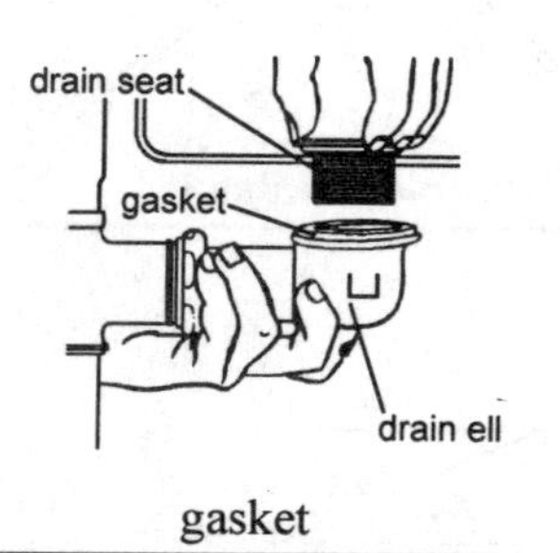

gasket

lower on the galvanic scale than steel, the zinc is eaten away by the galvanic corrosion, leaving the steel unharmed.

galvanic scale – a hierarchy of metals showing their relative susceptibility to galvanic action.

galvanize – to electroplate or hot dip a metal in zinc to provide a protective coating which inhibits corrosion or oxidation of that metal.

galvanized iron – iron coated with zinc.

galvanized steel pipe – steel pipe that has been internally and externally coated with a thin layer of zinc to provide corrosion protection.

galvanometer – a laboratory instrument used for measuring or detecting a small electric current.

galvanoscope – an instrument that can measure and determine the direction of an electric current.

gambrel roof – a roof style in which the rafters are at two different slopes from the ridge to the eaves.

gang – 1) to operate as a group; 2) a combination of similar instruments or tools arranged for the convenience of working together, such as a series of circular saw blades spaced on a shaft to make multiple simultaneous cuts.

gang box – a small electrical box made of sheet steel or plastic used for mounting electric switches or outlets in walls or joining cable or conduit. Also called handy boxes.

gang drilling – drilling holes in a number of stacked and aligned panels at the same time so that the holes are identically placed in each of the panels.

ganged – joined together in a group.

gang mandrel – a shaft designed to hold multiple pieces of work or tools for turning simultaneously on a lathe. The pieces are held in place by means of a nut threaded onto the end of the mandrel shaft.

gang milling – milling with multiple cutting heads or milling two or more pieces simultaneously.

gang operations – processing several identical pieces of work at one time, such as drilling through several sheets of stacked plywood at one time so that the holes drilled are all in the same location on each sheet.

gang saw – a power saw with multiple blades for making several parallel cuts simultaneously.

gangway – a temporary walkway providing access to a construction site or other temporary situation.

gantry crane – see crane, gantry.

garden bond – a brickwork pattern in which headers are placed at large intervals, three to five stretchers between headers in each course.

gargoyle – a grotesque statue of a man or animal used as a building decoration in some older styles of architecture.

garnet – a hard, red silicate mineral crushed and used as an abrasive material or set in jewelry as a semi-precious stone.

garnet paper – an abrasive paper coated with finely-crushed chips of garnet.

garret – an attic.

gas – the vapor state of a substance.

gas burner – the exit nozzle supplying fuel gas to an appliance.

gas cock – a plug valve designed to provide shutoff or flow control in a natural gas or other gas system. A gas cock is installed between the gas line in a building and each of the gas appliances connected to the gas line. This permits the gas line to a particular appliance to be closed off without having to shut off gas to the entire building.

gas cylinder – a cylindrical metal vessel used for the transport and storage of gas under pressure.

gas-fired furnace – a heating system that uses natural gas or propane gas as a fuel.

gasket – a compressible or conforming material used as a seal between two parts. The configuration and material used for a gasket is determined by the material and use of the two pieces that are joined. The gasket must conform to the mating parts, and the pressure, temperature and fluid conditions to which the parts are subjected. Rubber or other elastomer materials are often used in water service, but are not suitable for steam. Extreme temperatures, as in the case of steam, require that gaskets be made of mineral or fiberglass-based materials, or in some cases, metal or metal-clad materials.

gaslight – a lamp which utilizes burning gas as the source of light.

gas metal arc cutting (GMAC) – a cutting process, utilizing an external shielding gas, which melts metal with an electric arc produced between a consumable electrode and the metal being cut. A variable transformer is used to supply electrical current through heavy electrical leads. One lead is connected to the metal to be cut and the other is connected to the electrode. When the electrode is held close enough to the metal, an electric arc passes from the electrode to the metal, completing the electrical circuit.

gas metal arc welding (GMAW) – a welding process, utilizing an external shielding gas, which welds metals by heating them with an electric arc produced between a consumable electrode and the metals being welded. This process provides deep penetration of the weld and rapid deposits of weld metal.

gas metal arc welding, pulsed arc (GMAW-P) – an electric arc welding process that uses pulsed current. This offers the speed of gas metal arc welding, with reduced welding heat from the pulsed arc to avoid burn-through when relatively thin metal is being welded.

gas metal arc welding, short circuit arc (GMAW-S) – an electric arc welding process using repeated short circuits to deposit the filler metal.

gas pliers – a type of pliers, used for gripping pipe, that have a curved, toothed cutout in the jaws and two different ranges of jaw openings.

gas pocket – a void space in weld metal or a casting which weakens the work.

gas tungsten arc cutting (GTAC) – a process using an electric arc which flows from a nonconsumable tungsten electrode to the metal being cut. The electrode holder has a gas passage, or cup, around the electrode. An inert gas flows through this passage to isolate the molten metal of the cut from oxygen. This gas shield prevents oxygen from having detrimental effects on the hot metal. The gas is supplied via a hose connected to a tank or other inert gas supply.

gas tungsten arc welding (GTAW) – an electric arc welding process in which a nonconsumable electrode is used to maintain an arc surrounded by inert shielding gas with the weld wire fed separately into the weld.

gas tungsten arc welding, pulsed arc (GTAW-P) – an electric arc welding process using a tungsten electrode, an external shielding gas and pulsed current.

gas welding – a welding process in which the base material is melted by a fuel-and-oxygen-burning torch and the weld metal is hand fed into the weld.

gate – a closure for a fence or barrier which can be opened to permit ingress and egress.

gate post – a fence post on which a gate is hung.

gate valve – a valve with a wedge or disc that is raised and lowered to open and close the valve. When the valve is open, the disc or gate completely clears the opening allowing a straight line of flow. A handwheel is turned clockwise to move the disc into its seat, stopping the flow. Gate valves are used in applications where the valve is to remain either fully open or completely closed most of the time, such as a shutoff valve for a hot water heater.

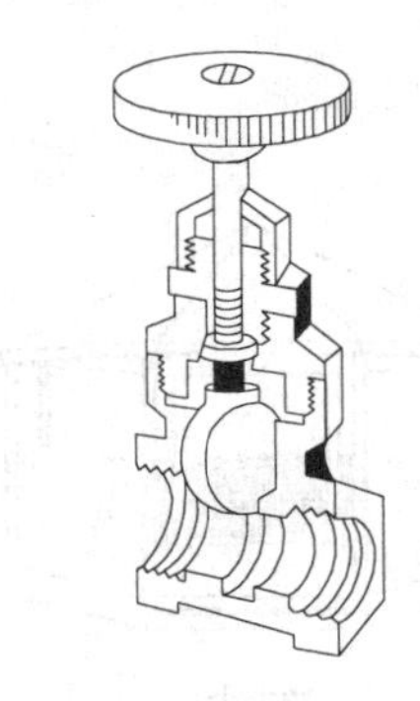

gate valve

gateway – a passageway with a gate.

gauge – 1) a measuring instrument; 2) the proportions of some materials as measured according to a standard for thickness, diameter or size, such as wire, thread or resilient flooring; 3) the horizontal distance between train rails or crane bridge rails.

gauge blocks – reference blocks cut or ground to a precise dimension, 3⁄8 inch by 1 3⁄8 inches by the desired dimension that the block is intended to represent. Gauge blocks are hardened and finished to within a few millionths of an inch accuracy when used at 68 degrees F. They are used for calibration of precision measuring devices, such as micrometers.

gauged brick – brick that has been cut and shaped to fit a specific location, such as in an arch.

gauged mortar – mortar to which plaster of paris has been added to speed up the hardening process.

gauging – cutting masonry units to a specific size.

gauging block – a block of wood used along a string line suspended between the corners of concrete forms. This block is used to align and level the forms.

gazebo – an open structure with a decorative roof supported by columns. Gazebos are built in many different sizes and styles. They can be raised several steps up on a platform like a small stage or they can sit on a ground level foundation or deck. They can be completely open or partially enclosed with lattice around the base. Most raised gazebos are either partially or totally enclosed around the base for safety. The more popular styles are the open hexagonal or octagonal shapes. They are usually found in gardens and used for outside entertaining. Large models are often located in public parks.

gazebo

gear – a toothed wheel which interlocks with other toothed or chain driven mechanisms to transmit power and motion as it is turned.

gearbox – an enclosure surrounding a gear train, which is a series of gears that can be meshed together in different combinations. The gearing combinations can change the mechanical advantage between the input and output shafts in the gear train.

gear ratio – the ratio between two gears of different sizes that are in contact.

gearshift – a means to change the gear combinations in a gearbox. The gearshift is the mechanism in a vehicle that selects the gears, such as first, second or reverse, in the transmission.

gearwheel – a gear.

gel – the top coat of resin applied in a layering of fiberglass mat or cloth and resin in a mold to form a fiberglass shape, such as a boat hull.

general contractor – the contractor who has overall responsibility for a construction project. Subcontractors work under the direct control of the general contractor who makes up the schedules, coordinates the tasks, and supervises the activities of everyone on the job.

generator – a rotating device, consisting of a conductor that is moved through a magnetic field, that converts mechanical energy into electrical energy.

geodesic dome – a structure composed of a formwork of triangularly arranged grid members whose finished appearance is a dome.

geodetic survey – a survey of large land areas which takes into consideration the curvature of the earth.

geology – the science of the history and composition of the earth.

geomagnetic – the magnetism of the earth.

geometry – a branch of mathematics dealing with the relationships of lines, angles, points and figures.

gesso – a plaster surface that has been coated with glue or other material which prepares it to accept paint.

geyser – a hot spring that builds up pressure and forces water and steam out of the earth.

ghosting – a defect on a paint surface which gives the finish a shadowed appearance, such as flat spots appearing on a semi-gloss surface.

gib – 1) a thin metal wedge used to control the movement of a sliding part, such as the moving table on a milling machine; 2) a part secured by a cotter pin; 3) a metal strap holding parts together.

giga – a prefix meaning billion, as in gigawatts or gigahertz.

gigahertz – one billion cycles per second.

gild – to apply a thin covering of gold.

gill – a measure of liquid volume equal to ¼ pint (American).

Gilsonite – a trademark name for a pure asphalt used in paints.

gimbal – a mounting consisting of two circular pivoting rings that permit movement in any direction. Used for such items as compasses, gyroscopes or nautical equipment that require self-balancing mounts. *See Box.*

gimlet – small wood-boring auger with a T-shaped handle.

gingerbread – ornate trim work on a Victorian style structure.

gin pole – a long pole rigged with guy wires and pulleys used for lifting objects or commodities and moving them to another location. The object can be hoisted up and swung around the base of the gin pole, and then lowered into the new position.

girder – a structural beam used to support concentrated loads at points along its length.

girt – 1) heavy timber corner post of a building; 2) a heavy beam.

girth – the measure of the distance around the outside of an object; its perimeter.

girt strip – a horizontal wood member used to support floor joists in balloon framing. Also called a ledger.

glare-reducing glass – transparent glass that is tinted, or translucent, to reduce light transmission.

glass – a rigid, noncrystalline material, usually transparent, made from fusing silica and alkali with other additives for stabilization. Lime, lead, barium, and various oxides are used as these additives, depending on the properties needed for the end product. Different additives produce different colors as well as different properties. The properties are also varied by the manufacturing process. Manganese oxide and selenium oxide are added to soda, lime and silica to make colorless window glass. Borosilicate glass and aluminosilicate glass are used for high temperature applications. Vycor, a type of glass with high silica content, can also be used for very high temperature applications. Fused silica glass has the highest continuous operating temperature and the greatest amount of chemical resistance. See also individual listings under plate glass, safety glass, tempered glass and wire glass.

glass blocks – building blocks of translucent glass used for non-bearing walls or partitions. The blocks are held together with mortar or a system of plastic strips and silicon sealant.

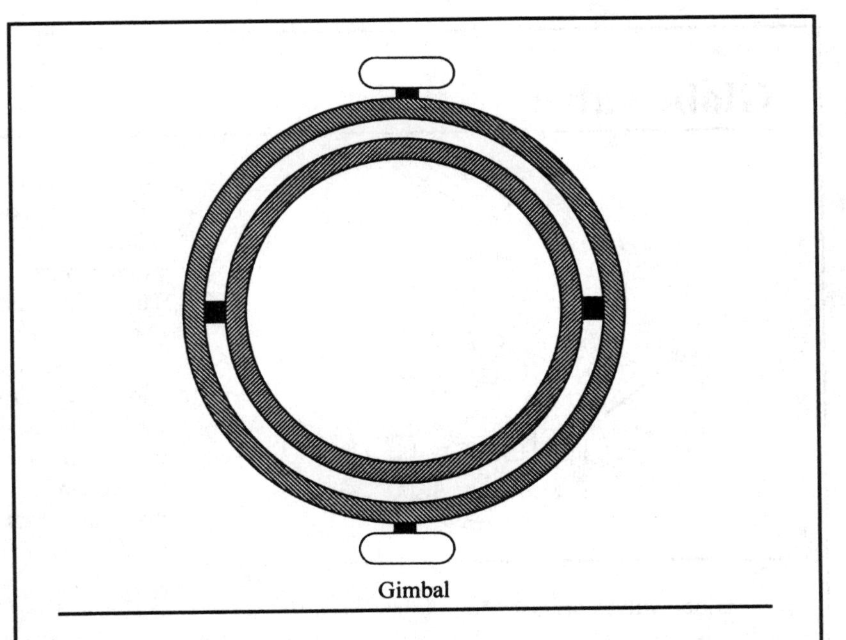

Gimbal

Gimbal

A mounting that permits movement in any direction. It consists of two rings, one inside the other. The outside ring is fastened to a mounting at two points that are opposite each other across the diameter of the ring. The outer ring can pivot on these two points. The inner ring is fastened to the outer ring at two points also across the diameter of the inner ring, but rotated 90 degrees to the points at which the outer ring is fastened. The inner ring can pivot about the two points at which it is fastened to the outer ring. The object to be supported by the gimbals, such as a compass, is mounted to the inner ring. This object can then remain level in all directions regardless of the movement or position assumed by the mounting to which the outer ring is fastened. Gimbals are commonly used to mount nautical equipment as well as surveying compasses.

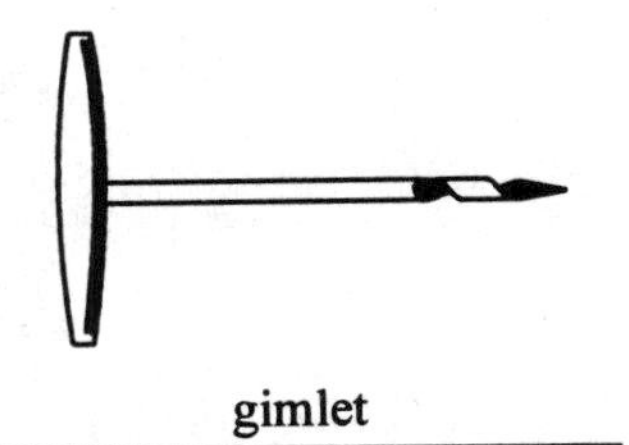

gimlet

Globe valve

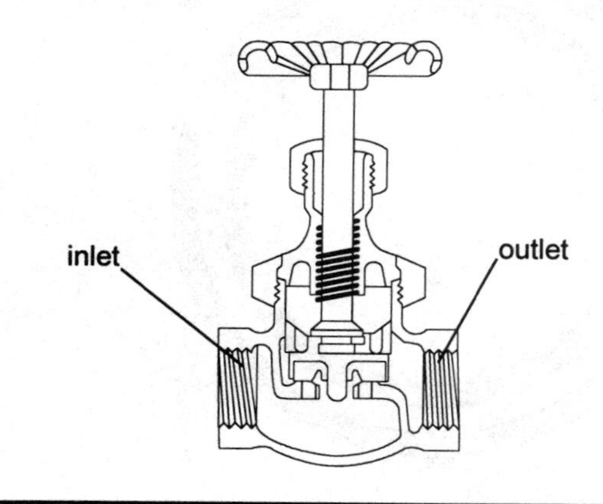

Globe valves are used to regulate fluid flow. The valve is opened or closed by raising or lowering a horizontal disc or plug from its circular seat. The most common globe valves are operated by turning a handwheel connected to the threaded valve stem, causing the disc to move up or down. To close the valve, the handwheel is turned clockwise until the disc presses firmly on the circular opening, which is the valve seat. When opening the valve, the volume of flow is roughly proportionate to the number of turns of the handwheel. The fluid must flow up and through the valve seat, changing directions as it passes from the inlet to the outlet. This change of direction slows down the flow, and causes a drop in the pressure. It also allows the fluid flow to be easily adjusted. Globe valves are used in applications requiring frequent operation, such as water faucets and hose bibbs. Because of constant use, the discs and seats are liable to wear. Most globe valves used in household water service are fitted with soft fiber or elastomer discs, called faucet washers, which can be replaced when worn. These are suitable for common service applications. High temperature, steam or high flow rate applications require a metal disc, which prolongs the life of the valve. Valve seating surfaces may be faced with a hard material, such as a chrome alloy, to increase wear resistance in severe duty applications.

glass cutter – a tool for scoring a line in glass along which the glass can be broken.

glass door, sliding – see sliding glass door.

glass drill bit – a diamond or carbide-tipped bit designed to drill holes in glass or ceramic tile.

glass mesh mortar unit – a cement and sand mixture cast into boards with fiberglass mesh reinforcement. It is used as a backing for ceramic tile installation. Also called cement board.

glass pliers – pliers with wide plastic or rubber faced jaws used for grabbing a section of glass to be broken off. Also called glazier's pliers.

glazed brick – brick that has been coated with a fusible material that creates a glaze on the surface when the brick is fired.

glazed tile – ceramic tile to which frit, a powdered glass material, is applied before firing. Firing fuses the frit to the tile creating a hard, glossy surface.

glazier – a glass installer.

glazier's pliers – see glass pliers.

glazier's point – small pieces of thin sheet metal, shaped like arrowheads, used to hold a pane of glass in a sash or frame while it is being installed. The glass is permanently held in place with glaziers putty which is applied around the glass edge-to-sash joint.

glazing – 1) a technique in which semi-opaque color is painted over a base color and then partially wiped away; antiquing; 2) installing glass in a structure, or the installed glass; 3) applying a special coating to a ceramic or masonry unit and firing it in a kiln to form a hard surface.

glazing compound – putty used as a seal between window glass and the window sash.

glitter – small shiny particles that can be spayed, painted or glued onto a surface to create a sparkling, reflective appearance or used as an additive in other substances to create that effect.

globe valve – a valve that is opened or closed by raising or lowering a horizontal disc or plug from its circular seat. *See Box.*

gloss – a high shine finish.

gloss oil – a varnish of limed rosin and petroleum thinner that dries to create a high luster finish.

gloss paint – paint that dries with a sheen to the surface.

glove – a protective covering for the hand.

glove box – 1) a protective box in which hazardous materials may be manipulated by the use of sealed gloves that are designed into the walls of the enclosure. The pressure inside the glove box is maintained at less than that of the atmosphere outside to preclude possible leakage from the box; 2) a small storage compartment in the cab an automobile or other vehicle.

glue – an adhesive.

glue block – a glued-in-place block of wood placed at the junction of two other pieces of wood to strengthen the joint.

glue gun – 1) a device that applies adhesive from a cartridge; 2) a heating dispenser used to apply hot melt glue.

glu-lam or glulaminated – a structural beam that has been fabricated by gluing and bonding several layers of boards together.

gnarl – the twisted section of tree trunk. When cut into boards, the wood displays an attractive swirl pattern and can be used for a natural finish trim or veneer.

goggles – safety protection for the eyes.

go no-go gauge – a gauge which can be used to judge if the size of an object is within an acceptable range or not. The gauge is made to the specifications of the upper and lower ends of the acceptable tolerance band and can be used without having to make a measurement. For instance, if a great number of fillet welds need to be made, a go no-go gauge can be made that will just fit over the minimum acceptable size weld, and then the gauge can be used to rapidly check weld sizes. Any weld that is undersize will show a gap between the weld and the go no-go gauge. This not only saves time, but increases the quality level of the end product by making inspections quick and accurate.

goodwill – good relations with customers, an intangible asset in a business.

Gothic arch – see major arch.

gouge – a cutting chisel with a U- or V-shaped cutting end used to make a groove in the material being chiseled.

gouging – rapidly removing excess material by any means appropriate to the material and its intended shape.

grab sample – samples taken over a short period of time, such as samples taken periodically from a newly laid pipeline to assess the quality of the water in the line.

grade – 1) the level of quality; 2) the ground level around a building site; 3) to change the natural ground slope to accommodate drainage or to lay pipe; 4) to level and prepare a surface for paving.

grade beam – a foundation beam set at grade and resting on pilings set in the soil.

grade break – the point at which there is a change in slope.

grade lath – lath that has been marked to indicate its grade.

grade level – the elevation of the ground surface at a construction site.

grade mark – a stamp put on lumber or lumber products which indicates the quality of the material.

grade, percent – see percent grade.

grade pin – a steel pin used to hold a string line along the course that a trench or other excavation is to be dug.

glue gun (2)

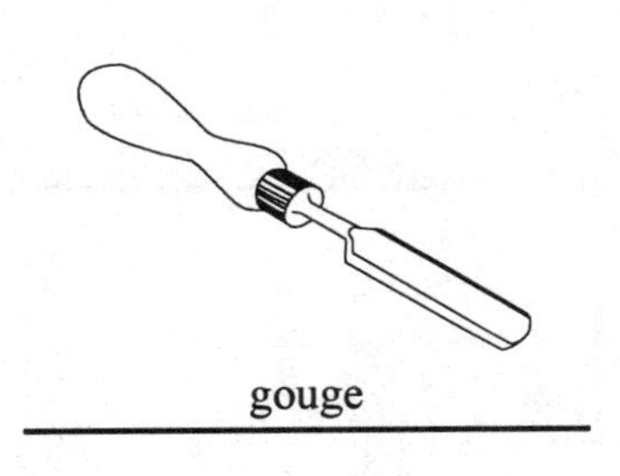
gouge

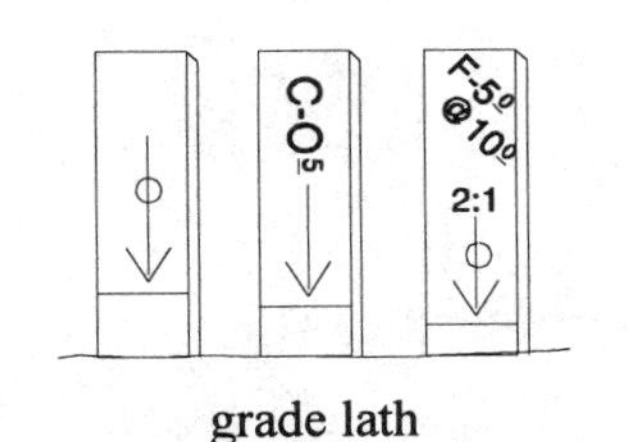

grade lath

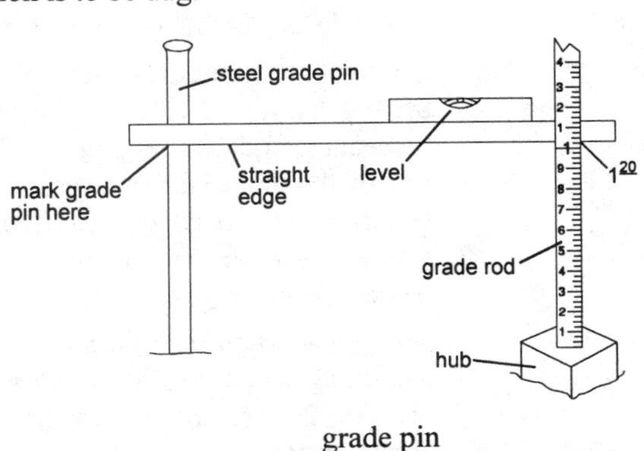

grade pin

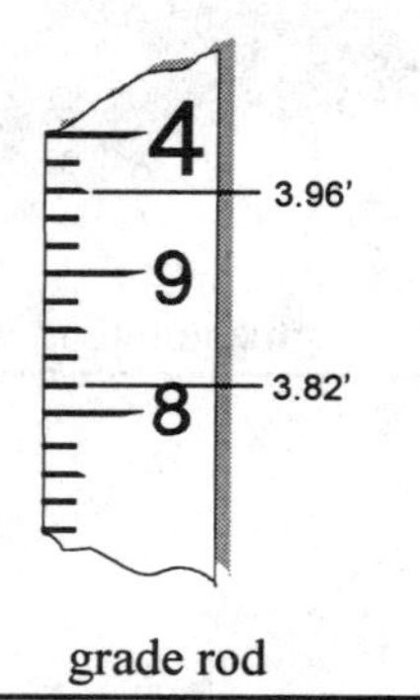

grade rod

end grain

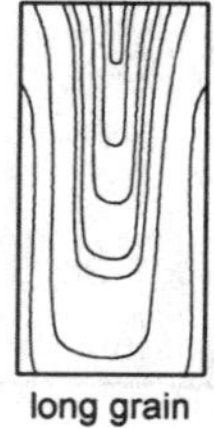
long grain

grain

grader – an earth-moving machine with a scraper blade for smoothing the ground surface.

grade ring – sections of cylinder placed on top of a manhole form to bring the top to near grade so that the manhole casting on the top will match grade.

grade rod – a measuring rod, marked in feet and decimal feet, used by surveyors. When viewed from a distance through a transit, the surveyor can determine the vertical change in grade between the transit and the point at which the grade rod is located.

grade stake – rebar stakes, cut to the proper elevation, that are set in a footing excavation where no forms are used. The tops of the grade stakes mark the correct height of the finished footing.

grade stamp – a mark or series of marks, placed on such items as lumber or steel, to indicate the classification or quality of the material.

grade stick – a stick cut to the proper length to measure the distance from an overhead structural member to the desired level of the top of a concrete surface, such as a slab being poured and leveled for the floor of a basement. The concrete is poured to the appropriate depth and the grade stick is used to check the measurement as the slab is floated. When the surface of the concrete is the same distance from the overhead structural members in several places and appears flat, it has been assured to be level. In some situations the use of a grade stick is the only practical means of determining if a pour surface is flat because there is no reference point available except the overhead members.

gradient – the amount of slope to a grade usually expressed as a percentage. The percentage is determined by dividing the rise of the grade in feet over a 100 foot length by 100. For example, a five foot rise in grade over a distance of 100 feet is a five percent gradient.

grading – 1) preparing the ground level of a building site to the proper elevation and contour; 2) determining quality levels of commodities.

grading plan – a plan view of an area showing the information necessary to shape the land for the proposed construction or use.

graduate – to mark in specific progressive increments.

graduated mortgage payment plan – a mortgage repayment plan in which the payments start lower than the normal rate for the size of the loan, and increase gradually over a period of time.

graduation – a mark showing a specific increment, such as the markings on a ruler indicating inches and fractions of inches.

grain – 1) the pattern of fibers in wood; 2) the arrangement of the lattice structure formed by the crystals created when liquid metal solidifies.

graininess – a paint defect in which the surface appears rough, as if it had sand mixed with it.

gram – a metric unit of weight equal to 15.432 grains or 0.0353 onces.

grandstand – stepped-up seating with more than three square feet of area per person.

granite – a very hard natural rock, composed of feldspar, quartz and other minerals, produced under intense heat and pressure.

granular – composed of small grains or having a grainy texture.

grape stake fence – a wood fence made up of 2-inch by 2-inch split redwood slats hung on horizontal rails.

graph – a diagram that plots one value against another and shows the relationship between them, usually over a period of time, such as days, months or years. This relationship can be shown in a variety of forms, such as a line graph, pie chart, or columns.

graphite – soft, fine carbon dust used as a lubricant in locks and other mechanisms.

grapnel – a multi-pronged hook with a line attached used for grabbing and holding like an anchor. The hook and line can be thrown over the top of a wall or other obstacle, allowing the hook to catch and take hold, thus securing the line to the top of the wall.

grate – a framework of evenly-spaced perpendicular metal bars used to cover openings, such as drains, and restrict the passage of large objects or materials into the openings.

grating – see grate.

gravel – a coarse aggregate of small rocks.

gravel roof – a built-up roof with a final covering of gravel to protect the roof covering from the elements.

gravel stop – a raised metal edging installed around a roof that is to be covered with gravel-coated built-up roofing. The edging forms a barrier against which the layer of gravel will end, preventing the gravel from rolling off the edge of the roof.

gravimeter – an instrument used to measure specific gravity.

gravity return condensate system – a return system operating solely by gravity. Steam heating systems can operate using one set of pipes with a gravity return condensate system. The water is heated in the boiler and turns to steam which rises, by gravity like warm air, through the pipes and into the radiators. When the steam cools, it condenses into water and drips back into the boiler through the same pipes it rose in.

gravity warm air heating system – a system that uses natural convection to circulate warm air through ducting on the principle that warm air, being lighter than cold air, rises.

gray paper – backing paper used on gypsum wallboard panels; the paper used on both sides of gypsum backing board panels.

grease – semi-solid lubricant.

grease interceptor – a device that traps grease from waste water and prevents it from entering the sewer system. Also called a grease trap.

grease trap – see grease interceptor.

green – uncured or unseasoned, such as wood from a freshly cut tree.

greenboard – see water resistant panels.

green concrete – concrete that has not yet set up.

Greenfield – see flexible metal conduit.

greenhouse – a structure, with controlled temperature and humidity, in which plants can be cultivated in natural sunlight admitted through glass or plastic panels to provide optimum growing conditions.

greensward – a grassy area.

grid – a pattern of regular lines laid out at right angles to one another.

grid system – a method of laying out in which a large area is divided into grids, or squares, in order to make it more manageable in size.

grillage – a steel rebar grid in a concrete foundation which spreads a load over a wide area.

grille – a grating; a protective open-mesh or lattice-like covering over an open space.

grinder – a powered device having an abrasive wheel which is used to remove surface material and sharpen tools.

grinding wheel – an abrasive wheel used to remove material or sharpen tools.

grip length – the length of the fastener, such as a nail shank, that penetrates the structural member.

grit – abrasive particles.

grit number – an indicator of the coarseness of abrasives; the higher the number, the finer the grit.

groin – a curved line that is formed where two vaulted structures intersect.

grommet – 1) a ring used as reinforcement for a hole in metal or fabric, such as the reinforcement around the holes in a tarp; 2) an insulator, such as the insulation required to protect electrical wiring as it passes through a hole in a metal plate.

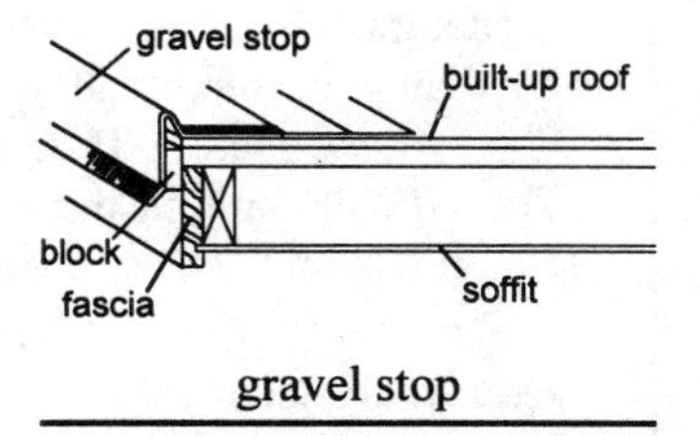

gravel stop

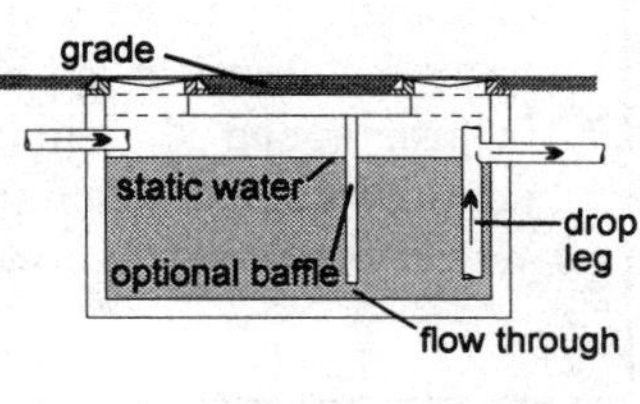

grease interceptor

Ground fault circuit interrupter (GFCI)

A circuit breaker designed to interrupt electrical current flow and protect people from electrical shock. Under normal conditions, the current in the hot wire and the neutral wire of an AC circuit are the same. However, if a fault in the circuit allows some of the current to leak to a ground, there will be a difference between them. The GFCI monitors the current flow and opens the circuit if it detects even a very small difference. It opens in about an ⅛ of a second, quick enough to prevent an accidental electrocution. The GFCI can be installed either at the panel or at the receptacle itself. Most building codes require a GFCI for all outside receptacles, and receptacles in indoor areas where people may be using electrical appliances near water, such as in bathrooms and adjacent to a kitchen sink. They are also required at construction sites where temporary electrical service is needed for power tools.

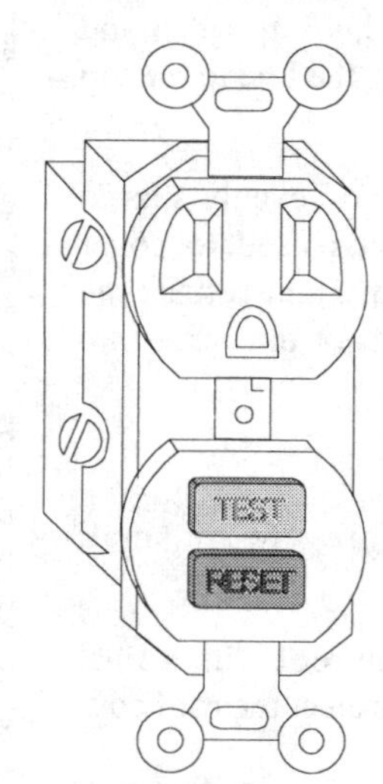

Ground fault circuit interrupter

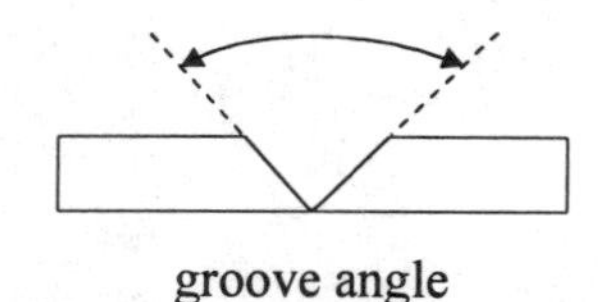
groove angle

groove – a shallow channel.

groove angle – the total combined angle created by the sides of weld preparation bevels on adjoining pieces.

grooved joint – a joint between a fence rail and a fence post in which a mitered-end rail is fit into a groove cut in the fence post.

groover – 1) a hand tool used to form an indentation or groove in wet concrete; also called a jointer; 2) a hand tool for setting sheet metal seams.

groove weld – a weld between two parts, the edges of which have been shaped so as to form a groove into which the weld metal is deposited.

gross margin – a percentage produced by dividing gross profits by total revenues.

gross profit – profit after direct costs are deducted and before expenses are deducted.

ground – the earth's surface.

ground coat – the prime coat of paint.

ground conditions – stability of the soil in a specified area.

ground cover – low-growing plants, such as ice plant, used for landscaping and/or erosion control.

ground, electrical – a conductive connection which provides a path for electrical current to pass from an electrical component into the earth.

ground fault circuit interrupter (GFCI) – a circuit breaker designed to protect people from electrical shock. *See Box.*

ground fault interrupter (GFI) – a type of circuit breaker designed to protect equipment from continued electrical current in case of a circuit fault. The current in the hot wire and the neutral wire of an AC circuit are normally the same. However, if a fault in the circuit allows some of the current to leak to a ground, there will be a difference between them. The GFI monitors the current flow and opens the circuit if it detects a difference of more than 4 to 6 milliamps. This sensing and interruption takes place in less than $1/10$ of a second, preventing damage to electrical equipment.

ground floor – the street level floor of a building.

ground glass – polished and shaped glass.

grounding clamp – a clamp which grips a metal pipe tightly, and has a slot and holding screw to secure an electrical grounding wire to the clamp.

grounding conductor – a wire that connects an electrical device to the electrical ground. Also called a ground wire.

grounding electrode system – a technique for grounding a building's electrical system by bonding together the metal underground water pipe and the metal building frame with a grounding ring and an electrode embedded in the building foundation concrete. The grounding ring should be a minimum of 20 feet in circumference and buried 2½ feet or deeper in the ground.

grounding plug adaptor – an electrical device with two prongs and a wire or metal loop for a grounding connection into which a three-prong (two conductor with ground) plug can be inserted. It serves as a means of adapting a three-prong plug to a two-slot outlet receptacle.

grounding rod – a copper-plated metal rod that is driven several feet into the earth to provide an electrical ground.

ground joint union – a pipe union with a brass grounding insert between the two halves of the union.

ground plan – the plan view of a plot of land showing the location of structures.

grounds – strips of wood, the same thickness as the lath and plaster to be applied, that are attached to walls around doors, windows and other openings to serve as plaster stops or for attaching trim.

groundwater – surface water in the ground that seeps into an excavation because of a high water table.

ground wire – a wire used in an electrical circuit to connect the circuit to the ground.

grout – a thin, fluid mortar made of a mixture of portland cement, fine aggregate, lime and water, and used for finishing mortar joints and filling voids.

grouted masonry – a multiple vertical unit masonry wall with the void between vertical rows filled with grout.

growth ring – the rings in a tree trunk that denote annual growth.

grubbing – clearing a land area of plant growth.

guarantee – to give assurance that an obligation will be fulfilled.

guard – a protective barrier.

gudgeon – a hinge pintle socket or gate hinge section consisting of a horizontal rod, secured on one end in a wall or gate post, with a hole for a vertical pin or pintle to pass through and pivot. The pin is part of an L-shaped unit, with one leg of the L connected to a gate or door. The gudgeon and pintle together make up the support and the hinging mechanism for the gate.

guide coat – a coat of the finish paint applied before the final coat.

guinea – a wood stake with a blue top used to mark grade. See also blue top.

guinea hopper – a worker who locates guinea stakes and signals to the grading machine operator to cut or fill as indicated by the marker.

gullet – the slot or depression between the teeth in a saw blade.

gum – viscous fluid from plants that swells when in contact with water and turns into a gelatin-like substance that is used in some paints.

gum arabic – a water-soluble gum used as a binder or as an adhesive in some paints.

gum spot – a concentration of gum in one spot on a board. Also called a gum streak.

gunite – sprayed-on concrete. Used in swimming pools and sometimes to reinforce an earth bank and prevent soil erosion.

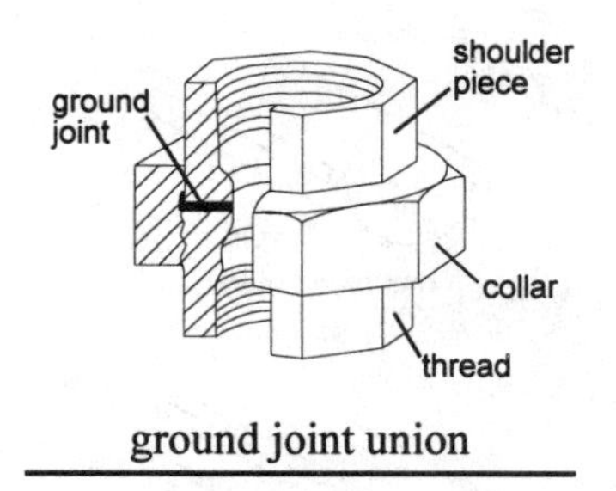

ground joint union

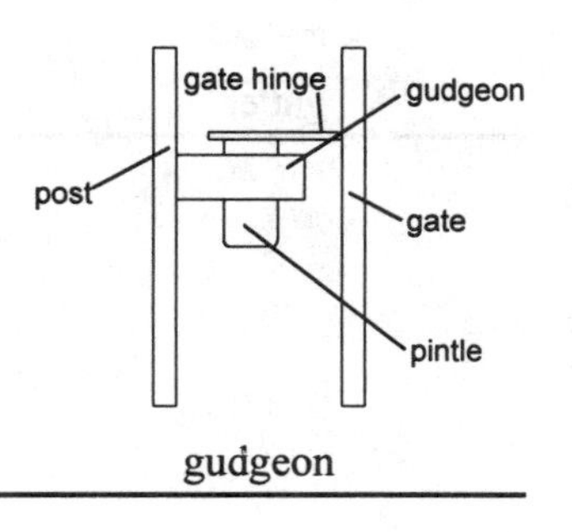

gudgeon

half-round gutter

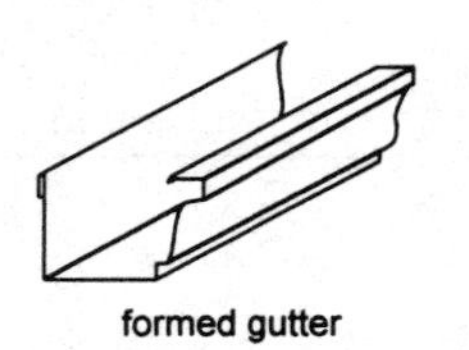
formed gutter

gutter

gunny – a coarse jute cloth material which is wetted down and used to cover concrete to prevent moisture loss during curing.

gusset – a plate of wood or metal used to fasten two structural members together and transfer the load between them.

gutter – 1) a trough at, or slightly below, the eaves of a building which catches and redirects the flow of rain water from the roof; 2) the water-gathering depression at the curb edge of a street.

guy – wire ropes placed at intervals around a structure, such as a mast, to provide lateral stability.

gymnasium – a building for athletic activities or events.

gyp board – a term for gypsum wallboard.

gypsum – hydrous calcium sulfate used as a core for drywall.

gypsum block – fire resistant gypsum building blocks that can be used in non-bearing applications.

gypsum board – wallboard that is manufactured with a gypsum core and paper or other facing.

gypsum concrete – a mix of gypsum, aggregates and water.

gypsum coreboard – gypsum panels, one-inch thick, designed for use in solid gypsum partitions. The panels have V-shaped tongue-and-groove edges.

gypsum formboard – gypsum panels deigned to be used as the forms for poured gypsum concrete roof decks.

gypsum lath – a gypsum wallboard with either a solid or a perforated face that is designed to be used as the backing for plaster.

gypsum partition blocks – two- to six-inch-thick gypsum blocks designed for use as the core material for solid gypsum partitions.

gypsum planks – gypsum roofing members with steel reinforcement in the form of galvanized wire matting cast into the planks. They are two inches thick with tongue-and-groove edges.

gypsum plaster – gypsum powder used as a base-coat plaster; plaster of paris.

gypsum sheathing – a water repellent exterior gypsum wallboard used as a base for finish exterior building siding.

gypsum wallboard – see gypsum board.

gyroscope – a rotating device that maintains equilibrium.

H: hacking to hygroscopic

hacking – stacking brick in a kiln for firing.

hacksaw – a wide, shallow, inverted U-shaped saw frame with a handle on one end and a fine-tooth blade stretched across the ends of the U. Used for cutting metal.

haft – a handle, such as for a knife or chisel.

hairline cracks – very fine cracks.

half bath – a bathroom that contains a water closet and lavatory, but no bathing facilities.

half-lap dovetail joint – a dovetail joint in which half the thickness of each member is cut away so that their surfaces are on the same plane when joined.

half-lap joint – a lap joint where half the thickness of each of the members to be joined is cut away so that the surfaces of the members are on the same plane when they have been joined.

half-life – the time required for 50 percent decay of a radioactive substance.

half round – a molding in the shape of a cylinder that has been cut in half lengthwise.

half round file – a file that has one flat face and its other face is a curved lengthwise half-section of a cylinder.

half section – see section, half.

half-timber – a type of building construction in which large wood members are left exposed on the outside walls and the wall surface between them is covered with another material, such as stucco.

halogen – any of the elements astatine, bromine, chlorine, fluorine, or iodine. They have varied uses. For example, fluorine, in the form of fluorocarbon compounds, is used as a refrigerant, lubricant, and in fire extinguishers; bromine is used in fumigants, dyes, and flameproofing; and chlorine is used for disinfecting and as a bleach.

hammer – a hand tool for striking, consisting of a head attached at a right angle to a handle. The head may be of various configurations to be used in different applications. For example, a carpenter's hammer has a two-sided head, one side for driving nails and the other side a claw, which may be straight or curved, for pulling nails. See also listings for individual hammer types.

hammer fracture – a tear in the facing paper of gypsum wallboard caused by overdriving a nail.

hand drill – a manually operated drill driven by turning a crank which is geared to rotate the drill bit at a faster speed than the crank turns.

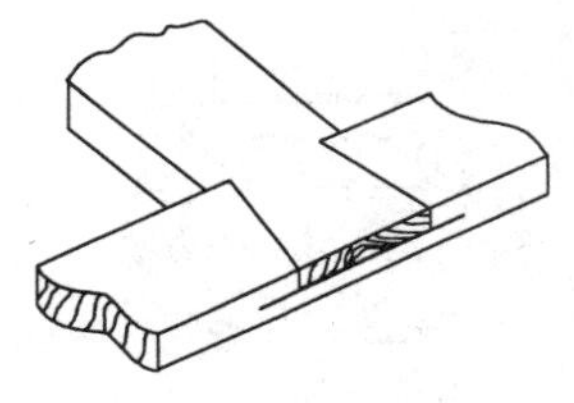

half-lap dovetail joint

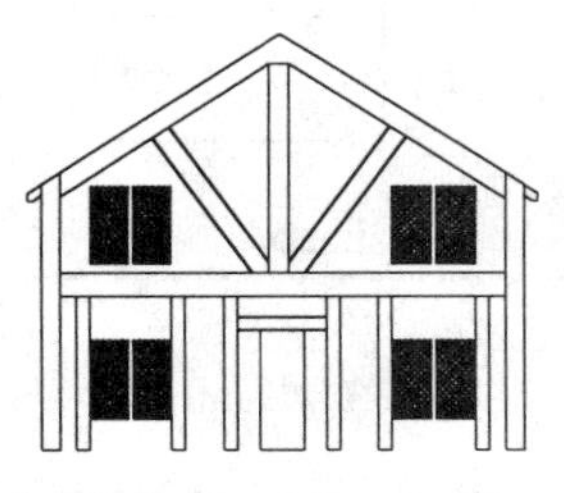

half-timber construction

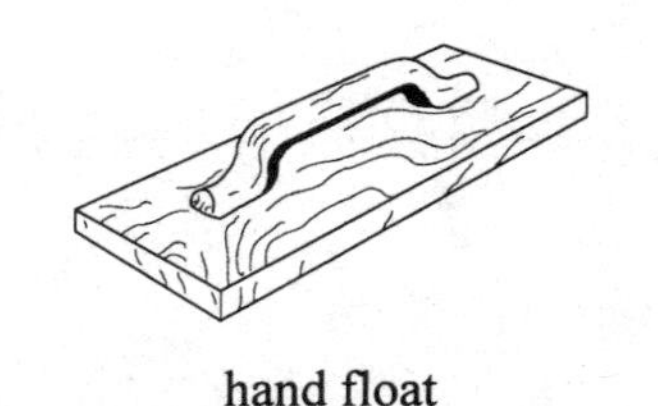
hand float

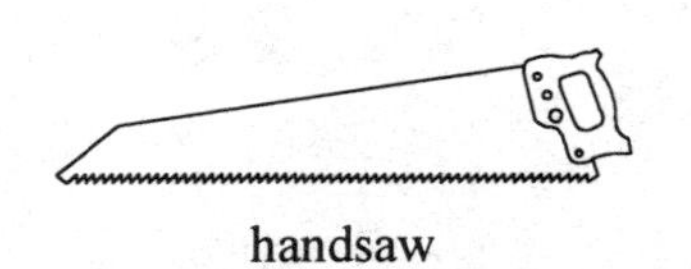
handsaw

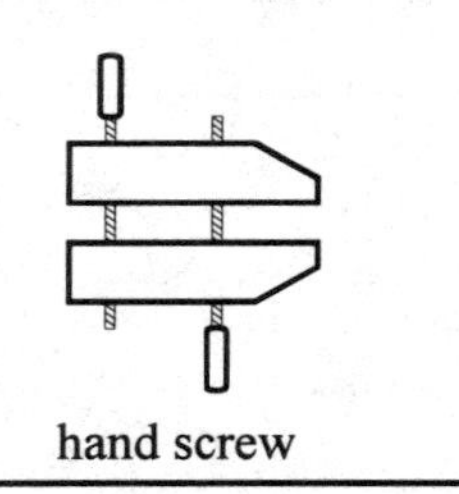
hand screw

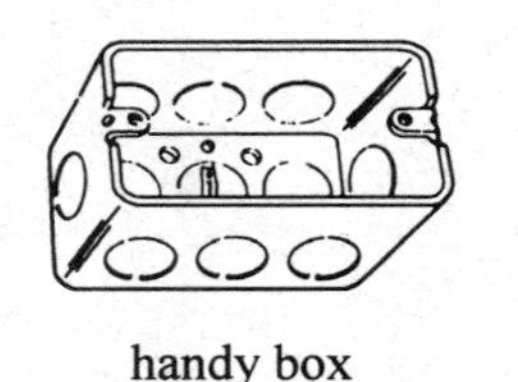
handy box

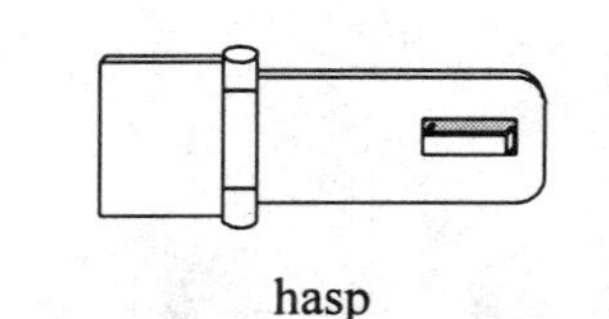
hasp

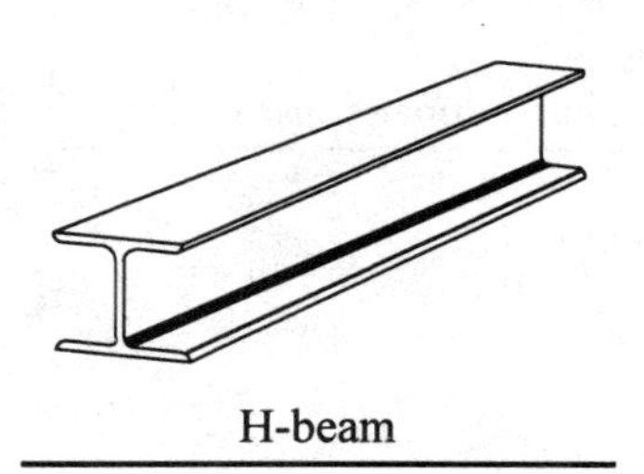
H-beam

hand float – a flat wood board with a handle meant to be gripped by one hand. It is used to smooth and level the surface of wet concrete or plaster.

handrail – a rail at about waist height that parallels the staircase stringers and is used as a guide and handhold while using the staircase.

handsaw – a manually powered saw. The term is generally used to refer to a manual wood saw.

hand screw – a clamp held together and adjusted by two screws.

hand tool – a tool that is manually operated rather than power driven.

handy box – a single electrical outlet or switch box.

hanging scaffold – a platform, suspended from a structure by cables, that provides access for work or maintenance to be performed on the exterior of multi-story buildings. The scaffold can be lowered and raised by manual cranks or electric motors.

hardboard – panels or sheet material made from wood fibers compressed under heat and pressure to a density of 31 pounds per cubic foot or more.

hard burned – clay masonry that has been fired at a high temperature. Hard burning makes brick or other masonry stronger and more durable.

hard edge – a strong section of the gypsum drywall core along the paper-bound edge which is designed to resist damage.

hardener – 1) any additive that makes a compound, such as concrete, set up harder; 2) the catalyst for epoxy resin.

hardfacing – a wear-resistant coating of one metal applied to a different and softer base metal. It is applied by weld overlay or by inserting a section of the hardfacing material into a recess in the softer metal. Hardfacing is commonly used on the disks and valve seating surfaces for valves in high pressure steam service. Hardened materials such as type 410 stainless steel or Stellite, which is a cobalt-chrome-tungsten alloy, are used in these applications.

hard hat – protective head gear that consists of a hard shell with a shock-absorbing webbing suspended inside which cushions blows to the head.

hard money – a term used to denote the actual cash exchanged in a loan.

hardness – a measure of the resistance of a material to dents or scratches.

hard oil finish – an interior varnish that dries hard to a moderate or high sheen.

hardpan – rock or very hard soil.

hard surfacing – see hardfacing.

hardware – hinges, nails, door knobs, latches and other small metal fasteners or accessory items used in construction.

hard water – water with a high mineral content.

hardwoods – wood produced from broad leaf, deciduous trees such as oak, walnut, ash, maple or birch.

hasp – a hinged closure with a slot, through which a U-shaped staple can be passed and a lock attached.

hatch – a removable cover over an opening, such as the entry to a cellar or the hold of a ship.

hatchet – a hand tool with a blade on one end of the head and a nail-driving surface on the other end of the head.

haul – to move or transport a load.

haul road – an access road to a work site.

haunch – a thickened section of a structural member.

hawk – a flat square with a handle on the bottom that is used to hold and carry mortar.

hazardous electrical location classifications – the National Electrical Code has established hazard classifications, including: Class I — areas which have flammable gases or vapors in explosive or burnable quantities; Class II — areas with combustible dust; Class III — areas with easily-ignitible fibers or particles.

hazing – a defective paint finish in which the surface appears cloudy or smokey.

H-beam – a structural member, usually of steel, with a cross section resembling the shape and proportions of the letter H.

head casing – the top framework for a window.

header – 1) a structural member running perpendicular to the studs that spans an opening for a door, window or stairway and adds support between studs. *See Box*; 2) a short section of brick or a brick laid so the end is to the wall surface; 3) a masonry unit that ties together different wythes. Also called a bonder.

header, blind – see blind header.

header, clipped – see clipped header.

header course – see heading course.

header, flare – see flare header.

heading course – a bond course of brick that is set in a wall so that the end of the brick faces outward. Also called a header course.

head jamb – the upper horizontal finish member in a window frame. Also called a yoke.

head joint – the vertical joint between adjacent masonry units. Also called a cross joint.

headland – a section of land projecting into a body of water.

head plate – see wall plate.

head pressure – the maximum pressure against which a pump can move fluid.

headroom – 1) the distance from a stair tread to the lowest overhead point directly above the tread; 2) any measure of vertical clearance.

headstock – the portion of a lathe that holds one end of the work and turns the work.

hearth – a fireproof section of flooring extending out from a fireplace opening. It can be raised or flush with the rest of the floor.

hearthstone – 1) a stone used for a fireplace hearth; 2) a soft stone used to clean a fireplace hearth.

heartshake – splitting that occurs across the growth rings of a log where rot exists at the center of the log.

heartwood – wood from the center of the tree to the sapwood, where the cells have hardened. Heartwood is preferred over sapwood because it is stronger. This is particularly true with a wood like redwood, which is used specifically for its durable qualities. Redwood heartwood is more rot and insect resistant than the sapwood.

heat-absorbing glass – glass which contains iron to absorb heat. This type of glass is used in windows and skylights where there is a need to reduce the heat transmitted through the glass.

heat-affected zone – the volume of base metal affected by the heat, but not melted, during soldering, brazing, cutting, or welding. These operations must be closely controlled so that the heat generated does not have a detrimental affect on the area surrounding the activity. Means of control include limiting the preheat, postheat and process time as well as limiting the temperature.

heat collector – see solar collector.

heater, baseboard – a heating unit with a low profile that fits along the baseboard.

heat exchanger – a device for transferring heat from one fluid to another in cooling systems. Heat exchangers are used in many familiar applications. Air conditioners use two different heat exchangers, a condenser and an evaporator. The condenser is located at the discharge of the compressor. It removes the heat of compression from the refrigerant by air flow across the unit. In the evaporator, the refrigerant is expanded to absorb the heat from the surrounding air. Steam and hot water radiators are also heat exchangers, even though they are used to produce heat. They impart heat from the steam or fluid contained in the system to the air in the space being heated. Nonresidential heat exchangers include condensers that condense steam exhausted from a steam engine into water, or radiators used to cool internal combustion engines.

heat gun – a device that converts electricity to heat and uses a blower to aim that heat at a surface to accelerate drying or to soften paint for removal.

heating cable – a cable, embedded in the plaster of a ceiling, which generates heat in a room by means of electrical resistance.

heartshake

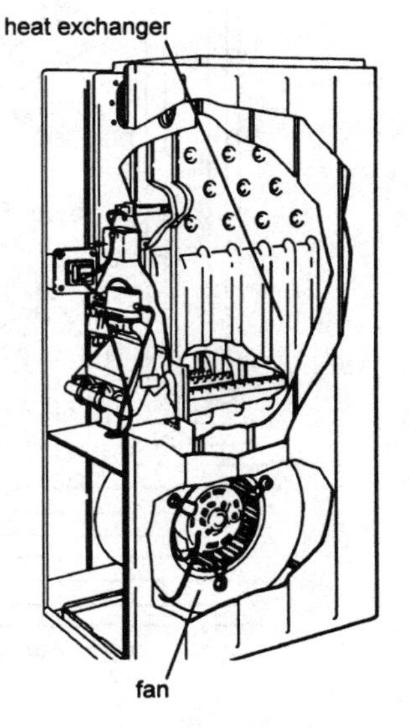

heat exchanger

Header

A header is a horizontal structural member, or a structural member placed perpendicular to studs, joists or rafters, to reinforce openings. The header boxes in the opening and adds continuity for the load transfer around the opening, compensating for the material removed. Headers are used to transfer upper structural loads to the studs alongside doors, windows, closets or fireplaces. They are also placed as reinforcing joists at right angles to floor or ceiling joists that have been cut for an opening, such as for a stairway or attic access. For normal light-frame construction, headers may be made of the same type of lumber used for the floor joists. For other than light-frame construction, specially-designed headers may be needed. As the size of the opening increases, the depth of the header must increase as well. Wide openings for bays or porches may need trussed headers. Headers may be made of wood, steel, or other structural shapes or a combination of these. Metal is more expensive, but stronger for a given dimension. If a design calls for a window that extends very near the ceiling in a load-bearing wall, a metal header will take up less space than a wood header. The metal can be drilled and sections of wood screwed to it to provide a nailing surface.

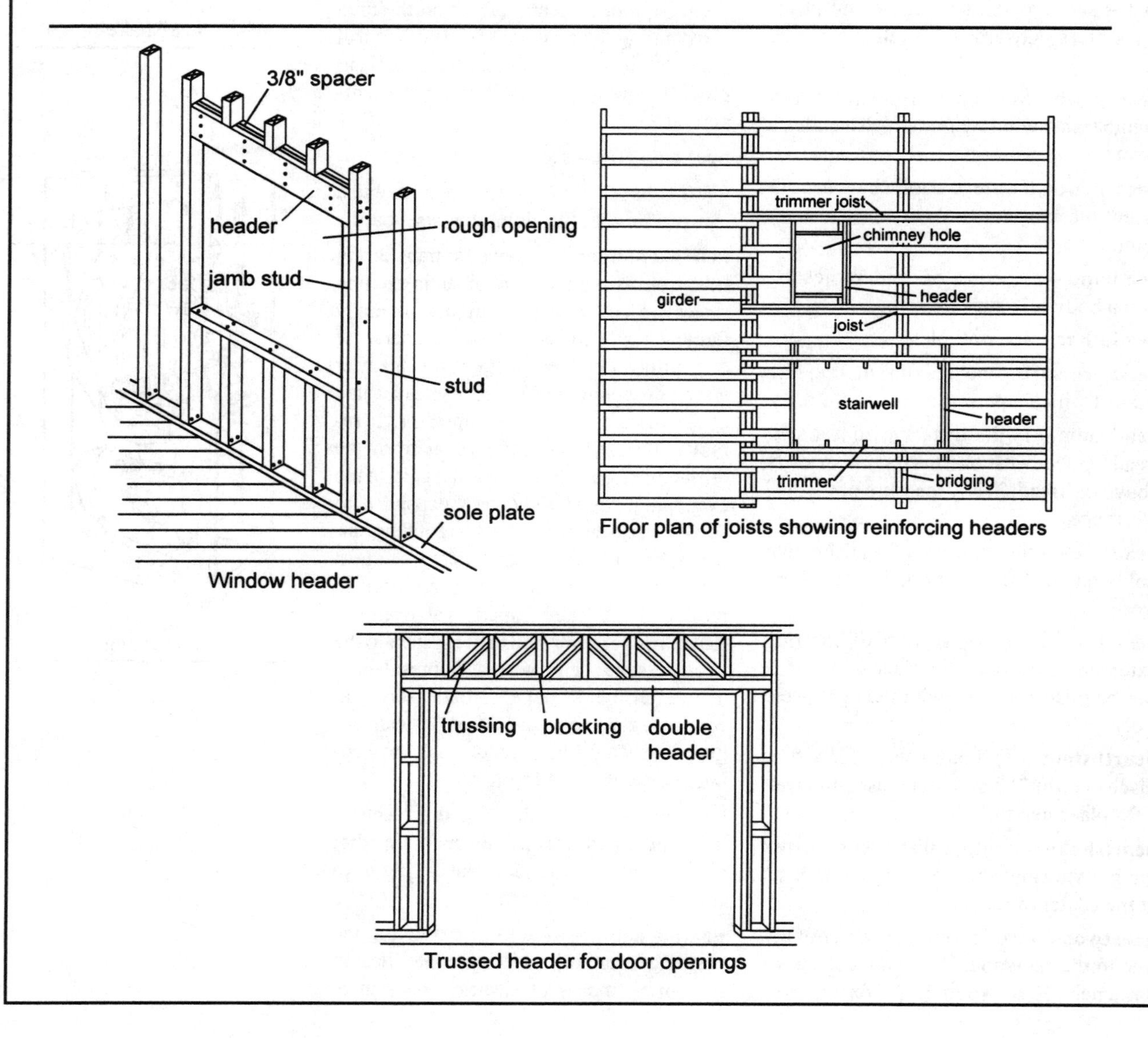

Window header

Floor plan of joists showing reinforcing headers

Trussed header for door openings

Resistance heating occurs when an electric current is passed through a conductor that has a high resistance to the flow of the current. This resistance converts some of the electrical energy into heat.

heating, ventilating and air conditioning (HVAC) – the system and process of regulating temperatures and atmosphere inside residential or commercial buildings.

heating, ventilating and air conditioning (HVAC) duct – large tubular piping fabricated from sheet metal (usually galvanized steel), vinyl or fiberglass. It is used in venting systems to circulate fresh air or to carry warm or cold air from heating or air conditioning units to room outlets in homes and buildings. The ducting comes in round, rectangular or square shapes with a variety of fittings to connect to heating, ventilating and air conditioning equipment.

heat of fusion – the amount of heat required to take a particular solid to its melting point. Tin, for example, becomes liquid at 450 degrees Fahrenheit, gold at about 2,000 degrees F and iron at about 2,730 degrees F.

heat of vaporization – the amount of heat required to turn a liquid into a gas.

heat pump – a device that can heat, or by changing its cycle, cool a space using a refrigerant, a compressor, and heat exchangers.

heat sink – a metal shape with good heat conductivity used to draw heat away from an area or component. Heat sinks are used widely in electronics to keep heat from sensitive circuit parts during soldering.

heat-strengthened glass – glass to which a ceramic glaze has been applied on one side, approximately doubling its strength.

heat transfer – the movement of heat from one place, fluid, or object to another.

heat treating – using controlled heat and cooling rates to enhance a material.

heat welding – 1) a process using heat to seal roofing membranes together; 2) a heat process for fusing plastic pipe and fittings together.

heave – 1) to cause to be lifted upward; 2) a displacement in the earth.

heavy timber construction – construction in which the main structural members are heavy timbers. The timbers are connected with bolting and metal plates at their intersections. Studs added to form partitions are non-load bearing for both interior and exterior walls, since all the structural load is carried by the heavy timbers.

heavyweight concrete – concrete made using dense, heavy aggregate. It is used for radiation shielding. The approximate weight is 400 pounds per cubic foot, compared to regular concrete which is approximately 150 pounds per cubic foot.

hectare – a metric unit of land measure equal to 10,000 square meters, or 2.47 acres.

hectogram – a metric unit of weight equal to 100 grams, or 3.527 onces.

hectoliter – a metric unit of volume equal to 100 liters, or 26.4 gallons.

hedge – a row of upright close-growing plants used as a border or barrier.

hedgerow – a fence of close-growing plants.

heel – 1) the outside corner of a framing square; 2) the juncture of paintbrush bristles and the handle; 3) the outside of a bend in pipe or conduit; 4) the bottom of a metal lathe cutting tool; 5) the bearing point of a joist or rafter.

heel hardened – an accumulation of dried paint at the heel or base of a paintbrush where the bristles join the handle.

heel post – the gatepost with the gate hinges.

height above plate (HAP) – the vertical distance from the outside corner of the rafter plate to the top surface of the rafter.

helical – spiral shape or form.

helical gears – gears with teeth cut in a spiral around their axes.

helicoid – spiral.

helmet – 1) a protective head covering; 2) a protective shield for the face and neck used during welding.

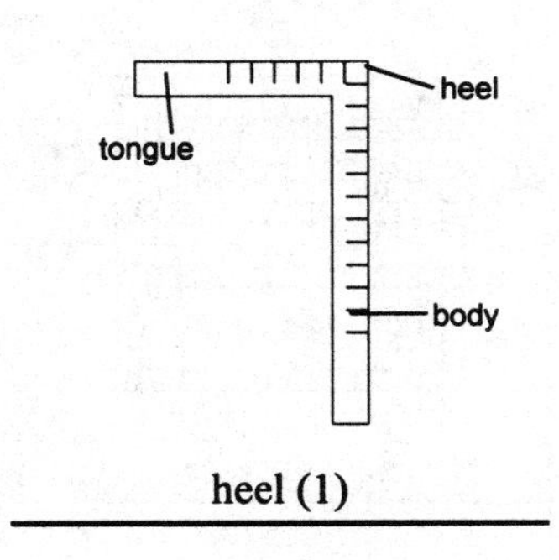

heel (1)

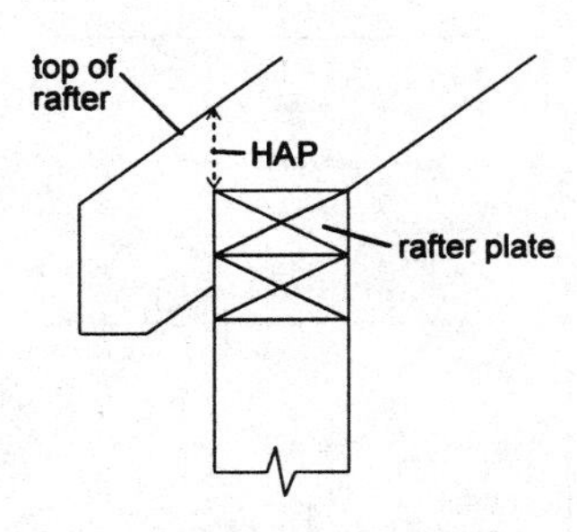

height above plate (HAP)

herringbone

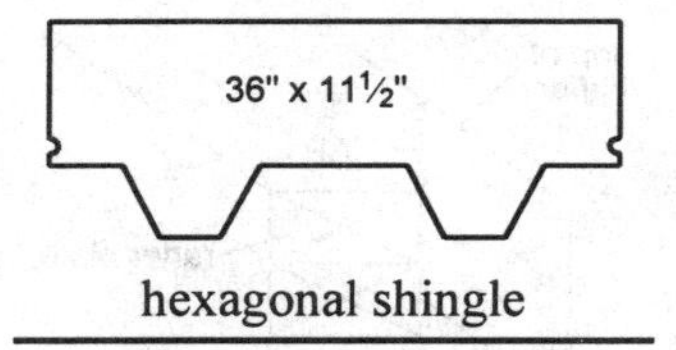

hexagonal shingle

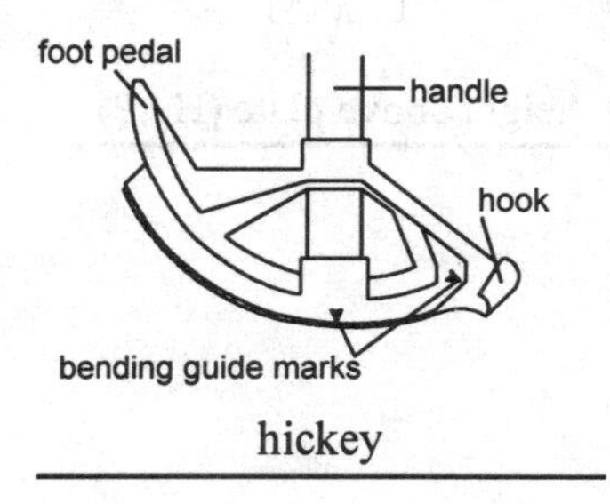

hickey

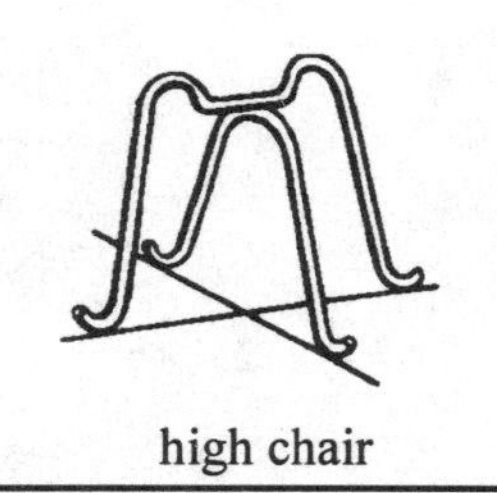
high chair

hemlock, west coast – a straight grain, long fiber wood that is light weight, a good nail holder, and free of gum and pitch. It is a good construction wood that gets harder with age, but does not darken. Of the softwoods, hemlock is most like the hardwoods in its qualities, such as finishing, color, texture and aging.

henry – a unit of electrical inductance.

heptagon – a plane figure with seven sides and angles.

heptahedron – a solid figure with seven surfaces.

hermaphrodite caliper – a caliper that has one straight leg and one leg with the end bent inward. This caliper is used to scribe lines parallel to a surface and for locating the center of a circle.

hermetic – sealed air-tight.

herringbone – a pattern in which bricks or other rectangular shapes are laid at alternating right angles to one another to form series of chevrons.

herringbone gears – cylindrical gears with teeth in a herringbone pattern.

Hertz (Hz) – cycles per second in electricity and acoustics.

hewn – wood that has been worked or shaped with a hand tool, such as an axe or adze.

hexagon – a plane figure with six sides and six angles.

hexagonal shingle – a type of interlocking strip asphalt shingle that is 36 inches long by 11½ inches wide with two half-hexagonal tabs.

hexahedron – a solid shape with six surfaces.

hickey – a tool used for bending pipe and conduit that grips the outside diameter of the pipe, and has an integral curved shoe that shapes the inside radius of the bend being made.

hidden lines – dashed lines on a drawing that depict a feature that is not visible, such as a hole bored through a part perpendicular to the viewer's line of sight.

hide glue – a non-waterproof glue, made from animal hides, that is very strong even with joints that are not well fitted. It is available in flake form, which is heated for use, or in liquid form.

hiding power – the ability of a paint to cover.

high centered – a condition where a tracked or wheeled vehicle has sunk so far into the soil that the vehicle chassis is resting on the ground, preventing further movement.

high chair – a metal support used to hold reinforcing steel horizontally in place while concrete is poured and cured. The high chairs prevent the concrete from pushing the reinforcing to the bottom. Also called bolster.

high early cement – a portland cement that sets more quickly than standard cement. It is used where the immediate strength of concrete is needed.

high frequency resistance welding (HFRW) – a process using resistance heating with high frequency AC (10 to 500 kHz) and the application of force to join the pieces after heating. Resistance heating occurs when an electric current is passed through a conductor that has a high resistance to the flow of the current. This resistance converts some of the electrical energy into heat.

high joint – an installation flaw in which a joint in gypsum wallboard is raised higher than the rest of the wallboard surface.

high lift grout method – installation of a masonry wall to the top of one story prior to pumping the grout in to fill the voids in the wall.

high rise – a multi-story building usually equipped with elevators. The number of stories that determines if a building is classified as a "high rise" is determined by local codes.

high strength bolts – bolts made of a steel alloy with high tensile strength.

high-tension – high voltage electricity.

high water mark – the highest recorded level that the water in a river, stream, lake or other body of water has ever reached.

highway – a main or direct public road.

hinge – a pivoting fastener used to attach a door to the frame or molding of the door opening and provide support and a pivot point for the door. There are many types of hinges, such as full and half mortise, full and half surface and thrust pivot.

hinge, butt – a pivoting fastener consisting of two rectangular metal plates joined to alternating barrel sections and held together by a pin.

hinge, concealed – a pivoting fastener that is not visible when the door which it supports is closed.

hinged pipe vise – see yoke vise.

hinge gain – the recess in a door edge or jamb for a hinge plate so that the hinge plate sits flush with the surrounding surface. See also gain.

hinge hasp – a hinged steel plate with a slot which fits over a staple so that it can be secured by a pin or padlock.

hinge, loose pin – a hinge held together with a removable pin.

hinge point – the location where a slope stops and a road grade starts. Also called a catch point.

hip – the external angle formed where two adjacent roof slopes rise and meet.

hip rafter, backing – chamfering or beveling the top edge of a hip rafter to the slopes of the adjoining sections of roof. The top of the hip rafter is cut at an angle along its length, sloping downward from a centerline on its top edge. These angled slopes match the slopes of the adjoining sections of roof.

hip roof – a style of roof which slopes on the ends as well as the sides, so that the eave line formed is constant on all walls.

hod – a trough, mounted on a perpendicular handle and balanced on a workman's shoulder, used to carry bricks and mortar on the job.

Hole saw

A hole saw is a small cylindrical saw blade that can be attached to a power drill to cut holes. It has an arbor or shaft extending through its center with a drill bit on the end. The drill bit contacts the work first and acts as a guide for the saw. The base of the arbor is held in a drill motor chuck which provides torque to the saw. The hole saw is used to cut circular sections from a material with less work than would be required of a drill bit. The hole saw cuts out or cuts around a plug of material rather than grinding out the entire width of the hole. The depth of the hole is limited by the depth of the hole saw. However, part or all of the plug can be removed periodically with a chisel in order to increase the cutting depth.

hog valley – a roof valley between two roofs sloping toward each other that is partially filled and sloped so that runoff water from the two roof slopes is directed toward a gutter.

hoist – a mechanism built to raise or lower a load.

holding grade – keeping the bottom of an excavation at the proper depth.

hole – an opening through an object, or a hollow or cavity in a surface.

hole gauge – a gauge for measuring the diameter of a small hole.

hole punch – a punch that removes a circular section of material leaving a hole in the remaining material.

hole saw – a cylindrical saw attachment used with a power drill. *See Box.*

holiday – a gap in a finish surface coating; a place that was missed when the surface coating was applied.

hollow – an empty space inside an object.

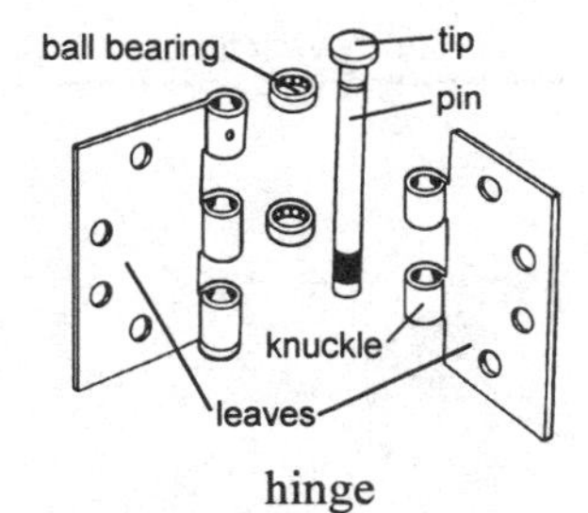

hinge

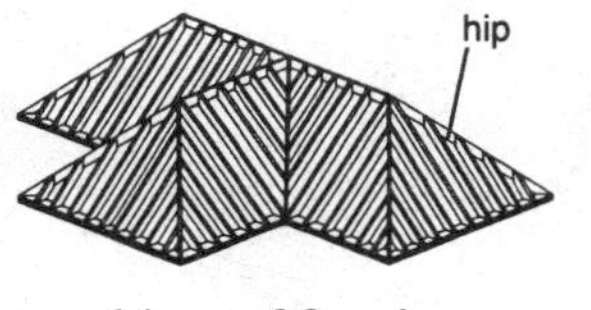

hip roof framing

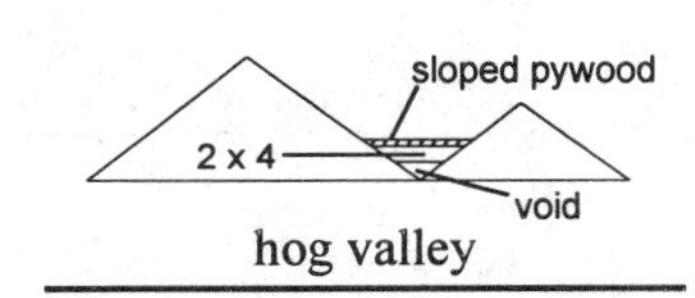

hog valley

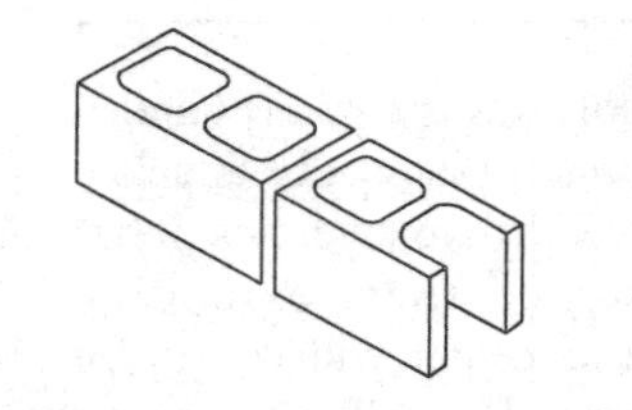
hollow concrete masonry unit

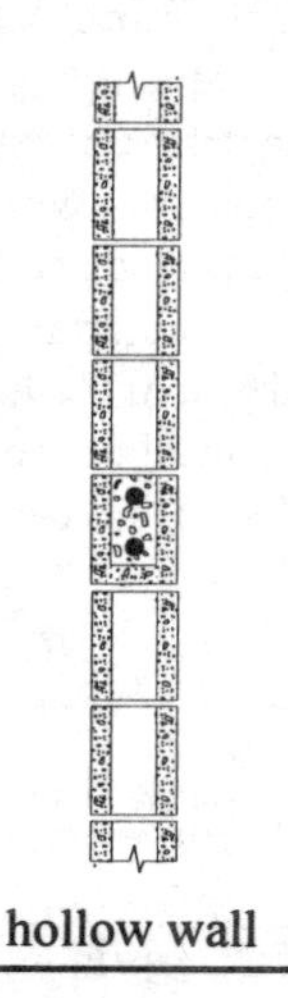
hollow wall

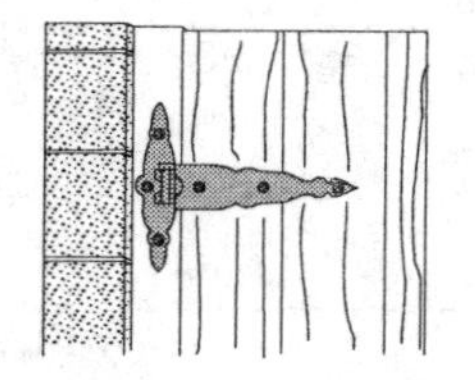
hook and eye T-strap hinge

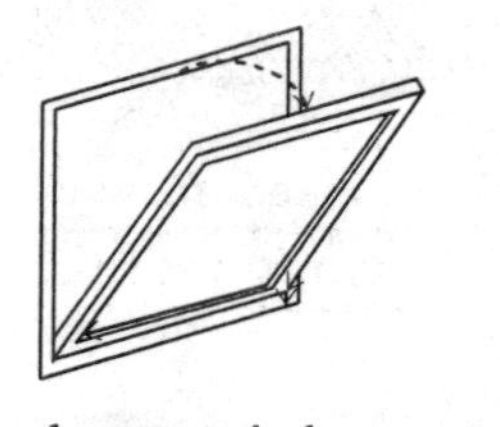
hopper window

hollow brick – a brick manufactured with one or more large holes through its core, showing an opening from side to side. The solid area remaining on the sides of the brick is less than 75 percent of the total surface area of those sides.

hollow concrete masonry unit – a concrete masonry unit manufactured with one or more large holes through its core, showing an opening from side to side. The solid area remaining on the sides of the brick is less than 75 percent of the total surface area of those sides.

hollow-core – a method of construction in which a framework, as for a door, is covered with thin plywood which may be supported by pieces of lightweight material in the center of the framework.

hollow ground – a method of grinding a circular saw blade, or other blade, so that the blade tapers from a thicker cutting edge to a thinner section toward the blade hub so that the blade doesn't bind in the material being cut. Also called taper ground.

hollow masonry – single-withe construction using concrete masonry units with the voids on the ends of the blocks filled with grout as each course is laid.

hollow wall – a masonry wall with an air space within the wall.

homogeneous – composed of parts that are all of the same type of material.

homogeneous beam – a structural beam composed completely of one material.

hone – 1) a very fine grinding stone; 2) to grind to a smooth surface or a sharp edge.

honed finish limestone – see limestone, honed finish.

honeycomb – 1) a concrete pour that does not completely fill the form, and is not totally consolidated by some means before it sets, leaving it weaker than it should be; 2) a lightweight, rigid structural material, whose cross-section resembles a honeycomb, used in aircraft/aerospace applications and in the manufacturing of products in which a high rigidity-to-weight ratio is needed or desired.

hood – a metal canopy with a ventilating exhaust fan that collects and traps fumes.

hook – the curved end member of a crane or hoist which connects to the load via rigging. A hook is roughly in the shape of the letter "U" with one leg longer, and with a means to attach it to the lifting cable.

hook and eye hinge – a hinge commonly used for gates in chain link fences. It consists of a strap around the gate post and a strap around the gate, joined together by a pin.

hook and eye T-strap hinge – a pinned-type door hinge that mounts on the exposed surface.

hook latch – a bridge across the opening of a hook which closes the opening and prevents a load from becoming unhooked.

hook rule – a ruler with an overhanging projection on one end which can be "hooked" onto the edge of something to be measured. This is a convenience to ensure that the end of the ruler remains at the proper point.

hook stick – a pole with a hook on the end used to turn an overhead busway switch on or off. Busways are often mounted overhead, and the switch to turn electrical equipment on and off is not always accessible unaided from the floor.

hook travel – see lift.

hoop – a circular ring used to hold barrel staves together and to strengthen and hold cylindrical wood water tanks together.

hopper window – a window with a sash that is hinged at the bottom and swings inward at the top.

horizontal – on a level plane with respect to the earth.

horizontal application – the installation of gypsum wallboard with the long edge perpendicular to the framing members and parallel to the floor.

horizontal branch – a lateral drain pipe from fixture branches to a soil or waste stack in a building drain system.

horizontal position – a welding position in which the work is approximately horizontal and the weld is on the top surface.

horizontal roof area – the area, taken in a flat plane, disregarding roof slopes, that is covered by a roof.

horizontal slat fence – a fence made of slats installed horizontally between posts.

horizontal sliding window – window with sashes that slide horizontally in tracks.

horizontal supply tapping – a water supply line for a boiler that is connected horizontally.

horizontal turret lathe – see turret lathe.

horsepower – a unit of power that equals 550 foot-pounds of work per second.

horsepower-rated switch – an electrical switch for use with small motor loads.

hose bibb – a faucet with a threaded outlet to which a hose can be connected.

hospital latch – a latch with push-pull levers which retract the bolt.

hot-dip galvanizing – a process in which the material to be galvanized, such as a nail, is dipped in molten zinc.

hot-melt glue – glue in the form of sticks which melt in the presence of heat. The glue is fed through an electrically-heated nozzle that melts it and directs its application. The glue sets rapidly as it cools.

hot pressure welding (HPW) – a welding process using heat and pressure, either in a vacuum or with a shielding of inert gas.

hot spray – spray painting with paint that has been heated. The paint flows more smoothly and adheres better to the surface when applied hot.

hot wall – a plastered wall that has a lot of free lime which is detrimental to oil-based paint.

hot wire – the high voltage electrical conductor in an electrical circuit.

hot-wire anemometer – see anemometer, hot-wire.

hot wire welding – an electric arc welding process using filler metal heated by resistance. Resistance heating occurs when an electric current is passed through a conductor that has a high resistance to the flow of the current. This resistance converts some of the electrical energy to heat.

housed – 1) enclosed; 2) stored.

house paint – exterior paint.

house rack – see service insulators.

housing – a protective enclosure.

hub – 1) the center of a circle; 2) the center of a wheel.

hub and spigot piping – see bell and spigot piping.

hub stake – a small stake driven in the ground to mark a point.

hue – a color.

Humboldt undercut – a common tree felling method used with large diameter trees. *See Box.*

Humboldt undercut

The humboldt undercut is a common method of felling large diameter trees using the natural lean of the tree. A horizontal face cut is made about a quarter of the way through the tree trunk, followed by an upward angled cut just below, which intersects the first cut. Removing this chock of wood unbalances the tree. A second horizontal cut is made on the back side of the trunk, just opposite, but slightly higher than the face cut. Using a wedge in the back cut, the tree is forced past its center of balance and falls forward.

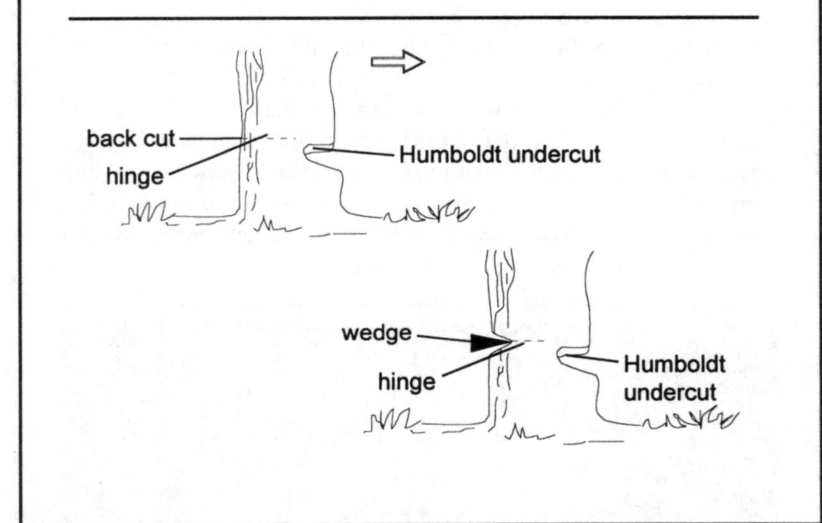

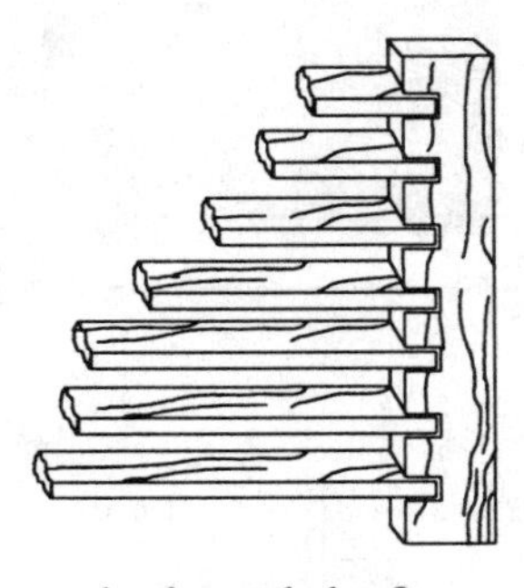

horizontal slat fence

Hydraulic jack

A jack that is operated by hydraulic pressure. It consists of a large piston in a cylinder and a small piston in a cylinder, each enclosed in a system containing noncompressible fluid. The cylinders are connected and have a one-way valve between them. Pressure applied to the small piston tries to compress the fluid. Since the fluid is noncompressible, the fluid transfers that pressure to the large piston. There is a mechanical advantage between the pistons because they are of different size. If the area difference between them is 4 to 1, the force applied to the small piston will be multiplied 4 times on the large piston. So a small amount of force applied to the small piston can be used to lift a large weight with the large piston. The one-way valve prevents the fluid from flowing back to the small cylinder. Another valve that bypasses the one-way valve can be opened to let fluid flow out of the large cylinder, allowing the large piston to be lowered. Hydraulic jacks are used to jack up building structural members for repairs or replacement of studs or other support members. They are also used in excavation shoring to hold the shore planks against the trench wall.

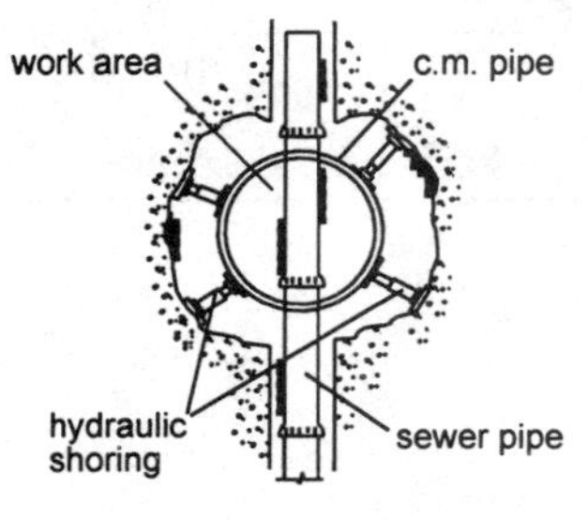

hydraulic shoring

humidifier – a device for adding humidity to indoor air, usually by simply heating and vaporizing water. It is used to bring the moisture in the air to a comfortable level for breathing.

humidity – the moisture in the air.

humidity, relative – a ratio of the moisture present in air to the maximum possible moisture content of air at the same pressure-temperature conditions.

humidity, specific – the weight of moisture per unit weight of dry air.

hundredweight – 100 pounds American weight, 112 pounds British weight.

HVAC – see heating, ventilating and air conditioning.

hydrant – a standpipe, set in the ground and supplied from a fire main or other water main, with connections for fire hoses and a valve to start and stop the flow of water.

hydrated lime – quicklime (calcium oxide) that has been mixed with water before packaging (factory-slaked).

hydraulic – 1) a machine or system that is powered by applying pressure to a noncompressible fluid; 2) the process of cement hardening in the presence of water.

hydraulic control – a machine controlled by a fluid pressure system.

hydraulic jack – a jack that functions by hydraulic pressure. *See Box.*

hydraulic pressure – pressure within a system that contains noncompressible fluid.

hydraulic shoring – excavation shoring that is made up of hydraulic jacks between shoring shoes that are placed against the sides of the excavation and held in place by the jacks.

hydrocarbon – a substance that is composed of hydrogen and carbon, such as gasoline, kerosene, plastics and other petroleum products.

hydronic – a forced hot water system.

hydropower – power generated by water flow.

hydrostatic design basis – the long-term hoop stress rating of plastic pipe, in psi, at a specified temperature. Hoop stress is a force that is exerted in an outward direction on an item with a circular cross section, such as pipe. It is the measurement of the item's resistance to pressure from the inside.

hydrostatic test – a pressure test of a system or component using water or other noncompressible liquid under pressure. The test pressure is equal to, or greater than, the maximum operating pressure that will be seen in service, depending upon the specification requirements. Many codes require a hydrostatic test be done at 1½ times the normal system operating pressure to ensure that the system is safe and leak-tight. Some systems are tested only before initial operation. Other systems of a more critical nature, such as steam boilers, are hydrostatically tested periodically to ensure the system has not deteriorated during service. If a system that contains a fluid other than water is hydrostatically tested using water, it will usually have to be thoroughly dried before returning to use.

hydrous – containing water.

hygroscopic – a material that takes up and retains moisture.

I: I-beam to isothermal

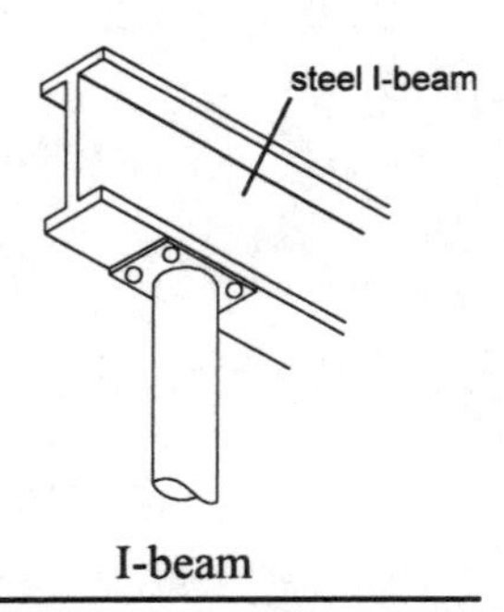

I-beam

I-beam – a structural steel member with narrow flanges and a cross section resembling the letter I. I-beams come in different values of weight per foot and width, the combination of which specifies the strength of the beam. Once the required load capacity is calculated and the length of the necessary beam is known, the next largest combination of length, width and weight that is available can be selected for the job. The I-beam should always be slightly larger than the calculated size to ensure that it will be strong enough for the weight it must support. I-beams are used for long spans, such as structural members in large buildings, or over wide wall openings, such as a double garage door opening.

idle – 1) not in use or not working; 2) a gear that transmits power but is not driven by or driving a shaft.

igneous rock – rock formed from the solidification of molten lava. Basalt and granite are two forms of igneous rock that are sometimes used in construction.

ignite – to start something burning.

illuminate – to shine light upon.

impact noise rating – a rating system that provides an indication of the transfer of noise through a floor and ceiling by such actions as moving furniture or simply walking.

impact resistance – the ability to resist damage by a sudden force. For example, impact-resistant steel can resist damage by impact during cold temperatures that would crack or fracture another steel.

impact wrench – a powered wrench that drives a socket which fits over a nut or the head of a bolt and applies an adjustable amount of torque to the fastener in rotary impact blows.

impedance – the total opposition to the flow of current in an AC electrical circuit. It is equal to the square root of the sum of the squares of resistance and reactance in the circuit. The flow of current in an AC circuit is equal to the circuit voltage divided by the circuit impedance.

impeller – a rotating wheel with vanes used inside a casing, such as in a centrifugal pump, to move a fluid. The fluid enters the center of the spinning impeller which flings it to its outer periphery by means of centrifugal force. This force imparts velocity to the fluid, the velocity being a function of the rotational speed of the impeller. The casing around the impeller directs the flow of the fluid through the pump casing discharge outlet. Centrifugal pumps with impellers are used for a wide variety of fluids, including water and gasoline. They are often used when a fluid must be moved uphill or when pressure has to be added to a system.

Imperial gallon – a liquid measure equal to 1¼ U.S. gallons or 160 ounces.

impermeability – the ability of certain materials, such as plastic sheeting or roofing felt, to resist penetration by water.

impervious – 1) not permitting the passage of a substance, such as water; 2) not capable of being damaged or harmed.

implement – 1) to put into use; 2) a tool or device with which to perform work or a task.

implosion – a bursting inward. For instance, a tube that contains a vacuum, such as computer monitor tube, will implode when it breaks because the surrounding atmospheric air pressure will push the tube walls inward.

impost – the load-bearing portion of a wall, pier, or column which supports the end of an arch.

impregnate – to fill all the pores or spaces of a material with another compound. For example, roofing felt is saturated or impregnated with an asphalt compound to add water resistance to the felt.

impression – 1) an inward deformation under pressure, such as the impression left in metal by a welder's identification stamp; 2) an imitation or representation of the features of an object or person.

improvement – a change that increases value, such as the addition of a structure to a piece of real property.

impulse resistance welding – an electric resistance welding process using pulses of electric current.

inadequate – 1) not suitable; 2) not meeting specifications or requirements.

inaudible – not within the range of human hearing.

incandescent – glowing or luminous with intense heat.

incandescent light – a bulb containing an inert gas and a conductive filament through which current flows. The current reacts with the gas, creating an intense heat that makes the filament glow.

inch – a unit of length measurement equal to 1⁄12 of a foot.

inching – moving a crane hook, trolley or bridge in short increments that are discontinuous. Also called jogging.

inch-pound – a unit of work equal to the force required to move a weight of 1 pound over a distance of 1 inch.

incline – a sloped section of ground.

inclined manometer – a sensitive measuring device that indicates slight changes in the liquid level in a tank. It consists of an inclined tube filled with fluid. One end of the tube is open to the atmosphere, and the other end is connected to the tank. Pressure created by changes in the liquid level in the tank will cause the fluid level in the tube to change in greater proportion, drawing attention to even minor changes in the tank level.

inclined position – a pipe welding position in which the pipe axis is inclined at about a 45-degree angle.

incombustible – not capable of being burned.

incomplete fusion – a weld in which the weld metal is not fused to the base metal over the entire weld area. Areas of incomplete weld fusion are weak and reduce the overall strength of the weld.

increaser – a plumbing drainage fitting, larger on one end than the other, used to enlarge the diameter of a straight-run line. It can also be used for the reverse purpose, to reduce the diameter of the line, in which case it is called a reducer.

increment – 1) a small increase; 2) regular consecutive additions, such as an increase in pressure in 1 pound per square inch amounts.

incumbrance – see encumbrance.

incuse – a stamped figure, as on patterned metal ceiling panels.

indemnify – 1) to provide security against damages due to loss or injury; insure; 2) to compensate for damages.

independent chuck – a four-jaw lathe chuck in which each jaw is moved independently. This enables the work to be centered exactly, using a dial indicator.

index of refraction – an indication of the speed of light in a medium.

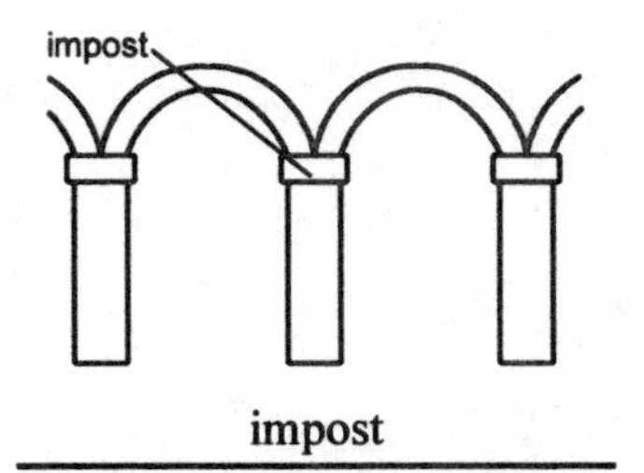

impost

incandescent light

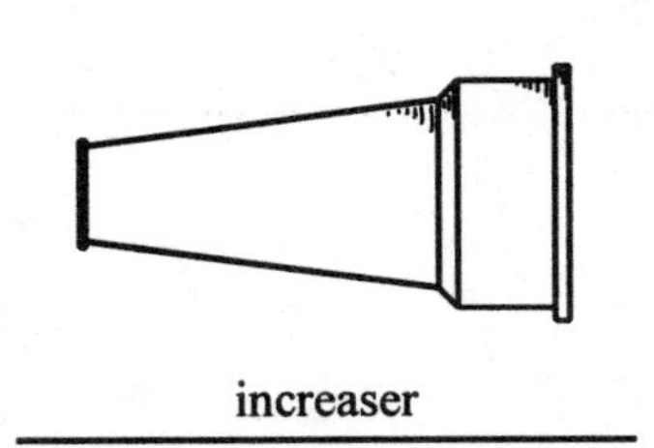
increaser

indicating gauge – a precision gauge that provides a visual indication of variations in a surface. It consists of a plunger with gear teeth cut along portions of its length which engages a gear on an indicator needle. The needle moves around a precisely calibrated dial to show variations in gradations as fine as one ten-thousandth of an inch. The indicating gauge is firmly clamped to a rigid surface and the plunger is adjusted so that it is in contact with the work to be measured. The work is mounted in such a way that it can be moved past the plunger in a controlled manner. This provides a good reading with all indications on the gauge being the result of variations in the surface of the work and not the result of irregular movement of the gauge. It is used for precision measurements, such as might be needed in centering a piece of work in an independent chuck on a lathe. Also called a dial indicator.

indicator – a pointer or other means of showing a value, status, or measurement, such as the needle on a pressure gauge which indicates the pressure.

indirect lighting – lighting that is reflected, often off a ceiling, before shining on an area or object. Indirect lighting creates fewer shadows and can reduce the glare in the area being lighted if the intensity level is maintained below 75 footcandles.

indirect overhead – overhead costs that are not directly related to a specific job, such as office rent, electricity and telephone expenses, but are a general cost of doing business.

individual vent – a separate vent for a plumbing fixture in a building drainage system.

induced current – electrical current created in a conductor by a fluctuating magnetic field around the conductor.

induced-draft cooling tower – a cooling tower with a suction fan that draws air through the tower to cool water.

induced soldering (IS) – a soldering process using heat generated by the electrical resistance of the material being soldered when an electric current is passed through the material.

inductance – the opposition to flow of electrical current. Inductance opposes a change in current in an electrical circuit. Since AC varies with time, electromotive force is generated, opposing the current flow. See also counter electromotive force.

induction brazing – a brazing process using electrical current through the material being brazed to create the heat required. This heat is generated from the electrical resistance of the material and the current passing through the material.

induction welding (IW) – a welding process which uses heat generated by the electrical resistance of the material being welded when an electric current is passed through the material.

inductive reactance – the counter electromotive force in the AC circuits of a coil which cause opposition to the AC flow.

industrial – pertaining to industry.

industrial park – a development for commercial or industrial uses.

industrial waste – waste produced through manufacturing processes.

industry – manufacturing or manufacturing activity.

inert – inactive chemically, such as helium or glass.

inert gas – a nonreactive gas used as an oxygen shield during welding operations. The inert gas prevents oxidation by displacing air or oxygen from the weld area. The presence of oxygen would be detrimental to the finished weld because oxidation action is greatly accelerated by the heat of welding.

infiltration test – a water line leak test used for a buried pipeline. The empty line is sealed and watched to see if any water leaks into the system from outside sources.

infinite – without limit.

inflatable plug – an inflatable ball-type plug used to seal off a portion of pipeline for pressure testing. *See Box.*

inflate – to increase or expand in size with air or gas.

information stake – a survey stake located near a hub or other reference point that gives grades at different distances from the hub.

infrared – long wavelength light outside the visible frequency range.

infrared brazing (IRB) – a process that uses infrared radiation to heat the material being brazed.

infrared soldering (IRS) – a process that heats the material to be soldered using infrared radiation.

infrared testing – a test that can be employed using either the natural infrared radiation of an object or by inducing heat energy into the object. The first method is passive, and the second, active. Both methods measure infrared radiation from the object. Variations of 0.01 degrees Fahrenheit can be detected, as well as variations within the material being examined. Infrared testing is used to test insulation materials and insulation installations. A thermal pattern can be developed over the surface of a material that shows the temperature variations and heat loss. This indicates where insulation is inadequate or nonexistent and should be improved upon or added to in order to minimize heat leakage.

infrasonic – outside the low frequency range of human hearing.

inhibitor – a chemical that retards or prevents a chemical reaction, such as the addition of starches or fatty acid compounds to a concrete mix during hot weather in order to retard its setup rate.

initial rate of absorption – the weight of water that is absorbed by a brick that is partially immersed in water for one minute in accordance with ASTM C67.

inlay – to set pieces of a different material within a base material in order to form a pattern, such as the patterns created by using different types of woods in furniture, particularly table tops.

Inflatable plug

An elastomer plug used in pipeline testing. The plugs can be inflated and used to seal off a portion of a pipeline. Pressure can then be introduced into the pipeline to check for leaks in the line. Usually the pressure must be held in the pipe for at least five minutes to ensure the pipe does not leak. This type of plug can be used to test pipelines before the manholes are built, or used in situations where the line cannot otherwise be sealed for testing. For example, if a line enters a tank and the tank cannot be exposed to the pressure that the line is to be tested at, the inflatable plug can be used to seal off the line before it reaches the tank. There are practical limits to the pressure that this type of plug can be exposed to before it loses its ability to seal. These limits are determined by the strength of the plug, its friction with the inside diameter of the pipe into which it is inserted, and the diameter of the pipe itself. The larger the pipe and the more area of the plug that is exposed to pressure, the weaker the seal will be under that pressure.

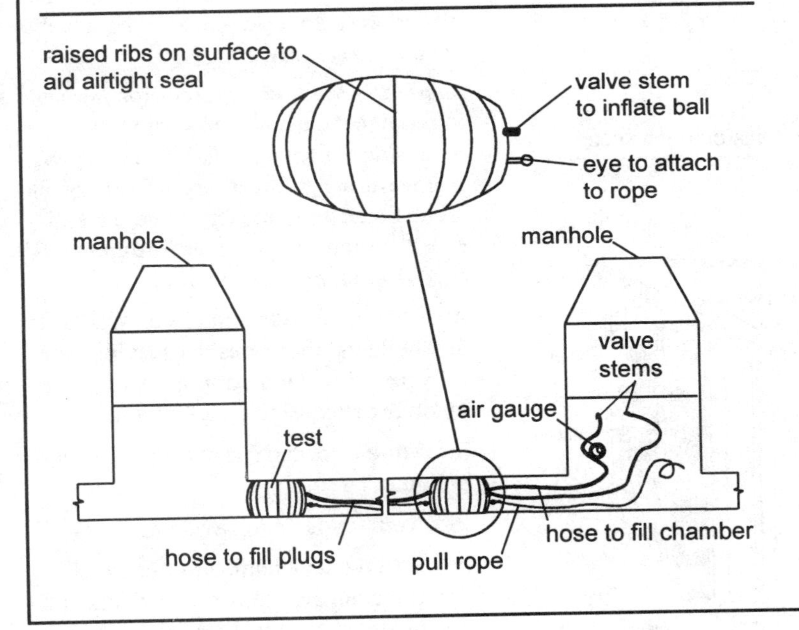

inlet – 1) a bay or water passage into a recess along the shore of a lake, sea or river; 2) an opening for the intake of water, fuel or other fluids.

in-line pump – a small pump mounted on and supported by the line through which it is pumping fluid.

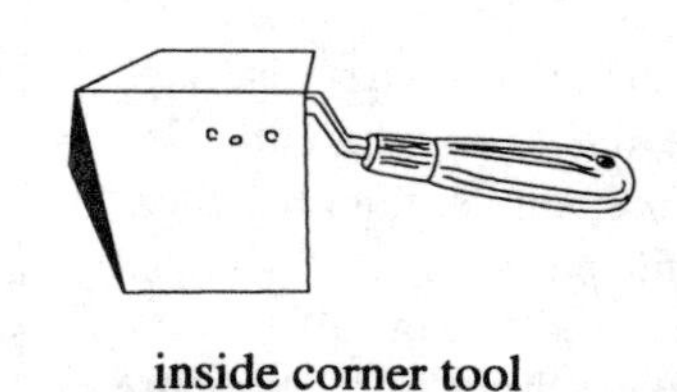
inside corner tool

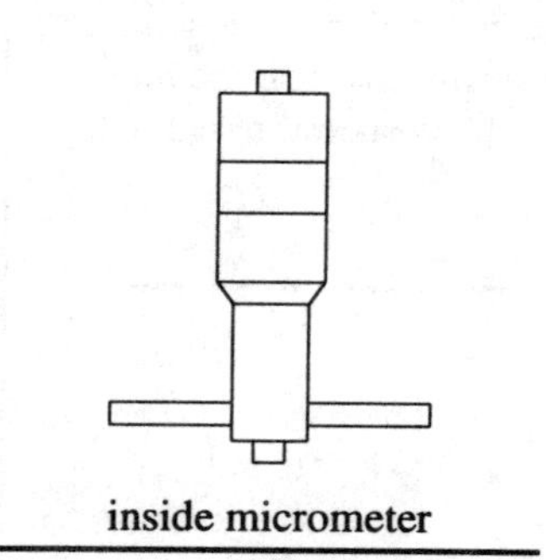
inside micrometer

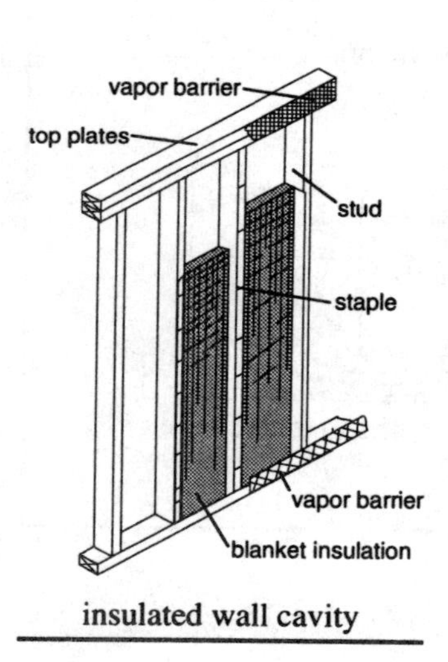

insulated wall cavity

inorganic – a compound or substance that does not contain hydrocarbons, such as animal or plant matter. An example of such a substance is inorganic zinc paint.

insert fittings – fittings used to connect certain types of flexible plastic and rubber tubing. These fittings are pushed into the tubing and have external annular rings to aid in gripping.

inset – to set within something.

inside calipers – calipers for measuring an inside dimension, such as the inside diameter of a hole.

inside corner – a corner that does not protrude into a room. Also called an internal corner.

inside corner tool – a type of metal drywall joint or taping knife with the blade bent longitudinally at slightly greater than 90 degrees so that the spring action of the blade forces both sides against both walls at an inside corner.

inside micrometer – a precision measuring tool that expands outward to measure the inside diameters of holes. It is capable of measuring to the nearest 0.001 of an inch. Some micrometers have a vernier scale which makes measurements of 0.0001 of an inch possible.

in-situ – undisturbed, in place. Architects and builders often consider leaving naturally existing landscaping in place, or in-situ, when designing a building site.

insoluble – not capable of being dissolved in liquid.

insolvent – unable to pay debts.

inspection – an examination in which an object is compared to an established standard in order to rate its quality. This is designed to ensure consistency in the finished product.

inspector – a person designated to perform inspections which determine the acceptability of a person, product or performance.

installment – 1) a periodic payment of money owed when a debt is divided into parts to be paid over time; 2) one of a series presented at intervals, as in a serial story.

instant start fluorescent light – a fluorescent light in which the ballast applies a high voltage between electrodes to produce an arc between them and activate the light. The filament is kept heated for an instant start, as opposed to the slower hot cathode fluorescent lights, in which the filament must first be heated before the fluorescence begins.

instrument – 1) a legal document; 2) a device to measure and/or record data; 3) a mechanical or electrical device used for navigating; 4) a means by which something is achieved; 5) an implement.

insulated cavity wall – a wall in which the cavity between framing members, or the space between the solid portions of a masonry wall, is filled with insulation material in order to inhibit thermal and sound transmission.

insulated window – a window, having two or more panes of glass with air space between them, designed to retard heat transfer through the glass area.

insulating concrete – see concrete, insulating.

insulating flange kit – electrically nonconductive gasket, sleeves, and washers for use on dissimilar metal flange joints to prevent galvanic corrosion. The sleeves are fitted over the bolts or studs used to hold the flanges together. This isolates the shanks of the bolts from the inside diameter of the flange bolt holes. The washers are placed under the nuts isolating them from the flange, and the gasket is placed between the flanges to remove the last path of electrical conductivity. This prevents the metal lowest on the galvanic scale from being corroded away by the adjoining metal.

insulating glass – multiple panels of glass separated by an air space.

insulating varnish – a non-electrically-conductive varnish, usually made of a synthetic material such as a plastic compound, that can be used to coat and insulate copper wire.

insulation – materials designed to retard the transmission of heat, sound or electricity. See also individual types.

insulation, batt – a flexible insulating material made of glass wool, cotton or wood fibers. Organic materials are treated to resist fire, decay and insects. Batt insulation is precut to fit between framing members and comes in 4- and 6-inch thicknesses for 16- and 24-inch joist spacing.

insulation, blanket – insulation that comes in rolls or packaged precut sizes to fit between 16- or 24-inch studs. The usual thicknesses are 3½, 5½ and 7½ inches. It is made of flexible materials such as rock or glass wool, cotton, or wood fibers, with the organic materials pretreated for fire, insect and decay resistance.

insulation flange – a flange along the edge of an insulation batt that is stapled to the studs and joists to hold the insulation in place.

insulation, flexible – see insulation, batt or insulation, blanket.

insulation rating – the number assigned to a form of insulation to denote its effectiveness. *See Box.*

insulation, reflective – a type of insulation material employing a surface that reflects heat. Aluminum foil, sheet metal and paper products coated with a reflective oxide compound are some of the reflective materials used to back insulating surfaces.

insulation, rigid – insulation material in the form of nonflexible panels made of expanded synthetic materials such as Styrofoam and urethane foam, as well as hard plastic materials such as polyvinyl chloride. Balsa wood, cellular glass, corkboard, processed wood and gypsum panels are also commonly used for both thermal and sound insulation.

insulation, thermal – materials used to minimize the flow of heat in or out of objects, such as steam pipes, refrigeration units or buildings. Heat transfer through an insulator occurs by conduction or radiation. Although no material can completely prevent this transfer, materials can be selected which give the best insulation for a specific application. The properties, shape and thickness of a material plus the heat-flow characteristics from a given source will help a builder or engineer decide on the proper choice of insulating materials. Thermal insulators are available in a variety of forms and materials, each designed for specific and limited ranges of application. Insulation foams are made of polyurethane or polystyrene. Gas-tight barriers are added to foam insulation to prevent an increase in thermal conductivity within the insulation as it ages. Loose fill insulators are made of such materials as diatomaceous silica, perlite, vermiculite, calcium silicate, plastic beads, cork, and charcoal granules. Rigid insulation materials are of polyurethane, PVC, balsa wood, urea, cellular glass, corkboard or polyethylene. Fiberglass is used to make batts, blankets and loose fill insulation. Other types of insulation include sprays, made of such materials as macerated paper or insulating concrete. Vapor barriers are installed on the warm side of insulators if the insulation temperature will go below the dew point, which would cause condensation. Reflective surfaces, such as aluminum foil, are used to decrease radiated heat transfer. Reflective surfaces turn back a large portion of radiant energy and can be combined with other materials to increase their insulation capabilities.

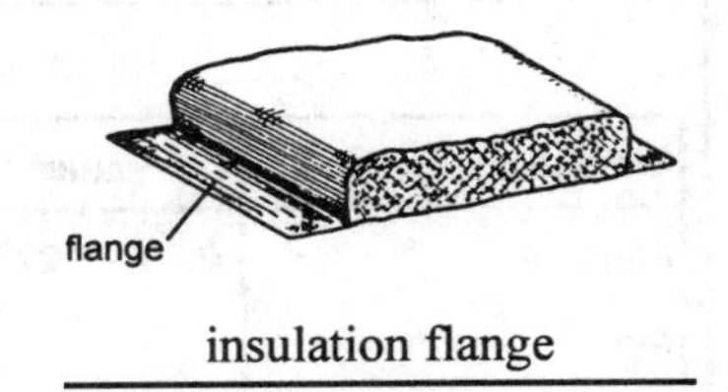

insulation flange

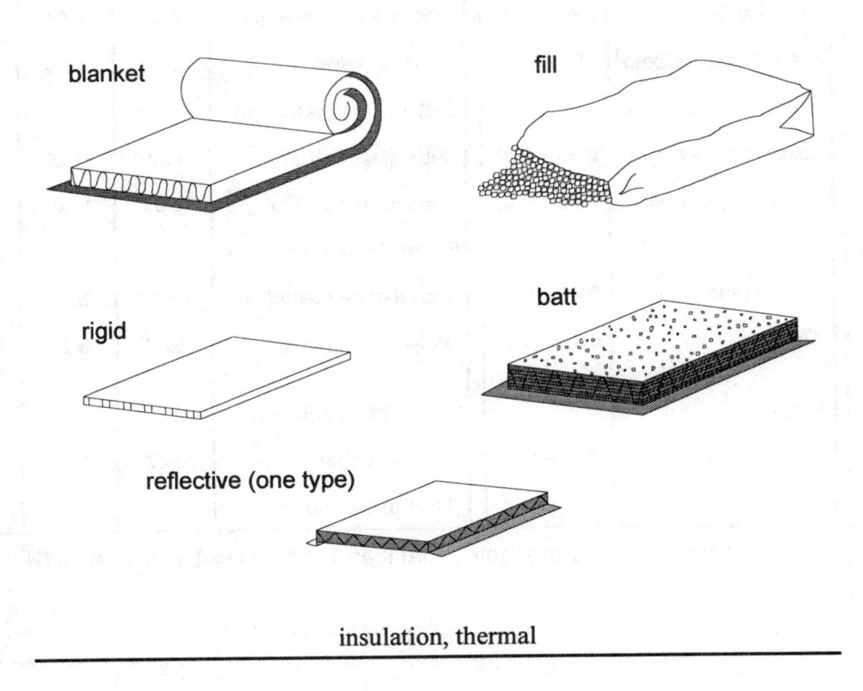

insulation, thermal

Insulation rating

Each type of insulation has its own advantages and disadvantages based on its thermal properties. These properties, plus established insulation ratings, help builders and engineers determine which type of insulation would be most effective in a given situation. The insulation rating indicates how well a material insulates. Insulation is rated in three different ways:

The **U-value** rating indicates the overall heat flow between air on the warm side and air on the cold side of a wall, floor or ceiling which is insulated with a given material. The lower the U-value, the more insulated a unit is against heat transmission.

The **R-value** measures the thermal resistance of the material. It indicates the total resistance of a material to the passage of heat or cold. The *higher* the R-value, the more effective its insulation properties. The R-value is equal to the U-value divided into 1 (R = 1/U).

The **K-value** is the measure of heat conductivity of a material. It is the equivalent of the U-value per square inch of thickness of the material. That is, the measure of the amount of heat, in Btus per hour, that will be transmitted through one square foot of material that is one inch thick to cause a temperature change of one degree Fahrenheit from one side of the material to the other. As with the U-value, the lower the K-value for a material, the better it insulates. If the K-value of a material is known, the R-value per inch can be determined by dividing 1 by the K-value (R-value per inch = 1/K-value).

Building codes set minimum R-values for insulation in various regions, or thermal zones, in the U.S. The tables indicate the insulation values of common building materials and the K-value range of insulating materials.

Insulation group	K-value
Flexible	0.25 – 0.27
Fill	
Standard materials	0.28 – 0.30
Vermiculite	0.45 – 0.48
Reflective (2 sides)[1]	
Rigid	
Insulating fiberboard	0.35 – 0.36
Sheathing fiberboard	0.42 – 0.55
Foam	
Polystyrene	0.25 – 0.29
Urethane	0.15 – 0.17
Wood	
Low density	0.60 – 0.65

Material	U-value	R-value
Single glass	1.11	0.90
Dual glazing		
¾" air space	0.55	1.82
High performance glass	0.26	3.85
Triple glazing	0.34	2.94
Batt or roll insulation		
Fiberglass batt 3½"	0.091	11.00
Fiberglass batt 5½"	0.53	19.00
Board insulation		
Polystyrene board / in.	0.28	3.57
Polyurethane board / in.	0.16	6.25
Loose fill insulation		
Cellulose fiber per inch	0.27	3.70
Mineral fiber per inch	0.34	2.92

Material	U-value	R-value
Building components		
Douglas fir / pine ¾"	1.06	0.94
Douglas fir / pine 1½"	0.53	1.89
Douglas fir / pine 3½"	0.23	4.35
Drywall ½"	2.22	0.45
Drywall ⅝"	1.78	0.56
Plywood ⅜"	2.13	0.47
Plywood ½"	1.60	0.62
Plywood ⅝"	1.29	0.77
Stucco ⅞"	4.38	0.23
Concrete masonry unit 8"	0.96	1.04

[1]Insulating value is equal to slightly more than 1 inch of flexible insulation. (R-value = 4.3)

insulator – 1) a material that does not conduct electricity. Electrical insulation is important in all products using electricity to control and direct the flow path of current. There are a wide variety of materials in common use as electrical insulators, ranging from varnish coating to Teflon. They are selected for particular applications by their performance under the conditions that the product is to be used, as well as economic considerations; 2) a material that resists thermal or acoustic transmission. Acoustic transmission, or sound energy, is absorbed to some degree by materials such as fiberboard, fiberglass, and plastic foam materials. Commonly-used thermal insulators are some types of plastic foams, wood and fiberglass.

intake – the suction device or opening of a machine or system that allows air, water or fuel into the system.

intangible – 1) not palpable; 2) not a material item. Customer good will and employee loyalty are both valuable intangible items.

integral lockset – a lockset for a door that combines a spring lock and a dead bolt lock in one assembly.

integral vanity top and bowl – a one-piece bathroom vanity top and wash basin unit.

integrated circuit – an electrical circuit contained within a single semiconductor chip.

interceptor – a device in a plumbing drainage system which separates out and traps hazardous materials, such as grease or oil.

interest – 1) a fee paid to a lender for the use of borrowed money; 2) a title, right or share in something.

interface – 1) the common boundary surface or meeting place of objects, substances, or bodies which do not combine with each other, such as the common surface between oil and water; 2) the means of interaction or communication.

interference – 1) an obstruction; 2) unwelcome contact between objects; 3) the undesirable mixing of two sources of energy.

interference fit – see drive fit.

intergranular stress corrosion cracking (IGSCC) – a type of corrosion and cracking which affects stainless steel. Under certain conditions, the carbon in austenitic steel can become concentrated between the metal grain boundaries, setting the stage for corrosion and separation to occur. This condition can be prevented by limiting the amount of carbon, by adding an alloying element such as molybdenum, and/or proper heat treatment when the steel is made to insure that the carbon is thoroughly integrated into the steel.

interior – the inside of something.

interior door – a door not designed to be exposed to the elements. It is used to close a doorway on the inside of a building.

interior finish – any material that covers the interior walls of a building and is generally not suitable for exterior use.

interior wall – a wall, both sides of which face an inside space within the building.

interlacing arcade – arches that appear to intersect.

interlock – 1) two or more objects which connect to each other by fitted parts, such as tabs and slots; 2) an electrical circuit that blocks an action until preset conditions have been satisfied. For example, if a pump is moving water from a well to fill a tank, there may be a float switch with an interlock at a low level in the tank that prevents the pump motor from turning on until the tank water level drops to a preset point. Electrical interlocks are widely employed in automatic and semi-automatic operations to ensure desired sequences of events are followed. They are also employed as safety devices to ensure an action will not take place until some prerequisite condition has occurred.

interlocking shingles – roof shingles with tabs that fit into, or interlock, one shingle to the next.

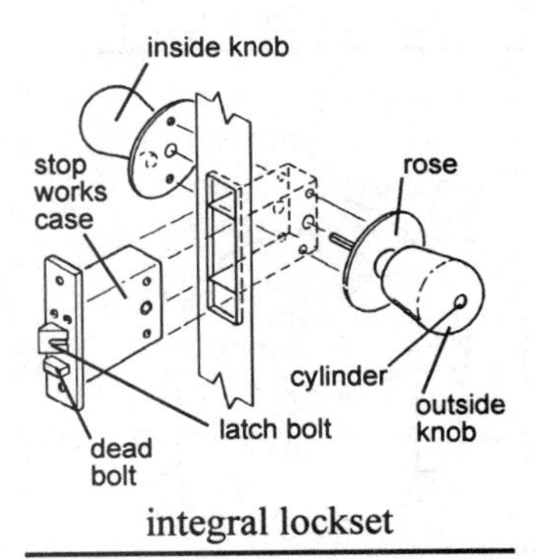

integral lockset

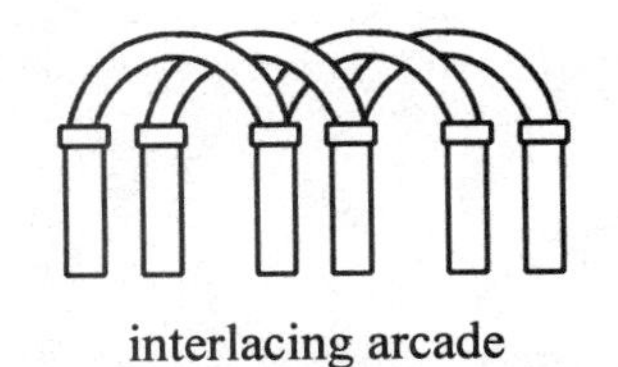
interlacing arcade

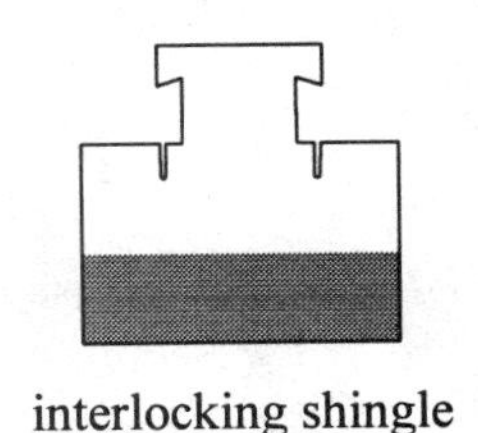
interlocking shingle

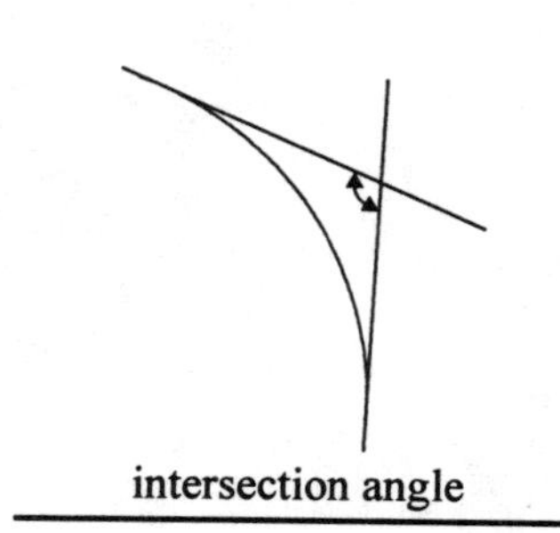
intersection angle

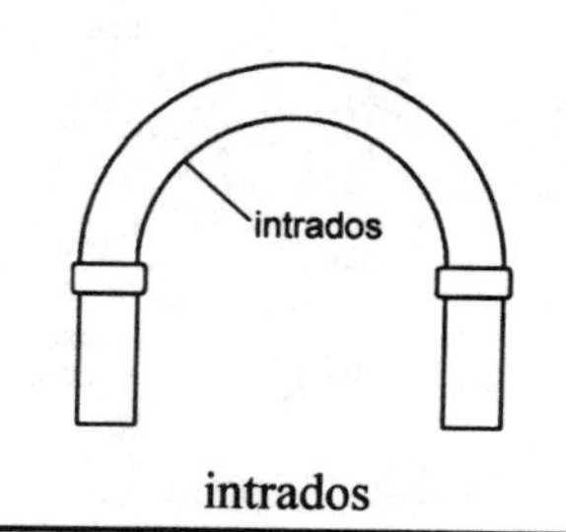

intrados

intermediate flux – a soldering flux that leaves a residue that is non-corrosive to the base metal. Some fluxes leave a residue that will eventually corrode the base metal. As a rule, all flux residue should be removed after soldering, even a non-corrosive one.

intermediate metal conduit – see conduit, intermediate metal.

intermittent weld – a non-continuous weld that leaves unwelded sections along the weld line.

internal combustion – a type of engine in which the power is developed by the combustion of a fuel within the engine. A gasoline engine, in which fuel is burned in a cylinder to power the engine, is an example of an internal combustion engine. A steam engine, powered by steam that has been created by burning fuel in a boiler rather than in the engine, is not an internal combustion engine.

internal gear – a gear that is hollow with teeth cut on the inside surface.

internal mix – a paint spray gun that mixes air and paint inside the gun before it exits as a spray.

internal threading tool – a tool for cutting internal threads in metal parts. An internal threading tool is used in a lathe to cut internal threads in large holes.

interpass temperature – the temperature the deposited weld metal should be before the next weld pass is deposited in a multiple weld pass application.

interrupter feed-through receptacle – a type of ground fault interrupter receptacle which can be installed in an electrical circuit between the power source and other standard receptacles in the circuit. The ground fault interrupter receptacle will protect all the receptacles on the circuit. Any appliance plugged into a receptacle on that circuit is protected just as if it were plugged directly into the GFI receptacle.

intersection angle – the external angle formed by the intersection of two tangents to a curve.

interstice – 1) a small space between objects, such as between boards in a fence; 2) a short space of time between events; the time between flashes of a flashing signal light, for example.

interval – a period of time between actions, such as the period of time required for paint to dry between coats.

in toto – on the whole (Latin); entirely. An example of a job bid "in toto" would be a complex bid that includes all facets of the work from clearing the land and erecting the structure to completion and occupation by the owner or tenants.

intrados – the inside curve of an arch or a vault.

intrusion – a trespass or wrongful entry onto the property of another.

intrusion detector – a motion, heat-sensing, or other type of detector that identifies a presence within specified boundaries. Used as part of a security system in both commercial and private applications.

inventory – a list of goods or assets on hand.

iron – a metallic element found in the earth; atomic symbol Fe; atomic number 26; atomic weight 55.847.

iron alloy – a mixture of iron and one or more metals which are combined to enhance the performance and/or reduce the cost of the resulting metal material. Alloy steel is an iron alloy whose characteristics are controlled primarily by the proportions and types of elements added to the iron and the heat treatment applied to the alloy. Chrome and nickel are added to make stainless steel, valuable for its corrosion resistance. The addition of chrome and molybdenum to iron increases the high-temperature strength of the resulting steel. Carbon steel is a type of steel whose characteristics are controlled primarily by its carbon content. It is stronger and harder than iron. Tools such as chisels and drill bits are made of carbon steel. They can be sharpened to a fine edge and are hard enough to retain that edge.

iron, cast – an iron, shaped by the casting process, that is relatively brittle and has a high carbon content. Both ductile iron and gray iron are sometimes referred to as cast iron, though the term can be applied to any iron shaped by the casting process.

iron, ductile cast – iron that has magnesium or cerium added while in a molten state in order to form the carbide into globular nodules. It is commonly used for pipe, fittings, valves and other components. It is less brittle than gray cast iron and has good corrosion-resistant characteristics. Also called nodular cast iron.

iron, gray cast – an iron with a controlled carbon content in the form of cementite or iron carbide and carbon also in the form of graphite flakes. For many years it was the material most used for water main piping and for sanitary waste piping within a structure. It was also used for pipe wrenches, valves and other components. It has high corrosion resistance and is relatively economical. Its disadvantage is its brittleness. Gray cast iron is often called cast iron.

ironing center – a small closet containing a built-in ironing board that swings into place for use, and folds up out of the way for storage.

iron, ingot – an iron that is low in carbon and other impurities.

iron, malleable cast – a form of white cast iron that has been heat treated to change the cementite to free carbon, making it more formable and less brittle. It is used to make pipe fittings.

iron phosphate coating – a cleaner and primer for industrial equipment.

iron, pig – a crude iron resulting from the refining of iron ore.

iron soldering (INS) – soldering using a soldering iron as the source of heat.

iron, white cast – a brittle iron with its carbon content in the form of cementite.

ironwood – a hard dense wood from the caprinus caroliniana or lyonothamnus floribundus trees found on islands off the coast of southern California and in some tropical locations. One common name for the wood is Catalina ironwood. It is not commonly used for building because of the cost and scarcity, but it is strong and may be used in a variety of applications, including furniture.

iron, wrought – an iron that has small amounts of slag added to and distributed within it while it is hot and partially solidified. The slag is added to the hot iron at a temperature at which it does not fuse to the iron. Wrought iron is used to make ornamental iron work, such as fences, railings and screens.

irregular brace table – the twelfths and sixteenths scales on a framing square, which can be used to calculate sides of right triangles with angles other than 45 degrees. A regular framing square brace table can only be used to calculate dimensions for right triangles with 45-degree angles.

isocyanate resin – isomeric isocyanic acid salt, a chemical binder used in paints.

isofootcandle graph – a graph showing lighting patterns and illumination data for outdoor lighting.

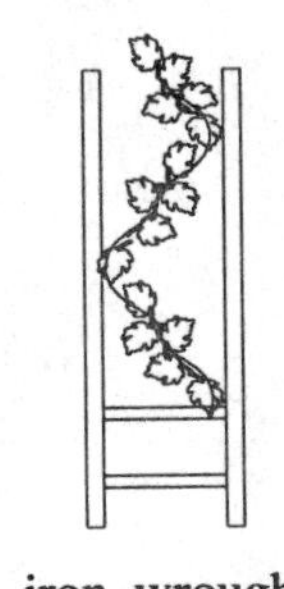

iron, wrought

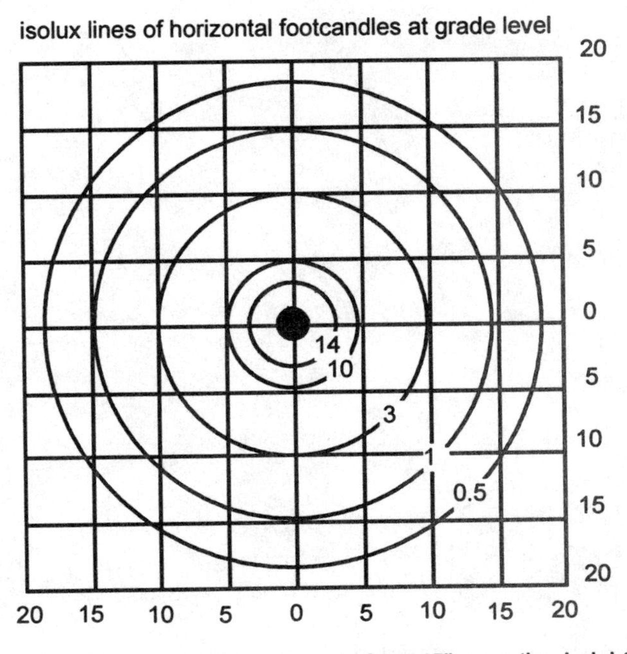

isofootcandle graph

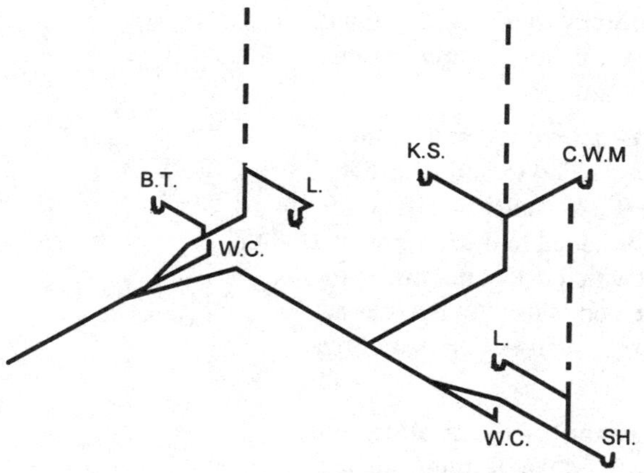

isometric drawing of plumbing layout

isofootcandle lines – the lines on an isofootcandle graph that illustrate lighting patterns. Also called isolux lines.

isolation – 1) quarantine; set apart from others; 2) separation from a sound source.

isolation joint – a joint between parts of a structure designed to prevent cracking and buckling when temperature changes cause the structure to expand or contract, or when wind, ground settling or earthquakes cause a degree of movement in vertical or horizontal directions.

isometric drawing – a drawing, often used for piping layout, which shows a three-dimensional view of the routing of the piping. The three-dimensional view is accomplished by means of the angles at which the lines are drawn. The piping may be to scale, or dimensioned, or both. Junctions are shown at 120-degree angles, rather than 90-degree angles, in order to provide a perspective of depth. Also called an isometric projection or an isometric.

isothermal – action taking place without a change in temperature.

J: jack to jute

jack – a lifting device placed beneath an object, such as an automobile or a floor joist, to raise it using a mechanical advantage. Most jacks are of the screw type or hydraulic type. Jacks are used as temporary supports while work or repairs are being done to the object raised.

jack arch – an arch with a horizontal or near-horizontal top surface. Also called a flat arch or straight arch. See also minor arch.

jack hammer – an electric or pneumatic device, with a chisel or pointed bit, used to cut or break up a surface such as asphalt or concrete.

jack plane – a hand tool with a long flat base plate (approximately 15 inches) and a sharp, angled cutting blade that extends through a slot in the base. As the base plate is moved along the surface of a piece of wood, the extended blade shaves off thin layers of wood. The extension on the blade is adjustable for shaving thinner or thicker layers.

jack post – a type of screw jack, or lifting device, with a wide base footing and a plate at the top to spread the load. A jack post can be used to raise and support a sagging horizontal structural member, such as a cracked floor joist. It can be left in place as a permanent part of the structure.

jack rafter – a roofing rafter that spans from a hip to a wall top plate, or from a valley to a ridge.

jackscrew – a lifting device that uses a threaded shaft which turns in an internally-threaded sleeve to adjust its height and the height of whatever load it is supporting.

jackshaft – a short, solid, cylindrical shaft that is used to transmit power, such as from a motor to a pump.

Jahn forming system – a trade name for concrete forming accessories which include clips, brackets and ties used together in a system to hold sheets of plywood in place as concrete forms.

jalousie window – a type of window composed of a number of rectangular panes of glass, or louvers, placed in slots

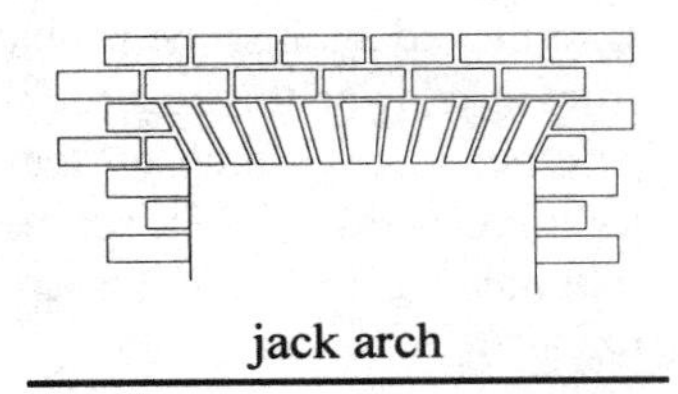

jack arch

jack plane

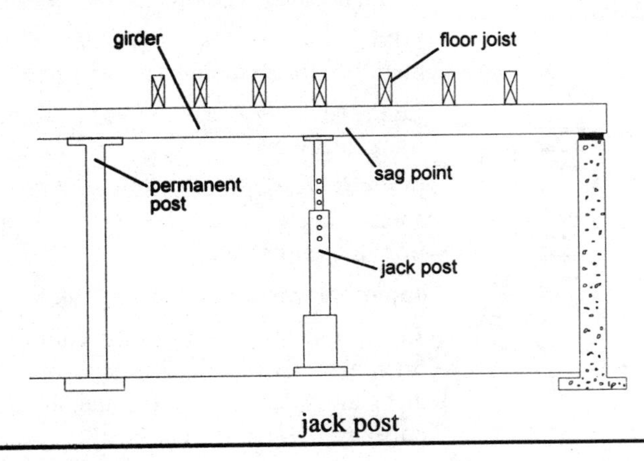

jack post

Jam nut

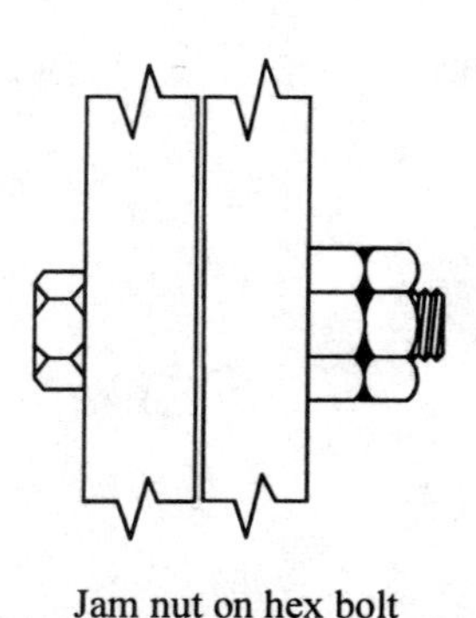
Jam nut on hex bolt

A jam nut is a nut that is tightened against another nut on a threaded shaft to keep the first one from loosening. The jam nut may be of the same type and size as the primary fastener nut, or it may be made of stamped sheet metal. When the first nut is tightened to the desired torque, the jam nut is tightened against it. (The first nut must be held against turning to keep its torque from changing.) The tightening of the jam nut against the first nut puts the section of the threaded shaft between the nuts in tension and distorts the threads a small amount. A stamped sheet metal nut becomes distorted itself when it is tightened against the primary nut. These forces prevent the nuts from moving on the threaded shaft and coming loose due to vibration or other forces acting on them. Jam nuts are used where fastener tightness is essential for proper operation and/or access to the nut for retightening is limited, such as for holding parts of machinery together.

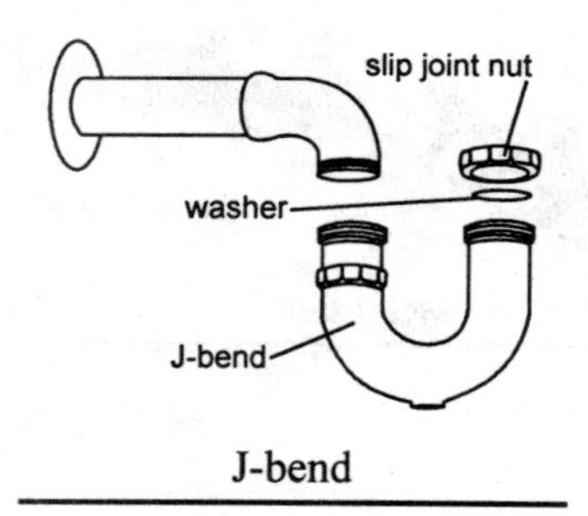

J-bend

in a frame, one above the other. The panes are held in place by a pinned connection that rotates, allowing the panes to be opened simultaneously to provide a flow of air. When closed, the panes overlap slightly, preventing rain from entering the structure. They do not form a tight seal however, so this type of window is recommended for use only in mild climates. Also called a louver window.

jamb – the side, top, or bottom framing members for a window or door opening.

jamb, head – the top window or door frame member.

jam nut – a nut that is tightened against another nut on a threaded shaft to keep it from loosening. *See Box.*

Japanese wood oil – see tung oil.

Japan wax – a wax from the fruit of an oriental sumac plant. It is used as an ingredient in some floor waxes and furniture polishes.

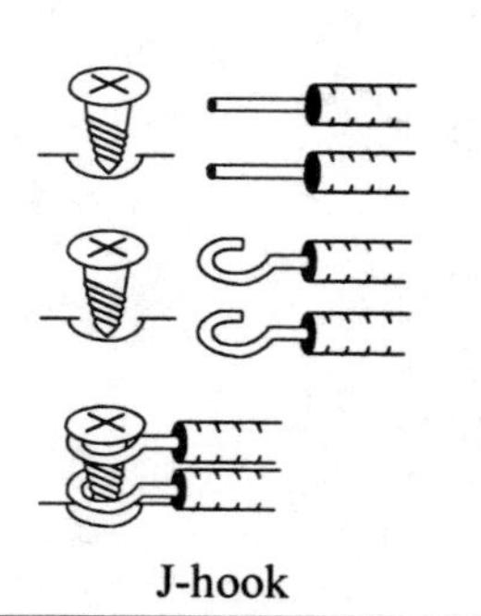
J-hook

Jarno taper – a standard taper for lathe headstock and tailstock sockets. The amount of taper is 0.600 inch taper per foot (tpf). The taper grips similarly-tapered objects, such as a tapered-end drill bit.

J-bend – the bottom or trap section of a multi-piece P-trap. There is a 180 degree bend in the fitting. Also called a return bend.

jetting – 1) an earth compaction method using water injected into the earth with a metal pipe. The water action consolidates the mass of the loose earth and helps in settling unstable ground or backfill; 2) a method for digging holes in the earth using high pressure water directed at a specific location.

J-hook – a hook-like bend in the end of an electrical wire shaped to fit around an electrical terminal.

jib – an arm projecting from the main body of a crane that can be swiveled about the body to reach and move loads.

jig – a device used to hold parts or a guide that ensures that parts are being cut or shaped within a particular set of tolerances that can be reproduced. *See Box.*

jigsaw – a power saw with a thin blade that moves up and down for cutting curved, irregular and intricate patterns.

jimmy – a short pry bar.

jitterbug – a hand tamper used in finishing concrete flatwork, consisting of a metal grille with long handles. Its purpose is to force large aggregate below the surface of the wet concrete, leaving the surface smooth.

job site – the location where construction activities are taking place.

jogging – see inching.

join – to fasten two or more pieces together, as in fastening or joining sections of gutter together for installation.

joinery – the craft of cutting and fitting woodwork using a variety of joint-making techniques.

joint – the juncture between two pieces of material, such as wood or drywall pieces.

Jig

A jig is a fixture fabricated to hold a part being worked on, and/or a guide or pattern designed to aid in shaping, drilling or cutting the part to ensure that each part produced of the same design is identical to those already completed. The use of a jig guarantees each part is made within the same tolerances. The jig is often fabricated along with the prototype of the mechanism or part with which it will be used. The design of the jig must accommodate precise positioning of the parts and yet permit easy access to the parts when fabrication operations are complete. An example of the use of a jig would be in a welding operation which employs a jig to hold two pieces of metal at the precise angle for welding. When the weld is complete and the pieces removed from the jig, they will be exactly the same as all the previously-welded pieces held in the jig. Simple jigs can be fabricated to aid in a variety of carpentry applications as well. For example, a jig may be made as a guide for drilling holes for installing latch mechanisms. The cost of fabricating a jig is made up in the time and effort saved in doing repetitious work. Jigs enable parts to be interchanged and reduce costs in production runs.

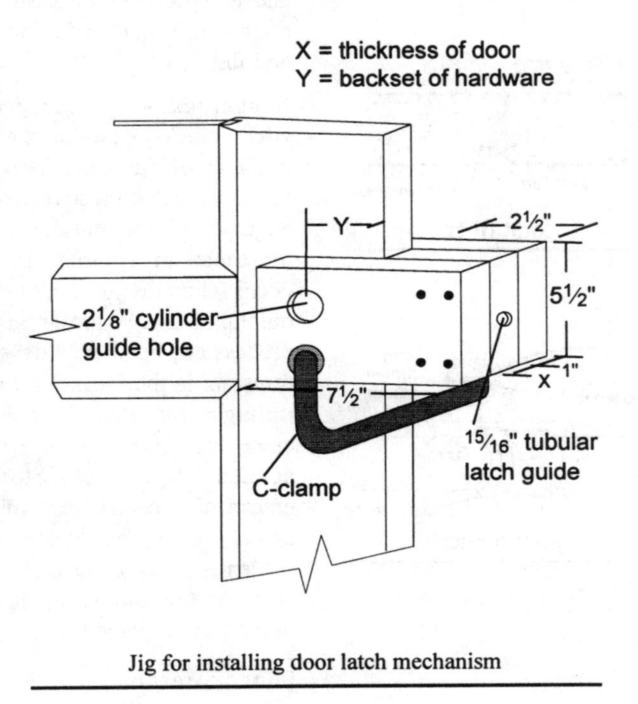

Jig for installing door latch mechanism

joint clearance – the gap between two surfaces to be joined, particularly with a weld. In welding, the gap helps to ensure that the weld filler will penetrate to the depth desired in the base metal.

joint compound – the putty-like material used to embed joint tape and bond it to the drywall. It is also used to cover fastener heads when working with gypsum wallboard. A water-resistant type is manufactured for use with exterior and water-resistant drywall. This compound requires more work in application than the standard type. It shrinks more and may require extra coats, as well as being harder to sand and smooth than standard compound. Both standard and water-resistant compounds come premixed or packaged as dry powders to be mixed with water.

joint compound, all-purpose – a combination drywall joint and topping compound with many of the smooth-spreading qualities of topping compound but with greater adhesion. It can be used for embedding tape, as well as for covering joints and fasteners. It comes premixed or packaged as a dry powder to be mixed with water.

joint compound debonding – a gypsum drywall defect in which the joint compound loses its bond with the tape or the wallboard. It is caused by a foreign substance, such as dirt, sticking to the drywall or tape during application, using dirty water to mix the compound, working with dirty tools, mixing the compound improperly, or using old compound. Debonded tape may be repaired by removing the loose tape and just enough of the dried compound to permit a new application of compound without leaving a raised section in the repair area.

joint compound, topping – a fine grain compound used for second and third coats over standard drywall joint compound. It

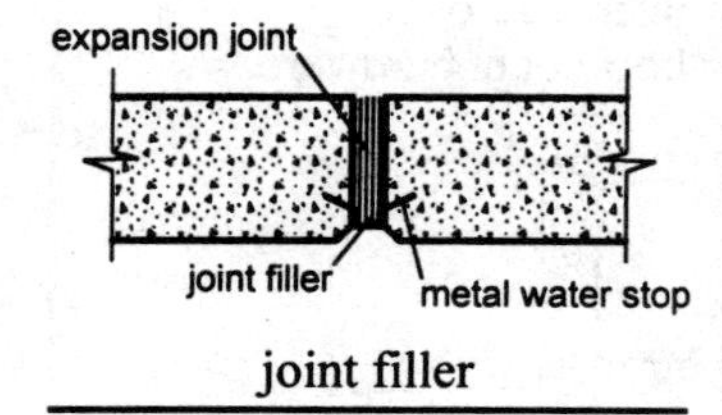

joint filler

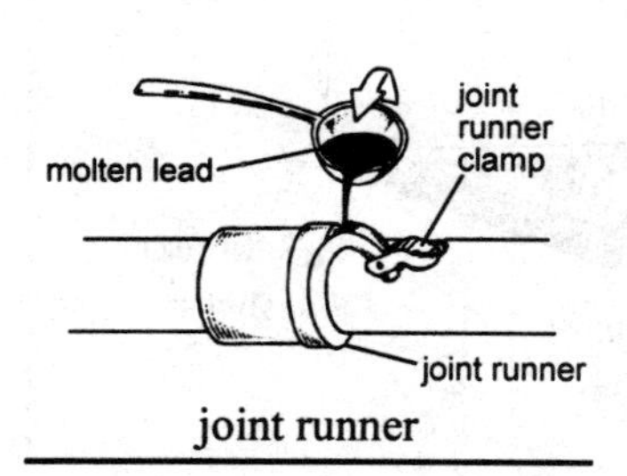

joint runner

is not suitable for embedding tape because it does not have the necessary adhesive qualities. However, it spreads smoothly and is easier to sand than standard compound, making it useful for covering joints and fastener heads.

joint cracks – a gypsum drywall defect in which cracks appear in the joint between two panels. There are two common types of joint cracks: those that form along the edges of a joint, and those that run along the center of the joint. Edge cracks can occur when the joint compound dries too quickly. During warm weather, the drying process can be slowed down by applying moisture in the form of a fine spray or by rolling a wet roller along the joints. Edge cracks can also occur if the compound is applied in layers that are too thick. Center cracks also occur from joint compound layers that are too thick or from building settlement. To repair the cracks, the torn tape and the joint compound must be removed and replaced.

joint depression – a defect in gypsum drywall application in which there is a valley in the joint between panels. A joint depression may occur because too little joint compound was applied, the compound was too thin, or the joint may have been sanded too deep. Joint depressions can be repaired by filling in with additional layers of compound, and sanding with a sanding block as required.

joint design – the configuration and dimensions of a weld joint.

joint discoloration – a defect in gypsum drywall application in which the joints discolor or turn lighter or darker than the rest of the finished wall. This may be caused by moisture being trapped in the joint when the joint was sealed, by the drywall being painted while there was high humidity in the air, or from the use of poor quality paints.

joint efficiency factor – a number that is equal to one or less, which represents the strength of a weld joint in relation to the base metal strength. A factor of one indicates that the weld strength is equal to that of the base metal joined by the weld. A factor of less than one indicates the weld strength is less than that of the base metal.

jointer – 1) a hand tool used to form a groove in wet concrete; also called a groover; 2) a power tool with rotating cutter blades used to smooth wood for precise joints.

jointer plane – a hand tool with a long base for smoothing surfaces and straightening edges on long cuts of wood.

joint filler – preformed cork or plastic material used to fill an expansion joint in a concrete structure.

joint penetration – a measurement of the depth of weld-metal penetration taken from the surface of the base metal.

joint photographing – a drywall defect in which the joint tape, or the shadow of the joint tape, shows through the gypsum wallboard finish. It can be caused by leaving excess joint compound under the joint tape, highly humid conditions which delay the drying of the second and third coats of compound, or by dry tape absorbing too much moisture from the joint compound. To correct the problem, the compound covering the tape must be sanded to feather it into the surface of the drywall, and the tape covered with thin coats of joint or topping compound. This condition is also called tape photographing.

joint reinforcement – 1) a variety of steel wire shapes used for reinforcing masonry unit joints; 2) weld metal added above that required for the weld. The extra metal increases the strength of the joint.

joint runner – a steel clamp that is used when making a horizontal lead and oakum pipe joint. The clamp fits around the pipe and keeps the molten lead from escaping as it is poured through a special opening in the clamp. The lead flows into the joint, which is packed with oakum, and hardens to make a seal. Once the lead cools, the joint runner is removed. Lead and oakum joints may be used with hub and spigot drainage pipe.

joint striking tool – a tool for smoothing and shaping masonry mortar joints before the mortar has set.

Joint tape

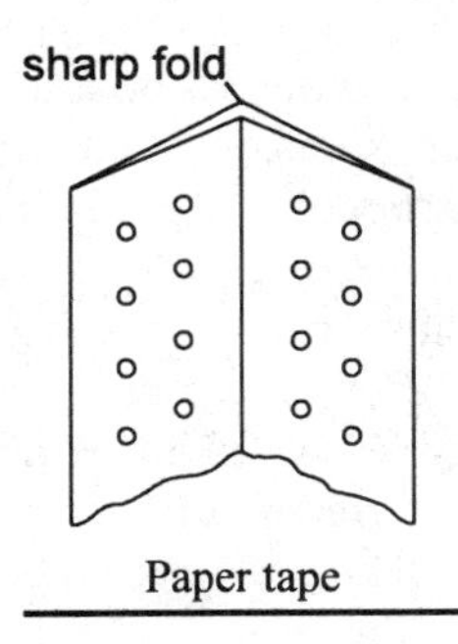

Paper tape

Joint tape is used to finish and reinforce joints in gypsum wallboard applications. It comes in paper or fiberglass mesh. Paper joint tape is about 2 inches wide and has a crease down the center. The crease is a guide for folding the tape when it is used in corner joints. The joint tape is embedded in joint compound to bond it to the wallboard. Paper tape usually has small holes in it to increase its gripping strength. The joint compound is forced through the holes in the embedding process. After the paper tape is pressed into the joint compound, a taping knife is used to smooth the tape into the compound. Excess compound is pressed out from under the tape so that it does not bulge. A second coat of joint compound or a topping compound can be applied when the first coat of compound dries. The finished joint is sanded smooth before the wall finish is applied.

Fiberglass tape has fiberglass strands woven into a coarse mesh. It is 2 inches wide, easy to apply and moisture resistant. It is available with or without an adhesive backing. When using fiberglass tape, there is no need for an application of joint compound before the tape is attached to the wallboard. The adhesive tape is pressed directly onto the wallboard joint. The non-adhesive fiberglass tape is stapled to the wallboard along the joint. The tape is then concealed by applying and smoothing successive coats of joint compound. Fiberglass tape is more expensive to use than paper tape but it can be used in any taping application. It is the only type of tape used where water-resistant compound is required.

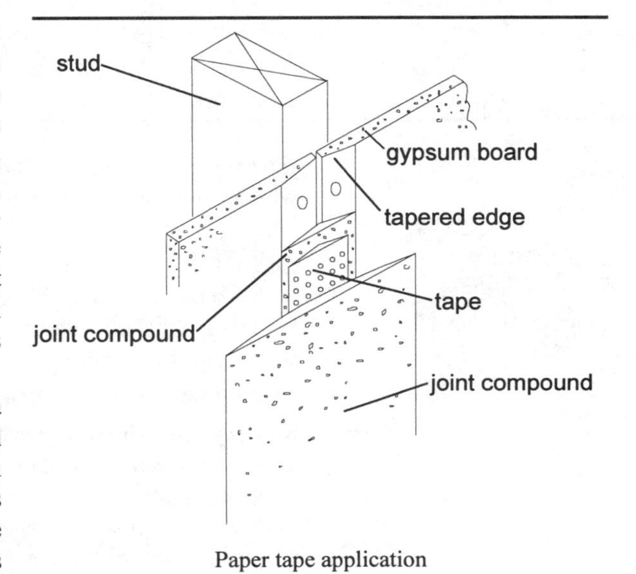

Paper tape application

joint tape – a special paper or fiberglass mesh tape used to cover and reinforce joints in gypsum wallboard. *See Box.*

joint tenancy – property held with equal ownership rights by two or more people. Upon the death of one, the ownership of the property transfers to the remaining tenant(s).

joint venture – an undertaking by two or more people or groups working together toward a common end, such as two contractors pooling their resources to build a house for profit.

joist – a horizontal structural member that supports the load of a floor or ceiling.

joist cleating – see cleating methods.

joist hanger – a preformed sheet metal fixture that is used to connect the end of a joist to a structural member that is at right angles to the joist.

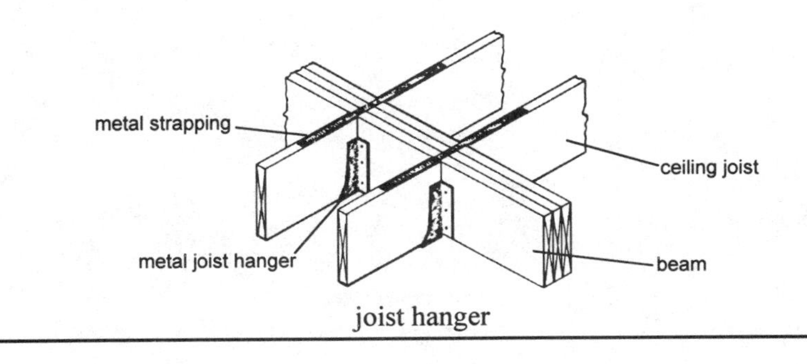

joist hanger

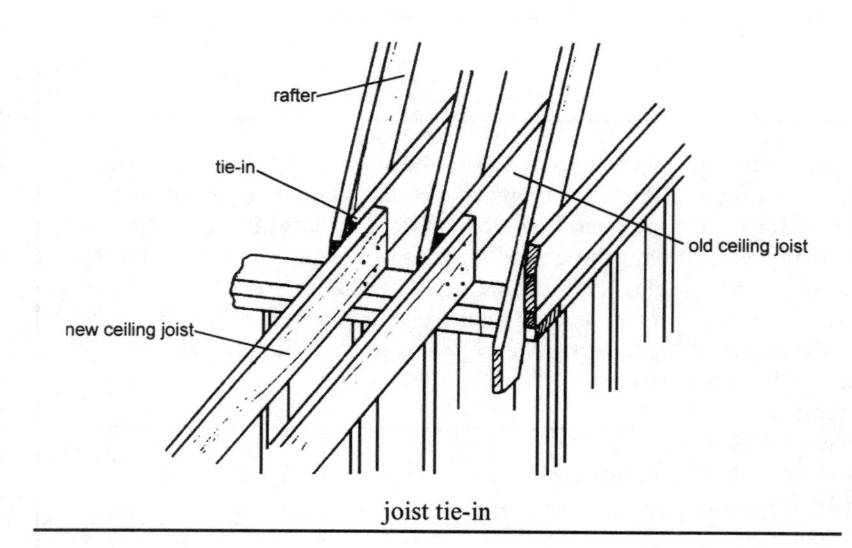

joist tie-in

junction boxes

joist tie-in – a method of adding joists in which the new joist rests on the existing sill.

journal – a portion of a rotating shaft within a bearing, such as the section of an engine crankshaft rotating in the main bearing of the engine.

journal box – a journal bearing housing.

journeyman – a worker who has served an apprenticeship and gained the experience required to work at a construction trade such as plumbing or carpentry, but who is still under the supervision of another.

jumbo brick – a brick measuring $3\frac{5}{8}$ x $2\frac{3}{4}$ x 8 inches.

jumbo brick, 6-inch – a brick measuring 6 x 4 x 12 inches.

jumbo brick, 8-inch – a brick measuring 8 x 4 x 12 inches.

jumbo utility brick – a brick measuring 4 x 4 x 12 inches.

jumper – a wire used between two points in an electrical circuit to make a connection and/or to temporarily bypass part of the circuit.

junction box – a metallic or non-metallic box, designed with knockouts in the sides and back, used to support and protect electrical wire connections or conductor splices. One or more connections can be made within one box. There are several sizes and configurations of boxes available for use with different wire sizes and quantities. The rectangular or square boxes are designed to fit into wall installations. Octagonal junction boxes may be the best choice for ceiling light fixture installations, and flat, round "pancake" boxes can be used for limited-space installations. A standard outlet box can be used as a junction box as well, and has the advantage of having a decorative cover that matches switch and receptacle cover plates. The *National Electrical Code* limits the numbers and sizes of wires that can be routed into each size of a box.

junior beam – a smaller, secondary structural member used as a brace or to transfer the load between two other structural members.

jurisdiction – the limits or territory in which legal authority may be exercised. Building codes vary depending on whether the city, county or state has the legal authority, or jurisdiction, over the area and type of construction being done.

jute – a strong coarse fiber, made from East Indian plants, that is used for rope, carpet backing and burlap cloth.

K: Kafer fitting to kraft paper

Kafer fitting – a cast iron building drainage pipe fitting with a threaded-on hub for connecting to existing cast iron lines. Also called an insertable joint.

kame – a geologic formation consisting of a short ridge of compressed sand and gravel usually caused by glacial movement.

kaolin – a fine white clay used to make porcelain.

Keene's cement – a hard, dense finish plaster.

kerf – 1) the cut or groove made by a saw blade; 2) a series of parallel grooves on the back of a masonry unit to aid in breaking the unit with a clean break; 3) a series of parallel grooves cut on the back of a member to allow for bending; 4) the width of any cut.

key – a square or rectangular cross-section piece of metal that fits into a groove or shaft to secure interlocking parts. A key would be used to secure a pulley or wheel to a shaft, for example.

keyhole saw – a hand saw with moderate-to-coarse teeth coming to a point at the end of the blade. The pointed blade allows the saw to be used to start an inside cut (one that does not extend to the edge of the material) from a hole that has been drilled in the material.

keystone – the center stone at the top of a masonry arch. It is called the keystone because it locks the other masonry pieces of the arch together.

key switch – an electrical switch that is operated with a key to guard against unauthorized use. It is simply a single-pole switch from which the toggle has been removed and replaced with a key slot. It can only be activated when the proper key is inserted. The controls for an electric hoist are often operated by a key switch.

keyway – the groove cut in a member into which a square or rectangular key can fit for the purpose of locking two members together, such as a gear to a shaft; 2) a groove or a channel formed into one concrete pour that is used to interlock another concrete structure poured at a different time, such as interlocking a wall to a footing.

K-factor – a constant factor used to figure system air flow in HVAC calculations. This factor is unique to each particular air diffuser design and model, and can be obtained from the diffuser manufacturer.

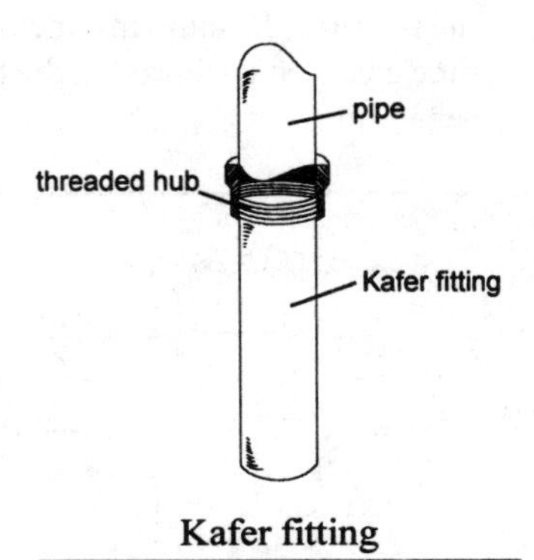

Kafer fitting

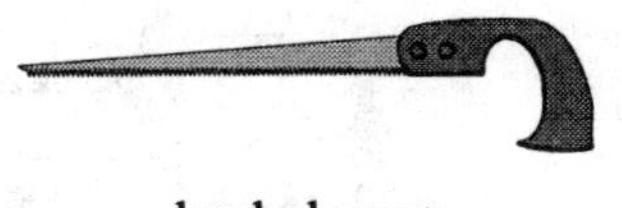
keyhole saw

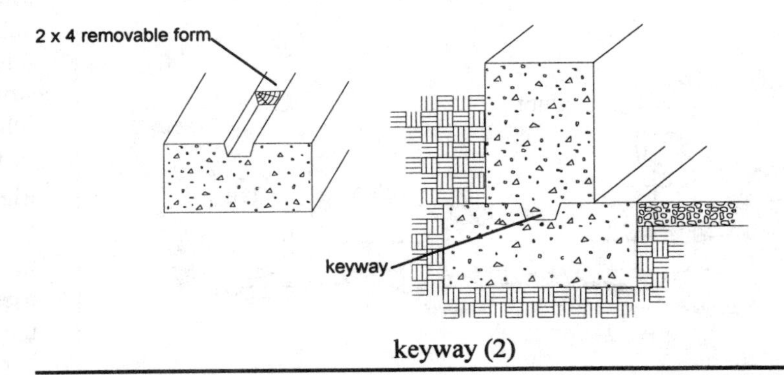

keyway (2)

Kicker block

A kicker block is a concrete mass that is placed to support hub (or bell) and spigot pipe joints in an underground system. Most buried piping has hub and spigot type joints that slip together and depend on a snug fit to ensure a watertight seal. This type of piping has the advantage of having flexible joints that can adjust to small amounts of earth movement. However, the thrust of the water moving through the pipeline puts constant pressure on the joints and connections. Kicker blocks must be placed to support pipe joints at each change of direction, elevation or pipe size. The concrete is poured beneath or behind the joint so that the connection cannot be pushed apart by the thrust of the water. These blocks are also placed to support the weight of valves or fittings, such as hydrants, that are joined to the pipe. The kicker block, or thrust block as it is also called, holds the pipe and fitting firmly in place, protecting it from separation due to water pressure or major shifts in the earth surrounding the pipeline. In some instances, metal straps are wrapped around the pipe and then anchored in the block to give added protection against separation.

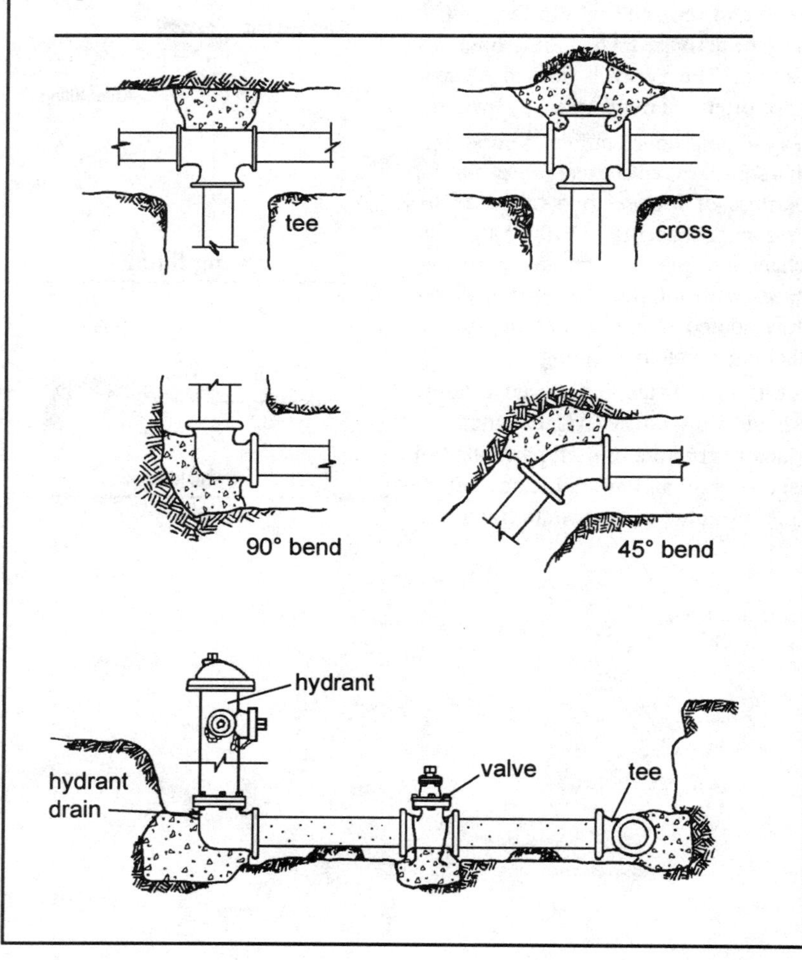

kicker block – concrete placed at a bend in an underground hub and spigot pipe system. *See Box.*

kick plate – a protective plate that is attached to the bottom of a door or other location likely to be damaged or worn from contact.

kiln – 1) a chamber for drying wood under controlled conditions; 2) a furnace for firing brick or tile.

kiln dry – to cure lumber, brick or tile in a kiln. This drying process removes moisture at a controlled rate, increasing the stability of wood and hardening bricks and tiles.

kiln run – a load of several pieces of wood, brick or tile fired and cured in a kiln at one time.

kilo – a prefix meaning one thousand.

kilocalorie – a kilogram calorie or 1000 calories.

kilocycle – 1000 cycles.

kilogram – a measure of weight equal to 1000 grams, or 2.205 pounds.

kilometer – a measure of length equal to 1000 meters, or ⅝ mile.

kiloton – 1000 tons.

kilovolt – 1000 volts.

kilowatt – 1000 watts.

kilowatt-hour – measurement of electrical energy usage equal to one thousand watts in one hour.

kinetic energy – the energy of a moving body, such as a hammer that is being swung toward impact with a nail.

kingbolt – a metal rod used as the central vertical framing member in a roof truss.

king closer – a brick that is cut down to half or three-quarters of a full brick length. It is used to fill the final opening in a brick structure where the opening is less than full length, but greater than half a brick length.

king post – the central vertical framing member of a roof truss.

king-post truss – a style of roof truss that uses a central vertical framing member as the main link between upper and lower horizontal truss members.

kitchen cabinet – a cabinet designed for the storage of food and/or food-related items, such as dishes, cooking utensils, and storage containers.

kitchenette – a small kitchen consolidated into a living area where space saving is of primary importance.

knee and column milling machine – a milling machine for tool and die making, called knee and column because of the shape of the casting that supports the table. It is used to make taps, reamers, ratchet teeth, fluted cutters, milling machine cutters, etc.

knee brace – diagonal bracing that extends from a mid-point on wall framing to the base of the wall or from the mid-point to the top of the wall. Used for reinforcing building framework where an opening near the corner of the structure prevents the use of diagonal bracing; 2) diagonal bracing from one structural member to another, such as between a cantilevered member and the main structure, to increase the load-carrying capacity of that member.

knee pads – padded protective covering for the knees used when working in a kneeling position. Used by brick layers, tile layers, and carpet installers.

knee wall – a short wall, usually about 4 feet high, such as might be found between a floor and a sloping roof.

knife – 1) a tool with a hard, flat metal blade sharpened on one or both edges for cutting; 2) a tool with a flat blade used to spread soft materials such as joint or texture compound.

knife switch – an electrical switch that opens or closes a circuit when one or more blades come in contact with one or more clips. The switch is manually raised or lowered by a handle to make or break the contact. Knife switches were formerly used in electrical panels, and can still be found in some very old existing panels.

knob and tube wiring – an obsolete form of house wiring in which the conductors are strung between porcelain standoffs (knobs) and porcelain tubes are used to line holes in structural members through which the wires pass.

knock-down texture – a type of sprayed-on texture applied to drywall that is created by spraying coarse and random droplets of texturing compound on the drywall and then troweling them flat.

knockout – circles of various sizes, factory-stamped into the sides of an electrical box, which can easily be punched out to form a hole through which wiring can be fed.

knockout plug – a plug that can be installed in a knockout hole of a metal electrical box when the hole is no longer needed.

knot – a hard, dense section of branch that runs through a piece of wood. Knots can come loose or come out of the board, leaving a hole in the wood.

knot hole – an indented space or hole left when a knot comes out of a piece of wood.

knurl – to make straight line or diagonal cross-hatch depressions in a metal surface to roughen the surface and/or to increase the diameter of the object. As the depressions are forced into the metal, the areas between the depressions are forced outward radially from the axis of the part being knurled, enlarging the surface area.

knurling tool – a tool consisting of two hardened steel wheels held in a fixture. Each of the wheels has an opposite-direction diagonally-embossed surface. It is used to create a rough surface finish for tool handles to aid in gripping and to knurl, or make depressions in, pistons to increase the piston diameter.

kraft paper – a strong, moisture-resistant, tan-colored building paper. It is reinforced with fibers bonded into it for added structural strength. In narrow rolls, it is sometimes call paper flashing. Paper flashing is used around rough window and door openings, overlapped from the top down, to shed moisture that might get through the exterior siding. Wider kraft paper is applied under the final exterior siding, such as stucco, to ensure the sheathing or framing under the siding remains dry under all conditions.

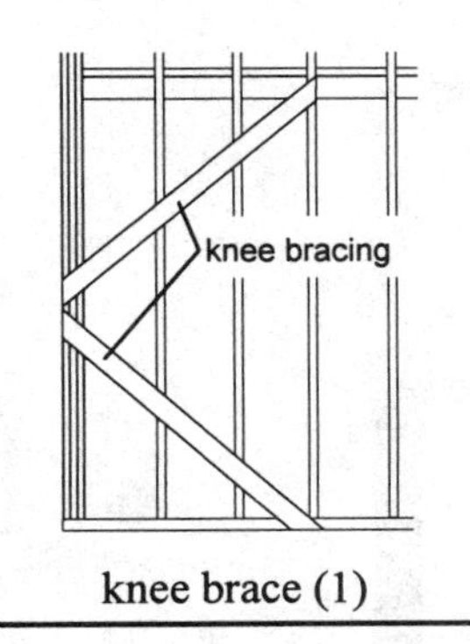

knee brace (1)

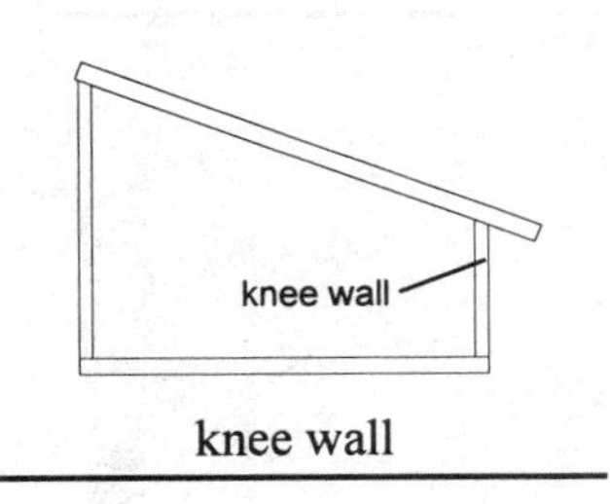

knee wall

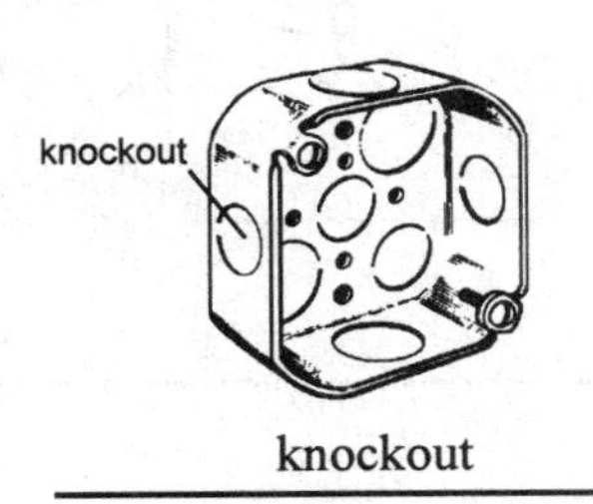

knockout

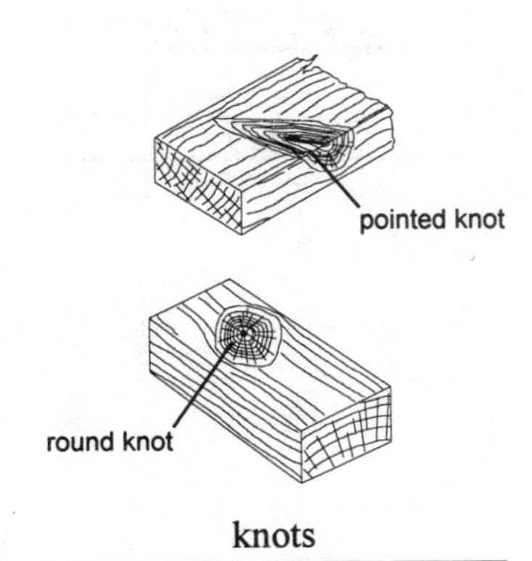

knots

L: labor burden to lux

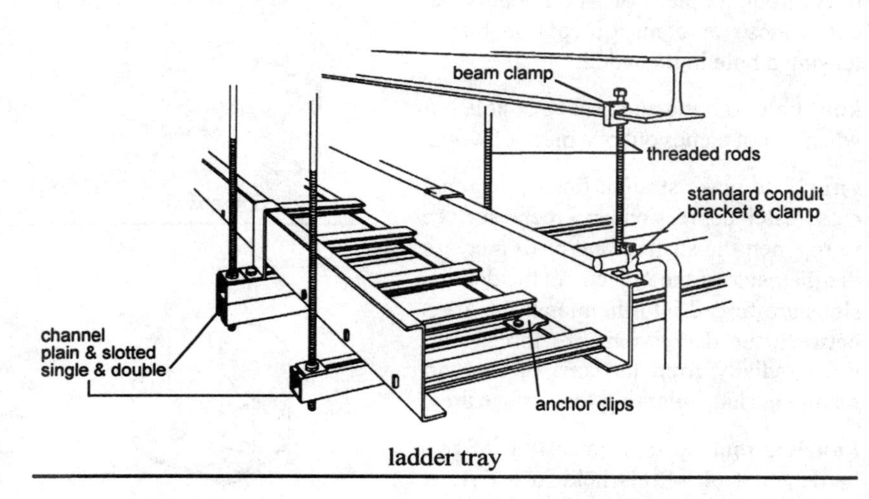

ladder tray

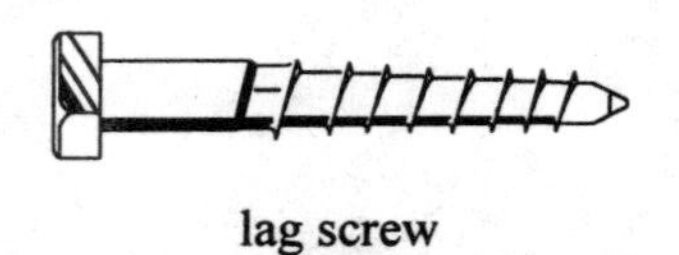
lag screw

labor burden – the cost to the employer for employee wages, taxes, insurance and benefits.

laches – failing to assert a legal right or privilege within a given time, such as neglecting to file a mechanic's lien within the legal time period.

lack of fusion – see incomplete fusion.

lacquer – 1) a synthetic coating made from cellulose acetate or cellulose nitrate in butyl alcohol or butyl acetate. It is commonly used as a finish for wood furniture because it dries quickly, minimizing the problem of dust particles adhering to a wet finish; 2) a resin found in plants (trees); 3) resin in turpentine.

ladder – a fixed or portable climbing structure consisting of two vertical rails with horizontal crossbars attached at convenient intervals for stepping.

ladder bar – a masonry joint reinforcement consisting of parallel wires with cross wires, similar in shape to a ladder.

ladder tray – a type of cable support with a shape similar to a ladder, having side rails spanned and connected by intermittent cross pieces.

lagging – thermal piping insulation used to minimize heat loss through the pipe wall.

lag screw – a fastener used with wood that has a hexagonal or square head and a threaded, pointed shank. Also called a lag bolt.

laitance – a white dusty film on the surface of cured concrete that remains when water evaporates from the surface. Laitance occurs when the concrete has not been worked properly during finishing or there is too much water in the mix and it is left to evaporate, rather than being removed. The concrete must be mixed in the proper proportions, and after placement, moderate vibrating or tamping is required to fill the forms. It should then be floated and troweled. In addition to being unsightly, laitance is an indication of weakness in the concrete.

Lally column – a structural column formed by filling a steel pipe with concrete.

lamella – 1) a thin flat membrane or layer; 2) short pieces of wood used in a roof framing method. The wood is laid out and joined in cross pieces creating a diamond pattern.

laminated – 1) in layers; 2) two or more layers of material joined with adhesive, such as in glu-lam beams or plywood.

laminated glass – multiple layers of plate glass with plastic sheeting between the layers. The glass adheres to the plastic sheeting and remains in a single piece, even when broken. It is commonly used for such applications as automobile windows, where there is considerable danger from flying glass in case of accidental breakage. Also called safety glass.

lampblack – a fine black, nearly pure carbon pigment made from soot and used as a coloring agent.

lamp cord – see SP wire.

lampholder – a simple lighting fixture, or holder for a light bulb.

lanai – a covered porch, gallery or veranda on one or more sides of a house.

landing – a level platform between flights of stairs to break the climb or allow for a change of direction in the stairs; a platform at the top or bottom of a flight of stairs.

landlord – the owner of real property that is leased to another.

landscape fencing – the use of shrubbery and trees to form a boundary or boundary demarcation.

landscaping – arranging shrubs, trees, and other plants on a specific site.

land use plan – a land development plan prepared by the developer. It details all the changes or improvements to be made on the property.

lane – 1) a narrow walk or roadway; 2) a marked parallel division on a highway which provides space for passing or side-by-side travel.

lane delineators – bright-colored cylinders or cones which are set up in roadways to mark temporary lanes or detours.

lantern – 1) a light or portable casing for a light; 2) the chamber of a lighthouse that houses the light; 3) open architecture that admits light or ventilation to an area below it, such as a dome window, skylight or glass roof; 4) a cupola; 5) the indicator light in an elevator lobby that signals the arrival of the elevator.

lanyard – 1) a short rope or cord used for pulling an object or from which an object is suspended; 2) a rope or cord attached to a safety belt on one end and a lifeline on the other; 3) a rope or line used to fasten down equipment or cargo on a ship.

lap – 1) to sand into a final shape, dimension, or finish using a fine abrasive compound; 2) a cylindrical metal rod, coated with an abrasive, used to smooth the inside diameter of a hole in metal. The lap is chosen to match the diameter of the hole to be lapped so that it is a tight turning fit in the hole.

lap cement – an asphalt-based cement used between the overlapped layers of roll roofing to make a tight bond between the layers.

lap joint – a joint in which the two pieces to be joined overlap each other.

lap siding – a technique for installing horizontal siding boards in which the bottom edge of one board is lapped over the top edge of the board below.

lap weld – a weld between overlapping members.

large scale integrated circuit (LSI) – an electronic chip comprising many semiconducting components.

laser – an acronym for light amplification by stimulated emission of radiation.

laser beam cutting (LBC) – using the concentrated beam of a laser to cut wood or metal.

laser beam welding (LBW) – using the concentrated beam of a laser for the welding heat to join metals.

laser level – see level, laser.

latch – a two-piece fastening mechanism designed to secure a door, window or gate to a jamb, sill or post for closing.

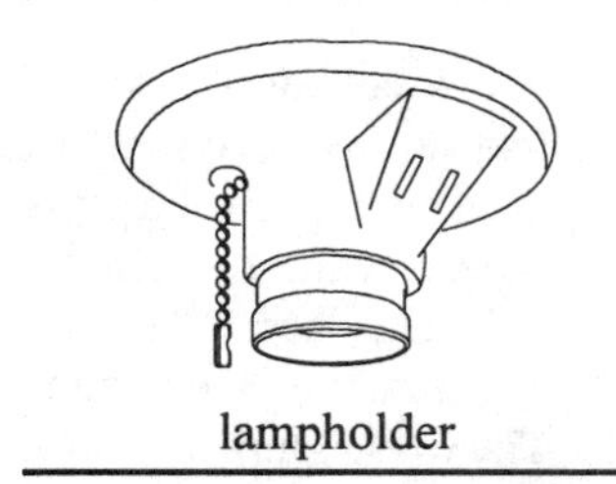
lampholder

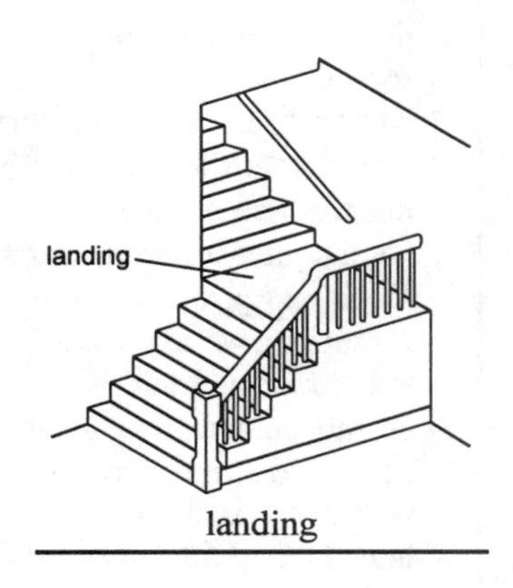

landing

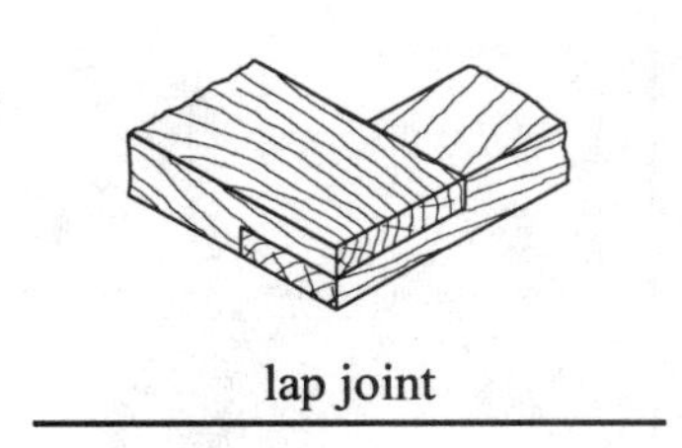
lap joint

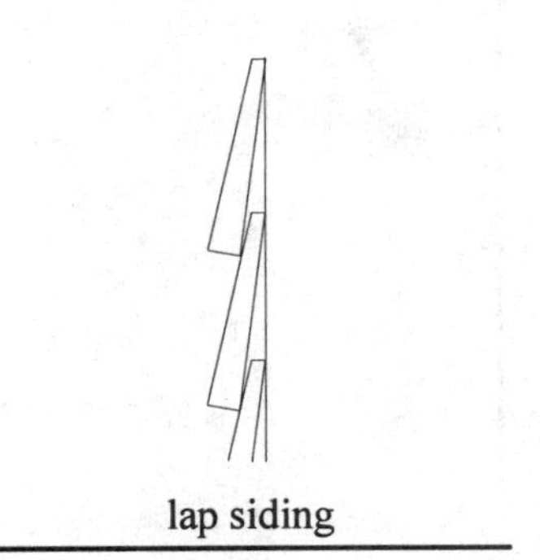
lap siding

Lathe

A lathe is a power tool which rotates a wood or metal object so that it may be symmetrically shaped with cutting tools. The lathe headstock grips the top of the object in a chuck which rotates the object, while the adjustable tailstock secures the other end of the object in place, but still allows it to rotate. The tailstock can be moved along rails on the lathe to adjust to the length of the piece of work. The work is turned at an adjustable rate of speed by the power source, usually an electric motor. The speed at which the lathe turns is mostly a function of the hardness of the material being worked. The harder the material, the more power is needed to shape it. There are lathes for shaping wood and lathes for shaping metal. They differ in the cutting tools used, the method by which both the tools and the materials are held, the number of automatic drive features, the power needed, and the size of the lathe structural members. A wood lathe has a lighter structural framework than a metal turning lathe because much less power and force are needed to shape wood. Lathe tools for turning wood are mostly hand-held. Metal lathe tools are usually held in a metal tool post firmly attached to the lathe. The tools can be moved into position manually or automatically adjusted by a power-driven feature. Wood lathes are used to make sculptured woodwork such as balusters, newel posts or furniture legs. A metal lathe is used to make items such as screws or valve stems.

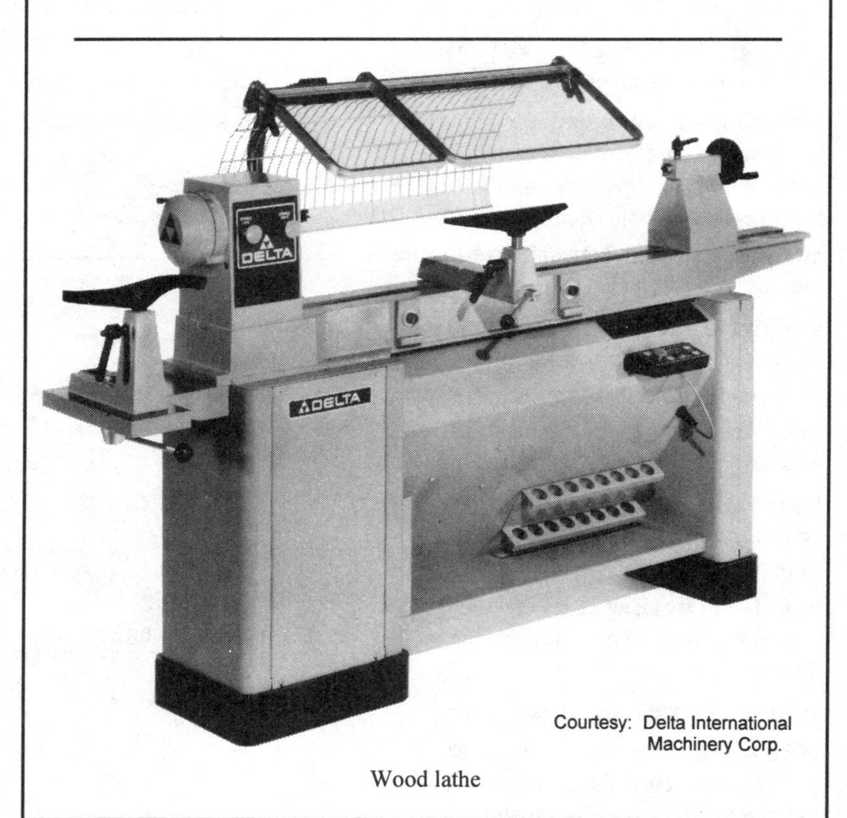

Courtesy: Delta International Machinery Corp.

Wood lathe

latch post – the fence post at a gate to which the gate latch is attached.

latent heat – heat involved in a change of state, such as the heat required to change a solid to a liquid.

lateral – acting to the side, or at a 90-degree angle to an object.

lateral reinforcement – horizontal reinforcement surrounding a concrete column.

lateral support – a structural support designed to resist a load or force applied to the side of a structure.

latex – a milky fluid which dries to a rubbery coating.

latex paint – a water-base paint with a natural or synthetic latex emulsion vehicle that dries by the evaporation of water. It dries very quickly, gives off minimal fumes, and can be used for both interior and exterior applications.

latex-portland cement mortar – a mixture of portland cement, sand and liquid latex. This mortar may be used to set ceramic tile on masonry, concrete, gypsum drywall, cement board, mortar beds and similar applications. It is moisture-resistant when allowed to fully cure before being exposed to water.

lath – the backing fastened to structural members onto which plaster is applied. Lath originally was strips of wood fastened to the framing members. Today however, gypsum or metal lath are the most commonly-used types of lath because they are inexpensive and can be installed in sheets which cover a large area very quickly.

lath and slat fence – a slat fence using vertical fence boards that are alternating 1 x 1 lath boards and wider slats, often 1 x 2s, with spaces between them.

lathe – a power tool that turns an object so that it may be shaped with cutting tools. *See Box.*

lath nailers – horizontal framing members added as fastening points for lath where the wall and ceiling framing join. Lath is the wood, metal or gypsum product fastened to the structural members to support the application of plaster to the walls.

latitude – 1) the angular distance north or south of the equator measured in degrees; 2) in surveying, the angular distance north or south from a fixed point, measured in degrees.

latten – thin sheet metal which can be used as a protective covering, a shim or a variety of other purposes.

lattice – thin wood strips fastened together at angles in a crisscross pattern to create an open basketweave effect.

lauan – Philippine mahogany or other related Philippine softwood, often used for interior doors.

laundry tub – a deep sink for soaking and/or washing laundry or other items.

lavatory – 1) a sink with water supply and drainage used for washing; 2) a room with a sink and water closet, with both running water and drainage.

lavatory, flush-mount – a lavatory, supported by a metal ring, which sits even and flush with the countertop around it.

lavatory, pedestal – a lavatory which sits on a pedestal base and is supported from a wall-mounted bracket.

lavatory, self-rimming – a lavatory with an integral rim that supports the lavatory on the countertop.

lavatory, undercounter – a lavatory which is mounted to the underside of a water-resistant countertop, such as marble.

lavatory, wall-hung – a lavatory supported from a wall-mounted bracket.

lawn sprinkler system – a water system for lawn irrigation which includes pipe, sprinkler heads, valves and other necessary parts used above or below the ground.

laying off – a painting technique in which a surface that has been freshly painted with a brush is brushed again with upward strokes to smooth the paint.

laying out – determining in advance how pieces of a system or structure should be arranged, usually involving a plan or design which indicates the relationship of one component to another.

laying overhand – building the face of a masonry wall from the side opposite the face of the wall.

lazy Susan – a circular or semi-circular shelf or platform that revolves in order to provide access to all sides.

leach – to remove or separate by percolation, as solids are removed from liquids by percolating through soil.

leach field – an area of land through which leach lines from a septic tank run. The lines empty treated liquid waste from the tank into the field. The soil in a leach field must be porous, like sand, so that the effluent can percolate down through it quickly, or the field must be constructed in such a way as to allow it to drain readily.

leach lines – effluent lines from a septic system to a leach field. As new sewage is added to the septic tank, the treated liquid is forced out of the tank into the leach lines and then carried to the leach field where the lines empty and drain into the soil.

lead – a dense, soft metallic element with the atomic symbol Pb, atomic number 82, and atomic weight of 207.19.

lead – 1) a section of a masonry wall that is built up, and from which successive courses are stepped back; 2) the linear distance a threaded object moves in one revolution. For example, a machine screw may move into a nut a linear distance of $\frac{1}{8}$ inch with one revolution.

lead carbonate – a poisonous salt formerly used as a white pigment. Lead is no longer used in most paint products.

lead drier – a drying agent made from lead and an acid that was formerly used in paint. The use of lead in paint products has been greatly reduced since 1972.

leaded zinc – zinc oxide and lead sulfates combined in a primer. At one time it was used as a coating to prevent rust. Lead is no longer used in most paint products.

laundry tub

lavatory, pedestal

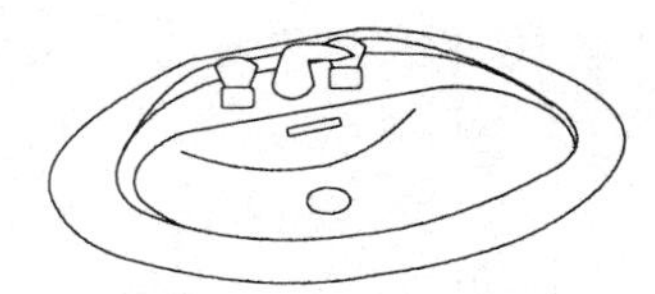

lavatory, self-rimming

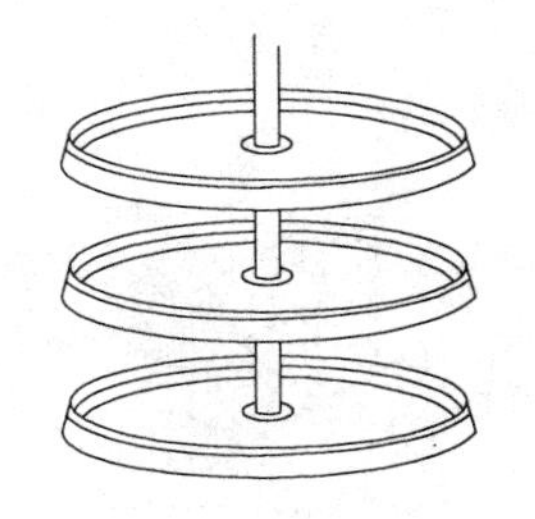

lazy Susan

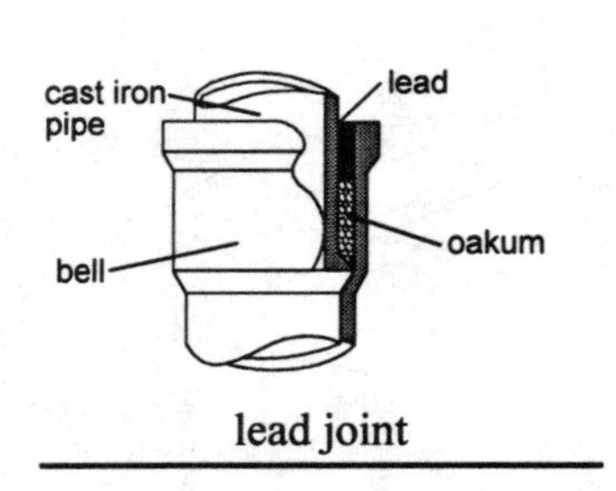

lead joint

leader – see downspout.

lead expansion anchor – a lead sleeve that is split lengthwise and threaded internally; the sleeve is installed in a hole, such as in a concrete floor, and a screw is tightened in the sleeve, expanding the sleeve against the sides of the hole, resisting pull-out. It can be used to permanently anchor an item to a surface.

lead joint – a cast iron, bell and spigot plumbing drainage pipe joint in which oakum, a rope made of coarse untwisted hemp, is tucked into the joint and molten lead is poured over the oakum to seal the joint. Also called a lead and oakum joint.

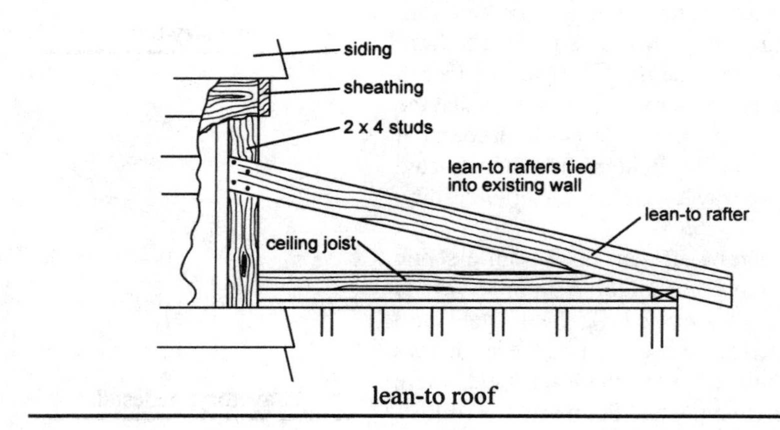

lean-to roof

lead oxide – a lead and oxygen compound formerly used as a paint pigment and a rust preventative.

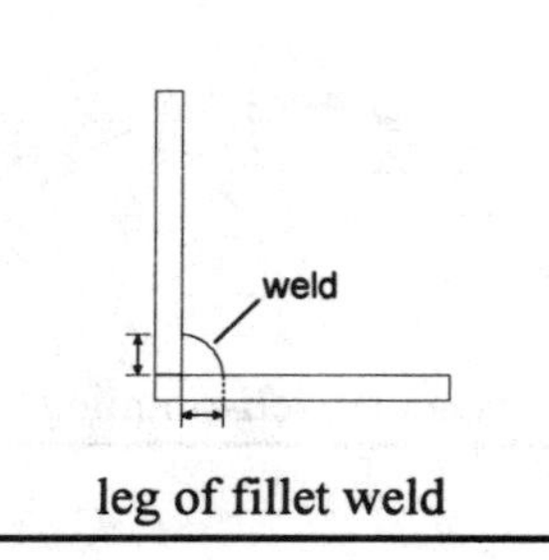

leg of fillet weld

leads – electrical wires connected on one end to a meter, instrument, battery, or other electrical/electronic item and on the other end to a circuit or component.

leak – a crack or hole that allows a substance, such as light, gas, electric current or fluid, to escape its boundaries.

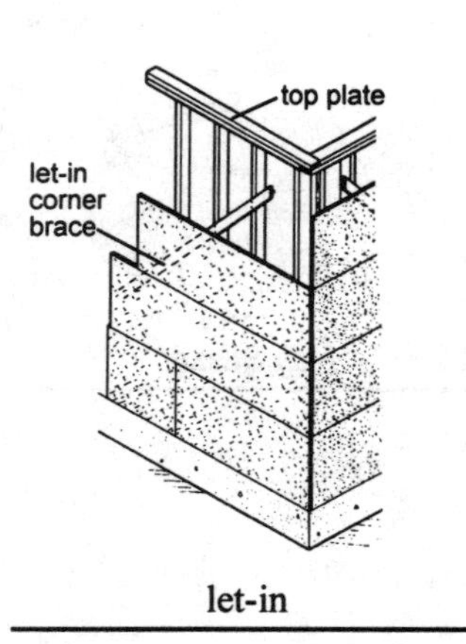

let-in

leaning edge – the factory taper on the long edge of a gypsum wallboard panel.

lean mortar – mortar that has insufficient cement in the mix. It results in a weak mortar that can be used as a soil stabilizer or a bed under pavement.

lean-to – a small building with a shallow roof that slopes in only one direction, similar to one-half of a gable roof.

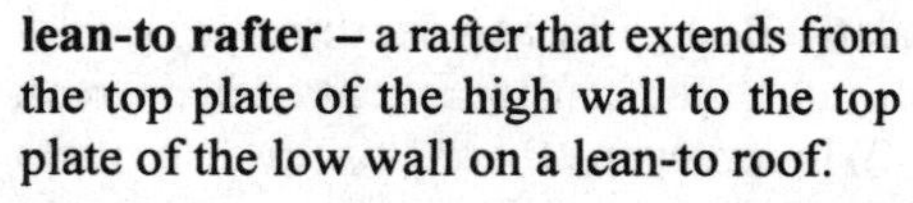

lean-to rafter – a rafter that extends from the top plate of the high wall to the top plate of the low wall on a lean-to roof.

lean-to roof – a roof that has only one slope, similar to half of a gable roof; the slope of this roof is often fairly shallow.

lease – a contract in which use of property is conveyed for a specific period of time.

ledge – a narrow projecting shelf.

ledger – 1) a record of financial transactions; 2) a horizontal support member fastened to a wall, on which joists rest. Also called a ledger board or ledger strip.

leg – a vertical support member.

leg of a fillet weld – the legs of the triangle formed by fillet weld. A fillet weld is a weld at the intersection of materials forming a 90 degree angle. The two legs join to form the right angle, with the weld forming the hypotenuse of the triangle.

length – a linear dimension along the longest side of an object or structure.

lens – a transparent, convex or concave shaped piece of material that is used to provide focus and/or magnification by changing the convergence of light rays passing through it.

lessee – a person who is granted a lease, such as a person renting a home.

lessor – the person who grants a lease, such as a landlord renting out his property.

let-in – a juncture of members where a section of one object is removed, or notched, so that a second object will fit flush with the surface of the first object. An example would be a section of stud being notched so that a let-in diagonal brace surface fits flush with the stud surface.

lettering guide – a mechanical drawing tool, used as a guide for lettering, that has holes or slots for drawing parallel lines.

letter of intent – a written declaration by a client indicating an intention to enter into a contract.

levee – an embankment or ridge used to confine water and prevent flooding, frequently used along riverbanks where water has a tendency to rise and flood after a rain.

level – 1) horizontal or aligned with the horizon; 2) a device, usually using a fluid-filled tube with an air bubble trapped within the fluid, that indicates when a surface is either horizontal or plumb by the position of the bubble within the tube.

level, builder's – a device consisting of a telescope, bubble level, and a movable mounting ring for the telescope that is used to establish straight work lines and verify elevations or the level of job site reference points in relation to each other.

level, dumpy – a type of builder's level consisting of a telescope and leveling vial mounted on a rotating plate. A sight taken through the telescope at a graduated rod some distance away can be used to find the point on the rod that is level with respect to the telescope. The telescope is leveled using the leveling vial, and its crosshairs are lined up on a point on the rod.

leveling – a desirable property of paint that causes the surface to flow, eliminating brush marks.

leveling bed – a bed of mortar in which pavers are set.

leveling course – a pavement layer or course put down to create a level surface before the final pavement course is laid.

leveling rod – a rod, marked with graduated measurements, used in conjunction with a transit or builder's level to measure elevation differences. They are available with graduations marked in feet, tenths, hundredths, meters, decimeters, and centimeters for both elevation readings or stadia readings. A target with graduations of a thousandth of a foot on a vernier scale may be attached to the leveling rod to provide even greater accuracy. They come in varying heights up to 25 feet. Also called a level rod.

leveling vial – a sensitive bubble level that is part of a surveying instrument, such as a transit or dumpy level.

level, laser – a leveling device that uses a laser transmitter and a laser receiver mounted on a vertical rod which has measuring marks along its length.

level, optical – a telescope, with cross hairs in the viewer, that rotates 360 degrees on its base. The base is mounted on a tripod which provides a stable position for viewing. The elevation of an object that is sighted and lined up with the cross hairs can be determined by turning the telescope and focusing on another object on the same horizontal plane whose elevation is known. The elevation of the object can also be compared to a leveling rod positioned within the horizontal range of the object. If the elevation of the telescope and leveling rod is known, the elevation of any object can be determined by comparing it to the leveling rod. Objects sighted will be at the same elevation, because the cross hairs remain at the same horizontal position as the telescope is rotated.

level protractor – a rotatable protractor, with a level in its center, mounted on a fixture with a base. Used to determine the amount of angle deviation from level.

level rod – see leveling rod.

level, string – a lightweight level with hooks so that it can be attached to a line stretched between points.

level, transit – a type of builder's level, with both a telescope and a leveling vial, that can take vertical angle measurements and horizontal measurements to establish elevations for residential building or for small construction sites. *See Box.*

lever – a rigid bar that acts against a fulcrum or pivot to gain a mechanical advantage, used to pry or move something.

leverage – 1) a mechanical advantage gained by using a lever; 2) using borrowed funds to invest, thereby gaining a greater percentage of return on one's own money.

lewis – a series of plates that are fit into a dovetail cut in a stone to aid in lifting the stone.

lewis pin – a metal dowel used to aid in lifting heavy blocks into position. *See Box on page 193.*

liability – a debt or obligation.

license – formal authorization granted by a legal agency to perform specific acts.

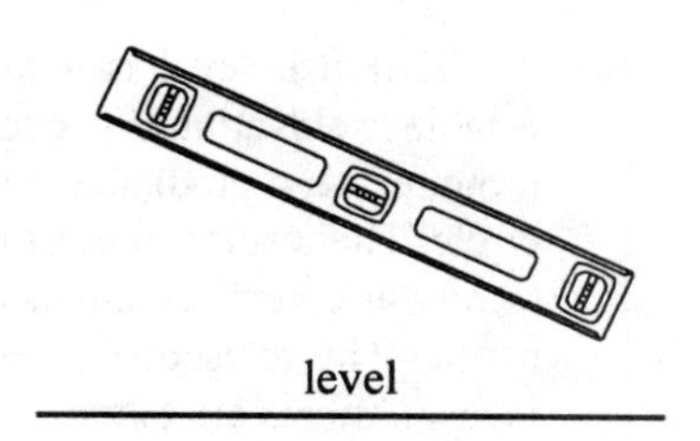

level

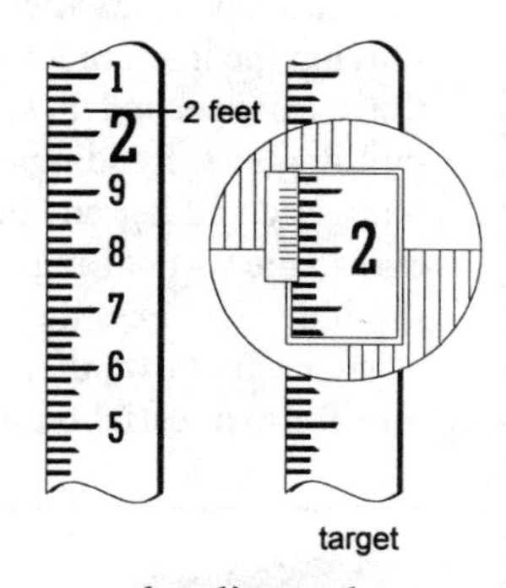

leveling rod

Level, transit

A transit level is a surveying instrument used by builders to establish elevations. It consists of a telescope and a leveling vial mounted on a graduated arc which rests on a rotating plate called a circle. The result is a transit and level which can move up and down to measure vertical angles as well as rotate for horizontal measurements. The telescope is leveled using the leveling vial, and then measurements are taken by lining up the telescope's crosshairs on a point on a level rod placed at a known elevation. This point can be used to find the difference in elevations for a point viewed through the telescope at another graduated rod some distance away. Without moving the instrument, elevations can be established with respect to the telescope and a known reference point at each station on a building site. Readings other than level can also be taken using the telescope and a graduated rod. The telescope can be tilted on its arc to read the vertical angle of sight from the telescope to the point being sighted. The versatility of the transit level makes it popular among builders. It allows them to make quick and accurate elevation readings for residential building or other small construction jobs.

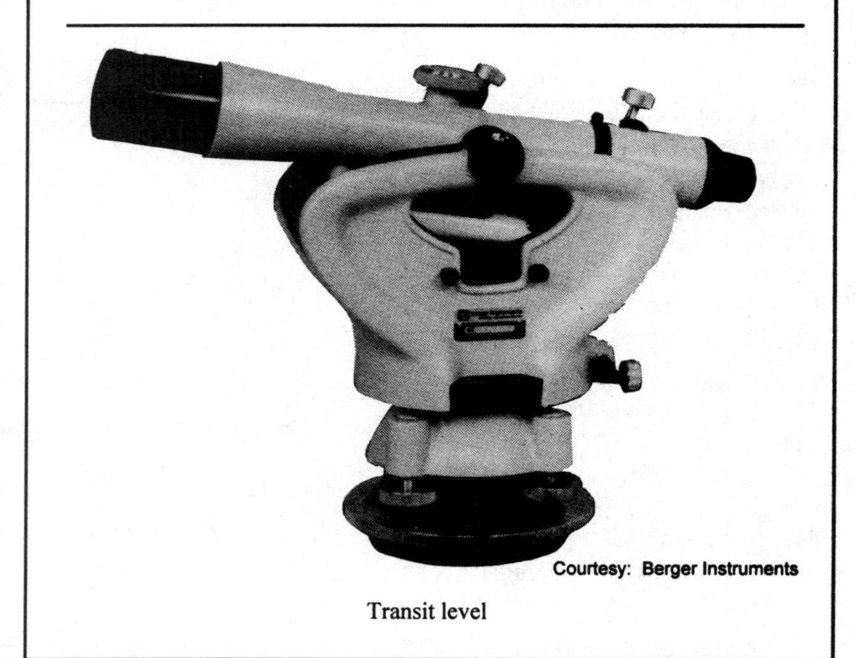

Courtesy: Berger Instruments

Transit level

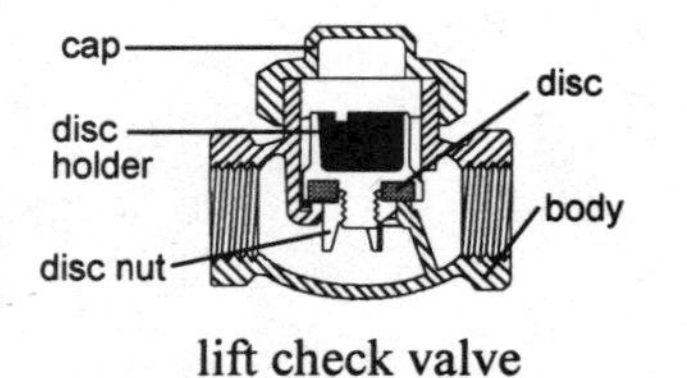

lift check valve

lien – a legal claim against real property for the satisfaction of a debt. A mortgage is an example of a lien, in which the property becomes security for the debt and clear title to the property is not obtained until the debt is paid. Some types of liens, like tax liens or mechanic's liens, result from the refusal or inability of a property owner to pay a debt, and can force the sale of the property to satisfy the debt. See also mechanic's lien.

lift – 1) to raise or elevate from a lower level; 2) the height or thickness of concrete that can be placed in one continuous pour, or a vertical layer of concrete as placed in a tall form; 3) a layer of spoil or backfill placed in a trench or excavation; 4) the amount of urethane foam that can be applied continuously; 5) the maximum vertical travel distance of a crane hook, also called hook travel.

lift check valve – a valve which permits fluid to flow in only one direction. When the valve is open, a disc is raised off its seat by the flow of the fluid in the direction that the disc is free to move. The disc closes by gravity, and/or a spring, when the flow tries to reverse. Lift check valves are often used in small diameter fluid systems.

lifting device – a lifting aid, such as a bucket connected to a crane hook. The lifting device is considered part of the load, as it must be lifted by the hook along with the load that it contains.

light or lite – 1) a source of light or illumination; 2) a medium, such as a pane of glass, through which light is admitted; 3) a small window, often placed in or alongside an opening, such as a door.

light bulb – an incandescent lighting element consisting of a filament heated by electrical resistance, converting some of the energy inside a glass envelope to light.

light diffuser – a translucent sheet of plastic or glass with facets that diffuse or spread light out, so that a concentrated beam of light can be expanded into a wide light pattern covering a larger area.

light framing – lumber, up to 4 inches wide, used to construct the skeleton of a structure such as a house, mobile home or other one- or two-story building. It can also be used to build portions of taller structures.

light handle switch – see switch, light handle.

lighting – 1) the illumination of an area; 2) the fixtures or apparatus that provide the illumination.

lighting maintenance factor – an estimation of how often cleaning and replacement of lighting devices is required. This information is used by maintenance scheduling personnel in commercial and industrial buildings.

light intensity meter – a device that measures the amount of light in an area. Photographers use these meters to take light measurements for correct shutter speed settings and aperture openings.

lightning – a high voltage atmospheric discharge occurring during certain types of storms. A charge builds up in a cloud to a point where the surrounding insulation, the air, can no longer contain it. The electrical charge travels through the air to the nearest conductor, often the ground or a conductive material connected to the ground.

lightning arrestor – a device that protects electrical equipment from high voltage surges, such as lightning strikes. The arrestor is connected to the equipment to be protected and then connected to a ground. It will divert a high voltage surge to the ground before it reaches the equipment.

lightning rod – a metal rod placed at a high point in an area or on a structure to attract and direct a high voltage atmospheric discharge (lightning) away from structures or objects which might be damaged if struck.

lightweight concrete – a concrete that weighs 90 to 110 pounds per cubic foot because of the use of lightweight aggregate. Regular concrete weighs approximately 150 pounds per cubic foot.

lightwood – a term often used when referring to pine, which is light in color. Pine is used for furniture, trim molding, flooring, and interior wall paneling.

Lewis pin

A lewis pin is a metal dowel or pin which is inserted into a stone block to aid in lifting the block into position. The pins are inserted into holes drilled at a slight angle into the top of the blocks. The holes are drilled so they slope away from each other. Each pin has an eye on one end through which rigging is threaded to connect the pins and balance the stress between them for lifting the stone with a crane or other large-capacity lifting device. Since the holes slope away from each other, the lifting stress does not pull the pins from the holes. When the lifting stress is discontinued, however, the pins can easily be removed individually. Stone blocks or panels can be ordered with predrilled lewis holes for ease of installation.

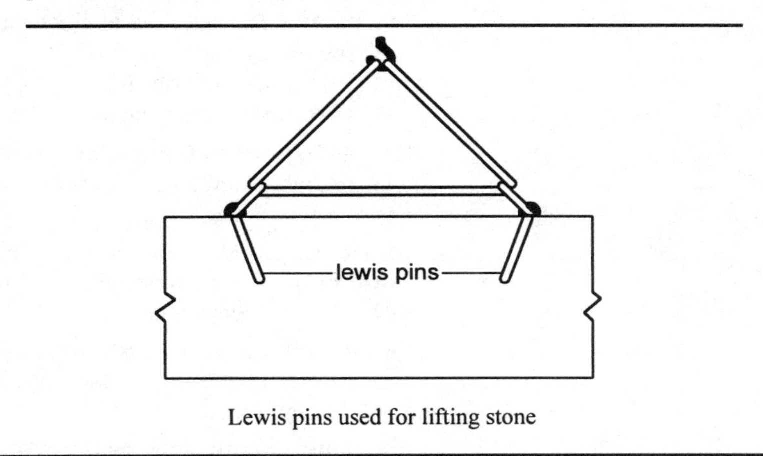

Lewis pins used for lifting stone

lignite – a brown, incompletely-formed coal, in which the texture of the original wood is still apparent. It is used as a fuel source in some areas.

lignum vitae – a hard, dense wood from either of two species of guaiacum (tropical American trees), used for such applications as pulleys or propeller shaft stern bearings on large ships.

limber – flexible, or capable of being shaped, such as a thin metal.

lime – a caustic form of calcium oxide used in cement, mortar and plaster.

lime putty – hydrated lime in a putty form. When added to mortar, it keeps the mixture workable by retaining water.

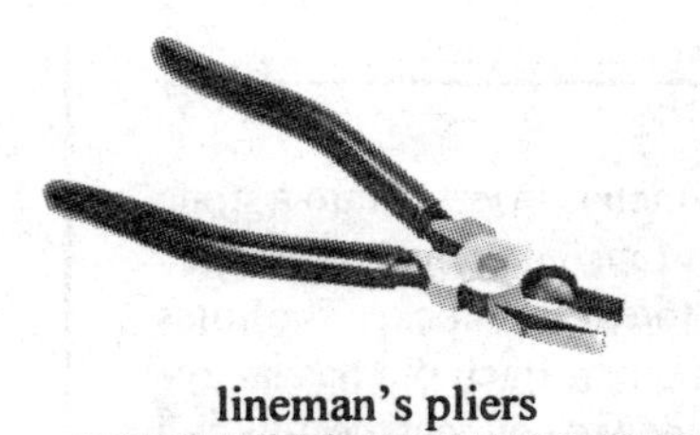
lineman's pliers

limestone – a building material of calcium carbonate. Limestone is a stable and attractive building material. It is used for building walls, columns, facings, and trim and is available in several different surface finishes.

limestone, Carborundum finish – limestone with a finely-ground surface. Used in building where a smooth surface appearance is desired. See also limestone.

limestone, honed finish – limestone with a very fine smooth finish. Used in building where a near polished surface appearance is desired. See also limestone.

limestone, machine-tooled finish – limestone blocks with four to eight parallel grooves per inch carved into the surface. The grooved finish is for decorative effect. See also limestone.

limestone, plucked finish – limestone with a rough textured surface appearance. See also limestone.

limestone, rough shot-sawn or chat-sawn finish – limestone with a coarse finish. See also limestone.

limestone, sand-sawn finish – limestone with a moderately smooth sawn surface. See also limestone.

limestone, smooth machine finish – limestone with a smooth finish produced by planing machines. See also limestone.

limestone, wet-rubbed finish – limestone with a finish rubbed smooth by using an abrasive. See also limestone.

limit – 1) a geographical boundary; 2) a restriction to the use of something; 3) a maximum amount allowed.

limit switch – a switch that turns off electrical power to a mechanism when a preset point is reached. The limit switch can be adjusted so that part of the mechanism contacts the switch. When the mechanism has reached the point of travel where it is intended to stop its motion, this contact actuates the switch and turns the power off.

line – 1) a mark of uniform width; 2) a term referring to a conductor of water or electricity, such as piping or electrical wiring.

linear – 1) in a straight line; 2) involving a single dimension.

linear foot – a foot of distance (12 inches) measured in a line along a surface, or the surface of a material such as conduit or piping. Also called a lineal foot.

linear indication – any sign of a potential problem, such as a flaw in a surface, found during a surface nondestructive examination (NDE), that requires further evaluation . The surface NDE can be visual, or use a magnetic particle or liquid penetrant examination. The flaw is called a linear indication if its length is at least three times its width.

line graph – a visual representation, by means of lines connecting points on a graph, showing how one quantity changes in relation to another or in relation to a constant factor, such as income from one period compared to another.

lineman's pliers – pliers with a tapered blunt nose and side cutters as part of the jaws. They are a multi-purpose tool, but used primarily for working with wire. The handles are covered with insulative vinyl to protect electricians and linemen from shock when cutting live wire. Also called electrician's pliers or side-cutting pliers.

lineman's splice – see Western Union splice.

linen closet – a closet with shelves for storing linens, such as sheets and towels.

line of sight – the visual path from the eye to a distant point.

line post – fence posts between corner posts and/or gate or latch posts.

liner – a material that serves as a lining, such as a rubber lining in a steel pipe.

lines of force – the lines of magnetic force that surround a magnet.

lining – 1) a material covering an object or the inside of a container to protect the surface or to protect the contents of the container; 2) a covering used to improve or alter the look or texture of the surface of a container or to alter the material that is placed against that covering. For example, a layer of material attached to the inside of forms can be used to alter the surface texture of the finished concrete poured into the forms.

Liquid penetrant examination

A liquid penetrant examination is a non-destructive examination which relies on a penetrating dye to seek out cracks or flaws in the surface of a nonporous material or part. The dye is applied to the surface and left for a prescribed period of time. The excess dye is then removed and an absorbent powder or foam, called a "developer," is applied. The developer draws the dye out of the defect, revealing its location and size. There are both colored and fluorescent dyes, the fluorescent being the most sensitive penetrant. This type of exam is only a "surface" examination and is limited to detecting defects that extend to the surface of the part being examined. It is used to check pressure-retaining parts subjected to high heat during manufacture, such as during welding, and to check items such as valve parts which have been hard-surfaced.

A type of liquid penetrant testing, called filtered-particle testing, can be used to examine surfaces that are so porous that the liquid penetrant is absorbed to the point that defects are masked. In this type of examination, dye particles are carried in a fluid that is usually oil-based. More of the liquid is absorbed into the defect than into the rest of surface of the material being tested. The dye particles in suspension in the liquid are filtered out, remaining as a deposit on the surface. By evaluating the difference in the absorption between the defective and defect-free surface areas, the defect location shows up, highlighted by a greater percentage of the dye. Fluorescent dyes are most often used with this type of test. Materials such as clay, concrete, ceramics and carbon can be examined with this method.

lining felt – 15 pound, or heavier weight, asphalt felt. It can be laid between a tongue-and-groove wood subfloor and the finished floor, to provide a cushion and smooth irregularities.

linkage – a series of mechanical connections used to transmit a force, such as the connections made between a gearshift lever and a transmission.

linoleum – a tile or sheet floor covering of ground cork, wood filler and pigment, held together by linseed oil or other binders which solidify, and then backed with felt.

linoleum knife – a hook-blade knife with a sharp edge ground into the inside curve of the hook. Originally developed for use in cutting linoleum, they are now used for cutting vinyl flooring, trimming drywall, and cutting detailed shapes as well.

linoleum varnish – a varnish that is formulated to be as flexible and elastic as linoleum, hence the name.

linseed oil – an oil made from the flax plant that is used as a drying agent and a vehicle for paint.

lintel – 1) a horizontal structural member that supports the load over a door, window, or other opening. Also called a header; 2) an opening in a masonry structure.

liquefaction – changing from a solid or semi-solid state to a liquid state, such as soil which has been saturated with water during an earthquake. The earth movement causes the saturated soil to lose its stability and ability to support weight and often intensifies the effects of the ground motion, increasing the resultant damage.

liquid – the state of a substance, such as water, in which its molecules tend to stay together, but move freely so as to assume the shape of the container holding them.

liquid membrane – roof sealing materials that can be applied in a liquid form which solidifies to form a membrane, such as an asphalt- or tar-based solution which contains glass fibers.

liquid penetrant examination (PT) – a type of test which uses a penetrating dye to locate cracks or flaws on the surface of a material. *See Box.*

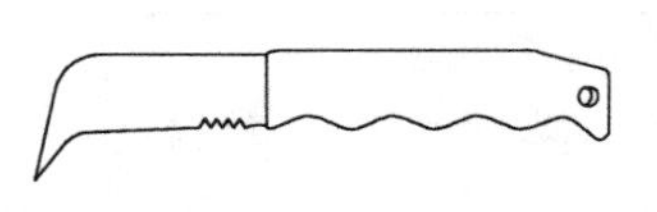

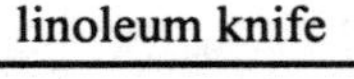

linoleum knife

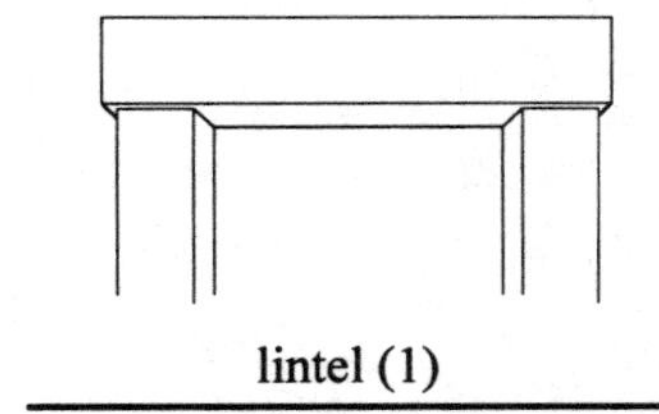

lintel (1)

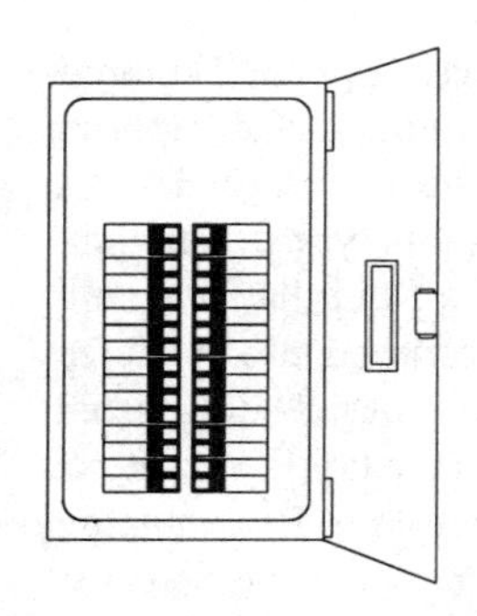
load distribution center

loader

liquid resin glue – a white, non-waterproof glue for wood. It is strongest when used with tightly fitted joints.

liquid sandpaper – a trade name for a chemical mixture that will remove many surface coatings in thin layers to provide an effect similar to that achieved with sandpaper. Its primary advantage over sandpaper is that it does not create dust.

liquid-tight flexible metal conduit – see flexible metal conduit, liquid-tight.

liquified petroleum gas – butane or propane, or a mixture of these gases, used for fuel.

list – 1) to lean to one side; 2) a series of words, numbers or items compiled for a particular use, such as a list of building materials needed to complete a project.

listing – an agreement between a real estate broker and property owner to sell a parcel of real property on a commission basis. The agreement usually involves placing a notice in the local real estate multiple listing service, making the property known to other real estate agents in the area, and some form of advertising to increase its market exposure.

list of materials – a list, usually on a fabrication or construction drawing, that shows the materials needed to build the item shown on the drawing. The list is made by the architect or designer.

litharge – lead monoxide, used as a component in vitreous enamel, and also used in making lead-acid batteries, ceramics and lead glass.

lithopone – a compound of barium sulfate, zinc sulfide and zinc oxide, which forms a white pigment. It is commonly used in white paints and as a coloring compound in linoleum.

litigation – a lawsuit or legal proceeding to settle a dispute.

litmus – a material from a lichen that turns red in acid solutions and blue in alkaline solutions. It is used to test the acidity or alkalinity of chemical combinations or body fluids, or to help maintain the pH balance in fluids such as the water in a swimming pool or hot tub.

littoral – pertaining to the shore of a body of water, especially that portion between the high and low watermarks.

live – an electrical circuit or connection with power applied.

live load – 1) the load on a structure caused by temporary or movable sources that are not part of the building, such as people, snow, or furniture. The basic capacity of a structure to carry a live load is predetermined and becomes part of the requirements for building the structure; 2) the load on any part of a crane that changes in amount and/or relative position to that part. For example, the live load on a crane trolley is the weight of the hook plus the added weight of the load on the hook.

live load deflection – the vertical deviation of a crane bridge caused by the weight of the crane trolley and the load being lifted. The crane is designed to compensate for this factor.

livering – a term used to describe paint that has congealed into a mass in the paint can after sitting for a period of time.

living unit – a single dwelling unit.

load – a weight or force on something.

loadbearing partition – an interior wall that carries part of the load of the structure.

load block – see block, load.

load distribution center – an electrical panel, containing circuit protection devices, from which circuits are routed throughout a house or building. Also called a service panel.

loader – a type of earth moving equipment used to pick up and move excavated spoil. A shovel-type bucket attached by movable arms to the front of the vehicle picks up the spoil, lifts and dumps it in a new location.

load factor – 1) an earth-moving factor that compensates for soil swell; the load factor equals the bank volume divided by the bank volume plus the soil swell percentage; 2) a comparison of the average load likely to be experienced by a system to the total load capacity of the system, expressed as a percentage.

load float – a crane control system that permits hoist operation without movement or stepping, in either direction, and allows the load to be held stationary with holding brakes released.

loading pump – a hand pump used to load drywall joint compound into application tools that have a reservoir for the compound.

lobby – an entrance area to a larger room or rooms within a building, often used as a public waiting area.

lock – a device for securing something against opening.

locked rotor torque – the minimum torque developed by a squirrel cage electric motor when voltage is applied and the motor shaft is not turning. A squirrel cage electric motor is the simplest and most common type of electric induction motor.

locking pliers – pliers with an adjustable over-center locking lever that can be opened and locked into the desired jaw opening position.

locknut – a second nut tightened against another to prevent the first nut from turning or becoming loosened from the bolt on which they are both installed. See also jam nut.

lock seam – a seam, between sheets of metal, that is folded over to lock the sheets together in a flat joint.

lock washer – a washer designed to ensure that a nut tightened against it does not come loose. Lock washers are either made of spring steel or with teeth. The spring steel washers are deformed, often with a split across one side. Tightening a nut against it flattens the steel and keeps the nut from loosening. Toothed washers dig into the nut and the part being fastened to keep the nut from loosening.

lock wire – stainless steel wire that is threaded through a hole in a bolt head or nut, twisted to a prescribed number of turns per inch, and then secured to an anchor point so as to resist the loosening of the nut or bolt to which it is attached. Lock wire is used primarily on aircraft, racing engines and rocket bolts.

loft – an open upper floor of a building, such as an attic.

log – 1) a length of tree trunk which can be used for the exterior structure of a building or cut into lumber or firewood; 2) a record of events.

loggia – a roofed room, arcade, gallery, or other enclosed area, open on one side to the outdoors, often an open court. It is different from a veranda or lanai in that its structure forms more a part of the main building.

logistics – the science of calculating the amount of materials, and coordinating their movements to the proper place in the proper sequence and time. Construction logistics involves figuring manhours, ordering materials, arranging for delivery, hiring workers and subcontractors and coordinating their arrival on the job site with the materials so that the work progresses on schedule.

longitudinal – along the axis or length of something.

long L – a type of stairway with two sets of steps. The first set ends in a platform landing, at which point one must make a 90-degree turn to continue up the second set of steps to the top. The second set of steps is often shorter than the first.

long-lead – a situation in which there is a long delay required between the placing of an order and the receipt of the ordered item. The delay is caused by the amount of time required to manufacture, assemble and deliver the item being requested.

long-nose pliers – pliers with long narrow jaws that taper to a near point, used for reaching and grasping items in tight locations.

long oil – a mixture of oil and resin, with the greater proportion being oil, used as a protective wood coating.

long oil varnish – a slow-drying varnish that remains flexible. It allows the surface to flex to some extent without cracking the coating, such as when a temperature change causes the surface to expand or contract.

long-term assets – see fixed assets.

locknut

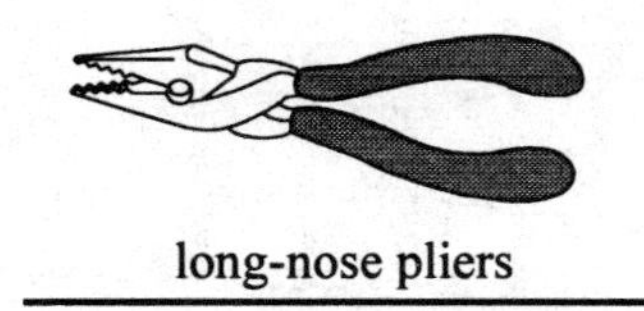

long-nose pliers

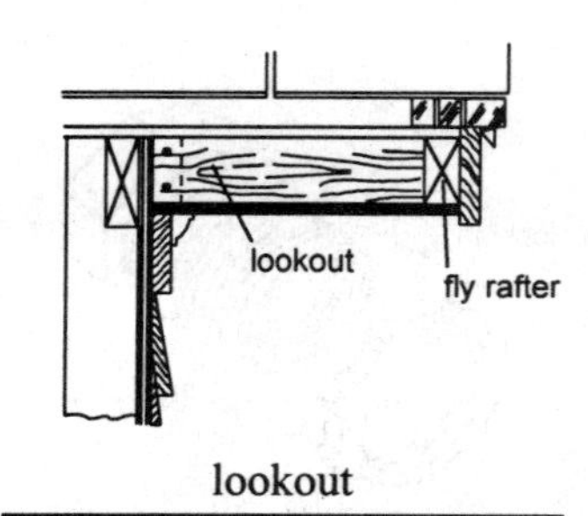

lookout

loop-top garden fencing

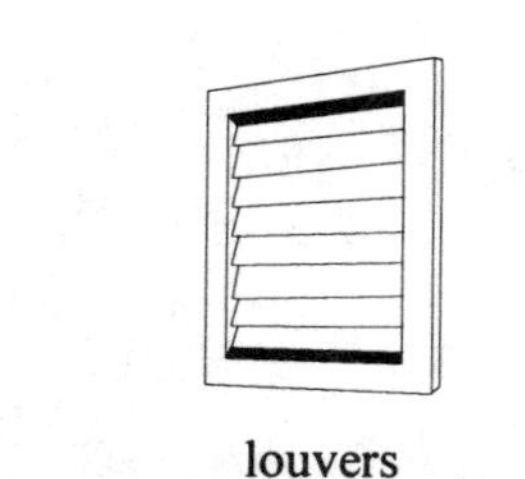
louvers

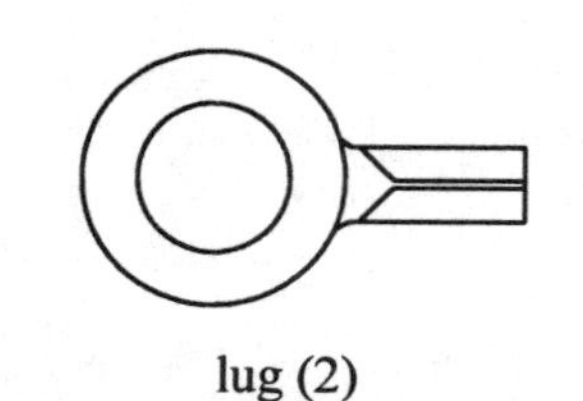
lug (2)

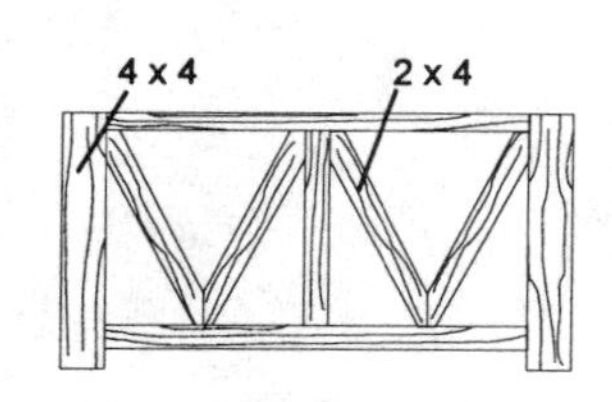

lumber fencing

long-term liabilities – monetary obligations that are payable beyond the next one-year period.

lookout – a structural extension that bridges the distance from the edge of a building to the fly rafter at the end of a gable extension. It provides support for the fly rafter and roof overhang. Also called a lookout block.

loop – a large section of metal heated to a temperature that makes it malleable for shaping and fabricating items.

loop-top garden fencing – a woven wire fence in which the wire is looped over to form rounded tops.

loosely-laid membranes – roofing membranes that are attached only around their perimeters and at penetration points where chimneys or vent stacks pass through the roof.

loose-pin hinge – a hinge with a pin that can be removed for disassembly.

lot – 1) a group of similar items; 2) a defined parcel of land.

lot line – a property boundary line.

louvers – slats placed at a downward slanting angle, in close proximity but not touching, so there are openings between them. They are used primarily for ventilation, allowing the passage of air and light but preventing rain from entering. Louvers are made of wood, aluminum, galvanized steel, or glass panes for louver windows.

low-density concrete – concrete combined with a lightweight aggregate such as expanded vermiculite or perlite, or foaming agents which create air pockets, resulting in a mixture that weighs 50 pounds or less per cubic foot, instead of 150 pounds per cubic foot for normal concrete. It is used where weight would be a detriment, such as for suspended concrete floors.

lower explosive limit (lel) – the lowest percentage of a flammable and/or explosive substance, in air, that can burn and/or explode.

low-lift grout method – grout poured into the voids of a concrete wall every few courses to add strength to the structure.

low water mark – the low water level, or the lowest level desired, in a water tank. A water tank may have a level switch set to turn on a tank fill pump if the water level drops to the low water mark.

L-screw – a thin metal rod, often made of brass, bent at a right angle on one end and with a screw thread on the other end. The rod can be screwed into a wall with the short end of the L protruding up, providing a support on which to place or hang objects.

lubricant – a material, such as oil, that is used to reduce friction between moving parts in a variety of machines or objects such as engines, locks, or hinge joints.

lubricate – to apply a lubricant, such as oil, to moving parts in order to prevent them from binding against one another.

lubricity – a measure of the lubricating properties of a lubricant.

Lucite – a trade name for a clear, tough plastic used for items such as skylights, windows and display cases.

lug – 1) a projection from a surface, such as a block welded to a length of pipe which provides a point of attachment for a support or restraint; 2) a terminal attached to an electrical cable that provides for quick connection and removal.

lumber – sawn and sized lengths of wood used for building.

lumber, appearance – a grade of lumber whose quality is based on its finished appearance. It is divided into industrial, architectural and premium grades.

lumber fencing – fencing made from 2 x 4s or 4 x 4s in various patterns.

lumber grade – grades for hardwood lumber range from "firsts," lumber that is 91½ percent clear on both sides, to "No. 3B common," which is 25 percent clear on one side only. "Sound-wormy" is a hardwood chestnut, a good grade of wood with worm holes that add to its decorative appearance. Grades for softwood lumber range from "select A," which is suitable for natural finishes that let the wood show, to "common No. 5," which is good only for filler lumber. The grading is outlined in the American Softwood Lumber Standard PS 20-70, with guidelines established by the National Grading Rule Committee, which is an industry sponsored organization.

lumber grade stamp – a marking on a piece of lumber by an official grading agency representing the National Grading Rule Committee, an industry sponsored organization. The mark identifies the lumber by grade for its intended use, and provides other information, such as the moisture content of the lumber at the time it was manufactured, the mill number, the species of wood and the mark of the inspection bureau that graded the piece. The lumber grades represented on the stamps are: SEL STR (select structural), NO. 1, NO. 2, NO. 3, CONST (construction), STAND (standard), UTIL (utility) and STUD (stud). The moisture content indicated by the stamps are: S-GRN (moisture content of 20 percent or less), S-DRY (moisture content of 19 percent or less), MC 15 (moisture content of 15 percent or less). Many grading agencies, such as the Western Wood Product Association, have their registered trademarks as part of their grading stamps.

lumbermill – a facility that saws logs into lumber.

lumberyard – a facility for the storage and/or sale of lumber.

lumen – a measure of the efficiency of a lamp. A lumen is the amount of light from a source of one candlepower that shines on a one-square-foot sphere, set at a distance of one foot from the source.

luminaire – a complete lighting fixture.

luminaire, direct – a lighting fixture that distributes 90 percent or more of its light downward.

luminaire, general diffuse – a lighting fixture that distributes 40 to 60 percent of its light upward.

luminaire, indirect – a lighting fixture that distributes 90 to 100 percent of its light upward.

luminaire, semi-direct – a lighting fixture that distributes 60 to 90 percent of its light downward and the balance upward.

luminaire, semi-indirect – a lighting fixture that distributes 60 to 90 percent of its light upward and the balance downward.

luminescent – emitting light at a lower temperature than that at which incandescence takes place.

luminous – giving off or reflecting light.

lump sum contract – a contract for a task to be performed for a set fee.

lux – the amount of light, equal to one candle intensity, on a surface that is everywhere one meter from the uniform point of light source.

lumber grade stamp

M: macadam to muntin

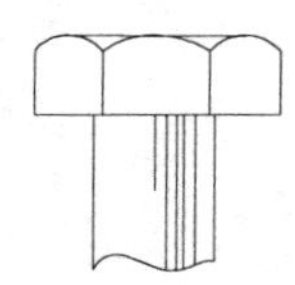

machine bolt

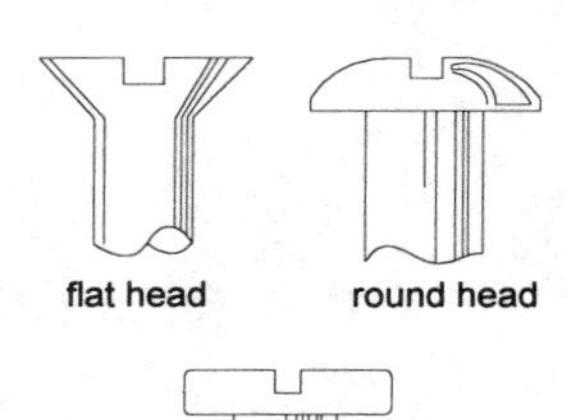

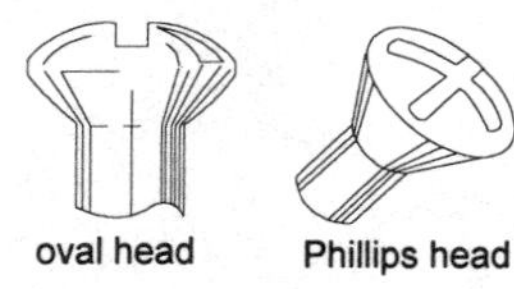

machine screws

macadam – a wear surface of compacted pieces of stone applied to walkways, driveways or roads. If an asphalt binder is added it becomes tarmacadam, a surfacing material that was more commonly used 60 years ago than it is today.

machine – a device or assemblage of parts that transmits force, motion or energy from one part to another in a manner designed to perform a task.

machine bolt – a threaded male fastener, usually having a hexagonal or Allen head, designed to fit in a nut or other machine threaded part in order to assemble members.

machine brazing – a process of bonding metals together using another metal that melts at a temperature higher than 840 degrees Fahrenheit, but lower than the melting point of the metals being joined. A bond is created between the molecules of the brazing filler metal and the metals being joined. Machine brazing uses an automatic or operator-controlled heat application and/or a brazing filler metal feed to perform the brazing operation. The advantage of machine brazing is the consistency and repeatability of this assembly-line type brazing operation.

machinery – 1) two or more machines; 2) the inner workings or parts of a machine.

machine screw – a threaded male fastener, designed to fit in a nut or other machine-threaded part, used to assemble members together. Machine screws usually have an indented head, often a slotted or Phillips head.

machine shop – a facility for making metal items using powered tools or machinery for cutting, grinding, shaping and smoothing the work.

machine tool – a powered tool, such as a drill press, grinder, milling machine or lathe, used for cutting, grinding or shaping metals or other hard materials.

machine-tooled finish limestone – see limestone, machine-tooled finish.

machine weld – welding is the fusing together of metals by melting them at the surface where they are to be joined. Filler metal is usually added during the process. Machine welding provides a degree of control that is consistent, less labor intensive, and sometimes faster than manual welding.

macro – involving large quantities or for use on a large scale. It is used as a prefix to another term to indicate the expanded definition of the term. An example would be macroetching, which is the process of chemically eating away at a large surface of metal to create a pattern or design for a building surface.

magnesium – a silver-white, strong, lightweight metal; atomic symbol Mg, atomic number 12, and an atomic weight of 24.312. Because of its light weight and strength, magnesium is sometimes used to make extension or very tall ladders.

magnet – a material that has the ability to attract iron and produces an exterior magnetic field of energy. Magnets are frequently used in electrical mechanisms, such as solenoids, electromagnets, motors, generators, loudspeakers and other items using, making or controlling electrical power.

magnetic cylinder lock – a lock cylinder that uses magnetic pins that are pushed into a plug to hold the lock closed. Since like magnetic poles repel one another, a key with magnetic inserts of the same polarity as the pins repels the pins, moving them out of their slots. This permits the plug to move in the cylinder and the lock to open.

magnetic energy – the power of a property to attract iron and compounds containing iron.

magnetic field – the area around a magnetic body in which the magnetic energy or force of that body can be detected.

magnetic particle testing (MT) – a nondestructive examination method for locating defects on or near the surface of materials containing iron. A magnetic field is induced in the material being tested, and fine iron filings are spread on its surface. A crack extending to the surface of the material will cause a discontinuity or interruption in the magnetic field. The filings will line up along the crack or flaw, indicating its presence. Half-wave direct current, alternating current or direct current may be used to generate the magnetic field for locating surface defects. Half-wave direct current has proven to be the most effective means for locating subsurface defects. There are also wet methods of magnetic particle testing. The wet method uses iron particles suspended in a low viscosity liquid, such as water, and is very effective for locating small cracks. Fluorescent wet particle examination is the most sensitive MT method.

magnetic starter – a starter or controller for an electric motor that uses an electromagnet to open and close the circuit. This type of starter requires a manual mechanical switch between the power supply and the magnetic starter in the event the magnetic starter fails in the closed position. The manual switch can then be used to interrupt power to the motor.

magnetic stud finder – a stud finder that detects the presence of nail or screw heads containing iron by means of the magnetic attraction between a freely-moving magnetic indicator and the nail or screw heads it passes over in the wall studs.

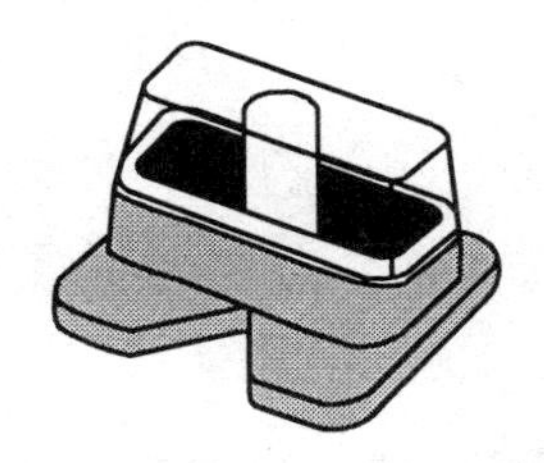
magnetic stud finder

magnetized hammer – a hammer with a head that has been magnetized to hold a nail, so that the worker can hammer the nail in with one hand while holding the material in place with the other.

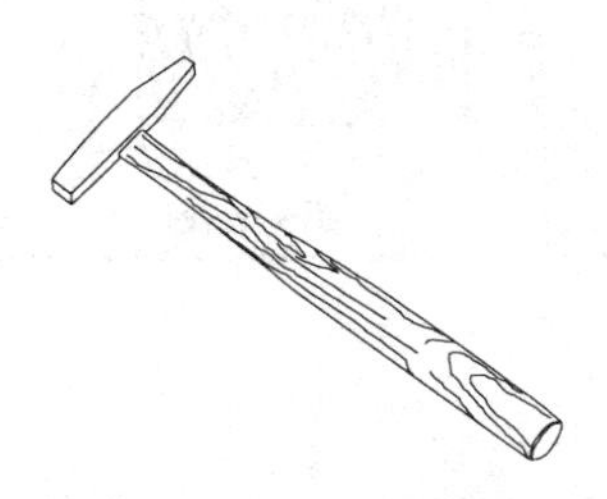
magnetized tack hammer

magneto – a small electric generator that uses permanent magnets to create a flux field. The flux field makes up the magnetic lines of force that are cut by a rotating armature to generate an electric current, such as that used to power spark plugs in an internal combustion engine.

magnet, permanent – a ferric or other type of material that has been magnetized and retains magnetism, as opposed to an electromagnet which requires electrical power for a magnetic field of significant density to exist.

mahogany – a strong, reddish-brown South American hardwood favored for fine furniture. It is used in both solid and veneer forms.

mahogany, Philippine – a tropical wood, somewhat resembling mahogany, but coarser and softer. It is used commonly as a veneer plywood for interior doors and also as solid wood for furniture and cabinets.

main – 1) the largest or most important; 2) the primary piping run in any application. The term is often used to refer to underground primary water supplies (water mains) or sewer lines (sewer mains).

major diameter

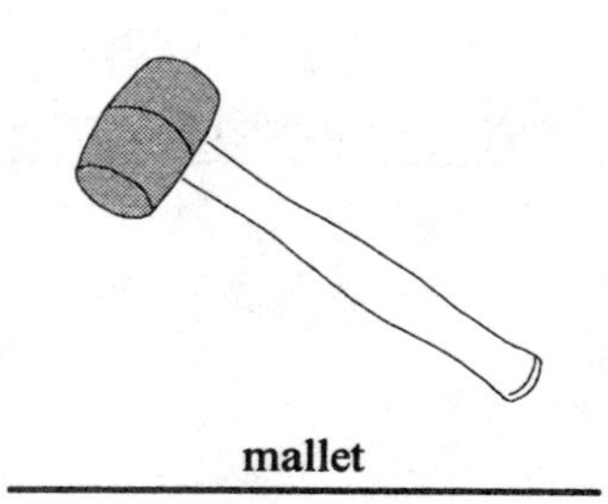

mallet

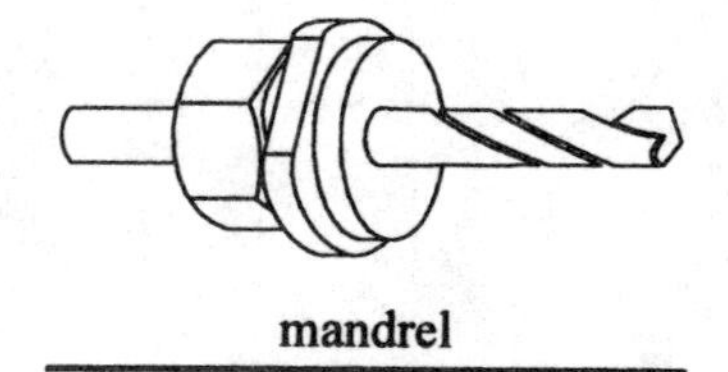

mandrel

main hoist – the primary hoist of a crane. Large cranes often have an auxiliary hoist for smaller loads, faster operation, easier access to difficult areas, or to assist in control of loads picked up by the main hoist.

main runner – the longitudinal metal support that runs the length of the room and holds the ceiling tiles of a suspended ceiling. The tiles are supported across the width of the room by cross tees which bridge the runners. The main runners are supported from the structure above them by lengths of wire fastened to the existing ceiling, ceiling joists or overhead beams.

maintain – 1) to keep as is and prevent deterioration; 2) to keep in a state of repair for proper operation.

maintain contact switch – an electrical switch that can be wired to run a mechanism, in two different modes, from each of two switch positions; a third center position opens the switch so that no current flows. For example, a blower may operate in high and low modes, with a switch set up so that the controls operate high/off/low.

maintenance bond – a warranty bond, with a specified time limit, that guarantees the contractor will make necessary repairs to anything that he/she installed for the customer during the warranty period without additional cost to the client. However, if an appliance fails through no fault of the contractor, it would be covered by the manufacturer's warranty rather than the contractor's maintenance bond.

main vent – the primary vent line in a ventilation system.

major arch – an arch with a span in excess of 6 feet with a rise-to-span ratio greater than 0.15 and with a load capacity of more than 1000 pounds per foot. Also called Tudor arch, semicircular arch, Gothic arch, or parabolic arch.

major diameter – the outside or largest diameter of the threaded portion of a threaded object, such as the largest diameter of the threaded portion of a bolt.

make-up air – replacement air added to a space from which air is being exhausted. If a room has an exhaust fan that extracts air from the room, and a fresh air intake to replace the air exhausted, the new air would be the make-up air. This process could take place on any size scale, from a bottle to an auditorium.

make-up water – water added to a closed system to replenish water lost from the system through leakage or evaporation. In a steam heat system, the boiler heats the water, turns it to steam, and the steam rises through the piping system to heat the building. Even though it is a closed system, steam escapes, so the water in the boiler must occasionally be replenished by a make-up water system.

male – external or projecting part, such as the projecting prongs on an electrical plug.

male fitting – a pipe fitting that is designed to be inserted into another fitting, such as bell and spigot piping.

male thread – a thread on an external shank.

malleable – workable; a material or metal, such as copper, which is soft and pliable enough to be pounded into a different shape without fracturing.

mallet – a hammer with a flat face used for striking another tool, such as a chisel. The face of the hammer is often made of a softer material than the rest of the hammer so that it will not mar the tool that it strikes.

management – 1) the planning, controlling, and directing of the actions necessary to accomplish a task; 2) those persons responsible for planning, controlling or overseeing a job or objective.

mandatory – required by order or command, as a license is required by law.

mandrel – 1) a steel shank, often threaded, for holding a cutting tool in a rotary driver, such as a drill motor; 2) a forming tool; 3) a steel shaft for holding work that has been bored, so that the work may be turned in a lathe.

mandrel, expanding – see expanding mandrel.

mandrel test – a method for testing the roundness of a pipeline by pulling a mandrel, a steel shank which has an outside diameter slightly smaller than the inside diameter of the pipeline, through the pipeline. The mandrel will be stopped or held up at any spot where the pipeline is out of round because the exterior dimension of the mandrel is so close in size to the inside dimension of the pipeline.

manganese – a gray-white metallic element; atomic number 25; atomic symbol Mn; atomic weight 54.938. Manganese compounds are used as coloring agents.

manhole – an opening with a removable cover that provides access from the surface to underground drainage piping, sewer piping, culverts or tunnels for maintenance, repairs or inspection. The opening is usually just large enough for a man to pass through.

manifold – a main header or run of piping from which auxiliary or branch piping is run to distribute liquids and gases to multiple locations. Headers are generally larger in diameter than branch piping.

manila paper – facing paper for standard gypsum wallboard. The facing paper covers the gypsum core on the side of the wallboard that is exposed after installation.

manometer – a U-shaped tube, containing fluid and marked with a graduated measuring scale, used for measuring slight changes in a low pressure system, such as an HVAC system, or for remotely measuring the level in a tank of liquid. One end of the tube is open and the other end is connected to the low pressure source. Slight changes in pressure cause the fluid in the tube to move or be displaced, and the change in the fluid level is read on the graduated measuring scale. The scale is calibrated to read directly in the desired units of measure. Also called a draft gauge.

manometer, dry – a measuring device used to determine changes in a low pressure system, such as an HVAC system, using a bellows and a pointer that moves across a scale in response to pressure changes. The flexible bellows is sealed, and the inside of the bellows is connected to the low pressure source. Changes in pressure cause the bellows to expand or contract, moving the pointer connected to the end of the bellows. A graduated scale under the pointer provides a means of reading the amount of change. Also called a draft gauge.

mansard roof – a type of roof with two different slopes around all sides of the structure, the upper of which may be nearly horizontal, and the lower nearly vertical.

mansion – a very large, elaborate house.

mantel – a shelf over a fireplace opening.

mantle – a fuel burner element in a gas lamp designed to give off a bright light. It is made of a coarse woven cloth in the shape of a small bag that is tied over the end of a gas pipe. The mantle is lit with a flame, which causes it to burn briefly and char, turning it into ash. The ash retains the shape and structural strength of the original material, but becomes brittle. The mantle serves to diffuse the gas and allow it to mix with air. The gas flowing through the mantle burns at the surface of the mantle, causing the mantle to glow brightly. It is used in decorative exterior lighting and portable lanterns.

manual – performed or operated by hand, such as a saw or sanding block which is manually, rather than power operated.

manual brazing – brazing performed and controlled by a person using a torch or arc welder.

manual motor starter – an on-off switch with a heating element designed to cut off power to the motor if the current is exceeded. The heating element is sized for the normal current required by the motor. If this is exceeded, the heater element temperature increases, tripping the switch to the open position and cutting off power to the motor. Manual motor starters are used for large motors such as large air compressors or deep well pumps.

manual welding – welding performed and controlled by a person using a torch or arc welder.

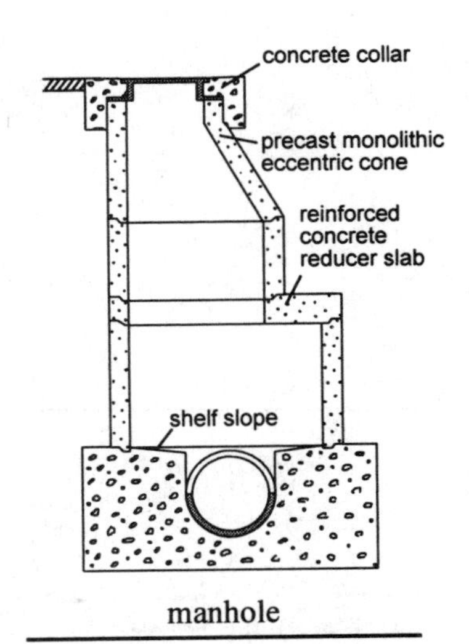

manhole

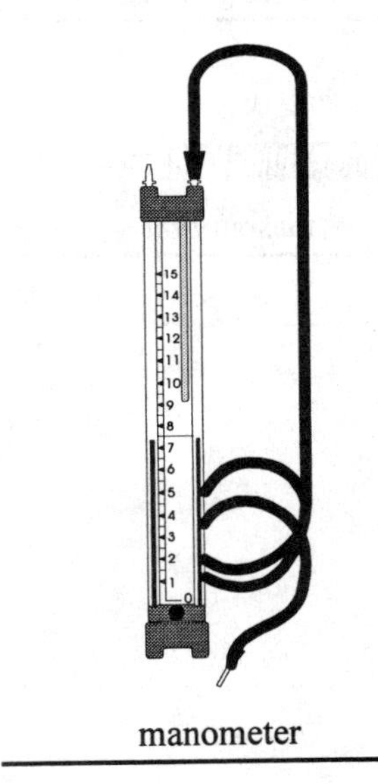

manometer

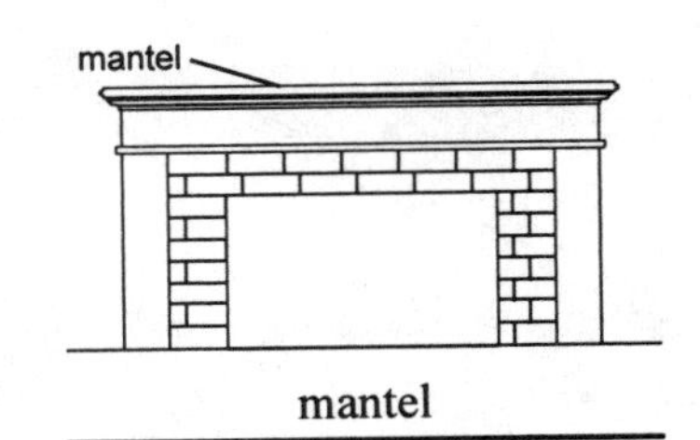

mantel

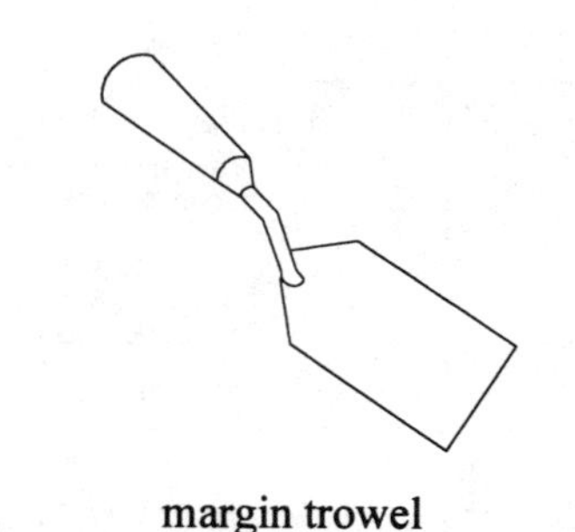
margin trowel

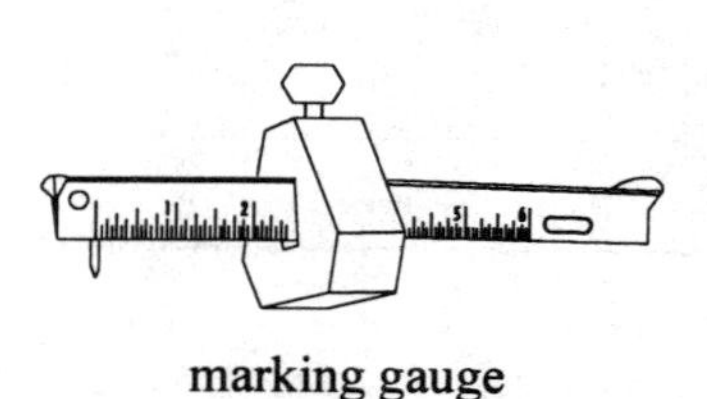
marking gauge

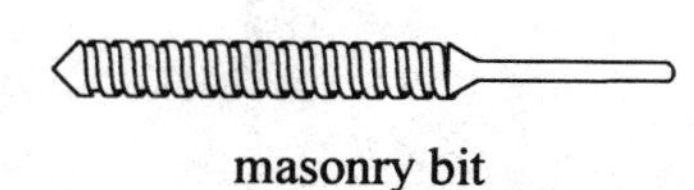
masonry bit

manufacture – to make something from raw materials, either by hand or machine.

manufactured gas – gas made from coal.

manufacturing defects – irregularities that occur in a product during the fabrication process at the factory.

manufacturing milling machine – a milling machine designed for general production work. It has a table with height and length adjustments and an adjustable height spindle head. During the cutting operation, the cutting tools move and the work being milled remains in place. The machine is semiautomatic and is used for repetitive milling work.

marble – crystalized limestone valued in architecture and sculpture for its durability and beauty. It polishes to a high luster and can be used for both exterior and interior surfaces. It is most frequently used in construction for fireplace mantels, countertops or floor covering, though the surface can become marred if used in high traffic areas. Marble is brittle and must be carefully installed and well supported in any application.

margin – 1) a capacity beyond normal operating range; 2) the difference between the cost and the selling price. Also called the profit margin; 3) a border or edge, such as the margin of a book.

margin trowel – a small masonry trowel that has a rectangular flat blade with a handle at one end.

marine – referring to the sea, sea life, or activities related to the sea.

marketing – advertising and selling a product or service.

market research – studies done to determine how well a product or service will sell in an area, how much it will cost to enter the market, and the extent of the competition. Businesses, such as construction or remodeling companies, may use this research before going into a new market area or expanding and/or changing their advertising approach.

marking gauge – a tool used for scribing a straight line on a surface parallel to the edge of that surface, or for transferring dimensions from one piece to another. It has a shaft with graduations marked in 16ths of an inch and a pointer projecting down from one end for scoring or marking a line. The shaft slides through a collar that locks into position along the shaft. To transfer a dimension, the collar is set at the dimension to be marked, such as a line 1 inch from an edge. The collar slides along the edge of the work while the pointer scribes a line parallel to the edge and 1 inch from it.

markup – the overhead, escalation, contingency fee, and profit that are added to a bid. The percentage of markup that a contractor uses varies with supply and demand, geographic location, costs, and other similar external influences.

marquee – a canopy or projection attached to the front entrance of a building, such as a theater or hotel. The marquee is often lighted to promote the business or display theater or entertainment information.

marsh – wet, swampy land usually located in low elevations near water.

mask – to cover or shield an object or person from the effects of an operation, such as paint spraying.

mason – a craftsperson who builds using unit masonry, such as brick, block or stone.

Masonite – trade name for a dense compressed board made from wood products and a binder. It is used as a nonstructural building material, such as an underlayment for vinyl flooring, or similar application where a smooth, flat surface is needed.

masonry – building materials made of stone, clay, cement, bricks, concrete, or other related materials. The term also refers to a structure built of masonry units and mortar.

masonry anchor – metal braces, embedded into the masonry, used to attach masonry to other structures or to attach additional members, such as door jambs and window frames, to the masonry. *See Box.*

masonry bit – a carbide-tipped twist drill bit designed for drilling holes in masonry.

Masonry anchors

Masonry anchors are used to attach masonry to a support, to other masonry, or to a floor slab, a spandrel beam or a column. Masonry anchors include masonry ties, used to hold masonry together, and fasteners which attach other building elements to masonry. Anchors are usually made of galvanized rods, wire or sheet metal, or occasionally of stainless steel or bronze. There are several different types of anchors used as a means of attaching to concrete. They may be placed as part of the formwork before the concrete is poured, or installed in the cured concrete. The types of anchors designed for these two methods of application vary.

Anchor bolts, such as J-bolts, are typical of the type of anchor placed either as part of the formwork or installed in the concrete while it is wet. A J-bolt has a curved hook on one end designed to anchor into the concrete as the concrete sets up. The other end of the bolt is threaded to receive a nut, and protrudes up from the surface of the concrete. Anchor bolts embedded in masonry walls are used to hold sill plates firmly enough to resist wind shear and seismic stresses. Embedded strip anchors are used to tie walls to foundations.

The embedded plate is a more sophisticated anchor used for heavier loads. It is a steel plate that has studs, with large-diameter heads or bent sections like the J-bolt, welded to its back. These studs anchor the plate into the concrete, enabling the plate to support substantial loads. The plate is placed in the form before the concrete is poured so that its exposed face will be flush with the finished concrete surface. Various steel anchoring devices can later be welded to the plate face, providing a practical method for attaching large loads to the concrete.

Flexible anchors, such as triangle ties, cavity wall ties and column anchors, are used in cavity walls and for attaching masonry veneer walls to steel, wood framing, concrete or other unit masonry. These ties provide lateral support but permit differential movements between the frame and the wall without causing cracking or distress. Wire mesh ties, also called metal lath, are used to bond intersecting walls at the joint.

Other types of anchor bolts are designed to be installed in holes drilled in cured concrete. They are called friction anchors or expansion anchors. A simple and commonly-used friction anchor is a tapered plastic sleeve that can be inserted in the drilled hole. A screw is inserted in the plastic sleeve and turned, causing the screw threads to bite into the plastic while the screw is drawn into the sleeve. This expands the outside circumference of the sleeve and forces it tightly against the inside of the hole, establishing a friction grip between the sleeve and the sides of the hole. The plastic sleeve probably has the least load-carrying capacity of any anchor. Friction anchors made of metal function in the same manner, but have higher pull-out resistance.

The split-lead anchor sleeve is made in two halves hinged together at the surface end. The sleeve is used with a lag screw. As the lag screw is turned in the sleeve's tapered internal thread, the two halves of the sleeve are forced to expand tightly against the inside circumference of the hole. Other types of friction anchors work on the same principle. Their relative strengths are a function of size, the material of which they are made, and the design and manner in which they achieve the needed expansion against the concrete.

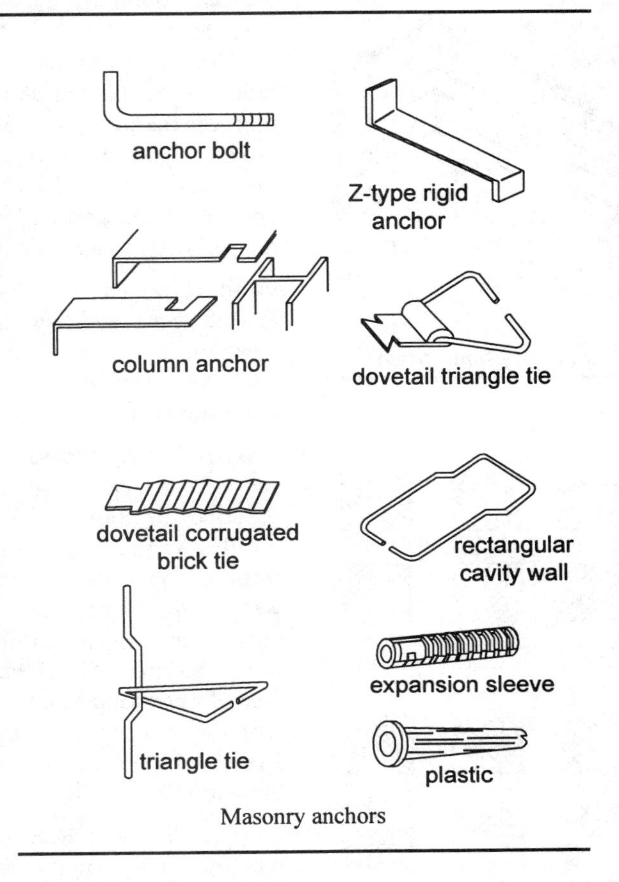

Masonry anchors

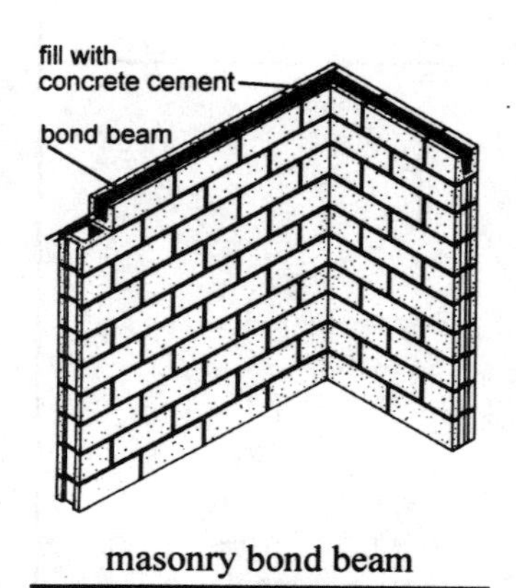

masonry bond beam

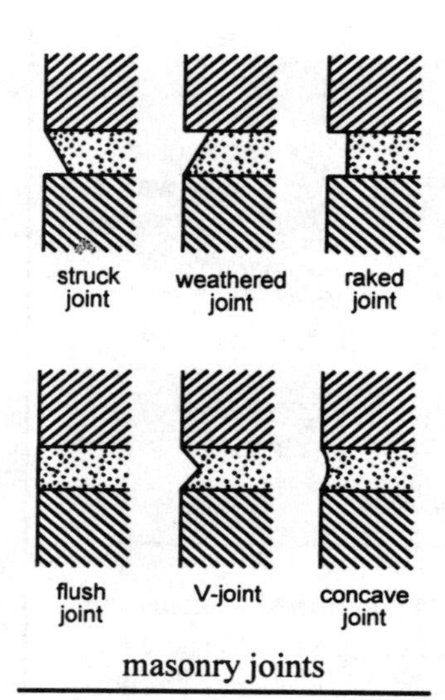

masonry joints

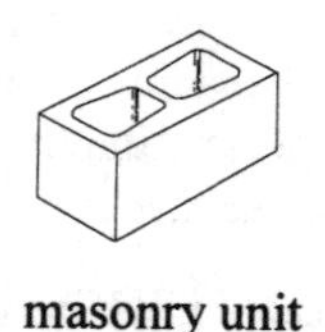
masonry unit

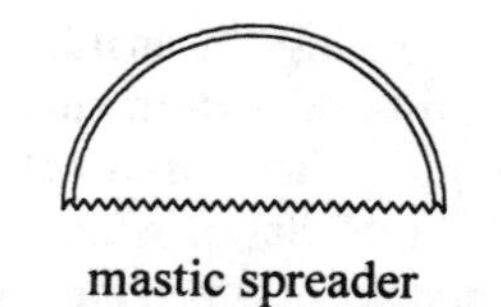
mastic spreader

masonry block – a hollow concrete building block.

masonry bond beam – a specially designed concrete block that is hollow on top which is used as one or more courses in the building of a concrete block wall. Rebar is placed in the hollow of the blocks horizontally along the length of the wall, and then the blocks are filled with concrete. The concrete beams add reinforcement around the perimeter of a building.

masonry cement – portland cement with air-entraining additives and inert limestone or hydrated lime added. These materials improve the workability, plasticity and water retention of the cement.

masonry joint – the joint between masonry units. Various terms are used to describe the mortar joint cross-section profiles: a struck joint has a tapered recess from the top to the bottom of the joint; a weathered joint has a tapered recess from the bottom to the top of the joint; a raked joint is squared off and indented from the face of the masonry; a flush joint is even with surface of the masonry; a V-joint has a recessed V-groove; and a concave joint has a rounded recess.

masonry lintel – a header for doors and windows made of precast concrete or block, mortar, and rebar. They are used in masonry walls so that the lintel material is compatible with the appearance and durability of the wall.

masonry paint – polyvinyl acetate or acrylic emulsion based paints manufactured specifically for use on masonry surfaces.

masonry primer – an asphalt base primer used to seal masonry pores and establish a surface onto which other asphalt-based materials, such as tar and a sealing membrane, will bond for additional moisture protection.

masonry reinforced wall – a wall constructed of stone, clay, cement, bricks, or concrete that has been reinforced with steel.

masonry symbols – drawing symbols that indicate different types of masonry.

masonry unit – manufactured or natural building units, such as block or brick, made of cement, stone, clay, or other related materials.

masonry veneer – a masonry facing for a wall as opposed to a wall in which the masonry serves as the structural support. An example of masonry veneer would be a single layer of brick applied to the exterior of a wood-frame house.

masonry wall – a wall constructed of stone, clay, cement, bricks, concrete, or other related materials.

masonry wall coping – see wall coping.

masonry wall ties – metal devices made of mesh, corrugated metal or heavy wire that are embedded in mortar between masonry units and connected to an existing wall.

mass coefficient (M) – a correction factor used in HVAC thermal calculations to adjust for real-world conditions.

mass spectrometer – a device for measuring leaks into a vacuum system. A mass spectrometer has a vacuum within it, and is able to sense the presence of the helium gas used to check for leaks around joints in vacuum systems. Helium is used because its molecules are very small, and can pass through microscopic leaks that would limit the vacuum.

mast – the heavy conduit on the side of a building from the service drop to the meter. The incoming electrical supply wires pass through the mast.

master antenna television system (MATV) – a central television antenna system for a multiple dwelling, such as an apartment complex.

mast head – see entrance cap.

mastic – 1) adhesive; 2) an adhesive composed of asphalt and nondrying oils that must be heated before using.

mastic spreader – a hand tool with a thin metal blade. The blade has a serrated edge for spreading mastic in ridges. The spaces between the ridges permit the mastic to spread out when the pieces to be joined are pressed together. Also called mastic trowel.

mat – 1) a smooth, flat layer of concrete, asphalt, or similar material forming a structural surface; 2) a surface with little or no sheen. Also spelled matte.

matchboards – boards milled to fit tightly together, such as those used to make a tabletop.

match line – lines used in mechanical drawing to permit matching a portion on one drawing with a portion on another drawing. These might be needed where portions of a building are too large to be shown on one drawing.

mat foundation – a reinforced concrete slab foundation for a building structure.

matrix – 1) an arrangement of rows and columns, as in an arrangement of circuit elements, for performing a specific function; 2) a material into which something is enclosed or embedded, such as cement paste or mortar in which aggregate is embedded.

matte finish – a low sheen paint finish, with less reflective quality than a gloss but more than a flat.

maul – a heavy hammer, often with a long handle, used to split wood or to drive wedges into wood to split it.

measure – 1) to determine the dimensions, capacity or value of something; 2) the basis or standard of comparison.

mechanical – 1) machine-like; 2) having moving parts; 3) produced or operated by machine.

mechanical advantage – a force which is multiplied by use of a mechanism or machine, such as a lever, pulley, gears or an inclined plane.

mechanical cleaning – a nonchemical process, often using abrasives, for cleaning surfaces such as the exterior of a metal pipe prior to painting.

mechanical drawings – building drawings that show the layout and details of plumbing, piping, and HVAC systems.

mechanical efficiency curve – a graph which shows the efficiency of a machine under different operating conditions. For example, the efficiency of a centrifugal pump will vary with its speed and output. By keeping records of such information and duplicating the conditions under which the pump operates most effectively, it can be utilized at its maximum efficiency.

mechanical equipment – the equipment, such as furnaces, hot water heaters, ducting, vents and piping, that make up the HVAC, water, gas, and other utility systems in a building or structure.

mechanical fingers – a device used to reach in and grasp items in tight places or locations that are too small or too distant for a worker to reach. It consists of a set of spring-loaded finger-like tines on a rod within a flexible housing. A pushbutton on the end of the rod opposite the fingers is pressed to extend the fingers out of the housing, and the spring forces them apart so they can extend around the object to be picked up. When the button is released, a spring on the button end of the rod pulls the rod back into the housing, closing the fingers around the object so that it may be moved.

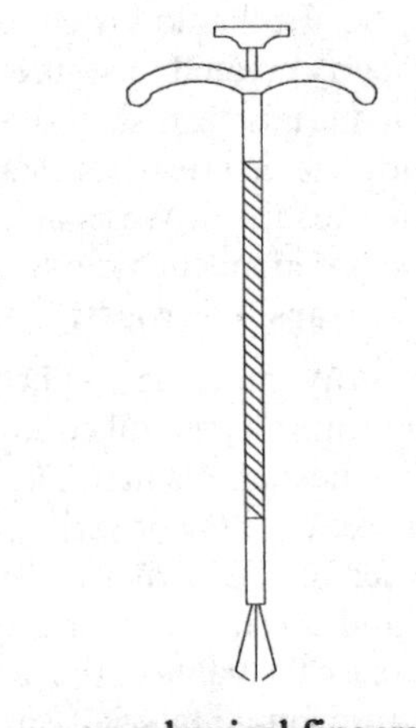

mechanical fingers

mechanical joint – a pipe joint that uses a gasket with a wedge-shaped cross section. The gasket is compressed against the outside diameter of the pipe by a pair of bolted ring fittings that partially encapsulate and squeeze the gasket as the bolts are tightened. One of the fittings is integral with the hub end of the length of pipe, and the other, a type of flange, slips over the end of the pipe to be joined to the first. Mechanical joints are commonly used with large diameter buried piping.

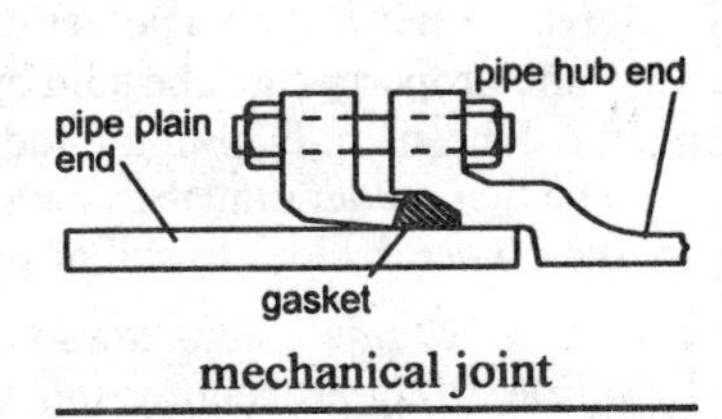

mechanical joint

mechanical pencil – a pencil that holds replaceable leads and feeds them in as the lead is used.

mechanical tamper – an engine-driven vibrating soil compactor for use in small areas and tight quarters.

mechanical taping tool – a tool that dispenses drywall joint compound and joint tape at the same time.

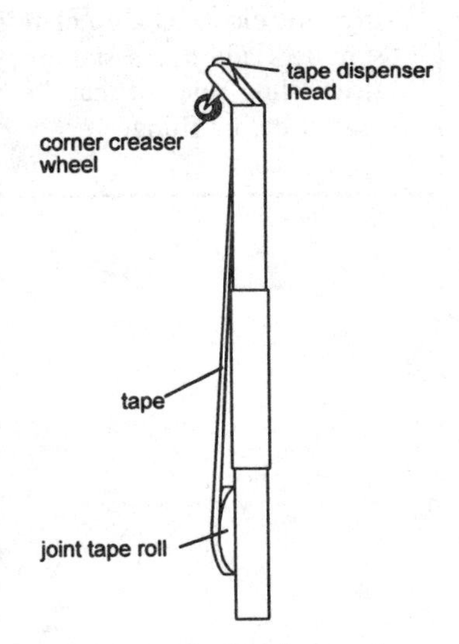

mechanical taping tool

mechanical ventilation – ventilation provided by an air-moving mechanism such as a fan.

Mechanic's lien

A mechanic's lien is a legal claim against real property created to secure payment for work performed or materials furnished while making improvements to land. The lien restricts clear title to the property until the lien is resolved. Most banks won't lend on property until a recorded mechanic's lien has been cleared. That usually means the property can't be sold.

Any tradesman, subcontractor (even if hired to do demolition), general contractor or material supplier can file a mechanic's lien. The lien must be based on a valid contract, written or verbal, between the claimant and the owner or the owner's general contractor. The theory behind the lien is that improvements to real property increase the value of the property. Those doing the work or furnishing materials should be guaranteed payment.

A mechanic's lien attaches to the property like a mortgage. But unlike a mortgage, most states do not give the lien holder a power of sale (although some do). If the property can't be sold by the lien holder to satisfy the debt, he may obtain a judgment against the owner. The lien holder may then have to wait for payment until the owner decides to sell or refinance.

Mechanic's liens laws vary from state to state. As a rule, filing procedures must be followed to the letter. Forms for filing a mechanic's lien are usually available from the city or county clerk's office, material or form suppliers and some stationery stores. Because of the strict filing rules, it may be good advice to consult a lawyer before filing.

Property owners may protect themselves from mechanic's liens by requesting lien releases from subcontractors, laborers and material suppliers. A lien release designates the materials and services furnished and certifies that the vendor or tradesman has been paid. Typically, the time limit for filing a mechanic's lien is 60 days from the time of completion for general contractors and 30 days for all other workers if the owner files a notice of completion and it is recorded within 10 days of completion of the project.

Owners who fail to request lien releases may have to pay twice for work done; once to the general contractor under the contract, and a second time to suppliers or subcontractors who weren't paid by the general contractor.

In most states there are two types of release forms, conditional and unconditional. Conditional releases are valid only after a condition has been met, such as that a check has cleared the bank. For example, a partial conditional release may be issued when a progress payment is received.

Contractors can protect themselves by paying suppliers, workers and subcontractors on time. All agreements with subcontractors, suppliers and owners should be in writing with copies to all involved. This is especially important when dealing with change orders, which are common causes of disputes between contractors and owners.

mechanic's lien – a legal claim against real property to satisfy a debt for supplies used or improvements made upon the property by a worker, subcontractor, supplier or general contractor. *See Box.*

mechanism – a device that uses moving parts working together to perform a task. A gasoline engine and a pulley are both examples of a mechanism.

median – an unused section on a roadway separating different-direction traffic. The median is sometimes left unpaved, or it can be paved as an emergency pull-off or raised and planted with shrubs or trees.

mediate – to bring about an agreement between parties using a neutral third party, or mediator.

medicine cabinet – a small storage closet for medicines or toiletries usually located above or near a bathroom lavatory.

medium – 1) intermediate in position, quality or degree; 2) a means or system of communication, such as radio or newspapers; 3) a base material into which other substances are added to make something, such as the liquid to which pigment is added to make paint.

medium curing cutback – asphalt which is thinned with kerosene to keep it liquid at lower temperatures.

medullary rays – fibers that run from the center or pith to the bark of a tree.

meld – to mix or blend a compound together.

melt – to change from a solid to a liquid state by the addition of heat.

melt-through – a welding defect in which the heat melts through the base metal.

membrane – a thin coating or sheet of material, usually water-resistant, such as the thin elastomeric sheets used to waterproof roof surfaces.

mercury – a heavy metal that is liquid above -38.9 degrees Fahrenheit. It has the atomic symbol Hg, atomic number 80, and atomic weight 200.59. Because it stays liquid at common atmospheric temperatures, mercury is used in scientific measuring devices such as thermometers and barometers.

mercury switch – a silent electrical switch in which the electrical contacts and mercury are enclosed in a bulb. When the bulb is in the appropriate position, the mercury bridges across the contacts, completing the circuit and permitting electrical current to flow.

merge – to combine into one, such as two lanes of traffic joining to become one.

merger – combining two or more businesses into one.

meridian – an imaginary circle around the surface of the earth that passes through the north and south poles.

merlon – one of the solid intervals between the open crenels of a castellated wall or notched surface.

mesh – material made up of spaced strands of fiber or wire in a weave or gridwork pattern which allows air or water to pass through it, filtering out larger matter. Mesh materials come with various size grid openings, from large openings used on earthen banks and for landscaping, to small weaves used for window screens or filters.

mesh tape – fiberglass joint tape for use with gypsum wallboard.

messenger-supported wiring – electrical wiring that is suspended by high-strength wire, such as in a service drop, or as used sometimes for outdoor lighting systems.

meta-anthracite – a coal with high carbon content that is very like graphite in structure and composition. It is difficult to ignite and burn.

metal – a class of chemical element, having a crystalline structure and the ability to conduct heat and electricity. Metals usually have a high luster and can be molded into various forms at high temperatures. Because of their durability, they have a variety of uses from jewelry to building materials. Gold, steel, copper, and iron are a few examples of metals.

metal arc cutting (MAC) – any process using an electric arc to provide the heat for cutting metal.

metal-clad cable (MC) – armored electrical cable, consisting of insulated inner conductors (wires) surrounded by spiral-wound metal sheathing. See also BX cable.

metal-clad switchgear – an electrical switch assembly in which all the equipment required to control a circuit is installed in a single metal cubicle.

metal corner – a building siding trim for corners which is made of metal and has the appearance of a mitered corner.

metal deck – a ribbed steel sheet used as a permanent form for concrete. It contributes some reinforcing strength to the concrete. Also called Q-deck.

metal drip edge – a preformed piece of sheet metal placed along the edge of a roof to encourage water runoff to drip off the edge of the roof rather than flow back under the shingles or eaves.

metal edge support clips – see plywood edge clips.

metal fencing – fencing made of ornamental iron or welded metal shapes, such as square tubing or pipe. Metal fencing may be decorative, used for security, or to

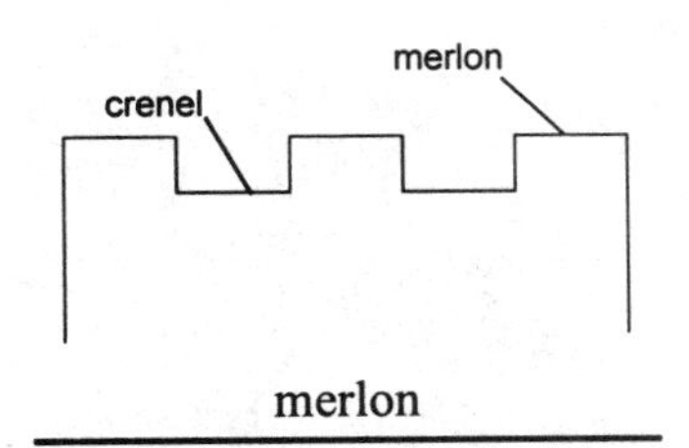

merlon

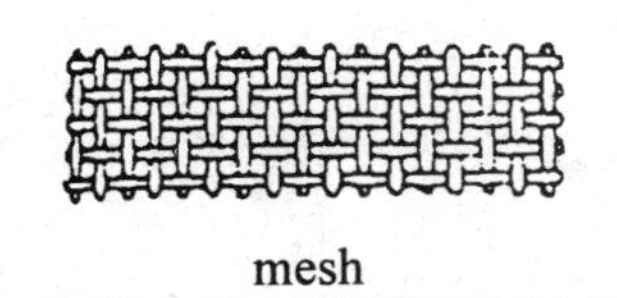
mesh

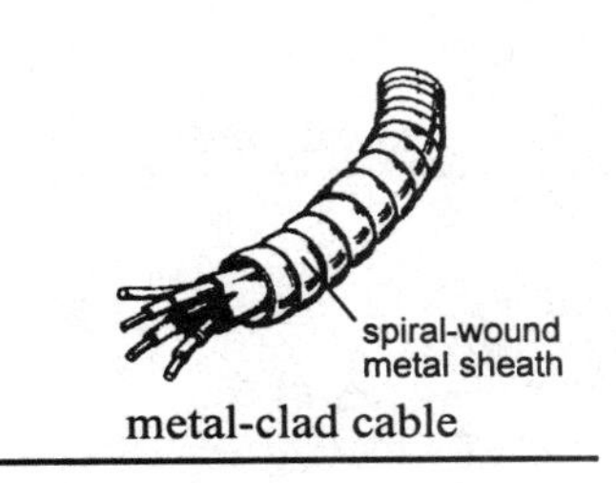

metal-clad cable

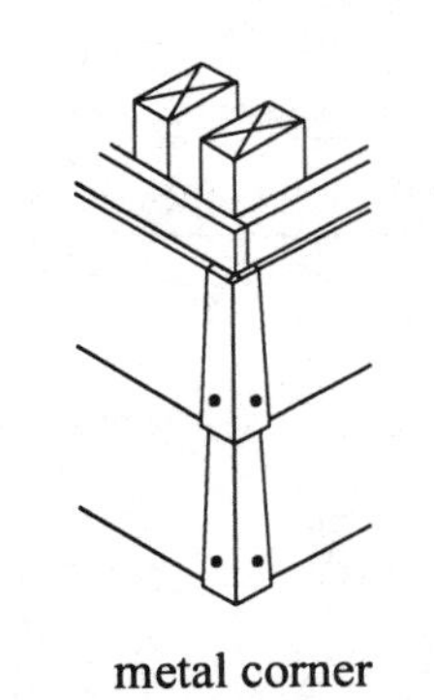
metal corner

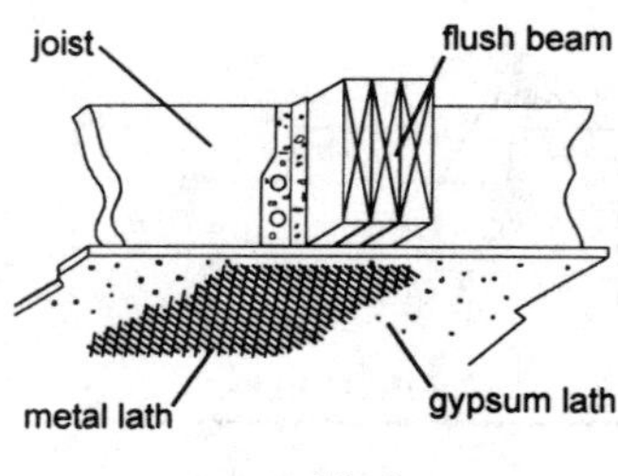

metal lath

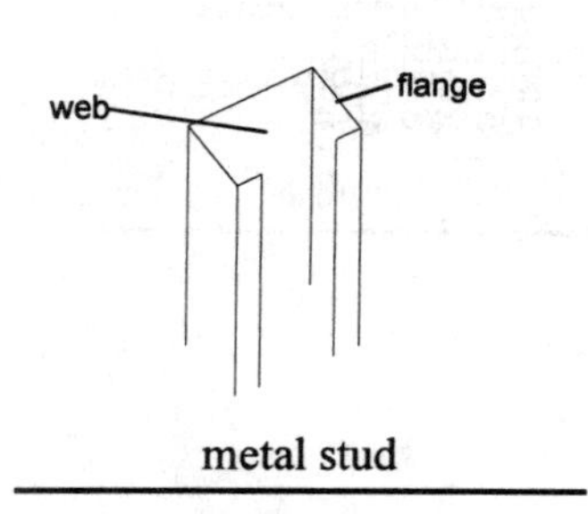

metal stud

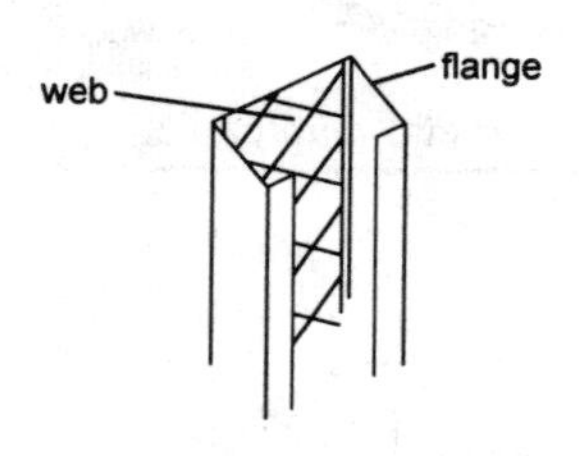

metal stud, combination

restrain large animals such as cattle or horses. Metal is also used for guardrails and safety rails or fences along highways.

metal lath – thin metal sheets that have been expanded or stretched and then stamped through with a slotted pattern giving them a screen-like appearance. They are used as backing and structural support for plaster and mortar.

metallic paint – paint containing metal powder, such as aluminum or bronze. The metal increases the protective ability of the paint and also makes the paint reflective.

metallography – the microscopic study of metal structure. Studies are done to examine the quality of weld cross sections in detail and also, on a larger basis, to study material failures to determine how and why they occur.

metallurgy – the science of working with, analyzing, and alloying metals. A metallurgist may be called upon to give advice in the selection or use of a material for a specific or unusual building condition or fluid service.

metal powder cutting (POC) – a metal cutting process using a torch to generate heat and oxygen, in combination with an oxidizable powder, to increase the chemical reaction to burn through the metal at a rapid rate.

metal-protecting paint – a paint with oil, or oil and alkyd, as a base and other compounds, such as zinc chromate, added to prevent rust or corrosion.

metal stud– preformed steel framing member that is protected against corrosion by galvanizing or coating it with aluminum or an aluminum-zinc compound. Studs are formed of sheet metal in different thicknesses for applications ranging from non-loadbearing partitions to loadbearing parts of a structure. Because they are made of steel, they permit the construction of a completely noncombustible wall. Metal studs are most commonly used in commercial buildings.

metal stud, combination – metal stud that has a wire that is bent diagonally to form the web of the stud and welded to rolled steel flanges. These studs are designed for the attachment of metal lath using clips or wire ties.

metal stud, screw – stud that is made from light gauge metal in the form of a channel with a turned-in edge or flange. The flanges are dimpled or serrated so that the metal screws won't drift as they are installed.

metal trim – U- and L-shaped edging designed to be installed on exposed edges of drywall to protect it and give a finished appearance.

metamorphic rock – igneous or sedimentary rock that has been changed in texture or composition while in a solid state as the result of extreme heat, pressure or chemicals. The deformation may in some cases create an interesting combination of different colored rocks that are layered, twisted and curved to form striking patterns in the newly formed rock. Cut sections of some forms of metamorphic rock are used as decorative siding and retaining walls as well as flagstone pavers.

meter – 1) a unit of metric measure, equal to 39.37 inches; 2) an instrument for measuring and recording units of measure, such as of electrical power consumed; 3) an indicating instrument, such as a volt-ohm-meter.

meter box – a box, usually metal, used to mount an electrical meter to a wall. The meter measures and records electrical usage in the building where it is installed.

meter socket – the receptacle into which the electric power meter for a building is installed.

metes and bounds – surveying terms which describe property lines or boundaries using landmark bearings, such as rivers and roadways, and the measurement of the distance between those landmarks.

methane – natural gas, commonly used for heating furnaces, water heaters, stoves and clothes dryers. It is readily available and a relatively inexpensive source of heating fuel in many parts of the country.

metric measure – a decimal system of measure, adopted by most of the nations of the world, which uses the meter for length and the kilogram for weight.

metric threads – machine threads that are in metric units, and commonly used in much of the world. As international trade increases, the availability of product forms using metric dimensions, such as threaded fasteners, will become more common in the U.S.

metric ton – a metric weight equal to 1000 kilograms or 2204.62 U.S. pounds. A U.S. short ton is 2000 pounds and the standard U.K. long ton is 2240 pounds. Metric tons are standard measure in most of the world and builders are increasingly more likely to encounter them.

metric wrenches – wrenches in metric unit sizes designed for use with metric fasteners. Metric tools are already commonly used in the U.S. for working on foreign cars, but there is also a growing market for imported construction products and equipment which require metric tools for maintenance and repair.

mezzanine – an intermediate level, which projects like a balcony between two floors, having less than ⅓ of the total floor area of any other level in the building.

M-factor – a number used to designate the relative heat retention properties of a mass. For example, one of the ways of storing heat energy in a solar heating system is to use a material of dense mass, such as a wall of water containers. The water containers are exposed to sunlight through a wall of windows and they absorb and store the heat. A large percentage of the heat gained from the sunlight is stored and given off slowly to heat the surrounding air when the sunlight is gone. The M-factor represents the thermal storage capacity of the water containers.

micro – 1) one millionth part; 2) very small or involving minute quantities.

microampere – one millionth of an ampere.

microcomputer – a small personal computer which uses a microprocessor.

micrometer – a precision tool which can measure up to the nearest $1/1000$ or 0.001 of an inch. It has a U-shaped frame and a spindle with fine threads (40 threads per inch). Each revolution of the spindle moves it $1/40$ or 0.025 of an inch. There are 25 equally-spaced graduations around the thimble, which is attached to, and encloses, the spindle. Making $1/25$ of a revolution of the thimble (and the enclosed spindle) moves the spindle $1/1000$ of an inch. Some micrometers have a vernier scale making measurements of $1/10{,}000$ of an inch possible.

microstructure – the microscopic structure of a material or cell, such as the crystalline structure of a metal as seen through a microscope.

mildew – a fungus growth that is enhanced by dampness. Mildew often shows up in damp places, such as bathrooms or the underside of a porch roof. Mildew must be killed on a surface and the surface cleaned before it can be painted. A mixture of trisodium phosphate, powdered detergent and water can be applied to the mildew to neutralize it, and vigorous scrubbing will remove the residue. Adding a mildewcide to the paint before it is applied will minimize the reoccurrence of the mildew.

mild steel – a steel with a relatively low carbon content and without special alloying elements. It is used in the production of nails, structural beams, brackets and other building hardware.

mile, nautical – a unit of linear measure equal to 6,080 feet used for international air and sea distances. Also called an admiralty mile.

mile, statute – a unit of linear measure equal to 1,760 yards, or 5,280 feet; the common U.S. linear surface measurement.

milestone – a significant event, such as completing a major goal within schedule.

mill – 1) a place of manufacture, such as a lumber mill where trees are sawn into lumber; 2) to shape a material into a pattern, as wood is shaped into molding or trim pieces.

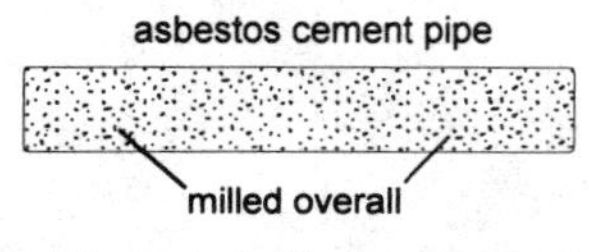

milled-each-end pipe (MEE)

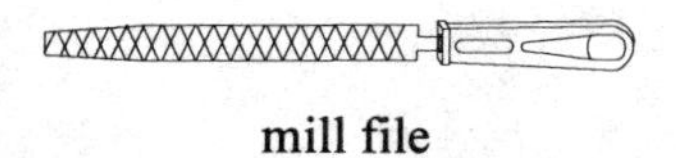

milled-overall pipe (MOA)

mill file

minor diameter

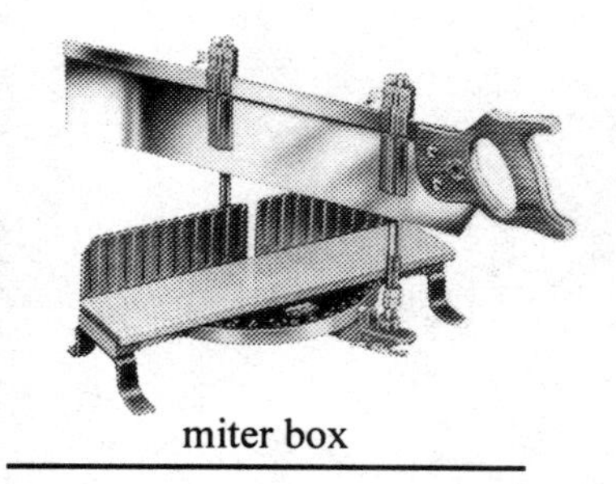
miter box

milled-each-end pipe (MEE) – sections of asbestos cement pipe that have been milled on each end to provide a smooth exterior surface for joints. The middle of the pipe is left in the rough, as-cast condition.

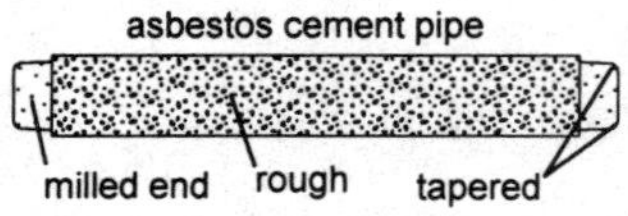

milled-overall pipe (MOA) – asbestos cement pipe that has been milled end to end to provide a smooth outside surface the entire length of the pipe. MOA pipe can be cut any length and joined with existing pipe without having to have the ends milled for a smooth fit.

mill file – a versatile file used for sharpening, smoothing and shaping metal. It is called a mill file because it was originally designed for sharpening saw teeth in lumber mills.

milli – prefix meaning 1⁄1000 or 0.001.

milligram – metric measurement for 1⁄1000 of a gram.

millimeter – 1⁄1000 of a meter, equal to 0.039 U.S. inches.

milling cutters – hardened steel or carbide-tipped cutting wheels, available in a wide variety of shapes, used in milling machines to shape and form material. The milling machine holds the work and moves it in an exact proximity to the rotating cutters to shape the work accurately and precisely. Milling cutters are used to shape items, such as gear wheels, or to make cuts, such as a keyway in a shaft.

milling machine – a machine that holds material and moves it in an exact proximity to rotating cutters to shape the work accurately and precisely. The cutters are also powered by the machine.

millwork – finished shapes, such as molding, jambs, gear wheels or any other solid material that is shaped into a form by the removal or cutting away of excess material by machine.

mineral insulated, metal sheathed cable (MI) – electrical cable in which the electrical conductors are insulated with compressed mineral refractory material and enclosed in a liquid- and gas-tight copper sheathing. This type of cable can be embedded, used on the interior or exterior of buildings in wet or dry locations, used in hazardous or explosive situations, or under any conditions except where the copper sheathing may be corroded away.

minor arch – an arch with a span of up to 6 feet, and a rise-to-span ratio of 0.15 or less, that can support loads up to 1000 pounds per square foot. Also called a jack arch, segmented arch, or multicentered arch.

minor diameter – the smallest diameter of the threaded portion of a threaded object.

mire – 1) a marshy area of land such as a bog; 2) deep mud or slush.

mirror hanger – an L-shaped fitting with a screw hole through it used to mount a mirror to a wall. The fitting supports the glass and is fastened to the wall surface with screws. This provides a secure means of holding up the mirror without the necessity of drilling holes in the glass.

miscible – capable of being mixed, as paint and thinner or plaster and water can be mixed.

mission stipple – a type of decorative texture created by dabbing a wet plaster surface or a wet joint compound surface with a sponge. This technique may be used on any room surface that does not require a smooth texture.

mist – 1) particles of water floating or falling through the atmosphere; 2) a thin or light spray used to dampen materials, such as that needed for curing concrete; 3) a liquid atomized into very small droplets.

miter or mitre – 1) a joint that is cut at an angle, most commonly 45 degrees; 2) a bend or change of direction in a length of pipe made by cutting the pipe at an angle and welding the pieces back together. This is an alternative to installing an ell fitting to make the bend. It is not as strong a fitting, but can be done quickly and inexpensively.

miter box – a device which holds a saw and the work to be cut at the proper angle for making accurate angle cuts.

miter clamp – a clamp designed to hold two pieces together at right angles to one another until glue at the joint between them has set up. A common use is holding pieces of picture frames, or other types of frames, together for gluing.

miter joint – an angled joint.

miter vise – a vise with a guide which holds two pieces at right angles to one another while a miter cut is made on them.

mix proportions – the proportions of all the different ingredients or components that make up a mixture of concrete, mortar, or any other compound.

mobile home – a dwelling unit on a chassis with axles and wheels which can be moved and mounted on a temporary foundation or anchored to a semi-permanent foundation.

mobility – 1) the ability to move; 2) the ability of a paint to flow well on a surface.

mock-up – a simulation or model built accurately to scale for study, display or testing.

MOD 24 – a framing system using 24-inch centers rather than 16-inch centers. The framing system aligns the joists, studs and rafters into a series of modules, or in-line frames. Plywood or fiberboard sheathing is used to add racking stability. The alignment of these structural members provides in-line transfer of structural loads through a strong and direct load path. This system consolidates a series of important construction advances made over the years, as well as saving in labor and material costs.

model – 1) a physical representation of an object, built to scale, but usually smaller than full-size; 2) a pattern, or something or someone capable of serving as a pattern.

model home – a finished living unit, usually decorated and furnished, which is used as a representative example of the homes offered for sale in a development.

modification – an alteration or change, such as changing the roofline of a building.

modified bitumen – an asphalt compound altered by the addition of polymers to improve its characteristics, such as its water resistance and toughness. It is used as a preservative to coat and saturate wood, especially wood that is exposed to the elements, such as fence posts and railroad ties. It is also applied as a water-resistant coating with some roofing systems.

modified epoxy emulsion mortar – a thin-set, non-chemical-resistant mortar for installing ceramic tile on interior and exterior floors and walls. It contains emulsion epoxy resins and hardeners which provide rapid curing and water resistance.

modify – 1) to make less extreme; 2) to change or alter.

modillion – an ornamental block or bracket extending outward from a wall to support an overhang or cornice.

modular – constructed of standardized modules or sections which can be easily assembled to complete the whole unit.

modular homes – houses that are manufactured, then factory-assembled into the largest practical sections or modules that can be transported to a building site. The sections are installed one at a time on a foundation at the job site to complete the unit. Wiring and plumbing are installed at the factory with the ends exposed at each module. The modules are designed to fit precisely together, with the connections between modules made as the modules are placed. This system ensures a consistency of product, and time-savings at the job site.

module – a unit, which may be complete in itself, but is usually intended to make up part of a larger assembly.

modulus – a numerical value representing a physical property of a material. A modulus is used in a calculation, such as a structural computation, to represent the way in which that material will react under a specific set of conditions.

modulus of elasticity – the ratio of the load or stress applied to a material, such as a structural member, in relation to the amount of deformation that will result from the application of that stress or load to the material. It indicates the ability of the material to recover its original shape after the removal of the load.

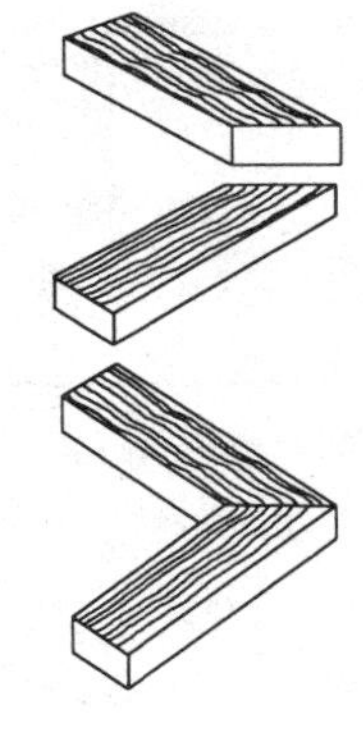

miter joint

miter vise

MOD 24

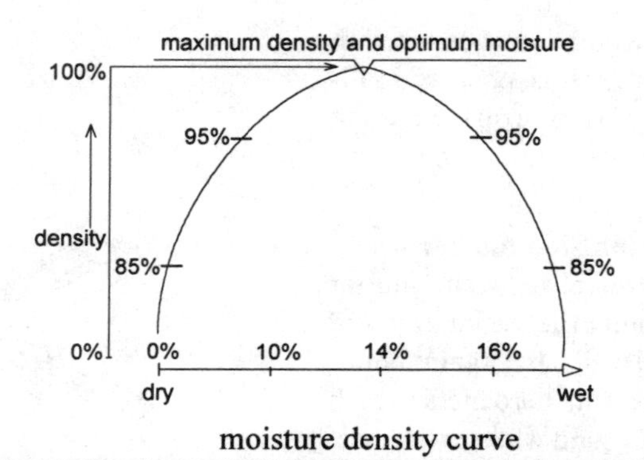

moisture density curve

moist – 1) damp; 2) having an atmosphere of high humidity.

moisture – water in small and diffused quantities.

moisture barrier – a membrane or coating designed to stop the penetration of moisture. They are made of materials, such as polyethylene, that prevent pass-through moisture. There are different types of barriers designed for different applications. For example, house wraps are applied to exterior house walls. They come in wide rolls that wrap around the framing and are stapled into place before siding is installed. Asphalt content moisture barriers, with or without plastic sheeting, are applied to the outside of masonry basement walls before earth is backfilled against the walls.

moisture content – a measure of the amount of moisture contained in something definable and specific. For example, lumber is available as S-Green, which has a moisture content by weight of more than 19 percent, or S-Dry which has a moisture content of 19 percent or less. S-Green lumber is milled green and must be milled to slightly larger dimensions than dry or seasoned lumber because there will be some shrinkage as the lumber dries. Lumber for framing should be S-Dry, seasoned or dried to the point that its moisture content is no more than 19 percent. In very dry parts of the country, only lumber with a maximum moisture content of 15 percent should be used for framing.

moisture density curve – a graphical representation of the soil moisture content compared to the soil density that can be achieved with that moisture content. Soil compaction depends to a degree on the moisture content of the soil. Soil compaction is desirable because it makes a given volume of soil more dense and increases its ability to support a load. For a particular type of soil, there is an optimum moisture content needed to achieve the greatest compaction. The moisture density curve indicates that optimum level.

moisture proofing – protecting against the absorption of moisture, such as applying an asphalt content moisture barrier to the outside of a basement wall.

moisture resistant panels – see water resistant panels.

moisture vapor transmission – a measure, in perms, of the amount of water vapor that passes through a material. A perm is one grain of water vapor per square foot, per hour, per inch of mercury pressure differential (0.491 psid). For example, one brand of 48 mil plastic roofing membrane has a moisture vapor transmission rate of 0.04 perms.

mold – 1) a form into which molten metal, or other liquid or plastic material, is poured for shaping while it hardens; 2) to form a material; 3) see mildew.

moldboard – a type of wood used for concrete forms.

molding or moulding – a length of material that has been milled into a pattern and is used to cover gaps, such as between the wall and floor, or provide a finished or decorative appearance. Molding is available in wood, wood-grained plastic, or paintable plastic or other synthetic materials. The plastic and synthetic materials are impervious to moisture and warping and are generally less expensive than wood. *See Box.*

molding head – rotary cutters mounted on a radial arm saw or table saw used to shape molding. The depth of the cut is adjustable and the fence can be set for the size of the material. The material is shaped by moving it past the rotating molding head. Also called a molding cutter.

molecule – the smallest particle of a substance, having one or more atoms, that contains all the properties of that substance.

molly bolt – an anchor assembly for use in drywall, between structural members. The anchor consists of a metal sleeve with internal threads at both ends, and a threaded bolt or machine screw which fits inside the sleeve. The sleeve is split lengthwise in four places around its center. A hole is drilled through the drywall, and the anchor sleeve is inserted. The end of the anchor assembly passes through the wall to the other side. As the screw or bolt is tightened inside the sleeve, it puts pressure on the center of the sleeve, causing it to collapse outward at the splits. The end of the sleeve is drawn toward the wall as the bolt is tightened, creating a flared end which anchors against the back side of the drywall.

monkey wrench – a type of adjustable wrench with flat jaws that remain parallel to each other.

monoplanar – a roof truss with all structural members of a single thickness to permit easy joining and joint reinforcement.

mop board – see baseboard.

morass – a section of marshy ground or swamp.

morro – a rounded hill or land projection.

Morse taper – a standard taper for drill press socket and tapered shank drills. The tapered shank of the drill is inserted into the tapered socket, making a snug fit and preventing movement between the two parts. The taper holds the drill securely but permits quick changes.

mortar – a mixture of portland cement, lime and sand used to fill voids in masonry units, bond them together, and add support. *See Box.*

mortar board – a flat square board, with a handle attached to the center of its underside, used to carry and hold mortar. Also called a hawk.

mortar box – a metal-lined mixing box for mortar.

mortar cleaning of concrete – the application of a cement and sand mortar to the surface of concrete after the forms have

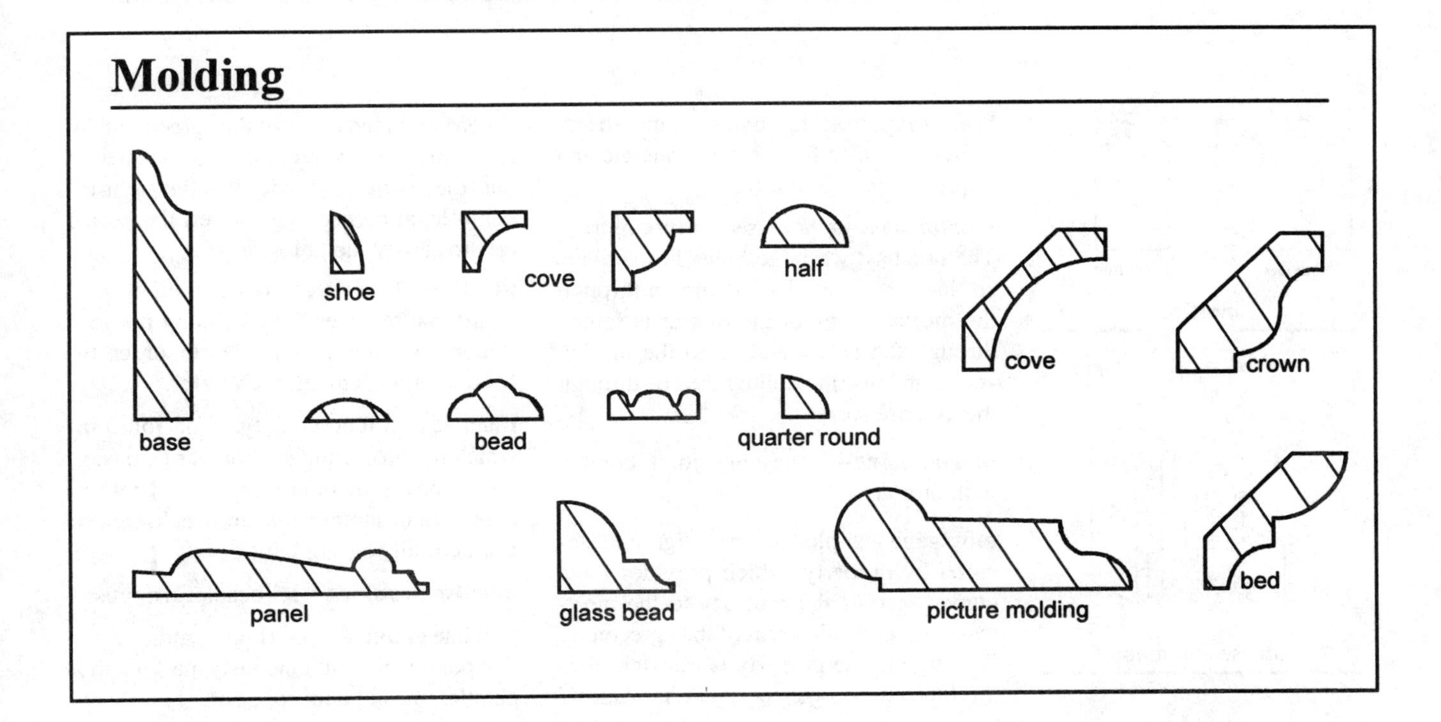

Mortar

Mortar is a mixture of portland cement, fine aggregate or sand, lime and water combined in proportions which will yield a plastic mixture that can be used to fill voids between masonry units, such as brick, block or stone. Mortar acts as a bonding agent to hold units together and add strength to the structure as a whole. The mixture may vary depending on the type of structure being built. Hydraulic cement or hydraulic cement with lime is used where greater plasticity and durability are desired than obtainable with portland cement alone. The different types of mortars are classified by the characteristics obtained in the mixtures:

Type K mortar has low compressive and bond strength and is used for non-bearing partitions and decorative work.

Type M mortar is a durable mortar with high compressive strength. It is used for reinforced brick masonry and is recommended for below-grade applications or other areas in contact with earth, such as foundations, retaining walls, sewers and catch basins. It is also used in areas that experience seismic movement.

Type N mortar is a medium strength, waterproof mortar used in areas above grade with high exposure to the elements, such as chimneys and exterior walls.

Type O mortar has medium-low compressive strength and is recommended for solid masonry non-bearing interior walls or in areas with limited exposure to the elements.

Type S mortar has medium-high compressive strength with high tensile bond strength, used for walkways, stucco, or loadbearing applications at or below grade.

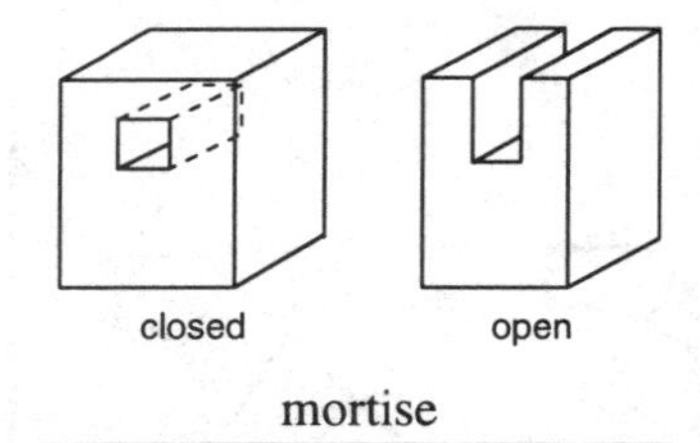

mortise

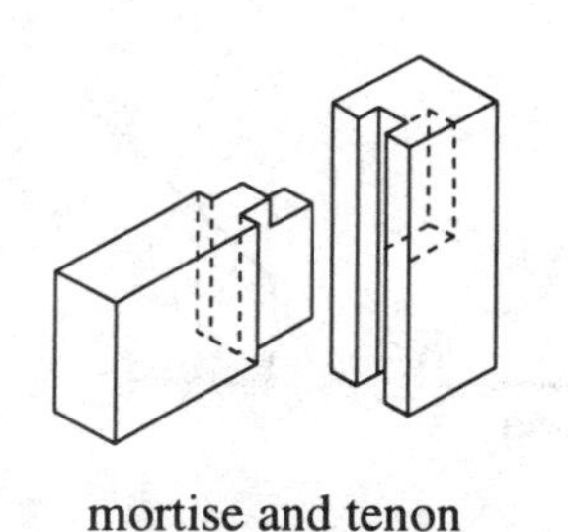
mortise and tenon

been stripped off. It provides a smooth and uniform surface for the new concrete and improves its appearance.

mortar hoe – a hoe used to mix mortar. The hoe has two large holes in the blade so that when the blade is drawn through the mortar, some of the mortar is forced through the holes, adding to the mixing action and making pulling the hoe through the mixture easier.

mortar joint – a masonry joint bonded with mortar.

mortgage – a pledge or obligation, secured by property, which provides for a conveyance of the property to the mortgage holder for the term of the agreement. Sole title to the property is transferred to the buyer when the mortgage is paid off according to the terms of the agreement. If the terms of the agreement are not carried out, the mortgage holder has the right to begin legal proceedings to sell the property to satisfy the obligation.

mortise – 1) a recess or slot cut into a board that receives the projecting portion (tenon) of another member in order to form a joint; 2) to cut such a slot.

mortise and tenon – a type of joint in which the projecting portion of one piece, the tenon, is fit tightly and glued into a recess cut in another, the mortise. It creates a structurally strong joint.

mortise deadlock – see deadlock, mortise.

mortise gauge – a marking gauge with two points for simultaneously marking the parallel lines of a mortise cut.

mortise lock – a type of lock installed in a mortise or recess cut in a door.

mortising chisel – a hollow, square-shaped chisel used on a drill press to make square holes. A drill bit is positioned in the center of the hollow chisel. As the drill bit bores a hole into the work, the chisel, which does not turn, cuts away the corners of the hole to form a square.

mosaic – a design created by the arrangement of small pieces of glass, tile, or rock in a pattern. The pieces are laid in a bed and mortared in place.

mosaic tiles – small colored ceramic tiles of the type suitable for use in a decorative mosaic pattern. The tiles are also used to tile any surface suitable for ceramic tile, such as bathroom countertops, flooring, patios or outdoor fountains.

motor – a prime mover that provides power for other operations and/or other devices. The energy for the mover can come from many different sources such as wind, steam, an internal combustion engine or electricity.

motor breakdown torque – the motor torque beyond maximum torque, or turning force, exerted by a wound-rotor electric motor when supplied with the rated voltage. As the load increases, the torque will increase to its maximum, then start to decrease. The point of torque decrease after maximum torque is the motor breakdown torque. An operator using a drill may experience motor breakdown torque when drilling a deep hole. The drill speed will begin to slow down when the maximum torque is exceeded under load.

motor control center – a magnetic electric motor starter unit in a single housing used for large electric motors.

motor, electric – a device that converts electricity into mechanical motion.

motor full-load torque – an electric motor running at a stated speed and producing the horsepower that it is rated for, such as one horsepower at 1800 revolutions per minute. Electric motors are rated at a stated horsepower at a particular speed.

motor pull-up torque – the minimum torque, or turning force, exerted by a wound-rotor electric motor. The pull-up torque occurs at a speed between starting the motor and reaching maximum torque.

mounted tile – ceramic tile that has been factory-assembled into sheets or sections using a facing, backing, or edge-mount material to hold the pieces of tile together. Sheets of tile are easy to work with and can be applied more rapidly than individual tiles as long as whole or partial sheets are used. If tiles have to be cut to fit a particular space, they must first be removed from the sheets, then cut and laid in the same manner as individual tiles.

mud – construction slang for mortar or gypsum wallboard joint compound.

mud pan – a rectangular container used for holding gypsum wallboard joint compound during application. The pan is long enough that a wide taping knife can be used to take the compound out of the pan.

mudsill – pressure-treated board or redwood board fastened to the top of a foundation and on which the rest of the building framing is erected. Also called sill plate.

mullion – a slender bar or divider between two window units.

mullion windows – two windows, side by side, with only a slender separation between them.

multicentered arch – see minor arch.

multiface fireplace – a fireplace having two or more sides open to a room or rooms.

multilayer – two or more layers.

multilite – a window sash having vertical and horizontal bars or muntins dividing the window pane.

multimeter – an electrical meter that has more than one function, such as a combination ohmmeter, ammeter, and voltmeter, which is known as a VOM or volt-ohmmeter.

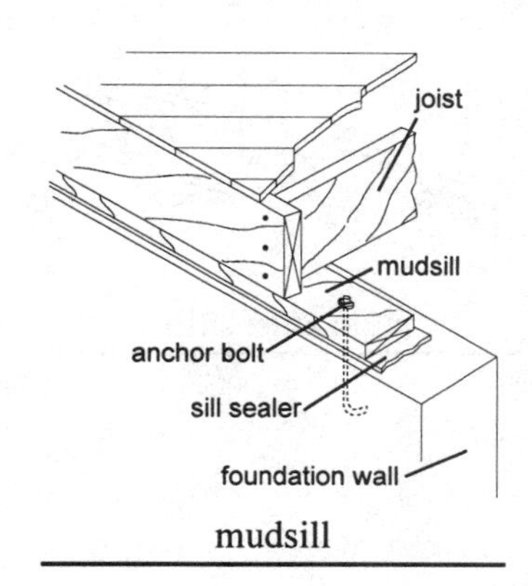

mudsill

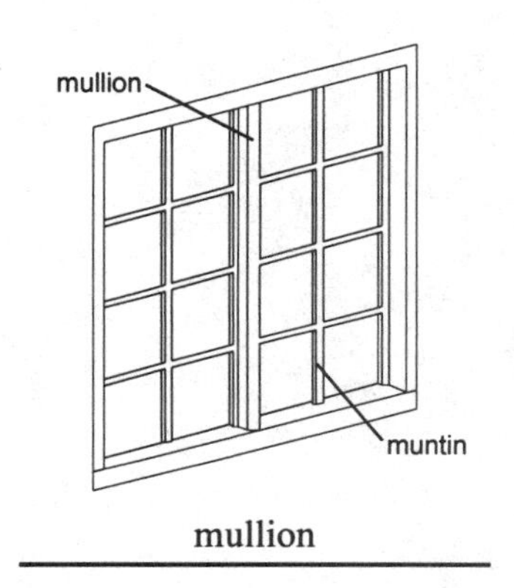

mullion

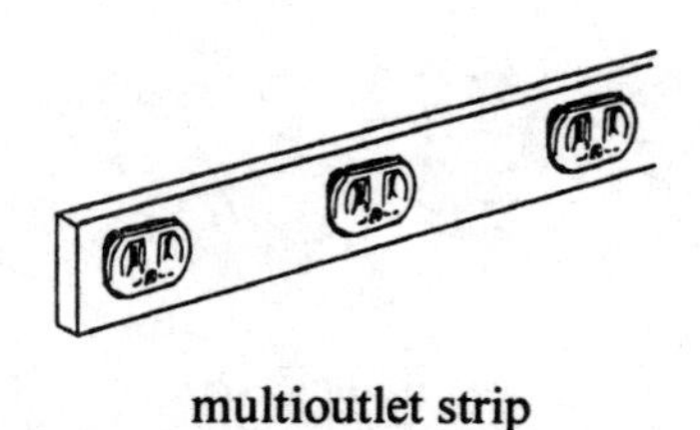

multioutlet strip

multimetering center – an electrical panel into which several electrical power recording meters can be inserted. They are installed for use in multiunit dwellings so that each unit has a separate electric meter.

multioutlet strip – an AC power supply strip housing several outlet receptacles. A multioutlet strip can be used when a number of low current draw components are grouped together and each requires an outlet, such as an entertainment center containing a television, VCR and surround sound decoder.

multiple dwelling – two or more dwellings that share a common roof and wall, such as a duplex or a condominium.

multiple switching – electrical switches that permit switch control from more than one location, such as a light fixture that can be operated from the top and bottom of a staircase.

multiple threads – a threaded section that has two or more thread starts located at different points around the circumference of the end of the part. This increases the total gripping surface of the threads while still providing a steep thread pitch for rapid installation. Type S drywall screws, used for steel studs, and Type G drywall screws, used for gypsum coreboard, are made in this style.

multi-ply construction – two or more layers of gypsum wallboard used to increase the fire rating of the structure.

municipal – relating to a municipality or local government.

municipal sewer – see public sewer.

muntin – bars that are used to divide window openings. Also called sash bars.

N: nail to nut splitter

nail – a slender metal rod with a pointed tip used to fasten materials together. There are a variety of sizes with different shaped heads for various applications. *See Box.*

nailer – 1) a piece of lumber installed in framing for the purpose of providing a nailing surface for the inside finish wall material; 2) a strip of wood fastened to concrete, masonry or steel to provide a nailing surface for the attachment of other materials.

nailing block – a block of wood fastened or set into a masonry structure to provide a surface onto which other members can be anchored with nails or screws.

nailing machine – a device that will hold a nail and direct it at the proper angle as it enters the wood. The machine is positioned at the location where the nail is to be driven, and then its plunger is struck with a hammer to drive the nail.

nail popping – a problem that appears both in decking and in gypsum wallboard finishes where the heads of nails pull or work themselves out of the framing members and "pop" through the surface. This can be caused by the flexing of the structure, force on the wallboard, or the use of too few nails, causing overload of individual nails.

nail puller – a bar with a built-in fulcrum point and slotted jaws that can grip a nail and pull it out of a surface. The jaws slide under the head of the nail, grip it, and lift it up as the handle of the puller is forced down.

nail set – a type of punch used to force the head of a nail flush with, or below, the surface of the wood.

nail spotter – an applicator used to smooth joint compound over the heads of fasteners when installing gypsum wallboard. It has a handle for reaching, a container/dispenser for the joint compound, and a smoothing blade for blending the compound onto the surface.

naphtha – a solvent distilled from petroleum. It is used to thin lacquer, paints and asphalt, and for removing the residue of these coatings.

narrow "U" – a stair run in which the first straight run ends at a landing, and the second run continues straight in the opposite direction after making a U-turn at the landing.

natural cement – a cement made by grinding and calcining a type of rock which consists of limestone and clay, with the clay comprising no more than 25 percent of the total volume.

natural draft cooling tower – a cooling tower with an internal design which utilizes natural air circulation to cool a fluid, usually water. This type of cooling tower

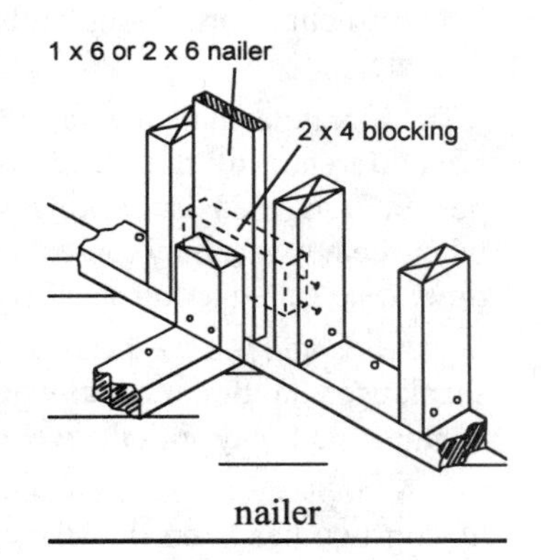

nailer

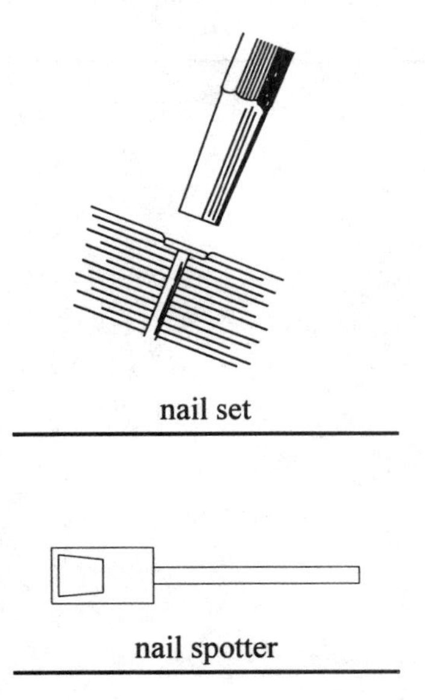
nail set

nail spotter

Nail

A nail is a slender metal rod with a sharpened tip used to fasten wood or similar materials together. It may have a smooth shank or be ribbed to aid in gripping wood or masonry. Nails have a variety of different head shapes and sizes, depending on their intended use. Most nails are made of steel, though they are also available in copper, brass and aluminum. Nails can be case hardened, galvanized or plastic coated for use with various materials. They are driven into the material using a manual hammer or a power tool, such as a pneumatic gun.

The common nail has a sturdy shank and flat head about twice the diameter of the shank. It is used mainly in construction for joining lumber, such as that used in framing.

An annular nail has raised rings forged into the circumference of the shank for increased gripping power. It is used in decking and other construction where heavy wear may cause the nail to work loose over time. Also called a ring shanked nail.

A box nail is a flat-headed nail with a thin diameter shank used on thin material where the thicker-shanked common nail may split the wood.

A duplex nail is designed for easy removal by having two heads on the shank, one above the other, with about ½ inch of shank between them. The nail is driven into wood until the first head is seated, leaving the second head sticking up above the surface. When it is time to remove the nail, the head can easily be gripped by a claw hammer or pry bar and pulled out. This type of nail is used for temporary installations, such as concrete formwork.

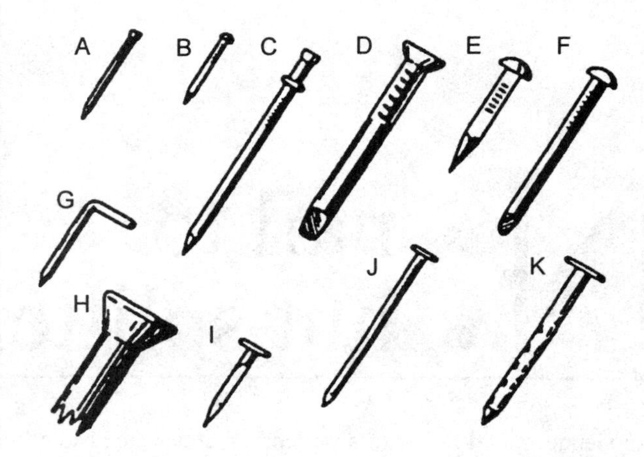

A - brad
B - escutcheon
C - duplex/double head
D to F - hinge nails
G - nail for metal lath
H - round wire spike
I - upholsterer's tack
J - shingle nail
K - asbestos shingle nail (barbed)

is used in power plants, refineries and chemical plants to cool water used in processing or cooling equipment.

natural finish – a finish, such as a penetrating oil, that protects the wood while permitting the grain and natural color of the wood to show through.

natural gas – methane, a gas formed in the earth in oil-bearing areas, that is used to fuel furnaces and water heaters.

natural ground – undisturbed land, before any grading or excavation is performed. Also called natural grade.

natural resin – an organic fluid from plants and trees used in making paints and varnishes.

natural ventilation – air flow or circulation that does not depend on a mechanical aid, such as a fan or blower.

neat cement – a mixture of cement and water, without aggregate.

needle – 1) a slender pointed metal rod used for stitching or sewing leather or fabrics together; 2) a structural beam used to support masonry that is in place while work is performed on other masonry below it.

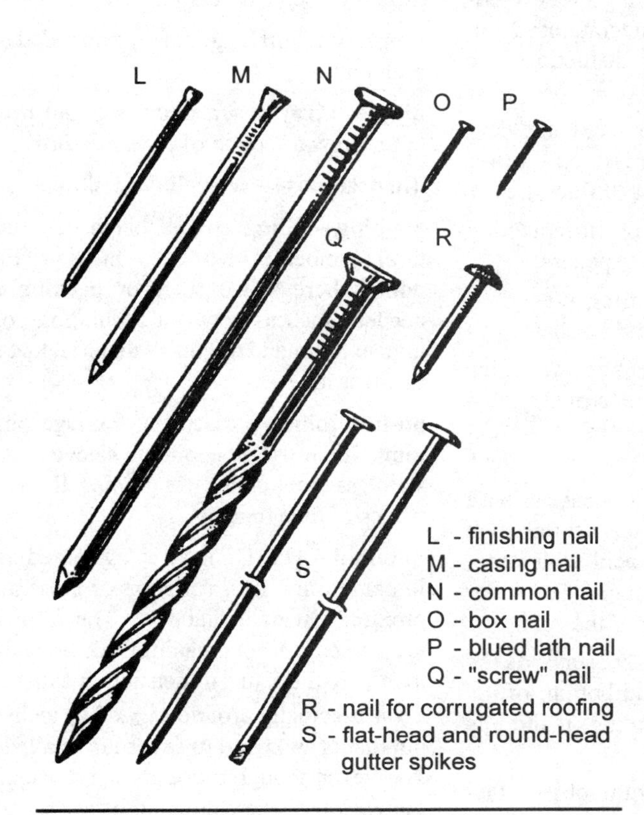

A finishing nail is a slender nail with a semi-spherical head only slightly larger than the shank of the nail. The head is driven flush with the surface of the wood, and then driven beneath the surface with a nail set. The resulting depression or hole at the nail head can then be filled with wood putty to conceal the nail. Finishing nails are used for trim work, molding, cabinetry and furniture. A small finishing nail is often called a fine nail. A brad is a very thin finishing nail, available with or without a head. Brads are used for delicate work, such as fasteners for picture frames.

Tacks are short, very sharp nails with shanks that taper from the head to the point, or at least a substantial portion of the shank length is tapered. Tacks have flat heads and are used most often for carpeting and furniture upholstery. Upholstery tacks may have ornate heads of brass or other material for use where they will remain exposed.

Roofing nails have heads that are ½ inch or larger in diameter. The large head prevents them from tearing the roofing felt or composition shingles on which they are used.

Drywall nails come with smooth or ring shanks, and with flat or concave heads. They are used to install gypsum drywall panels to wood framing.

Aluminum nails are made in a variety of sizes and shapes for use in situations where a nail of another metal may cause corrosion problems, such as in the installation of aluminum flashing. Aluminum panel nails have a large diameter head and an elastomer gasket or washer under the head. They are designed for use in contact with aluminum, such as in fastening an aluminum skylight rim to the skylight curbing.

needle file – a very slim, pointed file for use on delicate work such as small mechanisms like timers, locks or precision tools.

needle-nose pliers – pliers with tapered jaws that come to a point or near point. See also pliers.

needle valve – a valve in which the stem, or a disc attached to the stem, tapers to a point that fits into a seat in the valve body. The flow through the valve is regulated by moving the stem in or out of the seat. A needle valve is used for fine regulation of small flows, such as controlling the air for an HVAC system.

negative charge – 1) an electrical charge that is the same as the charge of an electron; it is the lowest potential charge to which electrons will flow; 2) an electrical charge that is minus.

negotiate – 1) to come to a settlement or agreement through discussion; 2) to bargain; 3) to maneuver or successfully travel through a difficult area.

neoprene – polymerized chloroprene synthetic rubber that is oil-resistant and flexible. It is used to make a seal or covering that is resistant to oil or other petroleum products.

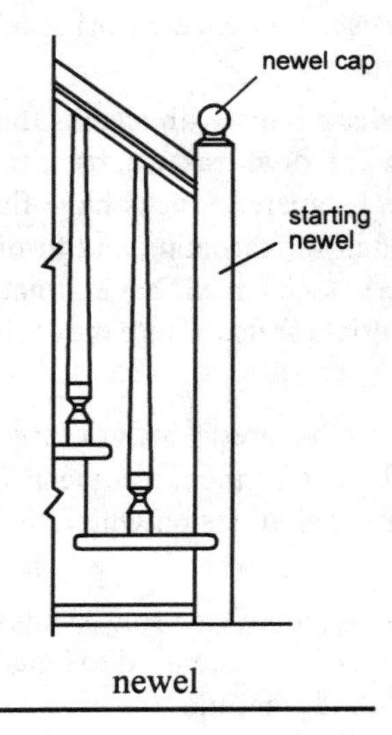

newel

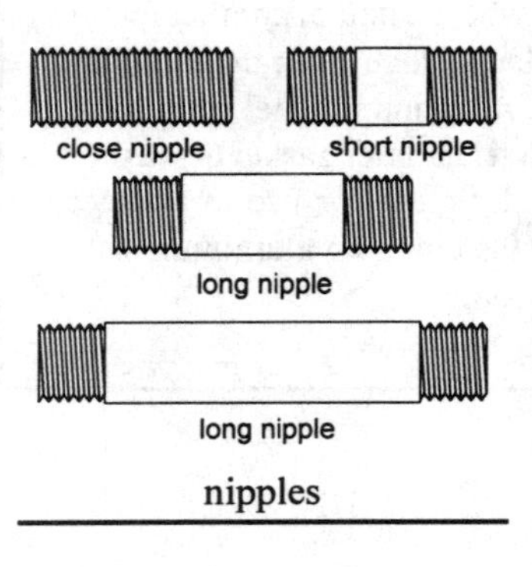

nipples

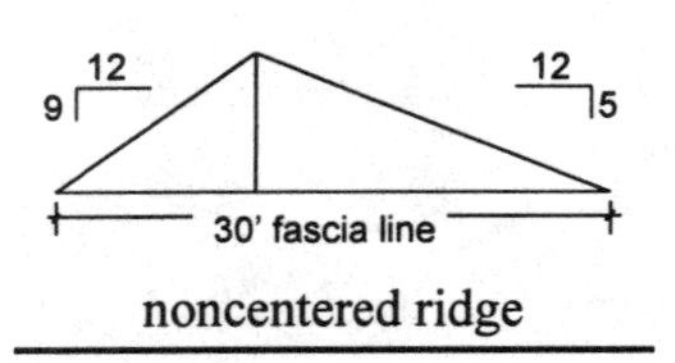

noncentered ridge

net – 1) an amount remaining after excluding all nonessential considerations; 2) an amount remaining after deductions are taken.

net floor area – the floor area not covered by partitions, stairs, and other construction features; usable floor area.

net income – income or profit remaining after deducting costs and expenses.

neutral – 1) not transmitting power; 2) not reactive; 3) neither acid nor base.

neutral conductor – the second wire (white) of a two-wire AC electrical system that carries the return electrical current from an electrical load.

neutralize – to add acid to a base or base to an acid until the solution is neutral, such as neutralizing muriatic acid with water after using it to clean masonry. The water stops the chemical action of the acid.

newel – the upright post that supports the stair railing at the top and bottom of the stair case, and at the landing, if there is one. Also called newel post.

newel cap – an ornamental object fastened to the top of a newel, or stair post.

nibbler – a sheet metal cutting tool that cuts away small sections at each stroke.

niche – 1) a recess in a wall designed to hold a decorative object, such as a statue or vase; 2) a position or employment for which a person or business is ideally suited. For example, a carpenter may discover that he really enjoys, and is good at, building circular stairs. He may become known and in demand for this skill, and make his entire business building circular stairs. This would be his niche.

nickel – a silver-colored metallic element; atomic symbol Ni. It is used to plate components, such as butterfly valve discs, for corrosion resistance.

niello – a black alloy of silver, copper, lead, and sulfur inlaid in ornamental designs cut into metal surfaces.

night seal – a temporary roof membrane seal made of synthetic sheet material used during construction. It is used to cover a roof opening until the opening is closed over with the finished roof material.

nippers – see wire cutters.

nipple – a short length of pipe threaded on each end.

nipple extractor – a plumbing tool used to unscrew a section of pipe. *See Box.*

nitrocellulose – see cellulose nitrate.

nogging – filling spaces between structural members with bricks. This is usually done where space filler or backing is needed, such as between a building column exterior and the finish material that is to cover it.

no-hub joint – a cast iron drainage pipe joint which uses a neoprene sleeve and a stainless steel clamp for joining the two pieces of pipe together.

nominal – 1) existing or being something in name only; 2) a variation or close approximation of actual size. The term is used when it is not meaningful or practical to refer to the actual dimension. A finished 2 x 4 is actually around 1½ x 3½ inches, but that is awkward to say, so it's called a 2 x 4 even though those are not its actual dimensions. There are published standards to which manufacturers adhere. Designers can plan actual dimensions based on these nominal sizes without having to measure each actual piece of material.

nominal pipe size – the approximate pipe size which, in some sizes, corresponds to the actual outside diameter of the pipe, but in other sizes it does not. For instance, the actual outside diameter of a steel pipe is larger than the nominal diameter in 12-inch sizes and smaller. A 4-inch steel pipe is actually 4½ inches in outside diameter, and a 10-inch pipe is actually 10¾ inches in outside diameter.

nonagon – a nine-sided polygon; not commonly used in building except occasionally in the design of a structure such as a gazebo.

noncentered ridge – a roof ridge that is offset with respect to the centerline of the outer walls of a structure.

noncombustible – a material that is fire resistant.

Nipple extractor

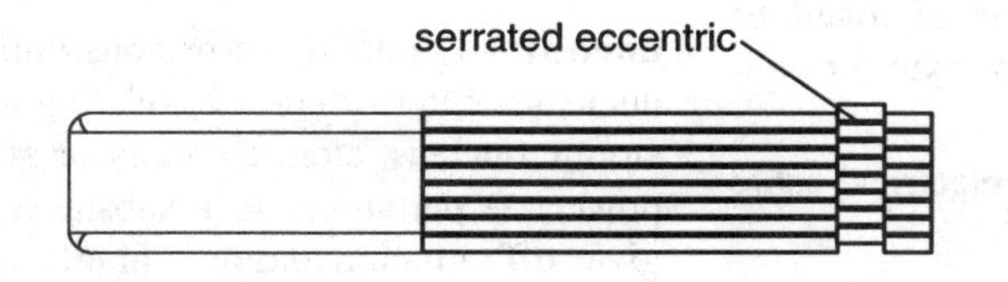

A plumbing tool designed to be inserted in the inside of a piece of threaded pipe, such as a nipple, and then turned with a wrench to unscrew the pipe. The nipple extractor grips the inside of the pipe, preventing damage to the outside of the pipe or to its external threads. It consists of a steel rod that is hexagonal on one end so that it can be gripped with a wrench, with the other end having serrations or grooves along its length. There is a groove machined near the end of the serrated section that is off center to the center of the rod, so that it is eccentric. A sleeve, also serrated along its length, fits loosely into this groove. When the extractor is inserted into the inside of a pipe and turned, the serrations on the sleeve catch on the inner pipe wall. This stops the sleeve from turning further, and forces it tightly against the wall of the pipe. The serrations on the extractor also grip the pipe wall. Further turning of the extractor transmits this turning force to the pipe, forcing the pipe to turn and unscrew. The extractor can be used to unscrew short lengths of exterior threaded or decorative pipe, such as chrome- or brass-plated shower head fittings, without marring the surface.

nonconductor – an insulator; a material that does not conduct much electricity or heat, or that does not conduct much sound or other vibration. Such materials are used for safety, to minimize heat loss or gain and to provide noise control.

noncorrosive flux – a flux and its residue that do not corrode the base metal. For example, chemicals are used which isolate heated metal from oxygen in the air, and so prevent, dissolve or aid in the removal of oxidation during welding, brazing or soldering. Preventing oxidation permits better metal flow in the joint and stops the formation of a corrosive residue on the base metal. A corrosive flux used with copper will leave behind a surface coating that will eventually oxidize the copper and turn it green.

nondestructive examination (NDE) – any of several methods of inspection to determine soundness or quality that do not damage or destroy the object being examined. Visual inspection, radiography, liquid penetrants and ultrasound are all examples of nondestructive examinations. They can be used singly or in combination to identify defects in materials.

nonferrous – not containing iron or relating to materials that do not contain iron. Some types of construction require the use of nonferrous or nonmagnetic materials, especially in situations where radio frequency interference must be considered.

non-loadbearing wall – a partition that does not carry any of the load of the structure; used only as a partition or divider. Such walls can be moved, removed or modified without affecting the structural stability of a building.

nonmetal – a material that contains no metal, and so meets required specifications for use under a particular circumstance, such as when some degree of chemical resistance is necessary.

nonmetallic sheathed electrical cable (NM) – cable used for electrical wiring within the building structure. It consists of individually insulated conductors within an outer protective insulating sheath. See also following definition for NMC.

nonmetallic sheathed electrical cable (NMC) – cable used for electrical wiring in moist areas. This wire consists of indi-

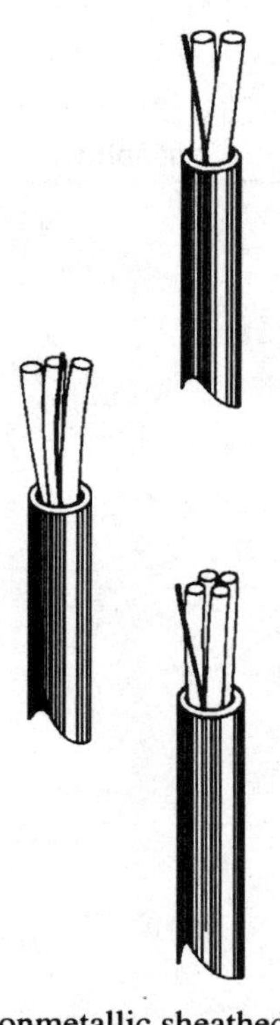
nonmetallic sheathed electrical cable

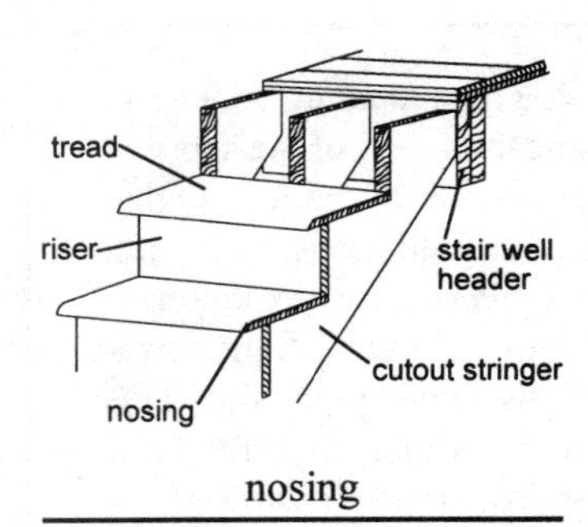

nosing

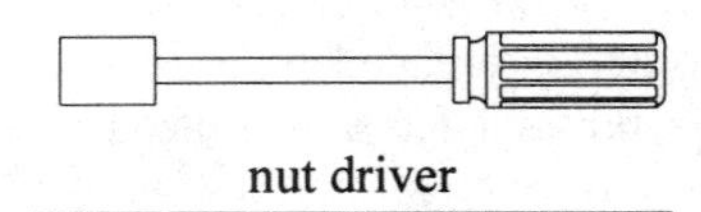
nut driver

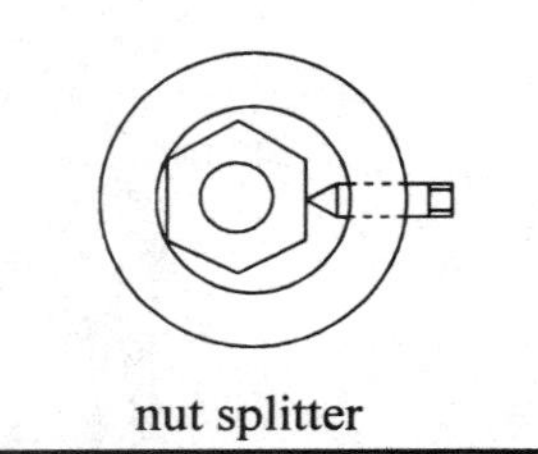
nut splitter

vidually insulated conductors within an outer protective insulating sheath that is cast tightly around the insulation of the inner conducting wires. This type of cable can be used inside in damp or wet areas or outside where it may be exposed to weather, such as wiring for exterior yard lights.

nonrigid – flexible, as in electrical cable that can be easily bent.

nonvolatile – 1) a material that cannot easily be made to vaporize; 2) the portion of paint that does not evaporate with the carriers and solvents.

normalizing – heating metal to the critical point, a specified temperature above which the metal begins to take on different characteristics. The metal is then allowed to cool in still air at room temperature. This process is used to alter the properties of the material.

Norman brick – a solid-core brick measuring 4 x 2⅔ x 12 inches. See also brick.

Norwegian brick – a solid-core brick measuring 4 x 3⅕ x 12 inches.

Norwegian brick, 6-inch – a solid-core brick measuring 6 x 3⅕ x 12 inches.

nosing – the rounded portion of a stair tread that projects beyond the face of the riser or step.

notary public – a person authorized to witness and authenticate writings or signatures on documents, affidavits or depositions.

notch – a cut or recess, often V-shaped, in the surface or edge of an object.

notch-sensitive – a metal that exhibits reduced strength at a stress concentration, such as a notch in the surface of an object made from that metal.

note – 1) a financial instrument, such as a promissory note for a loan; 2) a brief comment, instruction, or information on a drawing or other document.

notes payable – money due to a lender.

notes receivable – money or assets secured by a promissory note that is subject to call for payment by the lender.

nuclear – 1) relating to or constituting a nucleus, center or core; 2) relating to the atomic nucleus, atomic energy or atomic power; 3) pertaining to a substance that gives off radiation and/or is involved in a nuclear fission or fusion reaction.

nuclear density meter – a device that uses gamma radiation to measure soil density. It introduces a controlled amount of radiation into the soil. The meter then reads the amount of radiation scattered back from the soil and translates the radiation into density or moisture content based on its calibration from known conditions.

nuclear soil compaction test – a method of measuring soil compaction by measuring reflected nuclear radiation using a nuclear density meter.

null – 1) having no legal or binding force, such as an invalid contract; 2) equal to zero.

nut – a female fastener with a threaded hole used with a male threaded bolt or similar device. Some nuts have wings for easy installation by hand, other types have wrench flats around the outside.

nut driver – a tool, similar to a screwdriver, with a socket on the end of the shank which fits over a nut and turns the nut for tightening or removal.

nut splitter – a tool, with a collar that fits around a nut and a threaded rod with a sharp, hardened blade that can be forced against the nut to split it when the threaded rod is turned with a wrench. It is used to remove a nut that has frozen in place.

O: oak to oxyhydrogen welding

oak – a coarse-grained hardwood used in the construction of a variety of wood products including furniture, cabinetry, trim, and flooring.

oakum – long, untwisted hemp rope fibers used to pack the joint and retain molten lead in bell and spigot pipe joints.

oblique – not parallel or perpendicular to a plane or line; having no right angle.

oblique drawing – a drawing showing the shape of an object. It consists of the front of the object, with the body of the object shown as it would appear when viewed from an angle slightly off to one side. The side of the object becomes smaller from the perspective of the viewer.

obstruction – a blockage or obstacle which prevents passage, action or operation, such as a clog in a pipeline.

obtuse angle – an angle larger than 90 degrees but less than 180 degrees.

occlude – to close up or block off, such as to close up a duct.

octagon – a polygon with eight sides and eight angles.

octagon box – a box with eight sides used for supporting electrical fixtures in the ceiling or wall, or as a junction box in which wires can be joined.

octahedron – a solid with eight surfaces.

off-center – not centered; not located at an equal distance from edges or reference points.

off-center splice – a joint made between two structural members that are of unequal length, so that the splice is made at a place other than the center of the span.

offset – 1) a change in alignment from a single axis to two axes that are separated, as in one shaft geared to two other shafts; 2) something that serves to counterbalance or compensate for something else.

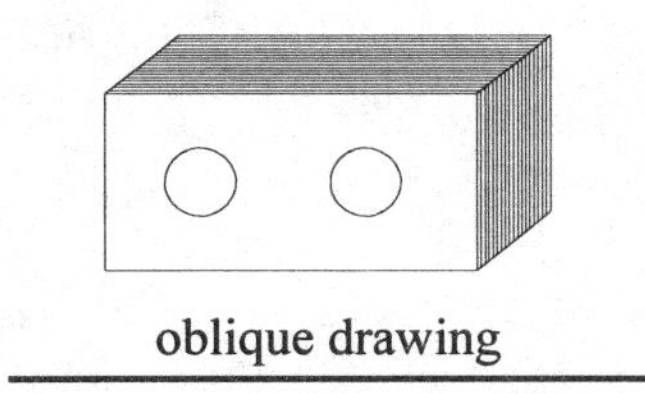

oblique drawing

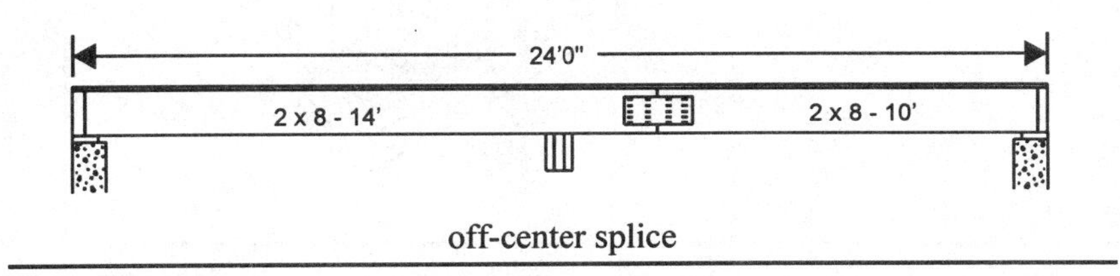

off-center splice

offset, 1/8-bend – a plumbing drainage fitting for routing a drain line around an obstacle.

offset nipple – a conduit connector with an offset centerline.

offset screwdriver – a screwdriver with the shank bent at 90 degrees near the blade tip to permit access to tight places.

offset section – see section, offset

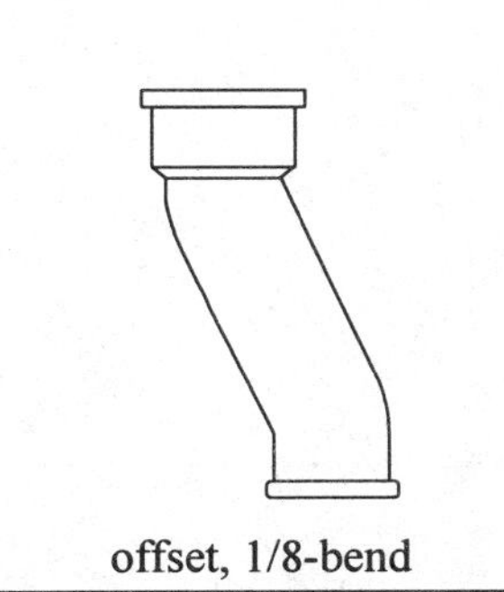

offset, 1/8-bend

Offset studs

A method of framing in which studs are staggered so that drywall panels on each side of a wall are fastened to a different set of studs. This is done by installing 2 x 4 studs on a 2 x 6 base plate in such a manner that alternating studs are flush with alternating sides of the plate. When the drywall is attached, there are no common studs from one side of the wall to the opposite side of the wall. This eliminates sound transmission from one wall, through the studs, into the next wall. When wall coverings on both sides of a wall are in contact with the same stud, sound is transmitted via vibration from one wall to the other, even if all the wall cavities are filled with insulation. Using offset studs with other sound elimination methods and materials can result in very effective sound isolation.

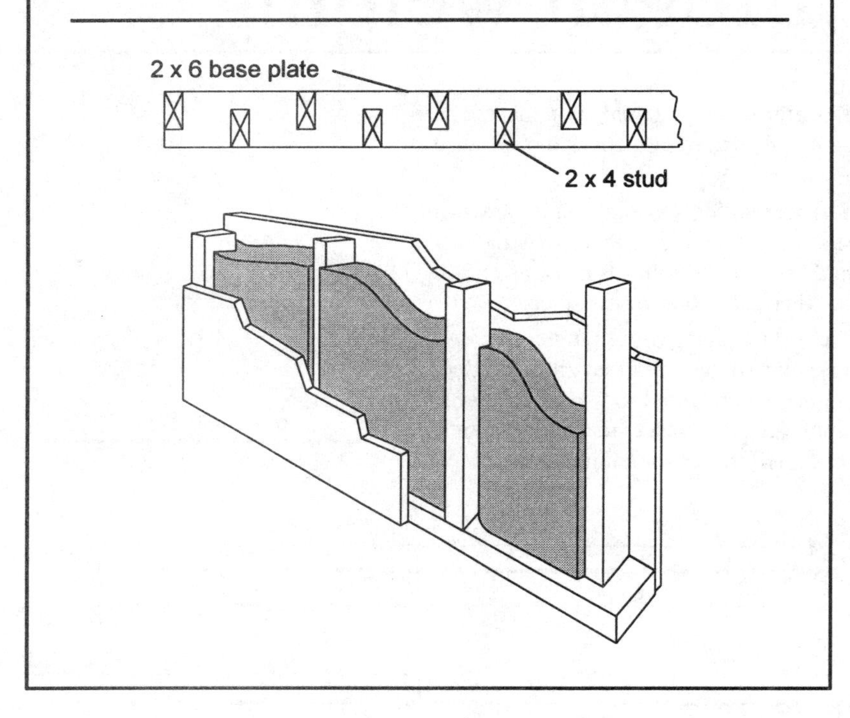

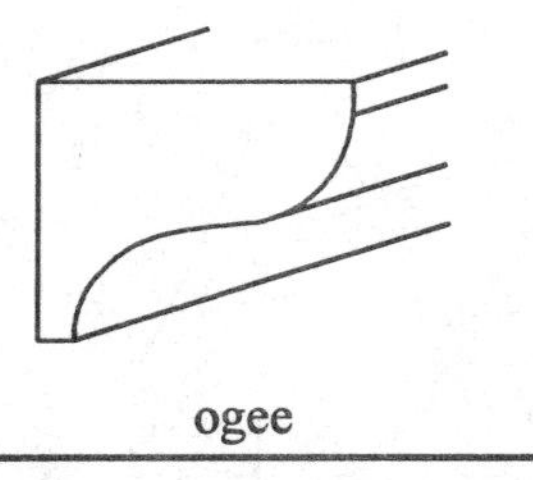

ogee

offset studs – a framing method in which the studs are attached alternately on one side of the plate or the other so that drywall panels on each side of a wall are fastened to different studs. *See Box.*

offset wrench – a wrench with the head offset at an angle from the axis of the handle, which permits it to be used to reach fasteners located in tight places.

offsite – at a location other than the construction site. For example, offsite electrical power would originate from a source near, but not on, the jobsite.

ogee – a compound curve shaped roughly like the letter S applied to a surface, such as molding or trim.

ohm – a unit of electrical resistance.

ohmmeter – a device for measuring electrical resistance.

Ohm's Law – a series of formulas describing the relationship between current, voltage, and resistance. Current (I) equals voltage (E) divided by resistance (R), or $I = E/R$. This relationship can be used to solve for any one of these three values if the other two are known. Ohm's law applies to all electrical circuits, and is therefore valuable in sizing circuit components and wiring.

oil – a smooth, greasy liquid substance, made from mineral or plant sources, that has a variety of uses such as a lubricant, a base for paints or a rub for wood finishes.

oil finish – a finish applied to wood in which oil is rubbed into the surface to add luster and preserve the wood.

oil of cedarwood – an oil, refined from cedar, that is used to oil cedar for restoration.

oilstone – a fine-grained stone used, with the aid of oil as a lubricant and cleaner, to sharpen tools. The oil prevents metal and stone particles from contaminating the stone surface and making it less effective.

oil well cement – cement formulated to cure at high temperatures. It is used to seal off oil and gas pockets found in drilling and repairing oil wells.

oleoresin varnish – see varnish, oleoresin.

on center (oc) – the distance between items as measured from the centerline of one item to the centerline of the next. For example, floor joists that are placed 24 inches on center, measure 24 inches from the centerline of one joist to the centerline of the next.

one-line diagram – see single line diagram.

one-way ribbed structural slab – a structural slab with hollow filler blocks made of lightweight concrete, clay tile, or gypsum tile arranged in rows, or with metal pan fillers between concrete joists. This results in a concrete slab with reduced dead load, as opposed to a solid slab that has equal load capacity. The depth of a one-way ribbed slab is greater than a solid slab, with two inches or more of solid concrete placed over the blocks to provide concealment space for piping and conduit, and to add strength.

one-way structural slab – a slab with a uniform depth and no filler material, that is reinforced in only one direction. This is an economical construction used for spans of up to 12 feet where heavy concentrated loads will be carried.

oolitic limestone – a type of limestone composed of very small round grains bonded together with occasional traces of ironstone or iron oxide. Blocks of oolitic limestone can be used for foundations, walls or similar construction. Also called oolite.

opaque – a material, such as paint, through which light cannot pass.

open – 1) not closed or obstructed to passage or flow; 2) not complete, so as to not permit flow, as in an electrical circuit or device.

open circuit – an electrical circuit that has a break, or is "open," so that the current cannot flow through.

open cornice – a cornice, or roof overhang, in which the rafter ends are exposed rather than enclosed.

open end wrench – a tool with a U-shaped head designed to fit a specific size nut or bolt head. The U or "open end" slides over and grips the fastener so it can be tightened or removed. The wrench may be single or double ended.

open grain – wood that has a coarse grain, such as oak. An open grain wood may need a filler or sealer before staining so that it will take the stain evenly.

open hearth furnace – a steel processing furnace with a shallow hearth and direct exposure of the steel to the heat source. Used in steel mills for ore smelting.

open listing – a contract with a real estate broker which allows other licensed brokers to show, and sell, for a percentage of the sales commission, the piece of property being listed.

open string stairway – a stairway with the treads visible from one or both sides.

open time – the time a mixture or compound may remain exposed to the air without impairing its usability. For example, when paint exceeds its open time, a dry film forms on the top of the paint. This film must be totally removed before the remaining paint can be used. Other products, such as drywall joint compound, become totally unusable when they exceed their open time. In this case, the compound forms a hard crust on top, which, even if removed, will leave small hardened particles in the compound, making it impossible to apply a smooth surface.

open valley – a type of roof installation in which the valley flashing is left exposed rather than being covered over by shingles. Water running off the shingles will be directed into the valley and run down the metal flashing and off the roof.

open web joist – a steel joist that is built up using lengths of thin, lightweight structural steel shaped into a lattice pattern. This method of fabricating joists makes a very strong, yet light, joist which contains relatively little steel for its size. These joists are easier to handle and less costly than solid steel joists and beams.

operating point – the airflow volume and static pressure produced by a fan in an HVAC system. Fans move a certain volume of air, but in order to do this they must build up pressure. The static pressure built up in the system and the volume of air moved constitute a combination called the operating point. The operating point is used to determine the effectiveness of the system in a building space.

operator – the person who controls or runs a process or machine.

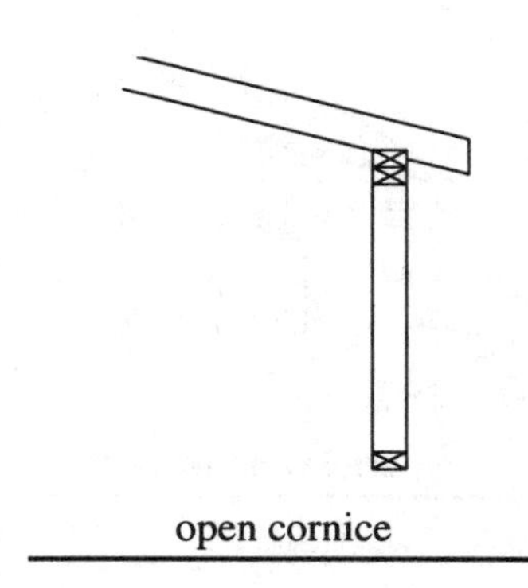

open cornice

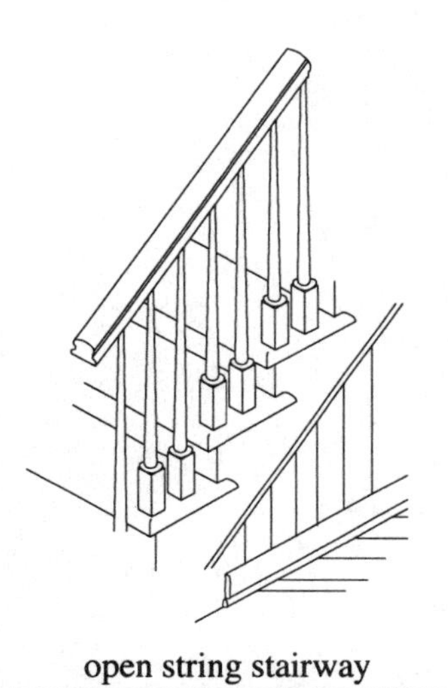

open string stairway

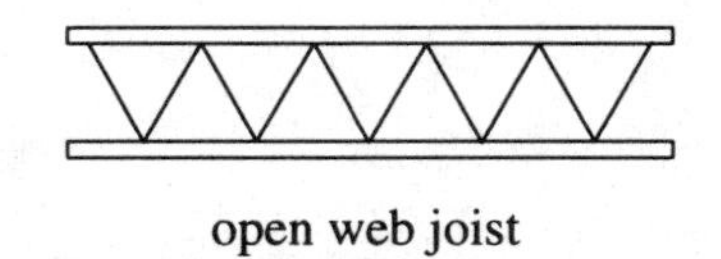

open web joist

oriel window

optical level – a telescope and level mounted on a tripod that has adjustments to lock the telescope in position. When it is locked level, it may be rotated and used to sight another point that is exactly level with the location of the telescope.

optical plummet – a prism device attached to a transit that allows the viewer to look down through the telescope and center the transit over a reference point. It is used instead of a plumb bob in locations where wind may affect an accurate reading.

optics – the science that deals with the origin and behavior of light and other phenomena associated with it.

optimize – to make something as well suited for its intended purpose as is reasonable or possible. For example, a water pump that is optimized for an application is one that is selected so that it runs at the combination of speed, discharge pressure and discharge volume that equals the highest efficiency for that pump.

option – 1) a right to choose among various possible selections; 2) an alternative; 3) a contract conveying a right to buy a particular commodity, such as a piece of property, at a set price during a specified period. For example, an owner may lease a piece of property for a set period of time and grant the lessee the opportunity to purchase the property at a set price during that period. If property values are increasing, it would be to the lessee's advantage to exercise his option to buy before the lease period runs out. Often, part of the lease payment is put toward the down payment, making it easier for the lessee to make the purchase. This is leasing with an option to buy, often called a lease option.

orange peel – a paint defect in which the finished surface has a texture resembling that of an orange peel. This can occur when the paint is not liquid enough, or has too much surface tension to flow smoothly. When a paint flows smoothly, brush marks and other surface irregularities disappear as the paint evens itself out.

orange shellac – a waterproofing agent, used on wood, made by dissolving lac in alcohol. In its normal state, shellac has an orange or orange-red color. It is called orange shellac to differentiate it from shellac that has had the color removed by bleaching.

orbital sander – a power sander that moves the sandpaper in an oscillating motion so that the work is sanded in all directions. Used for finish sanding.

order – 1) a rank, level, or classification; 2) a type of column and entablature, or horizontal members resting on the columns, forming a unit of an architectural style, such as Corinthian, Doric, Ionic, Tuscan or composite; 3) concentric rings forming an arch.

ordinance – a rule, regulation, or law set by governmental authority.

organic – relating to or containing carbon compounds.

oriel – a window that projects from the face of a building and rests on a structural extension, such as a corbel.

oriented strand board (OSB) – a three-ply wood product made by bonding mechanically oriented wood strands with resin under heat and pressure to make panels that are used for subflooring and sheathing.

orifice – an opening, mouth or constriction in a passageway. It can be reduced in size to a specific dimension or diameter to control flow, or to change the characteristics of the flow. For example, the nozzle of a propane torch has an orifice to control the flow of propane gas and to accelerate the gas so that it mixes better with air. The orifice provides improved combustion. An orifice may also be used to measure fluid flow passing through it by placing a pressure indicator across the opening.

original grade – the existing grade at a jobsite before any excavation or grading work has begun.

O-ring – a circular gasket designed to be placed in grooves to provide maximum sealing effectiveness.

ornate – elaborately styled or decorated, such as the elaborate trim on a Victorian house.

orthogonal – having right angles and or perpendicular lines and planes, such as a square or rectangular room.

orthographic projection – a drawing showing straight-on views of the different sides of an object, each side shown being projected at a right angle from the front view. Used for elevation drawings. Also called orthographic drawing.

oscillate – 1) to move back and forth or above and below a mean point or value in regular cycles; 2) to swing, vary or fluctuate from one extreme to another, as from fully on to completely off.

ounce – 1) a weight equal to $\frac{1}{16}$ of a pound avoirdupois or $\frac{1}{12}$ of a pound troy; 2) a measure of volume equal to $\frac{1}{32}$ of a quart.

outlet – 1) the discharge side of a fluid system or component; 2) an electrical receptacle.

outlet analyzer – an electrical device which, when plugged into an outlet, will indicate whether the outlet has been wired properly with the correct polarity and an open ground or an open common lead.

outlet box – a housing used to mount an electrical system or make electrical connections. The box itself is often metallic and rectangular in shape, but can be made in other shapes and of nonmetallic materials. Different sections of an electrical system can connected within the box.

outlet box cover – a cover for an electrical outlet box. It can be blank or have cutouts for various switches and receptacles.

out of plumb – not truly vertical, such as a wall that leans slightly. A plumb bob is used to determine if an object or surface is plumb, or lines up exactly with a string line and pointed weight (bob) dropped from a point parallel to it.

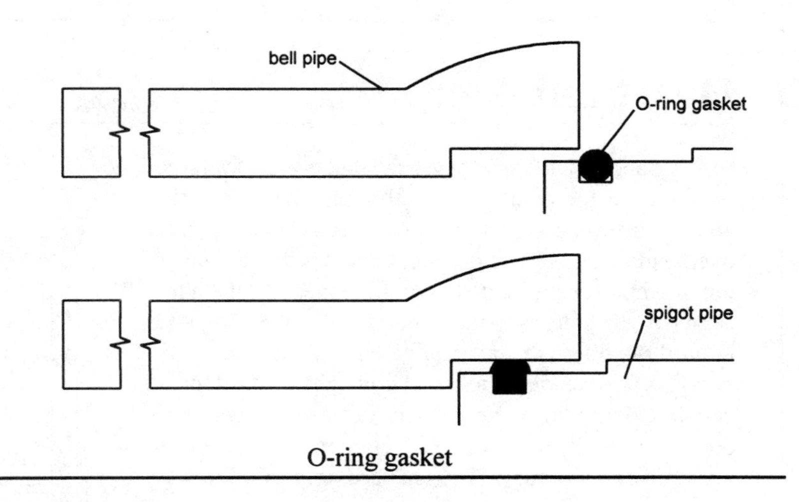

O-ring gasket

out-of-square – not square; a condition in which a true 90-degree angle does not exist, but should exist. During construction, floors, walls, openings, doors and forms for foundations should all be checked to ensure that they are square. Any square or rectangular object can be checked by measuring diagonals from one corner to the opposite. Diagonal measurements are equal only if the corners form 90-degree angles.

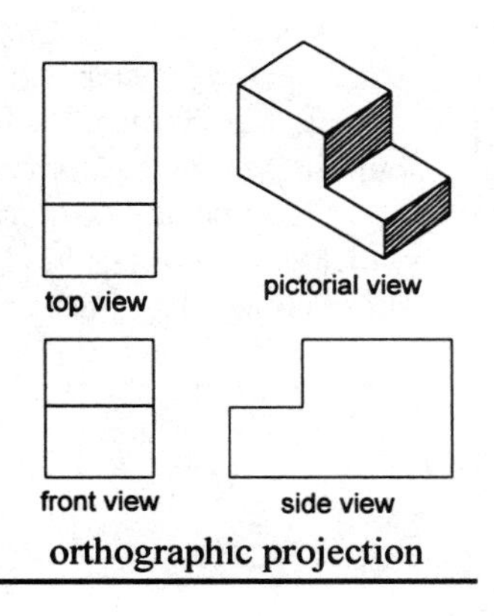

orthographic projection

outrigger – 1) the rafter extension and structural support for the roof overhang. This design feature provides shade, water runoff and some degree of weather protection for the building; 2) a horizontal beam used for supporting a lifting device or mechanism, such as a block and tackle; 3) stabilizing extensions on each side of a crane, or similar equipment, which provide a broader structural base for the crane and prevent it from becoming unbalanced when lifting a heavy load.

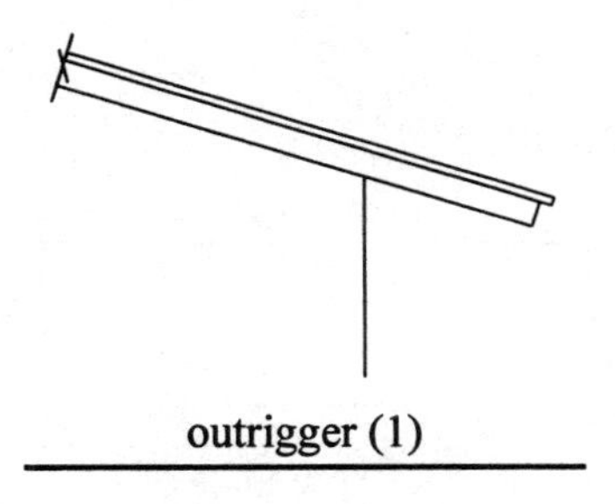
outrigger (1)

outside caliper – a caliper with ends that curve inward toward each other for measuring outside dimensions of an object.

outside corner tool – a drywall taping knife with a blade that is bent slightly less than 90 degrees along its length so the spring action of the blade bears against both walls at an outside corner. It is used to apply and smooth joint compound to both sides of a corner simultaneously, saving time and producing a uniformly smooth joint on each side of the corner. If the drywaller prefers, the compound can

Overhead door

A door which opens overhead that is used for large openings such as garages, warehouses, loading docks and manufacturing facilities. There are several types of overhead doors. The rigid one-piece overhead door is mounted on spring-loaded pivoted hinges. As the door is raised, the bottom of the door swings out and up and in until the door is balanced overhead inside the structure in a horizontal position. The weight of the door is counterbalanced by coiled springs in tension. One-piece doors are usually made of lightweight materials such as plywood or aluminum. They are used primarily for one- and two-car residential garages.

A sectional overhead door can be used for very large openings, such as service stations, car repair facilities, warehouses, loading docks and manufacturing buildings as well as for residential garages. This type of door is divided into horizontal sections, the ends of which ride in tracks via rollers attached to the ends of the sections. The tracks are nearly vertical at the door

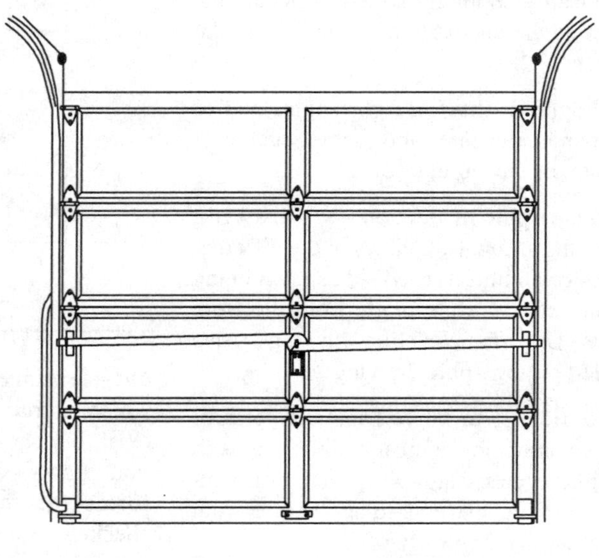
Hinges, lock and track system of sectional door

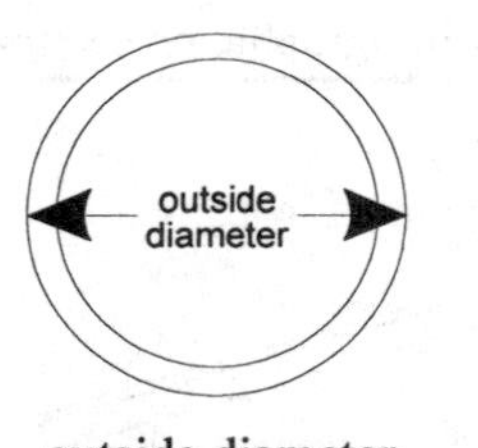

outside diameter

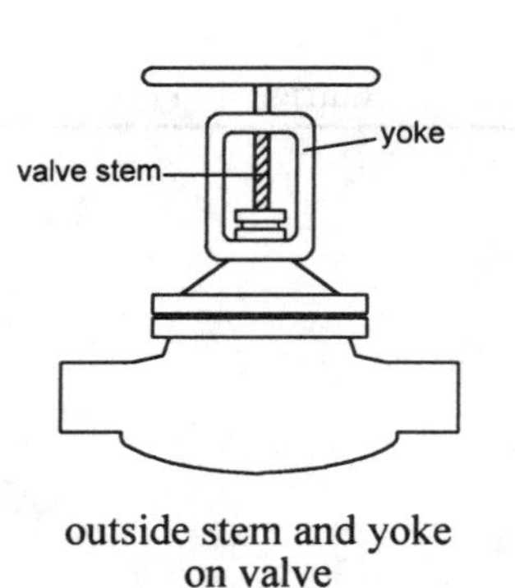

outside stem and yoke on valve

be applied with a straight taping knife and the corner tool used only to smooth the compound. Excess compound can be smoothed and feathered with the straight knife.

outside diameter (OD) – the largest dimension of a cylinder, sphere, or circle, measured across its diameter.

outside stem and yoke (OS&Y) – a type of valve with an exterior stem emerging from the valve bonnet which is supported by a yoke, an upward projection from the valve bonnet. Gate valves and globe valves utilize this design.

overburden – the soil that has to be removed to reach a deposit of rock, sand or gravel that is to be quarried or excavated.

overcoat – the finish paint coat.

overcurrent – electrical current that exceeds the limit of the equipment or amperage load of the circuit.

overcurrent protection – a device that interrupts an electrical circuit if the current, or current and temperature, in the circuit exceeds a preset amount.

overflow – 1) a flow of fluid in excess of the capacity of a container; 2) a means to control and direct excess fluid from a container, such as a channel which directs the excess water from a dam or a piping system and allows fluids to flow to another container when the first one is almost at its capacity.

overhang – 1) the projection of the second story of a building beyond the exterior wall of the first story; 2) the length a rafter extends beyond a building exterior wall; 3) any projection beyond the exterior wall of a building.

overhang ladder – a framework for extending a roof beyond a gable end.

overhaul – to repair and restore something to its design specification values.

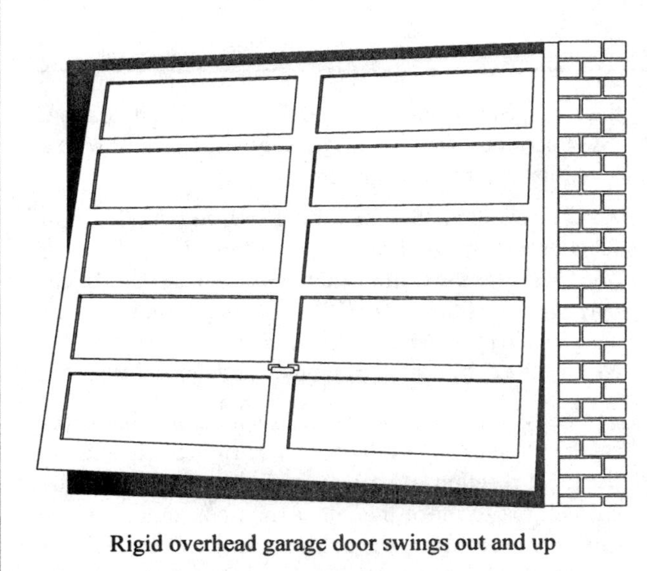

Rigid overhead garage door swings out and up

opening and curve up to a horizontal position away from the opening. As the door is raised, it moves up the tracks until it is suspended horizontally overhead. A cable and torsion, or twisted spring mechanism, counterbalances the weight of the door via a connecting cable. Sometimes a cable and weight is used as a counterbalance. The door sections can be made of a wide variety of materials including aluminum, fiberglass, insulated steel, and solid wood or wood products. Costs vary considerably with the materials used.

A third type of overhead door is the roll-up door. It consists of narrow interlocking horizontal metal sections that ride in tracks. These tracks spiral inward at the top inside of a cylindrical housing. As the door is raised, it rides up the tracks and rolls up inside the housing. Springs are used to counterbalance the weight of the door. This type of door is used for warehouse doors and loading docks.

overhead – the cost of doing business that is not related to a specific job, such as office rent, administrative staff and marketing. In construction, these costs are usually figured into bids and the expense spread out among all the jobs.

overhead crane – a crane that is mounted on overhead rails which are supported on trestle-like legs or on one leg and one wall of a building structure. This is usually a permanent installation. The crane bridge moves along the rails above the work area and has access to most parts of the work area under the rails and the bridge. Used for lifting and moving heavy materials or equipment. See also cranes.

overhead door – a large door that is raised overhead to open. *See Box.*

overhead position – a welding position where the weld is overhead and applied from the bottom, or underside of the joint.

overheat – to increase in temperature beyond the tolerance of the item being heated. For example, the heat exchanger in a forced-air furnace depends on air circulation provided by a fan to maintain the temperature within acceptable limits for the metal. The air absorbs heat from the metal surrounding the burning fuel in the furnace. If the air were to stop moving while the fuel continued to burn, the heat exchanger could overheat and suffer structural failure.

overlap – 1) the amount that one surface, such as a roofing shingle, is laid over the top of another; 2) to install a material with the bottom edge of one laying over the top edge of the next.

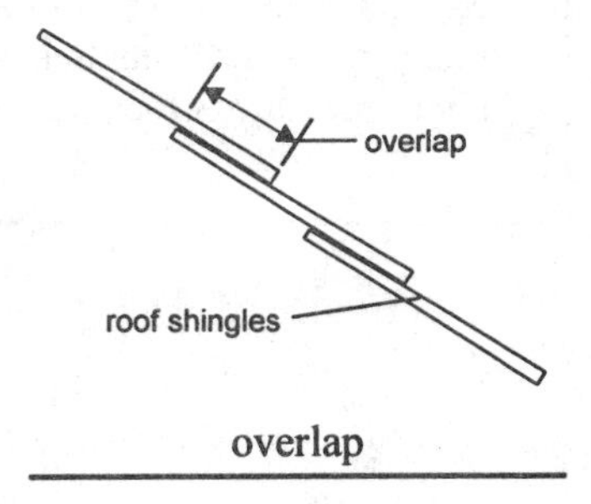

overlap

overlap joint – a fence rail joint in which the adjoining rails are fastened to the post, one above the other.

overlay – 1) a thin layer of material bonded to a panel or board for decorative or protective purposes, such as bonding a sheet of plastic laminate to particleboard or plywood to form a waterproof countertop; 2) a thin layer of metal deposited on a joint prior to welding to prevent the weld metal from being diluted while it is being deposited. For example, a stainless steel overlay on carbon steel that is to be joined to stainless steel provides a weld that will be fused with stainless steel on both sides.

Oxygen cutting

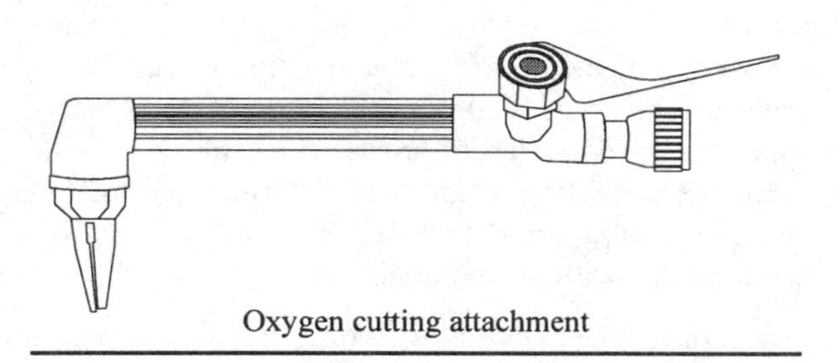

Oxygen cutting attachment

A method of cutting metal using heat and oxygen. The heat is generated by an electric arc or by burning a gaseous fuel, such as acetylene with oxygen in a torch. A stream of pure oxygen under pressure is then directed at the hot metal causing rapid oxidation which cuts the metal.

Oxyacetylene cutting (OFC-A) uses an oxyacetylene (oxygen and acetylene gases) flame to heat metal. An increase in the oxygen flow reacts with and cuts through the heated metal once it has reached a high-enough temperature. The acetylene and oxygen are usually contained in separate portable cylinders. The gases are fed through regulators and hoses into the torch, where they are mixed and burned. The resultant flame or electric arc is used to heat metal to around 1200 or more degrees Fahrenheit. A stream of pure oxygen under pressure is then directed at the hot metal through the torch, cutting the hot metal through rapid oxidation. The flow of oxygen is controlled by the operator with a lever that is part of the torch.

The same method of cutting is used with different fuels providing the heat source. *Oxyhydrogen cutting (OFC-H)* uses the combustion of oxygen and hydrogen; *oxypropane cutting (OFC-P)* uses the combustion of oxygen and propane; and *oxynatural gas cutting (OFC-N)* uses the combustion of oxygen and natural gas.

Oxygen-arc cutting (OAC) is a metal-cutting process that derives its heat from an electric arc and uses the chemical reaction of oxygen with the metal along the heated area to make the cut. A non-consumable electrode is used to strike and maintain an arc with the metal to be cut. This arc heats the metal. A stream of pure oxygen under pressure is then directed at the hot metal through the electrode holder, cutting the metal through rapid oxidation. The flow of oxygen is controlled by the operator with a lever that is part of the electrode holder.

Oxygen-lance cutting (LOC) uses oxygen supplied through a consumable pipe to heat and cut metal. When the metal reaches the proper temperature for cutting, oxygen under pressure is directed at the metal through the pipe, where rapid oxidation of the metal by the oxygen cuts the metal. The heat and oxygen tend to burn away, or consume, some of the pipe in the process.

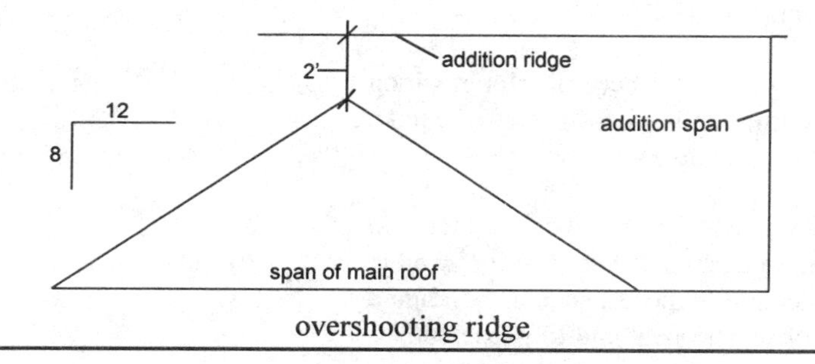

overshooting ridge

Without the overlay, that stainless steel weld metal would be diluted with the carbon steel, making a less satisfactory joint.

overload – 1) an electrical current demand that is greater than the rated current. An electrical overload causes protective devices to actuate, such as burning out a fuse or tripping a circuit breaker. These devices remove the electrical current from the circuit. If the devices fail to function, the wiring could melt from the overload; 2) excess weight or stress placed on a structural member. An overloaded structural member can sag, deform or fail completely.

overshooting ridge – an addition to a structure in which the new roof ridge peaks higher than the ridge of the existing roof structure.

overspray – paint that is sprayed on a surface other than the surface that is intended to be painted. Covering or masking adjoining surfaces as well as careful application and proper adjustment of the spray nozzle will prevent this from occurring.

ovolo – a convex trim molding with the rounded section equal to approximately one quarter of a circle. Also called a quarter round.

owner – 1) the person to which something belongs; 2) the person who has title to a piece of property.

oxidation – the chemical combination of a substance with oxygen. The combination of oxygen and a material that oxidizes can result in the deterioration of that material, or cause a discoloration. Carbon steel, unpainted and exposed to the elements, can rust away, completely destroying its strength and usefulness. In some instances oxidation can cause a protective coating to cover the surface, preventing deterioration. Copper and its alloys turn green over time and exposure to the elements. This green color is copper oxide, a protective coating that prevents further oxidation of the metal.

oxyacetylene welding (OAW) – a type of gas welding using oxygen with acetylene as the heat source.

oxygen – an element that exits freely as a gas in the atmosphere, with the atomic symbol of O, atomic number of 8, and atomic weight of 16. Oxygen combines with other elements to form water, rock, minerals and numerous organic compounds. It is active in physiological processes, and necessary for combustion to occur.

oxygen cutting (OC) – any of the metal cutting processes using the chemical reaction of oxygen with heated metal. *See Box.*

oxygen gouging – the cutting of a groove or bevel using heat and oxygen. A cutting torch is employed in the shaping of the material. It burns fuel in the form of a combustible gas, such as acetylene, to heat the metal to a high enough temperature that the introduction of excess oxygen through the torch cuts through the metal.

oxyhydrogen welding (OHW) – a welding process which uses the combustion of oxygen and hydrogen to heat the metal to be welded. This is done with or without filler metal and without the use of pressure on the pieces being joined. The weld is made by the heated edges of the two pieces of metal melting together and forming a fully fused joint.

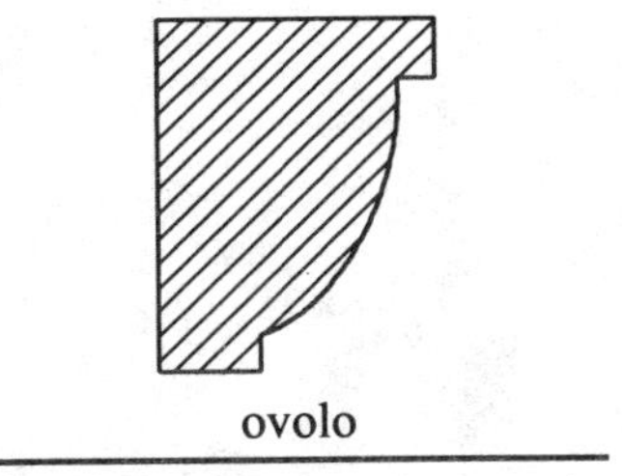

ovolo

P: package boiler to Pythagorean theorem

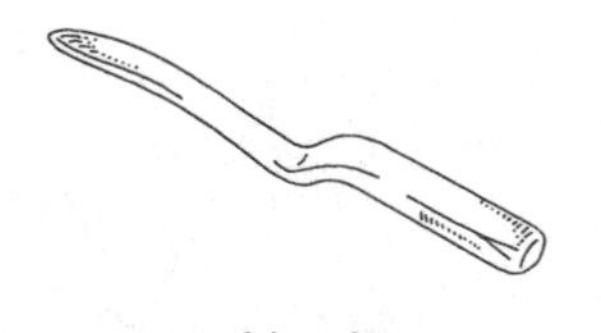

packing iron

package boiler – a complete steam boiler unit with all the required controls and auxiliary equipment ready to be connected to water, electrical and fuel systems for operation.

packaged equipment – a large equipment unit, such as a boiler system, delivered on site already assembled and ready for installation. *See Box.*

packing iron – a tool for packing oakum, rope fiber, into a cast iron drain piping joint. When the joint is tightly packed, molten lead is poured over the oakum to create a seal.

pad drum roller

padding – 1) the application of a stain or paint finish using a paint pad or folded cloth pad; 2) layering, cushioning or building up a surface; 3) a resilient underlayment for carpeting made of fibers, compressed foam-rubber pieces or waffle patterned rubber.

paddle alarm – an alarm on the riser of a particular type of wet-pipe fire sprinkler system that sounds a warning when water flows through the pipe. The paddle extends into the pipe and is moved by the flow of water, triggering the alarm.

paddle wheel scraper – see scraper, paddle wheel.

pad drum roller – a large machine-driven heavy metal cylinder, with short rectangular pads projecting from its surface, used for soil compaction.

paint – a coating, made up of a combination of pigments and a binder or vehicle to carry the pigment. It is designed to cover, color and protect the surface to which it is applied. Most paints have either an oil or latex base and can be mixed in an unlimited variety of colors. They can be applied as decoration to almost any type of surface, but are most commonly used on wood, metals, plaster and drywall to seal and protect them from moisture, heat or other types of damage. Vinyls, epoxies and urethanes are used to make paints that are also resistant to chemicals.

paint, bituminous – coatings made from asphalts or coal tar that are used to protect metals, such as buried pipe, and below-grade masonry, such as exterior cellar walls.

paintbrush – any of a variety of natural bristle or synthetic bristle brushes designed to apply paint. A natural bristle brush holds more paint and applies the paint more smoothly than a synthetic brush. However, they are not well suited for latex or water-based paints because the bristles absorb water and swell, softening the bristles. A brush with synthetic bristles, such as nylon, works well with latex and water-based paints and will generally outlast a natural brush. Paintbrushes come in different qualities, shapes and widths to suit specific applications.

paint, flame-retardant – a paint that absorbs heat, reducing heat transfer, while changing into a form that insulates the material it is covering. It may do this by forming insulation pockets between the surface and the flame, melting into a form resembling glass or producing nonflammable gases when exposed to high heat. There are many different types of flame-retardant paints. They usually contain silicones, polyvinyl chloride, chlorinated waxes, urea formaldehyde resins, casein, borax or other noncombustible substances which reduce the spread of fire on combustible materials. Some paints and paint binders have low flammability or will not act as a fuel in a fire, but they do not protect the material they cover.

paint, metallic – a coating made up of a mixture of varnish and very fine metal flakes. Aluminum paint contains flakes of aluminum and works well as a light reflector. It is also used for heat retention in hot vessels, such as thermos bottles. Zinc-rich primers contain enough zinc to enable the coating to conduct electricity, providing some degree of cathodic protection. Zinc acts as a sacrificial metal under galvanic corrosive conditions. It will be corroded away over time, rather than the base material that is being protected.

Packaged equipment

A complete equipment assembly mounted on a skid or other structural platform and ready to be installed in a system and operated. An example of commonly-used packaged equipment is an air compressor assembly, complete with compressor, electric motor, tank, regulator, cooler, interconnecting piping, pressure switch, and other miscellaneous parts, mounted and ready for connection to an air system and electrical power. Packaged equipment has the advantage of saving field setup and installation labor, which is generally more costly than shop labor. The savings in overall site construction schedule time can also be a significant factor in the decision to purchase packaged equipment. The disadvantage of packaged units can be their large overall size and weight. This may present shipping, on-site handling and placement problems. However, proper advance planning can overcome these difficulties.

Packaged equipment

paint pad – a sponge-backed material, with a short nap, cut in the form of a square or rectangle. The pad is held in a plastic handle so it can be easily dipped in paint and used like a brush. The paint is applied by wiping the pad over a surface. Paint pads work well on smooth or nearly smooth surfaces. They apply the paint rapidly and evenly. However, because they do not carry or hold much paint, they do not work well on rough or porous surfaces.

paint roller – a cylinder covered with an absorbent material and held in a handle which allows the cylinder to rotate freely.

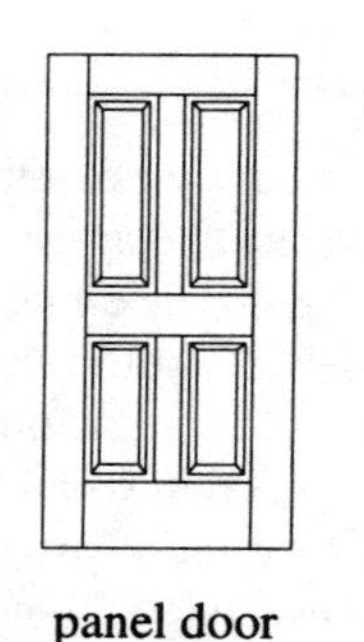
panel door

The roller is dipped in paint, the excess is squeezed out, and the paint is applied by moving the roller up and down over the surface to be painted. Rollers come in various sizes for painting convenience, and also with extensions for the handles so that paint can be applied to ceilings and walls without the need of a ladder. Rollers are also available which can be attached to a paint supply. They have a hollow handle to which a hose or a reservoir and a pump are attached. A continuous supply of paint is forced into the handle and onto the roller surface by pressure or pump. This gives the painter the freedom to paint without having to stop frequently and dip the roller into the paint supply. The advantage of using a roller is the speed and ease with which the paint can be applied over large surfaces. They do not work well with borders, trim or detail work.

paint thinner – mineral spirits, petroleum spirits or coal-tar products used to thin oil base paints and varnishes. They are also used as a solvents to clean oil base paint from painting implements, clothes and hands.

Palladian window – a set of three windows, the top of the center one forming an arch, with a rectangular window on either side. The design is a revival of the classic architectural style based on the work of Andrea Palladio.

pallet – a low, portable platform on which materials can be stacked for handling and storing. A pallet has openings in the base into which the tines of a forklift can fit for lifting and moving the pallet and its load.

pancake box – a flat, round, shallow electrical box for use where a deeper octagonal electrical box is not practical. Shallow boxes are needed for mounting in situations where the wall is very thin, such as a stucco wall with a plywood shear wall underneath. The stucco wall may only be about ¾ inch thick. In that case, a hole can be cut through the stucco and the box fastened to the shear wall beneath, making the mount flush with the stucco surface.

pane – a section of window glass.

panel – 1) thin, flat wood or wood veneer sections used for covering a wall; 2) any relatively thin, flat material used in construction, such as plywood panels for flooring, gypsum wallboard for walls, and acoustic panels for drop ceilings; 3) a metal surface on which electrical devices, such as circuit breakers, are mounted.

panelboard – an electrical cabinet in which circuit breakers are mounted and connected to wiring. In a residence, the three wires from the service entrance conduit enter the panelboard and are connected to three bus bars in the panel. Two of the leads are hot and one is neutral. The circuit neutral and ground wires are connected to the neutral bus bar in the panelboard. The circuit hot wire is connected to the circuit breaker. Circuit breakers are plugged into the panel and connect to a hot bus bar for 120 volt general service use, and to both hot bus bars for 240 volt service. The 240 volt service is used for ovens and clothes dryers.

panel bowing – a defect in gypsum drywall installation caused when a panel is forced into an area that is too narrow, preventing the panel from laying flat. The panel must be removed and the edges trimmed until it fits properly.

panel brick – a nominal 8-inch square (actual 7⅝-inch) or 12-inch square (actual 11⅝-inch) brick with a thickness of 3⅝ inches. Panel bricks are used for areas where their size and shape give the desired appearance. When used in areas where they will be under a load, such as walkways or patios, they must be laid evenly and well supported.

panel clip – a small piece of metal, shaped like the letter H, used to join the edges of two panels of plywood between structural supports.

panel door – a door that has exposed stiles and rails, with wood or glass panels installed in the spaces between them. The wood panels are often shaped in an attractive pattern.

panel fence – fencing made of panels of sheet material, such as fiberglass or plywood, held in a post and rail frame.

panel hoist – a device for lifting and positioning ceiling panels for installation.

panelized brick masonry – see prefabricated brick masonry.

panel lifter – a short, flat bar with a built-in fulcrum used to lift a panel, such as a drywall panel, into position for installation. One end of the bar is place under the bottom edge of the panel and foot pressure on the other end of the bar acts on the fulcrum to raise the panel so that it fits against the panel installed above it on a wall.

panel saw – a fine-tooth hand saw for cutting wood panels.

panel siding – a type of tongue and groove siding board for horizontal or vertical application.

pantile – 1) lengths of semi-circular clay roofing tile that resemble short sections of pipe cut in half lengthwise. They are designed to overlap the adjacent tiles and fit concave side up under adjacent tiles; 2) lengths of S or ogee shaped clay roofing tile laid to overlap each other at the edges.

pantograph – a drafting tool resembling two large X's laid side by side and connected at the points. The upper and lower unconnected points of one of the X's are fastened in place and a pen or pencil is affixed to the lower unconnected point of the other X. When the connected lower point of the two X's is used to trace a drawing, the pen or pencil on the unconnected lower point duplicates the drawing. Pantographs are also available for use with a router so that designs and figures can be duplicated in wood or other material.

pantry – a storage area for food and items used in the preparation of food.

paper, building – see building paper.

paperhanger – one who installs wallpaper or wall coverings on walls or partitions.

paper, sheathing – see sheathing paper.

parabolic arch – see major arch.

parallax – an apparent, but incorrect relation between two objects located one in front of the other when they are viewed from other than a straight-on view. Parallax causes distortion or incorrect interpretations of what the eye sees. For example, when a pressure gauge or other indicator with a pointer spaced away from a dial is viewed from an extreme angle, parallax causes one to make an erroneous reading. In such an instance, the line of sight across the pointer to the dial is directed to a spot offset from the spot actually indicated by the pointer.

parallel application – the installation of gypsum wallboard with the long edge running in the same direction as the framing members. Wallboard is usually installed with the long edge perpendicular to the framing members.

parallel circuit – an electrical circuit in which the loads are connected across the power source, rather than the power source passing through each load in series. In this manner, the continuity of the circuit is not broken if any of the loads become an open circuit. An example of a parallel circuit is a string of decorative lights in which one bulb can burn out and the others remain lighted.

parallel clamps – a set of clamp jaws with two sets of handscrews which allow the jaws to be adjusted to hold objects at various angles.

parallel lines – straight lines that remain the same distance apart along their lengths.

parallels – adjustable or non-adjustable gauges of hardened, precision lapped steel for measuring and laying out parallel lines.

parallel walls – two stud walls built parallel to each other, but not touching, in order to provide a measure of sound isolation. One of the ways in which sound is transmitted is by vibration from the surface wall of one room, through the wall, to the surface wall of an adjoining room. If the walls do not touch one another, the vibration of one wall does not directly result in the vibration of the other.

parapet – a low, protective masonry wall that extends above the edge of a roof, platform or bridge. It provides a protective barrier for people and also provides the structure with some protection from the elements, especially wind.

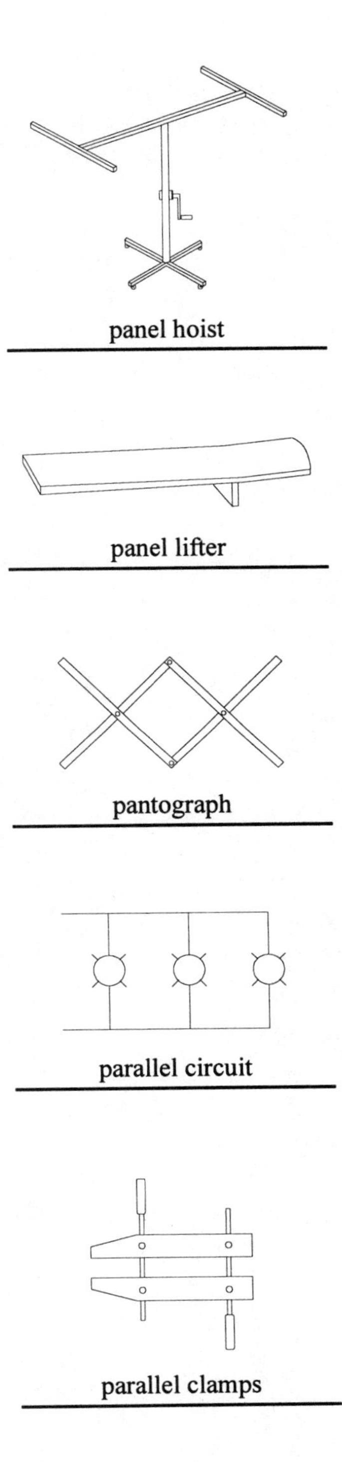

panel hoist

panel lifter

pantograph

parallel circuit

parallel clamps

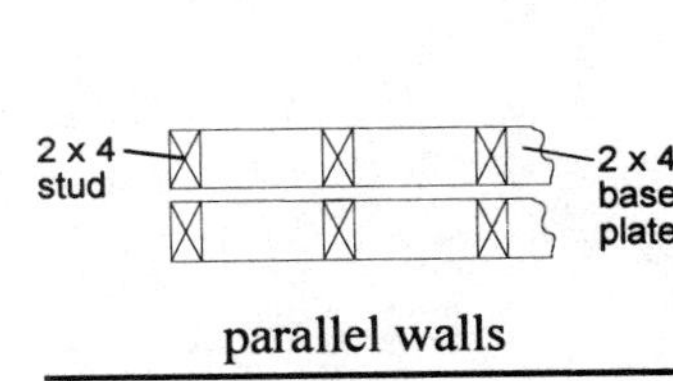

parallel walls

parcel – 1) a tract of land; 2) a wrapped bundle or package.

parent metal – see base metal.

parget – 1) a mortar or plaster containing lime used to coat masonry; 2) gypsum.

pargeting – applying a decorative coat of plaster to a masonry wall, particularly to a chimney or flue.

parging – 1) a waterproof coating for the outsides of foundation walls consisting of a ½-inch thick layer of cement-sand mortar covered with a tar-type layer, with or without a built-up fabric. The purpose of the parging is to establish a barrier that will prevent ground moisture from migrating through the exterior walls of the house. It can also be used on some interior walls, such as leaky basement walls that cannot be coated from the exterior; 2) a decorative plaster coating applied to masonry walls.

Paris green – a green colored paint additive made of copper acetate and arsenic trioxide used to control insects. It renders the material coated uninhabitable, killing insects that come in contact with it. Paris green is added by the painter before application.

parlor – 1) a room used for entertaining guests, receiving guests, or as a formal conversation area; 2) a place for conducting private business, such as a beauty parlor or funeral parlor.

parquet – small pieces of wood inlaid in a decorative pattern used for flooring or smaller areas such as table tops.

parquet flooring – wood flooring in which pieces of wood are positioned at angles to one another to form patterns. Solid wood tiles, prefinished in one-foot squares with interlocking edges, are available in many different patterns and wood types. They are laid on a subfloor or underlayment.

parquetry – a decorative pattern created with pieces of wood.

partial penetration – a weld in a joint that is designed to not go all the way through the joint.

particle board – a building material made by bonding wood particles together with resin. The result is a homogeneous material of uniform density with smooth sides that is relatively strong and inexpensive to manufacture. It is available in panels or sheets of varying thicknesses and sizes. Particle board is used for underlayment and cabinetmaking.

parting stop – a small vertical strip of wood positioned on each side of the window jamb of a double hung window to separate and maintain the relative positions of the sashes as they slide past one another when the window is opened or closed. Also called a parting strip or parting bead.

parting tool – a tool used to cut straight into a piece of work rotating on a lathe for the purpose of separating it from the lathe headstock. Parting tools are made for both wood lathes and metal lathes. A wood lathe parting tool is held in the hand. With a metal lathe, it is held in the tool holder or tool post. It is advanced into the work as the work spins, cutting completely through the work smoothly and squarely. Also called a cutting-off tool.

partition – 1) a divider that separates one area from another, such as an interior building wall. *See Box*; 2) to divide into sections or shares.

partition terminal – a metal edging used at the top of a plaster partition that does not extend at the top to connect with other construction. It serves as an expansion joint, permitting relative movement between the partition and adjoining surfaces.

partnership – a legal relationship between two or more persons who have contracted to become joint principals in a business or who have contracted to share the responsibilities in the accomplishment of a particular goal or task. A partner may be active or inactive. An active partner is involved in the day-to-day work or decision making for the partnership. Active partners may or may not contribute money to the partnership. An inactive partner is not involved directly in the work or decisions made in the course of the business, but usually contributes money to the partnership. Partners may have equal shares in the partnership or the shares may be apportioned according to each partner's contribution, or be divided by some other measure. Partnerships formed for the purpose of contracting to perform work should be clearly defined by legal papers drawn up for that purpose. The division of responsibilities and authority should be specifically defined. The idea is to preclude future conflicts between the partners. Not every situation can be addressed, but broad policies can be defined which can be used to deal with situations as they occur.

parts per million (ppm) – a unit of measure for particulate contamination, which is the unwanted presence of small or even microscopic solid particles in a liquid or gas fluid. The number of these particles in a relatively clean fluid is so small that they must be measured in parts per million or parts per billion.

party wall – a wall that is common between two separate living units, such as between apartment or condominium units, in which each of the owners shares the rights.

pass – 1) a layer of weld metal deposited at one time; 2) a single layer of a commodity or substance applied at one time when more than one coat or layer is required, such as the application of stucco, where each coat or pass must be allowed to set up before the next one may be applied; 3) a passageway or opening through a barrier.

passageway – an enclosed walkway connecting two or more areas, such as a hall.

passive – 1) not using an outside source of power to actuate, such as a one-way valve that permits fluid flow in one direction but not the other. The reversal or cessation of the fluid flow itself, plus gravity, causes the valve to close; 2) not active.

pass-through – a small opening in a kitchen wall for the passage of food to another room or an exterior patio area.

paste – 1) a thick adhesive, such as that used for hanging wallpaper; 2) a thick mixture.

Partition system

A partition system is an assembly used as a divider between or within rooms. Partitions can be loadbearing or non-loadbearing. Loadbearing partitions are an integral part of the building structure and carry part of the structural load. They cannot be removed or relocated unless suitable replacement load paths, such as structural beams or headers, are integrated into the structure to replace their function. Loadbearing partitions are constructed of studs, reinforced masonry or other structural components that serve as load paths. Non-loadbearing partitions do not carry part of the structural load and can be moved, replaced or eliminated as desired, without structural modifications to the building. Non-loadbearing partitions can be constructed of virtually any material suitable to the aesthetic and functional criteria of the partition itself. For example, if the partition must provide sound insulation, visual privacy, or establish a temperature-controlled area, those will be the primary criteria used in the design of the partition. Other considerations will be secondary. Some non-loadbearing partitions are designed to be easily relocated or removed, such as those that consist of top and bottom runners supporting decorated gypsum panels. These are often used to define office space in buildings or to create private meeting rooms in convention centers in such a way as to make it easy to expand or reduce the size of the space as it is needed.

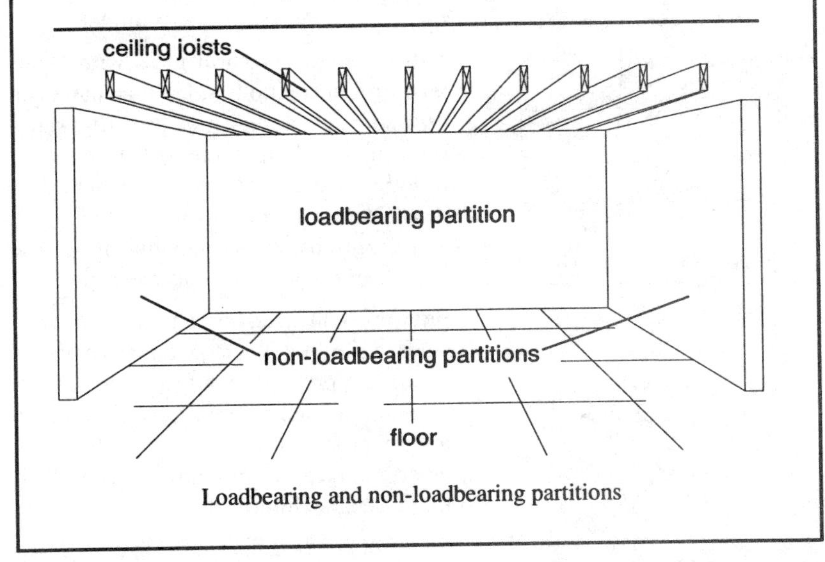

Loadbearing and non-loadbearing partitions

paste brush – a wide brush for applying wallpaper paste to wallpaper.

paste filler – a doughy wood filler material used to fill holes, gouges, defects, or cracks in poorly fitted joints in woodwork.

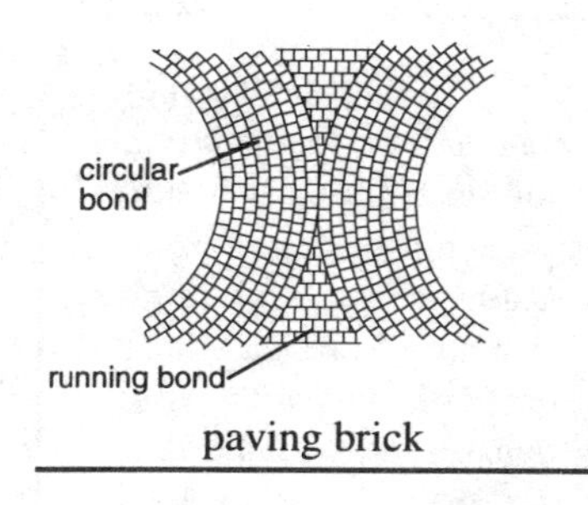

paving brick

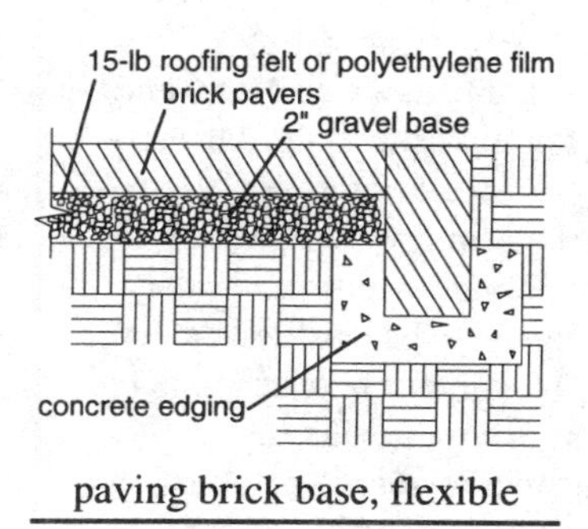

paving brick base, flexible

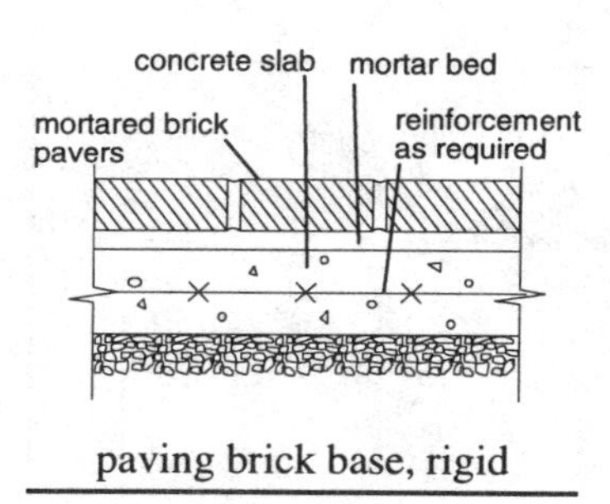

paving brick base, rigid

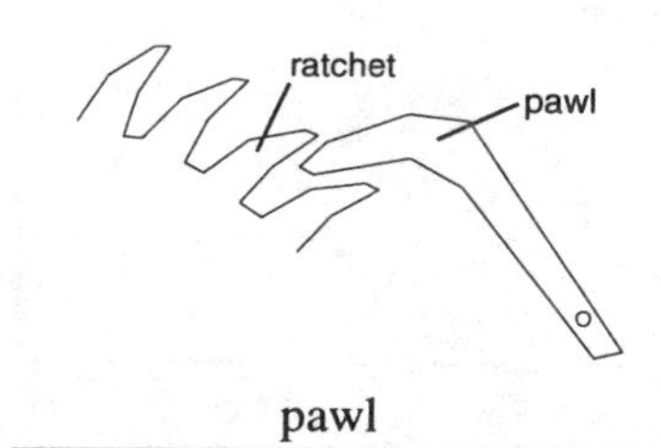

pawl

When the paste dries it can be sanded smooth and varnished, stained or painted to match the surrounding surface. Paste fillers also come in colors to match various stains and types of wood.

pastel – a light hue of a color, such as pale yellow or light blue.

patch – to repair a surface.

Patchal stick – a trade name for a type of putty stick used to patch defects in wood. The end of the stick is rubbed across the defect and the putty fills in scratches or indentations.

patio – a surfaced open area adjacent to a house or building used as an outdoor extension of the living area or a place to sit, eat or relax. Patios may be concrete slabs, wood decks or areas paved with brick, flagstone or other materials.

pattern – 1) a design; 2) a template for duplicating a design to ensure that the size and shape remains exactly the same from one location or duplication to the next. Such patterns are used for repetitive designs in building decoration or in marking and cutting woodwork; 3) a model.

patterned glass – sheet glass with a pattern on one or both sides that has been created by means such as pressing, sandblasting or etching. Patterned glass is used in light fixtures, decorative cabinet doors or as decorative accents in window lights and partitions. It is available in a wide variety of colors as well as clear glass.

pavilion – 1) a covered area with open sides; a canopy; 2) a large tent; 3) a part of a building projecting out from the rest; 4) a small ornamental building or structure in a garden or park area used for shelter or entertaining; 5) a temporary structure built to house an exhibition.

pavilion roof – a hip roof with each hip of equal length.

paving or pavement – road or parking lot surface intended to provide wear resistance and smoothness. Paving usually consists of a wear surface, such as concrete, over aggregate subbase material, such as gravel, placed over compacted subgrade soil. Asphaltic concrete, a common paving material, is a mixture of asphalt and gravel which is laid over a gravel subbase.

paving brick – brick used for outdoor paving applications. They are made of clay or shale and sometimes burned hard so they vitrify, providing a wear-resistant surface.

paving brick base, flexible – a paving brick base that can move easily, such as compacted sand or gravel. It is often used where brick is installed without mortar. It provides a stable base, good drainage and prevents moisture from working its way up from the soil beneath.

paving brick base, rigid – a non-moving base, such as a reinforced concrete slab.

paving brick base, semi-rigid – a base, such as asphalt paving, that provides some degree of flexibility for brick laid without mortar.

paving machine – an asphalt placing machine that spreads and levels asphalt to a predetermined depth. The paving machine rides on either movable tracks or tires.

pawl – a hinged and hooked device designed to engage a ratchet to prevent its turning. A pawl is used on a winch to prevent the load from pulling the cable back off the spool on which it is being wound. As the cable is turned, the pawl slips from one ratchet tooth to the next.

payment bonds – a guarantee that a contractor's bills will be paid so that the client is not liable for the payments. For example, if a client pays a contractor for materials, but the contractor buys the materials on credit and then fails to pay the material supplier, the material supplier may demand payment directly from the client. Without the payment bond, the material supplier has the right to file a mechanic's lien against the property to force the owner to pay.

peak above ceiling exposure limit – OSHA sets standards which limit the length of time workers may be exposed to certain hazards, such as pollution from hydrocarbon emissions. The peak above ceiling exposure limit allows workers to exceed that standard for a short period of time. For example, if the total hydrocarbon emissions from a particular process are limited to 50,000 part per million (ppm) per day, OSHA may permit those emissions to reach 75,000 ppm for one hour during a 24-hour period.

peak load – the highest electrical load during a time period. Public utilities that supply electrical power to communities measure usage to determine when peak loads occur. In most areas, the highest electrical loads occur during the daytime, the lowest at night. In order to keep generating enough electricity to meet growing load requirements, utilities may need to purchase power from outside sources, or use measures to store excess power not used during off-peak times. There are several methods of conserving power for use during the peak-load periods. One is by compressing air during the night and using it to power turbines to generate extra power during peak loads. Another method pumps water to an elevated reservoir during off-peak hours and then later uses the water to spin turbines connected to generators when extra power is needed. A third method involves charging batteries during off hours and then adding their power to the generating capabilities during peak-load periods. If a utility cannot meet the peak-load requirements, there will be brownouts or blackouts from lack of power to meet the demand.

pebble texture – a common type of drywall texturing in which small droplets of compound are sprayed on the wall surface and left as is, or flattened with a trowel.

pedestal – 1) a vertical support or column; 2) a support for an ornament or statue; 3) a single base supporting an object such as a table.

pedestal floor – a floor system in which removable floor panels are held on pedestals in order to provide a crawl space or a space for running wire, cable or ventilation or heating ducts under the floor.

pedestal floor, gridless – a type of pedestal floor in which the floor panels are supported at all four corners by pedestals, with no grid supports between panel edges.

pedestal floor, lay-in grid – a type of floor system in which the floor panels are supported on beams, which are in turn supported by pedestals.

pedestal floor, rigid grid – a type of floor system in which the floor panels are supported on beams which are mechanically attached to the pedestals which support them.

pedestal lavatory – a lavatory basin mounted on a pedestal stand.

pediment – 1) a gable pattern over a door or window; 2) a gentle slope of rock at the base of a steeper slope.

peeling – a paint defect where the paint debonds from the surface and peels off.

peel strength – the amount of force required to peel a bonded roofing membrane from a substrate to which it has been bonded. It is a measure of the adhesive or bonding strength of the roofing membrane and an indication of the overall strength of the roof covering and its ability to withstand this type of force.

peen – 1) the end of the striking head opposite the face of a hammer. The peen is usually shaped either like a wedge or rounded like a ball and used particularly for striking or shaping metal. See also ball peen hammer; 2) to apply a compressive force.

peening – striking metal to change its shape or surface characteristics, such as compressing the surface to increase hardness.

peg – a structural cylinder of relatively small diameter used to hold two or more pieces together by being fit into a hole drilled through those pieces, or driven into a hole with part left protruding on which objects may be hung. Pegs are usually made of wood, but they can be made of other materials as well. They are commonly used to lock joints together in the manufacture of furniture or in post and beam construction. Wood pegs add strength to the joints because the grain runs parallel to the axis of the peg making it resistant to shear loads. Also called dowels.

pegboard – a compressed fiberboard sheet with small, evenly-spaced holes, one inch on center, designed for insert-

pedestal lavatory

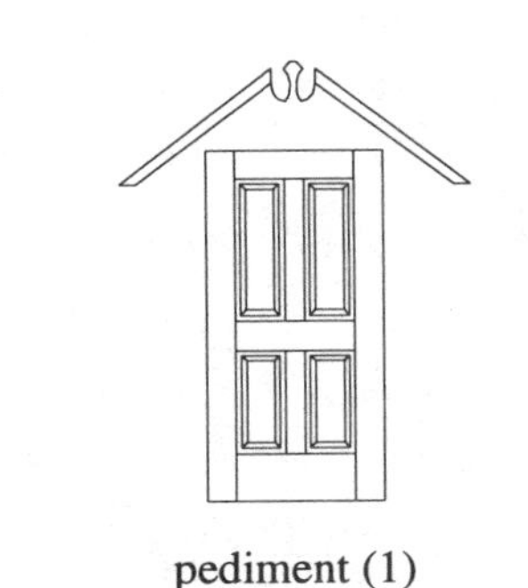
pediment (1)

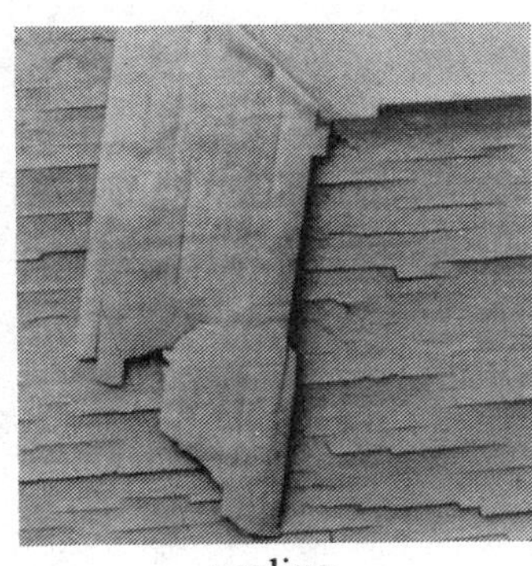
peeling

ing pegs, hooks or hangers. Pegboards are commonly used with hooks and hangers to store tools.

Pelton wheel – a water turbine wheel with buckets around the periphery. High pressure water is directed at the buckets, imparting a force that turns the wheel. The Pelton wheel was invented around 1890 as an improvement to the water wheel. It was used to drive electrical generators.

pendant – 1) a hanging decoration or decorative ornament that points downward. Pendant designs in plaster, wood or stone were used as decorations for Colonial and Victorian architecture. They were usually floral or plant patterns; 2) a decorative electrical fixture suspended from the ceiling.

pendentive – one of four triangular, concave braces joined together at the top to support a dome over a square structure. The braces form arches from one corner to another around the top of the structure.

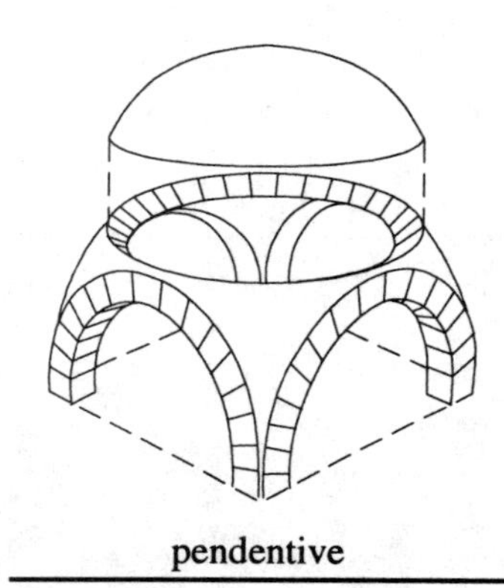
pendentive

penetrameter – a plate used in radiographic examination which shows the contrast available in the radiographic examination film. It has a constant thickness, which is selected as a function of the thickness of the object being examined. There are three holes in the plate, which are one, two and four times the plate thickness. The plate is located so that they show in the radiograph. Since the penetrameter is of a known thickness, shape and material, it provides a standard of reference for reading the radiograph, permitting the reader to evaluate details about the material being examined.

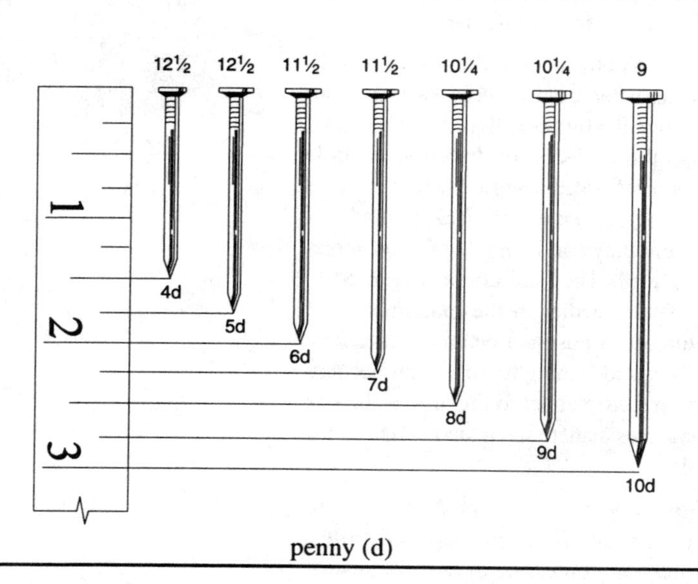

penny (d)

penetrating oil – an oil finish that sinks into the wood surface and hardens. It provides surface protection for the wood without an exterior coating that can be scratched off or damaged, or that will flake off over time. A penetrating oil finish darkens the wood and increases the contrast of the wood grain. It is popular because it gives the appearance of a hand-rubbed oil finish on properly prepared wood, without the labor that hand-rubbing requires. A penetrating oil finish is very forgiving. Most mars and scratches can be eliminated with another application of the oil, and the finish restored. It gives furniture a very warm look. When exposed to the elements, wood protected with penetrating oils will turn gray and take on a weathered appearance over time, but the structural integrity of the wood will be preserved. Many people like the appearance of weathered wood and use penetrating oils to achieve that look while protecting the wood's surface from being damaged by the extremes of outdoor exposure. The use of a penetrating oil will also prevent or limit redwood splintering, which occurs over time if the wood is used outdoors and not protected. To maintain the protection on woods subject to exposure, penetrating oils must be reapplied every one to two years, depending on the severity of the weather.

penetration – 1) in welding, the depth of the base metal to which the weld metal has flowed; 2) the depth that one material or commodity, such as conduit or piping, passes through the surface of another.

penny – a measure of nail size, often abbreviated with a "d" as in 4d, rather than writing out 4 penny. There are different stories as to the origin of this type of measure. One common history is that a given size nail cost that number of pennies per hundred, so a nail size that cost 4 pennies per hundred was said to be a 4-penny nail.

pentachlorophenol – a compound, often included as a component of shake and shingle stains and wood preservatives, that is toxic, repels water, and prevents decay and mildew. It is also used as waterproofing treatment for plywood.

pentagon – a five-sided plane geometric figure with equal sides and equal angles. Pentagonal shapes are used in construction to build a variety of structures, from garden gazebos to the famous Pentagon in Washington, D.C.

pentahedron – a five-sided solid figure.

percentage of completion accounting – a method of accounting in which cash received for unfinished work is designated as income only in the amount of work that is estimated to be complete at any time. When a payment is received for a given amount of work, say 80 percent, and the work is actually only 50 percent complete, then only 50 percent of the total job cost is shown as earned income. The remaining 30 percent received is shown as unearned income until that portion of the work has been completed as well. Using this method, the accounting books will reflect the actual progress of the work and the status of payments received versus work performed. Contractors are able to have a fairly accurate indication of what stage various jobs are in by looking at their books, as both underpayments and overpayments become clearly evident with this type of accounting.

percent grade – the rise or fall of the grade divided by the distance over which the rise or fall is measured. This calculation yields a decimal figure which is then multiplied by 100 to obtain the percent grade. The percent grade is important in evaluating the proposed slope of a parking lot, driveway, road or other passage route, because the steeper the grade, the harder it will be to use or traverse. Too steep a grade can make a roadway hazardous or impassable.

perch – 1) a unit of measure that equals a linear or a square rod; 2) a unit measure of stone equal to 16½ x 1½ x 1 foot, or 24¾ cubic feet. Stone of this size and shape may be used for stone stair steps.

percolate – fluid passing through a porous material, such as underground water passing through sand and gravel or the fluids from septic tank leach lines passing through and being absorbed by the soils through which they are run.

percolation test – a test, directed or ordered by the local building inspector, to determine soil absorption characteristics prior to locating a septic tank drain field. The test results are used to determine the size of the drainfield. The slower the absorption rate of the soil, the larger the drain field must be. Percolation tests are generally not required in areas with sandy, fast-draining soils. For the test, three or more holes are dug in the area of the drainage field, to a depth equal to that of the proposed field. A two-inch layer of gravel is placed in the bottom of each hole, and then the holes are filled with water to a level of at least six inches above the gravel. The water is monitored every 10 minutes for one-half hour. The time required for the water to drop one inch after the first 20 minutes have passed is the percolation rate. Three inches per minute is considered a very good percolation rate.

percussion welding (PEW) – an electric resistance welding process using a rapid flow of electric current with pressure applied to the joint during or after heating. The electrical current is passed between the two pieces to be welded, heating them to the proper temperature for bonding. A rapid application of force is applied to join the two pieces either during or immediately after the electrical current. This force bonds the two surfaces. The advantage of this process is that only a thin layer of metal is heated, so when the pieces are joined there is little or no excess molten metal forced out around the joint. This leaves a clean joint. The process can only be used on small flat surfaces because of the limited area that can be controlled and heated by the electrical current.

perfboard – see pegboard.

perforated – a surface with many holes.

perforated metal pipe – pipe with several equally-spaced holes drilled in rows along its length. One common use for this type of pipe is as a sparger or aerator pipe placed at the bottom of a tank filled with fluid. Air is passed through the pipe into the liquid in the tank to oxygenate the fluid and/or mix the fluids in the tank.

performance bonds – a bond, issued by a bonding company, that guarantees a contractor will perform work to the requirements of a contract. The contractor obtains the bond, and if he fails to perform the required work, the bonding company will have the work completed at no additional cost to the client. The contractor who fails to live up to the contract may not be able to get future bonds, or will have to pay a premium for them.

pergola – a horizontal trellis mounted on columns designed to be an overhead covering for an open-air structure.

perimeter – the outside boundary of a geometric shape; outside boundary of a building or piece of property.

perimeter drain – a drain placed around, or partially around, the outside of a foundation to carry away excess water.

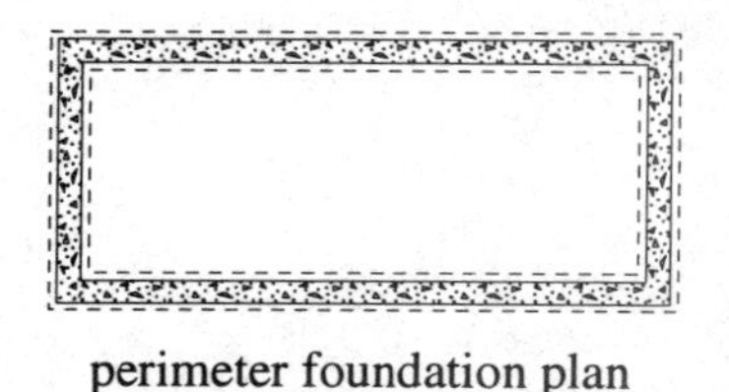

perimeter foundation plan

perimeter ducting – HVAC ducting that is routed around through the inside of a building's exterior walls. It may be located behind a false drop ceiling or disguised by some other method.

perimeter foundation – the portion of a building that is in direct contact with the soil and on which the building rests. Unlike a slab foundation that covers the entire area under the building, a perimeter foundation extends only around the perimeter, or the outline, of the building. In most modern buildings, it is constructed of concrete reinforced with steel bars. If the building has a cellar, the perimeter foundation usually comprises the cellar walls. If there is no cellar, there may be a crawl space under the building within the perimeter foundation.

perimeter heating – heating outlets around the inside perimeter of a building. This places the heat sources near those areas of the building where they can best counter cool drafts and heat leakages and allows a more even temperature to be maintained within the building.

perimeter heat loss – heat loss through the perimeter, or outside surfaces, of a building which are exposed to the exterior elements. Effective wall insulation, double-pane windows, and insulated doors can be used to help control this heat loss.

perimeter installation – an HVAC system with outlets located within the perimeter walls of a building. This puts the heating/air conditioning source near the areas where they can best counter exterior temperature leaks and aids in keeping the interior temperature stable. See also perimeter heating.

periphery – the outside edge or boundary.

perlite – a lightweight expanded mineral product used in insulation and as a lightweight aggregate.

perm – a measure of the amount of water vapor that passes through a material. A perm is one grain of water vapor per square foot, per hour, per inch of mercury pressure differential (0.491 psid). This measure is important in the selection of vapor barriers and waterproofing materials.

permanent magnet – a material, such as alnico and some iron compounds, that retain residual magnetism without outside influence. Permanent magnets are used for several applications, including doorbells and loudspeakers.

permeability – a measure of the ability of water to flow through a material, such as concrete.

permeance – the rate of water vapor transmission through a membrane in grains per square foot per hour per inch of mercury.

permit – a document that gives authority to conduct an activity or work, such as a building permit, or to use a particular machine or object, such as a gun permit.

perpend – a stone or masonry unit that passes completely through the thickness of a wall, such as a brick header extending from the interior surface to the exterior surface of a wall.

perpendicular – a surface or object at a right angle (90 degrees) to another surface or object.

personal computer (PC) – a small computer designed to fit on a desk top or similar location and be used by one individual at a time.

personal protective equipment – equipment worn to protect a worker against the effects of environmental hazards. See also protective clothing.

personnel – 1) persons; 2) a body of persons employed by a business.

perspective – a drawing that attempts to portray objects as they would actually appear to the eye of a viewer. Objects close to the viewer appear larger and those further away appear smaller. It is the diminishing dimensional perspective with increasing distance that aids in depth perception and gives the drawing its look of reality. Perspective drawings are used by architects and builders to give their clients a clear idea of how a proposed building project will look when completed.

petroleum – liquid hydrocarbons occurring naturally in the earth. They are used in many products from fuels and lubricating oils to asphaltic paving materials and wood preservatives.

petroleum refinery – a plant for the distiling of petroleum into its constituent parts, such as gasoline or lubricating oil.

pH – a measurement used to express both acidity and alkalinity on a scale running from 0 to 14. Substances with a pH less than 7 increase in acidity and over 7 increase in alkalinity; a pH of 7 represents neutrality. Many substances, such as boiler feed water and cleaning agents, may need to be checked periodically for pH in order to determine if they are within safe tolerances for use.

phantom line – a dashed line on a drawing used to indicate a surface that is hidden behind what is shown on the drawing.

phase – 1) part of a course, development or cycle; 2) to introduce or carry out in stages; 3) the relationship of two or more sine waves.

phased application – the application of materials at two or more different times, such as multiple coats of paint or plaster which can only be applied after the previous coat has dried.

phased building – work proceeding over a period of time in such a way that portions are completed in sequence or stages. An example would be the completion of the roof and exterior wall coverings of a building before the interior portions are done. The exterior work protects the interior work from the elements. On a large development, building is often completed in phases or sections to avoid having too much land under development at one time. This can be done for economic reasons or for erosion control. Also called phased development.

phenolic – a compound of formaldehyde and phenol which combine to form a resin. The resin can be combined with oil and used in protective paints, especially those used in the marine industry, or dissolved in solvent and made into a waterproofing varnish. It is also used to make a themosetting adhesive that bonds well with wood and paper and is used in the manufacture of plywood and similar wood products.

Phillips screw – a screw with crossed slots recessed into the center of the head. The slots do not extend to the edge of the head and so the screwdriver is automatically centered. This type of screw is easier and faster to drive than a single slotted screw because there is less tendency for the screwdriver to slip from the screw head.

Phillips screwdriver – a screwdriver that comes to a point on the tip and has cross-shaped edges raised in the tip end designed to fit into the perpendicular cross slots recessed in the head of a Phillips screw.

photoelectric cell – the light-sensitive device in a photoelectric or photo-control switch that generates the electrical current and causes the electrical circuit to remain open.

photoelectric switch – an electrical switch with a photovoltaic cell that generates an electrical current in the presence of

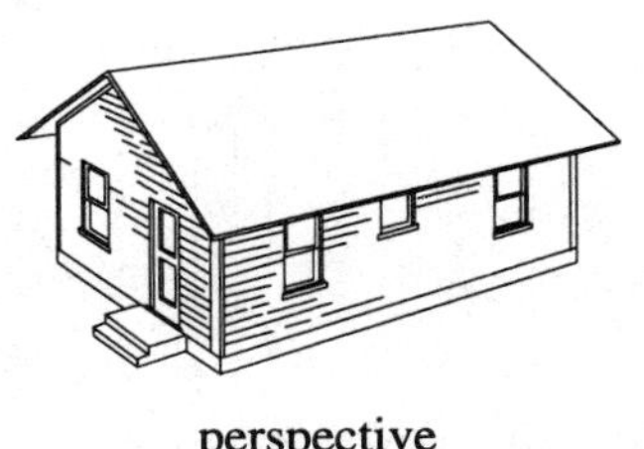

perspective

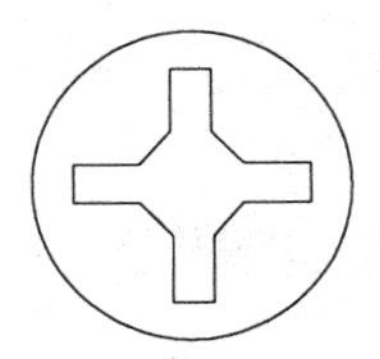

Phillips screw head

light; the current is used to hold the switch open until darkness causes the current to stop. When the current stops, the switch is activated. A photoelectric switch can be used in a variety of applications. They are commonly used for outdoor yard lights and street lamps which turn on automatically when the sun goes down, and for sentry devices or alarms which sound if the current is broken by a person coming between the light source and the switch. Also called an photo-control switch.

photogrammetry – the science of taking accurate measurements with the use of photography, especially aerial photography such as that used in surveying or map-making.

photovoltaic cell – a device that generates an electrical current when light shines on it. It is used in solar power systems to convert light to electrical energy.

piano finish – a high grade polished varnish or lacquer finish. It is used on furniture, and especially pianos.

piano hinge – a narrow-width pin hinge that extends along the lengths of the two parts that are joined. It is used in cabinetry to hinge large or heavy panels where the hinge load needs to be spread over a large area. Also called a continuous hinge.

pick and dip – a bricklaying method in which the brick is picked up with one hand and enough mortar to set the brick is simultaneously picked up (dipped) with a trowel in the other hand. It is a very fast bricklaying technique used by experienced bricklayers.

picket – 1) a narrow fence board, often shaped to a point at the top; 2) a stake or post that is driven into the ground.

picket fence – a wood fence made of pickets fastened to horizontal wood rails.

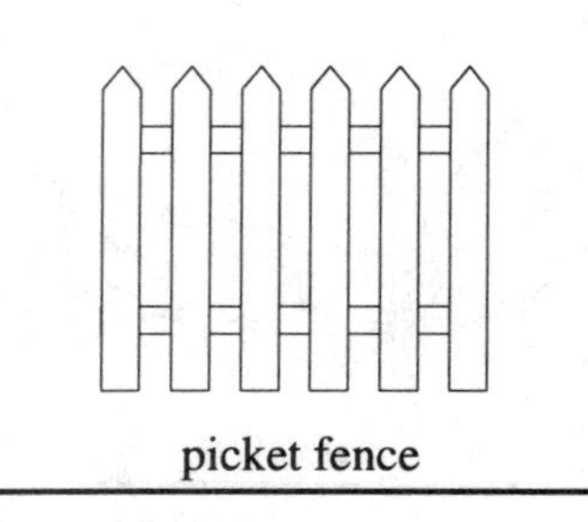

picket fence

picking up sags – rerolling or rebrushing a still wet painted surface to remove sags or drips in the paint.

pickle bar – a bar with a wedge-shaped forked end used for prying or separating, such as prying open crates or prying out studs during remodeling.

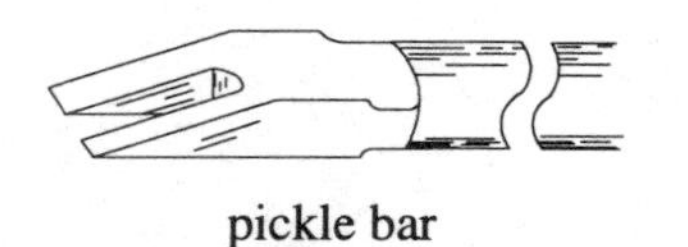

pickle bar

pickling – cleaning and passivating a metal surface with an acid wash to minimize the possibility of future corrosion. It is used to remove corrosion and restore old metal items, or to clean metal, such as the interior surfaces of carbon steel piping to be used for lubricating oil service.

pickup machine – a machine that picks up asphalt from a windrow on the ground and loads it into the hopper on the front of the paving machine. It is attached to the side of the paving machine and is operated or powered by the paving machine.

pico – a prefix meaning one trillionth. The most common application of the term is in the description of the capacitance of a small electrical capacitor, measured in picofarads.

pictorial drawing – a drawing that shows a three-dimensional view of an object. Perspective, oblique and axonometic drawings are three-dimensional.

picture framing vise – a vise that will hold pieces of frame together at a 90-degree angle for gluing.

picture window – a large, single-paned window designed to frame an exterior view. As a rule, the window is fixed and does not open.

pie chart – a diagram representing divisions of a whole, using a circle to represent the whole and wedge-shaped pieces to represent the divisions. Diagrams of this type are often used by businesses as a simple means to show expenditures for particular items in relation to the whole budget, such as advertising, rent etc. as a percentage of the total business expense. Also called a circle graph.

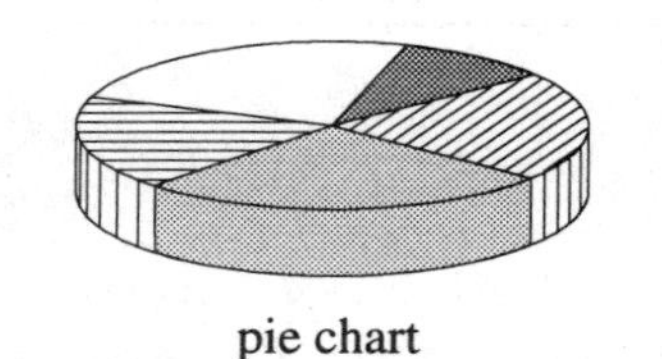

pie chart

pier – 1) a short vertical column used to support a foundation or deck. It may be wider at the bottom to increase its load-bearing area; 2) a wharf or dock extending out into a body of water.

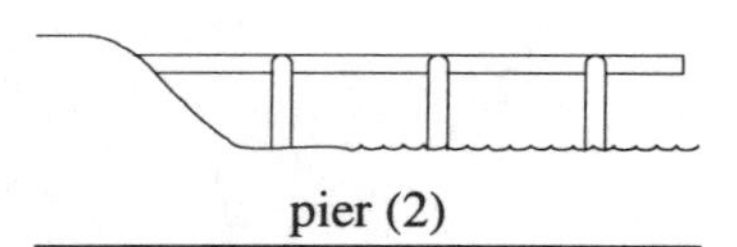

pier (2)

pier footing – a foundation footing for a pier or column.

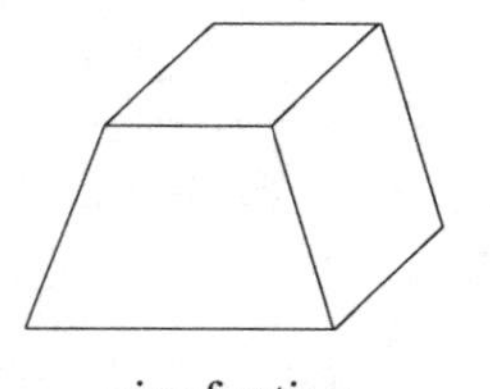

pier footing

piggyback breaker – an electrical circuit breaker designed to protect two electrical circuits. May be used only on circuits that are lightly loaded.

pig iron – crude iron, the direct product of the blast furnace, cast into a large section called a pig. It is ultimately refined into steel, wrought iron or ingot iron.

pigment – colored, insoluble particles used to add color to paint or stain.

pigtail – a length of electrical wire with a plug on the end attached to the electrical terminals of an appliance, such as a clothes dryer, to bring power to that appliance. Commonly called the cord or electrical cord.

pigtail splice – a simple electrical wire splice in which the ends of the wires are twisted together. It is used to connect new wiring to old wiring. The wires are insulated so that no bare wire is left exposed.

pilaster – a square support column, often projecting out from a wall, that is structurally attached to the wall to provide lateral support.

pile or piling – a vertical structural foundation member that is driven into the ground, and on which a foundation rests. *See Box.*

pile cap – a fitting placed over the end of a piling to transmit loads directly to the piling. Also called a piling cap.

pile driver – a machine for driving pilings or piles into the ground using a large weight and guides to hold the piles plumb as they are being driven. Pile drivers may be drop hammers, mechanical hammers, or vibratory types. Drop hammers are heavy weights that are raised by a cable and allowed to free fall, using gravity as the driving force. They are connected to leads which guide their fall. Mechanical hammers are power driven by steam, compressed air or diesel fuel combustion. The force drives the hammer into the soil. Vibratory hammers operate at 100 cycles per second and force the piles to vibrate, which causes them to alternately lengthen and contract by very small amounts. As the pile lengthens, it becomes smaller in diameter, momentarily diminishing the friction, which permits it to slip a little further into the ground. This method is often the fastest method of driving a pile. Extensions are used when driving pilings under water so that the pile driver is not under water.

piling – see pile.

pillar – a slender column or vertical structure of masonry or other material used to support a load.

pillar file – a file with no cutting teeth on one or both edges. The smooth edges allow it to be used to file work that is close to or against other surfaces. Also called a safe-edge file.

pilot flame – see pilot light.

pilot hole – a small diameter hole used as a guide for drilling a larger diameter hole. It is used when a large hole is to be drilled in a tough material, such as metal. It is easier to make a large hole if the hole is made in increasing increments of smaller holes so that the excess material is removed a little at a time.

pilot light – a small, constantly-burning flame used to light larger burners in a gas appliance. When the gas feed is turned on for the larger burner, the pilot light ignites the gas. Also called a pilot flame.

pilot light switch – a switch with a small light in it which indicates when the circuit controlled by the switch is energized. The light enables a person to tell at a glance (or from a distance) whether the switch is on or off. This type of switch is used primarily on control panels that have many different switches.

pin – a slender cylinder used to hold parts together, such as the two halves of a hinge.

pinch bar – a pry bar used for prying or pulling items apart. Also called a crowbar.

pinch dog – a heavy U-shaped staple, in sizes varying from 1/4 to 3 inches wide, which is driven into boards or timbers to hold them together.

pine, Arkansas soft – a short-leaf pine with soft fiber. The wood is very soft and lightweight, but it is tough, resilient, and has no heavy pitch. Used for door and window jambs, trim molding, paneling and flooring.

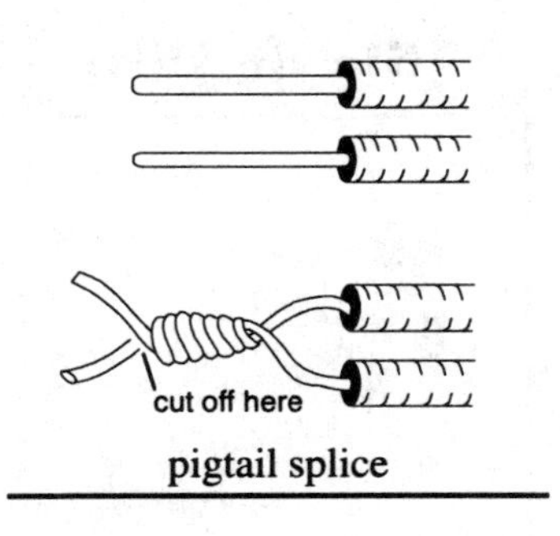

pigtail splice

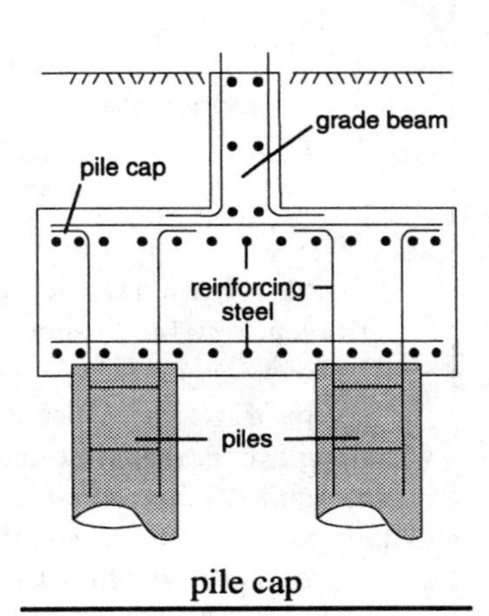

pile cap

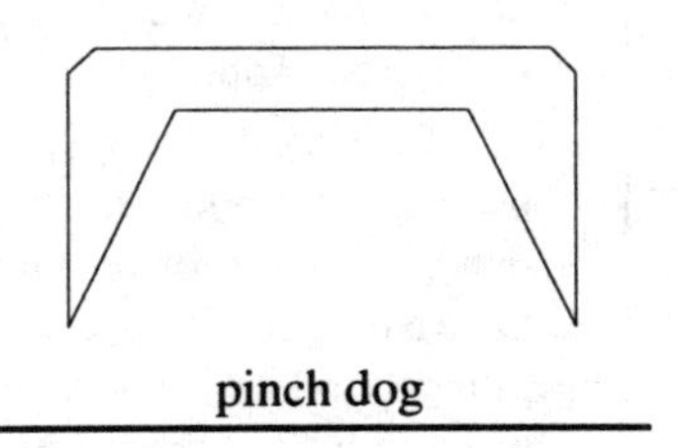
pinch dog

Pile or Piling

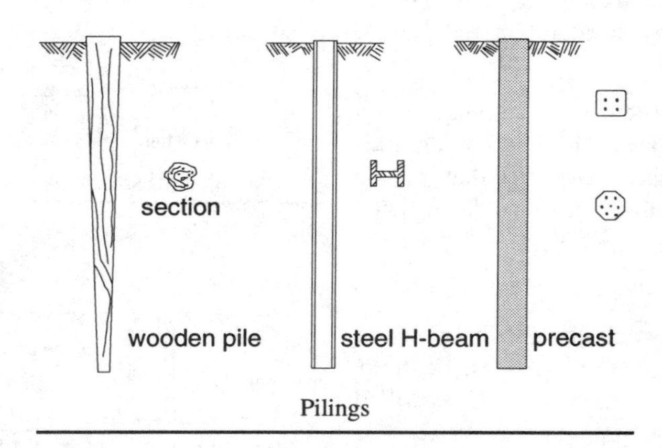

Pilings

Piles are used where surface soil conditions are not sufficiently stable to support the weight of a building. The load-carrying ability of a piling is determined by the type of soil into which it is placed. For example, clay soil will support approximately one ton per square foot, but granite, being more stable, will support about 30 tons per square foot. All pilings are driven or placed in the ground to a depth where they reach stable soil. The depth varies, depending on soil conditions in the area. Piles are made of wood, concrete, steel or a composite of concrete and wood, each offering different properties for various applications.

Composite pilings, made of concrete and wood, are used for foundation support where the groundwater level is 70 feet or less below the surface. Wood piles have a long life below water level, so they can be driven into ground soil where they will be under the groundwater level, but resting on stable soil. The concrete piles are then driven in on top of the wood piles. A foundation connects them structurally at the top.

Concrete pilings are driven or placed in the ground to support a building. Concrete pilings are either precast or cast in place. Precast and prestressed pilings are manufactured and then brought to the site and driven or water jetted into place. If they are to be water jetted in, longitudinal external grooves and a hole down the center are manufactured into the pile for the water to pass through. They are use in soft soil. Cast-in-place pilings are created by drilling a hole into the surface and then pouring concrete into the hole. These are of two types, shell and shell-less. The shell type are used where the soil is too soft for an uncased hole or too hard to compress. A steel shell is placed into the ground and left in place after the concrete pile is poured and cures. When the shell-less type is used, a steel pipe and shoe are driven into place. The pipe shell is removed before the concrete pour, but the shoe remains at the bottom.

Steel pilings, made of H-shaped or tubular steel, are driven into the ground to support a building. Steel piles can be driven into soils that would be too difficult for other types of pilings. The ends of the steel piles may be left above ground to be used as columns. The disadvantage of using steel is that it may be deflected by underground rock or bent with continued driving.

Wood pilings are vertical support members used for structures built over water or where the foundation members must be driven into the ground below groundwater level. Wood piles are expected to have a long life below water or below groundwater level, and they are lightweight, inexpensive, and readily available. They also have a greater skin friction than most materials. There are two main disadvantages in using wood pilings: they are subject to insect and fungi attack if not properly treated, and they may split while being driven.

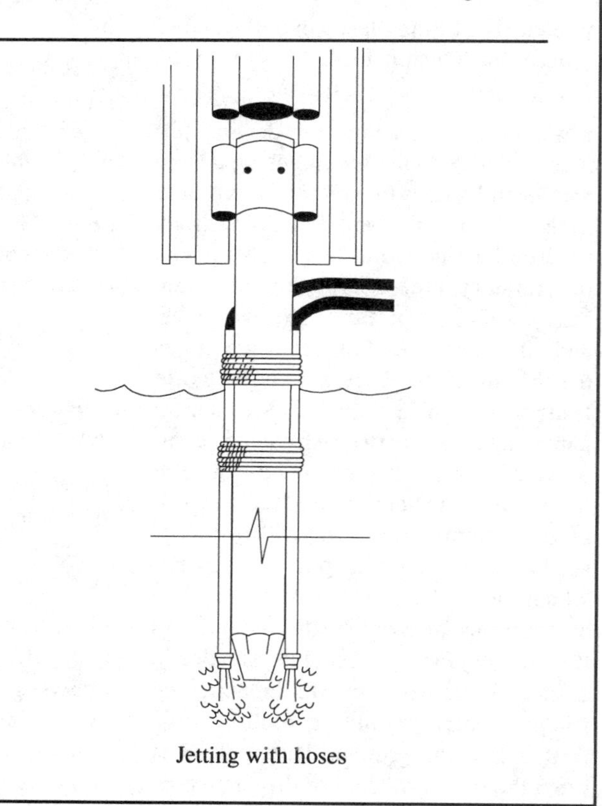
Jetting with hoses

pine, Idaho white – a wood with soft, light color and straight grain that is weather resistant and does not split easily. Used for exterior trim molding, siding, sheathing and shelving.

pine, ponderosa – a lightweight wood with a close, even texture. It is tough and usually free from warping and twisting. Used for doors, door cores, cabinetry, trim, window frames and the beds of trucks.

pine, southern or yellow – there are four types of this pine that are construction grade. The wood is strong and durable, it is stiff and has good fastener holding power. This type of pine is used both as structural members, such as beams, and in nonstructural applications, such as flooring and interior and exterior trim work.

pine, sugar – a lightweight, soft wood with an even, corky texture. It is strong and tough, with a low, uniform rate of shrinkage and it cuts easily in all directions. Used for carving and making patterns, interior and exterior trim, siding and doors.

pin-face spanner wrench – a wrench in the shape of the letter U with a handle. The two legs are joined at one end and each has a pin projecting through the other end at right angles to the axis of the leg. The pins can be inserted into holes or slots on a nut and torque applied to the nut.

pin hinge – a hinge that pivots on a pin that passes through both halves of the hinge. The pin hinge remains visible when a hinged door is closed.

pinhole – a very small hole in a material, such as might be made with a straight pin. A pinhole may be designed into a material, as in the pinholes in paper joint tape which allow compound to seep through, or they may indicate a flaw, such as a pinhole in a weld or paint surface.

pinholing – a paint defect in which pits appear in the finished surface. It is most often caused by air bubbles in the paint, and is more likely to occur when paint is sprayed. When the air bubbles in the paint dry, the trapped air creates a void. If the bubble breaks, it leaves a small hole in the paint surface. Pinholing can also be caused by slow evaporating solvents that are part of the paint mixture. Multiple coats of paint and/or adjusting the paint coat thickness and viscosity will prevent pinholing. Also called pitting.

pinion – a small diameter gear engaged with a larger gear. Used in gear sets, such as those in transmissions and speed reducers.

pinned file teeth – file teeth clogged with material particles.

pin punch – a device used to remove tapered pins from machine work or tools.

pintle – 1) a vertical pin-type peg that is used as a hinge pin for a gate hinge on a door or gate, or rudder on a boat or ship; 2) a metal or cast iron base for a wood post.

pin-tumbler lock cylinder – a type of lock cylinder that uses a series of pins that must be positioned by a key in order for the lock cylinder to be turned. A spring behind each pin holds the pin in place, bridging the small gap between the cylinder and the lock housing. The cylinder cannot be turned until the pins are pushed back into the housing by the insertion of the proper key. The pins are of different lengths so that only a specific key shape will simultaneously push all of them out of the cylinder.

pipe – a long, pressure-tight hollow cylinder that is used to contain and convey fluids. Pipe is available in a variety of materials, wall thicknesses and lengths for use under many different fluid and pressure-temperature conditions. Polyvinyl

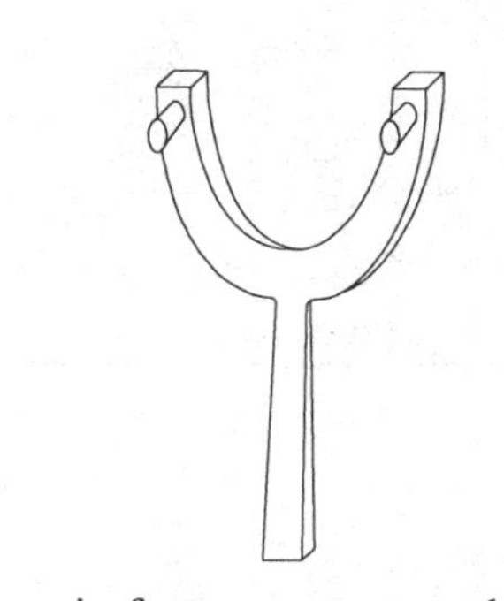

pin-face spanner wrench

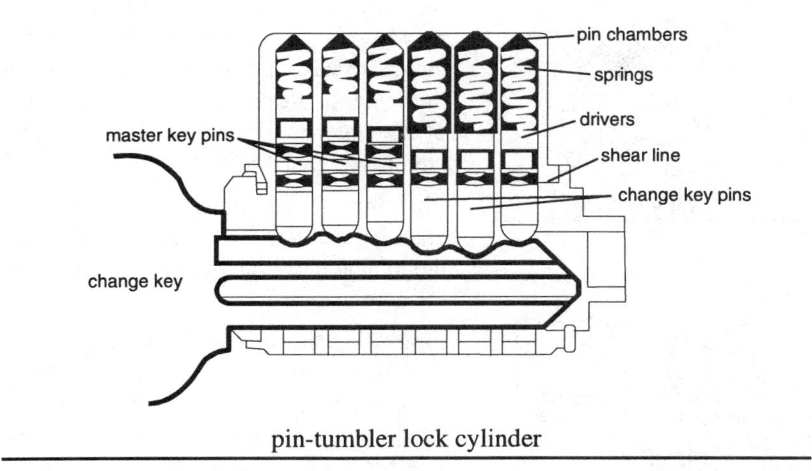

pin-tumbler lock cylinder

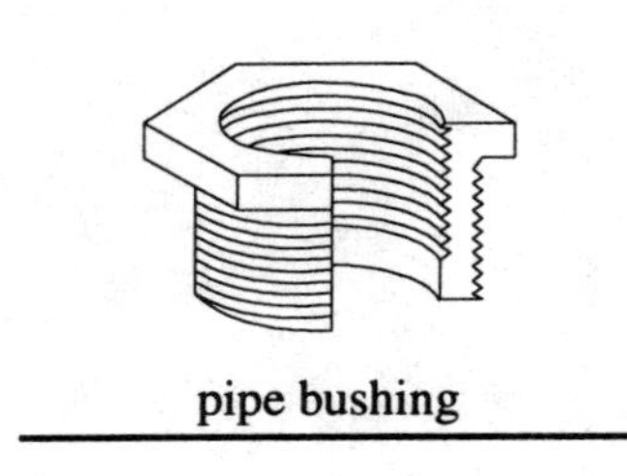
pipe bushing

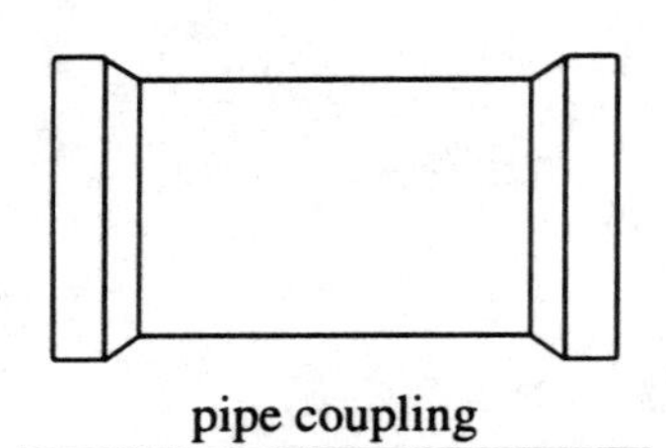
pipe coupling

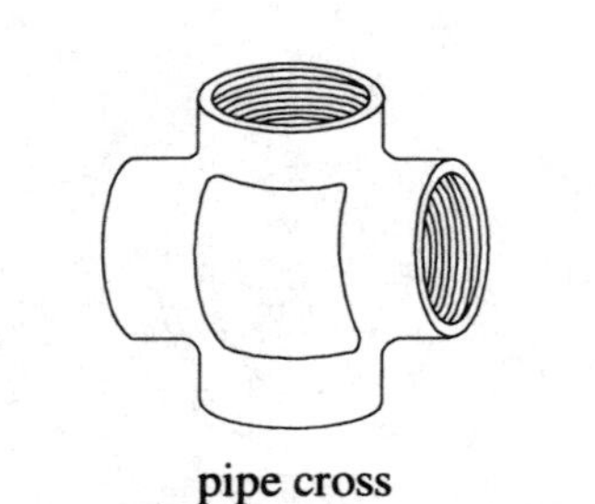
pipe cross

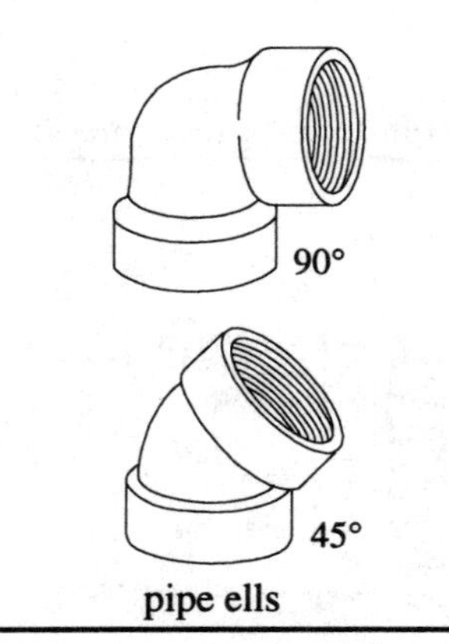

pipe ells

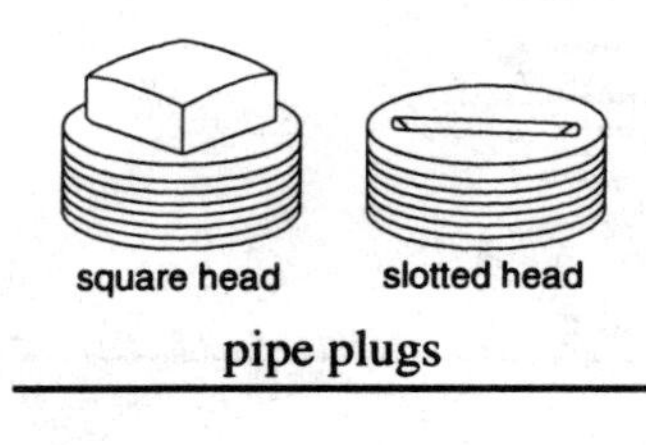

pipe plugs

chloride is used most often for water pipe and sewer mains. Carbon steel pipe is most often used within building for natural gas or for steam service up to about 750 degrees F. For higher temperatures, a chrome-molybdenum low alloy steel is generally required. Various stainless steel materials are used for pipe systems that have to be kept very clean, such as in food or beverage processing. Some stainless steels, such as grade 321 H stainless, are also suitable for use at very high temperatures. See also piping.

pipe bushing – a short length of pipe that is threaded both internally an externally and has wrench flats on one end for turning. A bushing is used to change pipe sizes in places where other size-change fittings, such as reducers, cannot be use.

pipe cap – a pipe fitting for sealing the end of a length of pipe.

pipe clamp – see bar clamp.

pipe coupling – a sleeve with the inside diameters of both ends threaded with pipe threads so that a length of pipe can be screwed into each end to join the lengths together.

pipe cross – a pipe fitting used to connect four lengths of pipe together at a single junction.

pipe cutter – a device with a set of rollers and a sharp cutter wheel that can be adjusted to clamp around the circumference of different sized pipes; the cutter is clamped in place and rotated around the pipe circumference, which causes the cutter wheel to make a groove in the pipe. The cutter wheel is advanced inward toward the center of the pipe with each revolution of the cutter until the pipe wall has been completely penetrated.

pipe dope – a compound, applied to pipe ends before they are joined, to lubricate the joint and/or seal the pipe threads once the pipe is joined.

pipe ell – a fitting with a bend in it used to change the direction of the pipe run. Two straight lengths are joined with the fitting, which bends, usually at a 45-degree or 90-degree angle. Also called an elbow.

pipefitter – a tradesperson who installs piping for steam, hot water heating, cooling and other systems.

pipe fittings – ells, tees, couplings, bushings, adapters and other parts used for joining pipe lengths and permitting bends or changes of direction in piping.

pipe flange – see flange.

pipe hanger – see pipe support.

pipe joint – the connection between two lengths of pipe or between lengths of pipe and fittings. Pipe joints can be threaded, flanged, welded, cemented or joined by other means, depending on the type of pipe material being joined.

pipe joint compound – see pipe dope.

pipe laser – an optical device with a laser beam used for aligning pipes. The beam is projected along the slope or grade of the pipe installation to a point some distance from the instrument. As the piping is installed, the pipe layer can check its alignment by looking through the viewer at a target placed at the end of the pipe series to see if it lines up with the light beam.

pipeline – a piping system for conveying fluids over distances, such as a water main system from a building to a main run, or a cross-country oil pipeline carrying oil from the source to a distribution point or refinery.

pipe nipple – a short pipe with male threads on each end used where there is room for a very short length of pipe.

pipe plug – a male threaded plug used with a female fitting to seal the end of a piping run. The plug may have either a square head with wrench flats or a slotted head.

pipe size – the approximate size of pipe from the smallest through 12 inches, above which point, the pipe size or *nps* (nominal pipe size) is the actual outside diameter.

pipe sleeve – a sleeve, that passes through a wall, which is made of a strong, durable material and is designed to permit other pipe to be inserted through it and the wall.

pipe street ell – a threaded ell with male threads on one end and female threads on the other. It can be used to connect a male end, such as a pipe, to a female end, such as a valve, and change direction, saving the need for a joint.

pipe supports – metal straps, clamps or other supporting devices used to restrain or anchor piping in order to control movement. The term includes all devices used to hold piping in place. Hangers are supports used to suspend pipe from above. Spring supports may be used to either suspend pipe from above or support it from beneath. Roller supports allow the pipe to move axially. Low friction plates, attached to the pipe by saddles, are sliding supports and can be designed to allow the pipe to move axially and sideways. Riser clamps fasten around a vertical pipe and support the weight of the pipe. There are many other types of pipe supports for various applications and systems. Pipe supports are considered an integral part of the piping system because they ensure that the piping remains intact and functional even during abnormal events, such as an earthquake. Also called pipe hangers.

pipe tee – a fitting used for joining three lengths of pipe at a single junction.

pipe threader, ratchet – a type of handle that can use a variety of die sizes and has a ratchet so that the die can be turned even when there is no room for a full revolution of the handle. It is used with a die to cut threads into the ends of pipe.

pipe threader, 3-way – a handle with three different sized pipe dies for threading pipe. The appropriate size die is fitted on the end of the pipe and then the threader is rotated to cut threads into the pipe end.

pipe threads – tapered threads that increase the sealing capability of pipe when the pieces are joined and tightened.

pipe vise – a vise with curved jaws or a chain for gripping the outside diameter of pipe without flattening the pipe. A pipe vise is used to hold pipe in place while cutting or threading it, or while fittings are installed on a length of pipe.

pipe wrench – an adjustable wrench with serrated jaw surfaces for gripping pipe. It has a movable jaw, at a slight angle to the fixed jaw, which pivots slightly so that as force is exerted on the handle of the wrench, the change in angle increases the jaws' grip on the pipe. Also called a Stillson wrench.

piping – the assembly of pipe, fittings, supports, valves and other control components that comprise a system for containing and transporting gas or liquid from one area to another.

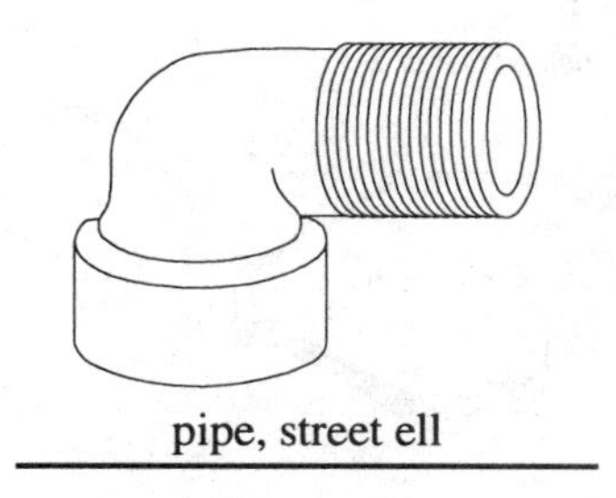

pipe, street ell

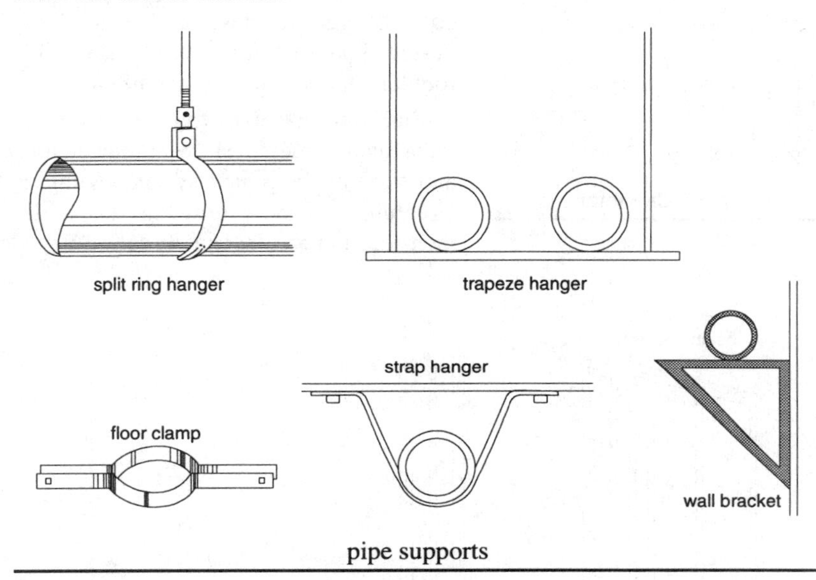

pipe supports

piping area drawing – a drawing to scale showing the routing of a piping system within a structure, using plan and section or elevation views. The drawing is done by the person responsible for the piping system layout.

piping codes and standards – local and state laws establish mandatory and optional codes which apply to piping and piping systems. Piping system materials are covered by ASTM (American Society for Testing and Materials) standards. Other societies publishing piping regulations are the American Society of Mechanical Engineers (ASME), American National Standards Institute (ANSI), American Petroleum Institute (API), American Water Works Association (AWWA), American Welding Society (AWS), Copper Development

pipe tee

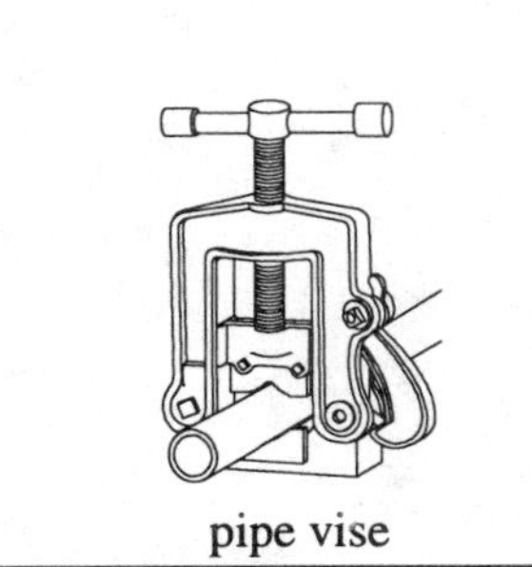

pipe vise

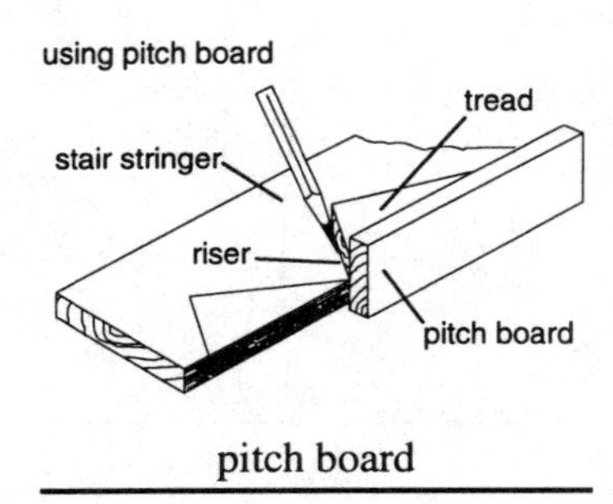

pitch board

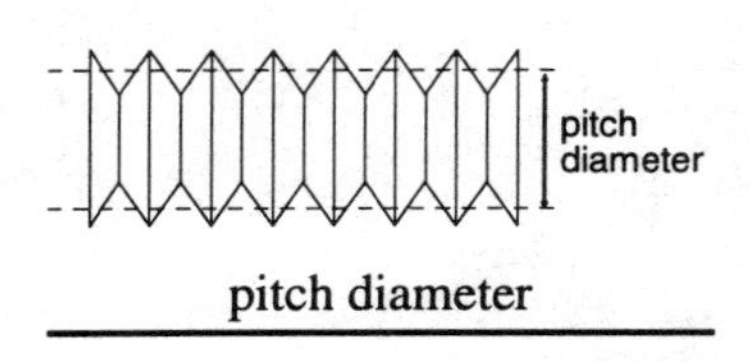

pitch diameter

Association (CDA), Manufacturers Standardization Society of the Valve and Fitting Industry (MSS), Uni-Bell PVC Pipe Association, and the Plastics Pipe Institute (PPI).

piping, copper – tubing made of copper which is used primarily for domestic water piping and compressed air piping within buildings because it is clean and corrosion resistant. It is available in three thicknesses, identified as Types K, L, and M, as drawn temper, which is rigid, and as soft annealed, which is flexible and comes in coils. Straight lengths are sold in 12- or 20-foot-long sections. Coils come in 100-foot lengths. See also copper tubing.

piping isometric drawing – a three-dimensional drawing showing the layout and sizes of the piping systems within a structure. Isometric drawings show dimensions but are not usually to scale.

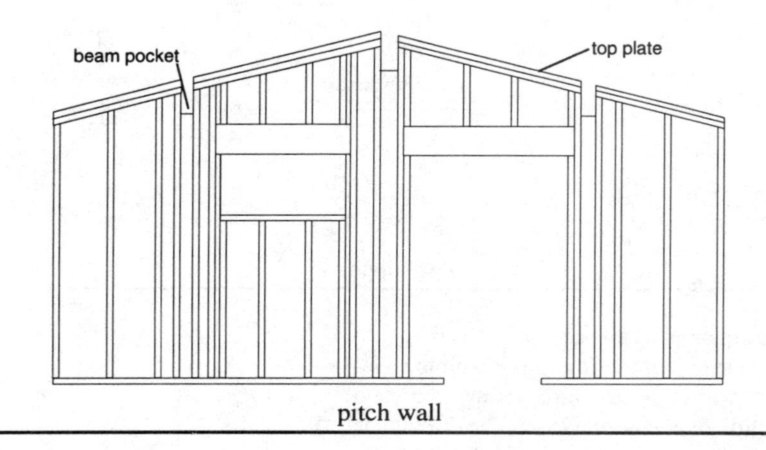

pitch wall

piping, PVC – pipe, fittings and valves made of polyvinyl chloride, a lightweight durable thermoplastic material used to carry a variety of fluids. The properties of various PVC compounds are defined in ASTM D1784. PVC pipe properties are defined in ASTM D3915. *See Box.*

piping schematic – a schematic drawing showing point-to-point routing of piping. Unlike an isometric drawing it does not show dimensions or piping lengths.

piston – a cylinder-shaped solid object within a cylinder that moves back and forth to pump or compress a fluid.

piston ring – a spring steel ring that fits in a groove around the circumference of a piston to provide a seal against the cylinder wall.

piston rod – a rod, with bearings at both ends, which connects a piston to a crankshaft for transferring power.

pit – 1) a small depression or indentation in a surface; 2) a hole or cavity in the ground; 3) an area that is sunken or lower than the adjacent floor level, such as an orchestra pit.

pitch – 1) a slope, grade or incline, such as the angle of the slope of a roof, which is expressed as a ratio of the rise to span; 2) an upward or downward movement; 3) the highness or lowness of a sound, voice or musical instrument; 4) the number of turns of a spiral per unit length; 5) the number of teeth or threads per inch; 6) the sap of a tree, particularly a conifer; 7) a resin substance from tar residues.

pitch board – a stair-building template cut in the exact size and shape of the required stair stringers and used to mark and make duplicate stringers.

pitch diameter – the diameter of a threaded cylindrical object at a point midway between the diameter at the root or base of the threads (the deepest cut to make the threads) and the maximum outside diameter of the threaded portion.

pitch seam – a wood defect in which a shake or check is filled with pitch.

pitch wall – a wall with an upper plate that slopes to match the roof slope.

pith – the soft inner core of a tree, branch or log. This part of the tree has little or no structural strength and is undesirable for use as framing or other structural wood members.

pitot tube – a fluid flow measuring device that compares the difference between dynamic and static pressure. It is used in various fluid systems to measure fluid velocity and monitor operating characteristics so that they can be controlled and optimized.

pitting – 1) a gypsum drywall defect in which small pits appear in the surface of the joint compound after it dries; 2) a paint defect in which pits appear in the finished

PVC pipe

PVC pipe is made of polyvinyl chloride, a lightweight, strong, resilient and durable thermoplastic material. Its discovery dates back to the end of the 19th century, but its practical application as a piping material began in Germany at the end of WWII, where the manufacturing methods were developed. Because the material has been shown to have an indefinite life span under most conditions, PVC is frequently used for cold water systems. It is also used for many chemicals because of its chemical-resistant properties.

There are two basic forms of PVC pipe; bell and spigot pipe, and solvent cement joint pipe. Bell and spigot pipe has a bell on one end with an internal elastomer seal into which the lubricated plain end of the next length of pipe fits. The elastomer seal in the bell makes a fluid-tight joint. Since the joints are slip fit, they must be restrained at the points where the pipe changes direction. Because of this requirement, this type of PVC piping is used primarily underground. The underground joints are supported with poured concrete thrust blocks or metal clamps which hold the joints together. The joints are secure but remain flexible and can easily accommodate underground earth movements.

Solvent cement joint pipe uses a solvent to join pipe ends. The mating surfaces of the joint are cleaned and then the primer is applied to soften the material's surface. The solvent cement is applied to the pipe end and the inside of the fitting end. The pipe is pushed into the fitting with a twisting motion to ensure an even spreading of the solvent cement. The cement cures quickly and fuses the joint together. The joint is rigid, and not well suited to underground applications in the larger sizes. The larger pipe is relatively stiff, and will not flex easily if the soil shrinks or swells. The joints do not need restraint, however, so solvent cement joint pipe is often used above ground. It must be protected by a latex-based paint if exposed to sunlight because ultraviolet rays will cause the material to deteriorate.

There are many different styles of valves and fittings available in PVC material for all the parts that are in contact with the fluid contained by the piping. The valves are available in sizes up to about 8 inches (nominal pipe size). They come with a choice of end connections, depending on the valve type. For example, ball valves are available with union ends or with flanged ends, and butterfly valves are designed to fit between flanges. When used with PVC pipe, the mating flanges will most often be of PVC that have solvent cement socket joint connections to the pipe. PVC flanges have flat faces and are rated for a maximum system working pressure of 150 psig at 73 degrees F. They are designed for use with full face soft elastomer gaskets.

End of 16 inch PVC pipe showing gasket in hub

PVC fittings are available for threaded or solvent cement socket joints. They are not generally available for use with hub and spigot PVC pipe. Cast gray or ductile iron fittings are usually used with hub and spigot PVC pipe. The use of threaded fittings with PVC pipe necessitates derating the allowable working pressure of the system by 50 percent because so much material is removed from the pipe for the threads.

PVC piping is used extensively for interior and exterior cold water distribution systems as well as industrial water and chemical use. It has rapidly gained popularity because of its freedom from corrosion and tendency to not lose flow capacity from interior changes, such as the growth of organisms on the inside diameter. It is also lightweight, easy to handle and install, has joint flexibility that accommodates ground movement without leaking, and has a long life with no required maintenance as long as it's protected from sunlight. It is available in a wide variety of sizes and wall thicknesses, with the uses of each size specified by ASTM and American Water Works Association ratings and guidelines.

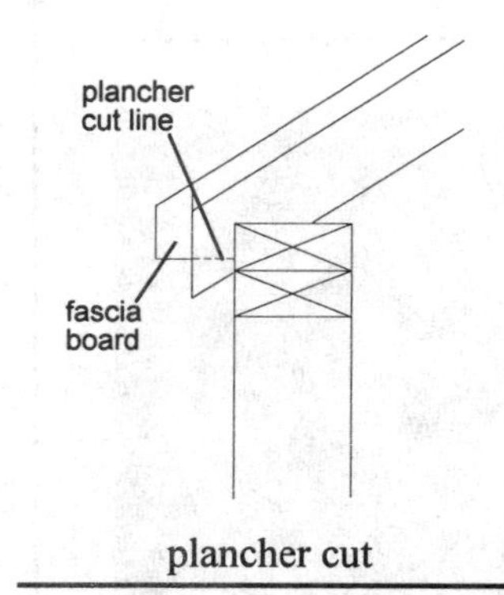

plancher cut

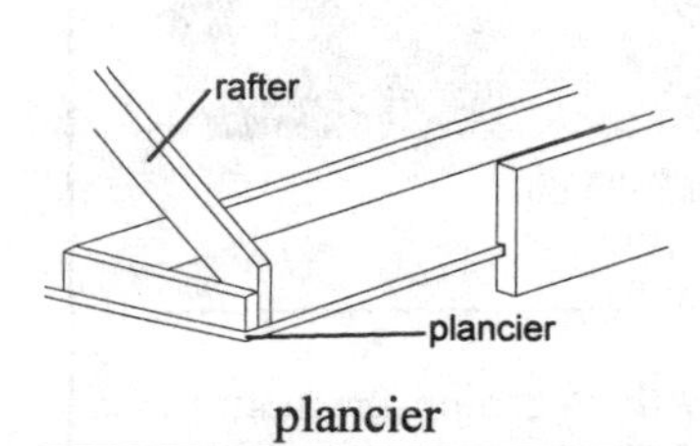

plancier

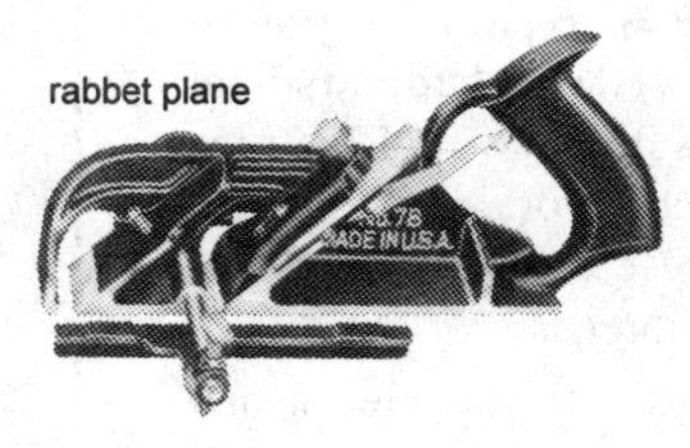

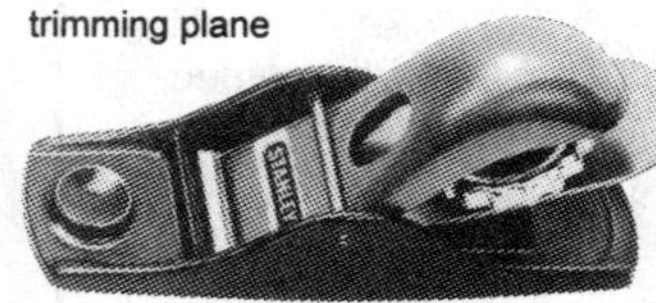

Courtesy: The Stanley Works

planes (1)

surface. They are caused by tiny bubbles in the paint that break as the paint dries, leaving little pits; 3) localized corrosion in metals.

pivot – a connecting or hinge point between two objects.

place – to pour concrete.

plain bibb – a faucet that is not threaded on the outlet end for a hose connection.

plan and profile sheets – a plan view, or view looking down on an area that shows the survey details of the area to be graded, and a profile, or view showing a vertical cut through the area to graded. All of the elevations and contours of the finished grade are shown on these two sheets, as well as the existing condition. The drawings are used by the crew doing the earth moving and grading to reach the final desired land shape and grade.

plancher cut – a horizontal trim cut on the tail of a rafter.

plancier – the underside of an eave or cornice; often a horizontal plywood surface.

plane – 1) a hand-held wood working tool used for smoothing, shaping and sizing wood. It has a flat base with a slot or mouth across the bottom through which a blade projects just enough to shave and smooth the wood as the plane is guided across the surface. The depth of the cut can be adjusted by adding more pressure to the tool, or by lengthening the adjustment of the blade. There are several types and sizes of planes designed for specific applications, such as a jointer, block or trimming plane; 2) a flat surface; 3) a two-dimensional shape.

planer – 1) a power driven wood plane, using horizontal blades mounted in a rotating horizontal cylinder which is adjustable for cut depth. The work is moved past the cutters while being held in a relatively constant position, so that an exact and even amount of wood is removed by the cutters; 2) a power driven metal shaping machine which uses a cutting tool mounted on a framework and a reciprocating table on which large pieces of metal can be moved back and forth for shaping and cutting.

planer gauge – see shaper gauge.

planer milling machine – a milling machine designed for heavy milling operations, such as cutting gear teeth or cutting keyways.

planer, rotary – a rotating blade cutter mounted on the end of a shank that can be installed in a drill press to plane the surface of wood.

planet gear – a gear arrangement in which one process is going on within another one. The planet gear engages and revolves around another gear, the sun gear, which is also revolving. Used in a differential gear.

plank – a rectangular cross section board that is more than 6 inches wide and more than 1 inch thick.

plank and beam framing – a type of framing in which the structural loads are concentrated on fewer and larger members than conventional framing. Used to create large, unobstructed areas in residential, commercial and industrial buildings.

plank flooring – random width and/or random length wood strip flooring.

planning grid – a grid used to lay out a structure so that the dimensions are in multiples of four. This gives the builder flexibility in locating openings and permits straightforward matching of vertical and horizontal surfaces of the building. Since most materials come in standard sizes, this type of layout is an advantage to the builder. Material take-offs are matched to the grid, minimizing wastage.

plano lens – a protective lens for the eyes used in nonprescription safety glasses.

plan view – the drawing of a structure or part of a structure as it would appear from a horizontal plane above the structure. A floor plan is a plan view.

plasma – a high-pressure ionized gas which is a good conductor of electricity and can be affected by a magnetic field. It is used in a cutting process which utilizes an electric arc to ionize gas and turn it to plasma.

plasma arc cutting (PAC) – a metal cutting process using an electric arc for heating the metal and high pressure ionized gas (plasma) to cut it. It can be used to cut thick and/or high melting point metals or any electrically conductive material, including nonferrous metals, which some other heat-cutting processes cannot. It is a high speed process and does not impart the physical force to the material that a mechanical cutting process would.

plasma arc welding (PAW) – an electric arc welding process, with or without pressure and/or filler metal, using shielding from hot ionized gas. The electric arc is either between the electrode and the metal being welded, or between the electrode and the gas nozzle. The gas coming from the nozzle serves as a shield. Plasma is a very concentrated heat source, existing at about 30,000 degrees F. Plasma arc welding can be used on most metals and in all welding positions. Because the heat is concentrated, there is a smaller heat-affected zone than with other processes, an advantage in minimizing the effects of welding on the base metal. The disadvantages are that this process requires costly equipment and is more complex to control than other processes.

plasma spraying (PSP) – a metal spray-on coating process that uses an electric arc between an electrode and a nozzle through which an ionized gas flows to melt the coating and spray it onto the work being coated. It is used to apply hard surfacing to metals and give then wear resistance, while retaining their desirable characteristics, such as toughness, ductility, and economy of the base material that has been coated.

plaster – a compound of gypsum or calcium sulfate and water mixed to a dough-like consistency and applied to walls and ceilings. It cures to a hard surface. See also plaster of paris.

plasterboard – see drywall.

plaster, exposed aggregate – a mixture of cement, sand and lime, commonly known as stucco, used primarily as an exterior wall covering.

plaster grounds – strips of wood temporarily attached to surfaces being plastered for use as guides by the plasterer to keep the plaster thickness consistent and the surfaces smooth and level.

plaster, gypsum – calcined gypsum powder and/or lime with sand or other additives that can be worked as a paste with the addition of water. It sets up hard when dry and is used to plaster interior surfaces.

plaster lath – see lath.

plaster of paris – calcined gypsum, a compound that mixes with water to form a paste which can be worked, shaped or spread and dries hard. It is used for fine work or castings.

plaster, portland cement – a plaster wall finish for interior or exterior use, consisting of portland cement binder with aggregates.

plaster ring – a steel or plastic extension for a recessed electrical box that is used to bring the face edge of the box out to the surface of the finished wall.

plastic – 1) any of a number of synthetic compounds, made from petroleum products, that can be formed into shapes, and that harden or solidify under the proper conditions; 2) moldable; malleable; easy to shape or manipulate; 3) a material that is pliable and workable, such as properly-mixed fresh mortar.

plastic anchor sleeve – a tapered sleeve of plastic that is inserted into a fastener hole in drywall. The sleeve lines the hole so that a screw inserted into the sleeve grips the hole tightly.

plastic cement – cement with plasticizing agents added to it which make the cement more workable.

plastic conduit – electrical conduit made from PVC or other plastic material. It is more economical and easier to work with than metal conduit.

plastic electrical boxes – electrical boxes made of polyester, fiberglass or PVC designed for use with nonmetallic cable.

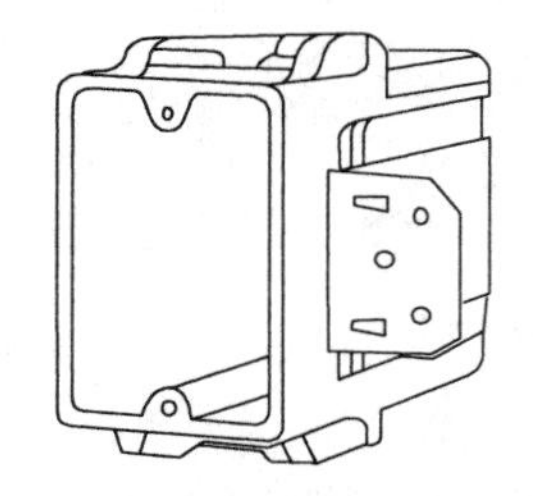

plastic electrical box

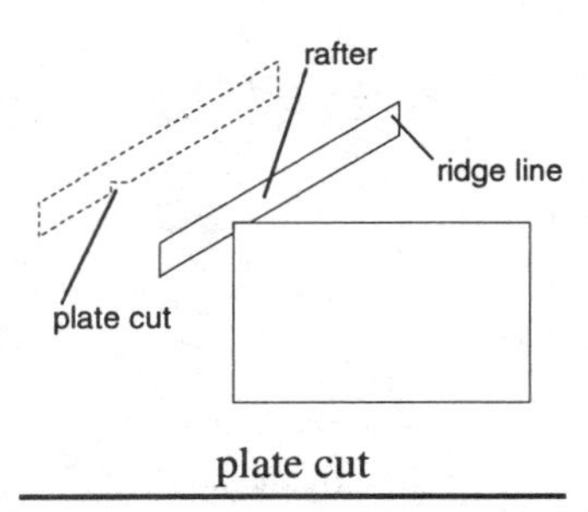

plate cut

plasticizer – a compound added to another compound to increase toughness and/or flexibility.

plastic laminate – a sheet material built up of fibrous material and resins. It is used for countertops and similar items.

plastic piping – see ABS pipe and piping, PVC.

plastics – plastics are materials of long-chain molecules, called high polymers, which contain carbon and also may contain hydrogen, nitrogen, sulfur, oxygen, silicone and halogens. Plastics are either thermoplastics, which melt with the application of heat and resolidify when cooled, or thermosetting, which do not melt once formed. Thermoplastics are easily molded into complex shapes. They are tough, low cost, have low thermal conductivity, are chemically resistant, flexible and can have good optical clarity. Thermosetting plastics are usually made in the form of composites, blending or laminating polymers with particles or fibers. Thermosetting plastics have good dimensional and thermal stability. They are used to make corrosion resistant pipe, pipe fittings, valves and other parts.

plastic solvent welding – fusing plastic parts at the interface where they join using a solvent that dissolves the plastic. Solvent cement is used to join plastic piping and pipe fittings, such as PVC piping. The cement is applied to the end of the pipe and the inside of the fitting socket into which the pipe is to be inserted. The pipe is then inserted into the socket with a twisting motion. The joint sets up in a matter of minutes, and is ready for use. The solvent cement softens and partially dissolves the plastic of the pipe and fitting so that they fuse together making an integral material.

plastic welding – fusing plastic parts using heat to melt the plastic at the interface where the parts are joined. A heating coil is used to bring the plastic surfaces to the proper temperature to be joined. The temperature will vary depending on the type of plastic being used (usually the manufacturer will provide this information). When the surfaces reach a near molten state, they are held in contact with each other until they fuse together.

Plastic Wood – a trade name for a putty-like substance used to patch wood. The material will accept stain and it is also available in a variety of colors to match different stained and unstained wood finishes. It hardens in place to a wood-like consistency that can be sanded smooth.

plate – 1) a horizontal framing member forming the top or bottom of a wood-framed wall. Studs attach to the bottom plate (which is attached to the subfloor), and rafters and joists to the top plate; 2) a material form that is relatively thin compared to its other dimensions, but is generally thick enough to have structural strength, such as the steel deck plates used on board ships; 3) to chemically, electrically or mechanically cover one material with a thin coating of another which adheres to, or bonds with the first material.

plate cut – a birdsmouth or angle cut out of a rafter to allow the rafter to sit into the top plate of a wall, rather than balance on top of it.

plate girder – a structural member built up of steel plates and angle iron fastened together to form a strong, yet relatively lightweight, section.

plate glass – glass manufactured in large, smooth sheets that are $\frac{1}{16}$ inch thick or more. It is ground and polished to increase its quality. Because of its thickness and strength, plate glass is used for large windows and other large expanses of glass.

platen – the part of an electric resistance welding machine that applies force to the parts being welded during the welding operation.

plate rail – the bottom fence rail of a fence which has a concrete foundation. The plate rail is supported by the foundation and is the wood member to which the rest of the fence attaches.

platform – a raised horizontal floor section commonly used as a stage area.

platform framing – a type of house framing in which the wall framing sits on top of the subfloor, and each story is built up as a separate unit. The other major type of framing is balloon framing, in which the studs run from the bottom of the first floor to the top of the second floor.

platform stairway – a stairway with a change of direction or landing between floors.

platinum – a heavy white corrosion-resistant metallic element with the atomic symbol of Pt, atomic number 78, and atomic weight of 195.09. It is used in some spark plugs and in catalytic converters, as well as for fine jewelry.

plat map – a map that shows land divisions within a specified area with township, lot lines, streets and improvements. It also includes major geographical features, mineral claims, elevation measurements, and survey bench marks.

play – free movement in a machine where no work takes place. For example, there is clearance between gears to allow for production tolerances and changes in temperature for smooth running. When movement is reversed, there is a small amount of free movement, to take up this clearance, where there is no force exerted on the driven gear by the driving gear. This clearance where there is movement but no force being imparted, is know as "play." Similarly, there is some play in most mechanical linkages when movement occurs across the clearance gaps and no work is taking place.

plenum – 1) a chamber or vessel that is designed to hold stored energy in the form of compressed gas; 2) an accumulation and distribution chamber to which ducts are connected in an HVAC system; 3) the space between the electrode and the inside of the nozzle through which ionized gas passes during a plasma gas welding, cutting or spraying process.

Plexiglass – the trademark name for methyl ethacrylate, a transparent plastic sheeting material used to make windows and the removable inner panes of double-pane storm windows.

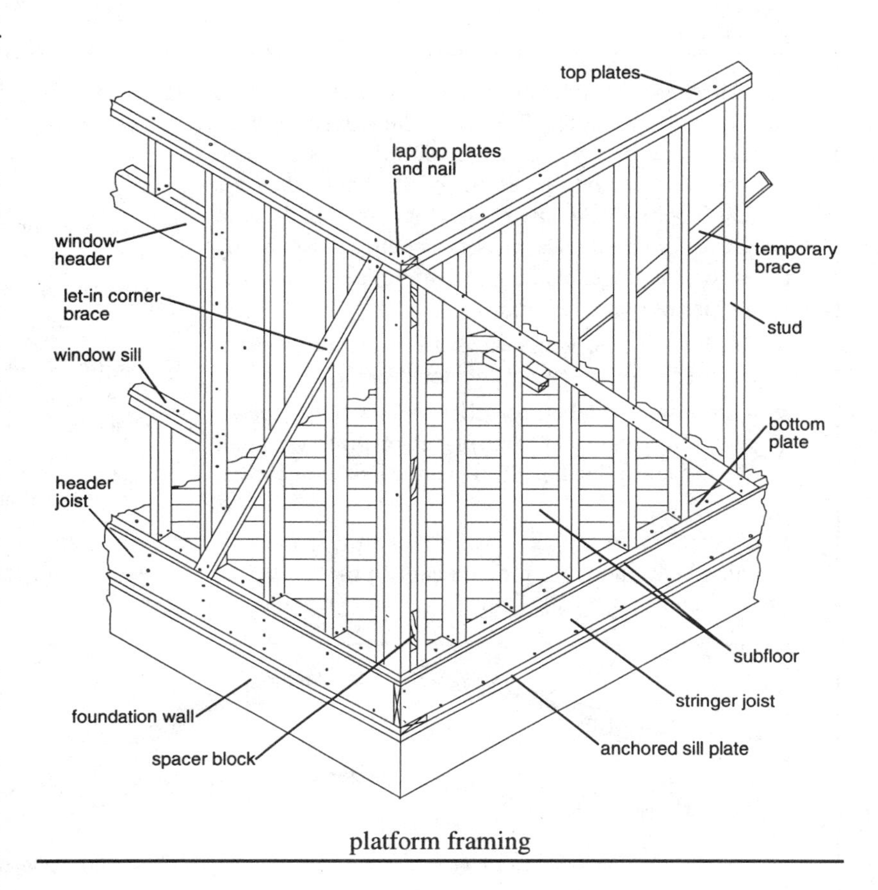

platform framing

pliable – flexible, bendable or moldable, as in products made to be pressed or molded into position for use, such as duct seal putty, weatherstripping and caulk.

pliers – a hand tool with a hinged jaw for holding or manipulating work. *See Box.*

plinth – 1) the square base of a pedestal or column; 2) the base course of a stone wall or rough stonework.

plot plan – a plan view drawing of a structure which includes the dimensions of the building site, the location of the structure in relation to the property boundaries, elevation of key points, existing and finish contour lines, utility services, and compass directions.

plough – cutting a groove, such as a dado, with the grain of a board.

plow – a wood plane designed to cut grooves, such as dadoes, in wood.

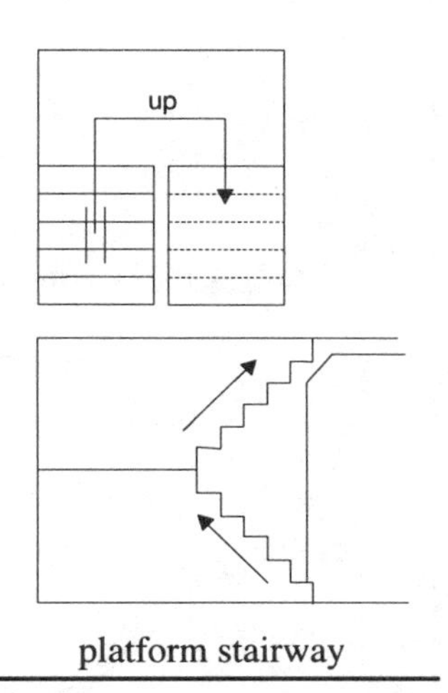

platform stairway

Pliers

There are many styles of pliers for general purpose and specialized tasks. The jaws often have serrated surfaces for a better grip. The handles are longer than the jaws and provide leverage for working. They are made of carbon or tempered steel and the handles are frequently covered with vinyl for a better grasp. Many types come in large, medium and miniature sizes. Common types of pliers include:

- slip-joint pliers, which have two-position serrated jaws for gripping larger or smaller objects;
- long-nose or needle-nose pliers, which have tapered jaws for working in tight places or gripping small objects, and a fine honed edge on one side for cutting wire;
- diagonal-cutting pliers, which have precision machined cutting edges for true cuts on most types of wire, and long handle grips for cutting leverage;
- lineman's pliers, with honed cutting edges for cutting most types of wire and serrated jaws which provide a strong grip and leverage for use in heavy applications;
- groove-joint pliers, with an angled head and adjustable jaws which can hold and turn work of various sizes; also called tongue-and-groove, adjustable or water-pump pliers;
- bent-nose pliers, with curved, tapered serrated jaws for gripping small objects in applications where the space is limited;
- end-nipper pliers, with finely honed cutting edges for gripping and cutting fine wires;
- flat-nose pliers, used for gripping and crimping, which have smooth jaws that will not mar the work surface.

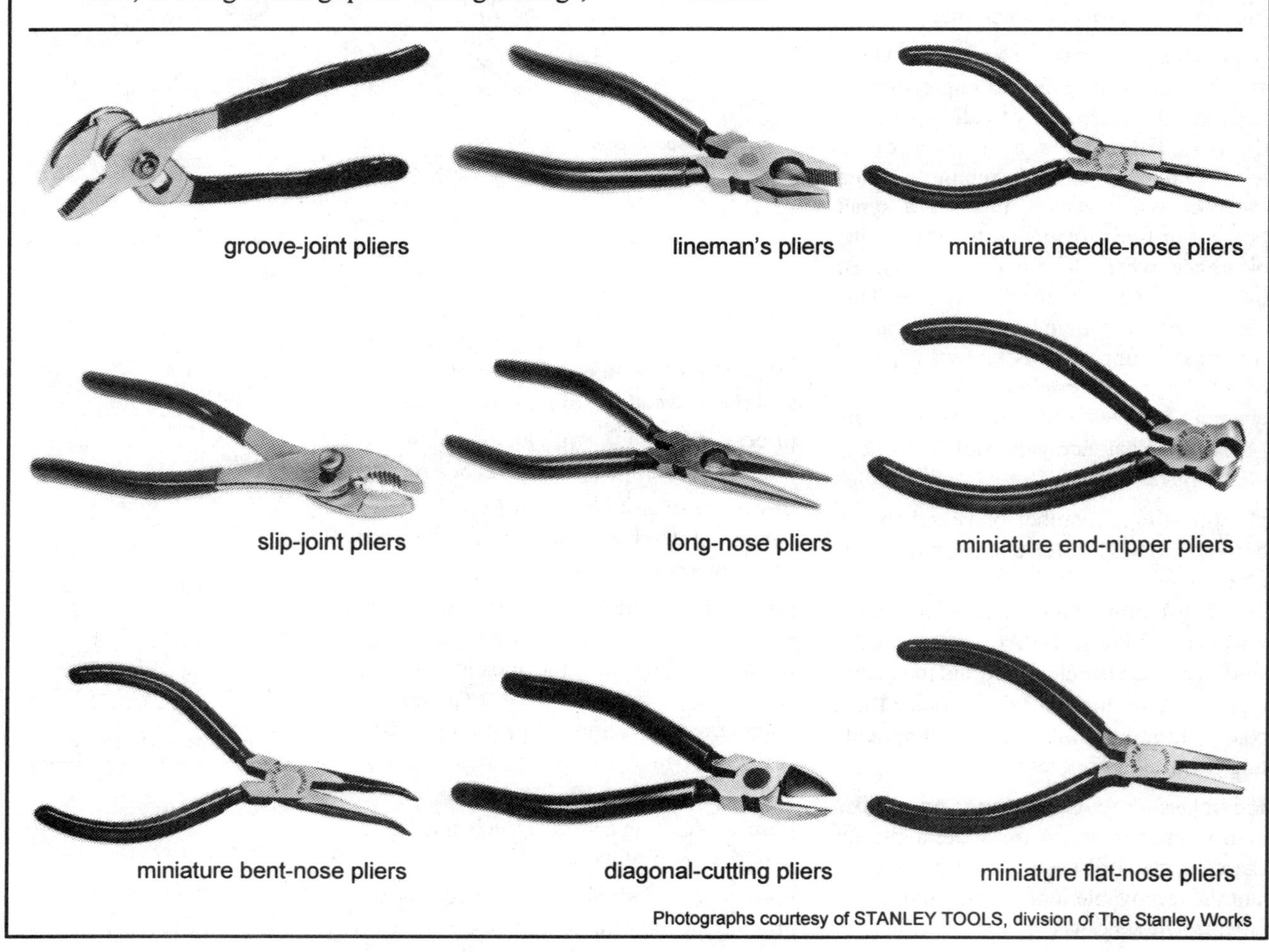

groove-joint pliers | lineman's pliers | miniature needle-nose pliers

slip-joint pliers | long-nose pliers | miniature end-nipper pliers

miniature bent-nose pliers | diagonal-cutting pliers | miniature flat-nose pliers

Photographs courtesy of STANLEY TOOLS, division of The Stanley Works

plucked finish limestone – see limestone, plucked finish.

plug – 1) a male fitting at the end of an electrical cord or cable that fits into a receptacle or other connector providing electrical power; 2) a piece of wood or other filler material used to fill a hole; 3) a threaded fitting that screws into the end of a female plumbing pipe fitting or valve to seal it off; 4) a stopper for a drain; 5) to change the voltage polarity or phase sequence to an electric motor before the motor has stopped, generating a force to retard the motor rotation.

plug and feather rock splitter – a rock splitting device consisting of a split plug with an internal taper and a wedge-shaped rod, called a feather. The plug is inserted in a hole drilled into the rock and the feather is forced into the plug, wedging it apart and splitting the rock. These are used in groups along a line of holes drilled into a large rock at the position where the rock is to be cut.

plug cutter – a hole saw, designed for use with a drill press, which cuts cylindrical wood plugs to be used to fill holes in wood.

plug gauge – precision inside diameter gauge for measuring small holes. It is an essential tool for precision machine work, and is also used for measuring hole sizes for dowels, drilling holes or selecting tap sizes.

plug-in bus duct – a bus duct run through an industrial building which provides power for pieces of heavy equipment, such as lathes and milling machines, via plug-in ports along the bus.

plug weld – a weld made through a hole in a piece of metal, either to fill the hole or to join the piece to another piece below the hole.

plumb – true vertical; to be, or cause to be, vertically aligned.

plumb bob – a weight, usually with a point at the bottom, that is suspended from a cord and allowed to swing free. The force of gravity causes the cord to hang in a true vertical plane. A plumb bob is used to align framing, walls, wallpaper and any other item that must be constructed or installed true to vertical. Also called a plummet.

plumb cut – a vertical cut that is exactly straight up and down. Plumb cuts are made at the top end of a rafter where it meets the ridge.

plumber – one who installs and/or repairs plumbing systems and plumbing fixtures.

plumber's friend – see force cup.

plumber's strap or tape – galvanized steel strap with a row of holes along its length. The strap is used for supporting small-size piping.

plumbing – piping and/or tubing used to carry and control fluids, particularly those used for domestic service.

plumbing bend – an elbow fitting used to make a bend at an angle to the main line. Used in drain and vent lines. See individual listing which follow for the various types of plumbing bends.

plumbing bend, closet – an elbow fitting that connects to the outlet of a toilet or water closet.

plumbing bend, long-sweep – a long-radius elbow.

plumbing bend, reducing sweep – an elbow fitting that is larger on one end than the other so that the pipe can change direction and adjust to a change in pipe size. The elbow or ell can be used in either direction to join with a pipe that increases or decreases by one size.

plumbing bend, return – an elbow fitting that makes a 180 degree bend.

plumbing bend, standard bend – a short-radius elbow.

plumbing double Y-branch – a plumbing drain or vent fitting with branches coming in at an angle on each side of the main to join the main run. Also called a double wye.

plumbing drawing symbols – shapes used on plumbing drawings to denote specific fittings, and other items.

plumbing fixture – a device that receives water and/or discharges wastes into a drainage system, such as a lavatory or a water closet.

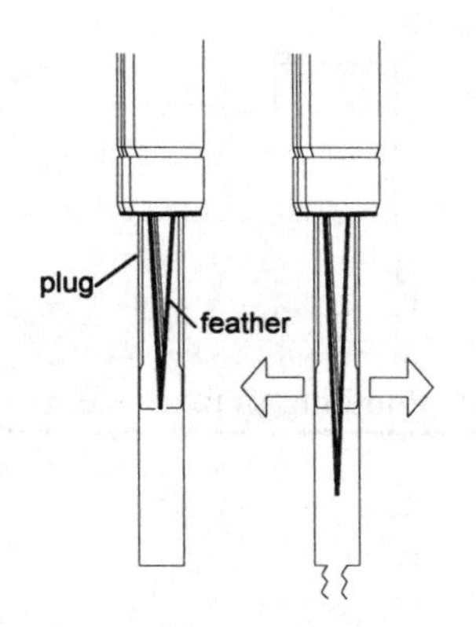

plug and feather rock splitter

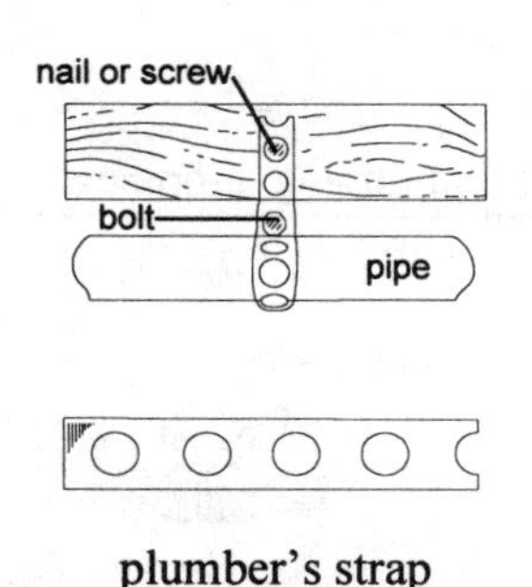

plumber's strap

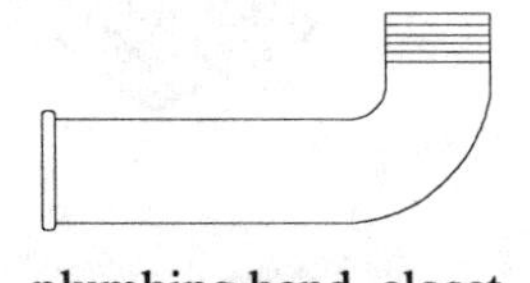
plumbing bend, closet

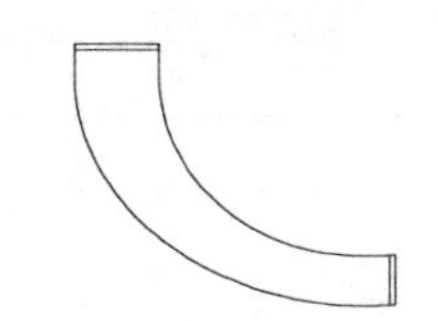
plumbing bend, reducing sweep

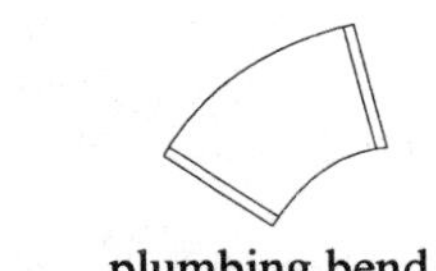
plumbing bend, standard bend

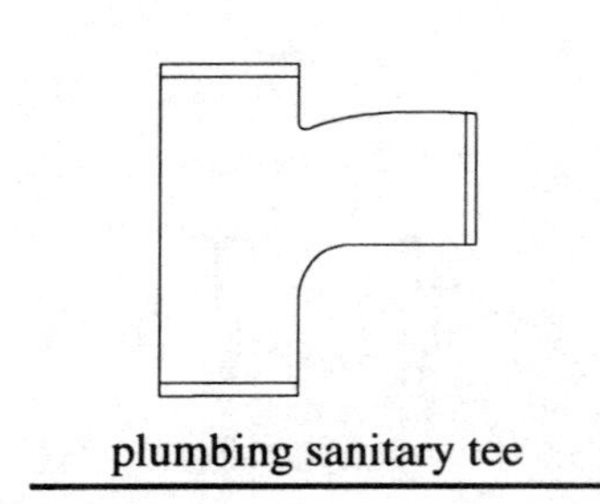
plumbing sanitary tee

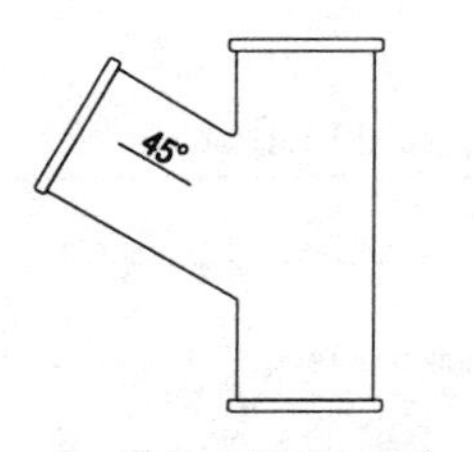

plumbing Y-branch

plumbing, rough-in – installation of the plumbing system in preparation for the plumbing fixtures, but not including the installation of the fixtures.

plumbing sanitary tee – a short-pattern 90-degree tee with the stand, the branch or right angle portion of the tee curving in to the run section in same the direction as the flow. This is so there is no abrupt change of direction, as there would be with a standard tee. This type of tee is used in sanitary waste lines. Also called the sanitary tee branch.

plumbing snake – a flexible, thin spiral-wound length of metal used for inserting down a drain. The snake is rotated in order to clear a clogged drain.

plumbing Y-branch – a Y-shaped plumbing drain fitting used to join a branch line to the main run. The fitting has one straight side, with the branch joining in at an angle. Also called a wye.

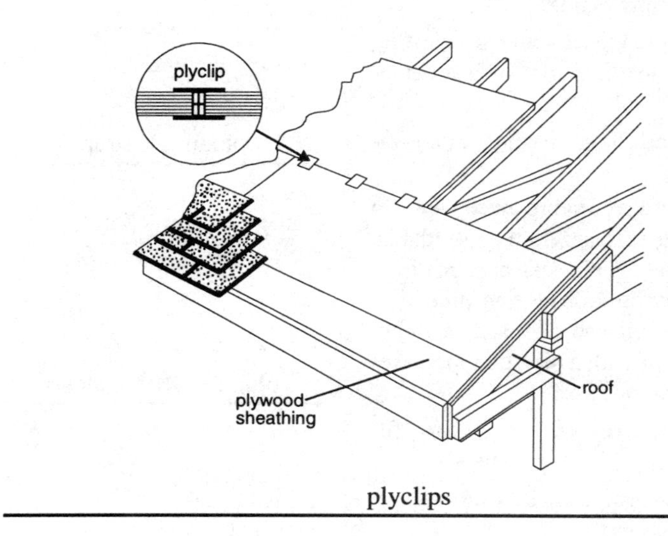

plyclips

plumb line – a string line with a weight on the end that hangs freely in a true vertical orientation when it is suspended from a point with the line fully extended. Used to determine and mark a vertical line on a structure or to identify a point directly beneath another point; 2) a line that is exactly vertical.

plumb rule – a narrow board with a bubble indicator that centers itself between two vertical lines when the board is plumb. A variation of this is a narrow board with a plumb bob suspended from the top which indicates when the board is plumb.

plummet – see plumb bob.

ply – a single layer, strand or thickness in a multi-layer material, such as plywood.

plyclips – metal clips that are used to bridge the edges of plywood installed as roof sheathing. See also panel clips and plywood edge clips.

Plyform – a trade name for plywood that is especially designed and treated for use as concrete forms. The plywood is made from veneer grades and woods chosen for long term durability. They are bonded with exterior grade waterproof glue. The plywood is sanded and oiled during manufacturing to reduce potential moisture penetration from wet concrete and minimize the tendency of concrete to adhere to the plywood as the concrete cures.

Plyform, HDO – a trade name for a high density overlaid Plyform, with a resin-impregnated material bonded to the face of the panels to improve longevity. Used where a longer form life is needed than standard Plyform provides, such as for a large, long term concrete job. They are also more expensive than standard Plyform, and so more likely to be used on larger, more costly jobs where the expense would be justified.

plytooth saw blade – a fine-tooth saw blade designed to minimize splintering when cutting plywood.

plywood – a building material made by laminating several thin layers of wood together. The grain of each succeeding layer is at right angles to the grain of the preceding layer to add strength.

plywood edge clips – metal clips that bridge between adjacent plywood panel edges to connect those edges. Also called metal edge support clips, plyclips or panel clips.

plywood, exterior – plywood, made with waterproof glue, used for exterior siding, roof sheathing and other applications where waterproof glue is essential to retain the integrity of the material.

plywood grades – ratings that indicate the quality of the plywood, ranging from Grade 1, which has matched grains and can be used for furniture or panels, to Grade 4, which has any type and number of defects, but is sound enough to be used in construction.

plywood, interior – plywood made with a water-resistant glue which can be exposed to occasional or moderate amounts of moisture, but is not suited to exterior applications.

plywood, lumber core – a type of plywood, sometimes used for cabinets, made with a core of boards encased in a top and bottom layer, each 1/16-inch thick, covered by two 1/28-inch facing layers.

plywood, veneer – plywood that is made of several layers of veneer glued together and in which the outer layers are of an expensive or higher quality wood, such as walnut or oak.

pneumatic – powered by, using, or filled with a gas, usually compressed air. Many tools are driven by compressed air, such as grinders, impact wrenches and saws.

pneumatic roller – a paving machine with wide, smooth, air-filled tires used to compact the surface by rolling over it. It is used for compacting clay soils because the clay does not adhere to the roller.

pneumatic tools – power tools driven by compressed air.

pocket door – a door that slides on an overhead track into a recess, or pocket, within the wall.

podium – 1) a low wall that is used for the base of an additional structure; 2) a short wall or footing comprising a fence foundation; 3) a raised speaker's platform, lectern or pulpit.

pointing – placing mortar into masonry joints after the masonry is laid. It may also include cleaning out old mortar and replacing it with new to restore the masonry.

pointing trowel – a small triangular mason's trowel used for finishing mortar joints.

point of curve (PC) – a surveying term defining the beginning of a curve. See also point of tangent.

point of intersection – a term used by surveyors or engineers to indicate the point where two tangents to a curve meet. Also called the vertex.

point of tangent (PT) – a surveying term defining the end of a curve. The curve extends from the point of curve (PC) to the point of tangent (PT).

points – a fee, expressed as a percentage of the loan amount, that a borrower pays the lender to originate a loan. One point equals one percent, therefore, if the loan is for $100,000, one point would be $1,000.

polarity – the direction of electric current flow through a circuit indicating the positive or negative charge of the current. In direct current, there are two poles, one positive and one negative. Current flow is from negative to positive.

polarized plug – an electrical plug designed so that it can only be inserted into a receptacle one way. This ensures that the "hot" side of the appliance is connected to the "hot" side of the receptacle. Usually the hot side of the appliance is operated by a switch, so when the switch is off, the power to the appliance stops at the switch.

pole – the number of hot wires that are connected to an electrical circuit breaker. For example, a single-pole switch has one hot wire and a double-pole switch has two hot wires.

pole gun – a paint spray gun with an extension handle that allows the painter to reach high or difficult areas without a ladder or scaffold.

polishing bonnet – a cloth or sheepskin bonnet that slips over a wheel on a motor, such as a sanding wheel, used to polish the surface of an object. It may be used alone or with a wax or a polishing compound.

polybutene tape – a non-hardening mastic which is pressed into place. Used for sealing window panes.

polyethylene (PE)– a thermosetting plastic used for a variety of applications, including sheeting, tubing, pressure pipe and pipe fittings. As piping or tubing, it

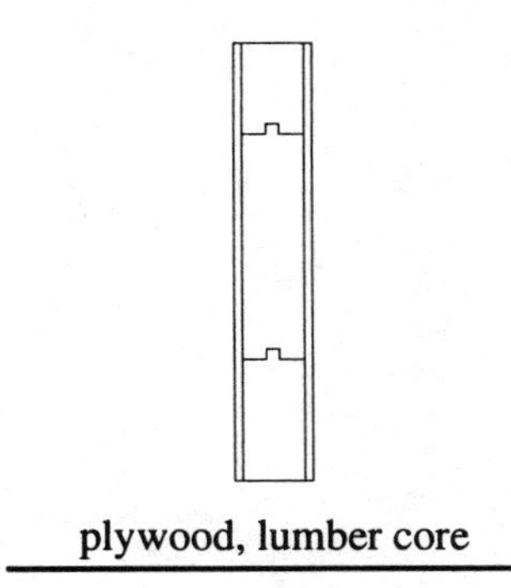
plywood, lumber core

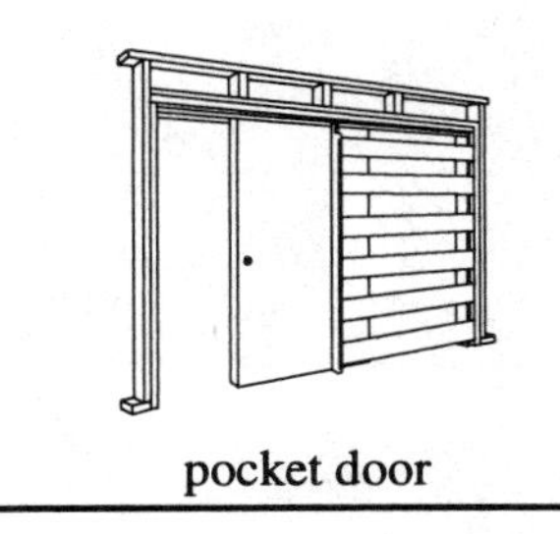
pocket door

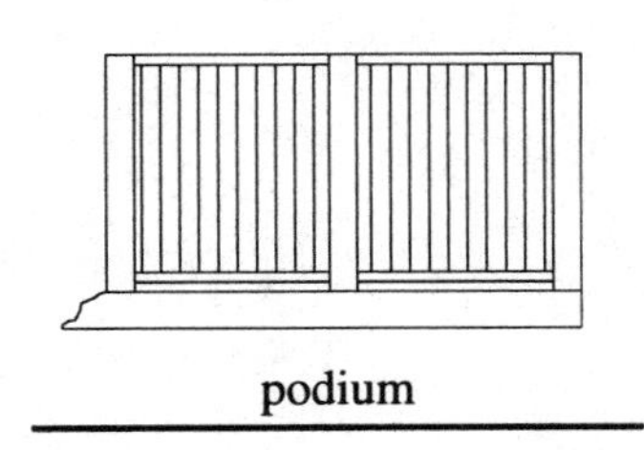
podium

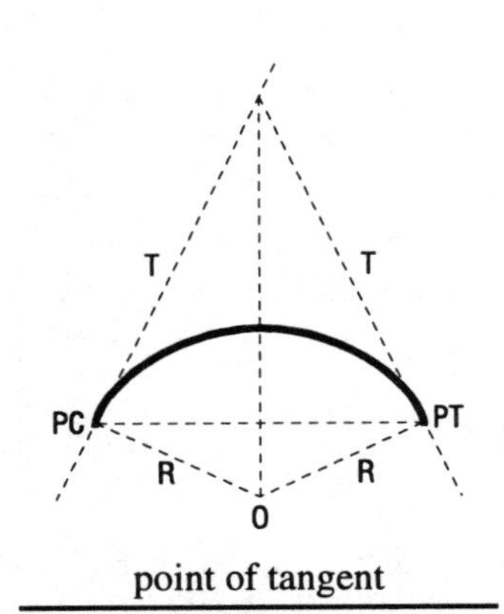

point of tangent

remains flexible and is easy to install. Polyethylene is corrosion resistant and not affected by many chemicals, which makes it an ideal material for use in containing or transporting chemical compounds. It will not deteriorate when exposed to ultraviolet light, so unlike PVC, polyethylene piping can be installed out of doors without the need for a protective coating.

polygon – a multi-sided figure.

polymerization – a chemical reaction that interlocks molecules. The process is used in the manufacture of polystyrene resins, acrylic resins, silicon resins and many more materials.

polystyrene – a plastic manufactured in the form of clear sheets, foam beads or foam boards. The foam boards are made by expanding polystyrene foam beads with heat in a mold. They are used as a rigid insulating material and have a high R-value, or high thermal resistance. They are manufactured in many different sizes and configurations. Insulating board can be extruded through a die to make any desired shape. Because of its insulating value, rigid board is often used as roof deck sheathing. Expanded polystyrene is also used to make styrofoam.

polysulfide – a flexible, synthetic rubber-like material used as a sealing material for products such as caulking or gaskets. It has good weather and chemical resistance and can be used in both liquid and solid forms.

polytetrafluoroethylene (PTFE) – a type of Teflon with good chemical and heat resistance. It is usable, under some conditions, up to about 450 degrees F. PTFE is widely used for pipe and valve lining materials, as flange gaskets in piping systems, and soft seats in valves, because of its good sealing characteristics. As a lining material, it is supported and strengthened by the metal body systems of the pipe or valve. It is very smooth and slippery, reducing the pressure drop through piping systems caused by some other lining materials. It is also resistant to biological growth which can cling to the inside of a piping system and reduce the inside pipe diameter over time. It has a very low coefficient of friction and can be used for bearing surfaces where low friction is a must, and the environmental conditions are suitable.

polyurethane – a polymer plastic used most often as a component in paints, varnishes and insulating foams. Polyurethane insulating foams can be bought in rigid board form or liquid foam. The rigid board insulation has high thermal resistance. The liquid foam comes in pressurized canisters that permit directing the foam into small crevices, such as around windows, where it will expand to fill up the space. The foam also comes in large quantities in a two part mixture of urethane and a catalyst. When mixed, the foaming action is activated, and the compound expands rapidly as gas is generated within the mixture. Foam used in this manner grows quickly in volume and molds to whatever shape is used to control the foam until it sets up hard. When used to insulate the outside of a container, it is often covered with fiberglass cloth and resin, after it has cured and the form or mold is removed, to add structural strength to the hardened foam and prevent it from being damaged. The foam is also available in preformed sections that can be clamped around piping and piping components as insulation.

polyurethane foam board (PU) – Insulating board with a high R-value, or high thermal resistance, used in construction when a higher R-value per inch than fiberglass batting is required and space is limited.

polyvinyl acetate (PVA) – a thermoplastic material that is a component of certain types of strong wood glue and some latex paints. It adds a hardness that is not brittle, as well as adding water resistance. It is durable and forms a gas-permeable film which helps to prevent blistering in paints.

polyvinyl acetate glue – a white glue for wood that forms a very strong joint, but because it is not waterproof, it is not suitable for exterior use or for use in very humid areas.

polyvinyl chloride (PVC) – a thermoplastic polymer, synthesized from vinyl chloride, which is used for a variety of products including plumbing pipes and fittings.

polyvinyl chloride (PVC) pipe – see piping, PVC.

ponding – 1) flooding a new concrete surface with water to aid curing; 2) the accumulation of water on a surface that does not completely drain. This is usually a surface problem caused by unwanted depressions or insufficient slope for water runoff, and can be cured by filling depressions and/or improving the slope.

pony trowel – a small power trowel, driven by a large drill motor, used for concrete finishing.

popcorn – 1) asphalt paving mix using primarily ¾-inch aggregate; 2) a slang term for a sprayed-on decorative acoustic ceiling texture that is made of expanded perlite or vermiculite in a bonding agent.

poppet – a flow control valve with the disc attached perpendicular to the stem. The valve opens and closes by being moved up and down on the end of the stem. This type of valve is commonly used in internal combustion engines, such as the gas or diesel engines that use reciprocating pistons in portable compressors and generators.

pop rivet – a rivet consisting of a metal sleeve, flanged on one end, and a center stem which fits into the sleeve. The stem is upset on one end so that it braces itself against the end of the sleeve. The sleeve and stem are inserted into the hole drilled through the two pieces to be joined until the sleeve flange is flush against the outside of the hole. The stem protrudes out of the flanged end of the sleeve. A riveting tool grabs the protruding stem while holding the sleeve in the hole, and then pulls the stem. The pressure on the sleeve created by the pulling force on the stem causes the sleeve to collapse and expand against the surface surrounding the hole. The pieces of material are gripped and held tightly between the flange on one side and the expanded end of the sleeve on the other. During the process of pulling, the stem breaks, or "pops" off inside the sleeve.

pop riveter – a tool designed to install pop rivets; see the definition of pop rivet.

porch – a part of the structure that projects out from the exterior walls to provide a covered entrance to a house or a covered recreation or seating area outside of the main structure.

pore – a microscopic opening, such as a wood cell.

porosity – 1) a weld defect in which there are small holes and voids in the weld filler metal. The holes lessen the weld strength and require changes in the welding process to correct; 2) microscopic and larger holes in a material through which fluids can pass.

porous – a material that has very small holes or pores, through which a fluid may pass. For example, wood has long tube-like cells, which, when cut in half, reveal the tube ends or pores of the wood. Oaks are porous woods with large, often visible, cells.

portable – an object that is easily moved about or transported, such as a circular saw.

portable electric tool cords – portable tools come with a variety of cords types which are selected by the manufacturers to withstand varying degrees of use, voltage and oil-resistance. All these are designed for use in damp locations, they include:

SJ – two, three, or four 18 or 10 gauge conductor wire with a thermoset plastic insulation and outer sleeve, designed for hard use and up to 300 volts.

S – two or more 2 to 18 gauge conductor wire with thermoset plastic insulation and outer covering, intended for portable and hanging fixtures. Can take very hard use and up to 600 volts.

SO – two or more conductors of 2 to 18 gauge with thermoset insulation and an oil-resistant thermoset plastic outer covering for portable and hanging fixtures and extra hard use.

STO – two or more conductors of 2 to 18 gauge with thermoset insulation and an oil-resistant thermoplastic outer covering for portable and hanging fixtures and extra hard use.

portal – an entryway, door or opening.

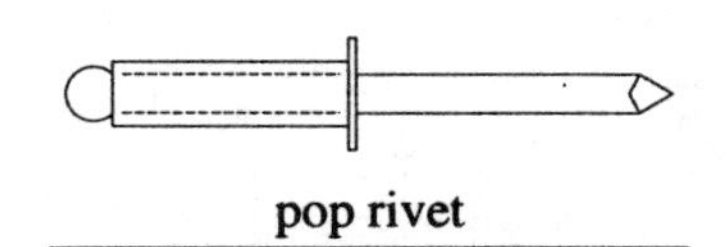

pop rivet

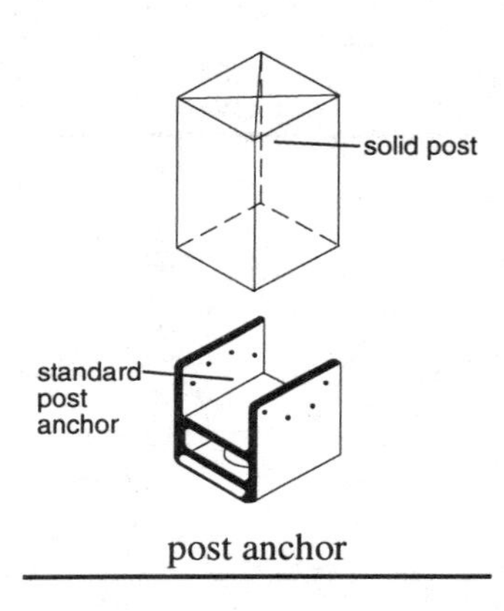

post anchor

portico – 1) a covered porch or walkway, supported by columns, usually extending out from a building; 2) a classical architectural form of covered colonnade or ambulatory at the entrance of a building.

portland blast-furnace slag cement – a type of cement, in accordance with ASTM C595, that has granulated blast furnace slag added to the mixture. Blast furnace slag is nonmetallic, consisting of the silicates and aluminosilicates remaining from iron ore. The addition of slag makes a lightweight concrete, with 25 to 70 percent of the total weight being slag. There are two types, the difference being the addition of an air-entraining compound to one. Lightweight concrete is used for fireproofing, fill and precast concrete.

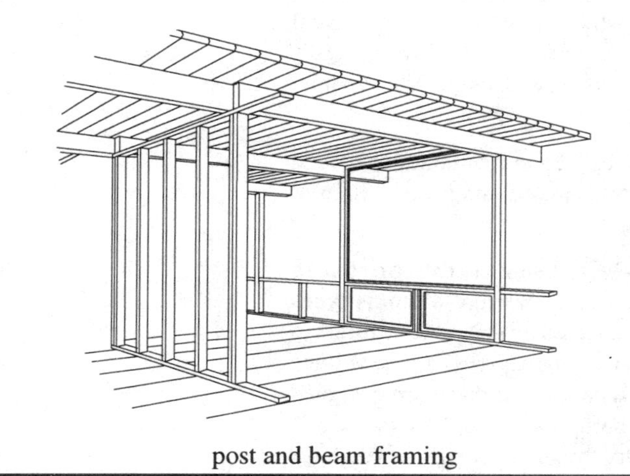
post and beam framing

portland cement – a ground and calcined mixture of shells, limestone, cement rock, clay, marl, shale, sand and iron ore. When mixed with water, it undergoes a chemical reaction resulting in a hard, strong homogeneous structural material. It is used as the base material for mortar and concrete.

portland-pozzolan cement – ASTM C595 cement with pozzolan (siliceous and/or siliceous and aluminous material) added. The pozzolan reacts, in the presence of water, with the calcium hydroxide in the cement to form compounds that bond together, adding strength to the cement. This also reduces the heat and volume change generated during curing. Portland-pozzolan cements are used primarily for dams and bridge piers.

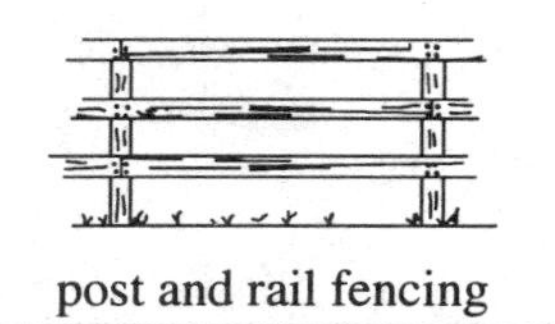
post and rail fencing

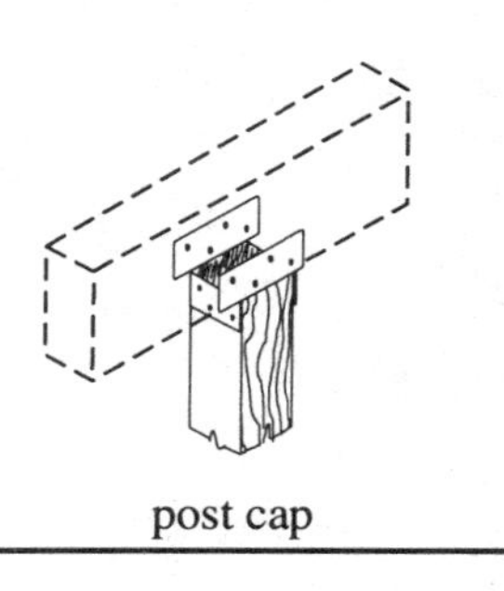
post cap

positioned weld – a weld in a joint that has been made from a favorable position for welding so that the welder has the best opportunity to make a good weld.

positive drainage – a roof design which ensures complete drainage by having the proper slope and the correct drain sizes and types in the proper locations.

post – a slender vertical support member.

post anchor – a preformed metal piece used to anchor the bottom of a vertical post to a foundation.

post and beam framing – a method of framing that uses heavy framing members that can be arranged far apart in order to create large, unobstructed open areas in a building. Heavy planks are used for the floor and roof, sized to accommodate the design load without being overstressed. Sometimes called plank and beam framing or post and girt framing.

post and board fencing – a fence made of horizontal boards spaced evenly between vertical fence posts.

post and girt – see post and beam.

post and rail fencing – fencing made of horizontal split rails between vertical posts.

post cap – sheet metal fasteners shaped to protect posts or to connect the tops of posts to horizontal beams.

postern – a private entry, other than the main entry.

postheat – heat applied to a welded metal joint and the heat-affected zone after the welding, soldering or brazing is completed. The additional heat improves the condition of the metal.

post hole – a hole dug for a fence post.

post-tensioning – a building method in which steel cables are drawn through holes in a concrete slab or other concrete structure and then tension applied to the cables after the concrete has reached a specified strength. The tension load is applied to the concrete by means of plates attached to the cables. Concrete is strongest in compression and this increases the strength of the overall concrete structure by putting it under compression stress.

pot – see potentiometer.

potable water – water that is safe to drink.

potential difference – see electromotive force.

potentiometer – a variable resistor that is used for control, such as the volume control on a radio. It divides the voltage in a way that is proportional to its resistance. It consists, in one form, as a wiper or slider that can be moved across fixed contacts, which varies the resistance. The greater the resistance, the less electricity flows through the potentiometer. To one extreme of the wiper's movement, there is maximum resistance and no electricity flow. At the other extreme, there is no added resistance, and electricity flow is at its maximum for the circuit in which the potentiometer is located.

pot life – the length of time a liquid mixture, such as an epoxy, adhesive or plaster, remains workable and usable before it hardens or cures.

pound – a unit of weight measure equal to 16 ounces avoirdupois measure or 12 ounces troy measure.

pounds per square inch (psi) – pounds per square inch is a measure of pressure, a load applied by one commodity on another. Fluid pressure, which includes both liquid and gas, is measured in terms of either pounds per square inch absolute or pounds per square inch gauge. Atmospheric pressure at sea level is 14.696 psi. See also pounds per square inch absolute and pounds per square inch gauge.

pounds per square inch absolute (psia) – a measure of pressure, such as the air pressure within a tire or the water pressure within a pipe, measured on a pressure gauge, plus the atmospheric pressure, which at sea level is 14.696. So, a pressure of 30 psig is equal to 44.696 psia. See also pounds per square inch and pounds per square in gauge.

pounds per square inch gauge (psig) – fluid pressure, which includes both liquid and gas, measured on a pressure gauge. It does not include atmospheric pressure. See also pounds per square inch and pounds per square inch absolute.

poured concrete wall – see concrete wall, poured.

poured in place – mortar or concrete that is poured and cured at its permanent location and position. Perimeter and slab building foundations are examples of poured-in-place concrete. They are poured in forms at the building site. By contrast, concrete building veneer is precast and brought to the building site for installation. Poured in place is also called cast in place.

powder flame spraying – melting and applying a powdered metal coating using a spray method. Many hard-surfacing metals, such as chrome-cobalt alloys are brittle by themselves, but are effective at preventing surface wear. When these are melted and sprayed onto a softer and more ductile metal, such as a stainless steel, they are well supported, not as brittle, and provide a hard and wear-resistant surface. This technique is often used with items, such as the seating surfaces of valves, that will be used in high temperature service.

powdering – a gradual deterioration of varnish into dust. This can occur when a varnish is exposed to conditions for which it is unsuitable, such as a varnish used on an exterior surface that does not include an ultraviolet protection formulation. When varnish begins to powder, the old varnish must be removed and a new more suitable coating applied.

power – 1) the capacity of an electrical circuit expressed in watts, a watt being equal to the voltage multiplied by current; 2) energy, or the rate at which work is done; 3) strength, or the quality of an object or person to exert force.

power and control cable – electrical cable that has multiple conductors. Used for most electrical applications, such as for power tools and a wide variety of machines.

power buggy – a diesel or gasoline powered cart with a wheelbarrow type bin used to transport wet concrete short distances at

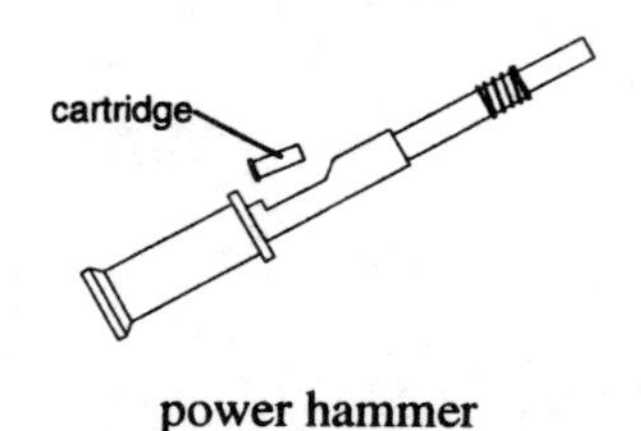

power hammer

a job site. The bin has a rocker attachment which allows steep dumping when handling stiff mixes. They come in varying sizes with the capacity to carry from 9 to 12 cubic feet of concrete mix.

power factor – a measure of electrical efficiency expressed as a ratio of the true electrical power (the amount of power remaining after it is reduced by losses in a circuit) to the theoretical (apparent) power in the circuit. Reactance in an AC electrical circuit can cause the current and voltage to be slightly out of phase. When this occurs, multiplying volts times amps, will not give the true average wattage of the circuit. This value must be multiplied by the power factor to obtain the true wattage, which is a significant factor in electrical circuit design. Without this value, the circuit may be undersized for the intended electrical load.

power float – a gasoline engine or electric motor-powered device used to compact wet concrete. It consists of a number of flat blades tilted at a slight angle to the horizontal and attached at one end to a rotating hub. The hub and blades are turned by the motor and worked across the surface of the uncured concrete to compact it.

power hammer – a gun, powered by blank cartridges, used to drive fasteners into concrete. Each fastener requires a new cartridge to generate the force to drive it. The explosive force, powered by igniting the powder in the cartridge, gives the hammer considerable power and speed. It can be used to drive fasteners in situations where a nail gun could not generate the proper force.

power miter box – a powered circular saw and miter box combination. The saw is mounted with a hinge on a table in such a way that it can be set to cut at various angles, from 45 to 90 degrees. On some units, the saw blades can be tilted at different angles to the horizontal. A power miter box makes a lightweight and very portable substitute for a radial arm saw. It is frequently use to cut trim and molding on the job.

power of attorney – the documented legal authority to act on behalf of another.

power plane – a portable, power-driven plane for smoothing wood, and/or rapidly reducing the size of a piece of wood.

power plant – a plant in which electricity is generated.

power pulling equipment – a powered winch or other means of pulling electrical cable through conduit. The equipment is used to install long runs of conduit.

power ratchet – an air or electric powered, hand-held ratchet for driving a socket to turn a bolt or nut.

power riser diagram – a drawing, made by the electrical circuit designer, that shows the service entrance components. It can be used by the electrical contractor as a reference when installing the service entrance components and wiring.

power saw – any of a variety of saws that are power driven, either by electricity or compressed air. These include such saws as saber saws, circular saws, table saws and radial arm saws.

power screed – a motor-powered screed consisting of two screed boards connected to a framework on which the motor is mounted. The screed boards are worked back and forth across the uncured concrete to level it.

power screwdriver – a battery powered (cordless) or electrically powered screwdriver with interchangeable, replaceable bits. Used primarily for driving fasteners into panels and wallboard, or with drill bit attachments for drilling holes.

power trowel – a gasoline or electric powered device with revolving blades for smoothing and finishing large areas of wet concrete. Troweling is the next step after floating the concrete.

preassembled brick masonry – see prefabricated brick masonry.

precast architectural concrete – a type of precast concrete that is used for finished parts of a building. *See Box.*

precast concrete – concrete that is cast in forms at a location other than its final placement. Precast concrete parts, such as panels, columns and beams, are often fabricated in factories and delivered to the site.

precipitate – 1) a substance caused to separate from a solution or suspension by chemical reaction; 2) to condense from a vapor and fall, such as rain falling from clouds.

precise – accurately conforming to a strict pattern or standard. It implies working to precise tolerances as in finish carpentry or machining parts.

precoat – the application of a coating to a metal surface before soldering or brazing. A flux coat protects the metal from oxidizing during the soldering or brazing process, dissolving existing oxides and acting as a temperature indicator. It melts at a point near the right temperature for the brazing or soldering operation.

pre-engineered steel building – a steel framed building covered with sheet steel. The building parts are manufactured in a factory and shipped in pieces to the site. There they are erected and assembled on a foundation, forming a complete structure.

prefabricated – modules, parts or units made or constructed at a location, often at a factory, apart from the final place of installation. The prefabricated modules or units make assembly of the finished product easier, faster and more uniform.

prefabricated brick masonry – brick masonry sections that are put together at one location and moved to another location for installation. They are reinforced and limited in size to permit handling and retain their structural integrity during transit and placement. Prefabricated brick units reduce the on-site labor time and costs. Also called preassembled, panelized, or sectionalized brick masonry.

preformed ductwork – see ductwork, preformed.

pregrouted ceramic tile sheets – a factory assembly of ceramic tile into sheets that are grouted at the factory with a flexible silicon grout. The tile is fastened to a mesh backing to provide structural integrity during handling. The flexible grout permits handling of the sheets without damage to either the tile or the grout. The installer uses adhesive to bond the sheets to the wall and then applies a silicon grout, of the same type used at the factory, to the joints between the sheets. The sheets are used for large areas such as showers, saving both time and labor costs.

preheat – applying a prescribed amount of heat to a joint before welding, soldering, brazing or thermal spraying in order to condition the metal and improve the quality of the weld to be made.

Precast architectural concrete

Precast concrete was developed to be used as part of the structural framework of a building. However, improved design and manufacturing control techniques have resulted in the ability to produce architectural precast concrete for exposed applications, such as building siding, as well. It adds weather protection and decorative design to the buildings. Factory-made units have the advantage of consistent quality control and design freedom for shape and size variations, including interior finishes and services. There are a number of different precast wall units available for use as either curtain or loadbearing walls, shear walls, wall support units and exterior forms. The curtain wall units are designed to carry only wind loads, while the loadbearing units carry floor and roof loads. Shear wall units transfer lateral or sideways forces. Exterior form units are used for exterior siding and can also be used to form cast-in-place columns and structural walls. The steady supply of factory-made precast units enable builders to erect structures rapidly. They can also be used for such items as fences, planters and screens. Architectural concrete that is cast at the jobsite is known as cast-in-place or tilt-up construction.

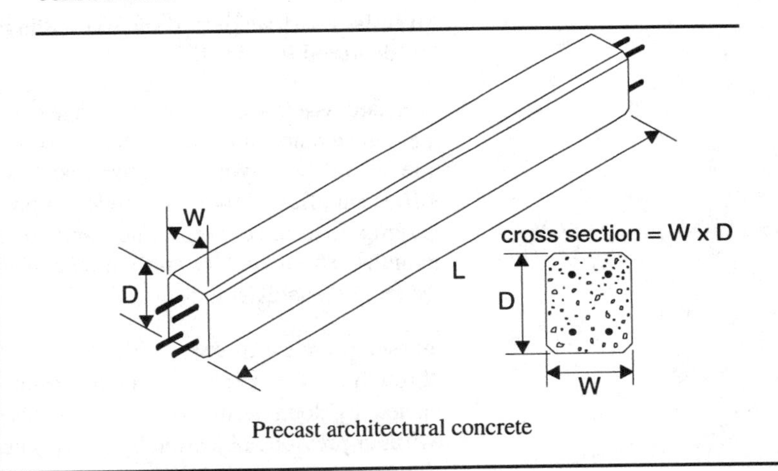

Precast architectural concrete

preheat temperature – the temperature specified for preheating a material before welding, soldering, brazing or thermal spraying. Preheating minimizes the potentially harmful effects of the high heat input on the material at the time of welding. The preheat temperature for carbon steel pipe that is less than one inch thick, for example, is 50 degrees F. Carbon steel pipe that is one inch thick or greater should be preheated to 175 degrees F. before welding.

prehung door – a door, prefabricated with frame, hinges and trim members, that is intended to be installed as a unit. Prehung door units come in two types, both save time on fitup and installation. One type has the door hung on the hinge-side jamb, which is installed in the opening first, followed by the top and latch side jambs. The other type has the door and complete jamb assembly installed as a single unit. Both are commonly used.

premises – a tract of land and its buildings as identified in a deed.

prepaid assets – expenses that have been paid in advance of the due date. An example would be buying inventory, such as office supplies, before it is needed or prepaying insurance premiums for future months. The inventory and the insurance become prepaid assets.

prescriptive easement – a right to pass through another person's property established by long term precedent and use without previous objection by the owner. When an owner allows passage through his property over time without objection, common law may consider that a right to continue such passage has been established.

preservative – a chemical used to inhibit rot, deterioration or infestation of wood products or other materials. Many copper-bearing compounds are used as preservatives to paint or soak wood that is exposed to conditions that may cause rot or insect infestation. Copper napthenate and pentachlorophenol are common preservative compounds.

pressed brick – a brick molded under pressure for uniformity and homogeneity of the product. The resulting brick is strong because it is of the same density throughout.

pressure – the force exerted by one commodity or object on another. For example, the weight of a vehicle exerts pressure on the ground or pavement at the point of contact, and compressed gas in a container exerts pressure on the walls of the container.

pressure drop – a decrease in pressure from one point to another due to friction losses in a fluid system, such as a water system.

pressure gas welding (PGW) – an oxy-fuel welding process that heats the entire surface to be welded and then fuses the parts under pressure without filler metal. This process can be used in automated welding operations. The parts to be welded are heated to the proper temperature and then forced together with sufficient pressure to make a good weld.

pressure gauge – a device for measuring the amount of pressure exerted by a fluid, which includes both liquids and gases. Pressure gauges have many applications. They are essential in systems that have dynamic operating characteristics and the constantly changing conditions need to be monitored by the system operator. They are also useful tools for making system adjustments, such as to the water system pressure regulator setting for a building supply.

pressure loss – see pressure drop.

pressure rating – the non-shock pressure at which a component or system may be operated continuously. In a fluid system containing a noncompressible fluid, such as water, a sudden pressure shock, as from a check valve closing suddenly, sends a pressure wave through the system. The predictability of pressure shocks in a system is imprecise. The pressure rating of a component is given as a "non-shock" rating because each shock is different and shocks vary with every system. The system designer evaluates the potential for pressure shocks, estimates their probable magnitude and then selects components with a pressure rating adequate for the anticipated operating conditions. A Class 150 pound rated steel flange may be acceptable for use at 275 psig at 100 degrees F, but if system pressure surges that exceed 275 psig are predicted, even for a short period of time, a higher rated flange would be needed. In that case, the designer would select a Class 300 pound rated flange.

pressure rating, working – see pressure rating.

pressure-reducing valve – a control valve that can be set to reduce pressure between its inlet and outlet, and maintain the set outlet pressure with fluctuations in inlet pressure. A common use for a pressure-reducing valve is in the water system entrance to a house or other building. If the water main pressure is 150 psig, it is too high for sprinkler system valves, washing machines or other household uses, most of which are designed for 40 psig. The pressure-reducing valve will decrease the 150 psig main pressure to 40 psig.

pressure-treated wood – wood that has had a preservative fluid forced into it under pressure. Because of the pressure, the fluid goes deep into the wood, protecting it against rot and insect attack.

pressure washing – cleaning mortar with a stream of water that is pressurized to between 200 and 600 psig. This is a labor saving method of cleaning a masonry wall to improve the appearance of the wall or prepare it for painting.

pressurize – to increase a fluid pressure internal to a container, such as pumping a hydraulic jack to increase the pressure on the hydraulic fluid in the jack.

prestressed concrete – see pretensioning.

prestressing – see pretensioning.

pretensioning – stretching steel cables or wires, used in prestressed concrete, in the forms to put tension on them before the concrete is placed. The tension is kept on them until the concrete has set up. After the concrete hardens the steel remains in tension, adding to the strength of the concrete.

prick punch – a punch that tapers to a slender, pointed shank. It is used to make small indentations along the length of a line scribed on a piece of metal. The indentations make the line more visible.

prima facie – at first look (Latin); a fact that appears self-evident or legally sufficient unless proved otherwise. For example, if a contractor named Jones was driving a truck that had "Jones Contracting" written on the side, it would appear that the truck belonged to Jones. That would be prima facie evidence that the truck belongs to Jones.

primary coil – the first coil in a transformer through which the electrical current passes.

prime contractor – the contractor, contracted directly by the client in writing, who is responsible for overseeing the work to be performed by his employees and any subcontractors that he employs to complete the job.

primer – 1) a paint especially formulated to seal a surface and provide a base for subsequent coats of paint or other finish materials. Primers dry with little or no gloss; 2) an explosive device used to set off a larger explosive charge. A primer is more sensitive and easier to ignite than the secondary charge, so for safety, it is usually very small. A primer requires very careful handling. A small amount of a primer compound, such as fulminate of mercury, is placed in a capsule in intimate contact with dynamite. The primer can be easily ignited from a distant source, such as an electric current that is conducted through wires from a remote location. The ignited primer will then cause the dynamite to explode.

principal – 1) one who authorizes another to act for him and from whom this agent's authority derives. For example, a principal may hire an architect and give him power to act in his behalf in the design and building of a project; 2) a capital sum of money to be returned with interest, as in a loan or investment; 3) first, main, or highest in order.

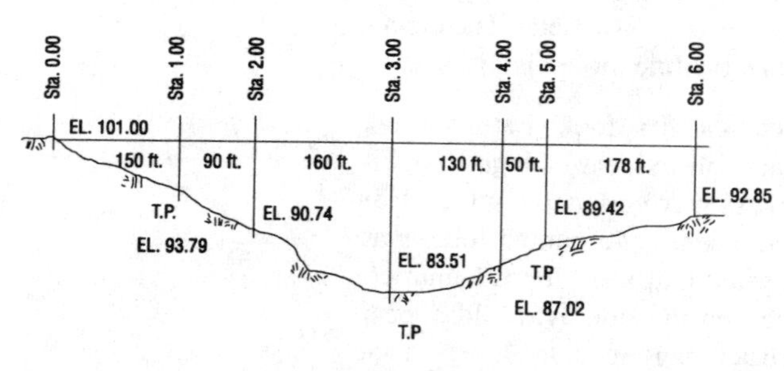

profile leveling

print-free – a term indicating that paint is sufficiently dry so that it won't be marked or damaged by touching.

prism – 1) a transparent solid, with parallel corners, that will break a beam of visible light into the spectrum of colors; 2) a small scale replica of a structure, using the masonry units and mortar to be used in the full-scale structure, built to predict the strength of the full-scale structure.

private property – property held in private ownership.

private sewer – the sewer, from a building to the public sewer, belonging to the property in which it is installed. Also called a building sewer.

procedure – a routine or operating system set up as a means of producing a specific end result, such as a welding procedure which defines the correct method of producing a particular type of weld. It would include all the variables that affect the outcome of the weld, including the welding current setting, filler materials, etc.

procedure qualification – a demonstration intended to prove a procedure suitable for its intended purpose. An example would be a welding procedure qualification which proves that the variables controlled by a welding procedure result in a specific weld suitable for a specific application or condition. Each welding procedure must be qualified to be valid and acceptable for use.

procedure qualification record (PQR) – a record indicating that a welding procedure has been qualified as suitable for specific applications or conditions. The record is kept by the contractor or welder using the welding procedure.

process time – the time required to perform a task or operation to completion. For example, the process time for a concrete job includes the concrete placement, finishing and curing time.

Proctor test – a test, usually performed by a soils engineer, to determine the optimum moisture content of soil for proper compaction. A sample of soil from the site is dried in an oven and weighed. Then small amounts of water are added gradually and the soil compressed. The weight to volume ratio is measured each time water is added. This value increases as water is added, up to a point. When too much water is added, the weight to volume ratio begins to decrease. The amount of water needed to obtain maximum soil density is determined by this method.

procurement – the function of purchasing materials or services. For example, a general contractor may procure materials and the services of subcontractors for the customer. This may include obtaining and evaluating competitive bids, placing and administering a purchase order, inspection during fabrication and/or testing, arranging for transportation, expediting orders, and receipt inspection at the building site.

profile – a shape in outline, such as a cross section of a piece of molding.

profile duplicator – a series of small diameter rods held snugly at their centers in a handle. The rods slide in and out to follow the shape of virtually any profile so that the shape can be transferred to another piece of work. It is used most often to mark the shape and size of vertical molding around a doorway. The shape is then transferred to vinyl flooring so that the outline of the molding can be precisely trimmed out of the vinyl before it is laid.

profile leveling – to survey a parcel of land and determine elevations at a number of different points in order to map out the contours of the land.

profile machine – a self-powered machine that can mill the surface off an asphalt roadway to prepare it for resurfacing. Also called a profiler.

profile view – see section view.

profilograph – a straightedge that is mounted on a wheeled carriage, used for checking the profile of a roadway during paving. An inspector compares the paved surface to the straightedge to identify areas that need more smoothing or filling.

profit – 1) something gained; 2) the excess of returns over expenditures in a transaction or series of business transactions, as in the net income of a business; 3) the return on an investment.

progress chart – a graphic visual indication of the state of completion of a task or project.

projected centerline grade – the planned grade down the center of a roadway to be built. Roads are designed to be higher at the center, or to peak, so that water will run off to the sides of the roadway.

project evaluation and review technique (PERT) – a chart used as a job scheduling and management tool. The work to be done is identified and analyzed, and then broken down into individual tasks, with an estimate of the time required and cost for each task. The tasks are then arranged in sequence, showing parallel tasks, with arrows indicating the flow of work from start to finish. A time line with calendar dates is shown along the top of the chart. Three completion times are estimated, the best possible time, the expected average time and a worst case time. These times are used to determine a "normal" time for each required task. One method of doing this is to add the best and worst time and four times the average time, then divide the total by six to obtain a weighted average. The chart clearly shows the processes and sequences that must be completed separately or at parallel times. In this way it can be an effective tool for planning and scheduling crews and material orders.

profile machine

projection – 1) the distance that an object projects beyond a plane or surface; 2) the horizontal distance that the eave projects beyond the exterior wall or face of a building.

projection welding (RPW)– an electric resistance welding process which uses projections built up on the parts being joined to make simultaneous localized welds with heat produced by electric resistance heating. Projection welding enables pieces to be welded to one another at several points at one time. This is a common method of attaching spot welded brackets to a surface. The bracket mounting plate is made with several projections. The welding cable lead is attached to the bracket and the grounding wire to the other sur-

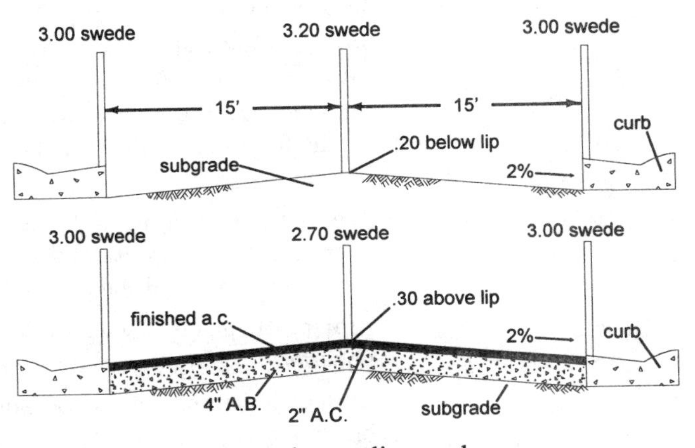

projected centerline grade

pry bar

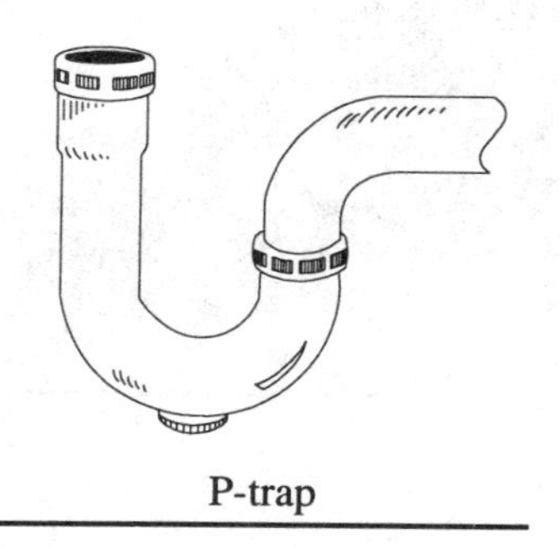
P-trap

face. When the two parts are brought into contact, each projection on the bracket is welded to the other surface by the electric current-generated arc between the bracket projections and the surface to which it is being welded.

promissory note – a written promise to pay a specified sum of money on a fixed date, or over a period of time, to an individual or lender.

proof – evidence that compels acceptance of a truth or fact.

propane – a flammable natural gas used as fuel in devices such as soldering torches.

propane torch – a hand-held torch consisting of a burner nozzle screwed on to a bottle containing pressurized propane gas. The gas is ignited and the nozzle used to direct and control the burning gas for preheating, soldering, brazing and other applications that require moderate temperatures. Propane torches are used by plumbers to solder copper tubing joints.

propeller – a device with shaped blades radiating from a rotating hub. It is designed to move fluid axially as the hub and blades rotate, propelling the object it is attached to either forward or backward.

propeller fan – a fan that uses propeller blades rather than a squirrel cage rotor to move air.

property – 1) something owned or possessed; 2) real estate.

property line – the boundary of a specified piece of land.

proportional – sized with relation to other adjacent or nearby objects, such as the windows in a house must be proportional to the walls in which they are located.

proposal – 1) making a suggestion for consideration; 2) an offering to perform work under specified conditions. *See Box.*

protect – to guard or shield from damage or harm.

protected membrane – a roofing membrane with a protective surfacing on top. For example, 90-pound roofing felt intended as the exposed layer in a built-up roofing system is protected with fine gravel.

protected membrane roof assembly (PMRA) – a design in which ballast, such as gravel, and insulation are applied over a roof membrane.

protective clothing – clothing that is designed to minimize or prevent injury while performing a task, such as welder's goggles, aprons, steel-toed shoes or hard hats.

protective coating – a coating, such as a paint, designed to provide protection against environmental elements, such as rain, sun or pests.

protective life – the effective life of a protective coating, such as the effective life of exterior latex paint applied to wood siding, which is approximately ten years.

protractor – a gauge or tool used to measure angles. It is circular or semicircular, with degrees marked off around the edge.

protractor and depth gauge – a graduated ruler that can be slid through a pivot on a protractor so that the actual depth and angle of a hole can be measured using the ruler.

proximate – 1) near or very close; 2) preceding or immediately following.

pry bar – a steel bar with one or both ends flattened, curved and notched. One end may be curved more than the other for wedging between materials and pulling them apart. Pry bars are used for prying and pulling nails or other materials or fasteners. Also called a crowbar.

P-trap – a plumbing device which maintains a water seal in a drain line to prevent sewer gas from entering the building through the line.

public address system (PA) – a loudspeaker system used to broadcast announcements throughout a large area, such as a stadium or park.

public sewer – main sewer system to which private sewers can be connected. Also called municipal sewer.

Proposal

A proposal to perform work may include several items, depending on the size and complexity of the job and the customer's requirements. Some customers require that the contractor's qualification to perform the work be included, along with a detailed list of similar jobs that have been performed within recent years, descriptions of those jobs, and the names and addresses of the respective customers. They may also request information on key subcontractor personnel and their qualifications to perform the work.

The job being proposed may be described in differing levels of detail, depending on the type of job and the client's objectives and preferences. It may be as simple as stating that a building will be constructed for some particular purpose in accordance with the drawings and specifications. Or, it may have detailed design criteria to be met, with production of the requisite drawings and specifications to be part of the proposed job to be performed.

The details of a proposal must be tailored to fit the situation, so each may be different, even for similar types of jobs. Intermediate milestones or completion points may be defined, such as foundation completion, rough framing, etc. with a payment schedule and an estimated date for each of those points. Or, it may simply have a projected completion date as part of the proposal. The proposal should always be in sufficient detail to clearly define, either directly or in terms of referenced documents, the work to be performed and the costs associated with the job.

Proposal and Contract

For Residential Building Construction and Alteration

Date AUGUST 7 19 94

To MR. SAM JONES

Dear Sir:

We propose to furnish all material and perform all labor necessary to complete the following: A TWO-CAR GARAGE ADDITION ADJOINING THE FRONT OF THE EXISTING GARAGE AND CONVERSION OF THE EXISTING GARAGE TO A RECREATION ROOM IN ACCORDANCE WITH DRAWINGS SJ-1 THROUGH S-J-6, ATTACHED.

Job Location:

All of the above work to be completed in a substantial and workmanlike manner according to the drawings, job specifications, and terms and conditions on the back of this form for the sum of

Dollars ($ 83,800.00)

Payments to be made as the work progresses as follows: IN ACCORDANCE WITH THE ATTACHED CONSTRUCTION MILESTONE SCHEDULE;

the entire amount of the contract to be paid within 30 days after substantial completion and acceptance by the owner. The price quoted is for immediate acceptance only. Delay in acceptance will require a verification of prevailing labor and material costs. This offer becomes a contract upon acceptance by contractor but shall be null and void if not executed within 5 days from the date above.

By I.M. BILDER

"YOU, THE BUYER, MAY CANCEL THIS TRANSACTION AT ANY TIME PRIOR TO MIDNIGHT OF THE THIRD BUSINESS DAY AFTER THE DATE OF THIS TRANSACTION. SEE THE ATTACHED NOTICE OF CANCELLATION FORM FOR AN EXPLANATION OF THIS RIGHT."

You are hereby authorized to furnish all materials and labor required to complete the work according to the drawings, job specifications, and terms and conditions on the back of this proposal, for which we agree to pay the amounts itemized above

Owner ____________

Owner ____________ **Date** ____________

Accepted by Contractor ____________ **Date** ____________

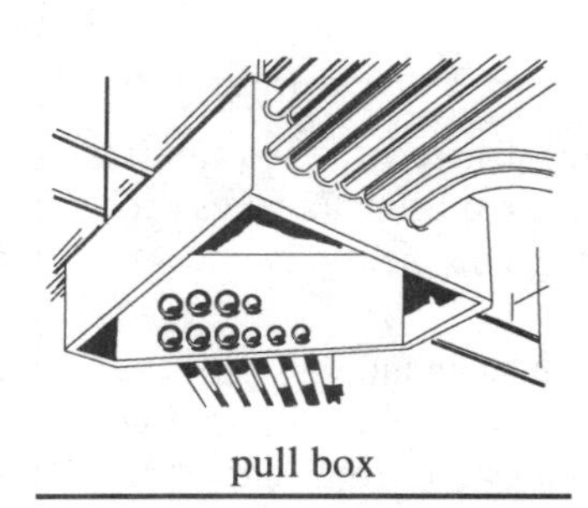
pull box

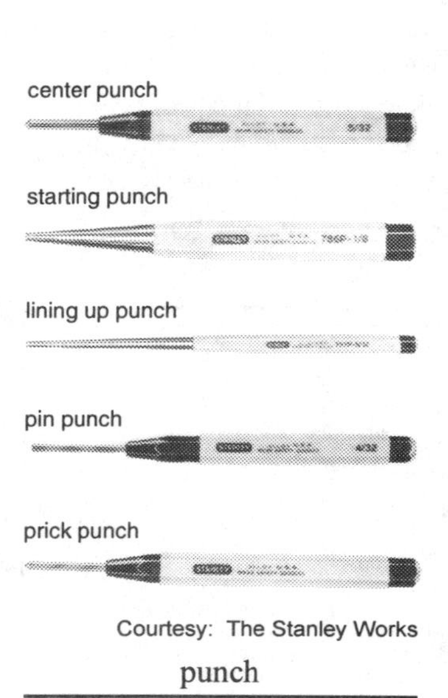

Courtesy: The Stanley Works
punch

puddle – a molten weld on metal, created when heat is applied during the welding process. The puddle is where the metals fuse together.

puddling – 1) a soil compaction method in which loose backfill is flooded with water to saturate it, then allowed to dry and settle; 2) a paint defect where the painted surface looks very wet in spots. It is caused by applying the paint too thick.

pugging – a coarse mortar used to fill cavities in masonry walls in order to deaden sound transmission.

pull box – a box installed at a corner or other junction where two conduits meet, especially at a sharp angle to one another. The box has a removable cover which provides access to the conduit runs. The cable can be pulled straight through one conduit into the box, and then reinserted and pulled through the next section of conduit. The box supports the conduits and relieves much of the strain on the wires, making the pulling, change of direction and connection easier. Long runs of wire should not be made in one pull, so pull boxes are also installed on long runs where support may be beneficial.

pulley – a rotating wheel with a groove around its rim into which a belt or cable fits. It can be used to transmit power or lift and move heavy loads.

pull switch – an electrical fixture that has a pull cord to operate the switch, turning the fixture off and on. Used most commonly for overhead fixtures, such as basement or closet lights, where the cost of running wire through the wall or conduit and putting in a wall switch is not considered a necessary expense.

pulsed power welding – an electric arc welding process using programmed short pulses of electrical power from a computer controlled welding machine.

pultrusion – pulling an object through a die to shape and size it. Many fiberglass materials are shaped by the pultrusion process.

pulverize – to break into small particles by crushing, grinding or beating. Gravel is a commodity that is pulverized to a desired size for specialized uses.

pumice – a gritty volcanic material used as an abrasive, usually in a powdered form, for smoothing and polishing.

pump – a device that raises, moves or compresses fluids by pressure or suction.

pumping – the rippling movement of water-saturated soil as heavy equipment rolls over it. Pumping indicates the instability of the soil.

punch – a cylindrical metal tool with one end tapered to a point or near point that is used to make holes or indentations in a piece of work. The punch point is placed against the work and the other end is struck with a hammer. Punches are used to drive a nail head flush with or below the surface of the work, to mark and align holes, or for other similar applications. Specialty punches are designed for particular tasks.

punching shear resistance – the ability of a roofing membrane to resist damage from normal expected concentrated loads, such as foot traffic.

puncture – 1) to pierce or penetrate the integrity of a membrane or other seal; 2) a hole, wound, or penetration made by a pointed instrument or object.

puncture resistance – the ability of a material to withstand localized forces which might otherwise puncture it. This quality is important if a puncture would compromise the ability of the material to perform its intended function, as with conduit, pipe and waterproofing membranes. Puncture resistance is determined by measuring the force required to puncture the material with a standardized instrument. The instrument will vary with the material being tested. Testing is done by the manufacturers and/or by independent testing laboratories.

purchase order – an authorization to supply materials and/or services at a specified rate. A purchase order may be one page or many pages, depending on the quantities being ordered and the nature of the items being purchased.

purge – to use an inert gas or other fluid under pressure to displace fluids within a system or area. Nitrogen can be used to purge gas lines containing flammable or explosive gases prior to working on the lines. Purging the lines and filling them with an inert gas removes potential hazards for the workers.

purlin – a horizontal structural member used to support rafters between rafter end points. A purlin runs at right angles to the rafters.

push drill – a drill that looks like a screwdriver but works by pushing the handle in the direction of the work. There are steep helical grooves in the shank and pawls in the handle that ride in the grooves. As the handle is pushed toward the work, the pawls force the shank and drill bit to turn. A return spring returns the handle to its starting position. Used to drill small holes, especially for drilling holes through a hinge or other hardware while it is held in place.

push stick – a stick used to push work through a power saw, shaper or other power driven equipment, so that the operator's hands are kept away from blades. It is a safety device designed to prevent injury.

push welding – see projection welding.

putty – a pliable sealant made from a variety of different materials. The choice of sealant material depends on the intended use and temperatures to which it will be exposed. Several plastic and synthetic materials are used, many of which are formulas owned by the manufacturer who produces a particular type of putty. Putties are used to seal woodwork, plumbing, HVAC ducting, to retain glass in windows and doors, and many other functions.

putty coat – a smooth final coat of plaster applied to the surface of a wall or ceiling.

putty knife – see drywall knife.

PVC pipe – see piping, PVC.

pylon – 1) a massive gateway or a monument at the entrance or approach to a bridge; 2) a tower or post supporting electrical power lines or other overhead electrical equipment.

pyramid – a multi-sided solid shape with triangular sides that converge at a point at the top.

pyroxylin – see cellulose nitrate.

Pythagorean theorem – the square of the hypotenuse (c) of a right triangle is equal to the sum of the squares of the other two sides (a and b), or $c^2 = a^2 + b^2$.

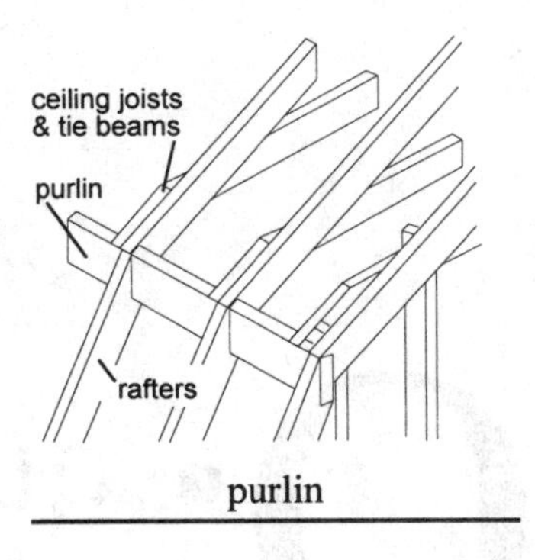

purlin

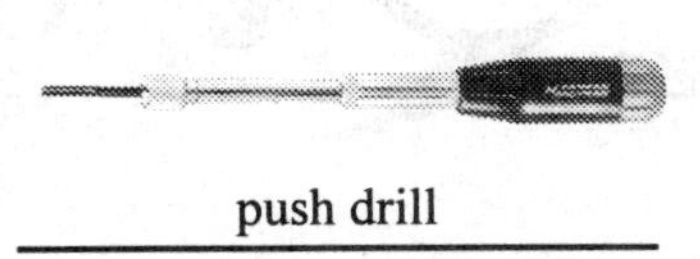
push drill

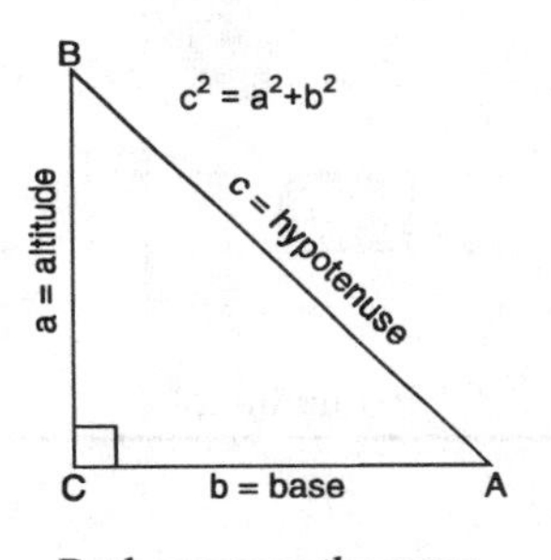

Pythagorean theorem

Q: quadrangle to quoin

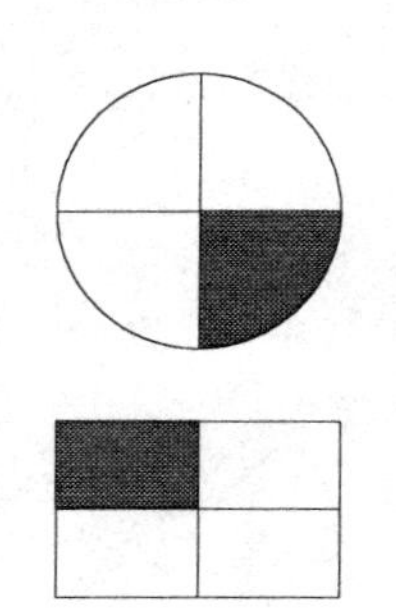
quadrant

quarter round

quadrangle – a four-sided open courtyard, usually surrounded by buildings.

quadrant – 1) one of four equal quarters created by dividing an object or area with lines that intersect at right angles; 2) one-fourth of a circle.

quadrilateral – any figure defined by four lines.

quality assurance (QA) – a program of actions and procedures that are implemented to achieve and maintain a desired level of quality. *See Box.*

quality control (QC) – a system designed to ensure that a desired level of quality is maintained in a product being produced under a defined standard. The process involves periodic inspection of the manufacturing procedures, materials, tools and instruments used, as well as the products produced, to ensure that minimum standards are being met.

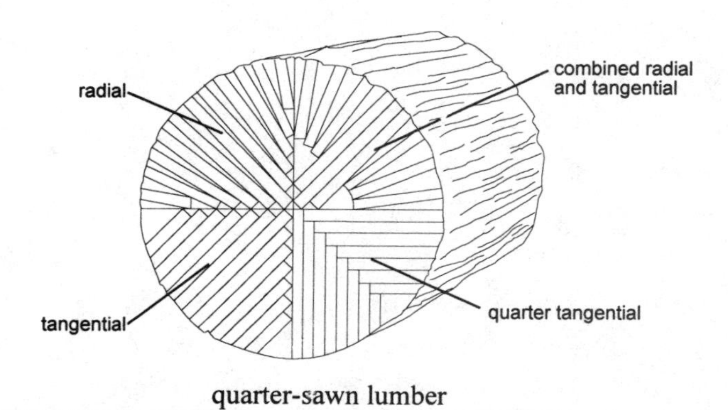

quarter-sawn lumber

quantity – 1) a measured or estimated amount; 2) the total amount or number.

quantity take-off – a listing of the total number of items needed to complete a building project as shown on a drawing or blueprint. For example, if the drawing is of a piping system, the quantity take-off would be a list of all of the pipe, fittings, valves, and supports needed to completely assemble that piping system.

quarrel – a diamond-shaped pane of glass used in a leaded window.

quarry – 1) a deposit of rock strata from which stone, slate, limestone, marble or other rock pieces can be removed for use in building; (2) to break up and shape rock pieces or sections.

quarter crown – the area of a roadway between the centerline and the edge of the roadway.

quarter round – a trim molding that is one quarter of a circle in cross section.

quarter sawn – wood that has been cut from a log quartered lengthwise and at an angle of at least 45 degrees between the board face and the annular growth rings. This method of cutting lessens the warpage and shrinkage of the lumber.

quarter section – an area of land equal to 160 acres that is ½ mile on each side, or a quarter of a square mile.

quatrefoil – an ornamental figure, often a flower or group of leaves, with four equal lobes connected at a common center point. Used as exterior building decorations; not often seen today.

queen closure – a brick that is split vertically along its length, so that it is the full length of a brick, but only 2 inches wide rather than the normal 4 inches wide. It is used to create an "English" corner in a brick wall. The queen closure is placed next to the last brick in the row.

queen post – one of two secondary vertical tie posts in a truss.

quenching – rapid cooling of metal using water, oil or cold air.

quick assets ratio – the ratio of current assets without inventories (liquid assets) to current liabilities. It is a balance sheet or financial ratio used to quickly ascertain the financial health of a business. It is a measure of a business' ability to pay current debts. The ratio is considered acceptable if it is 1:1.

quick drying – a paint or other coating that dries rapidly. Solvent-based paints dry by oxidation or polymerization. Driers, which act as oxidation or polymerization agents, are often added to oil-base paint to speed up the drying process. Such driers are soaps that are the result of a reaction between a metal oxide, such as cobalt, manganese, zinc or calcium, and an organic acid, such as petroleum-based naphthenic acid. Driers are soluble in the paint vehicle.

quicklime – calcium oxide (CaO), a caustic lime that is mixed with water to make a fine putty-like substance used in finish coats, such as plaster. See also lime.

quick set – a fast-setting drywall joint compound that has chemical accelerators added to it to speed the reaction and drying time. It is used under conditions when normal drying would be too slow, such as in cold, damp weather, or in situations where it is necessary to complete taping and finishing a drywall job as quickly as possible. Fast-setting drywall compound is used only as necessary because too much accelerator can shorten the working life of the compound to an impractical amount, as well as cause shrinkage and cracking problems.

quick setting asphalt cement – a roof cement, made with a fast-evaporating solvent, used in cold weather or when it is necessary for the cement to set up quickly.

quick valve – a valve that opens or closes with a quarter turn (90 degree rotation) of the stem. *See Box.*

quiet switch– an electrical switch, such as a mercury switch, that makes no noise when moved from one position to another. One type of switch consists of a glass bulb with wires running into both ends, but not connected, and a small amount of mercury. The mercury is electrically conductive. When the bulb is tilted into a horizontal position, the mercury connects the two wires, making a path for the electricity to flow. When the bulb is in a

Quality assurance

A program intended to guarantee a standard level of quality, beginning with the design and following through the manufacturing of the product or the completion of the service to be performed. The program establishes, as a goal, a set minimum acceptable quality level for all products produced or services rendered. It is then put into operation to ensure that the goal is reached or bettered. This is done through training, observation and evaluation of work or manufacturing procedures, and the implementing of controls and improvements as necessary to enable the workers to meet these goals. Good supervision and product monitoring enable flaws to be detected and negative variations in quality to be corrected before the work is completed. Periodic evaluation of the quality program results also helps in identifying and correcting problems within the program. To be meaningful, the quality assurance program must be separate from the production arm of the company, reporting directly to senior management. There was a time when formal quality assurance programs were limited to a few industries, such as aerospace and nuclear power plants. However, with the growing number of clients insisting that design and engineering/construction firms have ISO (International Standards Organization) certification, formal quality assurance programs are becoming increasingly widespread. Certification requires demonstration that such a program is in effect.

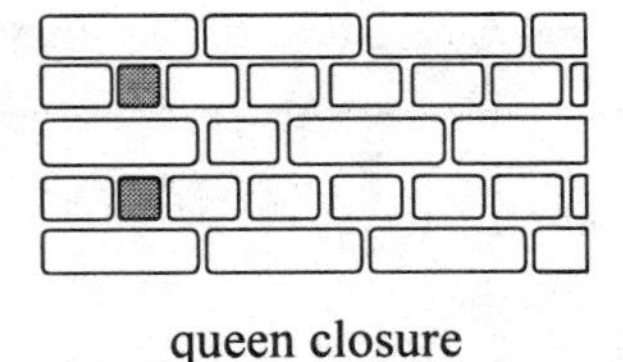
queen closure

Quick valve

A quick valve is a valve that can be fully opened or closed with just a quarter turn of the stem. These valves are available in a wide range of materials, configurations, and pressure ratings for different service applications. Common varieties of quick valves are plug valves, ball valves and butterfly valves. Some types are totally lined with materials that can withstand harsh chemical environments. The actuating mechanism on smaller valves is usually a lever, sometimes furnished with detents to hold the valve in intermediate positions between open and closed. Larger- and/or higher-pressure valves require too much torque to open or close for a directly-coupled lever, and require an intermediate gear reduction system, often with a handwheel, to generate the requisite torque. These tend to become very expensive. Besides manual operation, electric motor, pneumatic, or hydraulic remotely-controlled actuators may be used with this type of valve, sometimes with gear reduction where required by the valve operating torque. This type of valve, when designed for the application, can be used for throttling or flow control as well as for on-off service.

A quick valve is frequently used in gas and other fuel lines, as well as in many other services. The ability to close them quickly is an asset in many systems, in the event of changing conditions or an emergency situation. Most designs have the added benefit of very good flow characteristics; that is, they do not represent much of a flow restriction in the line. This feature results in a lower pressure drop, and sometimes a smaller diameter pipe can be used, which is a cost-saving feature. Some designs do not have any pockets that can trap fluid or particles, making them ideal for situations where the fluid contains particles or is radioactive, as there will be no collection of undesirable products in the valve. In the case of a radioactive fluid system, this prevents the valve from becoming a hot spot. Some quick valves are made with metal seating surfaces for high-temperature applications, and may be the best choice for very high temperatures. These are usually quite expensive. Most quick valves, however, are the soft-seated type, designed for temperatures of 400 degrees F or less.

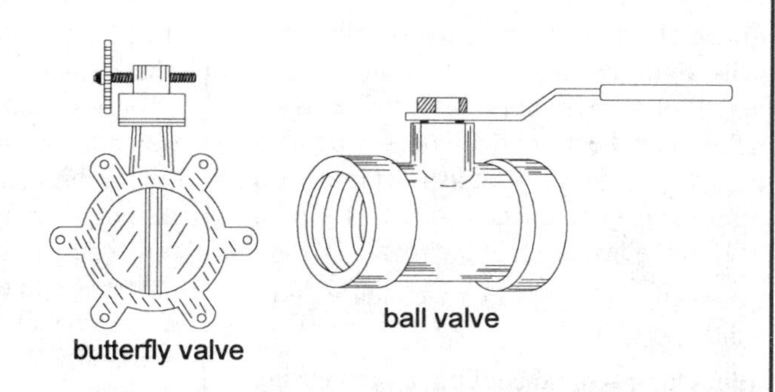

vertical or near vertical position, the mercury touches only one of the wires, breaking the circuit. The bulb is positioned by the switch toggle.

quitclaim deed– a deed releasing one person's rights, title and interest in a property to another without warranty of title.

quoin– 1) a building corner material, such as a block, that is different from the material used for the rest of the walls; 2) a masonry unit that is finished on both the face and the end; 3) a right-angle masonry corner that projects from the wall; 4) a building cornerstone; 5) an arch keystone.

R: rabbet to R-value

rabbet – an L-shaped cutout at an edge or along the side of a piece of material. These grooves are useful in joining materials so that the pieces joined are flush, such as fitting a back on a bookcase so that it is flush with the sides of the bookcase. Also called a rebate.

rabbet plane – a hand-held wood-cutting tool with a blade designed to cut an L-shaped groove along the edge of a piece of material.

race – 1) a ring or groove in which ball bearings are retained; a ball bearing often comprises an inner and an outer race. The balls fit between the races, with the races held by the two surfaces which rotate in relation to one another. The balls provide a low-friction bearing. For example, the outer race may be held in a vehicle wheel while the inner race may fit on the axle shaft. When the wheel turns, the inner and outer races rotate in relation to each other, riding on the bearing balls; 2) a strong current in a natural body of water, such as a stream.

raceway – 1) an enclosed channel, usually made of metal, designed to hold electrical wiring or cables installed in a building. They form a strong, protective, usually fire-resistant, means of routing wiring. Raceways are required where wiring may be exposed to physical damage; 2) a chute or other form designed to guide the flow of fluids or objects, such as a mill raceway which directs water to the water wheel.

raceway, one-piece – a channel for electrical wiring shaped to resemble decorative wood molding. It is designed to blend with the surface where it is installed.

raceway, overfloor – a channel for electrical wiring designed to be mounted on a floor. It has gently sloping sides that minimize the chance of someone tripping over it.

raceway, two-piece – a galvanized steel channel, 3/4 inch high by 1 1/4 inches wide, used for surface run electrical wiring. It has an open, U-shaped channel which holds the wiring, and an inverted U-shaped top channel which fits over the first channel to enclose the wiring.

rack – 1) a framework designed for temporarily holding or permanently storing materials or products. Some racks are manufactured to protect particular tools or electronic gear, or to store tools in relation to one another; 2) a flat bar with gear teeth on one side. The teeth are cut at 90 degrees to the longitudinal axis of the bar. It is part of a gear set used in conjunction with a pinion gear that engages the gear teeth on the rack. As the rack or bar moves, the pinion gear rotates, or vice versa. Such gear sets are used in some vehicle steering gear boxes.

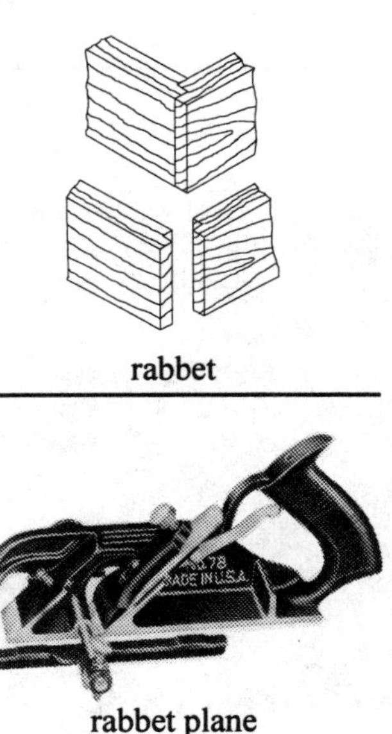
rabbet

rabbet plane

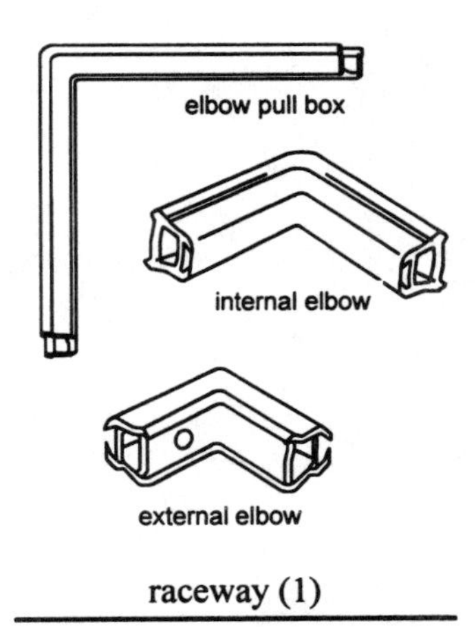

raceway (1)

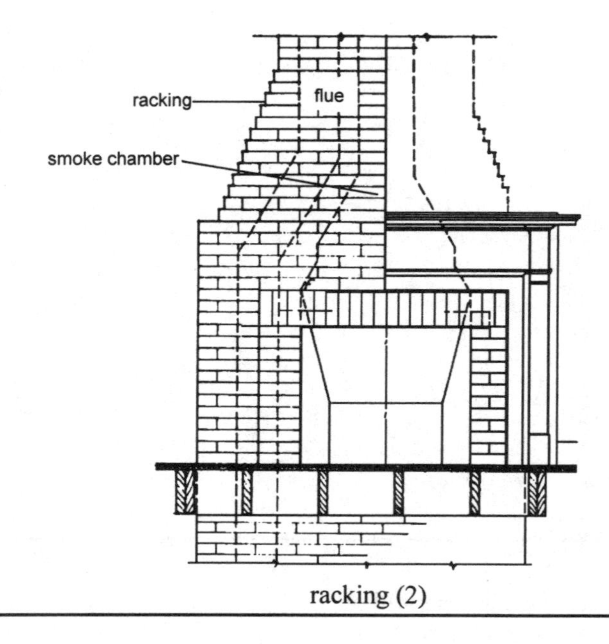

racking (2)

radial blade fan

racking – 1) a sideward force tending to move a structure out of plumb or out of parallel. This could be caused by high winds, an earthquake, ground settling or similar phenomena; 2) the angle transition used to join the different widths of a chimney and fireplace. This transition makes the structure gradually smaller as it rises from the hearth to the roof; 3) a method of installing masonry so that each successive course is stepped back from the preceding course.

radial – moving outward in a straight line from the center of a circle.

radial-arm saw – a circular power saw mounted on an arm above the saw table. The arm can be raised or lowered or swung about a pivot point so that the blade can make cuts at various angles or be tilted with respect to the work. The blade and motor assembly slide in tracks along the length of the pivoting arm.

radial blade fan – a type of axial flow fan with blades that extend out like arms from the fan hub.

radial drill – a drill press with a radial arm that can swing to various positions on the work table. The radial arm is pinned at one end so that it can be pivoted about a central point.

radial ducting – a heating, ventilating or air conditioning system that has ducting running outward from a centrally-located heating/cooling source.

radial surface – the surface of a log that has been cut lengthwise down its center.

radian – an arc of a circle equal in length to the radius of the circle.

radiant floor heating – a heating system in which pipes carrying hot water, or electric wiring, are embedded in a concrete slab floor. The heat from the hot water or wiring radiates upward, warming the living space.

radiant heat – a method of supplying heat by transmitting heat energy created by hot water pipes, steam pipes or electric coils installed in the floors, walls or ceilings of a building.

radiant heater – an appliance that generates heat by radiating the heat toward the object or person being heated. It is a heat transfer, and can be accomplished by conducting the heat through a material (such as heated coils in a floor), by convection (in which heated gas transfers the heat), or by radiating the infrared wave from the source to the object (such as an electric resistance wire heater in which the wire glows red and gives off heat). Almost all heating sources use radiant heat, including a fireplace.

radiation – the transmission of energy by particles or waves, such as the heat transfer from a fire to a person close to the fire.

radiator – a device for heating or cooling which utilizes a source, such as hot or cold water or steam, to heat or cool the air around it.

radiographic examination (RT) – a type of nondestructive examination which uses penetrating radiation and photo-sensitive film to perform a volumetric examination of objects. The film is placed on one side of the object to be examined and the radiation source, which may be X-rays, gamma rays or other penetrating radiation, on the other side. Defects in the volume of the material will alter the path of the radiation, showing up on the developed film as a contrast to the nondefective material. The sharpness of the image is determined by the focal size of the radiation source, the distance between the source and the object, and the proximity of the film to the object. The area being examined should be centered on the film. Exposures at different angles may be necessary to show the size, type and orientation of the flaw. Special calibrated plates and exposures with known defects are used to establish guidelines and required exposure times for various thicknesses and materials being examined. Tables of test data are also available for use in setting up radiographic examinations to give the most meaningful result. These tables are based on performance obtained under varying conditions. This type of NDE is known as a "volumetric" examination because it shows flaws through the volume of the material being examined.

radius – one-half the diameter of a circle, or the linear distance from the center of the circle to its circumference.

radius gauge – a gauge for checking the regular radii of convex and concave surfaces. The tool is made up of a series of blades of different radii. By comparing the appropriate blade to the surface and reading the radius from that blade, the unknown radius can be identified. Also called a fillet gauge.

rafter – the main structural support members for the roof of a building, extending from the peak of the roof to the eave line. *See Box.*

rafter anchor – a sheet metal fastener used to attach the rafter to the top plate of the wall on which the rafter rests.

rafter bearing – the amount of the rafter that bears on the top of the wall, the measurement of which is equal to the length of the rafter seat cut.

rafter, fly – a common rafter that is parallel to a gable end and projects out from the gable to create an overhang. Also called a floating rafter.

rafter overhang – the end of the rafter extending beyond the outside of the building exterior walls. Also called the rafter tail.

rafter pattern – a template used to mark rafters for cutting.

rafter plate – the top plate of the wall on which the rafters rest.

rafter plumb cut – the vertical cut made in a rafter where it joins the barge board; the vertical portion of the birdsmouth cut that permits the rafter to mate solidly with the top plate.

rafter seat cut – the horizontal cut which, with the plumb cut, forms a "birdsmouth," the cutout that permits the rafter to sit firmly against the top plate.

rafter shortening – shortening the length of a rafter to compensate for one-half the thickness of the ridge board. The opposing rafter on the other side of the ridge board is shortened a like amount.

rafter tables – tables printed on a framing square that have the information needed to calculate the lengths and angles of rafters for different roof designs, such as those cuts required for common, hip, valley, and jack rafters. The cuts can be marked using the square. Rafter tables are also available in books and manuals published as references and guides for roofers, but are not as easily used for common cuts as those printed on the square.

rafter tail – the portion of a rafter that extends beyond the exterior wall of a building. Also called the rafter overhang.

raggle – see reglet.

railing – a horizontal member used as a safety barrier, or a member used as a handhold running parallel to a stairway or elevated walkway.

rails – 1) the top and bottom portions of a window sash; 2) the cross members in a panel door structure; 3) long sections of steel or metal, anchored to a floor or roadbed, forming the tracks on which steel-wheeled vehicles move.

rain cap – see entrance cap.

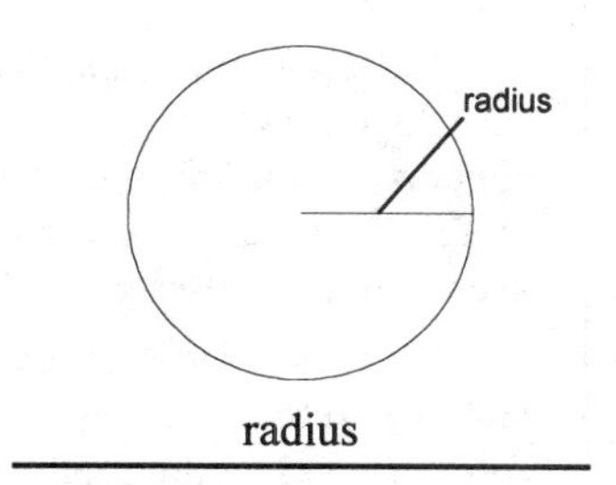

radius

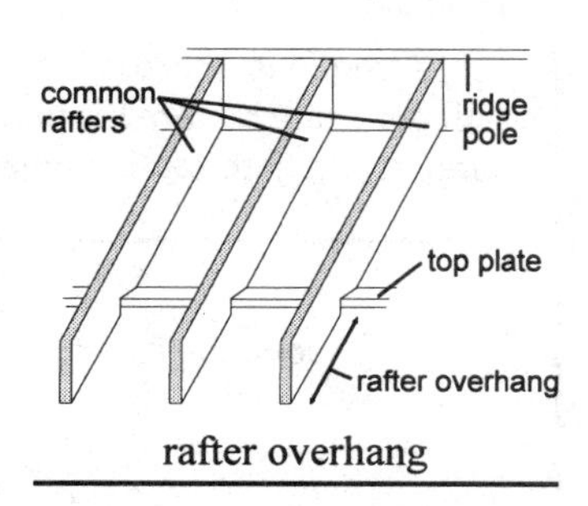

rafter overhang

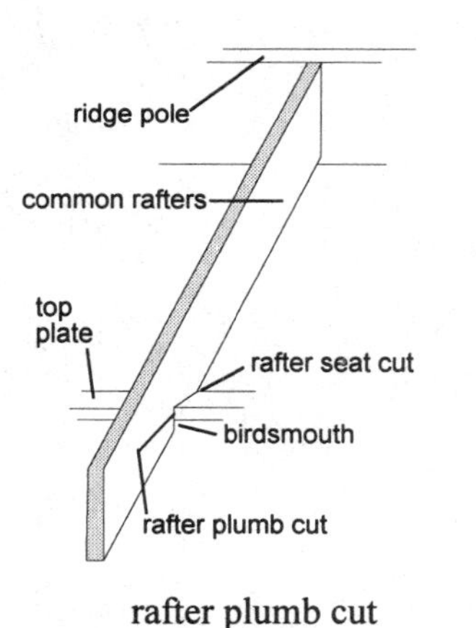

rafter plumb cut

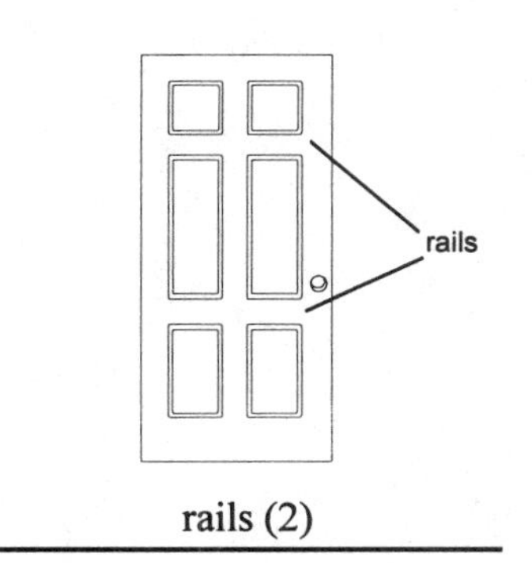

rails (2)

Rafters

Rafters are the main structural support members for the sloped roof of a building. They are designed to support the roof sheathing and covering as well as roof loads. There are several different types of rafters to accommodate different roof styles.

Most roof styles utilize the *common rafter*, often in addition to the other types. The common rafter extends from the peak of the roof to the eave line, and is perpendicular to both the ridge and the wall plate to which it is attached.

A roof with more than two slopes, such as a hip roof, will generally require hip and valley rafters. A *hip rafter* supports the intersection where two roof slopes slant up to meet at an external angle. The hip extends diagonally from the ridge to one of the outside corners of the wall. The inside intersection of two roof slopes is supported by a *valley rafter*. The valley rafter extends from the ridge to an inside corner of the exterior wall.

Jack rafters are the short rafters that span the distance from a hip to the wall top plate, or from a valley to a ridge. These rafters are often referred to as hip jacks or valley jacks to identify their location. *Cripple jack rafters* run between two other rafters, such as between a valley and a hip rafter, and do not bear on the ridge or wall plate. Like other jack rafters, their location is designated by their names, valley cripple or hip cripple.

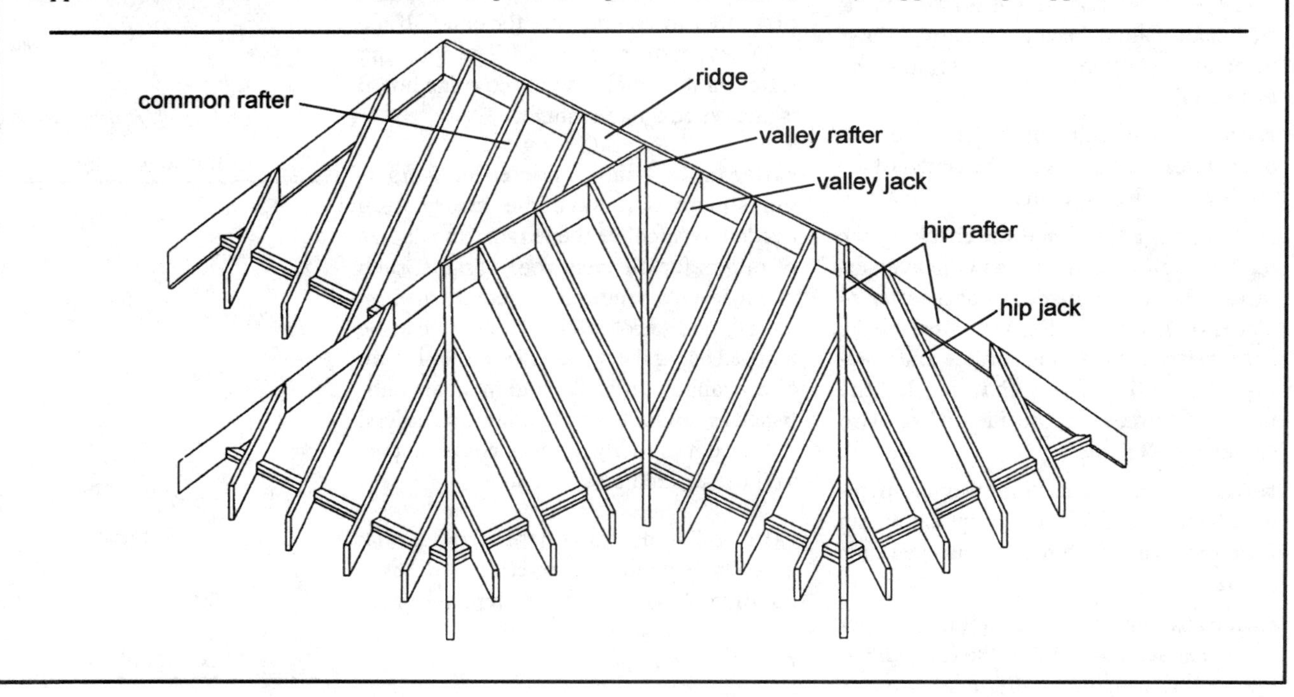

raised grain – wood grain that has been wetted with water, causing some of the wood fibers to swell and raise from the surrounding surface. Raising the grain in this manner and then sanding the wood flat is one way of achieving a very flat and smooth surface. If water-based stain is to be used, this can be an important step in ensuring that the stained wood does not end up with a raised-grain, uneven surface.

rake – 1) the overhang of a roof on the gable end of a building; 2) an angle to the vertical or the horizontal, such as the angle a rafter makes with the adjoining wall; 3) an angle between any two objects; 4) a long-handled tool with fork-like tines arranged in a row at the end of the handle. Used for gathering debris, such as leaves or weeds, or smoothing and leveling, as in smoothing out a small area of newly-placed top soil.

rake cornice – the junction of the eaves on a gable roof end and the walls of the building.

raked joint – a mortar joint between masonry units in which the mortar has been removed, forming a groove at a set depth from the face of the masonry.

rake wall – the wall of a building that has a sloping top plate, such as a structure with a shed roof.

ram – a piston in a cylinder that is pneumatically or hydraulically powered for applying force or driving a weight, such as a pile hammer.

ramp – 1) a sloping surface used to change elevations, such as a wheelchair ramp; 2) the sloped portion of a staircase handrail.

ramset – see stud driver.

ranch-style architecture – a single-level house with a low-pitched roof and informal design.

ranch trim – a type of window and door trim with a simple curved outer surface.

random – 1) not uniform; consisting of a variety of sizes, lengths, or shapes; 2) not planned, ordered or sorted.

random access memory (RAM) – the portion of a computer memory which stores information for retrieval.

random rubble veneer – a stone veneer facing for building walls made from stones of various sizes, shapes and irregularities.

random sampling – the periodic, and sometimes unannounced, sampling of a product to test for quality.

range – 1) the operational limits of a machine or object, such as the minimum to maximum speeds of a variable speed drill; 2) a cooktop, often part of a unit with an oven.

range, drop-in – a cooktop/oven combination designed to be installed as a built-in unit. The unit is supported on each side by the kitchen base cabinets, with its weight resting on the countertop.

range hood – a metal shield installed above a cooktop which contains a fan, venting system and usually a light, and is designed to conduct cooking fumes and vapors to the outside via a duct. The hood also shields cabinets above the cooktop from flames if a fire should occur.

range poles – long poles used to establish midpoints between points in a survey that are not visible from one another. The range poles mark the location of the points and several midpoints between them. Since the range poles are long enough to be visible, sighting along the poles permits the survey to be made.

range receptacle – an electrical receptacle for 50 amp service at 240 volts, designed primarily for electric ranges.

rapid curing cutback – asphalt which is cut or liquified using gasoline. The rapid evaporation of the gasoline speeds up the curing of the asphalt when it is laid.

rapid-start fluorescent light – a fluorescent light with extra ballast transformer windings which keep the electrodes (filaments) in the tubes hot. This type of fluorescent light offers the advantage of coming on almost instantly, like an incandescent light, rather than flickering on slowly.

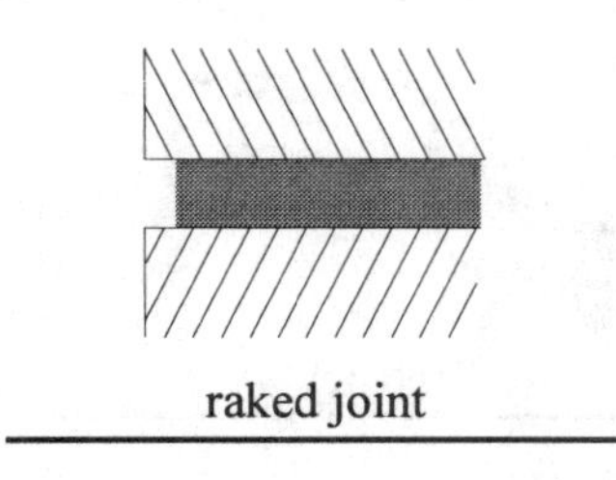
raked joint

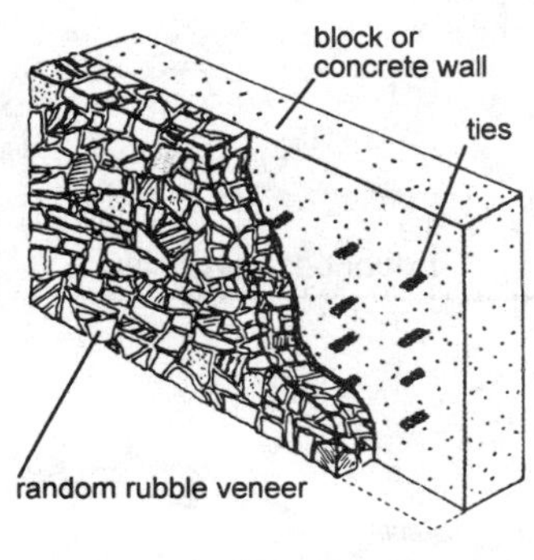

random rubble veneer

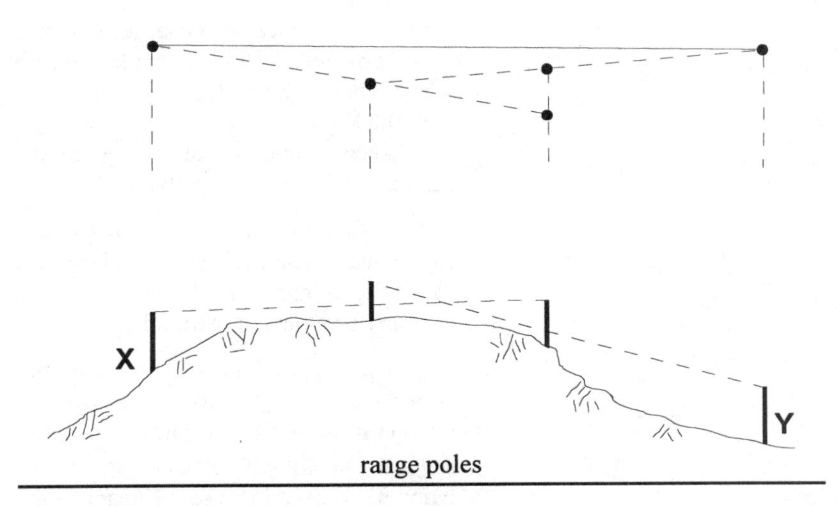

range poles

rasp – a coarse file used for rapid shaping of wood or other soft materials. The file teeth are sharp and staggered.

ratchet – a mechanism, used with braces, wrenches and lifting devices, that permits free movement in one direction but locks in the other direction.

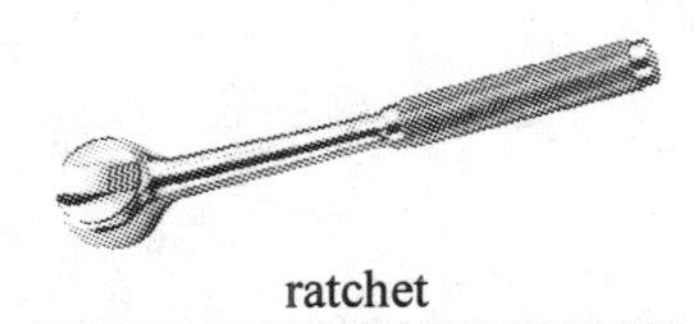
ratchet

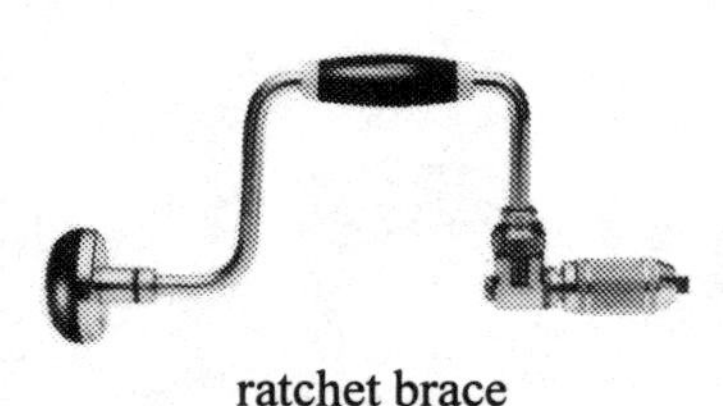
ratchet brace

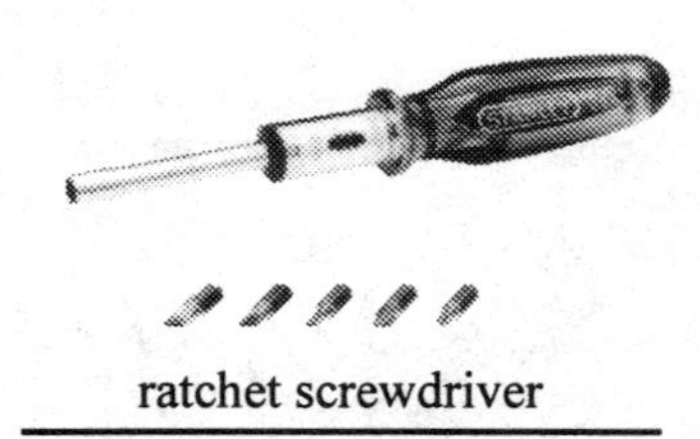
ratchet screwdriver

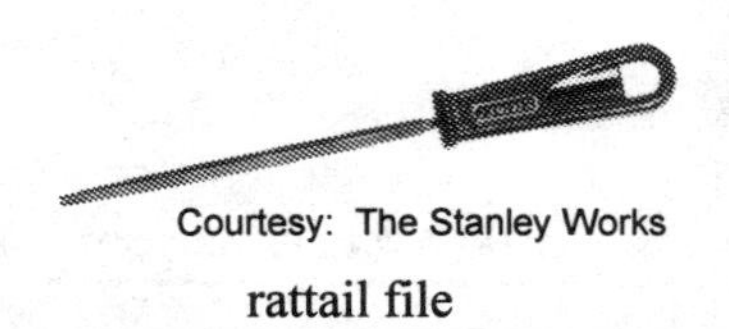
Courtesy: The Stanley Works
rattail file

ratchet brace – a brace which holds an auger or other drill bit and has a ratchet device which permits the brace to turn the bit with a back and forth movement when a full swing of the brace is not possible.

ratchet hoist – a manual lifting device with a ratchet operated by a lever which provides a mechanical advantage. The lever is pumped up and down, lifting on the downstroke and locking into place, before going on to the next stroke. This prevents the load from slipping backwards between strokes. A come-along is one example of a device that utilizes a ratchet hoist.

ratchet pipe threader – see pipe threader, ratchet.

ratchet screwdriver – a screwdriver with an integral ratchet in the handle, allowing the user to turn the tool with a flick of the wrist, without releasing their grip.

ratchet wrench – a wrench that incorporates a ratchet so that the object can be grasped by the wrench and turned, even in close positions where the wrench handle can only move a short distance.

rate – 1) a fixed value or charge; 2) a fixed ratio or degree, such as the angle at which a valve body tapers down to fit the pipe wall thickness to which it will be welded. The taper is the rate of change of the section thickness of the valve body.

rating – a numerical value or quantification based on comparisons or studies, such as the fire ratings or the thermal insulation ratings (R-values) of materials.

ratio – a proportional relationship between one size, quantity or number and another described in the form of an equation by dividing one by the other, such as using $5/1$ to describe a set of gears which operate at a ratio of five revolutions to one revolution. One gear revolves five times while the other gear revolves only once.

rattail file – a file that has a tapered cylindrical shape that comes to a rounded point at the end, bearing a superficial resemblance to a rat's tail. It is used to smooth and shape holes or curves.

raveling – a separation over time of the coarse and fine aggregate on a road surface, with the larger aggregate raised and exposed above the surface.

rawhide – untanned leather.

rawhide mallet – a mallet with a metal head having rawhide facings.

raw linseed oil – oil from the flax seed that is unprocessed or only slightly processed.

raze – to demolish or tear down to the ground.

react – a movement or change in response to an action, force or chemical.

reactance – a counter voltage or force that appears in AC circuits because of fluctuating supply voltage. This counter voltage works against the supply voltage, causing power losses in the circuit.

reaction flux – a soldering flux that deposits metal by chemically reacting with the metal being soldered when heat is applied. The flux, which is a paste that cleans the metal to be soldered as well as serving as an indicator when proper soldering temperature is reached, also prevents oxidation of the hot metal during the soldering or brazing process.

read only memory (ROM) – the preprogrammed portion of memory in a computer that allows information essential to the operation of the computer to be retrieved, but not changed.

ready-mix – materials or compounds, such as joint compound or concrete, that have all essential elements combined at the factory. Some ready-mix materials are ready to use, others may require an additional ingredient, such as water, before application.

ready-mix concrete – concrete that has been mixed before delivery to the jobsite. All ingredients, including water, have been combined and the concrete arrives at the jobsite ready to be poured in place. This is a time and labor savings, and is the most common form of concrete used for all but very small pours.

ready-mix truck – a truck designed to agitate pre-mixed wet concrete en route to a jobsite and then dispense it.

real estate – land, the structures and improvements affixed to the land, and the air, water and mineral rights above and below the surface. Also known as real property.

Realtor – an agent licensed to sell real property who is a member of the National Association of Realtors.

ream – to enlarge the diameter and/or smooth the inside of a hole or pipe.

reamer – a cylindrical tool, sometimes tapered, with straight or helical teeth along its main length, used to enlarge and/or smooth round holes.

rebar – see reinforcing bar.

rebar ties – wires used to connect two or more lengths of rebar so they form a continuous length, or to tie lengths of rebar together where they cross.

rebate – 1) see rabbet; 2) a reduction in price, or partial refund offered by the owner or manufacturer after the product sells.

receivables – amounts of money that are due to be paid to a business by customers or clients.

receiver – a pressure tank installed on the outlet of an air compressor. The tank is pressurized by the compressor and serves as a reservoir of compressed air.

receptacle – 1) a female connector into which electrical equipment, fixtures or appliances may be connected to the power source. Also called an outlet; 2) a container.

recess – 1) a cavity or indentation; 2) to create a cavity or indentation so that an object, such as a light fixture, which is installed in that cavity, is flush with the surface around it.

reciprocal leveling – a means of finding the elevation difference between two points by taking the average of two survey sightings, each taken from one of the two points.

reciprocating compressor – see compressor, reciprocating.

reciprocating saw – a power-driven saw with a blade that moves back and forth along the axis of the blade.

reclaim – 1) to change land or material from an unusable condition, such as swampland or waste, to a usable state; 2) to obtain the return of property.

reclaiming machine – a machine that pulverizes old asphalt roadway with added asphalt oil to prepare the road surface for refinishing.

reclamation – restoring land to a usable state.

recoat time – the time required between applications of paint or other coverings. The first coat must set up sufficiently that a second application does not destroy the first. Manufacturers provide recommended recoat times on the product containers.

recoil – 1) a backward reaction to a force, such as the reaction that occurs with the use of a jackhammer or pneumatic nailer; 2) to react to a noise, shock or fright by moving backward suddenly.

reconciliation – 1) to bring something into balance, as in balancing a financial account; 2) to bring into harmony.

record – 1) to file a deed or other document with an authorized government agency; 2) to register permanently by mechanical means, as in recording an earthquake on a seismograph; 3) to set down in writing or on a document.

rectangle – a plane geometric figure having four sides and four right angles.

rectifier – a device which changes alternating current to direct current by allowing the current to flow in only one direction. Rectification can be achieved by different means. One example is an electrolytic rectifier which uses an aluminum and lead plate immersed in a sodium phosphate or ammonium sulfate solution. Current flows only from the solution to the aluminum. Reverse current causes an insulating oxide to form on the aluminum, preventing current flow. Current reversal dissolves the oxide coating. This allows

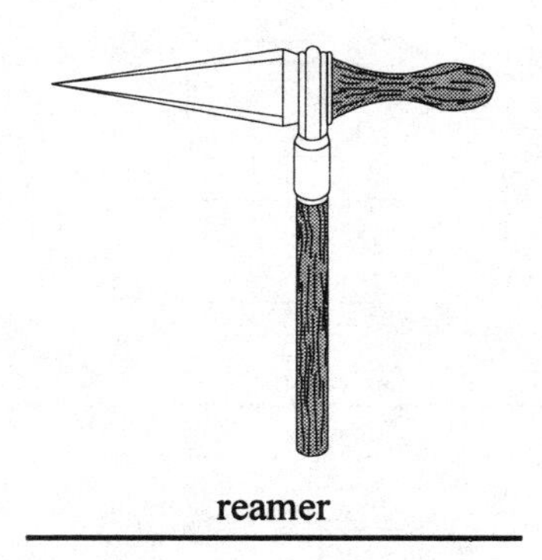

reamer

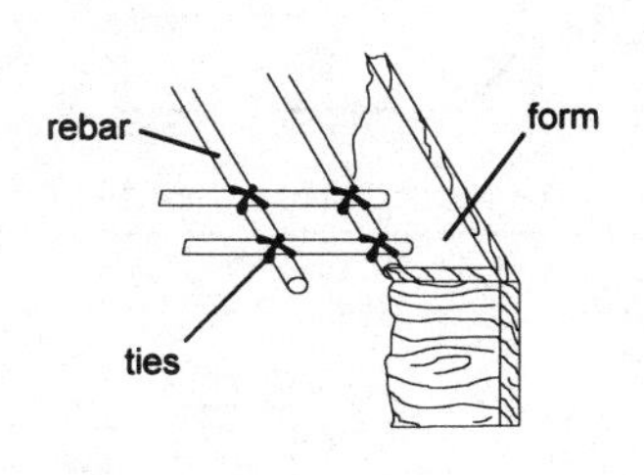

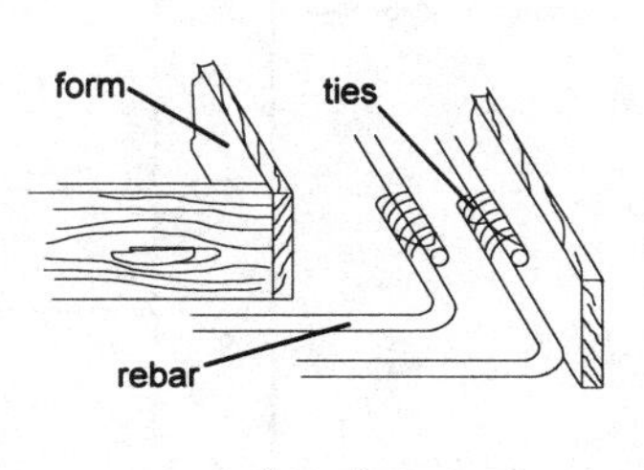

rebar ties

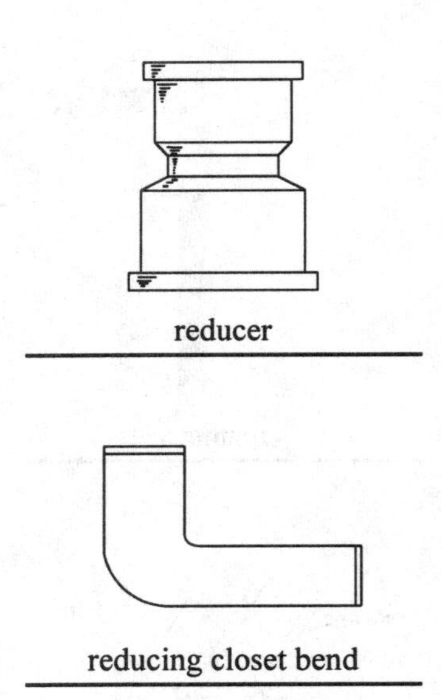

reducer

reducing closet bend

the current to only flow in one direction. Another type, a copper oxide rectifier, allows current flow from copper to a copper oxide on the copper, but not in the opposite direction. Rectifiers can be mechanical, electrolytic, metallic or gas-discharge types.

red cedar – an American juniper tree that produces a fragrant red rot-resistant wood commonly used for wood shakes and shingles.

red lead – a toxic primer containing a lead-oxygen compound which inhibits rust. It is no longer in general use, having been replaced by red oxide.

red oxide – a rust preventative used as a paint primer.

reducer – a fitting, with one end larger than the other, used to adapt one size of pipe or tubing to another. A reducer works either direction, depending on the need.

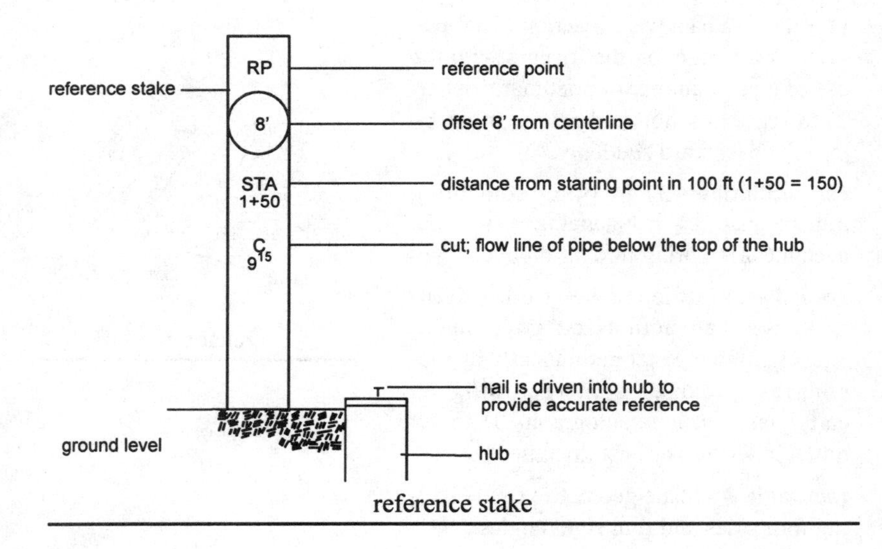

reference stake

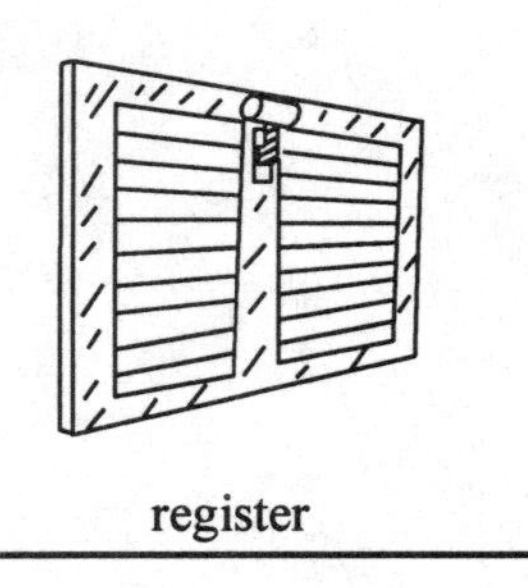

register

reducing atmosphere – a chemically-active atmosphere created to deoxidize metal oxides and bring them to their metallic state. This condition occurs at elevated temperatures, such as during the metal refining process.

reducing closet bend – a plumbing drainage fitting which connects to a water closet outlet. The fitting is bent at a right angle with the fitting diameter below the bend smaller in order to join with the pipe.

redwood – a durable and very workable wood which is resistant to decay and insect infestation. Widely used for outdoor decking and siding.

reed fence – a fence made of bamboo stalks fastened to fence rails.

reeve – to pass a rope through or around an object, as between blocks in a block and tackle system.

reference stakes – stakes placed adjacent to hub stakes in laying out excavation and earthmoving work for pipelines. The reference stakes show the elevation and location of the hub stakes with relation to the pipeline to be installed.

refine – to perfect, purify or extract.

reflective -1) the property of redirecting or throwing back sound or light, such as reflective paint used to mark roadways.

reflective insulation – a type of insulation employing a surface that reflects heat, such as foil backing. Used in wall insulation.

refraction – the bending of light or energy waves as they pass through a medium, such as water.

refractory materials – materials, such as various types of brick, that are resistant to heat. *See Box.*

refrigerant – a fluid used in cooling systems which absorbs heat by changing to a vapor, then gives up the heat as it condenses back into a liquid.

refrigeration system – a cabinet, apparatus, or appliance used for cooling, such as an air conditioning system, refrigerator or freezer.

regenerative braking – a method of braking an electric motor by feeding energy generated by the motor, acting as a generator, back into the power system. Regenerative braking is used on power saws.

register – an outlet, usually a grille or louvered panel, in an HVAC system, through which air can be released and/or directed.

reglet – a slot or groove used to hold or guide a panel, or to receive roofing or flashing. Also called a raggle.

Refractory material

Refractory materials are materials that are capable of enduring high temperatures without structural failure. They are used in construction to make fire-resistant bricks which have a variety of applications in both the home and industry. They are used in homes in the construction of fireplaces and chimneys, and in industry as building and lining material for the fire boxes of boilers, kilns and furnaces. They are also used as building insulating materials.

Refractory materials are fired to create ceramic bonds between the particles of material. Glass often forms as a part of many refractories. However, if there are large amounts of glass in the material, it may soften at higher temperatures and deform if exposed to a load. By controlling the purity of the raw materials used in the production of refractories, the amounts of glass-forming materials are limited. The higher the temperature a refractory will be exposed to, the more importance must be placed on the purity of the raw materials.

Another desirable characteristic of a refractory material is its ability to store heat. This is proportional to its bulk density, which is the ratio of its mass to its volume. The higher the density, the greater the heat storage. High density also aids in its ability to resist penetration by gases and foreign matter, such as slag.

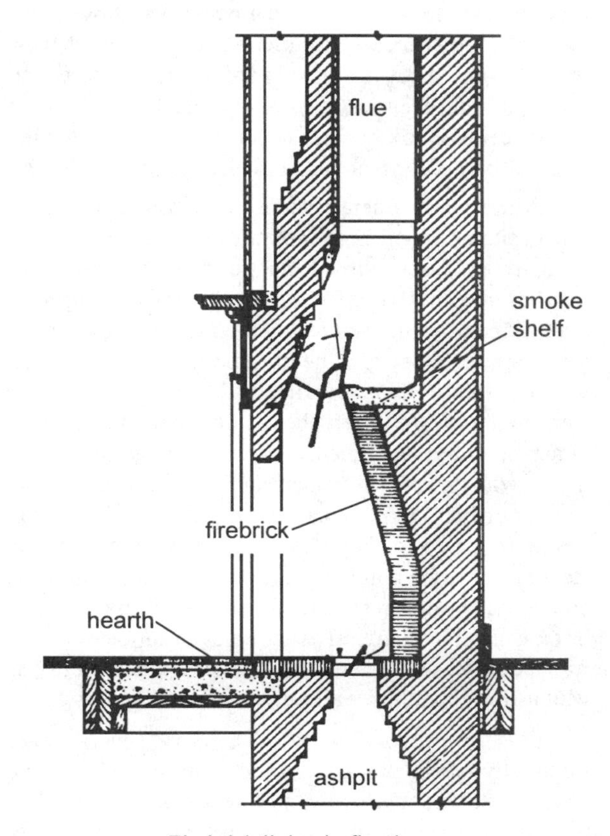

Firebrick lining in fireplace

Fireclay brick: The most commonly-used refractory materials are made from fire clay, a hard flint clay which is mostly hydrated aluminum silicate with small portions of other materials. Fireclay brick consists of flint and kaolin for its refractory properties, calcined clay to control shrinkage, and plastic clays to provide molding properties and bond strength. The characteristics of the brick are controlled by varying the proportions of these components. The chemistry of the gases to which the refractory material will be exposed is also an important factor in selecting the composition. The clays are molded into bricks and burned in kilns at temperatures between 1200 degrees F and 2700 degrees F. The American Society for Testing and Materials (ASTM) identifies five classes of fireclay brick: superduty, high duty, medium duty, low duty and semi-silica.

Fireclay bricks have greater stability than most refractories, with some formulated to have cracking and spalling resistance under rapid temperature changes. They are a good choice for use where frequent temperature cycles will be incurred. They are often used in glass furnaces, aluminum melting furnaces, carbon baking furnaces, malleable iron furnaces, enamel frit furnaces, blast furnaces, kilns, incinerators and arches.

High alumina brick: High alumina bricks are made from materials such as diaspore and bauxite, which have a rich alumina content. The bricks are manufactured to contain from between 45 to 99 percent alumina. They can be used in temperatures up to 3300 degrees F, with their refractory capacity being proportional to the alumina content. Alumina bricks are used in blast, foundry, forging and refining furnaces, induction furnace linings, glass melter walls, phosphate and electric furnaces, industrial incinerators, coke and ceramic kilns, rotary lime kiln linings, sulphur burners, hot metal cars, continuous casting sleeves and ladles for high temperature metals.

Silica brick: Silica bricks are made from crushed rock containing about 97 percent silica. Lime is used as a bonding agent and the bricks are fired in a kiln for several days at temperatures between 2700 and 2800 degrees F. They have good refractory properties combined with good mechanical strength and rigidity. They are resistant to spalling at temperatures above 1200 degrees F, with nearly no thermal expansion at these temperatures. They are used in applications where thermal shock and chemical resistance is needed, and in coke ovens, ceramic kilns, glass tanks, chemical reactor linings, carbon black reactors, and electric furnace roofs.

Basic brick: Basic brick is made from dead burned magnesite or magnesia, sometimes with the addition of chrome ore. Dead burned magnesite is the crystalline form of magnesium oxide. Firing the brick changes the material bond from chemical to ceramic. A ceramic bond gives bricks better load-bearing properties and greater volume stability than chemically bonded bricks of the same material. However, chemically bonded brick have better spalling resistance. Tar or pitch is sometimes mixed with the refractory material in making basic brick. The tar or pitch turns to carbon in use, inhibiting penetration by gases or slag. Basic brick is also available in steel cases which bond to the refractory under high heat and provide good resistance to surface damage. Basic brick is used in many different types of furnaces and in furnace parts, such as hearths, sidewalls and bottoms, as well as for glass tank walls and vacuum degasers.

Other refractory bricks: There are several other frequently-used types of refractory materials which have specialized uses. *Dolomite*, a mineral with a crystalline structure consisting of calcium carbonate and magnesium carbonate, is used in making bricks for furnace bottoms. *Chrome brick* is made from chromite ore, which is calcined. (Calcining is heating to remove all the moisture.) Chrome bricks are resistant to slag, and are used in metal processing furnaces. *Silicon carbide bricks* are made of alumina, clay and a flux to provide a ceramic bond when fired. They are abrasion resistant, have good dimensional stability under high temperature conditions, and are not affected by rapid temperature fluctuations. They are used in incinerators, ceramic kilns and nonferrous foundries. *Zircon bricks* are acidic and resistant to spalling and expansion under high heat up to 2900 degrees F. They have high density and a low porosity and are used in glass regenerator tanks, continuous casting nozzles and parts of aluminum furnaces. *Chemical resistant brick* is a nonabsorbent vitreous brick that is well suited for use in damp and corrosive situations. It is used in chemical processing tanks, gas scrubbers and on the floors of dairies and canneries.

Insulating refractories: Insulating fire bricks are made from porous clay or kaolin, a fine white clay used in making china. Their insulating properties and light weight come from the small bubble-sized air pockets they contain. They are used in industrial heat-treating furnaces where temperatures exceed 1600 degrees F. Insulating refractories are also used as backing for other refractory brick that is exposed to heat. They improve fuel economy because of their insulating capability.

Refractory concrete: Refractory concrete, made with calcium-aluminate cement and refractory aggregate, is suitable for use at high temperatures, such as in boilers and furnaces, and is used as mortar with refractory bricks. It is also used as an insulating material because of its low thermal conductivity.

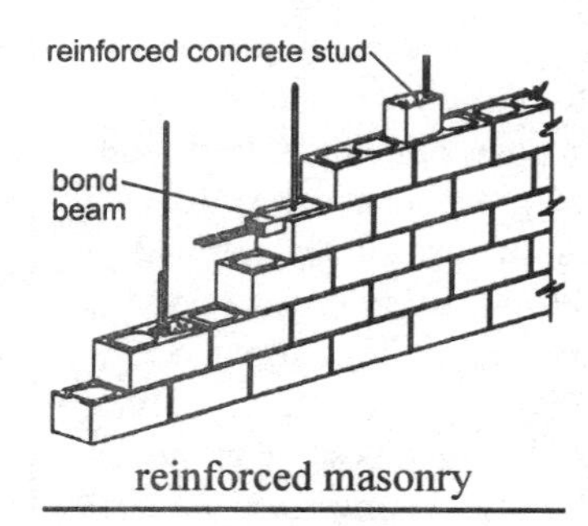

reinforced masonry

regular alternate saw tooth set – saw teeth that are set alternately to the left and right sides rather than straight down the center of the cutting edge of the saw. Used to make saw cuts with a wider kerf than the thickness of the saw blade.

regulator – a device used to control voltage, flow, or pressure.

reinforce – to add structural strength to a material, such as steel rods embedded in concrete for reinforcement.

reinforced concrete – concrete to which reinforcement has been added in the form of steel rods, bars or mesh to increase its strength and resistance to cracking.

reinforced faced masonry – walls of two widths of masonry units, each of different materials, such as concrete masonry units with a glazed masonry unit facing bonded to it.

reinforced masonry – a structure of hollow or drilled masonry units that is reinforced with steel and mortar. Also called reinforced filled cell masonry.

reinforcement – structural strengtheners, such as steel reinforcing bars, used to add stiffness and/or strength to materials.

reinforcement of weld – adding a greater amount of filler metal to a weld than is needed to fill the joint. This technique is used where the pipe or other material thickness at the weld is too thin for the pressures or stresses to which it will be subjected.

reinforcing bar – steel bar designed to be placed in concrete for reinforcement. Concrete has compression strength; reinforcing bar adds resistance to breaking when other types of forces are applied. The bars have a patterned, or textured exterior, which allows them to bond or interlock with the concrete mix. They come in a variety of sizes, starting at ¼ inch diameter and increasing in size in increments of ⅛ inch. Also called rebar.

relative – having a relation to, a connection with, or a dependance on another object, process or person.

relative humidity – a measure of the amount of moisture in the air with respect to the temperature. It is the ratio of the moisture present to the maximum amount of moisture the air can hold at that temperature.

relay – an electrical switching device that opens and closes a circuit by means of a remote source of current operating a solenoid or similar device.

reliction – land that is made usable by the gradual recession of water over time, as in the receding water line around a lake.

relief – 1) a sculptured figure or design that projects from the background surface; 2) a sharp or distinct outline of an object appearing in contrast with its surroundings; 3) a change of elevations in land contours; 4) a cut or removal of material made to prevent binding between parts, such as between rotating and stationary parts in machinery.

relief valve – a valve that is set to automatically open at a predetermined pressure, preventing the pressure inside a tank or container from building up beyond the safe tolerance of the container. The pressure or fluid within the system must be vented to another location where its discharge will not damage other equipment or personnel. The relief valve should be set to relieve at a value that will limit the maximum accumulated system pressure to a point below the allowable working pressure of the weakest component in the system. Many valves have a lever which permits manual opening of the valve at a pressure below the set pressure at which it opens automatically. Relief valves, such as those used on water heaters, may also be set to open at a given temperature as well as pressure.

relief vent – a plumbing pipe which is placed between a building drainage system vent stack and drain line to provide air circulation. Also called re-vent.

relieving arch – an arch that is built over a lintel or other arch to distribute the load over the opening. Also called a discharging arch or safety arch.

remodel – to alter or restore an existing structure.

repair – to mend or return to a usable state.

repair clamp – a device used to repair a leak in a pipe by tightly surrounding the pipe and sealing the leak.

replace – to remove an item and install another item in its place.

replacement cost – the current cost to replace an item with an identical item.

repose – the greatest angle from a horizontal plane at which an object will lay in a stable condition. For example, the steepest angle of a slope on which loose dirt will rest and remain stable without sliding. See also angle of repose.

repousse – a figure in relief on metal, made by hammering from the reverse side.

require – 1) to need or demand as essential, such as a minimum size of wiring required by the *National Electrical Code* for use in a given situation; 2) to ask for by right of authority, such as an operator's license required by law for use of certain types of heavy equipment.

requisition – 1) a document for the purchase of items; 2) a formal request.

resaw – 1) to saw again, usually into thinner sections; 2) to make two rip cuts, one from each side, on a board that is too thick for the saw blade to cut completely through from one side.

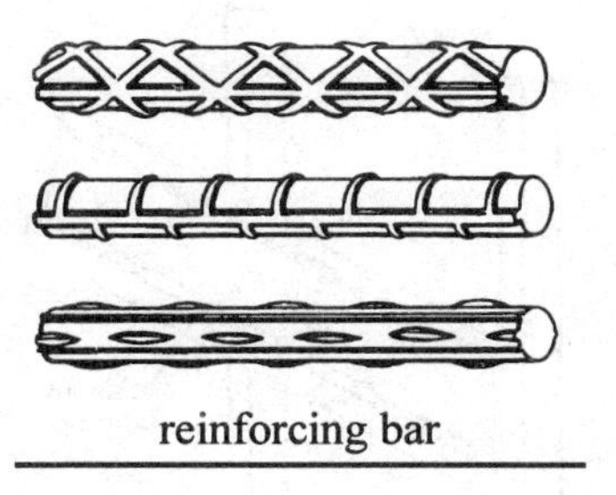

reinforcing bar

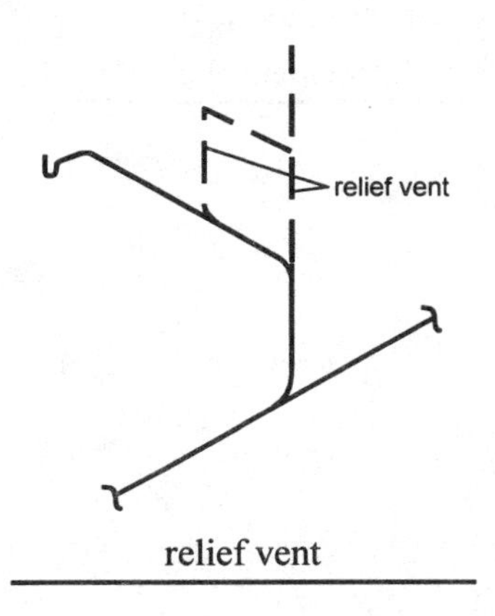

relief vent

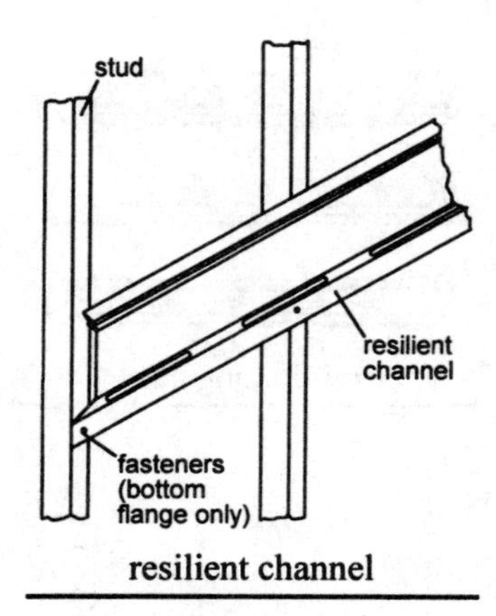

resilient channel

rescind – 1) to take back; 2) to void.

reseat – to renew a valve seat surface by grinding or by using a special cutting tool to remove part of the seat material, leaving a like-new surface.

reservoir – 1) a collection and storage area, such as water storage in a man-made lake; 2) a fluid container used to replenish a fluid system.

reshores – temporary shoring placed under concrete to help support it after the original forms have been stripped.

residence – a dwelling place or home.

residential – 1) pertaining to residences or homes, as in residential construction; 2) areas zoned for homes or living quarters, as in residential neighborhoods.

resilience – the ability of a material to be deformed under a load and return to its former size and shape when the load is removed.

resilient channel – a preformed section of sheet metal, ½ inch deep by 2½ inches wide by 12 feet long, installed between wallboard panels and framing to reduce sound transmission through walls. The resilient channel is attached to the framing studs by a flange which runs along the bottom of the channel. When the wallboard is attached to the channel, the channel acts as a hinge, with the weight of wallboard pulling the channel slightly out and away from the framing studs. By preventing the wallboard from laying against the studs, the channel inhibits the transmission of sound through the framing. The channels can only support a single layer of wall panel and still retain their resiliency.

resilient tile – floor tile made of vinyl, rubber, or other resilient material. It is available in many sizes, colors and patterns.

resin – a gummy sap secreted by certain plants and trees. Various resins are used in making varnishes, printing inks, rubber, plastics and medicines.

resistance – the opposition offered by a body or substance to the passage of a steady electric current through it. Internal friction slows the passage of electrons within a conductor and causes heat. The larger the diameter of electrical wire, the less friction there is per foot.

resistance brazing (RB) – a brazing process which uses electric resistance for heating. Electrodes with high electrical resistance are fed an electric current. Because of their high resistance, the current is converted to heat and the electrodes become very hot. The electrodes are placed in contact with the work to be brazed, and heat is transferred by conduction from the electrodes to the work. When the work reaches the proper temperature for brazing, the brazing material is applied to the joint. The brazing material is melted by the heat and flows into the joint, making a strong durable joint.

resistance seam welding (RSEW) – a welding technique that uses the heat generated by resistance when an electric current is applied to the parts to be welded, heating them to welding temperature. When the proper temperature is reached, the parts are brought into firm contact along the joint, creating a seam weld along the joint.

resistance soldering (RS) – a technique which uses electrical resistance to produce the heat for soldering. Electrodes with high electrical resistance are fed an electric current. Because of their high resistance, the current is converted to heat and the electrodes become very hot. The electrodes are placed in contact with the work to be soldered, and heat is transferred by conduction from the electrodes to the work. When the work reaches the proper temperature for soldering, the solder is applied to the joint. It melts and flows into the joint.

resistance spot welding (RSW) – a technique used to make spot welds to hold two pieces together. The welds are made using two electrodes and an electric current to produce heat. The electrodes are placed at the locations where the spot welds are needed. Because of their high electrical resistance, the electrodes convert the electric current to sufficient heat at the weld points to make the welds.

resistance welding – a welding process in which the heat required for welding is produced by an electrical current passing through material of high electrical resistance. This causes the current to be converted to heat.

resistor – an electronic circuit device that is calibrated to provide a specific number of ohms in a circuit, ohms being the unit of electrical resistance. These devices are used in circuit design to limit electrical flow to fixed amounts.

resonate – to vibrate in response to a stimulation. Each solid object has a resonant frequency. If a force is applied to an object, such as by a blow of a hammer, the object will vibrate or resonate at that frequency. A common resonating device is a tuning fork.

resorcinol glue – a type of adhesive that is moisture and temperature-resistant. It comes in a powder form to be mixed with a liquid for application. The glue is dark in color, and is 100 percent waterproof. It is used in exterior applications and for bonding laminated timbers.

respirator – a filter worn over the nose and mouth that is designed to remove harmful particles from the air and protect the respiratory tract. These filters are used by workers in many areas of construction, especially in painting and sanding.

retaining wall – a wall that is subjected to lateral loads other than wind loads, such as a wall used to hold back earth or water.

retaining wall

retard – to slow down.

retardant – an additive to cement, concrete or mortar which slows the rate of setting. It is generally used only in high heat conditions, where too-rapid curing would result in weak concrete.

retemper – to add water and remix concrete, stucco or mortar to increase its plasticity after it has stiffened from sitting too long before use. This is commonly done to maintain a proper working consistency when using these materials over an extended period of time.

retention clause – a clause in a contract that permits the client to retain some percentage of progress payment money that is due to be paid to a contractor. It is designed as an incentive for the contractor to keep the job on schedule and/or to guarantee the work is done satisfactorily.

reticule – the glass in a transit that has cross hairs and stadia hairs used for sighting and focusing in on a target area.

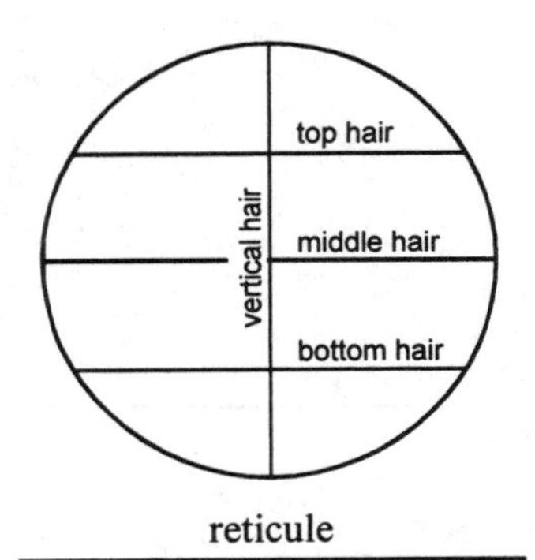

reticule

retrofit – to furnish with new parts or equipment in order to bring up to current standards, such as replacing a two-wire electrical system without a ground with a two-wire system with a ground while rewiring a building.

return – 1) a surface that turns back on itself or away from the primary or main surface, such as some types of decorative moldings; 2) air in a conditioning system that is routed back to the conditioner; 3) a plumbing or piping fitting with a 180-degree bend.

reveal – the exposed side of a door or window frame between the face of the frame and the surface of the surrounding wall.

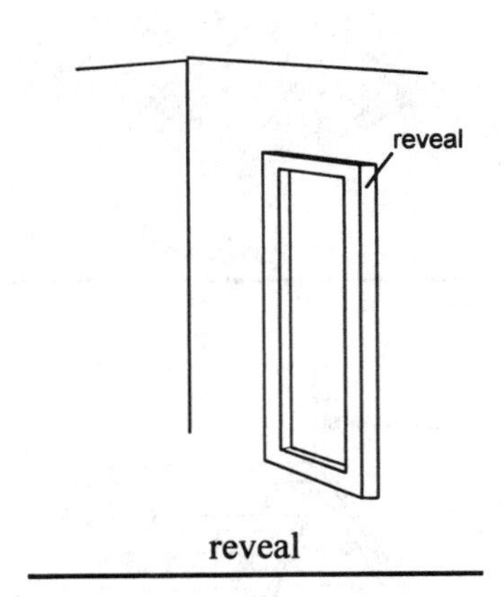

reveal

re-vent – see relief vent.

reverberate – to reflect or rebound, as in a series of echoes.

reverse mortgage – an equity mortgage, used especially with older persons, which provides for monthly income payments to the property owner rather than a lump sum loan which must be paid back. When the payments have been completely made, usually over many years, the equity in the property belongs to the mortgage holder. The object of this type of loan is to allow property owners to live off the equity accumulated in their homes for the stated period of the contract. In most cases, the contract period is longer than the life expectancy of property owner. Upon the death of the property holder, the property is sold to clear the debt, unless it is paid off by the heirs. In these days of longer life expectancy, cases have occurred in which the property owner outlived the mortgage period, and had no remaining equity and no place to live.

revetment – a facing, such as stone or cement, used to sustain an embankment or sloping surface.

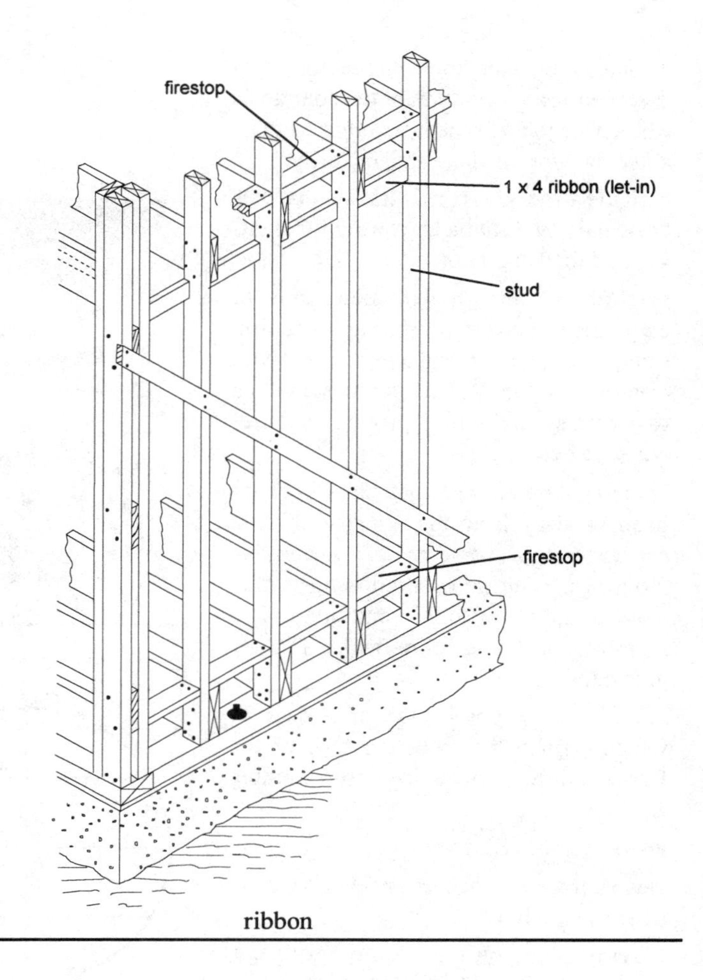

ribbon

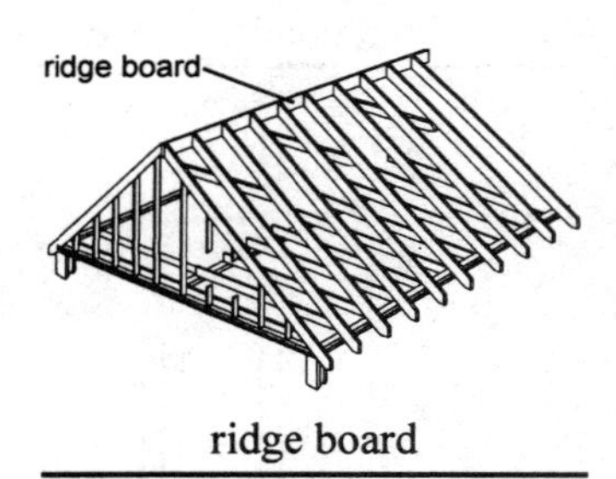

ridge board

ridge course

revision – a change, improvement or alteration to the original, such as a revision to a contract.

revision block – a record on a drawing, such as an architectural drawing, of the numbers, nature and dates of revisions made to that drawing.

revoke – to annul by taking back or recalling, as in revoking a business licence.

revolution – 1) a rotation or turning about an axis; 2) a drastic change.

revolved section – see section, revolved.

rewire – to remove existing electrical wiring and replace it with new.

rheostat – a control for regulating electrical current by means of varying resistance. When used instead of a standard light switch, it allows the brightness of the light to be adjusted rather than just having the light on or off. In this use it is also called a dimmer.

rhodium – a corrosion-resistant metallic element with the atomic symbol Rh, atomic weight of 102.905, and atomic number 45. Used to plate steel instrument parts.

rib – a reinforcement that projects from a surface. Ribs are often used with metal containers and structures to increase the stiffness of the unit without requiring large amounts of extra metal, keeping the thickness, weight and cost of the item reduced.

ribbon – a piece of 1 x 4 lumber that is installed horizontally into recesses cut or "let" into the studs in balloon framing. The second floor joists rest on this ribbon. Also called a ledger or girt.

rich mortar – mortar with a high proportion of cement. It is stronger and harder, but also more expensive to use.

rider – 1) a brace used at the corner of a fence; 2) an addition to a document, often included as a separate page at the end, which modifies or amends requirements or adds new terms to the agreement.

ride the brush – a painting term used to denote bearing down too hard on the paint brush.

ridge – the highest point of a sloped roof; the roof peak.

ridge board – the upper horizontal support member of a sloping roof against which the top ends of the rafters rest. Also called a ridge pole.

ridge cap – a roof covering along the ridge of a roof, often of continuous metal.

ridge course – the layer of shingles or other roofing material along the ridge of a roof.

ridge cut – the cut at the ridge end of a rafter.

ridge pole – see ridge board.

ridge reduction – a reduction in rafter length, equal to one-half the thickness of the ridge board, which allows the rafters to retain their proper slope and relation to each other when a ridge board is used.

ridge vent – a continuous vent along the ridge of a roof.

ridging – a defect in gypsum wallboard installation that shows up as a raised line along the surface of a finished joint. There are several causes for this defect, requiring different types of repair. High humidity or poor ventilation may result in expansion and contraction of the framing and drywall panels. This can be repaired by installing back blocking or multiple layers of drywall. A ridge can result during installation if the panels are forced together, squeezing out the joint compound. This problem can be corrected by cutting away a sliver of drywall at the joint, creating a small gap. The gap should be taped and smoothed with joint compound. Too much compound applied during installation can also cause a ridge. The ridge should be sanded smooth and a finish coat of compound applied. A flashlight held at an angle to the area will show whether the ridge has been smoothed away.

rigging – the slings, lines and other attachments necessary to move or lift loads with a crane.

right angle – an angle of 90 degrees.

right-angle drill – a drill with a gearset attachment which permits drilling holes at a 90-degree angle to the axis of the drill motor. It is used for drilling holes in tight places, such a through installed studs, as the position of the drill motor is less of a factor with this type of drill.

right-of-way – 1) an easement or strip of land that permits work crews access to an area for maintenance or construction; 2) a legal right of passage over another person's property.

right triangle – a three-sided geometric figure with one angle equal to 90 degrees.

rigid – not flexible.

rigid base diaphragm – a reinforced concrete slab placed on the ground.

rigid brick paving – masonry units laid in a mortar bed, with mortar joints between the units, installed for use as a patio, walkway or driveway trim.

rigid metallic conduit – see conduit.

rigid nonmetallic conduit – see conduit.

rim – an edge, such as the edge at the overflow point on plumbing fixtures.

rim lock – a door lock that mounts on the inside surface of the door and locks by turning a thumb-turn to secure a latch, bolt or pin through a mated section mounted on the door jamb.

ring shank – nails with rings around the shanks to increase the gripping power of the nails.

rip – to cut wood parallel with the grain. Rip cuts are made to cut boards to width (thickness). A board has the most strength perpendicular to the grain, so rip cuts make maximum use of the board's strength. Cuts to length are crosscut, or across the grain.

riparian – located on the bank of a stream, lake or other body of water.

riparian right – the right of the owner of riparian property to access the shore or use the water.

rippable rock – rock that can be easily broken up and removed mechanically, such as sandstone or granite in a fractured condition.

ripper – 1) narrow strips of gypsum wallboard used to fill in gaps or as reinforcement; 2) an attachment with hardened steel teeth used with earth-moving equipment to cut through rock and very hard soil.

ripple – an uneven or wavy surface or finish on a cabinet, wall, floor or other flat surfaced area.

riprap – 1) loose, irregular stones or crushed rock used as fill; 2) large rocks or stones piled along river banks or shore lines to prevent erosion.

ripsaw – a hand saw designed to cut wood parallel to, or with, the grain. The faces of the teeth are alternately bent to the right and left of the blade so that they are at an angle to the direction of the cut. This permits rapid cutting with the grain. This angle would cause the teeth to bind if used to cut across the grain. Crosscut saw blades have teeth beveled in relation to the direction of the cut so they can cut across the grain without tearing the wood.

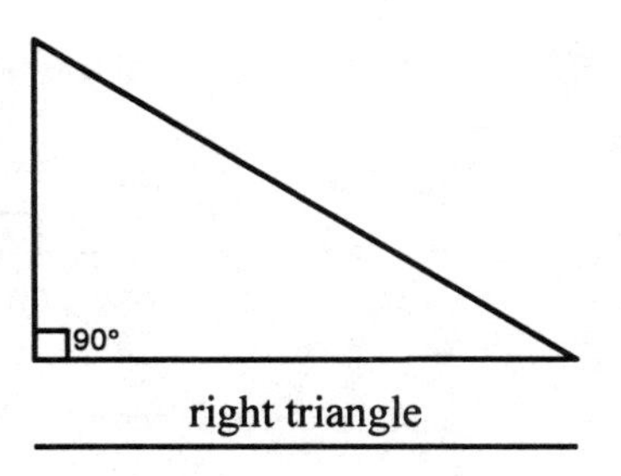

right triangle

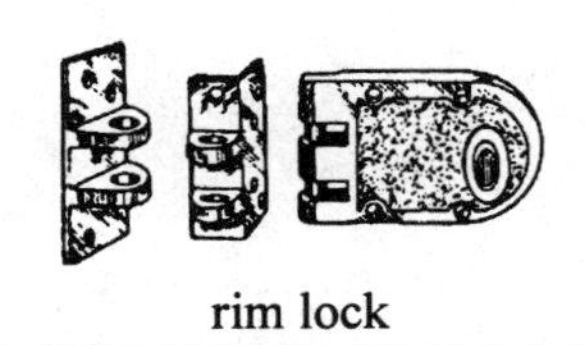
rim lock

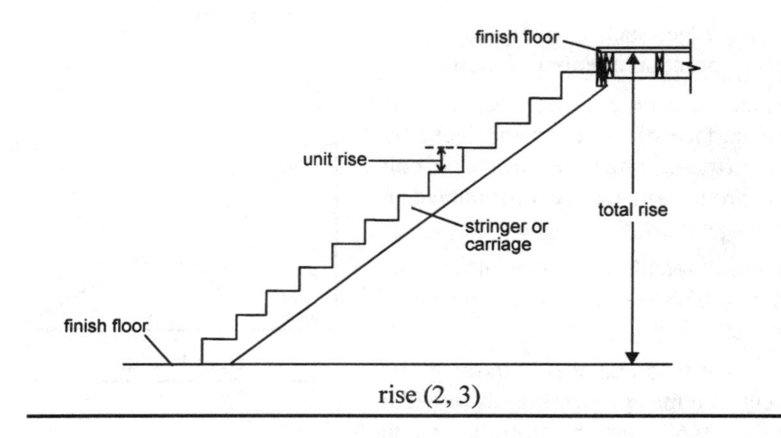

rise (2, 3)

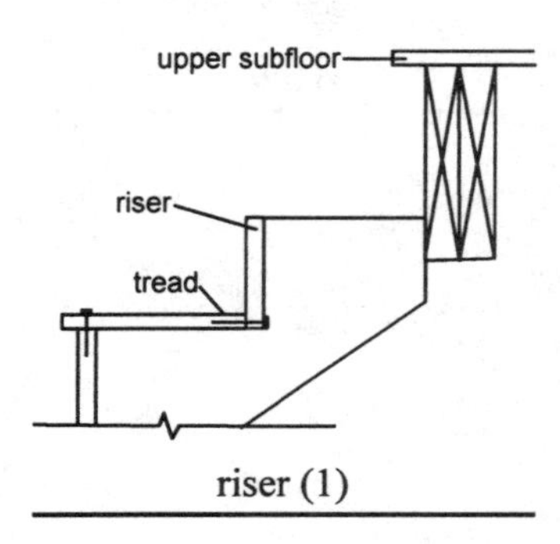

riser (1)

rise – 1) the vertical distance from the top of the wall to the ridge of the roof. See also roof rise; 2) the vertical distance that a flight of stairs rises; 3) the vertical distance that one step in a staircase rises.

riser – 1) vertical boards between stairway treads; 2) vertical water supply pipe in a fire sprinkler system. Also called a riser pipe.

riser diagram – a simplified drawing that shows electrical equipment and terminals and simplified wiring. The diagram is drawn by the electrical system designer for use by the installer.

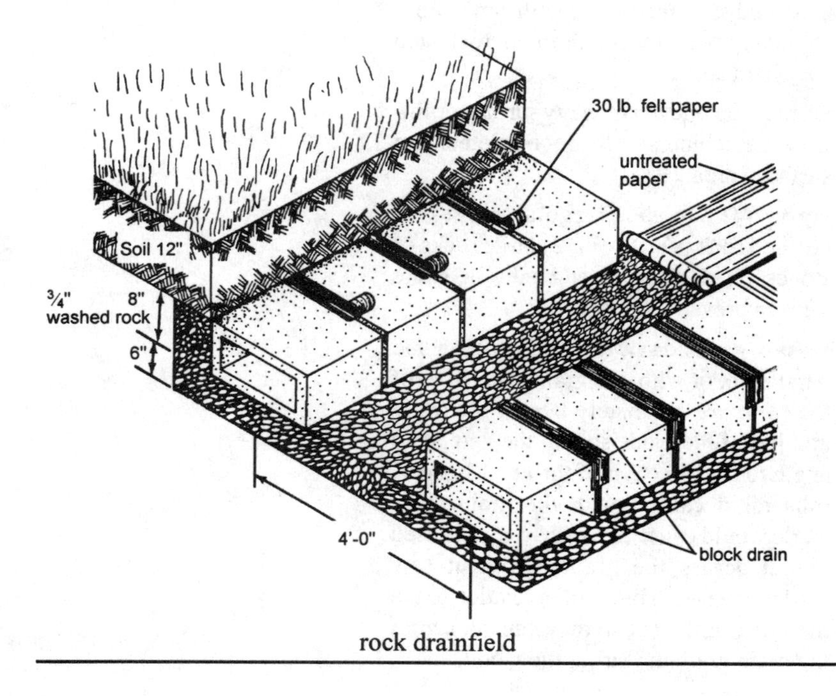

rock drainfield

rivet – a metal rod used to hold two pieces of material together. The rivet is placed in a predrilled hole with the head on one end bracing the back of the material. The shank end is then rounded over with a special tool, called a rivet set, permanently fastening the material between the two ends of the rod. Rivets are inexpensive, fast and easy to install, and will not work loose like a nut and bolt.

rivet hammer – a hammer designed to upset or deform the ends of rivets to clinch them tightly against the material they are holding.

rivet set – a metal tool which is placed against the shaft end of a rivet and struck with a hammer to upset or deform the end, fastening the rivet into place.

roadbed – the graded base over which pavement is laid for a road, or over which rails are placed for train or trolley lines.

rock drainfield – a drainfield, used for the leach lines from a septic tank, composed of ¾-inch rock with no more than 10 percent of the rock passing a ½-inch screen.

rocker arm – a pivoted lever used to actuate a device. Many internal combustion engines use a rocker arm to actuate the valve mechanism. The rocker arm is moved by the camshaft, either directly or through a pushrod. The other end of the rocker arm pushes on the intake or exhaust valve to open the valve.

rock holes – shallow pits that are dug at spaced locations along the side of a pipeline trench and filled with bedding material. The stored material is used as the bedding for the pipeline when it is laid in the trench. The holes provide easy access to the material during the installation process.

Rockwell hardness test – a test that measures material hardness by the depth of the impression that a ¹⁄₁₆-inch steel ball (used for softer materials) or a 120-degree diamond cone (for harder materials) makes on the material when it is struck with a calibrated force.

rock wool – a type of insulating material, loosely resembling wool, that is made by blowing steam through molten rock or slag. It is used for insulating attic spaces and is usually blown into place with air pressure or poured from bags.

rod – 1) a slender cylinder, often of metal; 2) a unit length measurement equal to 16½ feet.

rod saw blade – a slender metal rod coated with abrasive material, such as corborundum or diamond dust. The blades fit in a coping or hack saw frame and can be used to cut virtually any material.

rod tightening wrench – a wrench designed for tightening concrete form tie rods.

roller – a cylinder, often with a length greater than its diameter, used as a wheel for supporting (as in a bearing), moving (as in a conveyor) or compressing (as in a pavement roller).

Rollerbug – a tradename for a type of wet concrete tamper designed to force aggregate below the surface or to embed exposed aggregate in concrete.

roller catch – a type of cabinet door latch which has two spring-loaded rollers mounted on the cabinet, and a projecting piece on the door. The projecting piece is held between the rollers when the door is closed.

rolling door – see door, rolling.

rolling mill – a manufacturing facility which uses large rollers to produce plate or sheet products and bars.

roll-over gasket – a soft ring gasket used in bell and spigot piping joints. The gasket is fitted around the spigot end of a length of pipe and the spigot is inserted into the bell end of the next length of pipe. The gasket is compressed between the bell and spigot to make a seal.

rollover mortgage – a mortgage with a fixed interest rate for a period of time, such as five years, with a buyer's option to renew the mortgage at the end of that time. The interest rate adjusts to the prevailing rate, up or down, at the time of renewal.

roll roofing – an asphalt roofing material, available in 3-foot wide, 36-foot-long rolls, that can be rolled out onto a roof in horizontal lengths, overlapping downhill to prevent water entry.

roll welding (ROW) – a welding technique which uses a heat source and pressure applied by rollers to join the surfaces.

rolok – see rowlock.

Roman brick – a solid masonry unit measuring 3⅝ inches x 1⅝ inches x 11⅝ inches. See also brick.

Romex – a trade name for nonmetallic sheathed electrical cable.

roof – the covering for the top exterior surface of a building or structure. *See Box.*

roof deck – plywood, hardboard panels or spaced boards laid on top of the rafters or other roof support framing, onto which the roof covering, such as shingles, is fastened.

roof drain – a drain, installed at the low spot on a roof, which prevents water from standing on a flat or nearly-flat roof. The water drains into a drain pipe and is carried to the ground or into a drainage system.

roof drain strainer – a strainer at the inlet to a roof drain designed to trap debris and prevent it from entering and clogging the drain pipe. The strainer is usually a domed wire cage or casting of metal or plastic that fits over the top of the drain pipe. Since it projects above the roof surface it is less likely to become clogged than a strainer that is flush with the roof surface.

roofer – one who installs roofing material or makes repairs to existing roofing.

roofer's knife – a knife that has a triangular blade with a hooked cutting surface on one edge designed especially for cutting asphalt shingles, linoleum, roofing paper and similar materials.

roofing nail – a nail with a ½-inch-diameter head used with composition shingles. The large head spreads the holding capacity over a wider area and minimizes the chance of the nail tearing through the shingle material.

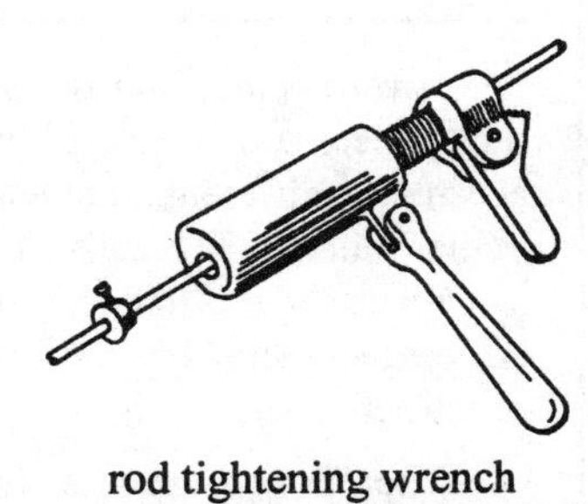

rod tightening wrench

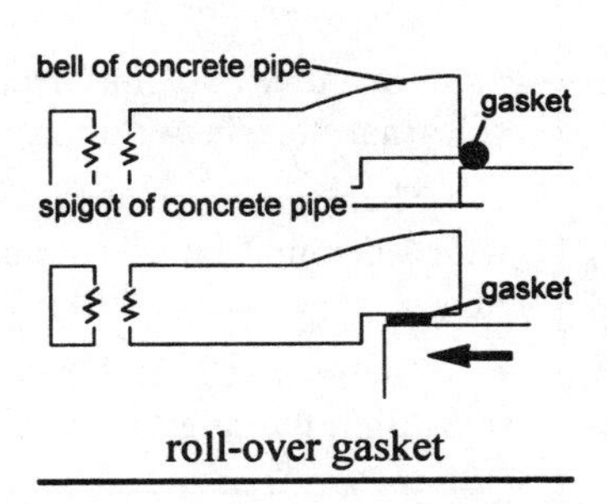

roll-over gasket

Roof

A roof is the exterior covering of the top of a structure. There are many different styles of roofs which accentuate different architectural types or provide a particular function or creative form. Geographical location, as well as style, often determine the type of roof used in building. Several types of roofs are commonly used in construction today:

The *flat roof* has no slope or just enough to allow for drainage, no more than a 10 percent incline.

The *shed roof* has one slope.

The *gable roof* has a gable at one or both ends and sides that slope in two opposite directions from the ridge.

The *rainbow* or *whaleback roof* has a pitched roof with sides that have a slightly convex slope from the ridge.

The *gambrel roof* is sloped in two opposite directions from the ridge, with each surface divided into two sections with different pitches. The lower section has a steeper slope to the outside walls. This type of roof is also called a barn roof.

The *mansard roof* has two slopes on all four sides of the structure, with the lower slope being the steeper of the two.

The *hip roof* has four sides sloping from a center ridge, with hip rafters that extend from the ridge to the outside corners of the structure, creating sloping, triangular ends.

The *intersecting roof* has two or more ridges meeting at different angles, creating hips and valleys. Also called a hip-and-valley roof.

The *dormer roof* has a projection built out from a sloping roof, often with a window in it. The projection provides added interior space, and may vary in form, having a gable roof, shed roof or another type that complements the rest of the building.

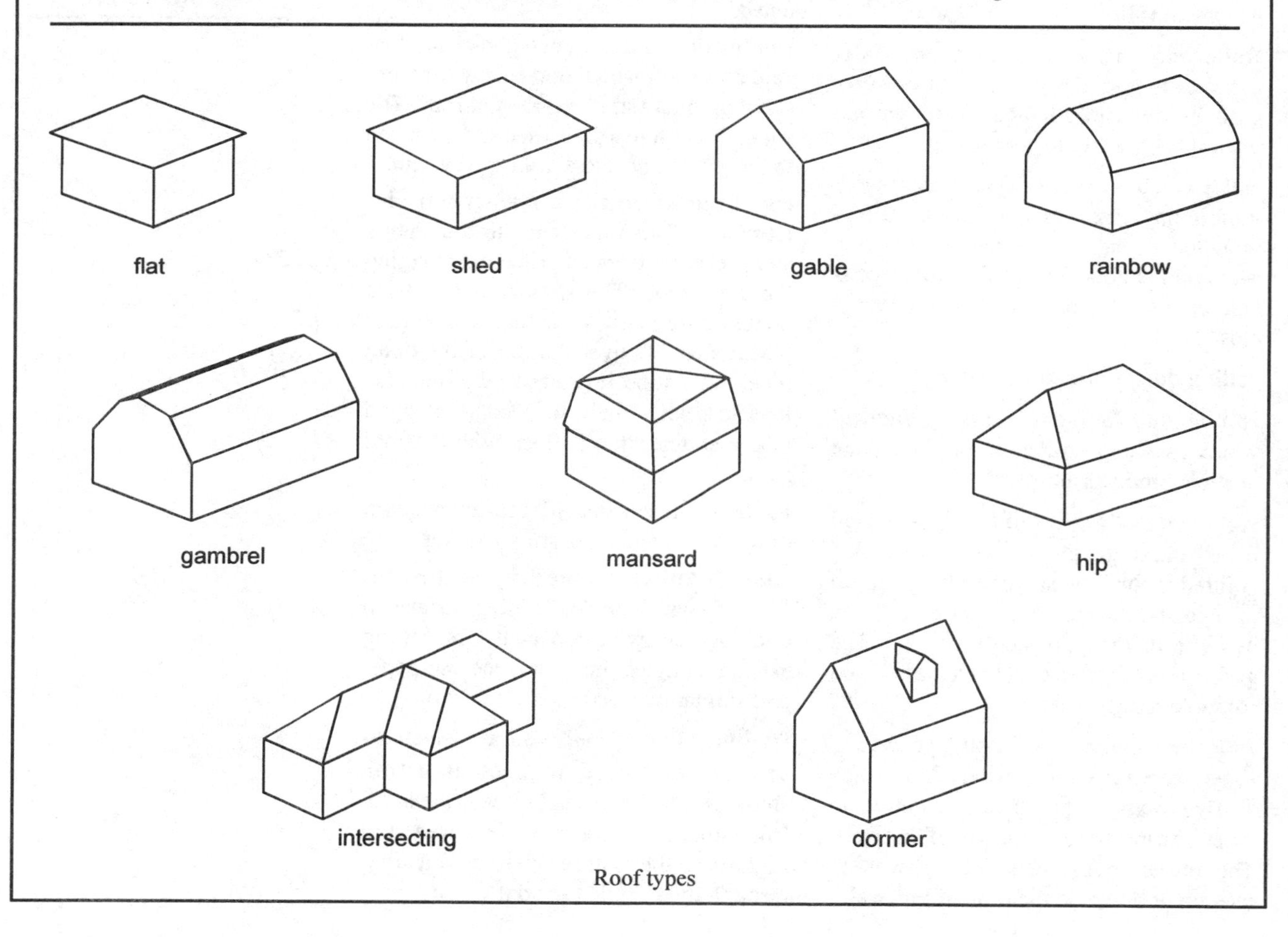

Roof types

roofing tape – asphalt-saturated cloth roofing material used with roofing cement as reinforcement for roof patches and for flashing.

roof jack – a specialized piece of flashing with a sleeved hole which fits over a vent stack that protrudes through a roof. The flashing seals the opening and diverts rain around the vent stack, preventing leaks at that point in the roof.

roof pitch – see roof slope.

roof rise – the height or vertical rise of a roof measured from the wall plate, on which the rafters rest, to the peak of the roof ridge.

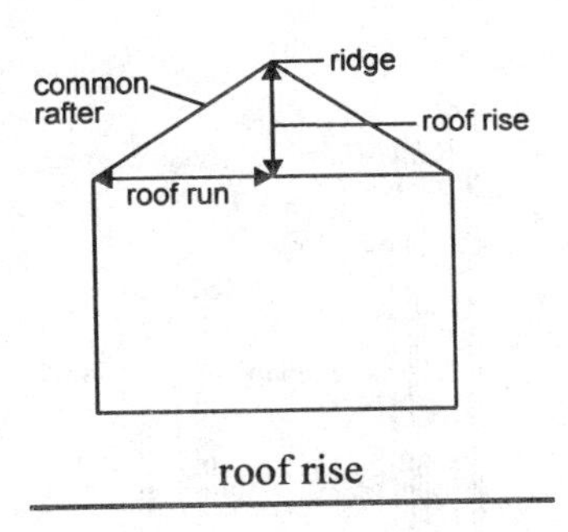

roof rise

roof run – the horizontal distance covered by the roof from the top plate of the wall to the midpoint, equal to one-half the span of the roof.

roof sheathing – the structural covering of the rafters or trusses, usually plywood or hardboard panels or closely-spaced boards. The roof covering is attached to the sheathing.

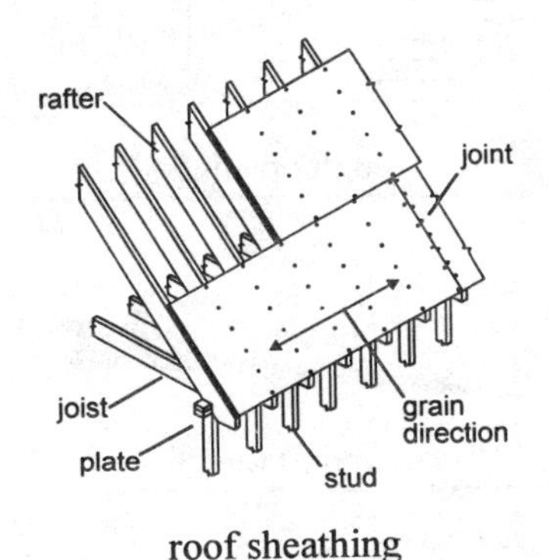

roof sheathing

roof slope – the angle of a roof expressed as the rise per 12 inches of horizontal distance, such as 3 inches of rise per 12 inches of horizontal distance.

roof tile – tiles of clay, cement, or other material designed for use as a roof covering.

roof truss – a rigid framework, made up of triangular sections, that spans across the walls of a building and which is used, in combination with other trusses, to support the roof. Roof trusses are usually prefabricated and delivered to the building site ready to install. This is especially true for large jobs involving several structures. On a small single-home construction job, trusses may be built on the job site.

roof truss

root crack – a crack in the root of a weld or the first weld pass that is made.

root gap – see root opening.

root of weld – the first weld pass or the deepest point of penetration of weld metal into the base metal.

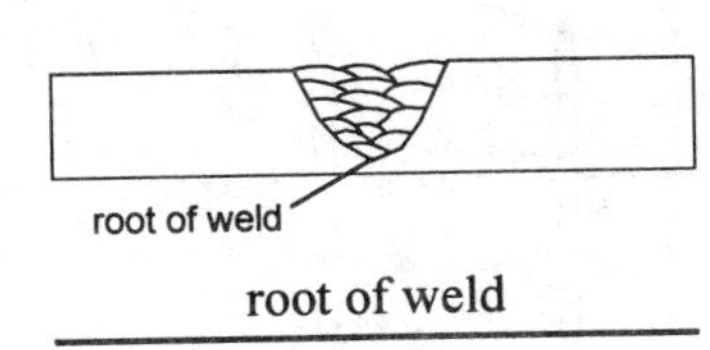

root of weld

root opening – the space between the pieces of base metal at the point the root pass of weld metal is to be laid. Also called the root gap.

root pass – the first layer of weld metal.

root penetration – the depth of weld metal penetration into the root opening at its centerline.

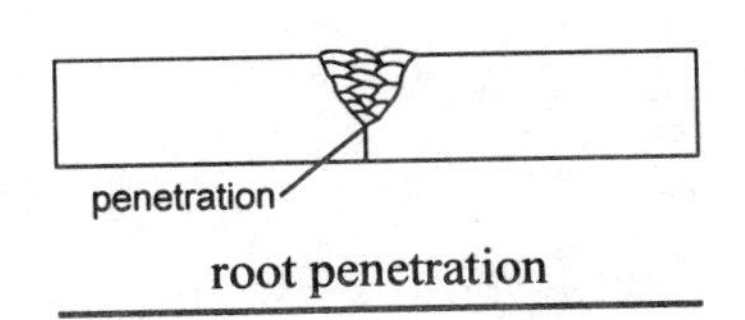

root penetration

rope – plant fibers or metal strands twisted into long lengths that can be used to secure or support objects.

ropey – a paint defect in which the finished surface dries with ridges or brush marks, rather than having a smooth or leveled finish.

rosewood – a reddish tropical hardwood with a distinctive grain pattern that is used in making expensive furniture.

rosin – an amber, gum-like substance distilled from pine tree resin. Rosin is the residue left after the turpentine has been evaporated from the resin. It is used as soldering flux and as a drying agent in paints and varnishes.

rosin core solder – a non-acid solder, with a hollow core filled with rosin soldering flux, used with electrical and electronic circuits.

rot – the deterioration and breaking down of a material caused by dampness and/or bacterial action.

rotary planer – see planer, rotary.

rotate – to turn about an axis.

rotated section – see section, revolved (rotated).

rotating vane anemometer – see anemometer, rotating vane.

rottenstone – decomposing limestone, or other friable stone, used as a fine abrasive for smoothing wood and soft metals.

rough-cut joint – see flush joint.

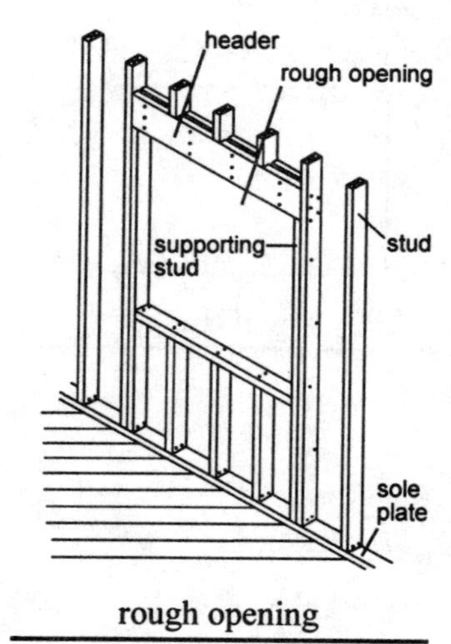

rough opening

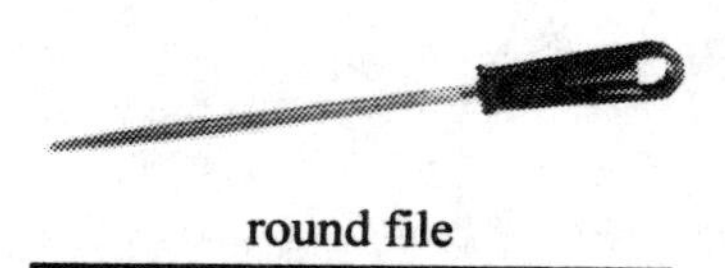
round file

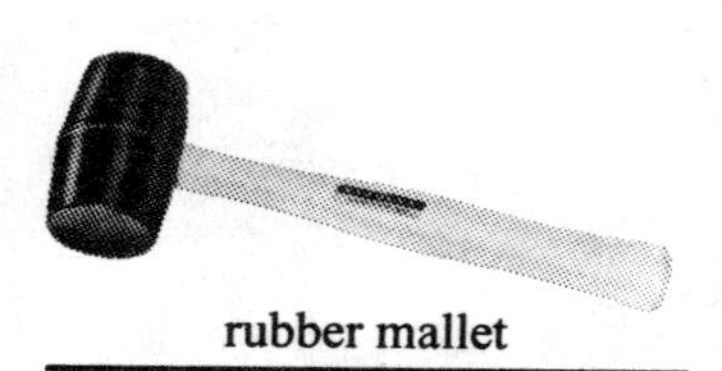
rubber mallet

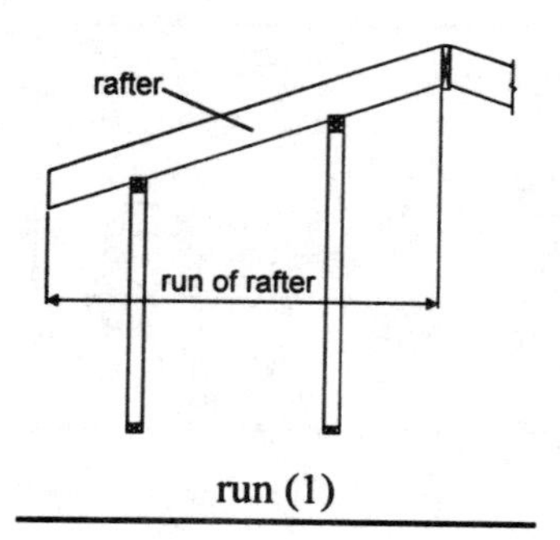

run (1)

rough file – a coarse file with about 20 teeth to the linear inch used to remove material quickly.

roughing-in – the installation of a system, such as electrical or plumbing, that will be behind finished surfaces and not show when the structure is complete. In an electrical system this would involve installing the boxes, and the conduit and wiring to the boxes; in a plumbing system it would involve the installation of the entire system except for the fixtures.

rough lumber – rough-sawn, full-dimension lumber with an uneven, unfinished surface. Also called rough-hewn lumber.

rough opening – a rough-framed opening in a building wall intended for the installation of a window, door, or other such device.

rough shot-sawn finish limestone – see limestone, rough shot-sawn.

round file – a slender cylindrical tool with cutting teeth around its circumference used to smooth and shape curves and holes in wood and other materials.

round key lock – a pin/tumbler type of lock that has vertical pins within a cylinder into which a key with a shank in the form of a hollow cylinder is inserted. Notches cut in the edge of the key cylinder are at the proper depth to move the pins just the right amount to align them.

rout – 1) to enlarge a crack in preparation for filling or patching it; 2) to work with a router.

router – a high-speed, portable power tool used for shaping wood by means of a spinning cutting bit that can be adjusted for depth. It is used for making grooves or for shaping board edges.

router bit – a cutting bit used in a router. Router bits come in a variety of types for making different cuts and shapes.

row houses – a series of dwellings which share common side walls.

rowlock – a course of bricks laid on edge with the ends of the bricks showing. Also called rolok or rowlok.

rubbed finish – a concrete finish made by rubbing the surface of cured concrete with carborundum stones to make it smooth and uniform. This type of finish is most often used on exposed walls.

rubber-base paint – a solvent-thinned paint containing chlorinated rubber, styrene-butadiene, vinyl tolune-butadiene or styrene-acrylate for use in damp areas, such as basements, to seal and waterproof the surface.

rubber flooring – a natural or synthetic rubber floor covering available in sheets of different thicknesses and sizes. It is used in labs and chemical plants because of its chemical resistance as well as in areas where an electrically-insulated floor surface is needed.

rubber insulated wire – wiring used for commercial and residential applications which is insulated for heat and/or water resistance. *See Box.*

rubber mallet – a mallet with a rubber head used on surfaces where a blow from a conventional hammer would leave unwanted marks.

rubber test plug – an expandable plug that is inserted into piping to seal off a section in order to test it for leaks.

rubber tile – see tile, rubber.

rubble masonry – 1) a masonry structure made of irregularly-shaped and -sized stones; 2) uncut stone used as backing material for masonry structures.

rule – a measuring device marked in increments that are actual dimensions, such as inches and fractions of inches. Also called a ruler.

Rumford fireplace – a shallow single-face fireplace with a high opening and flared firebox sides.

run – 1) the horizontal distance covered by a rafter; 2) width of a step; 3) horizontal distance covered by a flight of stairs.

rung – 1) the horizontal crosspieces or footholds of a ladder; 2) crosspieces on a chair which support or brace the legs.

runner – tracks or strips placed along ceiling and floor, usually as an attachment point for a partition or a sliding door.

Rubber insulated wire and cable

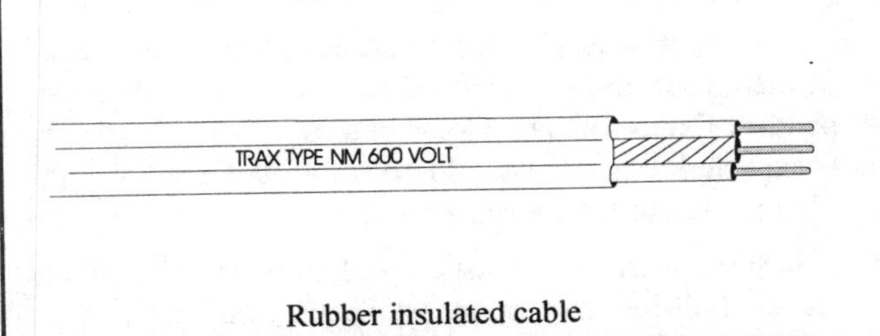

Rubber insulated cable

Insulated wire or insulated cable (two or more wires sheathed in a single protective jacket) is used for virtually all residential and commercial electrical wiring applications, as required by the National Electrical Code. At one time rubber was the main type of insulation used, and the code letter R signified rubber insulation. The term "rubber insulated" is intended here to include all insulated electrical wiring, whether the insulation material is natural or synthetic.

Most in-building wiring is Type NM (nonmetallic) cable because it is economical and works well for permanent in-wall circuit wiring. The insulation is rated at 600 volts for 140 degrees F (60 degrees C). Type AC (armor cable) is also used, particularly where the wiring may be exposed to potential damage, but its use is limited to dry locations only. The following are other insulated wire and cable types used commonly in construction. The code letters identify the type of insulation on the wire, such as H for heat resistant, R for rubber, T for thermoplastic, and W for water resistant, and sometimes its use, as in U for underground, F for feeder or SE for service entrance cable.

MTW - a moisture-resistant, oil-resistant, flame-retardant, and heat-resistant (up to 140 degrees F) thermoplastic insulated wire intended for use as machine tool wiring in wet locations.

R - a single conductor with moisture-resistant, flame-retardant, nonmetallic covering over the insulation. It is used indoors at operating temperatures up to 140 degrees F.

RH - a single conductor with heat-resistant insulation with moisture-resistant, flame-retardant, nonmetallic covering over the insulation. It can be used in dry or damp locations at a maximum operating temperature of 167 degrees F.

RHH - a single conductor with heat-resistant insulation with a moisture-resistant, flame-retardant, nonmetallic covering over the insulation. It can be used in dry or damp locations at a maximum operating temperature of 194 degrees F.

RHW - a single conductor with moisture- and heat-resistant insulation with a moisture-resistant, flame-retardant, nonmetallic covering over the insulation. It can be used in dry and wet locations at a maximum operating temperature of 167 degrees F.

RW - a single conductor wire with moisture-resistant insulation and moisture-resistant, flame-retardant, nonmetallic covering over the insulation. Its maximum operating temperature is 140 degrees F.

SA - a single conductor with silicone rubber insulation and a moisture-resistant, flame-retardant nonmetallic covering over the insulation. It can be used in dry or damp locations and special applications requiring conductor operating temperatures above 194 degrees F.

SIS - a single conductor with synthetic thermosetting heat- and moisture-resistant and flame-retardant insulation. Intended for use as switchboard wiring only.

T - a thermoplastic covered, flame-retardant wire for use in dry locations at a maximum operating temperature of 140 degrees F.

TF - a thermoplastic covered solid or stranded fixture wire designed for use at a maximum operating temperature of 140 degrees F.

TFF - a thermoplastic covered flexible-stranded fixture wire designed for use at maximum operating temperature of 140 degrees F.

TFFN - a thermoplastic covered, heat-resistant flexible-stranded fixture wire designed for use at a maximum operating temperature of 194 degrees F.

THHN - a thermoplastic covered, heat-resistant, flame-retardant wire designed for use in dry and damp locations at a maximum operating temperature of 194 degrees F.

THHW - a flame-retardant, moisture- and heat-resistant thermoplastic covered single conductor wire designed for wet and dry locations at maximum operating temperatures of 167 degrees F for wet locations, and 194 degrees F for dry locations.

THW - a flame-retardant, moisture- and heat-resistant thermoplastic covered wire designed for use in wet and dry locations at maximum operating temperatures of 167 degrees F (wet locations), or 194 degrees F (dry locations).

TW - a moisture-resistant, flame-retardant thermoplastic covered wire designed for use in wet and dry locations at a maximum operating temperature of 140 degrees F.

USE - heat- and moisture-resistant underground service entrance cable with a single conductor designed for use at a maximum operating temperature of 167 degrees F.

XHHW - a moisture- and heat-resistant thermoplastic covered wire for use in wet, damp and dry locations at maximum operating temperatures of 167 degrees F (wet), and 194 degrees F (damp and dry locations).

The following are insulated wiring used in portable power cords for use with power tools and lights:

S - thermoset plastic insulation with an outer jacket of thermoset plastic, designed for use with portable or pendant fixtures in damp locations where extra hard use is expected. It is rated at 600 volts for 60 degrees C and is called hard service cord.

SJ - thermoset plastic insulation with an outer jacket of thermoset plastic, designed for use with portable or pendant fixtures in damp locations where extra hard use is expected. It is rated at 300 volts for 60 degrees C and is called junior hard service cord.

SO - thermoset plastic insulation with an outer jacket of oil-resistant thermoset plastic, designed for use with portable or pendant fixtures in damp locations where extra hard use is expected. It is rated at 600 volts for 60 degrees C and is also called hard service cord.

STO - thermoplastic or thermoset insulation with an outer jacket of oil-resistant thermoplastic, designed for use with portable or pendant fixtures in damp locations where extra hard use is expected. It is rated at 600 volts for 60 degrees C and is also called hard service cord.

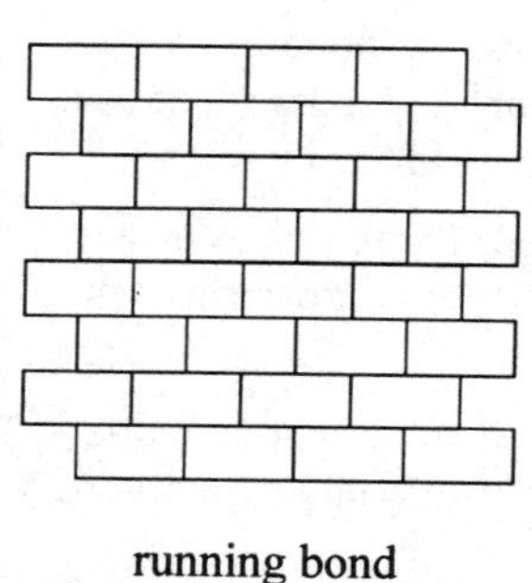

running bond

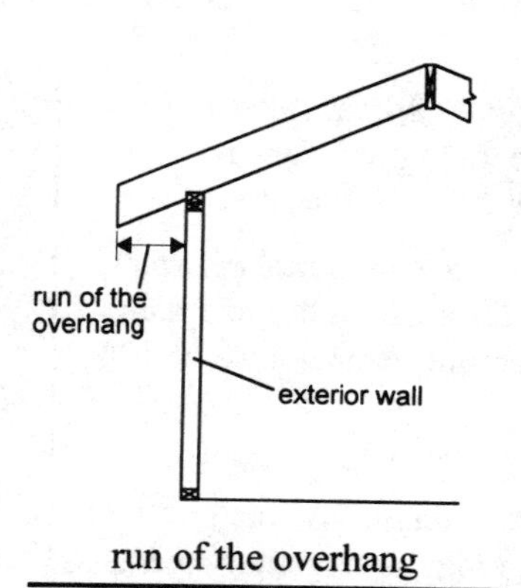

run of the overhang

running bond – a brick-laying pattern of all stretcher courses, each course offset from the next by half a brick or sometimes one-third, depending on the style of running bond.

run of the overhang – the horizontal distance, taken on a level, that the rafter overhangs the wall of a building.

runway rail – the rails on which a crane bridge travels.

rust – a reddish-brown coating formed by the oxidation of ferrous materials. Rust can be prevented by isolating the iron or steel surface from oxygen with a coating of paint or similar protective covering.

rustic – 1) rough-finished, uneven or irregular; 2) country-style; plain or unsophisticated in fashion or design, such as a simple wood cabin.

rusticate – to finish a wall surface and give it a rustic appearance.

R-value – a unit of measure of thermal resistance. The higher the value, the better the heat-insulating capabilities of the material. For example, an 8-inch lightweight concrete block has an R-value of 2.00 and a ½-inch sheet of plywood has an R-value of 0.63. The concrete block has far better heat-insulating properties than the plywood.

S: saber saw to system ground

saber saw – a portable power saw with a narrow reciprocating blade which can be changed for use with different materials. The blade is installed in a vertically reciprocating collar that grips it using set screws. Blades of different configurations and number of teeth are available for use with various materials, including metals. Most of the saw blades are only about ¼ inch to 5⁄16 inch wide. Such narrow blades make cutting around curves and rounded corners relatively easy. Attached to the saw is a metal shoe which rests on the work and adjusts to hold the blade either perpendicular to the work or at some other angle. Special features include variable speed controls, needed for cutting different materials, and scrolling, which permits turning the cutting direction of the blade independently of the saw body. Scrolling allows close control of direction changes, particularly in tight quarters.

sabin – a unit, equal to 1 square foot of material that absorbs sound perfectly, used to measure the sound-absorbing capability of materials.

sacking – to repair a finished concrete surface by rubbing a mixture of sand and cement into it using a coarse material, such as burlap sacking. This covers over minor blemishes, yielding an even surface.

saddle – see cricket.

saddle tap – a method for tapping into a main line, including one under pressure, and installing a corporation stop, or corp. *See Box.*

saddle-type repair clamp – a sleeve clamp that slips around a split pipe to make a repair. Bolts draw the clamp tightly around the pipe. There is a hole in the sleeve to slip over a branch connection fitting. This type of clamp is primarily used to repair pipe where a corp has torn out.

safe-edge file – see pillar file.

safety arch – see relieving arch.

safety factor – see factor of safety.

safety glass – two panes of glass with a strong plastic film sandwiched between. If the glass is broken, the fragments adhere to the plastic film, which tends to retain its integrity. This prevents the glass from fragmenting and causing injuries. One of its primary uses is for car windows.

safety goggles – protective eye covering, made with safety glass or impact-rated plastics, designed to protect workers from eye injuries while operating power equipment or during other work activities.

safety lamp – a lamp that is sealed to prevent it from becoming a source of ignition in the presence of flammable or explosive gases. This type of lamp is used around fuel storage areas, in mines, and in similar locations where an explosive atmosphere might exist. Also called a safety light.

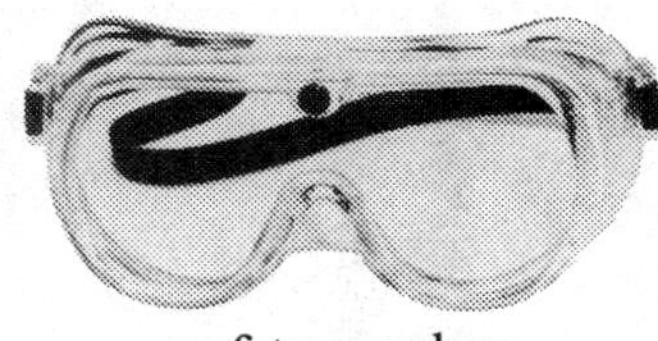

safety goggles

Saddle tap

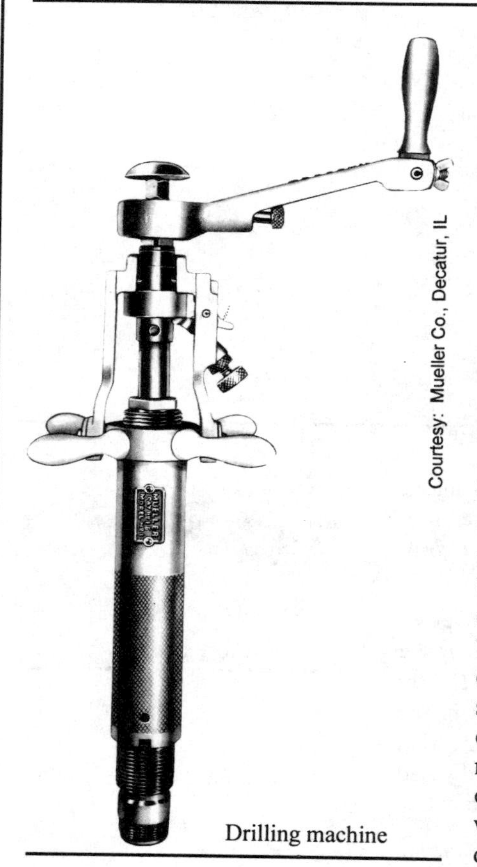

Courtesy: Mueller Co., Decatur, IL

Drilling machine

A saddle tap is a method used to tap into a main piping system line, including one under pressure, to install a corporation stop. It is used where the main line walls are too thin to permit drilling and tapping. A service clamp is placed around the main line and a corporation stop is threaded into the clamp. A drilling machine is mounted on the stop using an adaptor nipple, and the stop is opened. The drill taps into the main through the stop, without water escaping. When the boring bar is retracted, the corporation stop is closed. The stop now controls the opening. The corporation stop is threaded into the clamp, so threads in the main are not required, eliminating the stresses that would otherwise be introduced into the pipe wall. The clamp absorbs any bending load from the corporation stop, eliminating those potential stresses from the pipe wall as well. Once a corporation stop has been installed in the main line, and the drilling machine removed, a service line can be attached. Opening the stop activates the service line.

sailor

safety switch – an emergency shutoff switch for an electrical device, a circuit or group of circuits. A safety switch is the first device in the circuit so that when it is turned off, there is no power to anything else in the circuit. Safety switches are frequently installed in factories and machine shops where it may be important to quickly turn off all power with a single switch to prevent injuries.

safety valve – a pressure-relieving device, used to limit the pressure in a vessel or system to within a safe value. A safety valve is designed to open fully when it opens, rather than opening to an amount proportional to the flow and pressure. This is accomplished by means of a second disc larger in diameter than the disc that fits against the valve seat. When the pressure is sufficiently high to raise the primary disc slightly off the valve seat, the second disc (on the same valve stem) is then exposed to the pressure. This pressure acts on the larger area of the second disc, forcing the disc wide open. Pressure is thus rapidly relieved through the large opening. The valve reseats at a predetermined value. Also called a safety relief valve.

sailor – a brick placed upright with its length and widest face showing.

salinometer – an instrument that measures the salt content in a liquid, used on board ships or in other locations where fresh water is made by the evaporation of sea water. The salinometer can monitor the quality of the fresh water produced, either continually or on a sampling basis.

salmon brick – a salmon-colored brick that is underfired and relatively soft as a result of uneven or low temperature areas within the kiln. This problem was more prevalent in the past, as today's kilns produce a much higher quality brick. Salmon bricks are found among old salvaged bricks and can only be used for nonstructural or low wear applications.

salt glaze – a gloss finish on masonry created by firing the masonry in the presence of salt vapors. This causes a thermochemical reaction of the salt with the silicates in the masonry clay, producing a smooth and relatively impervious finish on the surface. The finish is easier to clean and less likely to absorb chemicals or gases than a standard finish. These bricks are used in extreme environments or in areas where they may be subject to exposure to chemicals.

sandblaster – a device that uses sand or other grit in a high-pressure air stream to wear away a surface or coating. The air stream containing the sand strikes the surface, causing the abrasive to wear away the surface. The spread and force of the stream can be varied by adjusting the air pressure and the size, shape and opening of the nozzle. The abrasive material and particle size can also be varied to suit the task. For instance, a soft abrasive such as ground walnut shells can be used to remove paint from a delicate missile or aircraft skin, while sand is used to remove paint from a masonry surface.

sand cone test – a test to compare the densities of a sample of soil from the field with a Proctor test sample for the same soil. The Proctor test is used to determine the correct proportion of water needed to be present in soil for optimum soil compaction. The Proctor test sample weight with 100 percent density is the known factor.

sander – a powered device that vibrates or otherwise moves sandpaper along a surface to smooth the surface or wear away excess material. Examples are belt, orbital, and straight-line sanders.

sander, belt – a power sander that moves an abrasive-coated revolving belt along the surface of the work. The belt removes material very rapidly and so must be kept in motion on the surface of the work so that it doesn't sand itself into a groove of its own making. Because it is a straight-line sander, it must be used with the grain of the wood to avoid scarring the wood by tearing.

sander, orbital – a vibrating-type sander that moves the sandpaper in an orbital motion along the surface of the work. This orbital motion eliminates sanding marks with finer grades of sandpaper, and ensures a flat, smooth surface.

sander, straight-line – a vibrating-type sander that moves the sandpaper in a straight back-and-forth motion on the work. The motion of this type of sander mandates sanding in the direction of the grain to avoid sanding marks. Material removal is not as fast as with a belt sander.

sanding – smoothing with sandpaper or a fine abrasive.

sanding a trench – using a loader to distribute sand evenly along the bottom of a trench. The sand provides a smooth, uniform and conforming surface on which to lay pipe that is to be buried. The pipe is then supported properly along its length within the trench and is cushioned against injury from sharp rocks and other damaging materials.

sanding block – a hand-held block with an abrasive material attached to it. Used for sanding flat surfaces or as an aid in maintaining the flatness of a surface.

sanding drum – a cylinder with sandpaper around its circumference, which is turned by a motor, such as a radial arm saw motor or a drill motor. It is used to sand curved shapes and contours.

sanding sealer – a sealer that is applied to wood to harden the surface wood fibers and fill the wood pores in order to yield a smooth surface when sanded. Sealers are used primarily on open grain woods, such as oak and pine. They are applied to a variety of surfaces, from furniture to floors, to ensure that the wood takes a stain or other coating evenly.

sanding wheel – a metal disc that is mounted on a power saw in place of a circular saw blade. Adhesive-backed sandpaper is affixed to the surfaces of the wheel. It is used to sand wood, to sharpen chisel blades and other tools, and to shape and smooth the cut edges of tile. A sanding wheel can be used with some precision when it is mounted on a fixed arbor, such as on a radial arm saw.

sandpaper – paper or other thin and flexible material with an abrasive coating made from crushed garnet, flint, aluminum oxide

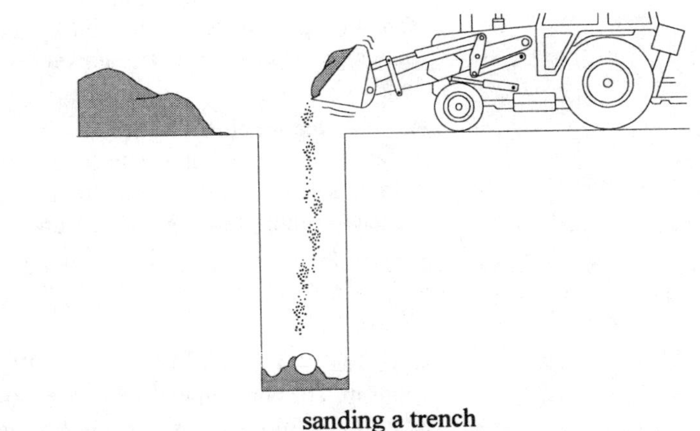

sanding a trench

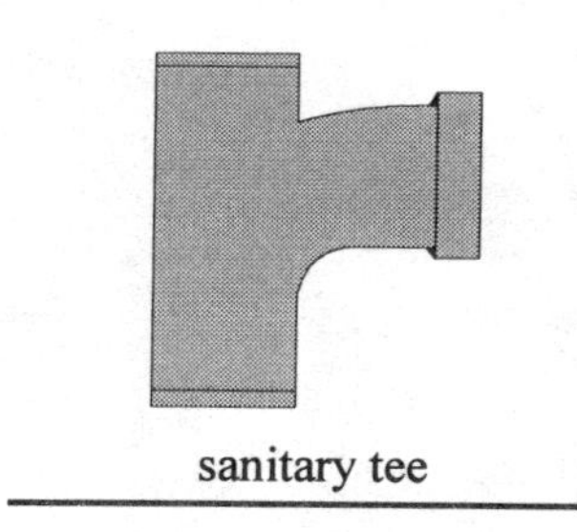
sanitary tee

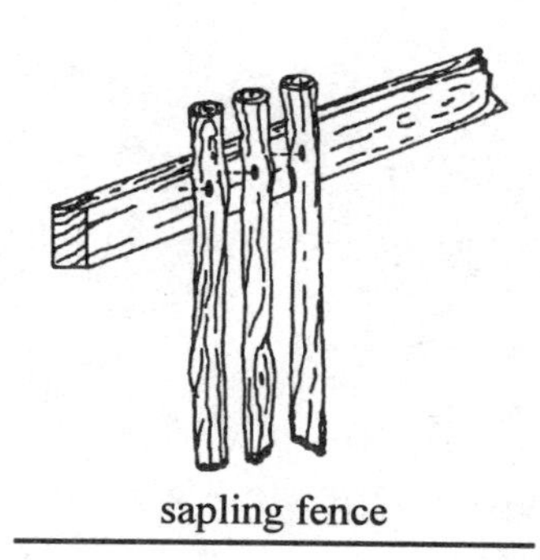
sapling fence

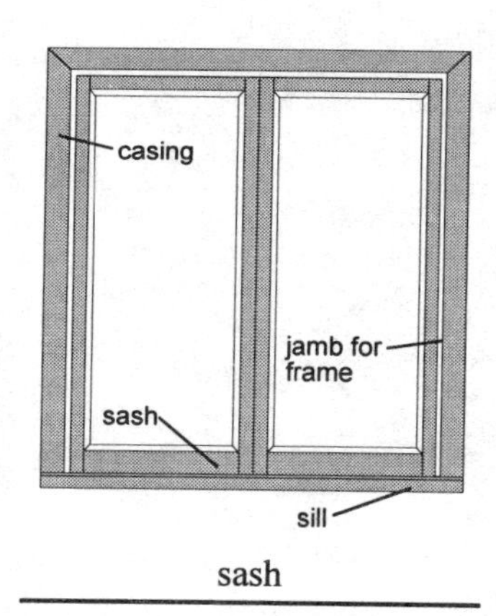

sash

or similar abrasive glued to one side. Sandpaper is made in a wide variety of grades, ranging from coarse to very fine. It is used to smooth surfaces or wear away small amounts of excess material.

sand-sawn finish limestone – see limestone, sand-sawn finish.

sandstone – a sedimentary rock consisting of quartz particles held together by silica, iron oxide and/or calcium carbonate. It has been used as a building material in the past, but now is most often used as a decorative facing.

sandwich braze – brazing two dissimilar materials with a brazing material in the form of a shim placed between them. Brazing involves heating materials to a temperature where the brazing material will bond with the base material. Most materials have different rates of expansion as they are heated. The shim, or flat sheet of brazing material, can take up some of the differential movement between dissimilar materials caused by different rates of expansion as they are heated.

sandwich panel – any panel that is composed of two or more materials with one or more other materials "sandwiched" between the different materials. The desirable or undesirable characteristics of one material can be offset by bonding it to another. For example, bonding sheets of metal to a center core of honeycomb plastic material creates a composite that is very strong, yet very lightweight.

sanitary sewer – the building or residential sewer system which carries sewage, but not storm, surface or ground water. It is the flow path and containment for waste matter from sinks, toilets, showers and baths.

sanitary tee – a plumbing drainage fitting in the approximate shape of the letter T but with a smooth, curved transition from branch to run in the direction of flow.

sapling fence – a fence constructed of sapling trees cut, trimmed, and nailed to fence rails.

sap streak – a section of a board filled with sap. The board may still be usable for a variety of applications, including structural ones, depending on the size of the streak and the requirements for the application.

sapwood – the wood directly under the bark of a tree, which is softer than the heartwood located at the center. Sapwood is less expensive than heartwood because it is not as strong, but it can still be used in many applications. Redwood and cedar sapwood is also less rot resistant than heartwood of the same type.

sash – a movable window frame containing one or more panes of glass.

sash balance – any means of offsetting the weight of a window sash so that it may be easily opened and left in an intermediate position. Often weights or springs are used for sash balances.

sash bar – see muntin.

sash block – a masonry building block with a ¾-inch groove in one edge to hold a window or door frame.

sash lift – a handle attached to the bottom rail of a double-hung window to aid in raising and lowering the lower sash.

sash lock – a lock to prevent opening the window.

sash weights – weights that are suspended from ropes attached to a window sash and strung over pulleys. The sash weight counterbalances the weight of the window sash so it may be easily raised, and can be positioned to remain in any intermediate location desired.

satin finish – a finish with a dull luster, having a similar sheen to satin cloth. Metal fixtures, such as bathroom faucets, towel bars and shower door sliders, are often manufactured with a satin finish.

saturated – soaked to capacity.

saturated felt – felt that has been impregnated with an asphalt compound to make it water resistant. It is used under shingles or other roof coverings, as part of the moistureproofing behind a shower wall, or in other construction applications where water resistance is needed.

saturated steam – steam that is at a temperature corresponding to its pressure such that any drop in temperature or rise in pressure will cause some of the steam to turn to water.

saw – 1) a hand or power tool with a toothed or serrated blade which can cut through materials such as wood or metal. See also individual saw types; 2) to cut.

saw arbor – the shaft on which a power-driven circular saw blade is mounted. The arbor is turned by a motor. The motor may be connected directly, through a gearbox, or through a V-box.

saw, back – a fine-tooth wood cutting saw with a stiffening rib along the back of the blade to prevent the blade from bending. It is used with a miter box to make straight and angle cuts, but can also be used independently of the miter box.

saw blade – the cutting portion of a saw, usually made of metal with a toothed edge.

sawbuck – two sets of 2 x 4s or similar-sized lengths of wood cut and fastened into two X-shaped sections which are hinged at the points where the legs cross. The two X-shaped sections are then connected from center joint to center joint by a fifth member so that they can support themselves and stand upright on their legs. The sawbuck provides a brace to hold a length of wood for cutting.

saw, chain – a power saw with a cutting chain rather than a blade for cutting. The cutting chain consists of a number of links pinned together so that they are flexible and free to travel around a guide bar. Cutting teeth are spaced along the outside length of the chain. Tangs along the inside periphery of the chain ride in a groove along the length of the bar guide to hold the chain in place. The chain saw is used for rough sawing wood.

saw, circular – a portable, hand-held power saw with a flat, circular blade. The blade is held on an arbor that is spun by the saw motor. A series of teeth around the outside diameter of the blade provide the cutting action. It is widely used in construction.

saw, coping – a saw with a narrow, fine-toothed blade stretched between the ends of a U-shaped holder. A handle attached to one leg of the U is used for gripping the saw. It is used for cutting curves and coped joints.

saw, crosscut – a saw with teeth designed to cut against (across) the grain of wood. The teeth are set with every other one bent slightly to the right or left to make fast cuts. See also crosscut saw.

saw file – a file expressly designed for sharpening saw blades.

saw horse – a portable bench or trestle used to support work at a comfortable height for cutting.

saw kerf – a cut that is the width of the saw blade; a groove made by a saw.

saw, keyhole – a hand saw with a blade that is tapered to a point. The point can be inserted into a hole that has been drilled in the material in order to make a cut that does not extend to the edge of the material.

saw, reciprocating – a power-driven saw with a blade that moves back and forth along the axis of the blade.

saw, rip – a saw with teeth designed to cut with the grain of wood. See also ripsaw.

saw, saber – see saber saw.

saw set – a tool for setting the teeth of a saw at the proper angle to the plane of the blade so that the teeth cut properly without binding.

saw teeth – the cutting points of a saw blade.

saw vise – a vise designed to hold a saw blade firmly, without damaging the blade, while it is being sharpened.

scab – a piece of lumber used to tie two other pieces together, providing support and strength at the joint.

scab ledger – a horizontal framing member that is nailed across the last complete set of common rafters on a gable roof, forming the base of a shortened gable end. It also serves as a stopping point for the rafters of a hip roof section in which the hip rafters do not reach all the way to the ridge, but are interrupted at the gable end, creating a Dutch gable roof end.

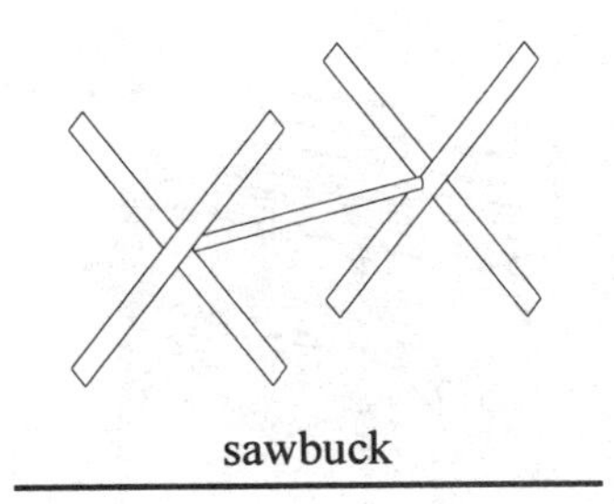
sawbuck

saw set

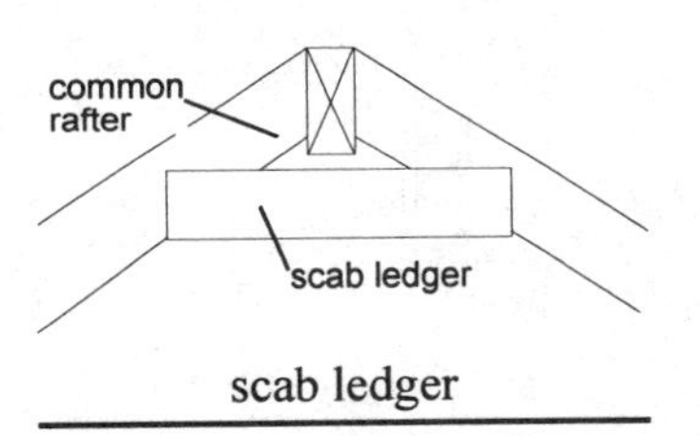

scab ledger

scaffold

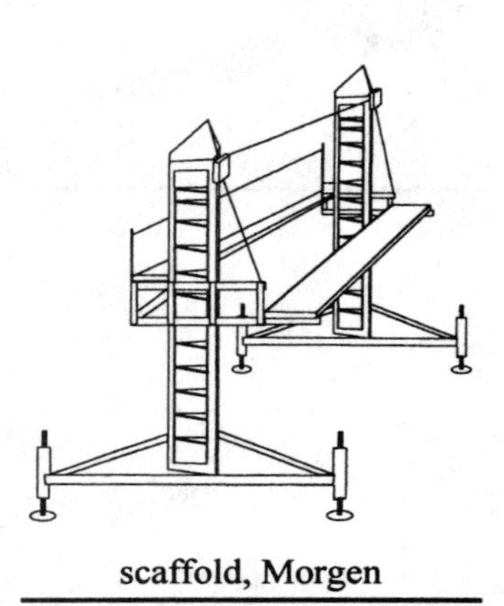
scaffold, Morgen

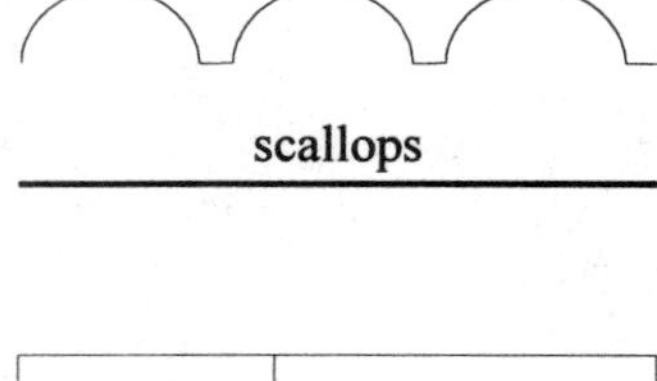
scallops

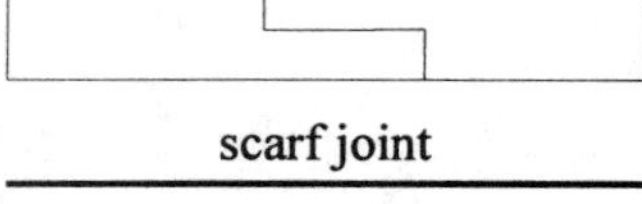
scarf joint

scaffold – a temporary wood or metal structure erected in or around a building to permit access to work being done above the first floor level during construction or for maintenance purposes. Also called scaffolding or staging.

scaffold bracket – a triangular structural form attached to a wall to provide a base over which scaffold planks can be laid. Also called a staging bracket.

scaffold, hanging – suspended scaffolding that can be raised and lowered; for use on high-rise buildings when exterior work or maintenance must be done.

scaffold, Morgen – freestanding triangular steel scaffold towers which support a platform section between them. The platform can be raised and lowered to access the work. This type of scaffold does not depend on an attachment to a wall, making it ideal for use in erecting masonry walls. The basic tower height is about 38 feet, but the height can be varied with extensions as needed when the towers are erected.

scaffold, sectional – scaffolding that is composed of metal sections that can be fastened together. The sections can be stacked as high as needed to access the work on buildings and other structures that are several stories high. At certain heights or locations it becomes more practical to use a more mobile type of scaffolding, such as the hanging type, rather than to continue stacking sections higher.

scaffold, swinging – scaffolds suspended from wire rope for drops of up to 200 feet. See also scaffold, hanging.

scale – 1) the relationship of the dimensions of the lines on a drawing to the actual dimensions of the structure, such as 1⁄4 scale equals 1⁄4 inch to 1 foot; 2) a measuring device that shows the relationship of numbers with actual dimensions, such as 1⁄4 inch = 1 foot; 3) a device for weighing or comparing weights; 4) loose peelings or flakes on a surface, such as paint scalings.

scaling – concrete that has thin sections or pieces sloughing off its surface.

scallop – rounded arches along an edge.

scantling – a small piece of lumber, such as a 1 by or 2 by, often fastened to a heavier member.

scarf joint – a lap joint at the ends of timbers where part of each timber is cut away at an angle so that the two timbers join on the same plane.

scarifier – 1) a powered roller with hardened steel cutters used to groove cured concrete; 2) a wide comb-like tool for scoring the surface of the first coat of stucco or plaster to provide a gripping surface for the next coat.

schedule – 1) an agenda or plan for organizing work and accomplishing tasks; 2) a list on a drawing showing information about the drawing and/or a list of sizes and types of commodities such as windows, doors, fixtures and materials.

schedule, fixture – a list on a drawing that identifies the details of electrical fixtures to be used at various locations.

schedule, lighting fixture – a listing on a drawing of the lighting fixtures needed for the job that is represented in the drawing. The fixtures are identified so that their respective locations are evident.

schedule, material – a listing on a drawing of the materials needed for the job that is represented in the drawing.

schedule, symbol – see symbol schedule.

schematic – a diagram of a plumbing system or electrical wiring routing or circuit.

scissors roof truss – a roof truss with a steep exterior slope and a bottom chord that angles up in the middle providing sloped interior ceiling surfaces.

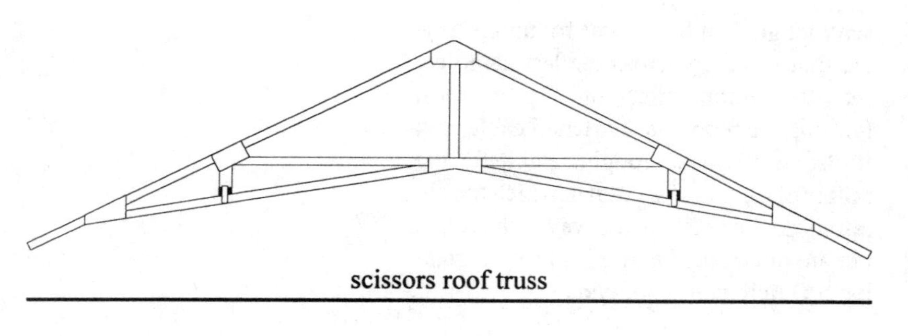
scissors roof truss

Scleroscope – a machine that tests the hardness of a material, such as a metal, by measuring the amount of rebound produced when a diamond-pointed hammer is dropped through a glass guide tube onto the material. The hardness of a material is related to wear resistance and/or brittleness, information often essential in designing metal components for severe duty service.

sconce – a wall-mounted lighting fixture.

score – 1) to scribe a line along a surface using a knife or other cutting tool; 2) to scratch or rough a surface in preparation for another coat of surface material.

scored block – a masonry unit with a groove across the midpoint of the block designed to give the appearance of a half block.

Scotia – a style of molding with an irregularly curved concave surface.

scraper – 1) a tool with a hardened edge for scraping surfaces clean; 2) heavy earthmoving equipment used for grading and leveling building sites and roadways. A cutting blade mounted under the machine scrapes soil from the surface and deposits it into a carrying bowl.

scraper, paddle wheel – earth-grading equipment with a paddle-wheel blade which rotates around, scraping off the top of the soil and moving it into a hopper. Used for fine trimming.

scratch awl – a sharp pointed tool used to scribe lines on surfaces.

scratch coat – the first surface coat of plaster or stucco which is roughened to ensure the bonding of the second coat.

screed – a rigid, straight piece of wood or metal used to level concrete or other materials that have been poured into forms, and to remove the excess material.

screed auger – an auger that is part of a paving machine. The auger spreads asphalt along the length of the screed to evenly distribute it, and the screed levels the asphalt on the roadbed.

screed, power driven – a screed, consisting of a framework of two boards held just over a foot apart and an engine that vibrates the boards, which is used to strike off and compact concrete.

screen – a fine mesh, usually wire or nylon, used as a covering, barrier or filter. Frame-mounted screen material is commonly used over windows, doors and other openings in buildings to keep out insects and debris. Screening material is used to filter large pieces of material out of mixes, such as sifting large pieces out of fine aggregate or sand before adding it into a cement mix.

screen block – a masonry building block with an open latticework structure, usually used for fencing or screening off a patio area.

screen wall bond – a pattern of brickwork in which the ends of the stretchers used in each course are about ½ brick apart to create a latticework effect with openings between each brick.

screw – a fastener with a raised spiral ridge down the shank, and a flared head slightly larger in diameter than the shank. The head is slotted so that a screwdriver can be inserted for placing or removing the screw. *See Box.*

screwdriver – a tool with a handle on one end of a shank and blade on the other end used for installing or removing screws. The end of the blade is shaped to fit into the slot on a specific type of screw. There are a variety of screwdriver designs, of both the manual and power-driven types. Various handle designs have been developed to improve grip, torque and convenience of use, and to provide insulation from electrical current. Some have interchangeable bits to accommodate different screw head styles and sizes. Power-driven types are available in cord and cordless styles.

screw extractor – a tool for removing broken-off bolts or screws. The tool has a left-hand steep spiral threaded shank so that when it is inserted in a hole drilled in the center of a bolt or screw that has a right-hand thread, the extractor exerts a loosening force on the screw when it is turned.

screw eye – a screw with a pointed, threaded shank on one end and a circle or loop on the other end instead of a head. It is screwed into wood, leaving the eye exposed. The eye can be used to hold wire, such as antenna cable, in place along its

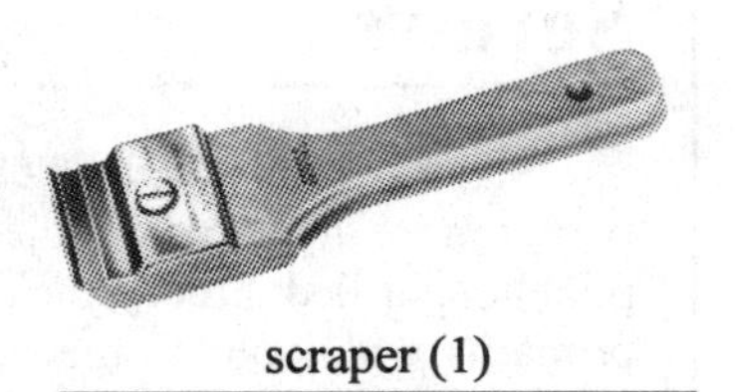
scraper (1)

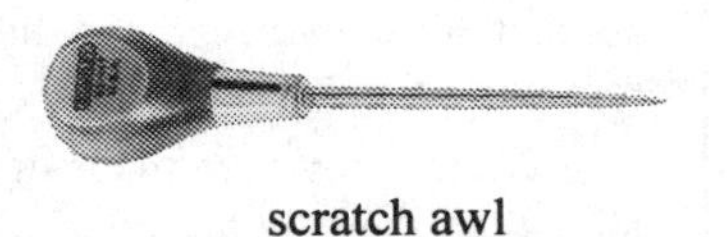
scratch awl

screen block

Screws

Screws are threaded metal fasteners designed to join two or more materials or objects by being turned into a pilot hole drilled through the materials. Screws are used primarily with wood or metal, though they can be used with other materials, such as drywall, as well. Screws have flat, oval, pan or rounded heads that are slightly larger in diameter than their shank. The head is slotted so that a screwdriver can be inserted for placing or removing the screw.

Wood screws are made of steel, brass or other metal alloys. As they are turned into the material, their threads cut into the wood and grip the sides of the hole. Their ability to grip gives them an advantage over fasteners with smooth shafts, such as nails. Within limits, the withdrawal load of a screw varies directly with the depth of penetration of the threaded shaft into the wood and the diameter of the screw. The longer the screw and larger the gauge (the diameter of the shank), the higher the withdrawal resistance. Some screws are self tapping. They have a pointed tip and sharp threads that spiral all the way to the head. They cut and thread their own hole as they are turned into the material. Self-tapping screws have a higher withdrawal resistance than other wood screws.

Machine screws are manufactured with a series of different pitches of even and regular threads. The threads fit into matching grooves cut into the interior of the hole into which they are turned. The pitch is the number of threads per inch. For example, a screw with a pitch of ¼ -20 signifies a ¼-inch diameter with 20 threads per inch, and one that is ¼-24 has a ¼-inch diameter with 24 threads per inch. There are also different designations used to identify threaded fasteners so that matching items, such as bolts and nuts, can be specified. These designations are:

- UNC - Unified National Coarse
- UNF - Unified National Fine
- UNEF - Unified National Extra Fine
- 8N - Coarse threads for diameters over 1 inch
- 12N and 12UN - Fine threads for diameters over 1.5 inches
- 16N and 16UN - Extra fine threads for diameters over 2 inches

Both wood and machine screws are sized by gauge and length (up to size 12). Common gauge sizes generally run from 0 which is 1⁄16 inch to 12 which is 7⁄32 inch, through they do come larger. The length size is the distance from the tip of the screw to the plane of the material's surface.

Drywall screws are designed specifically for use with gypsum drywall panels and provide up to 350 percent more holding power than nails. They have flat heads which flare out slightly from the shank and twist into the facing paper as they are tightened. The twisting increases the holding power of the screw as well as preventing tearing of the paper. Drywall screws have sharpened points that eliminate the need for predrilled holes, and they are threaded at a steeper angle than wood screws, reducing the number of turns required to drive the screw into the material. The flat head is easily concealed with joint compound, providing a smooth finished surface.

There are three common types of drywall screws. Type W screws are used for fastening drywall to wood framing. They are 1¼ inches long for use with ½-inch drywall and penetrate about ⅝ of an inch into the wood. Type S screws are designed for attaching drywall to sheet metal or metal studs. They have a smaller diameter shank than a Type W and come in a variety of lengths. They have a slotted drill-type point that easily penetrates the metal, eliminating the need for predrilling holes. The Type S screw should penetrate at least ⅜ of an inch into the metal. Type G screws have double threads and are used to attach drywall to gypsum coreboard or a base layer of drywall. The double threads allow the screws to grip the gypsum firmly, resisting pull-out. They should penetrate at least ½ inch into the drywall base or coreboard.

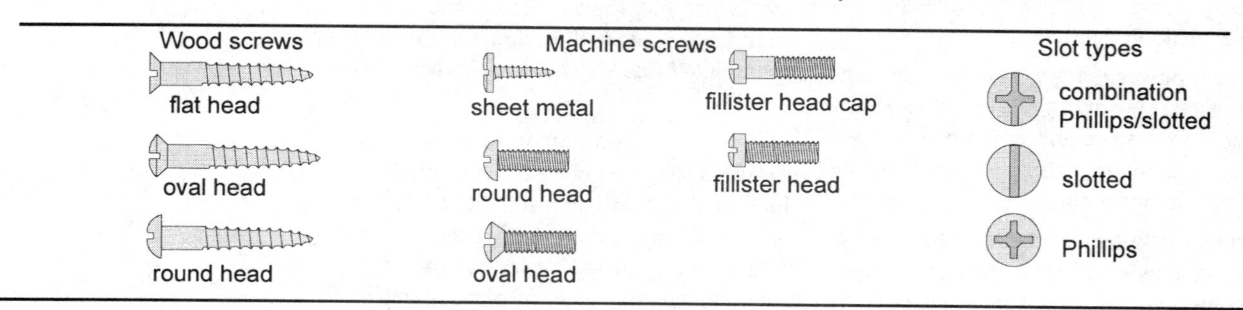

routing, for light-duty anchor points for pull cords, or as part of a two-piece fastener, the other piece being a small hook.

screw gauge – a gauge for measuring the size of a screw.

screw gun – a power screwdriver, usually having the features of a clutch, magnetic screwdriver bit, and adjustable screw installation depth. The clutch is combined with a depth adjustment so that when the screw is driven to the predetermined depth, the drive disengages from the screwdriver bit. This permits a series of screws to be driven to a consistent depth.

screw machine – a metal lathe designed to cut machine screw threads.

screw pitch gauge – a gauge used to check the number of screw threads per inch on a screw.

screw thread micrometer – an instrument that permits precise identification of the diameter of a threaded rod or shaft, often to the closest thousandth of an inch. It is used to make measurements where great precision and close tolerances are needed.

screw threads – the raised spiral ridges cut into the shaft of a screw. As the screw is turned into a hole in a material, the threads fit into matching grooves cut into the interior of the hole. In soft materials, the ridges or threads of the screw can cut their own grooves into the interior of a hole as the screw is turned into the material. See also screw.

scribe – 1) to cut or etch a line along a surface; 2) a sharply pointed and hardened steel tool used for marking a surface to be cut. The width of the marking tool can determine the precision of the cut. A sharply pointed scribe can be used to mark a very exact line or cut on a surface.

scribing – the fitting of woodwork to an irregular surface. The edge of the irregular surface can be held against the piece to be cut to match, such as a trim piece, and the contour of the irregular surface marked along the trim piece. The trim can then be cut to match the irregular surface contour and fit precisely.

scroll – a curved or spiraling ornamental design.

scroll saw – a power or hand saw designed to permit the cutting of larger pieces of work than a jigsaw will accommodate.

scroll work – ornamentation containing curved and spiral designs.

scuff – to scrape, causing minor damage to a smooth surface.

scuff sand – to sand lightly and break the sheen on a glossy surface to provide a surface to which new paint will adhere.

scupper – a drain on a roof or other deck or through the side of a building to allow rain water to run off.

scuttle – a small opening in a roof or ceiling which permits access to an attic or rooftop.

sealant – a caulking compound designed to form an airtight or waterproof bond at a joint, while remaining flexible.

seal bond – the pressure-tight mating at nonmetallic pipe joint surfaces.

sealer – a coating applied to a surface to close off the pores and prevent penetration of liquids to that surface, such as a coating of dilute varnish or shellac applied to a wood surface.

seal-off fitting – a conduit fitting that prevents the leakage of gases or spread of fire from one section of conduit to another.

seal weld – a weld bead that is laid for the purpose of preventing a leak. For example, a weld bead laid around the junction where a pipe is screwed into a threaded fitting.

seam – a taped wallboard joint, or a closure between two pieces of material, such as a sheet metal joint.

seam weld – a weld along joining or overlapping pieces.

season – to age or cure; to remove excess moisture, as in seasoned lumber.

seat angle – 1) a section of angle iron on which a beam rests; 2) the angle of a valve seat.

screw gun

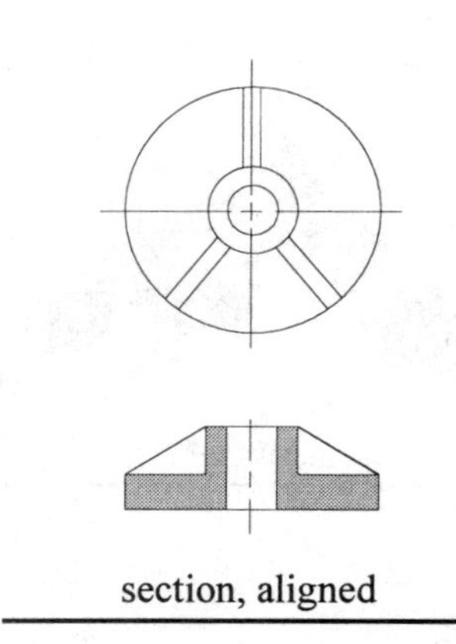
section, aligned

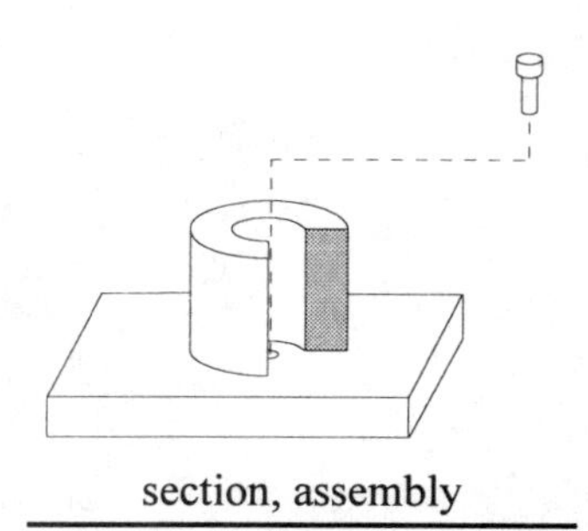
section, assembly

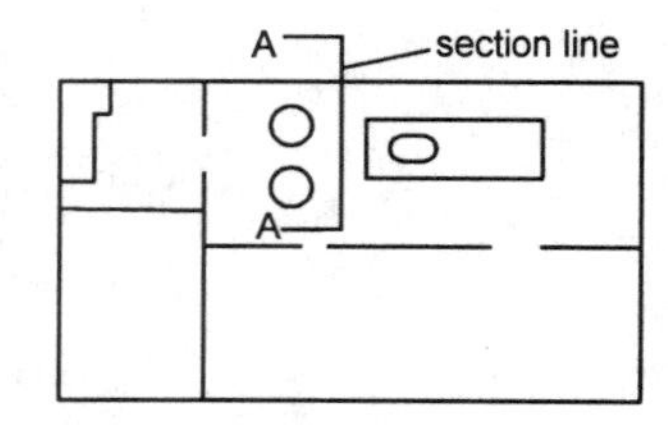

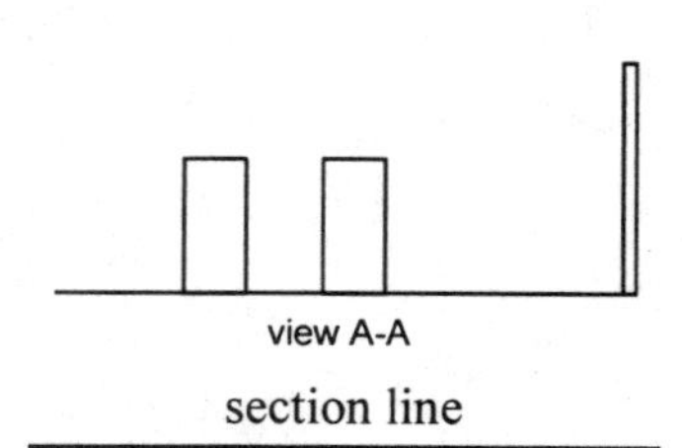

section line

seat cut – see rafter seat cut.

seawall – a barrier or embankment along a shoreline designed to act as a breakwater and prevent erosion and water damage.

secondary coil – the second coil in an electrical transformer, which is the output side of the transformer. Transformers can be used to drop voltage in order to provide 24 volts for a furnace gas solenoid valve or a doorbell.

second mortgage – a mortgage taken out against a piece of property which already has an existing mortgage. The holder of the first, or pre-existing mortgage, has priority in recovery over the holder of a second mortgage. For this reason, the second mortgage is generally for a smaller amount than the first in order to limit the risk of loss.

seconds – a lower grade material, or a material or product that in some way does not meet the standards set for the product. Seconds can be used where appearance or deviation from dimensional standards are of little consequence. For example, cabinet doors that have minor blemishes may not be acceptable in a kitchen, but could be used for garage cabinets.

section – 1) one of the 36 divisions of a township equalling one square mile, or 640 acres; 2) a part or division; 3) a cutaway view. See also section view.

section, aligned – a cut-away view into the interior of an object or building (as it would look if part of the exterior were removed), shown on a drawing lined up with the full view from which it is taken.

sectionalized brick masonry – see prefabricated brick masonry.

sectional switch boxes – electrical boxes with removable sides to permit ganging two or more boxes together. The boxes are held together with screws.

section, assembly – a section view that shows how parts of an object or building will fit together when completed. This is done by showing the parts in an exploded view, with dashed lines from the parts to their respective locations.

section, broken-out – a section view of a small portion of the whole object or building to show a detail. The section is enlarged with the detail called out.

section, full – a view of a whole object or building drawn as though the viewer is looking completely through it.

section, half – section view taken halfway through the object being depicted. The other half of the object is shown from the exterior of the object. Thus, the object is shown with half cut away to expose the inside.

section line – a line, often dashed or with arrows, drawn on a plan or elevation view drawing to indicate where an enlarged section view is taken.

section modulus – a value used to indicate the ability of a structural member to resist bending stress. The higher the section modulus, the more resistance the member has. The section modulus varies with the size and shape of a structural member. For example, an I-beam with a greater weight per foot will have a higher section modulus than one with the same overall size with a lower weight per foot. The heavier beam will be stiffer.

section, offset – a view cut through two or more planes through an object to show special features within the object. The offset section shows portions of the inside at different depths with relation to one another.

section, revolved (rotated) – a section view that is shown revolved (or rotated) through 90 degrees from the object from which it is taken.

section view – a cut-away view into the interior of an object or building as it would look if part of the exterior were removed. It is used to show interior features and add clarity to the drawing details.

secured loan – a loan that is made in exchange for an interest in a tangible item of value, such as a piece of property.

security – 1) the quality or state of being free from danger or want; 2) precautions taken to prevent or limit risk of loss or physical danger; 3) property or items of value offered in the event of default on a loan; 4) a financial instrument, such as a bond.

sedimentary rock – rock formed from sedimentary materials, such as sand, rock fragments, shell remains and silt, which have been compressed over time into a solid mass. It is generally homogeneous throughout, and in the absence of faults or cracks, is of approximately constant strength throughout. Limestone, sandstone, shale and gypsum are examples of sedimentary rock used in construction or in making various building materials.

sediment interceptor – a strainer device for the water outlet of a swimming pool designed to filter out hair, lint and sediment before the water is recirculated back into the pool.

seedy – a newly-painted surface that has specs of dirt, old paint, or other debris that was floating in the environment, marring the finish. Paint should be allowed to dry in a clean, dust-free environment whenever possible.

seep – an often undesirable fluid flow through porous material or small cracks in a surface, such as through the unsealed concrete wall of a basement.

segment – 1) a small piece or part divided from an object; 2) a line or portion of a line defined by two points.

segmented arch – see minor arch.

seismic – of or relating to an earthquake, or earth vibrations from a source other than an earthquake, such as an explosion.

seismic analysis – a scientific study of the strength of a structure or other object to resist damage cause by an earthquake.

seismic joint – a joint designed to permit movement during an earthquake in order to limit stress on a structure. Building code requirements vary depending on the geographical area and its history of seismic activity. The need for seismic joints and other requirements to limit seismically induced damage is determined by the structure's design and location.

seizure – the unintended bonding together of movable parts so as to prevent the designed movement. This may be caused by lack of proper lubrication between parts in a machine or from excess heat or cold. Also described as seizing or freezing up.

self-flashing skylight – see skylight, self-flashing.

self-rimming lavatory – see lavatory, self-rimming.

self-supporting partition – a gypsum wallboard partition consisting of one or more layers of wallboard fastened to gypsum coreboard without internal framing members added. It can support only its own weight.

self-tapping screw – a screw that has a pointed tip and sharp cutting edges on its steeply pitched threads to permit easy starting of thread grooves in an unthreaded hole. Commonly used with light metals and some nonmetal materials.

selvage – 1) the edge of a material that has been finished in such a manner as to protect the integrity of the body of the material from unraveling or coming apart, such as binding the edge of carpeting; 2) the edge or border of a material which has been finished in a manner which allows it to be easily joined with another piece; 3) the edge plate of a lock that has been bored for the passage of the lock bolt.

semi – 1) a prefix meaning half when used with a word that indicates bisection, such as semicircle; 2) a prefix meaning partially or not completely, as in semidarkness.

semianthracite – a coal that is softer than anthracite, easier to ignite, and which burns clean. It is used for heating, as in coal-burning boilers or furnaces, in the same manner as anthracite.

semiautomatic arc welding – electric arc welding with an automatic feed of metal filler to the weld, but with other operations controlled manually by the operator.

semiautomatic brazing – brazing with a machine that uses an automatic feed of filler metal, but with the braze speed manually controlled by the operator.

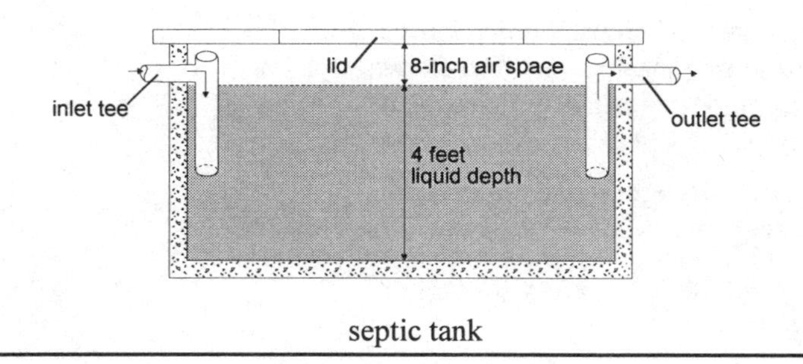

septic tank

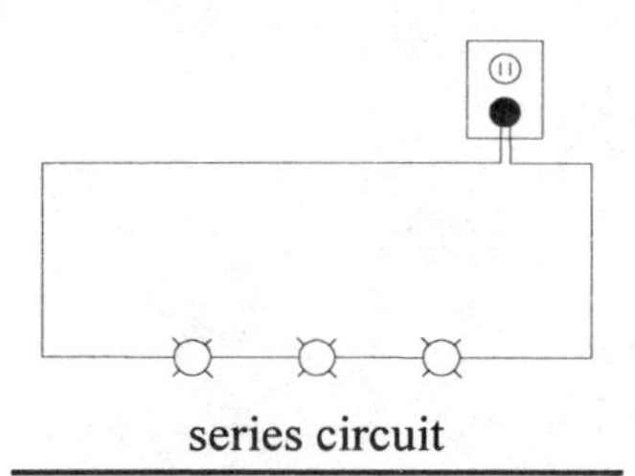

series circuit

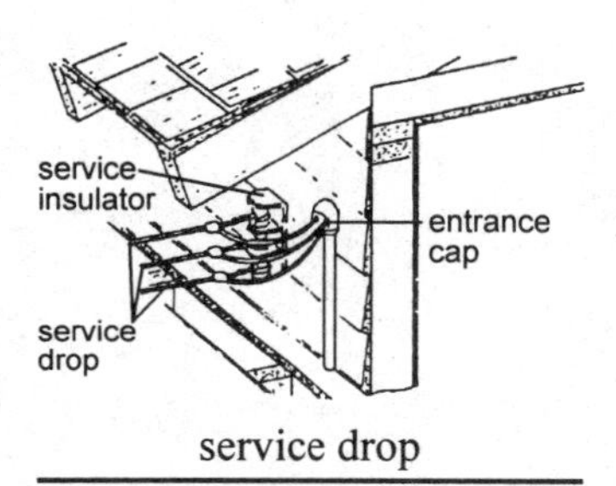

service drop

semiautomatic welding machine – a welding machine with an automatic feed of filler metal, but with weld application manually controlled by the operator.

semicircle – one-half of a circle.

semicircular arch – see major arch.

semigloss – a paint, varnish or other coating that dries with a sheen that is between a flat and a gloss.

semirigid continuous base – a road pavement made of asphalt that is firm but not as rigid as concrete. This type of base is better able to maintain the integrity of a road surface where there is some ground movement, such as where a road is laid over adobe soil which swells and shrinks with the absorption and loss of moisture.

sepia print – a print or drawing reproduction in a reddish-brown color, with dark lines on a lighter background. The print is made on semitransparent paper and can be used to make blueline prints. It is gradually being supplanted by the use of faster and better quality reproduction techniques.

septic tank – a water-tight tank for collecting sewage, and in which anaerobic bacteria decompose solid matter. The resulting gases are vented, and the remaining liquid is drained out through leach lines into a drain field or outlying soil beds.

sequence – an order of events or tasks. Sequencing events is an important step in scheduling and ordering for most businesses, and is especially so with the construction industry. An example of task or event sequence in house construction might be: excavation, foundation layout and pour, subfloor installation, wall framing, roof framing, and rough plumbing and electrical, etc. Some tasks can be done simultaneously, but others depend on the completion of one event in the sequence before the other can begin. Planning the sequential order of events saves time and money.

sequential firing – a method of blasting in which several charges are laid and then fired one at a time in series. This method is used effectively to break up rock underground in a trenching operation. Sequential firing controls the explosive force and direction of the blast.

series circuit – an electrical circuit in which the loads are connected so that the current passes through one load before it can go to the next. This type of circuit is used in perimeter alarm systems to alert against intruders. Breaking any part of the circuit, such as opening any window or door, triggers the alarm.

series submerged arc welding (SAW-S) – a welding process using two consumable electrodes that are positioned above the work. An arc is generated between the electrodes, creating the heat to make the weld. As the ends of the electrodes melt, they are deposited on the heated base metal to make the weld. The base metal is not part of the electrical circuit.

series wound – an electric motor in which the field of the motor and the armature of the motor are connected so that the electrical current passes through one before the other.

serrated – 1) notched or toothed on the edge, or having teeth directed forward or toward an apex; 2) a surface lined with ridges or marks.

service drop – the electrical wires from the power utility's overhead lines to a building. Also called service.

service entrance – the interface of the power utility's lines with the building's wiring. The utility service may come from an overhead service drop or underground cable connection.

service entrance cable – factory- assembled electrical wiring consisting of two or three insulated conductors with a stranded, uninsulated neutral conductor wrapped around them. It is similar to nonmetallic sheathed electrical cable.

service entrance distribution box – the circuit breaker or fuse box located at the service entrance that is the distribution point for the main circuits in a building.

service insulator – the anchor for the service drop where the utility's electrical wires are run to a building. Also called a house rack.

service lateral – underground power lines from the power line transformer to the meter box in a building.

service marker – a wood stake, placed at the capped or plugged end of a length of buried pipe, which extends above ground to mark the location of the pipe end. The marker is used in situations where there may be a need for additional pipe extensions later, such as for service lines to a housing tract that is being built one phase or section at a time.

service mast – the vertical section of conduit through which the service wires to a building pass before reaching the service entrance breaker box.

service panel – an electrical panel containing circuit breakers and serving as a distribution point for the electrical service entering the building.

set – the initial hardening or drying stage, as applied to such items as concrete, plaster, adhesives or paint. Also called set up.

setback – 1) the distance that local code requires a building to be from the street; 2) the amount a valley rafter must be raised so that the edges of the valley rafter match the tops of the rafters that abut it.

set retarder – a compound that can be added to concrete to slow the time it takes for the concrete to set up, such as on a large pour in warm weather, where it may be necessary to slow the setup time in order to ensure adequate working time to complete screeding, floating and troweling the surface.

setscrew – a screw used to hold two otherwise movable parts in position so that they do not move, such as to hold the blade in place in a saber saw or reciprocating saw.

setting boots – driving 4-foot long lath stakes, marked with the surveyor's cuts and fills rounded to the nearest foot, behind the surveyor's hubs. The stakes are easier for the grader to see than the surveyor's short hubs.

settling – 1) drooping or sinking down into to the earth due to the effects of gravity. Also called settlement; 2) solids, that were suspended in a liquid, falling to the bottom of the container or mold into which they are placed.

set-tooth blade – a saw blade with the teeth bent alternately to each side. The saw kerf width produced with this type of blade is greater than the blade thickness, preventing the blade from binding in the work. Many types of saws, such as rip saws and cross-cut saws, have blades with this feature.

sewage – liquid waste containing animal, vegetable or mineral matter.

sewage ejector – an ejector pump used to pump out sewage receiving tanks which collect sewage from appliances and fixtures installed lower than the adjacent public sewage main.

sewage lift system – a system using a pump to lift sewage from a building drainage system that is lower than the public sewage main to which it is connected.

sewer – drainage system used to convey sewage wastes to a treatment or disposal facility.

sewer brick – an abrasion-resistant brick with low moisture absorption, which works well in wet applications, such as storm sewers.

sewer gas – gas from the decay of organic materials conveyed through the sewer. The gases may be poisonous or combustible.

shade – 1) a color or a variation of a color; 2) shadows or areas in which sunlight is blocked or obscured; 3) a device for blocking or diverting light.

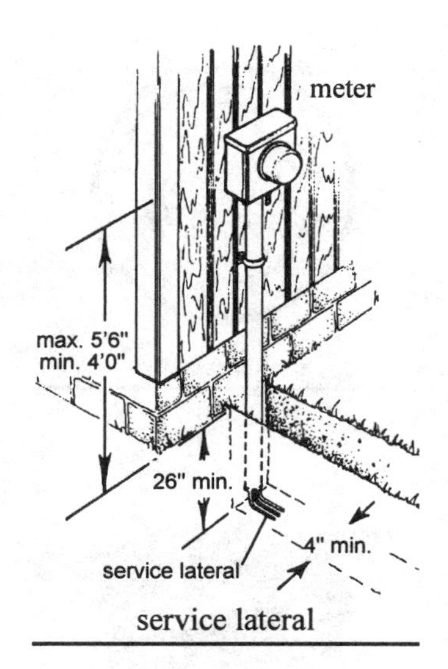

service lateral

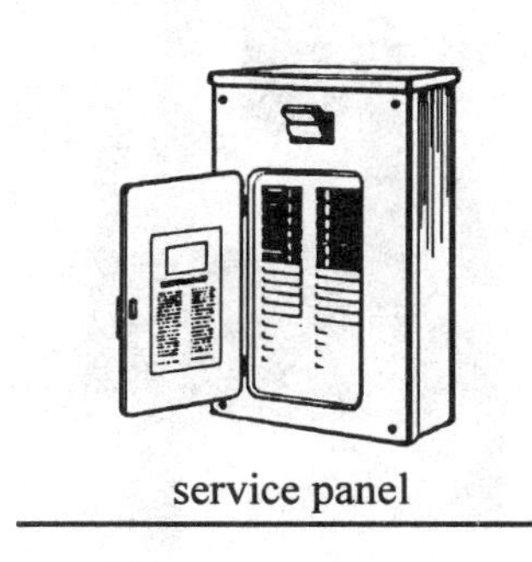
service panel

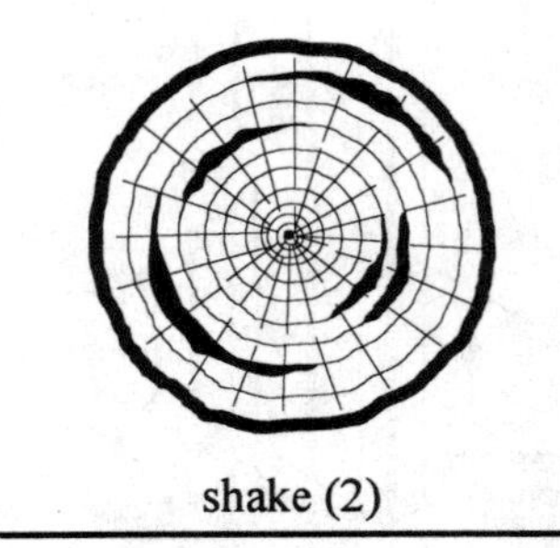
shake (2)

Courtesy: Delta International Machinery Corp.
shaper

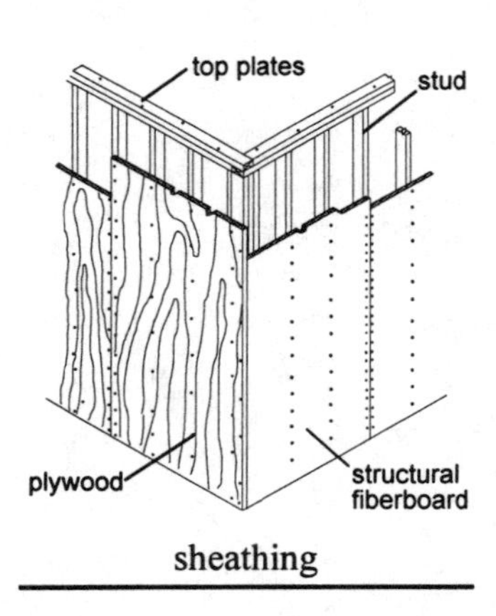

sheathing

shadowing – 1) a finished wallboard joint defect in which the edges of the joint tape are visible; 2) a paint defect in which the painted surface appears different when viewed from different angles. Shadowing may be caused by the paint itself, or as a result of the paint spray gun not being held perpendicular to the surface being painted.

shaft – 1) a narrow vertical opening or passage through the floors of a building, such as a ventilation shaft or elevator shaft; 2) a cylindrical rod used to support rotating pieces or transmit power or motion by rotation; 3) a straight handle on a tool or instrument.

shaft wall – a wall of an enclosure built for stairways, elevators, electrical wiring, piping or similar systems.

shake – 1) a thick wood shingle split from a log; 2) a separation in uncut wood along the grain of the wood (between annual rings).

shake roof – a roof covered with wood shake shingles.

shale – a rock formed from hardened, finely stratified clay, mud and silt.

shank – 1) the portion of a tool, such as a screwdriver, between the handle and the end of the blade; 2) the body of a nail; 3) a projection on an object by which it can be attached or connected to another.

shape – 1) to create or form an object out of raw material; 2) to adapt or make fit.

shaper – 1) a power rotary tool for cutting moldings and other shapes in wood. Cutter blades are mounted in a rotary head spun by a motor. The work is guided past the head and, depending on the shape of the blades in the head, the wood may be shaped into various moldings; 2) a metal working machine in which a cutting tool is moved across the work by a reciprocating driving arm. It is used to make metal parts such as gear teeth.

shaper gauge – a precision measuring instrument used to check slot widths, and for setting the blades of a shaper or planer to the proper depth. It consists of two inclined planes that are adjustable with relation to each other. The planes can be moved so that the assembly becomes thinner or thicker. Once they are adjusted to the precise dimension desired, they can be locked into place and used as a gauge to set gaps. Also called a planer gauge.

sharpen – to hone and improve the cutting edge on a blade.

shear – 1) to cut off at right angles; 2) a scissor-like tool used for cutting sheet metal. Also called tin snips; 3) resistance to sideways movement; the ability of an object, such as a screw, to keep two items from sliding against one another. A cabinet screw that keeps a cabinet from sliding down a wall is in shear, or in resistance to, sideways movement.

shear force – a force acting perpendicular to a plane. For example, the force of gravity acting upon a wall cabinet is perpendicular to the axis of the screws holding the cabinet in place. Gravity, in this instance, is a shear force.

shear plate – a plate incorporated into a structure that is designed to resist shear. The plate is attached by fasteners or by welding to structural members in such a way that it acts to prevent a shear force from moving the members in relation to each other. A plywood panel that is mounted on a wall from the bottom plate to the top plate with all the edges blocked is a shear plate. It provides shear strength and is nailed in place according to a shear schedule provided in the plans.

shear wall – a wall designed to resist horizontal loads. In wood frame construction, this is done by mounting plywood sheathing from the bottom plate to the top plate and nailing it in place according to a shear schedule provided in the plans. The plywood sheathing adds considerable stiffness to a stud wall, giving it the ability to resist sideways or horizontal loads.

sheathing – 1) the outer covering of a building, often structural, such as exterior plywood made from fir or redwood; 2) a protective outer covering, or insulation, for electrical cable.

sheathing paper – a moisture-resistant building paper that is applied to roofs or walls as a weather barrier before the final covering is installed.

sheathing stripper – a tool for cutting the sheathing from nonmetallic electrical cable.

sheave – a pulley or wheel with a groove around the circumference of the rim into which a belt or cable can be fit.

sheave beams – beams used as overhead supports for an elevator and the sheaves on which it operates.

shed – a small structure for storage or other purpose.

shed dormer – a dormer with a shed roof.

shed roof – a roof style in which the slope from the peak to the eave line is in one direction only.

sheen – the degree of gloss.

sheepsfoot roller – a heavy machine-driven roller with numerous short, blunt spikes projecting from its surface. It is used for soil compaction.

sheer – 1) extremely steep; 2) transparent fabric.

sheet – a thin, flat section of material, such as metal or wood.

sheet metal – metal that is in the form of thin sheets. Sheet metal generally ranges from 0.006 inch to 0.249 inch thick. It has several uses in building, such as in the manufacture of flashing, vents and ductwork.

sheet metal locking pliers – pliers with wide, flat jaws and an adjustable over-center lever locking mechanism. The flat jaws spread their force over a wider area than the jaws of other pliers. Used for bending and forming sheet metal.

sheet metal punch – a punch for making holes in sheet metal.

sheet metal worker – a tradesperson who shapes, forms, and installs components, such as flashing or ducting, made of sheet metal.

sheet piling – pilings made of wide sheets of corrugated steel or a combination of metal sheets and wood boards. They are used to hold back earth at the sides of an excavation.

sheet siding – building siding material in sheet form, such as plywood, as opposed to individual siding boards.

shelf – a flat level projection on which objects can be placed.

sheepsfoot roller

shelf angles – a metal anchor for masonry veneer walls that provides intermediate support along horizontal planes. The shelf angle bolts into the structural wall behind the veneer wall and provides a shelf on which the next horizontal course of masonry rests. These supports are generally used at each floor or at every other floor if the shelf angle can carry the load of two stories.

shelf life – the maximum time a product or material can be stored before it begins to deteriorate or lose its effectiveness.

shellac – a solution of lac dissolved in alcohol. Lac is the resinous secretion of the lac bug.

shell-and-tube evaporator – a cooling system evaporator in which refrigerant is expanded into tubes that are surrounded by water contained in an outer shell. The expansion of the refrigerant absorbs the heat, cooling the surrounding water. The chilled water is pumped to heat exchangers where it is used to cool air.

shield – a type of shoring made up of two large sheets of metal held apart and stiffened by a framework. The shield is lowered into an excavation to prevent it from collapsing because of unstable soil conditions. It surrounds and protects the workers in the excavation.

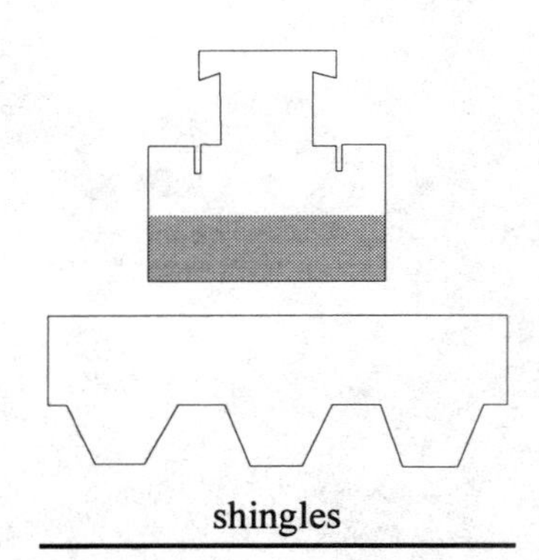
shingles

shielded metal arc cutting (SMAC) – a process using heat from a covered metal electrode to melt the metal being cut.

shielded metal arc welding (SMAW) – a welding process using a shielding gas which keeps oxygen away from the molten weld metal.

shielding gas – a gas used in cutting or welding to displace the oxygen in the air and prevent it from contaminating or weakening the weld through oxidation of the weld metal.

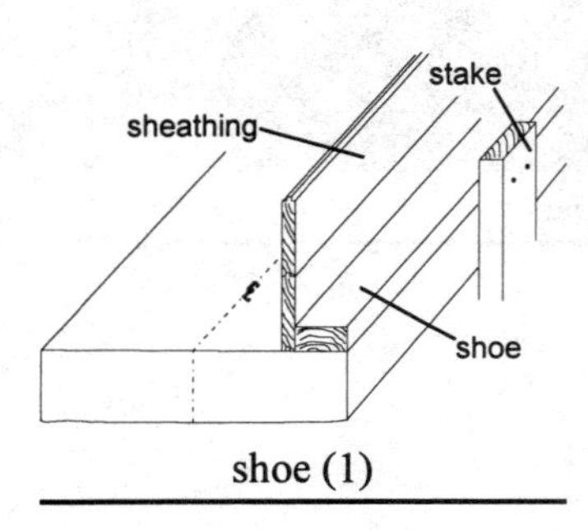

shoe (1)

shim – to use thin tapered pieces of material, such as wood, to level or plumb a structure, or part of a structure or surface. The shim is placed between two members, such as a wall and a cabinet, to fill in uneven areas and make a level surface for the cabinet installation.

shimmy – a thin wood shingle or other thin piece of material used to fill gaps between members, such as door or window frames and the rough opening, in order to make the installation level or plumb.

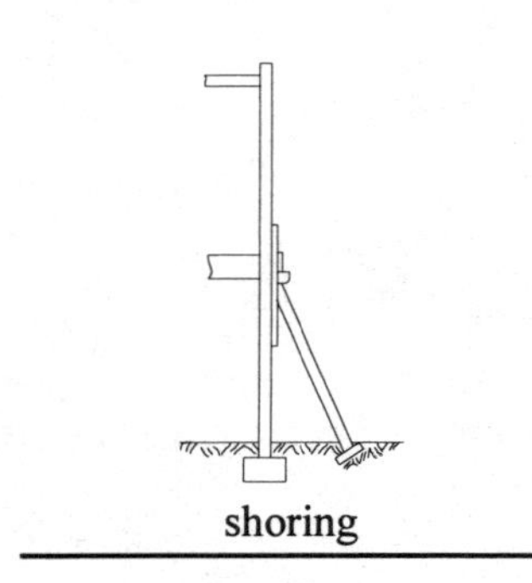
shoring

shiner – 1) a paint defect in which a glossy spot appears on a dull surface. It is caused by a difference in the absorption of the material which has been painted, often because of spot priming or overlapping paint; 2) a brick laid horizontally so that both its width and length show.

shingle – a unit of material intended for use as a weatherproof roof covering. Shingles come in a variety of shapes and styles and are made of several different materials, such as asphalt, fiberglass, cedar, slate, clay, and metal.

shingle hatchet – a hatchet used for nailing and trimming wood shingles. It has a head with a cross-hatched face for driving nails on one end, and a sharp-edged blade for cutting on the other end. The blade end also has a series of holes and a peg that can be used as a measuring gauge during installation. The peg is placed in the hole that will provide the correct distance between shingles when it is hooked over the end of the lower shingle. The next shingle is then butted up against the face of the hatchet. This feature allows the roofer to quickly measure the exposure of a shingle as the next course is laid on top. The butt end of the handle has a hole drilled in it with a thong through it for looping around the user's wrist.

shingle, interlocking – a shingle with tabs and notches that fit together and interlock as they are laid, making a uniform continuous roof covering. The interlocking tabs provide reinforcement for the shingles.

shingle nail – see roofing nail.

ship auger – a bit with a spiral-twisted shank used in a hand brace or power drill.

shoe – 1) a board, 2 x 4 or larger, used as the bottom member in a concrete form. The studs that are a part of the form rest on the shoe; 2) quarter-round molding used at the base of a wall.

shop weld – welding performed in a shop in a controlled environment.

shoring – temporary vertical or horizontal supports.

shoring jacks – jacks that exert side force on the shoring installed along trench walls to prevent a cave-in. The jacks hold the shoring in place, while putting pressure on trench walls.

short circuit – an unintentional or accidental failure in an electrical circuit in which a portion of the current is diverted to a conductor that is not a normal part of the ground or point on the circuit. For example, if a lamp wire has a break in its insulation and the bare wire touched the metal base of the lamp, the metal base would become a conductor. The electricity would follow the path of least resistance to the lamp base, creating a short in the electrical circuit.

shortened valley rafter – a rafter that extends from the top plate of an inside corner to the supporting valley rafter.

short paint – paint that is defective or inadequate for the purpose, such as an interior paint, without UV protection, that is applied to a surface exposed to the elements and deteriorates rapidly.

short ton – a unit measurement of weight equal to 2000 pounds.

shotcrete – pneumatically-applied concrete using air pressure to force the concrete through a nozzle. This method permits the spraying of concrete on walls and curved surfaces.

shoulder – 1) the edge surface along a roadway; 2) an increase in the thickness of a part which provides a bearing surface or limit stop.

shoved joint – a mortar joint made by pushing one brick against another with a bed of mortar between the two bricks. The pushing action forces the mortar between the ends of the two bricks.

shovel – 1) a manual digging implement consisting of a metal scoop affixed to the end of a long handle; 2) an attachment or an integral part of a type of motorized earthmoving equipment. The shovel is used to scoop up and move dirt or rock to another location.

shower enclosure – the walls and bottom pan that confine the water within a shower. Shower enclosures may be tiled or paneled in marble, laminated plastic or other types of waterproof wall coverings, or they can be made up of a complete fiberglass molded unit. The opening is covered with a waterproof curtain or a fitted glass door unit.

shower head – the nozzle that sprays water in a shower.

shower pan – the base or floor of the shower, with the drain for the water.

shrink – to contract or be reduced in size or volume.

shrinkage – a reduction in the volume of a material, such as concrete, often occurring during a curing process.

shrinkage void – a weld defect in which a cavity is formed by shrinkage of the weld metal. It is caused by incomplete fusion from inadequate heat or from insufficient fill metal.

shunt – a parallel electrical circuit used to achieve varied operating characteristics.

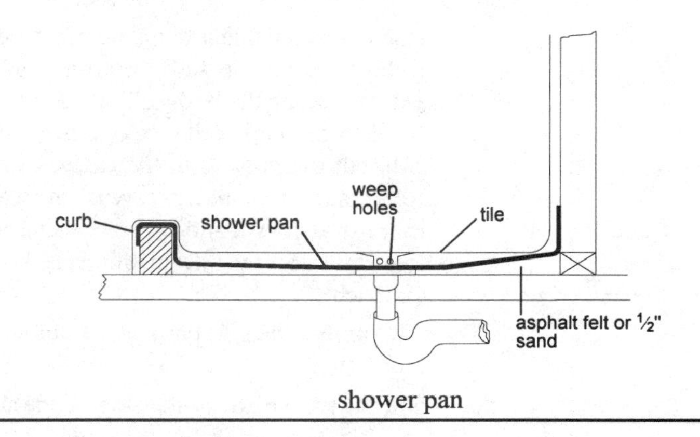

shower pan

shunt wound – an electrical motor in which the field and armature are wired in parallel.

shutters – 1) a covering for a window, consisting of two panels mounted on hinges on each side of the window. They can be swung open or closed with a lock at the center. Shutters may be made of wood, metal or plastic. They are usually louvered or slatted to provide ventilation; 2) decorative pieces fixed to a wall on either side of a window that give the appearance of being closable.

side clearance angle – the amount, measured in degrees, that the front edge of a metal lathe cutting tool bit deviates from plumb. A cutting tool bit must be contoured to provide the most efficient and cleanest cutting shape. When looking at the cutting end of the bit, the leading side of the cutting end is at an angle with respect to the vertical. This provides clearance from the work and enables the bit to make a clean cut.

side cut – 1) a method of felling a tree with an extreme lean, where cuts are made on each side of the tree part way through the trunk before the back cut is made; 2) a chamfer cut. See also cheek cut.

side cutters – plier-shaped cutters for use with nails and wire. The cutting edges are on the side of the head, parallel to the handles.

side hook approach – the distance between the hook and the wall at the closest point that a bridge crane hook can reach to the wall. On a bridge crane, the hook is

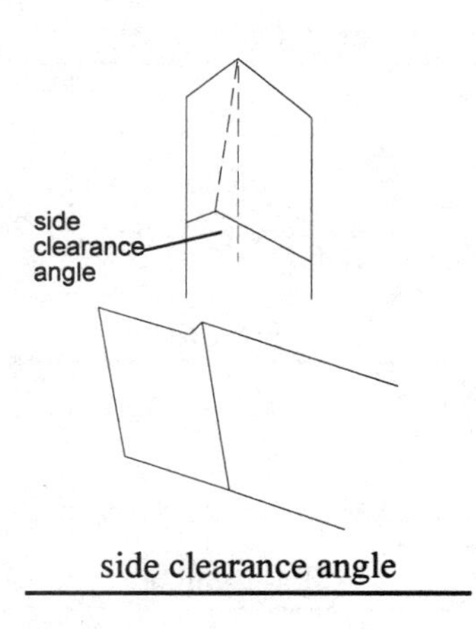

side clearance angle

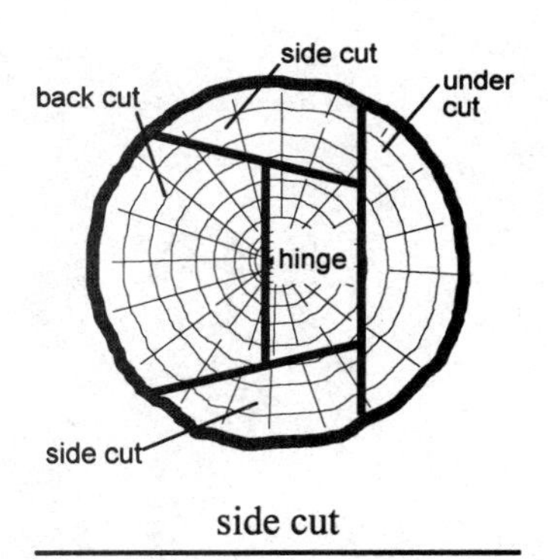

side cut

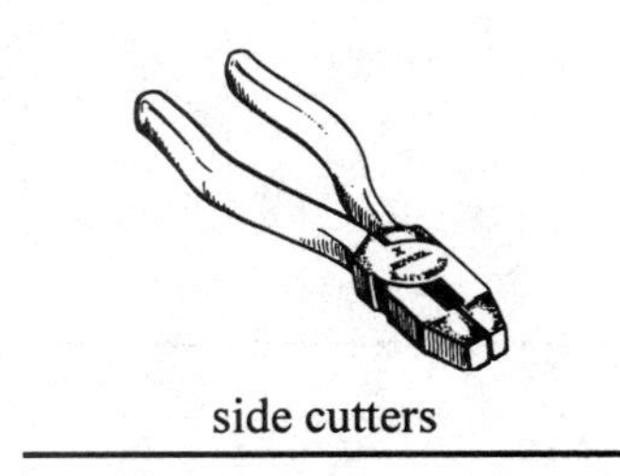
side cutters

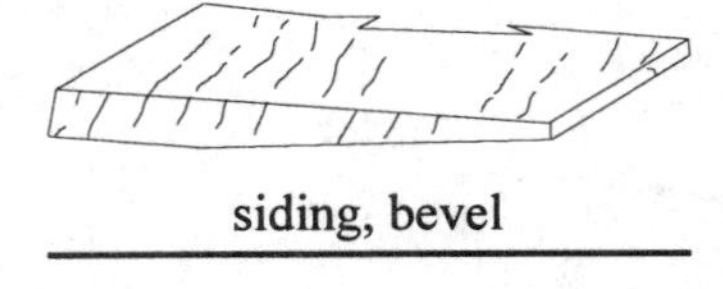
siding, bevel

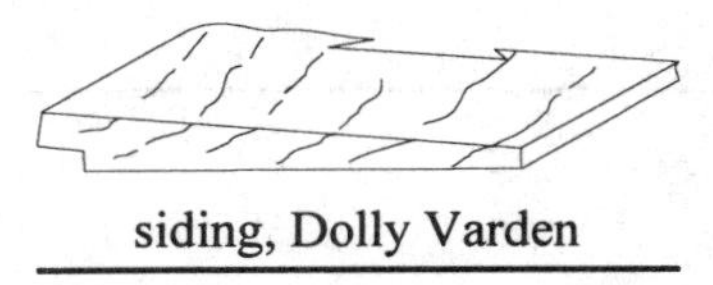
siding, Dolly Varden

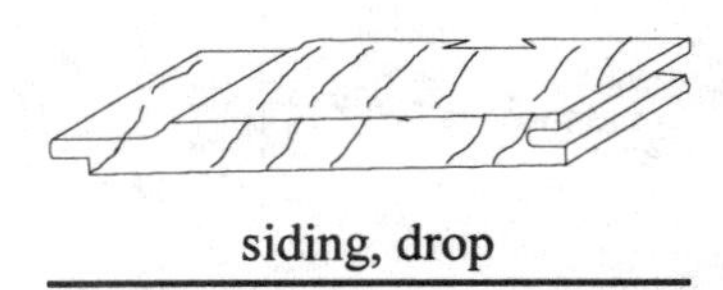
siding, drop

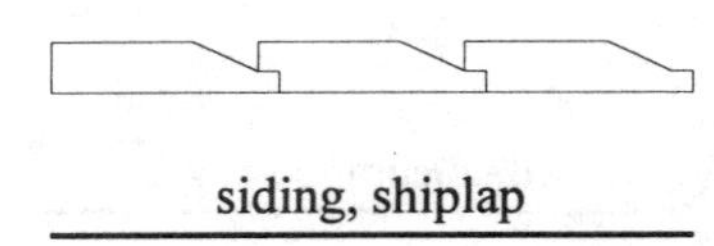
siding, shiplap

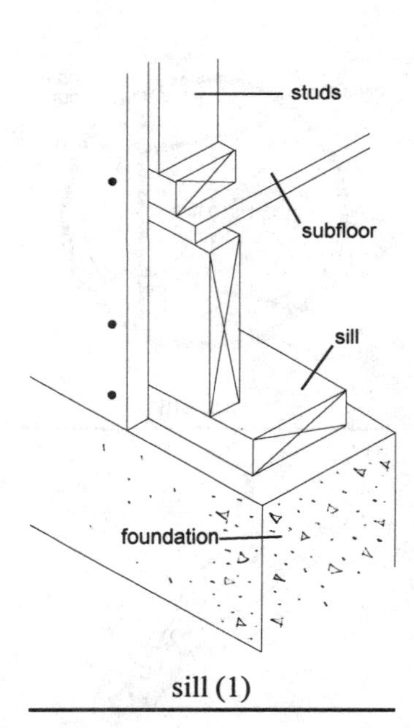

sill (1)

attached by a cable around a drum on the trolley, and travels with the trolley back and forth along the bridge. The trolley can travel to the end of the bridge next to a wall, but the hook is in the center of the trolley, and so, cannot quite reach the wall. There is always a short distance that the hook fails to cover from this side hook approach.

side jamb – the side parts of a window or door frame.

side rake – the horizontal slope at the cutting edge of a lathe tool bit, taken perpendicular to the axis of the tool.

sidewalk – a paved walkway, usually placed alongside a street.

sidewalk light port – glass panels that are set into a sidewalk to let light into a basement below. Sidewalk light ports are installed where a basement work area extends beyond the perimeter of the building.

siding – an exterior covering for a building. May be in the form of sheets, shingles or boards and made of wood, asphalt, vinyl, aluminum or other weatherproof materials.

siding, bevel – wedge-shaped siding boards used by lapping each board over the edge of the board below.

siding, Dolly Varden – a type of bevel siding with a rabbet that overlaps the top of the board below.

siding, drop – a tongue-and-groove type joint siding that may be milled into a variety of patterns.

siding, shiplap – a type of exterior siding board that is rabbeted at the top and bottom so that each piece of siding overlaps and fits snugly against the adjoining piece.

signal – 1) an indicator, often a light, used to visually alert users that a mechanical device is in operation; 2) a visual or audible device used to indicate the presence of a problem or danger.

signal systems – devices used to alert people to a variety of situations, including traffic crossings, safety warnings, burglary in progress, or telephones. Signal systems may be comprised of lights, sirens, horns, loudspeakers, beacons, chimes, bells, buzzers or a combination of these.

silica – ground quartz, frequently used as a component in the manufacture of a variety of refractory brick.

silica gel – a moisture-absorbing compound placed inside packing to absorb unwanted moisture.

silicon – an element, with the atomic symbol Si, atomic number 14, and atomic weight of 28.086, that makes up more than 25 percent of the earth's crust. It is used in alloying, glass and semi-conductors.

silicon bronze conduit – a corrosion-resistant metallic conduit made of silicon bronze metal.

silicon carbide – a hard abrasive used in the manufacture of abrasive papers, grinding wheels and some cutting tools.

silicon controlled rectifier (SCR) – a small semiconductor device used to convert AC current to DC current by suppressing the reverse direction of current flow in an AC circuit (called half-wave rectification), or by arranging the circuit so that both halves of the AC current flow in the same direction (called full-wave rectification). It is highly efficient, can operate at relatively high temperatures and is inexpensive.

silicone – silicon-oxygen polymers, used in construction in the manufacture of lubricants and caulking and sealing compounds.

silicone rubber grout – a silicone rubber compound designed for grouting ceramic tile. It remains flexible when cured and resists moisture and mildew.

sill – 1) the lowest member of a structure; it is attached to the foundation and provides support for the other framing members. Also called the sill plate; 2) lowest horizontal member in a window opening.

sill course – see belt course.

silver – a white metallic element with the atomic symbol Ag, atomic number 47, and atomic weight of 107.87. Silver is often used to plate electrical contacts.

silver brazing – brazing with an alloy containing silver. It is more costly than other brazing materials, but provides a fine finish for decorative work. Silver brazing has a higher melting point than solder, and is sometimes used in electrical connections to provide maximum conductivity through a soldered joint.

silver solder – a solder that contains a small percentage of silver. It is sometimes used in electrical connections to provide increased conductivity through a soldered joint.

simple cornice – a frieze board, or horizontal board placed flat against a wall to form a cornice.

simulate – 1) to copy or give the appearance of; 2) to duplicate for purposes of testing.

single coursing – the application of siding shingles in a single layer, with a small amount of overlap along the edge of the course below.

single-cut file – a metal file with a series of diagonally-cut rows of teeth running in one direction only across the face.

single-face fireplace – a fireplace with a single opening into a room.

single-hung window – a window with a fixed top sash and a moveable bottom sash. The bottom sash can slide up and down to open and close.

single-line diagram – an electrical drawing which shows the routing of electrical circuits, using only one line to depict the wiring, regardless of how many connectors it actually has. Also called a one-line diagram.

single lite – a fixed or stationary window sash with a single, undivided pane of glass.

single phase – AC electricity with one phase. Electricity is generated in three phases of power, 120 degrees apart.

single-pole, double-throw switch (SPDT) – an electrical switch that selects one of two positions. It can switch one wire or circuit two different ways. If two of these switches are placed on a circuit controlling lights, either switch can turn the lights on or off. This type of switch is commonly installed at both ends of a long hallway, or at the top and bottom of a staircase, so that the light can be turned off or on from either end. It is also known as a three-way switch.

single-pole, single-throw switch (SPST) – an electrical switch that connects or interrupts one conductor. This type of switch is used to control a light or an outlet from a single location. It is commonly used in residences and office buildings.

single-wall siding – a building exterior wall that has a single siding covering, as opposed to siding over sheathing. Most sidings can be installed in this manner.

single weld – a weld applied from one side only. This is common practice when the weld is performed where there is limited access.

siphoning – the process of transferring fluid through a tube from a higher to a lower level by means of suction and gravity. This method can be used to empty a tank, vessel or sump without a pump when it cannot be picked up and tilted to empty.

site – the physical location of a construction job; the location of proposed construction.

site plan – drawings that show compass directions, the property lines and dimensions, lot elevations and the locations of the building or buildings to be constructed on the property. Other details, such as the outside utilities and landscaping, may also be included. Also called a plot plan.

size of weld – the portion of the weld, in contact with the base metal, that can be considered the part that contributes strength to the weld joint. It is also called the effective size of the weld.

sizing – 1) a coating applied to porous material to seal the surface; 2) to cut or shape to size; 3) estimating the pipe sizes and components needed for HVAC or duct systems.

skeleton – the framework of a building or other structure.

sketch – a rough drawing.

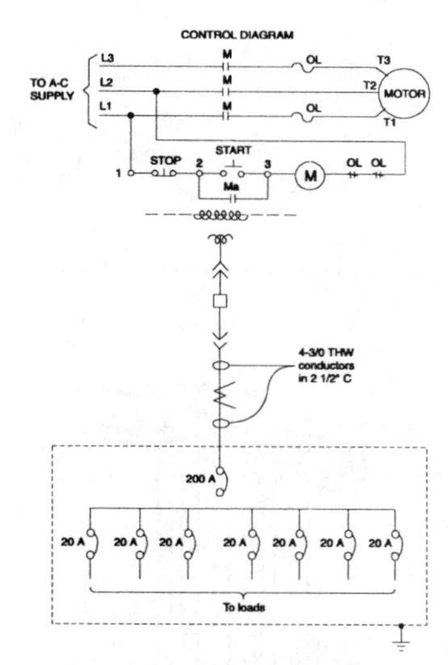

single-line diagram

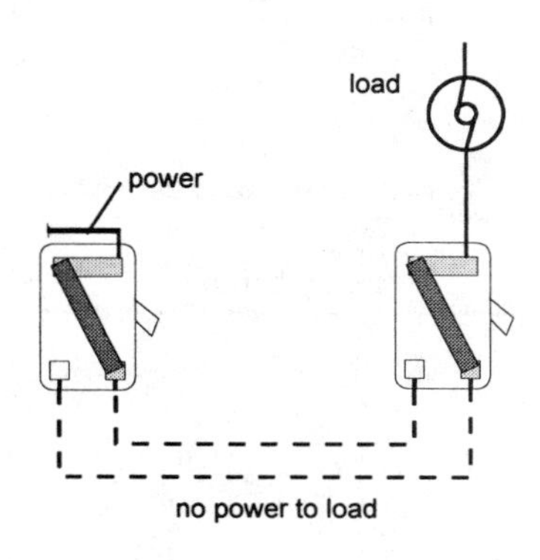

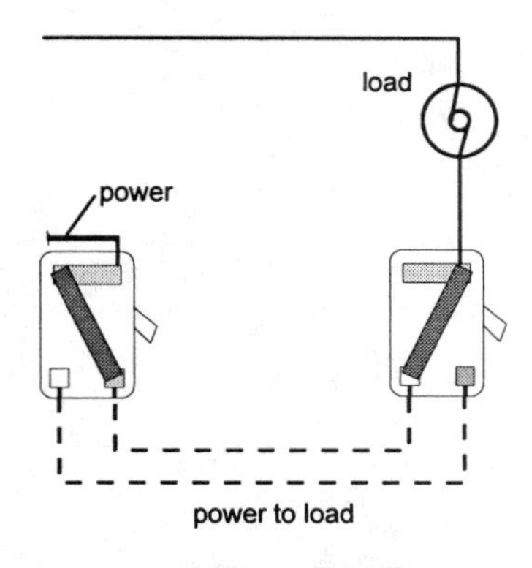

single-pole, double-throw switch

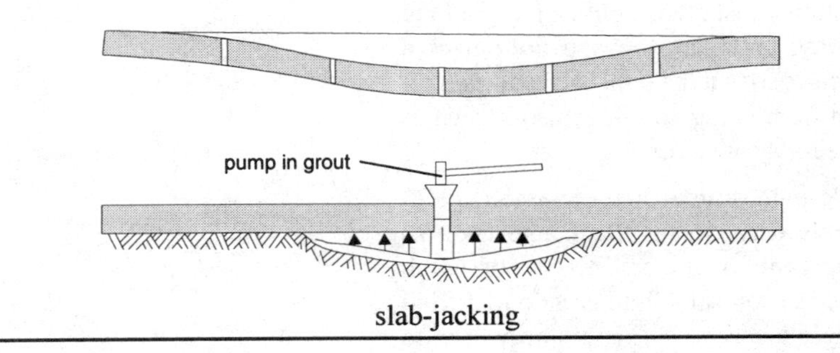

slab-jacking

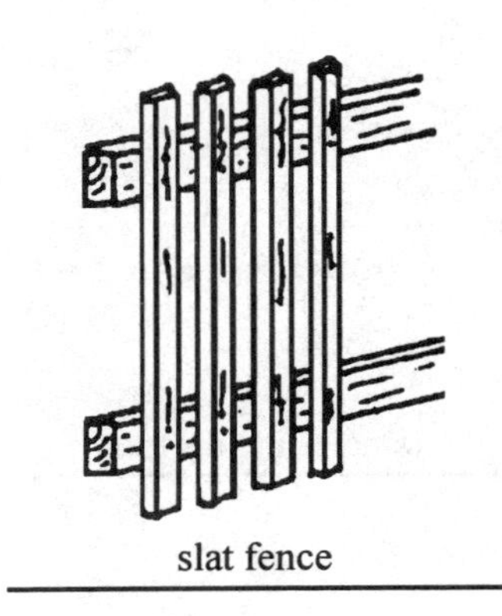
slat fence

sledgehammer

skew – 1) turn out of alignment or at an angle; 2) a type of chisel with the cutting edge at an angle other than perpendicular to the axis of the blade. Used primarily with a wood-turning lathe to make quick cuts.

skewback – a sloped surface that supports the end of an arch.

skewed – out of square with relation to a standard reference; not parallel, not level, not plumb; at an angle.

skewed beam – a structural member which intersects another structural member at other than a 90-degree angle.

Skil saw – a trademark name for a brand of portable circular saw.

skim coat – a thin coat of drywall compound.

skin – 1) an exterior layer or surface coating; 2) the dry surface that first forms on adhesive, caulk, paint or other coatings.

skintled brickwork – 1) a brick wall in which the bricks are not flush to the face but have been set in different vertical planes in order to create a rough and irregular wall surface; 2) a brick wall with a rough face caused by squeezing mortar out between the joints and allowing the mortar to set up that way.

skippy – a term used to describe a defective paint that leaves patches of uncoated surface.

skip trowel – a troweled-on texturing technique designed to produce flat, but slightly elevated areas as a finish coat on gypsum wallboard or plaster walls.

skirt – 1) siding at the base of a structure, such as a mobile home or porch, which extends from the bottom of the structure to the ground, or nearly so; 2) a trim molding, border or edge.

skirting – baseboard around the base of an interior wall.

skylight – a roof opening covered with glass or plastic designed to let in light and sometimes ventilation. *See Box.*

slab – 1) a relatively large horizontal concrete pour, such as one used for a house foundation; 2) a large flat piece or slice of material, such as a stone.

slab-jacking – forcing cement grout under a cured concrete slab to raise the height of the slab.

slack – 1) the sag or swag in a rope or cable; 2) free play in a mechanism; 3) excess time in a schedule.

slag – 1) refuse material resulting from the melting of metals or processing of ores. Used as aggregate in lightweight concrete; 2) impurities.

slag concrete – a lightweight concrete in which furnace slag is used for aggregate.

slag inclusion – a weld defect in which a nonmetallic material is trapped in the weld.

slake – hydrating or adding water to a material, such as lime, to cause it to disintegrate or crumble so that it can be used in mixtures.

slate – a fine-grain rock that readily splits along thin planes to form sheets. Used as a roofing material, for decorative flooring, or as an exterior wall facade.

slat fence – a fence made from 1 x 2 or 1 x 1 wood, or both, in an alternating pattern (called lath and slat), that is attached vertically to fence rails.

sledgehammer – a heavy long-handled hammer with two faces used for heavy pounding such as driving stakes into the ground or breaking up rock or concrete.

sleepers – 1) wood members embedded in poured concrete to be used as a nailing surface for wood framing; 2) treated wood or decay-resistant timbers laid across a concrete slab to support flooring; 3) wood

Skylight

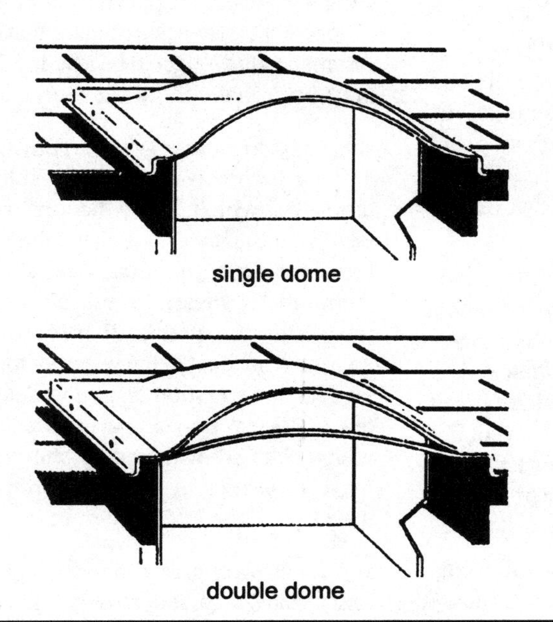

Skylights are glass- or plastic-covered roof openings designed to provide an inexpensive and attractive means of adding natural light to both homes and offices. It takes only about 1 square foot of skylight to provide light for 20 square feet of floor space. Skylights have long been used in studios and workplaces where natural light is preferred, but they are now becoming more and more popular in homes and commercial buildings as well. Skylights can be installed on slanted or flat roofs, and a simple skylight bubble can even be installed high up on a vertical wall to provide additional light at a particular time of the day.

Skylights may be flat or dome-shaped, single or double paned. A double dome or pane creates an insulation barrier to reduce the transfer of heat and cold. Skylight domes come in clear, frost or bronze tint. Clear gives the most intense light, while clear with a white diffusion panel is a good choice to light a large area without a harsh glare. The bronze eliminates glare and adds a gentle tint to the area below. Hatch-type skylights are designed to open and provide ventilation as well as light. They may be operated by remote control or a hand crank. They may be installed individually or clustered to provide a distinctively modern look.

Rafter spacing is the first consideration when selecting a skylight. The units are commonly manufactured to fit roofs with rafters 16 and 24 inches on center, although 4 x 6-foot and larger units are available for industrial needs. They come in self-flashing or curb-mounted types. The curb-mounted skylight fits into a curb that is installed around an opening in the roof. A self-flashing design installs directly into the slope of the roof and is preferred where a less conspicuous profile is desired. Attic skylights often require the installation of a light shaft. The shaft may be straight to provide direct overhead light, or angled to distribute light over a broader area. Once installed, skylights require almost no maintenance. Normal rainfall should keep them reasonably clean.

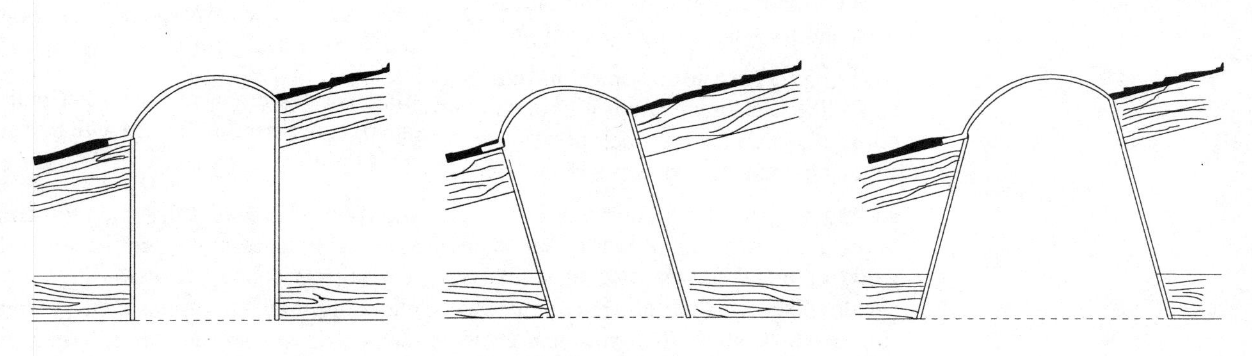

Angle of shaft controls light direction

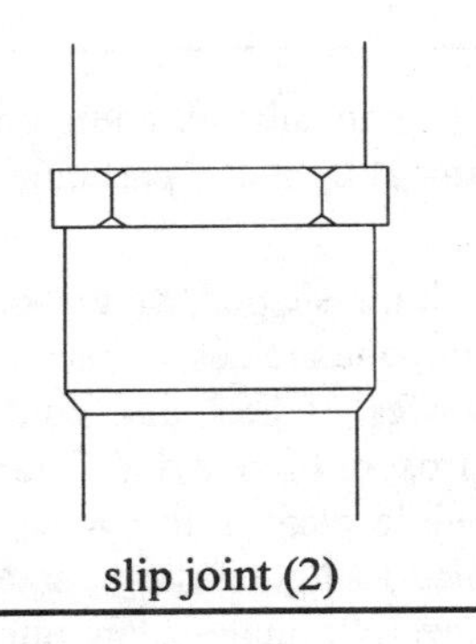
slip joint (2)

braces used in concrete formwork to help support the bottom of a wall form. Sleepers are placed at right angles to the wall form and project outward from it, forming a base to which other bracing can be nailed; 4) railroad rail supports.

sleeve – 1) a hollow cylinder through which another commodity can pass, such as the metal sleeve installed in poured concrete to allow for the passage of piping or conduit; 2) pipe or tubing that fits over another member.

sleeve anchor – a sleeve, with interior threading, which is inserted into a hole drilled in a surface. A screw is threaded into the sleeve, and as it is tightened, the sleeve is forced against the inside diameter of the hole, anchoring it into the wall, floor or surface. The sleeve anchor and screw assembly can be used to fasten a ledger, shelf, cabinet or other item to a surface.

slenderness ratio – the ratio of the length of a structural member to the thickness of the member. This provides information on the stiffness of a member, as stiffness is proportional to length and thickness.

slick – a wood chisel with a blade that is more than 2 inches wide. Used to smooth a wood surface, or to make wide cuts that are smooth and even.

slicker – 1) a metal smoothing tool for use on mortar joints; 2) a flat wood tool used for smoothing plaster.

sliding door – a door that is mounted in tracks or suspended from overhead tracks on rollers that allow the door to slide easily back and forth across the opening.

sliding gate – a gate that opens and closes by sliding horizontally on rollers that ride on a rail. The gate may be operated mechanically, electrically or manually.

sliding glass door – a door consisting of two (or more) glass panels mounted in wood or metal frames, one of which is usually fixed and the other mounted on rollers which allow it to slide back and forth along a track to open or close. The glass panels are used in a large door opening, providing light as well as easy access. They are commonly used for patio doors.

sliding window – a window with one or more panes mounted in tracks that slide open and closed.

sling – a hoisting support, such as a cable or rope with an eye attachment, that can be fastened to the object to be lifted and to the hoist hook that will lift the object.

sling psychrometer – a framework, holding both a wet and a dry bulb thermometer, which has a handle used to swing the framework, with the thermometers, through the air. The dry bulb thermometer measures the air temperature, and the wet bulb thermometer measures the heat in the water vapor in the air. This information is used to determine the humidity by comparing the results with a chart showing temperature/humidity relationships. A sling psychrometer is used to balance HVAC systems.

slip form – a concrete form for high structures that is raised slowly as vertical concrete pours are made and set up. The form is raised by means of a jacking mechanism and placed into position for the next pour which is made on top of, and extending, the portion of the wall or column that has already set.

slip-form concrete machine – a machine that forms and pours a shape as it travels. It is used to efficiently pour continuous forms, such as a continuous concrete barrier dividing traffic lanes on a highway.

slip-form paver – a machine that pours and forms a concrete roadway as it travels.

slip gasket – an elastomer gasket with a tapered cross section for use with bell and spigot pipe.

slip joint – 1) a joint designed to permit axial differential movement between the two objects connected at the joint; 2) a plumbing drainage piping joint in which one pipe slips inside the enlarged and threaded section of the next pipe; the joint is sealed with a slip-joint nut and an elastomer washer placed on the smaller pipe before the joint is made.

slip ring – an electrically conductive ring around a shaft for conducting current from a rotating part to a nonrotating part. Used in many electric motors and generators in place of carbon brushes.

slip sill – a sill that slopes away from the building face so that water, such as rainwater, runs away from the building.

slip stone – a sharpening stone that is shaped for sharpening a concave cutting surface, such as a concave wood lathe cutting tool.

slobber bits – steel guides, attached to each side of the bowl of a road scraper blade, which confine material to the front of the scraper bowl cutting edge.

slope – the angle of deviation from level. A slope may be expressed in degrees of angle, as a ratio of horizontal distance to vertical drop, as a decimal, or as a percent of vertical change to horizontal distance. See also pitch.

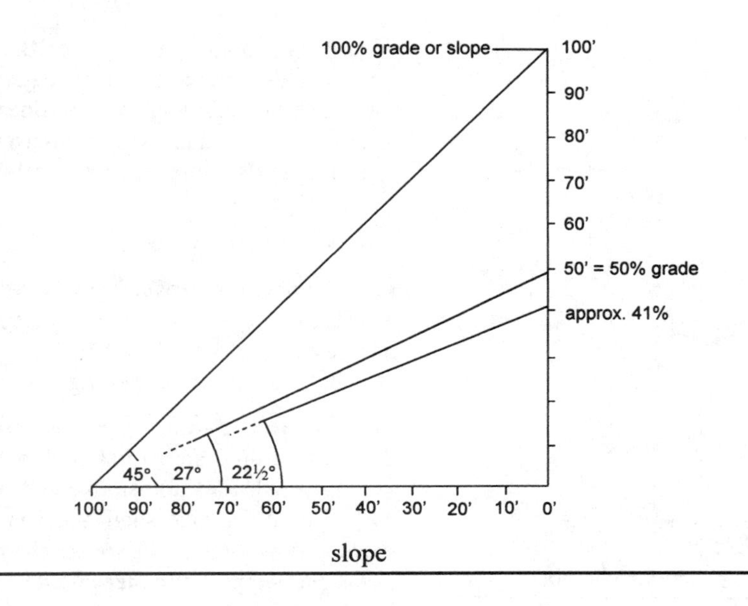

slope

slope bar – a dozer attachment used for cutting slopes where the soil is rocky. The bar is set to the proper angle for the slope that is to be cut back, and the dozer operator takes small cuts with each pass of the slope bar until the slope angle is correct.

sloping – widening a trench across the top so that it has sloping sides in order to minimize the chance of a cave-in.

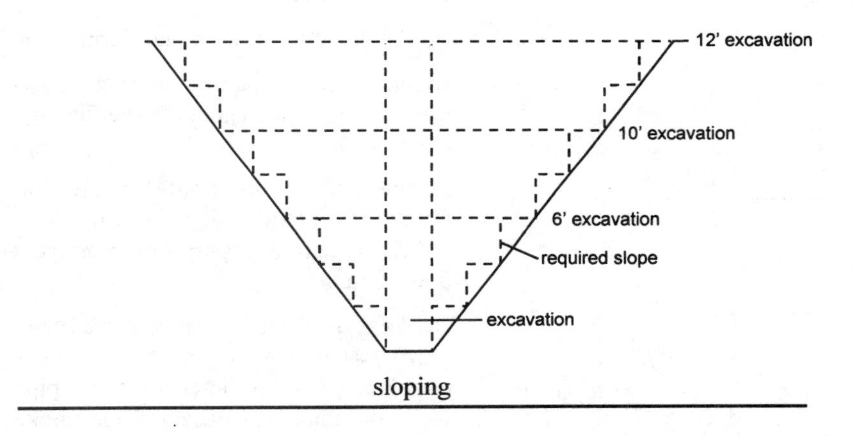

sloping

slot – an elongated hole.

slot dozing – to move material in an area the width of the dozer blade so that it is operating in a slot within the surrounding material. This limits the amount of blade load that is lost.

slot weld – a weld made through a slotted opening in one material to join it to a second material below. This is sometimes the most practical or economical method to join two overlapping pieces of material.

slow-blow fuse – see time-lag fuse.

slow-curing cutback – asphalt thinned with diesel fuel, which evaporates slowly, retarding the curing time of the mixture.

slump – 1) a test to measure concrete consistency by measuring the amount that a quantity of concrete of specified shape and size will slump due to the effects of gravity when unsupported on the sides; 2) the stiffness of a mix of concrete.

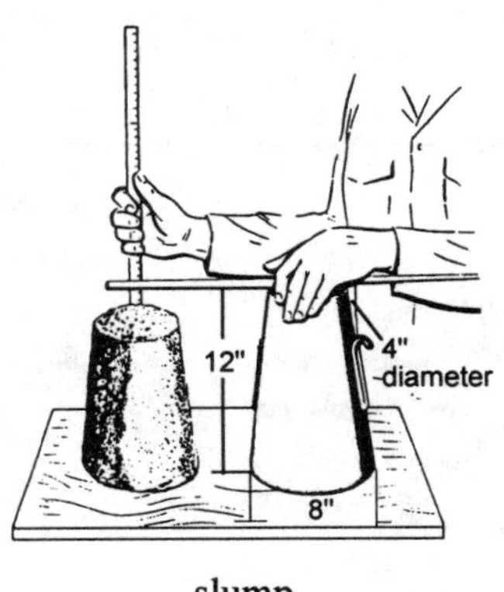

slump

slurry – a mixture of liquid and solid particles in which the particles are suspended in the liquid. Pumped concrete is an example of such a mixture.

slushed joint – a vertical joint in a masonry wall that is filled after the masonry units are installed by throwing the mortar into the joint with a trowel.

Small Business Administration (SBA) – a U.S. Government agency established by Congress in 1953 to assist small businesses. The SBA provides a series of free and low-priced publications that are available to anyone with an interest in starting

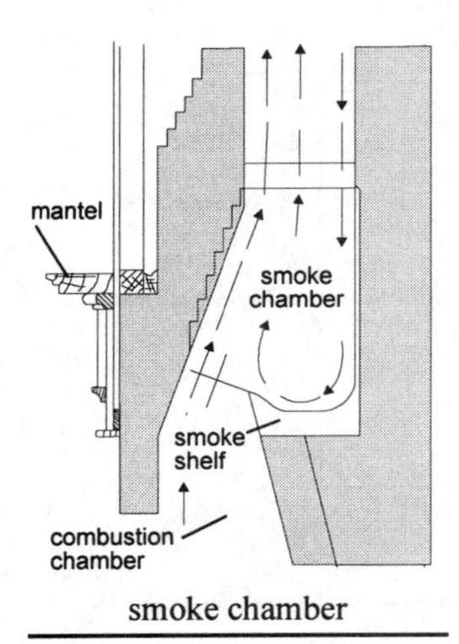

smoke chamber

snake fence

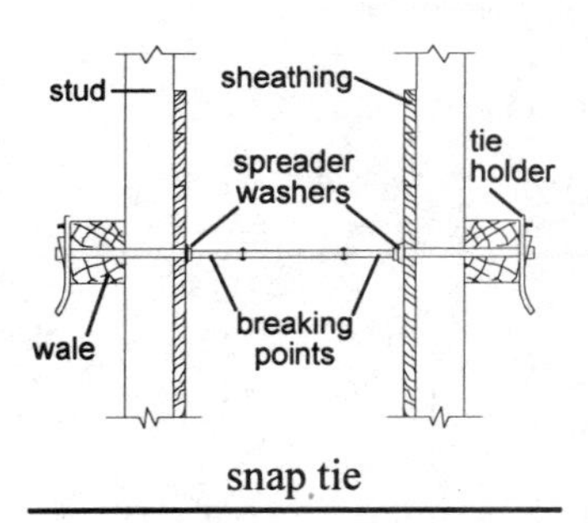

snap tie

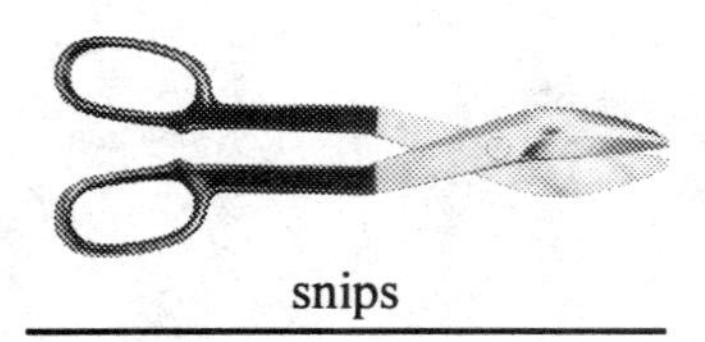
snips

or running a small business. The SBA can also provide financial and management assistance to qualifying small businesses. Requirements for qualifying for assistance or a list of SBA publications can be requested by mail.

smelt – to refine an ore by heat.

smoke alarm – a sensing device which detects the presence of smoke in a specified density and sets off an audible signal, providing an early fire warning.

smoke chamber – the open area above the combustion chamber in a fireplace assembly which traps the smoke and, with the aid of the smoke shelf and damper, prevents downdrafts. The smoke chamber forms the support for the chimney flue which rises above it.

smoke detector – see smoke alarm.

smoke pipe – a pipe from an oven or furnace combustion chamber to the flue of a chimney.

smoke shelf – a horizontal shelf that forms the bottom of the smoke chamber and fireplace flue; its purpose is to prevent downdrafts.

smoke test – a test sometimes used with gravity-rated plumbing in which smoke is introduced into the interior of the pipe while the joints are observed for smoke leakage.

smooth coat – the final coat of plaster.

smoothing brush – a wide, relatively stiff-bristled brush used for smoothing wallpaper on the wall during installation.

smoothing plane – a hand plane with a long base used for smoothing, flattening and finishing wood surfaces. The base covers a large area and helps ensure that the cutting action of the plane is smooth and even. Since the base is spread out, occasional high spots are leveled to the height of the surrounding surface. The cutting blade can be adjusted so that only a small amount of material is removed with each stroke.

smooth machine finish limestone – see limestone, smooth machine finish.

snake – a long, flexible plumbing tool made of tightly coiled wire used to clear drains and remove obstructions.

snake fence – a fence of rails that intersect and rest upon one another at angles less than 180 degrees, so that the fence zigzags back and forth along its length. Because they require no fence posts and are suitable for very rocky soil, snake fences were commonly used in the past for boundaries around farm or ranch lands. Today they are used only for aesthetic effect or to recreate a specific time or style of architecture. Also called a zigzag fence.

snap switch – an electrical switch with a spring mechanism to make or break contacts quickly in order to avoid arcing between the contacts. Standard household switches and circuit breakers often use this type of switch.

snap tie – a factory-made concrete form tie used as a combination tie and spreader. The tie holds the concrete forms in position a specified distance apart. After the concrete is set and the forms are stripped, the tie can be broken off below the surface of the poured concrete, leaving only a small hole to be patched in the concrete surface.

snecked masonry – rubble walls made from somewhat square, irregular-sized stones that are not set in any particular course.

snips – a type of shears used for cutting sheet metal.

snub gable – a type of roof design with a short hip extending from the ridge part way down the gable end.

soap – a masonry unit that has a nominal thickness of 2 inches.

socket – a wrench that is cylindrical on the outside, with the inside machined in a hexagonal or other configuration designed to grip a nut or fastener head on one end and to be driven by a driver having a square shank, such as a ratchet, breaker bar or torque wrench, on the other. Also called a socket wrench.

socket adapter – an adapter that permits a drive of one size to be used with a socket made for a drive of another size.

sod – a section of grass growing in a thin layer of soil which is just thick enough to cover the roots. The sod is cut into strips and laid on prepared soil into which the roots will grow. It is used to create a fully-grown lawn quickly.

soffit – 1) a covering over the space under the eaves of a structure; 2) a covering extending from the top of cabinets to the ceiling.

soffit vent – ventilation holes or openings in the soffit under the eaves of a building.

soft burned – clay masonry units that have been fired at low temperatures leaving them soft and unsuitable for use in other than ornamental situations.

soft face hammer or mallet – a plastic-faced hammer used to strike a surface without marring it. They are used on metal fabrications in machine shops.

soft mud brick – a brick made with clay having a 20 to 30 percent moisture content.

software – programs and associated instructional materials used in a computer system.

soft water – water that is free of most minerals, especially calcium and magnesium. It makes washing easier because it leaves no soap scum or calcium deposits on surfaces when they are rinsed. Water can be softened by running it through a tank filled with salt, which filters out the unwanted minerals.

softwood – wood from a conifer or evergreen tree. Softwoods are used for all types of construction in buildings, including framing, flooring, sheathing and trim, and for making cabinetry and furniture.

soil cover – plastic sheeting laid over bare earth in an under-structure crawl space to minimize moisture transfer from the soil.

soil density – the amount of compaction of soil. Soil density is often measured by a soils engineer using a Proctor test, which determines the optimum moisture of soil for proper compaction.

soil pipe – pipe used to convey the discharge of water closets or similar fixtures to the building drain and sewer system.

soil stack – a vertical vent stack that receives the discharge of water closets and is connected to the building drain.

soil structure – the mixture of varying types of earth that make up the soil in a section of land, such as sand, clay, loam or mineral particles.

soil swell factor – a factor that compensates for the swell in soil volume when it is loose as compared to when it is compacted. This factor is used to adjust the volume of soil to be handled so that the additional volume of loose soil is accounted for in determining the work to be done. Also called soil swell load factor.

soil test – a series of bore samples taken to determine soil composition and suitability for load bearing.

soil vent – see stack vent.

solar – pertaining to the sun or to energy from the sun.

solar collector – a device used to capture heat from the sun and turn it into energy. Solar energy captured by collectors is most commonly used for home water or pool heating. The collector may consist of a variety of materials in different forms, such as a coil of copper or plastic tubing, straight lengths of tubing, or flat plastic panels with passages through them. The coils or panels are exposed to the sun's rays, either in an insulated box under glass, or exposed to the elements. Water is pumped through the tubes or panels, absorbing their heat. The hot water in the tubes or panels may then be stored in an insulated tank for later use. Also called solar panels.

solar heating – using solar energy to heat a structure or to heat the water to be used in a structure or swimming pool.

solar panels – see solar collector.

solar storage – the storage of heat or heated water generated by solar energy. Hot water is stored in an insulated tank for use as needed. Other forms of storage may be a block wall system or a block floor system which absorbs solar heat and passes it, in the form of hot air, through the concrete block core to a thermal storage area, often in the walls or the roof area. Fans and ducting direct the hot air back

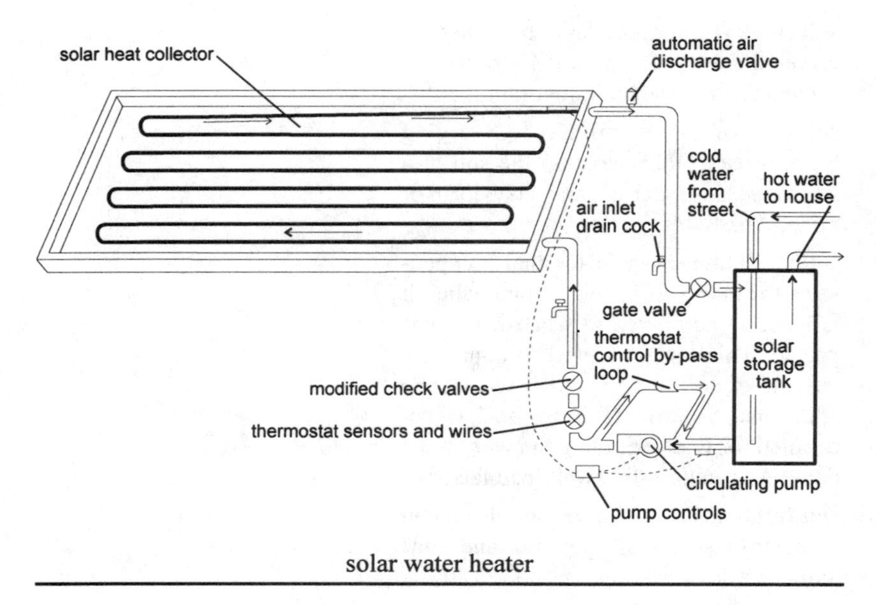

solar water heater

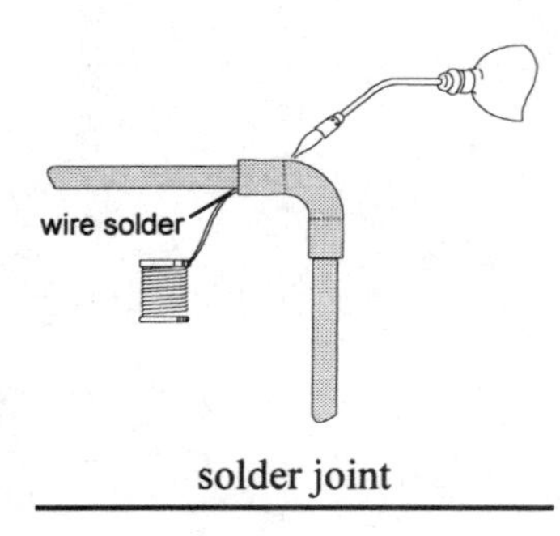

solder joint

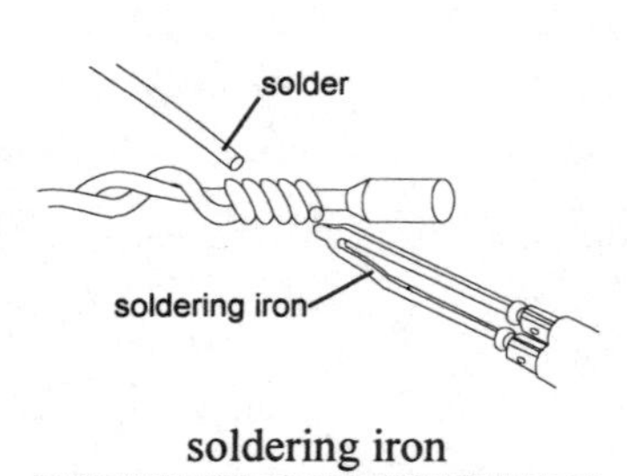

soldering iron

down through the cores in the walls or floors to provide heat to rooms as the outside temperature begins to drop.

solar water heater – a water heater that uses heat from the sun as its energy source. A solar water heater system includes a solar panel which absorbs the heat to warm the water in the system, an insulated storage tank to hold the heated water, a pump to circulate the water, a drain valve to empty the solar panels in cold weather, piping and insulation around the piping, and temperature sensors in various parts of the system to monitor and control the system operation. The system may be either direct or indirect. In a direct system, water flows into the system, is heated in the solar panels and then stored in the storage tank until needed. In an indirect system, another fluid in a closed recirculation system is circulated through the solar panels and then through a heat exchanger to heat the potable water which is stored in the insulated storage tank. Solar water heaters can be used to supply hot water for households or businesses, or may be installed for limited use, such as for hot tubs or swimming pools.

solder – metal alloys with relatively low melting points used for joining metals. The liquid point of solder is 450 degrees centigrade (840 degrees Fahrenheit) or less.

soldering – a process of joining metals with the aid of solder, or metal alloys, and heat. The melted solder creates a bond between the two metals being joined.

soldering gun – an electrically-powered tool shaped like a handgun with a shank and soldering tip in place of the gun barrel. The soldering tip is heated by resistance heating and is in turn used to heat the solder to the temperature at which the metal alloy melts and flows.

soldering iron – a soldering tool with a straight shank. It has a handle on one end and a soldering tip on the other. The soldering tip must be heated for use, either electrically or by another source, such as a torch. Also called a soldering copper.

solder joint – 1) a joint made by placing two or more pieces of metal or wire in contact, applying heat, and melting solder onto the joint; 2) a copper tubing joint in which the tubing end fits into a socket fitting and solder is melted into the joint. The bond created between the tubing end and the inside diameter of the socket forms a pressure-tight joint suitable for fluid service.

solderless connector – a plastic connector used to join two electrical wires together without the need for soldering. See also connector, pressure or wire nut.

solderless wire connector – see wire nut.

soldier – a brick set vertically on end.

soldier course – a course of bricks set vertically on end with the narrow edge exposed.

solenoid – a linear electric motor consisting of a shaft inside a coil which is moved when an electric current is applied to the coil. They are used for a variety of purposes, such as to open or close valves.

sole plate – the horizontal board to which the bottoms of wall studs and other framing members are attached. Also called the bottom plate, sole or sole piece.

sole proprietorship – having legal right or exclusive title held by one person.

solid brick – a masonry unit with a bearing area (the area against which a load can rest) of 75 percent or more of the total face area.

solid bridging – solid blocks of wood placed between floor joists to distribute concentrated loads.

solid-core door – a door that is made of solid pieces of material with no voids in the door structure.

solid masonry wall – a masonry wall made of units that are 75 percent solid material with mortar at all joints.

solid square – see try square.

solid state welding (SSW) – a welding process that fuses pieces of base metal together by applying pressure and heat, without the addition of filler metal. It is used most often in production line manufacturing.

solid wire – electrical wire in which the conductor is a single conductor, as opposed to a stranded wire of multiple conductors.

soluble – capable of being loosened or dissolved, as salts in water.

solution heat treatment – a process of heating austenitic stainless steel at a given temperature for sufficient time for the carbon contained in the alloy to become fully in solution. It is then rapidly quenched to below a specified temperature, usually 800 degrees F, to keep the carbon in solution. This process increases the corrosion resistance of the steel. Also called full solution annealing.

solvent – 1) a liquid chemical compound that can be used for dissolving, cleaning, or thinning a compatible substance. *See Box*; 2) a financial state in which one is able to pay any debts that are owed.

solvent weld – joining plastic parts, such as plastic pipe, using a solvent that dissolves the plastic surface at the interface between the parts so that they fuse together.

soot pocket – a space at the bottom of a chimney flue where soot can accumulate and be removed through a cleanout door.

sound insulation – a material, such as fiberglass batting, that resists the transmission of sound.

soundproof – 1) impervious to sound; 2) to insulate against the transmission of sound.

sound transmission – the ability of a gas, liquid, or solid to conduct sound.

space heater – an individual heating unit, usually electric, designed to heat a small area.

spacer – 1) small concrete block on which rebar can be supported to maintain the rebar at a specified height off the bottom of the concrete pour; 2) small cross-shaped pieces of plastic used to maintain consistent spacing between tiles during ceramic tile installation.

spacer strip – a metal strip placed in the root opening of a weld as a backing for the weld metal, and to maintain the root opening during the welding operation.

space standards – minimum dimensions needed in a building for various activities. These standards may be established by law, such as minimum aisle widths required to permit escape in the case of fire or other disaster, or by social convention, experience or design practices. Enough space must be allowed in a given area, such as a bedroom or kitchen, for people to move around comfortably. Different cultural and individual needs often dictate building design practices.

spackling compound – a patching compound formulated for use in filling small holes, cracks and other irregularities in walls. It can be smoothed and painted.

spade bit – a wood boring bit that has a flat head on a shank. The end of the head has a point for centering and guiding the bit in the wood, and the edges of the head on either side of the point are ground to a beveled edge. As the shank turns and the point enters the wood, the ground edges start to shave off the wood in a circular section equal in diameter to the width of the head of the bit. Designed for use with a power drill.

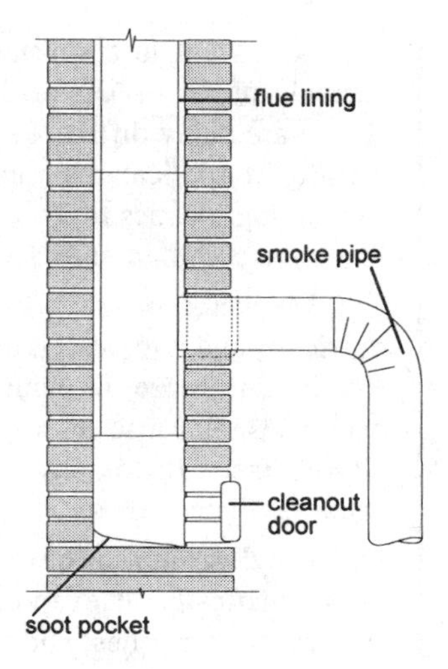

soot pocket

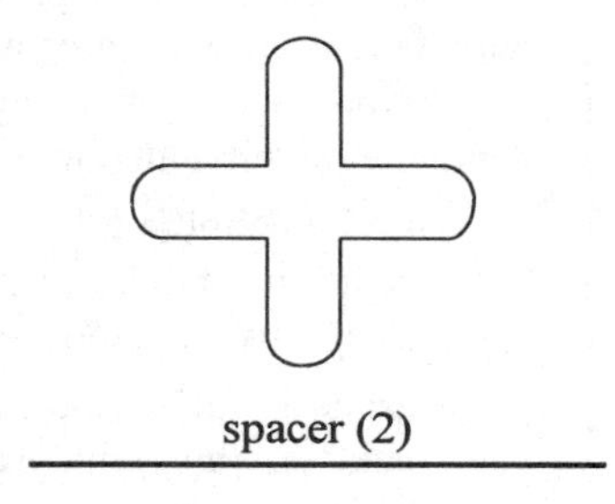
spacer (2)

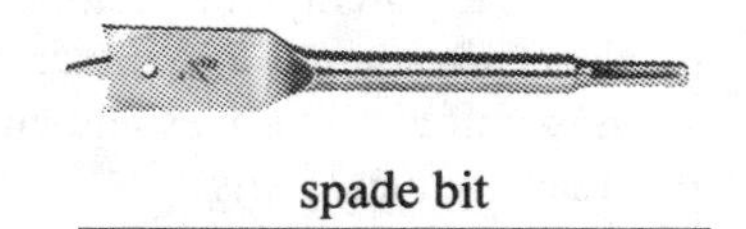
spade bit

Solvents

A solvent is a substance, usually in liquid form, capable of dissolving or dispersing another substance. There are many different solvents available for use in a variety of applications, ranging from thinning paints to cementing plastics together. The following are some of the more common ones and their uses.

Methanol, or methyl alcohol, is usually made synthetically and is toxic. It is used as a component in cooling system antifreeze, in paint removers, and in stains and varnishes. Because it evaporates quickly, it is a good drying agent. Its toxicity requires that it be used only in well-ventilated areas.

Ethyl alcohol can be made naturally by fermentation, or synthetically. It evaporates quickly, and is used in antifreeze, varnishes, and as a shellac solvent. In denatured form, it is toxic if ingested. Ethyl alcohol is denatured by adding in a toxic substance, often methanol, which makes it unfit for drinking without impairing its usefulness as a solvent.

Isopropanol, or isopropyl alcohol, is a petroleum derivative used as rubbing alcohol, a lacquer thinner and in the manufacture of cosmetics. It is an inferior solvent compared to ethyl alcohol.

Butyl alcohol is a flammable alcohol derived from butanes. It is used in lacquer, synthetic resin compounds, penetrating oils, in some paints, and as a metal cleaner.

Carbon tetrachloride is a clear, nonflammable solvent used on fats, oil, grease, wax and resins. It is hazardous to breath or come into contact with this solvent for any extended period.

Trichlorethylene is a synthetic compound, with a slower evaporation rate than carbon tetrachloride, used in manufacturing dyes and chemicals and for vapor-degreasing metal parts.

Tetrachlorethylene, or perchlorethylene, is another synthetic compound used in the same manner as trichlorethylene and carbon tetrachloride.

Ethyl acetate is a synthetic compound that evaporates very quickly and can be used to dissolve nitrocellulose, oil, resin, gum and fats. It is also used in coatings for paper, cloth and leather.

Butyl acetate is a synthetic compound that dissolves oils and synthetic resins, and is used in the manufacture of artificial flavors and perfumes.

Amyl acetate is a synthetic compound used in lacquers. Also called banana oil because of its scent

Aromatic hydrocarbons are distilled from coal tar. Benzene and toluene are used in paints and lacquers, paint remover, ink, asphalt and coal tar. Xylene is used in paints and lacquers and to make dye and synthetic chemicals. Coal tar naphtha is used in lacquer and paint.

Petroleum base solvents are benzine, kerosene, and mineral spirits. Benzine is used as part of the vehicle in paints, lacquer, rubber cement, ink and varnish remover. Kerosene is a solvent for petroleum compounds, and mineral spirits are used as a solvent for dry cleaning and in oil-base paint.

The ketones are acetone, methylethylketone (MEK), and glycol esters. Acetone is a solvent for many different materials in different forms (solids, liquids and gases), and is used in making dyes and lubricants. MEK is used as a general solvent and in various cellulose ester compounds. Glycol esters are solvents for cellulose esters, dyes, varnishes and lacquers.

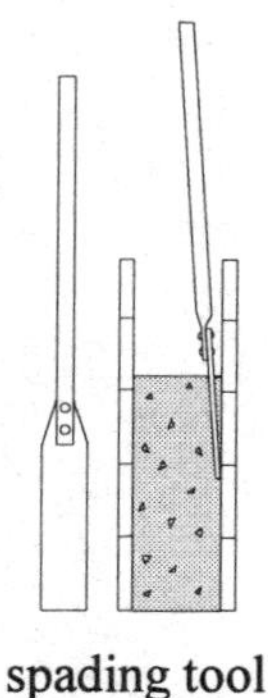

spading tool

spading tool – a special tool with a long flat blade attached to a handle, used to consolidate newly placed concrete in the forms.

spall – 1) flaking or chipping of a concrete surface caused by weathering, expansion, or a blow; 2) a small fragment broken from a masonry structure; 3) to shape by chipping off flakes from a surface, particularly stone or masonry.

span – 1) the distance between structural supports; 2) the width of a building.

span, allowable – the maximum permissible distance that a structural member may span between supports.

spandrel – 1) an area between and above two connected arches, or between one arch and the ceiling; 2) the area between the top of a window in a steel-framed building and the sill of the window directly above.

spandrel wall – the section of wall spanning the area from the top of one window to the sill of a window directly above in the next story of the building.

Spanish swirl – a style of heavy texture, used with drywall, plaster or stucco, that produces circular swirls. Also called a sponge twist.

spanner – a special wrench which usually has projecting teeth which fit in a groove or notch on a collar or particular nut.

spark arrestor – a screen or expanded metal covering on the outlet of an exhaust or a chimney which allows smoke to pass through but prevents sparks from exiting and creating a fire hazard.

spark lighter – a lighter for a gas torch which operates by moving a flint across a piece of rough steel, producing sparks.

spar varnish – a lacquer or liquid plastic-based varnish for exterior use.

spatter – pieces of molten metal that adhere to the surrounding base metal during welding. The spatter must be cleaned off or removed when the welding is complete.

spatula – a tool with a flat flexible blade such as a putty knife or taping knife.

specifications – detailed requirements to be met in the manufacture or purchase of an item; 2) written requirements which provide additional details or description included with the drawings or blueprints of a construction project. The specifications give technical standards to be met during construction. Also called specs.

specific gravity – the ratio of the density of a liquid or a solid to the density of the same volume of distilled water at a specific temperature, or a ratio of the density of a gas to an equal volume of hydrogen at a specific temperature. Also called relative density.

specific heat – the number of calories required to raise 1 gram of a substance by 1 degree centigrade, or the number of Btu needed to raise 1 pound of a substance by 1 degree Fahrenheit.

specular gloss – a very smooth, very high-gloss finish, such as the finish on a piano.

speculation – a business transaction with relatively high risk in anticipation of high returns. For example, a contractor may build a custom house, or spec house, at his own expense in anticipation of selling it quickly on completion for a large profit.

speed square – a three-sided measuring tool, similar to a builder's square except with a third side forming an open triangle. It is used to lay out angles or draw perpendicular lines on boards to be cut.

spigot – 1) faucet; 2) the straight end of a pipe designed to fit inside the expanded hub or bell end of another pipe.

spike – a large nail, 3 inches to 12 inches long, used for fastening heavy timber.

spindle – a slender cylinder of wood, especially one that has been turned to a shape on a lathe, such as a newel post.

spiral columns – concrete columns with a continuous spiral winding of reinforcing metal around the steel core.

spiral staircase – stairs with treads that curve around a central point as they ascend.

spire – a tall and slender tower with a pointed roof.

spirit level – a level with a bubble trapped in a fluid-filled glass tube. When the level is held in true vertical or horizontal alignment, the bubble is centered within the tube.

spirit stain – a wood stain with a spirit vehicle (a volatile liquid vehicle that readily evaporates) which enables the stain to dry quickly. Also called spirit varnish.

splash block – a masonry or other moisture-resistant block laid at the outlet of a downspout to route water from the downspout away from the building, and to protect the immediate area from the impact of a strong flow of water.

splatter paint – paint applied in droplets over a dry base coat which gives the finish a bumpy or mottled look. The splatter paint may be the same color as the base, or one or more different colors, depending on the desired effect.

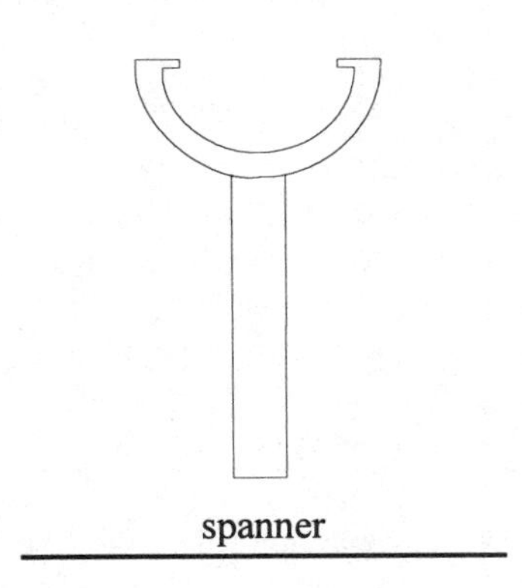
spanner

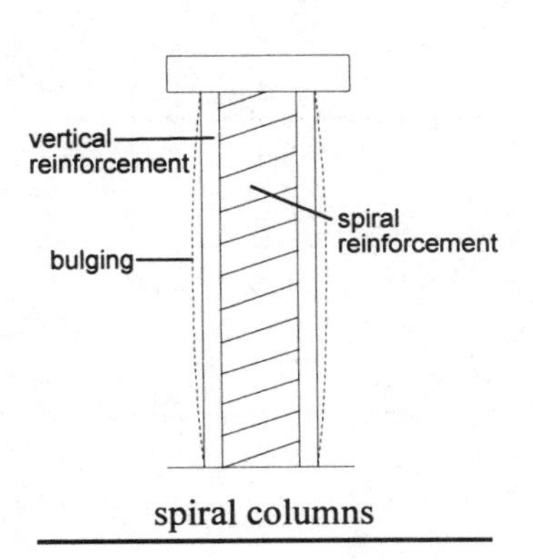

spiral columns

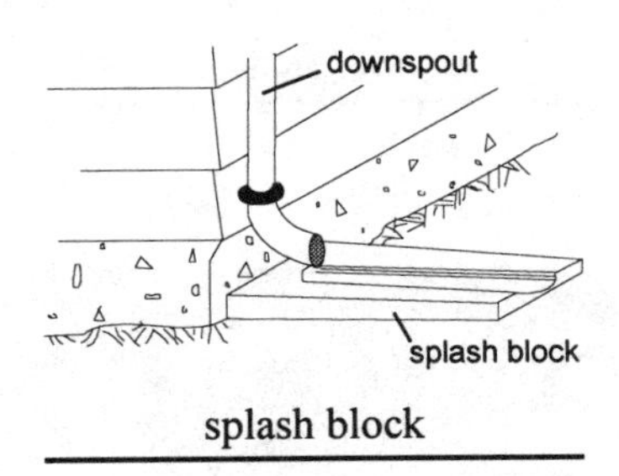

splash block

splay

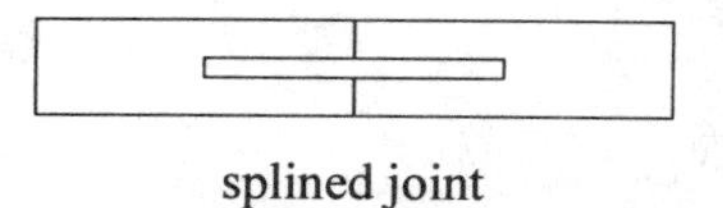
splined joint

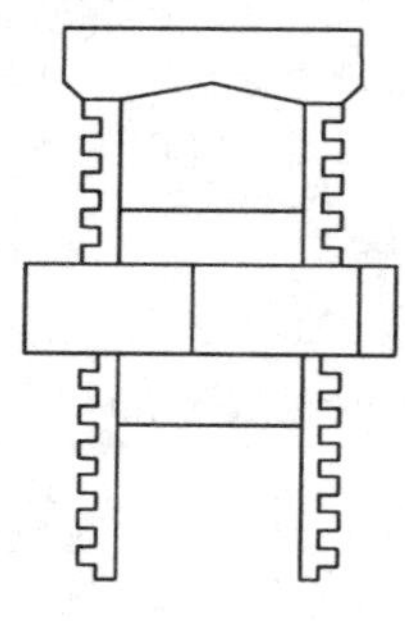
split-bolt connector

split-faced block

splay – 1) a door or window opening that flares out from the door or window across the thickness of the wall in which the door or window is located; 2) flaring outward; 3) a beveled edge.

splice – to join two pieces together.

spline – 1) a thin strip of wood used to join and strengthen two pieces of wood. The spline fits into a groove or wedge cut into the two adjoining pieces; 2) raised ridges that are evenly spaced around the circumference of a shaft, such as a drive shaft or axle. A male splined end fits in a mating female splined end of another shaft, wheel or other device. The splines permit some axial movement while maintaining rotational contact and load transfer.

splined joint – a butt joint in which a groove is cut in each of the members being joined and a spline fitted into the adjoining grooves to add strength to the joint.

split-bolt connector – an electrical connector used for joining wires when they are too large to be joined with wire nuts. The connector consists of a threaded bolt with a slot through the threaded portion. Wires are placed in the slot and a nut with an attached solid piece of metal which also fits in the slot is threaded onto the bolt. The nut and solid metal piece are tightened against the wires to join them in the slot.

split-faced block – a masonry building block with a rough textured face on one side.

split-level house – a house with floors at different levels less than one floor apart. Whole rooms may be at different levels, or a portion of a room may be at a different level from the rest of the room, such as a sunken "conversation pit" around a fireplace.

split rail fence – a fence made of logs that have been split lengthwise and fastened together with spikes or a threaded rod. The fence is laid out in a zigzag pattern. See also zigzag or snake fence.

split receptacle wiring – a method of wiring electrical receptacles in which one half of a double receptacle is wired to one circuit and the other half to another circuit. With this system, two high-current-draw appliances can be plugged into the two halves of a receptacle without overloading a circuit. If one circuit is lost, there is still power at one receptacle.

split-ring repair clamp – a pipe repair clamp designed to repair a leak at a gasket in a bell and spigot joint. One ring fits on the bell side, and the other on the spigot side of the joint. They are held together with interconnecting bolts that span the joint. A gasket is fitted around the spigot end of the pipe and the bolts are tightened. This draws the rings together and forces the gasket into the bell, sealing the leak.

splitting – a paint defect in which the top coat of paint cracks and splits. This is a result of applying the paint to a base coat that has not been allowed to dry completely.

spoil – material excavated from a trench.

spoil box – a container that is used to hold the material, or spoil, excavated from a trench. The spoil can then be used to refill the trench, or be easily removed if another material is to be used for backfilling.

spoil site – a temporary site used for depositing spoil, the material removed from a trench excavation.

spokeshave – a two-handled curved blade used for shaving wood to a curved surface. The blade is drawn toward the craftsperson across the surface of the wood being worked. It is used to make items such as wooden spokes for wagon wheels.

sponge twist – see Spanish swirl.

spotting – covering fastener heads in gypsum wallboard with joint compound in order to produce a smooth uniform finish.

spot weld – electric resistance welding that fuses the metals to be joined at spots, rather than along the entire length of a seam. Spot welding is fast and yields an adequately strong weld for many types of work, such as welding car bodies or other items in which a continuous weld seam isn't required to retain pressure.

spray can – a pressurized container in which the liquid contents are forced out through a nozzle, creating a fine mist. This type of container can be used to apply thin coats of paint in small quantities.

spray gun – a mechanical painting device which uses compressed air to force paint or other liquid coatings through a nozzle for application by spraying. This produces a fine, even coating over the surface. Spray painting is faster than painting by brush or roller, but the covering is thinner and the application less accurate. Care must be taken to prevent surrounding areas from being sprayed.

spray transfer – molten weld filler metal that is transferred across the gap from the end of a consumable electrode to the base metal during electric arc welding. The filler metal travels with the arc as is jumps between the electrode and the base metal.

spread – the width, in feet, of a paint spray pattern at the farthest possible point from the nozzle at which the paint can still be effectively sprayed on a surface.

spreader – 1) wood or metal spacer used in concrete formwork to space forms the proper distance from each other; 2) a metal device that rides in the kerf of a power saw cut to keep the kerf from closing and binding the blade as the work is pushed through the saw.

spreader bar – a stiff bar or beam used with the rigging when lifting large objects with a crane hook. The spreader bar separates the rigging that is attached at different points to the object to be lifted. This ensures that the rigging remains vertical or near vertical at each lift point, even though the object is to be lifted by a single hook.

spreader box – a paving machine attached to the back of a truck, used for spreading asphalt.

spread footing – a foundation footing with a wide base designed to add support and spread the load over a wider area. Used for foundations built on weak soils.

sprig – 1) a headless brad used to attach wood trim pieces; 2) a small dowel used to fasten together two pieces of wood, commonly used in cabinetry and furniture.

spring – a resilient metal coil that may be deformed, and then will either return to its original shape with no load, or will store energy under load to be released when it is allowed to return to its original shape. Commonly used on doors, especially overhead garage doors which use springs to counterbalance the weight of the door.

spring clamp – a hinged clamp with jaws that are held closed by spring tension. Force applied to the handles at the opposite ends of the jaws opens the jaws and allows them to grip an object.

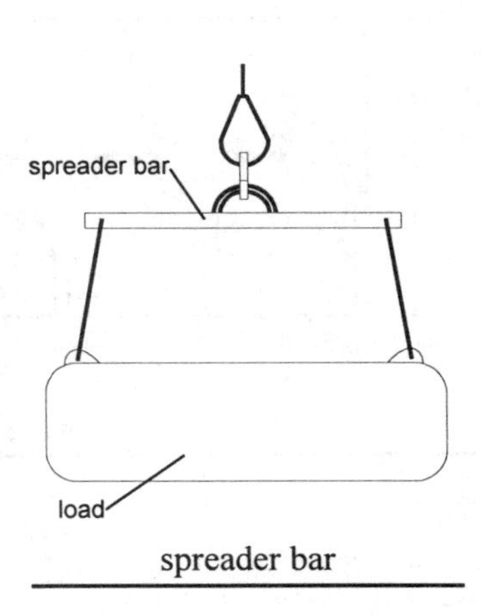

spreader bar

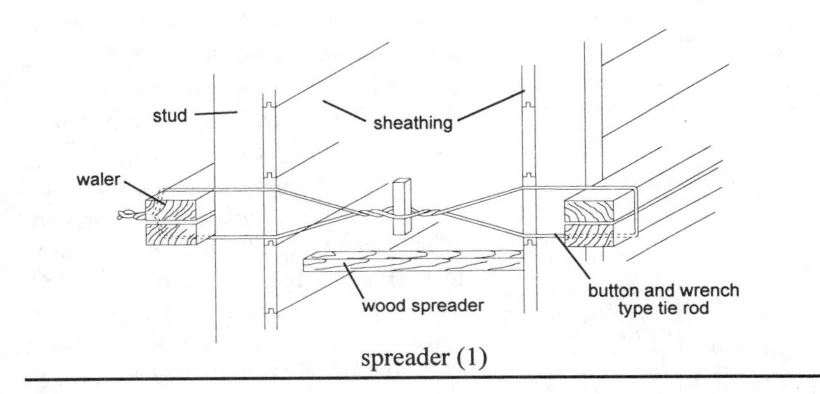

spreader (1)

spring steel – special iron and carbon alloys which are heat treated to develop the characteristics required for making springs. Carbon steel is used in making small springs, and chrome-vanadium or silicon manganese steel is used for large springs.

spring clamp

sprinkler – a water pipe fitting that disperses water in a spray. Sprinkler heads are used as hose attachments and exterior lawn and garden watering systems as well as for interior fire safety systems which operate with heat-activated valves.

spruce – a light colored, structurally-sound wood used in construction.

spruce, Sitka – a lightweight, moderately hard, very tough and strong wood from the largest variety of spruce tree.

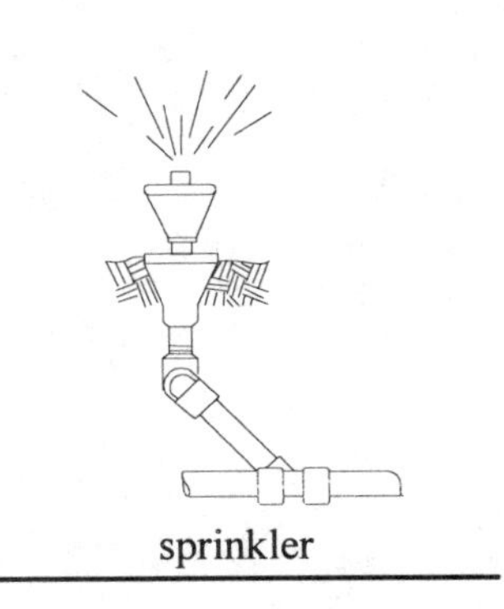
sprinkler

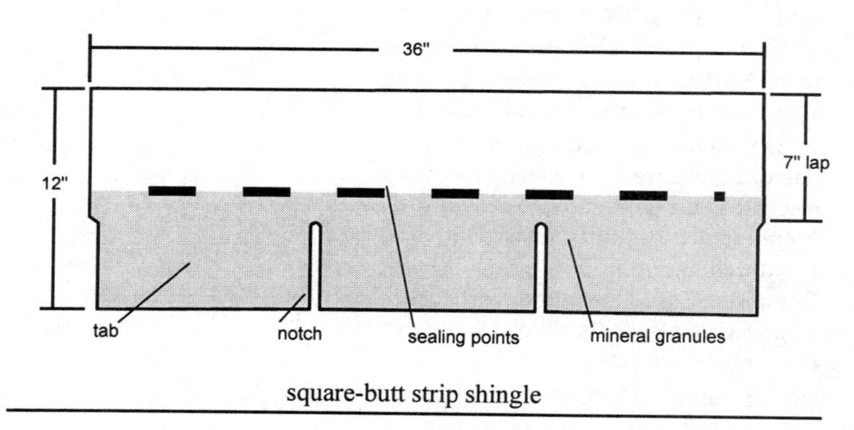

square-butt strip shingle

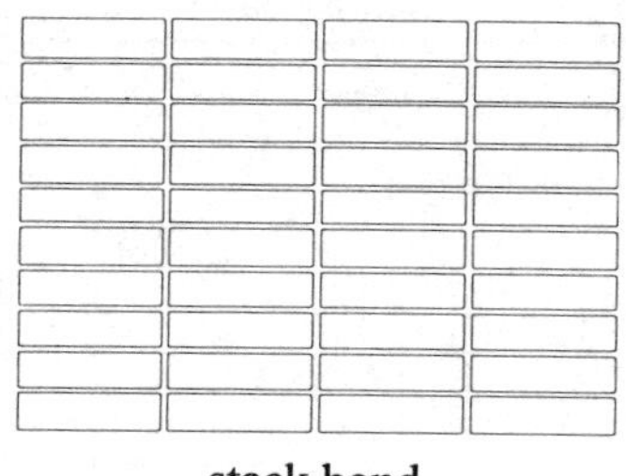
stack bond

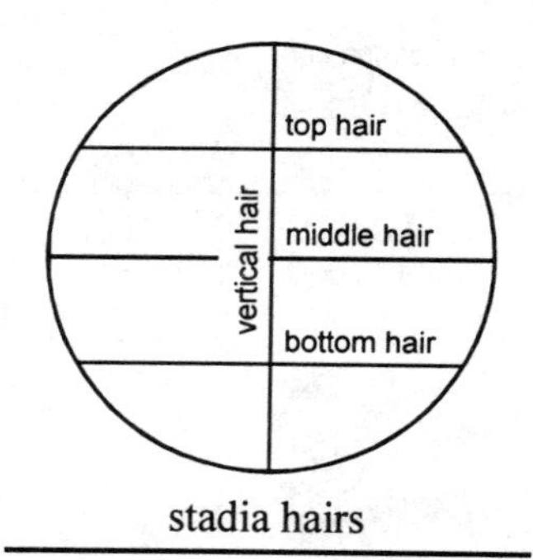

stadia hairs

spur gear – a gear with straight cut teeth which are parallel to the axis of the shaft on which the gear is mounted. They are used in transmissions and other gearboxes.

SP wire – small diameter stranded wires covered with rubber insulation. Also called SPT wire, zip cord or lamp cord.

square – 1) a four-sided geometric figure, with all sides equal and all angles 90 degrees; 2) a hand tool with two straight edges meeting at a 90-degree angle used to determine or mark right angles on work; 3) a unit of area, as in a square foot or square mile; 4) a construction reference indicating an area of 100 square feet; in particular, the amount of roofing or siding material required to cover 100 square feet; 5) a number multiplied by itself, as in the square of 5 equals 25.

square boxes – electrical junction boxes that are square in shape.

square-butt strip shingle – a commonly-used asphalt roofing shingle that is 3 feet long by 1 foot wide. It has three tabs formed by two slots cut into the shingle from one edge and extending part way across the shingle width.

square, combination – see combination square.

square drive screw – a screw with a square recess cut into the head. It is used when a non-slip tightening recess is required, such as when a set screw has to carry a load.

square file – a file with a square cross section used for shaping and smoothing.

square joint – a joint in which the members are fitted together at exactly 90 degrees.

square measure – a measure of area.

square up – to make square.

squaring shaft – see cross shaft.

stab rod – a metal rod with a movable collar stop; the stop is adjusted to the desired depth and the rod is pushed (stabbed) into freshly-laid asphalt to verify the asphalt thickness.

stab saw – a hacksaw blade held in a small handle from which the blade protrudes so that cuts may be made in tight places.

stack – 1) a chimney or vertical vent; 2) vertical piping, most often soil, waste, or vent piping.

stack bond – a brick laying pattern in which each course of stretchers is laid directly in line with the adjacent courses and all vertical joints are aligned.

stack cutting – cutting several stacked pieces of material, such as lumber, at one time.

stacked window – a large glass area that is formed by combining several awning, hopper, or casement windows together. Stacked windows are used in buildings as an architectural design feature as well as to provide light.

stack vent – 1) the portion of a soil or waste stack that rises above the highest horizontal drain that empties into the stack; 2) the extension of a soil or waste stack through the roof of a building.

stack venting – fixture drains that are connected independently to a vent stack in a building drainage system. Each fixture drain is connected directly to the vent stack rather than being connected to a header that is connected to the stack.

stadia hairs – horizontal lines in a transit lens that are set above and below the center hair. They are used in conjunction with a stadia rod to determine distances by relating the hairs to marks on the stadia rod. Also called stadia lines or stadia wires.

stadia rod – a rod, with graduated measurements in feet and tenths of a foot, used with a transit to determine distances in a survey. The stadia rod is located at the point of interest and sighted through the transit.

stadia surveying – determining distances using a stadia rod and a transit with stadia hairs or lines. The stadia rod is sighted through the transit, and the distance measurement established by observing the graduations on the rod as they are aligned with the stadia lines on the transit lens.

staggered intermittent welds – intermittent welds that are alternated from one side of a joint to the other.

staggered stud partition – a wall structure with plates of wider material than the stud material, and in which the studs are alternated from one side of the plate to the other; for example, 2 x 4 studs alternating to either edge of the 2 x 6 plates. This installation increases noise insulation, as vibrations are not passed through the walls by means of common studs. Also called offset studs.

staging – see scaffold.

stain – 1) a discoloration; 2) a natural or man-made coloring composed of dyes in a drying oil or water base used as a decorative or protective covering for a variety of materials.

stained glass – colored glass used to make decorative windows.

stainless steel – a corrosion-resistant steel alloy containing nickel and/or chrome. *See Box.*

stair – one step in a flight of stairs

stair carriage – the angled structural member on which the treads or walking surfaces of the stairs rest. Also called a stringer.

staircase – see stairs.

stair gauge – metal clamps, used in pairs, which fit onto a framing square and are tightened in place to allow repeated identical measurements to be made against a stair carriage. Used to lay out stairway stringers for the treads and risers.

stair riser – see riser.

stair run – see run.

stairs – a series of steps that allow passage between floors of a building, or between two different levels.

stair step – one stair tread.

stairway, closed-string – a stairway with walls on both sides.

stairway, housed – a staircase in which the treads, risers and stringers are connected with rabbet joints which are both glued and nailed to fit tightly together. This makes a staircase that won't squeak or allow dust and dirt to get into the joints.

stairway, open-string – a stairway with a wall on one side and (usually) a rail and balusters on the other.

stairwell – the vertical shaft, through one or more floors, that encloses the stairway in a building.

stake – a stick, board or peg, usually sharpened on one end, which is driven into the ground to mark a location.

staking an area – setting out survey stakes to indicate the location of excavation work to be done.

stamp – 1) a device used to make an identifying mark or impression; 2) a mark or impression made for grading or identification; 3) a sticker glued on items which indicates a tax has been paid or postage collected.

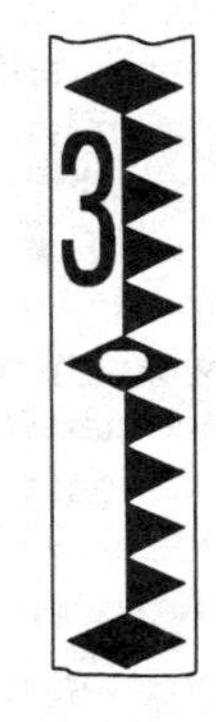

stadia rod

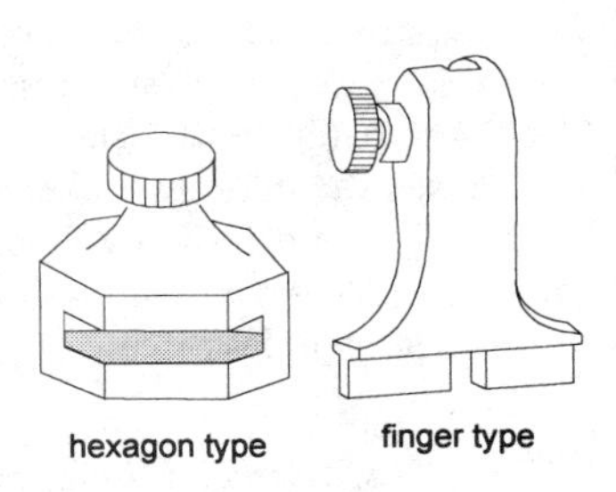

stair gauges

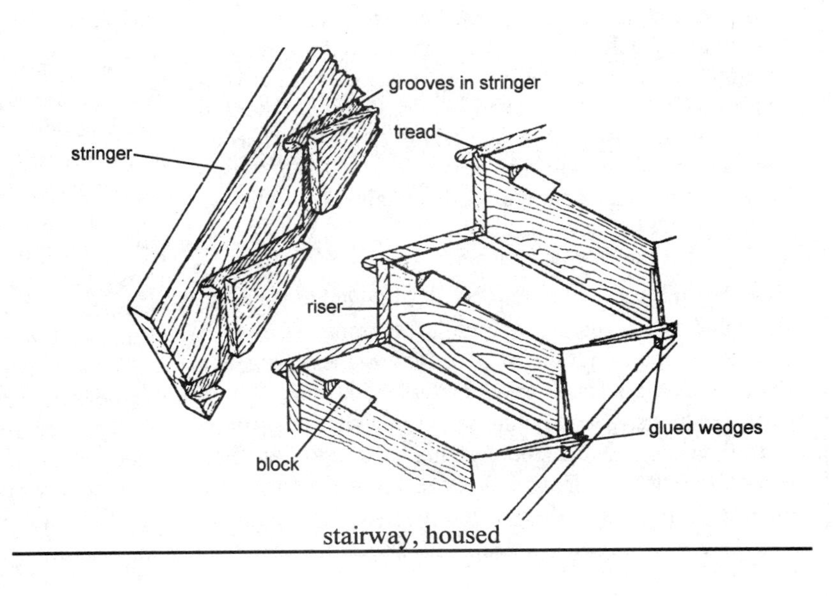

stairway, housed

Stainless steel

Stainless steel is a corrosion-resistant steel alloy containing nickel and/or chrome, often in combination with small amounts of other elements. A thin oxide coating forms on its surface which protects the metal. This coating is self-healing, since it is formed by oxidation. A chromium content of more than 11.5 percent provides greatly increased passivity. Stainless steels are high on the galvanic scale, and so do not usually become the sacrificial metal in a galvanic cell. A galvanic cell is created when two different metals are in electrically conductive contact in the presence of a conductive fluid. In this situation, the metal that is lower on the galvanic scale is eaten away or corroded by the electric current flowing through it.

The performance characteristics of the different stainless steels vary with their structure and the combination of alloying elements used in their manufacture. The American Iron and Steel Institute (AISI) has assigned class numbers to the different metallurgical structure groups of stainless steels. Group I encompasses the hardenable martensitic stainless steels. Group II are the nonhardenable ferritic stainless steels. Group III covers the austenitic chrome-nickel (with the addition of manganese in some alloys) stainless steels. There is a further classification by series number. These include 300 and 400 series stainless steels.

Group I, Martensitic Stainless Steels

Martensitic stainless steels are part of the 400 series stainless steels. Martensitic is derived from martensite, a metallurgical structure designed to increase strength and hardness that is developed by heat treating the steel. Martensitic steels contain between 12 and 17 percent chrome, they are abrasion and corrosion resistant, and can be formed hot or cold. They have a coefficient of thermal expansion less than that of carbon steel and thermal conductivity about half that of carbon steel. When used in assemblies that combine both metals, these characteristics must be taken into consideration to ensure that the intended function is realized.

The strength and toughness of martensitic stainless steels makes them useful for high-stress applications such as machine parts, jet engines, turbine blades, propeller shafts, shear blades, marine hardware, springs, valve parts, bushings, gears, pump parts, precision measuring instruments, hardware and fasteners, including screws and bolts. Their hardness and corrosion resistance make them a good choice where a sharp edge must be maintained, such as for knife blades, scissors, tools and cutting implements, as well as dental and surgical instruments and equipment. Type 440 has the highest hardness of any stainless steel. It is used to make ball bearings and other items that require a high degree of wear resistance, such as cutlery, dies, molds, nozzles, valve parts and surgical instruments.

Group II, Ferritic Stainless Steels

Ferritic stainless steels are also part of the 400 series. Their metallurgical structure is mostly ferrite, which is soft. They are low carbon, high chrome alloys that cannot be hardened except by working them using a process called cold working. They have better high-temperature corrosion resistance than the martensitic stainless steels and they are more easily welded.

Ferritic stainless steels can be easily machined and are used in making heat exchanger tubing, refinery equipment, corrosion-resistant linings for tanks, valve and pump parts and hardware. They are also used as a general purpose stainless steel for making fasteners, tubing, zippers, interior building trim, automobile trim and kitchen implements and hardware. Some ferritic alloys have up to 25 percent chrome, giving them the greatest corrosion and scaling resistance of the straight chrome stainless steels. These can be used in high-temperatures applications such as for furnace parts, heaters, heat exchangers, mufflers, valves and valve parts, oil burner parts, stack dampers and glass molds.

Group III, Austenitic Stainless Steels

Austenitic stainless steels make up the 300 series stainless steels. They are also known as 18-8 stainless steels because they contain about 18 percent chromium and 8 percent nickel, although some types have much higher percentages of these elements. They also contain various other alloying elements to develop particular performance characteristics. They have high corrosion resistance to atmospheric influences and also to many chemicals, including organic and oxidizing acids, such as acetic and nitric acids. They are not as resistant to mineral acids, such as sulfuric acid, nor halogen acids, such as hydrochloric acid. They can be hardened by cold working, and are readily formable. They are nonmagnetic when annealed, but may become magnetic to a degree from cold working. These stainless steels have good high-temperature properties, retaining strength and oxidation resistance. They also have good low-temperature properties, remaining ductile and having good impact resistance. They do not conduct heat as well as carbon steel, but have a higher rate of thermal expansion (they expand more for every degree increase in temperature).

The 300 series stainless steels are used for a variety of parts and equipment, with their use dependent on the particular qualities that are developed using additional alloying elements and cold working. Because they are strong and ductile and have varying hardnesses, they are used in a wide variety of metal products including: building trim, flashing, sheet metal products, fasteners, bushings, shafts, pipe, valves, flanges and fittings, plumbing fixtures, pump parts and roof drain parts. They are also used in high-temperature applications such as annealing boxes, furnace, boiler, oil and gas burner parts, heat treating equipment, jet engine parts, stack liners, oven linings, exhaust manifolds, air heaters and welding rods. Their corrosion-resistant applications include chemical process equipment, refinery equipment, forgings, marine trim, pipe, fittings, valves and other items which must be welded.

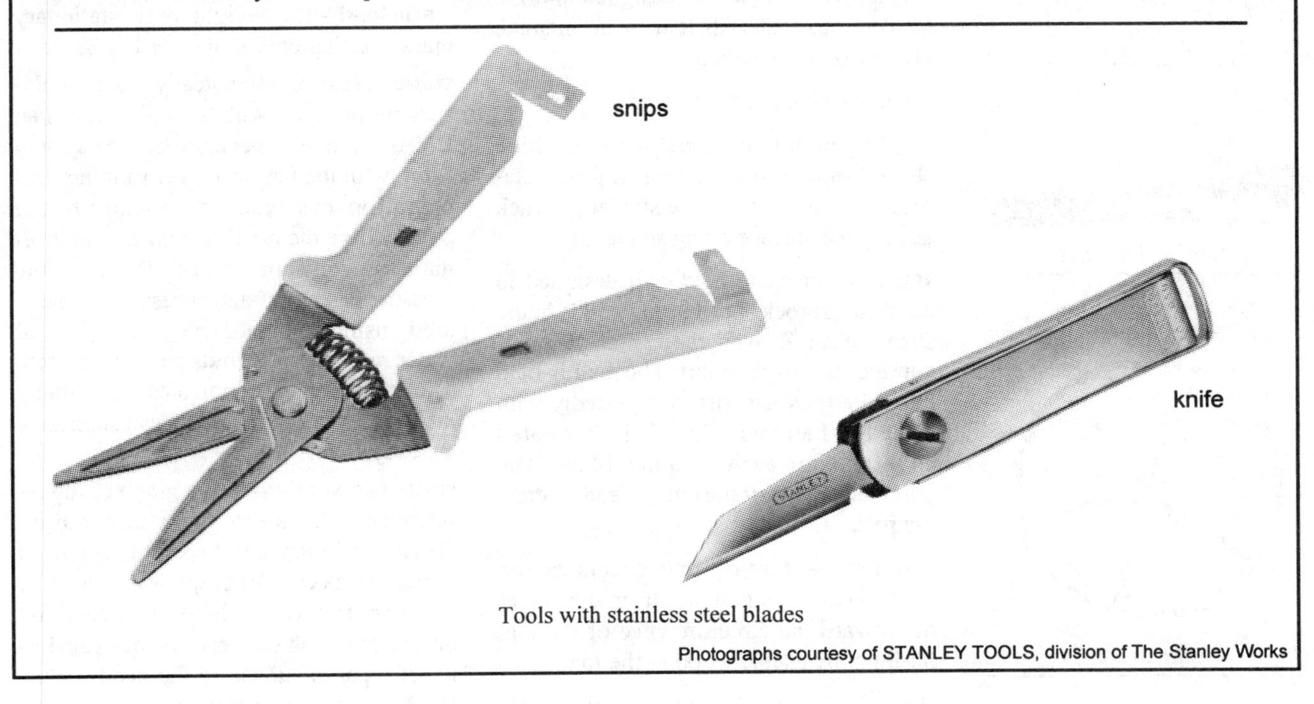

Tools with stainless steel blades

Photographs courtesy of STANLEY TOOLS, division of The Stanley Works

standard – a predetermined set of values, useful to ensure that minimum requirements are met.

standard brick – a brick with dimensions of 3⅝ x 2¼ x 8 inches.

standard conditions – those parameters that are established for use as a repeatable point of comparison. For instance, standard conditions for HVAC (heating, ventilating, and air conditioning) are taken as 70 degrees F, 50 percent relative humidity and 29.92 inches of mercury atmospheric pressure. HVAC systems are set using these conditions as a reference.

standard modular brick – a brick with dimensions of 2⅔ x 4 x 8 inches.

standby – a condition of readiness to operate or perform.

standby power generator – an electrical generator powered by an internal combustion engine used to supply electricity in the event of a failure in the primary electrical supply.

standing seam metal roof – a roof covering made of metal panels joined with interlocking seams that fold over at right angles to the plane of the roof panels. Some types of panels come with seams

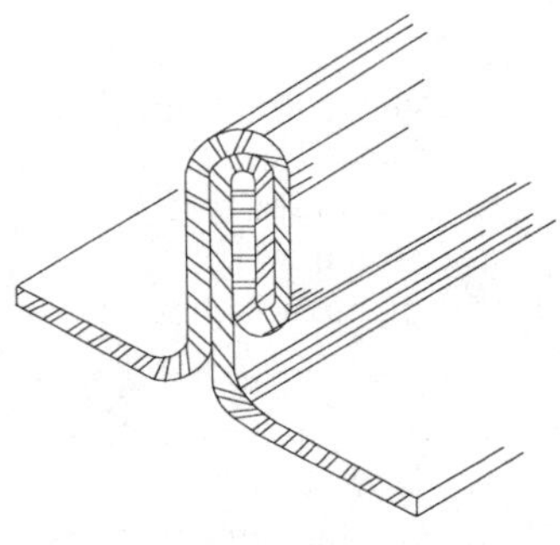

standing seam metal roof

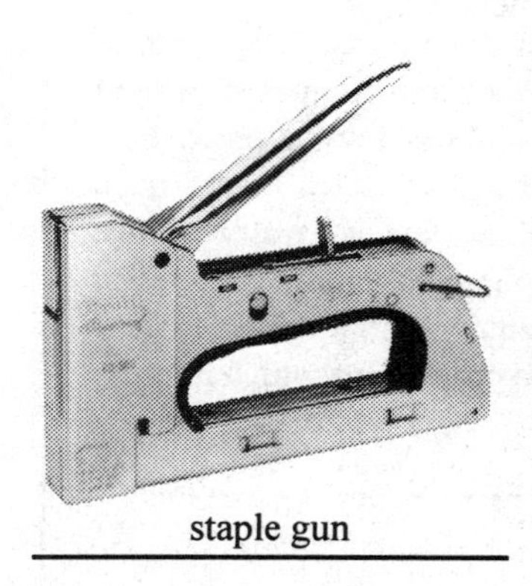
staple gun

stapling hammer

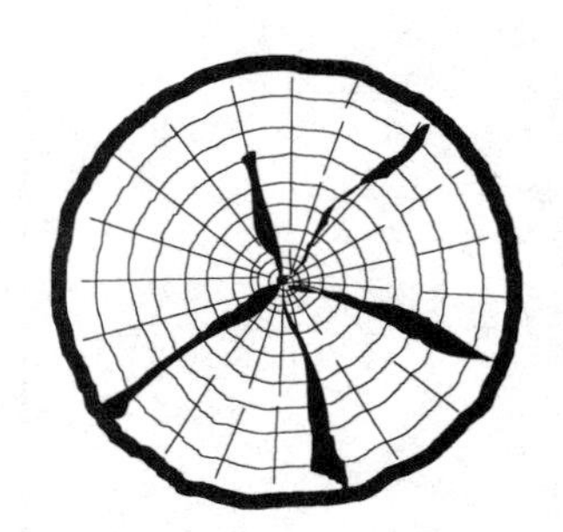
star shake

already folded which slide together for a tight fit; others are folded and crimped together during installation.

stand oil – a heat-thickened vegetable oil, such as linseed or tung oil, used to finish furniture.

standpipe – 1) a vertical run of pipe used as a water supply to a fire protection system; 2) an open vertical pipe into which waste water from an automatic washing machine is discharged.

staple gun – a machine designed to drive staples into material. It may be manual, electric or air powered.

staples – U-shaped fasteners.

stapling hammer – a stapling machine shaped similar to a hammer. It places staples when the head of the stapler is struck against the surface being stapled.

star drill – a hand-held drill designed to cut through rock. The cutting end is in the shape of an X, with the center of the X coming to a slight point. The drill is held against a rock and struck repeatedly with a sledge hammer. The drill is rotated slightly with each hammer blow. The point and shape of the cutter head shatters the rock.

starshake – splitting that occurs across the growth rings in a log, from the center out toward the circumference of the log, limiting the usable wood in the log.

starter – 1) an electrical control and switching device used to start and stop a motor; 2) the automatic switch that triggers the ballast in a fluorescent lamp.

starter course – 1) the first row of shingles at the eave line; 2) the first row of brick laid.

starter strip – 1) a strip of wood nailed on the outside of the framing against which the siding will rest; 2) the first row of roofing at the eave line in the application of asphalt shingles. These may be strip shingles that are properly trimmed, or mineral-surfaced roll roofing that is 7 or more inches wide. The starter strip is laid, and then the first course of shingles is laid on top of the starter strip.

starting tap – a tap with a tapered end to make starting the cutting of internal threads easier. The tapered end is inserted in the end of the hole to be tapped and the first threads on the tap make shallow starting cuts.

static – 1) showing little change or movement; 2) electrical interference or atmospheric disturbances affecting radio or television reception.

static electricity – a charge of electrons gathered on the surface of a material.

static load – the weight of a stationary mass exerting only downward pressure.

static pressure – the steady state, or unvarying pressure within an air duct in an HVAC system. It occurs when the system is on, with the fan or blower running, and operation has reached an equilibrium point where the air flow and pressure remain at constant values. Because the pressure is so low, static pressure is measured, using a manometer, in inches of water rather than pounds per square inch. The measurement indicates how many inches of water the pressure can support at a steady state.

static pressure curve – a graph of the relationship between the static output pressure of a fan and its output volume at a specific speed. The graph is provided by the manufacturer to help engineers and designers match the performance capability of a particular fan to the needs of the HVAC system in order to ensure optimum system performance.

static pressure regain – an increase in system static pressure in an HVAC system as the forward moving velocity of air in a duct decreases, such as when a damper in the system is closed.

stationary window – a window that is fixed in place and does not open.

station numbers – numbers assigned to stakes that are positioned as distance markers along the ground in an excavation or grading layout or other related building activity.

statute – 1) a law; 2) a regulation enacted by a corporation intended as a permanent rule of procedure.

stay-in-place forms – concrete forms designed to remain in place as a permanent part of the building structure.

steam cleaning – the use of steam under pressure to clean, such as to clean oil or grease off of concrete.

steam hammer, double-acting – a pile-driving hammer that is raised by steam and accelerated on the drop by steam. A double-acting steam hammer exerts more force than a single-acting steam hammer.

steam hammer, single-acting – a pile-driving hammer that is raised by steam and dropped by gravity.

steam heating – using steam, created by heating water in a pressurized boiler, to circulate heat through pipes and into radiators where it can be dispersed into the surrounding areas. The steam cools in the radiators and becomes water again, traveling back down the pipes and into the boiler to be heated again and recirculated.

steam shovel – a steam-powered earth-moving machine.

steam trap – a device that drains water off a steam line, but prevents the flow of steam through the trap.

steel – a strong alloy of iron and carbon with the addition of other metals, which may include manganese, chrome, molybdenum or nickel. Carbon steel is a steel whose characteristics are controlled primarily by its carbon content. Alloy steel is steel whose characteristics are controlled primarily by alloying elements other than its carbon content. The alloying elements are added to increase its high-temperature strength and corrosion resistance. See also stainless steel.

steel beam – see I-beam.

steel construction – building in which the main structural members are of steel. Steel is used for its strength in building very large or tall buildings that could not be supported by wood members.

steel-framed structure – a building whose framing members, including studs and rafters (where applicable), are made of shaped structural steel. The structure can be very strong, while maintaining a degree of flexibility. This provides earthquake resistance. If gypsum wallboard and other materials which do not support combustion are used in the construction, the structure can be made to be essentially fireproof as well.

steel-framed structure

steel wool – a pad of steel fibers which loosely resemble wool fibers. The steel fibers are available in different degrees of coarseness for use in cleaning metal surfaces and smoothing wood surfaces to a high gloss.

steeple – a tall tapered tower structure on top of a building. Often used in church construction to house a bell or chimes.

Stellite – a trade name for a metal alloy of chromium, cobalt and tungsten that is very hard, but brittle. It is used in areas of potentially high wear, such as the seat and/or disc of a valve in steam service. The parts may be made of Stellite, or Stellite may be welded to the areas where wear resistance is desired.

stemming a hole – backfilling a blasting charge hole with broken pieces of rock after the charge is placed in order to direct the blast.

stencil – a sheet with letters, numbers or designs cut out. The sheet is placed on the object to be marked and paint is applied over the cutouts, leaving the shapes or letters of the cutouts marked on the surface

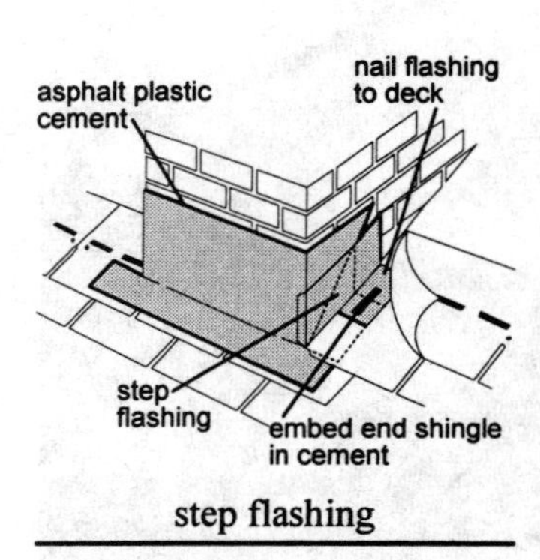

step flashing

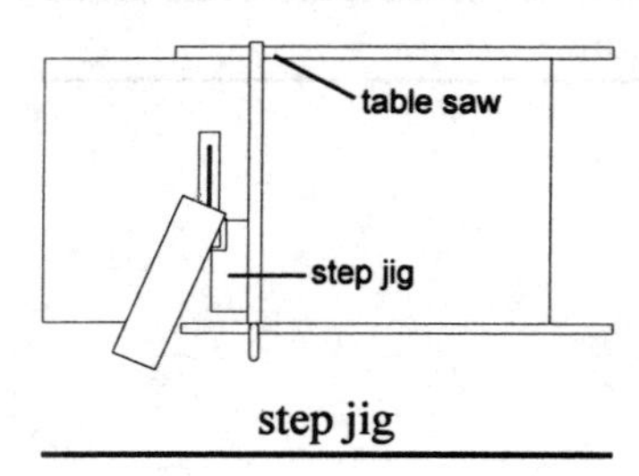

step jig

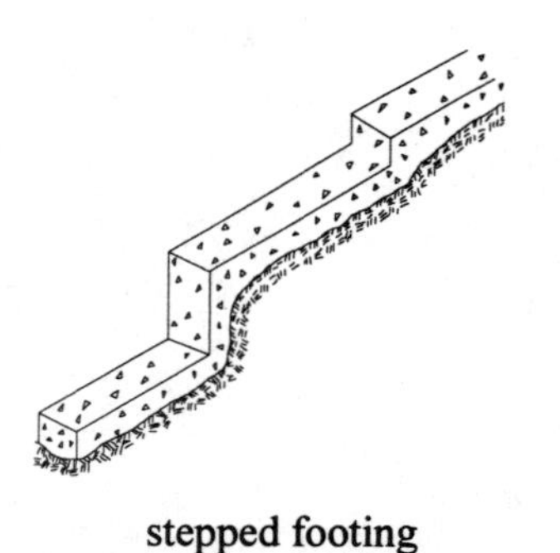
stepped footing

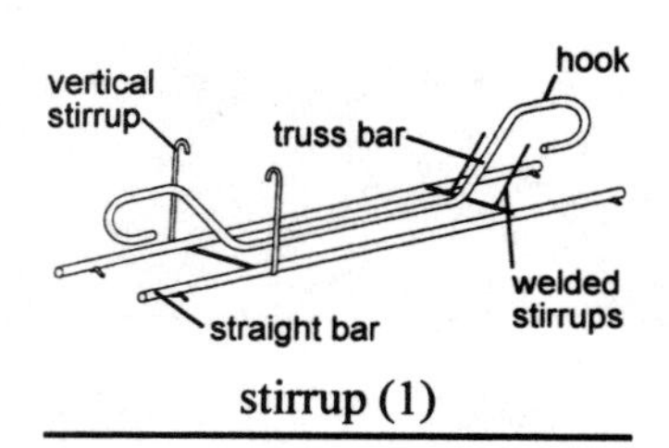

stirrup (1)

of the object. The stencil can be used over and over to create a uniform design on a number of items, or a different arrangement of the letters and numbers can be used to create different markings.

step brazing – brazing at different locations on the same part with brazing materials of progressively lower melting temperatures so that braze joints already made are not affected by later brazing operations on that part.

step flashing – flashing used where a roof slope abuts a vertical wall or chimney. It consists of a series of short sections of flashing bent at right angles across the center so that half of the flashing is on the roof and half up the wall. These pieces are placed along the junction of the roof and the wall, with each section of flashing overlapping the upper edge of the section below it in the same manner that shingles overlap. This prevents leaks between the wall and the roof.

step jig – a jig, consisting of a board with steps cut in its edge, used for cutting a taper on a table saw. The straight side of the board rests on the rip fence. The corner of the board to be cut on a taper rests in the notch formed by the step on the jig so that it is held at an angle to the rip fence. As the jig is moved along the fence, the work, which is at angle to the fence, is also at an angle to the saw blade, and so is cut at an angle.

step ladder – an A-frame shaped, self-supporting ladder, consisting of two sets of hinged supports with rungs across one or both sets for stepping. When in use, the two supports meet at the top and are held together at an angle by collapsible hinges. When not in use, the sides of the ladder collapse together for easy carrying and storage.

stepped flashing – see step flashing.

stepped footing – a building perimeter foundation in which the footing descends in step-like sections. It is designed for use on hillsides.

stepped ramp with risers – a walkway designed for use on a steep hillside. Sections of walkway are laid level, or approximately so, to form long shallow steps. Each section rises up to 6 inches above the one before it.

step soldering – soldering different locations on the same part with solders of progressively lower melting temperatures so that solder joints already made are not affected by later soldering operations on that part.

stick-built construction – a structure built on site from individual structural members, as opposed to prefabricated sections that are delivered to the site.

stick-built roof – a roof built on site from individual structural members rather than with prefabricated sections, such as trusses, which are manufactured away from the site and delivered for installation.

stick welding – electric arc welding using an arc produced between the work and a welding rod that is the electrode. See also shielded metal arc welding.

stiff mud brick – a brick made of clay having a 12 to 15 percent moisture content.

stile – the vertical members of a door or window sash.

Stillson wrench – see pipe wrench.

stilts – 1) posts used to support a structure, such as a dock or a dwelling raised above ground level or over water; 2) poles with footrests used to give workers, such as a plasterers, access to elevated work.

stinger – 1) a welding electrode; 2) a concrete vibrator.

stipple finish – a type of texture finish used on wall and ceiling surfaces consisting of closely-spaced raised bumps, often achieved by applying texturing compound with a paint roller or a special brush.

stirrup – 1) metal bar used in concrete reinforcement to resist diagonal tension; 2) a metal supporting strap designed to hold the end of a beam or joist and connect it to another structural member.

stock – 1) material on hand; 2) material that is standard-sized or unchanged from the original manufactured condition.

stockade stake fence – a fence made from peeled straight conifer branches with the top faces rounded and fastened vertically to fence rails.

stone anchor – a metal device used to attach or anchor stone to masonry.

stonemason – a tradesperson who works with stone and other masonry materials.

stool – the flat member between the jambs at the bottom of a window opening against which the bottom sash rests.

stoop – a raised, usually covered, porch or platform at the entrance of a house or other building.

stop – 1) a molding fastened to the inside of a window jamb to secure the sash; 2) a structural member designed to limit the travel of a machine or object, such as a sliding door, at a predetermined point.

stop-and-drain fitting – a corporation stop with a side opening used to shut off water flow and allow water to drain back from the isolated portion of the line. Used in cold weather areas where the pipes must be drained to prevent freezing.

stopcock – a valve with a tapered plug that opens or closes to control the flow of gas or water.

stopoff – material that is applied adjacent to a solder or braze joint to contain the flow of filler metal.

stop work order – a written statement from an owner or building inspector to halt work on a project until stated conditions are met.

storm door – a door installed on the outside of an existing exterior door to provide insulation and weather protection. The door may have an interchangeable window and screen to provide ventilation during warm weather.

storm sewer – a conduit or piping system which conveys ground water, rainwater and other surface runoff to a channel, river or other collection point.

storm window – a window placed on the outside of an existing window to provide insulation and weather protection. Usually installed for winter and removed in the spring.

story – a floor or level in a building.

story pole – a board or rod that is marked off in graduated increments along its length. Used to take repetitious measurements, such as the height of each course of brick or block installed in a wall.

stove bolt – a square- or round-headed steel fastener with a threaded shaft. The steel is not of a high strength alloy and so is limited in use to low stress applications.

straddle milling – milling opposite sides of a piece of work at the same time.

straight arch – see jack arch.

straight claw hammer – a claw hammer with a very shallow curve to the claws. A straight claw will fit more easily under a nail head in tight quarters. It can also be used to hold the head of a nail with the nail shank pointing outward from the claw so that the nail can be driven into a surface using only one hand. The first stroke sets the point of the nail in the surface. The hammer head can then be turned around to drive the nail the rest of the way in. This is an advantage when driving nails into surfaces that can only be reached by extending the hammer the full length of the arm.

straightedge – 1) a rigid piece of wood or metal that is straight and true, used as a guide for marking or checking straight lines; 2) a straight level tool for screeding plaster or concrete to a smooth finish.

straight-grain wood – wood with a grain that is fine and parallel, or nearly so.

straight polarity – direct current electric arc welding in which the work is the positive pole and the electrode is the negative.

straight-run staircase – a staircase leading directly from one floor to the next without turns or landings. Also called straight-flight stairs.

strain – a deformation or change in the shape of an object caused by an applied force.

strainer – a device used to trap and remove solids from fluids that are passed through it.

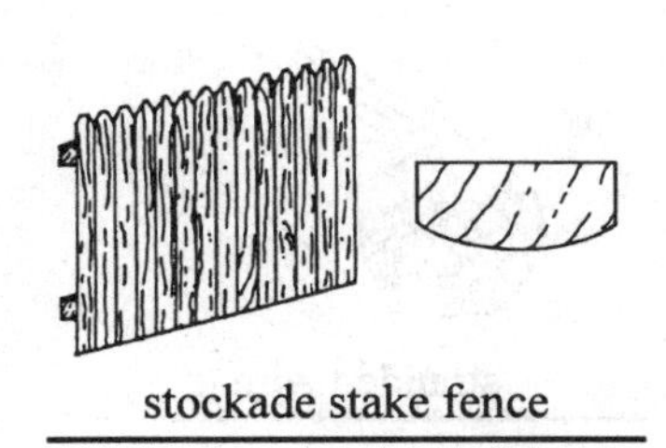
stockade stake fence

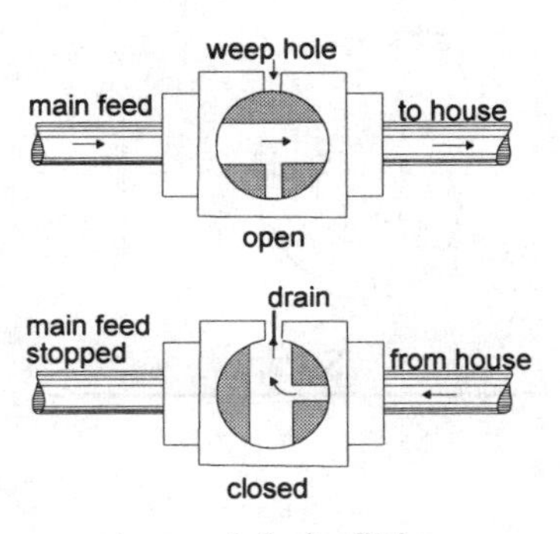

stop-and-drain fitting

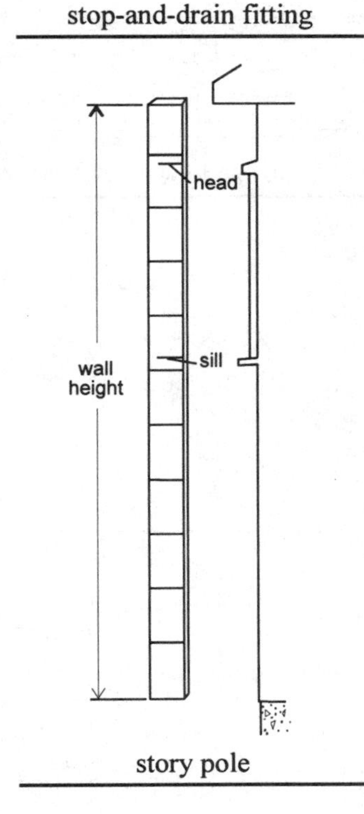

story pole

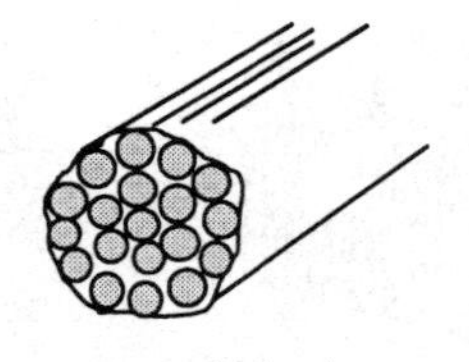

stranded wire

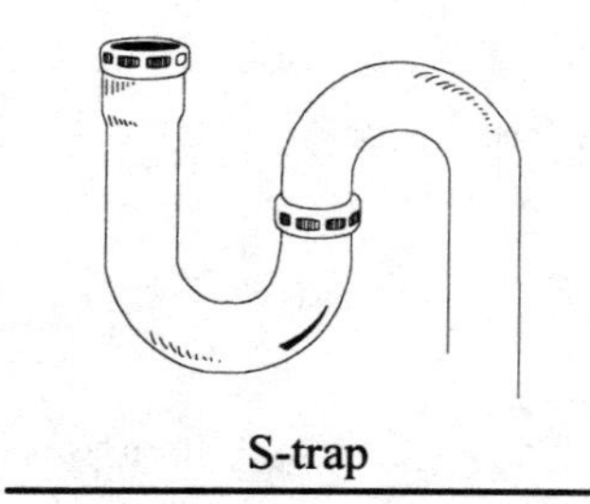

S-trap

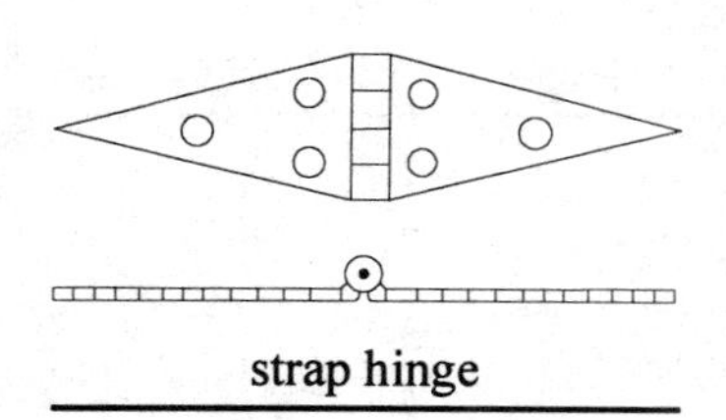

strap hinge

stranded electrode – a welding electrode made up of several strands of material, or strands of different materials used together to improve the weld.

stranded wire – electrical wire in which there are multiple conductors bunched or braided together.

S-trap – a plumbing fixture shaped like the letter S which is installed in a manner which obstructs the pipeline, preventing sewer gases from rising up into living areas.

strap – a piece of sheet metal or thin metal plate used to tie timbers or boards together at a joint.

strap hinge – a hinge designed to be installed across the face of a door. Used with heavy doors and in rough framing for barns, sheds and other structures where exposed hinges are appropriate.

strap wrench – a pipe wrench that utilizes a heavy fabric strap fastened to a handle to provide gripping force. The strap is wrapped around the pipe and force applied with the handle, tightening the strap. The resulting friction between the tightened strap and the pipe grips the pipe tightly to turn the pipe (or keep it from turning). This type of wrench is particularly useful on pipes where external scratches that mar the pipe would be objectionable.

stratify – to separate into layers. Fluids, such as oil base or latex paints, stratify if left standing for a long period. The pigments settle to the bottom and the lighter liquids rise to the top. Gases, such as air, also settle into layers, with the warm air rising and the heavier cold air settling close to the ground, floor or tank bottom.

street stake – a stake that marks the first road cut where there will ultimately be numerous streets, such as in a subdivision. Street stakes are placed down both sides of the street in order to establish the grade and mark the excavation to be made.

strength – the ability to carry a load, such as the ability of a loadbearing structural member to hold up the weight imposed on it without bending or breaking.

stress – the action of forces on an object or system which tend to cause deformation, strain or separation of adjoining members.

stress, allowable – the maximum permissible stress that may be placed on a structural member or a pressure-retaining component. The safety tolerance is determined through testing, and published in codes related specifically to the material in question.

stress corrosion cracking – a fissure in a piece of metal caused by a load imposed on the metal in a situation where the metal is exposed to a corrosive medium to which it is sensitive. This is most likely to occur in stainless steel grades 304 and 316 when they have been heated to and held at a temperature range where the carbide present in the metal precipitates at the grain boundaries of the steel, and then the steel is placed under a load in a corrosive atmosphere. An example would be a weld in a stainless steel pipe that contains internal pressure and is filled with a corrosive fluid, such as chlorinated water.

stress cracking – see crazing.

stressed skin – plywood sheathing fastened or bonded to framing in a manner that makes the combination work as a unit to resist loads.

stress relief cracking – cracking in the weld metal or heat-affected zone during or after exposure to high temperatures. When a metal is heated and worked, as during welding or bending, and then cooled, stresses build up in the metal and cause cracking when the material is used. This is especially true if the material is under a load, such as the internal pressure of a piping system. Controlled heating and cooling can relieve these stresses and prevent such problems from occurring. Codes and industry standards mandate stress relieving for certain combinations of materials and thicknesses.

stress relief heat treatment – exposure of metal to an elevated temperature for a given period of time to permit stresses to relax.

stretcher – a brick laid horizontally with a narrow side of the length exposed.

strike – 1) to hit; 2) to ignite by the heat of friction; 3) to find or locate, as in locating oil when drilling a well; 4) a work stoppage organized by a trade union.

strike board – a board used as a guide for screeding concrete and cement.

strike-off rod – a rigid, straight piece of wood or metal used to remove excess concrete and to level concrete that has been poured in forms. See also screed.

strike plate – a metal plate, recessed flush with a door jamb, into which a lock bolt latches.

striking an arc – initiating the arc between the electrode and the base metal in electric arc welding. The work to be welded is grounded, a source of high current is connected to an electrode and the electrode is brought close enough to the work for the electricity to jump from the electrode to the work, creating the electric arc. The electric arc generates the heat to melt the metal being welded.

striking off – running a straightedge back and forth across the tops of concrete forms to remove excess concrete and to level the concrete.

string course – see belt course.

stringer bead – a straight weld laid with very little weaving across the width of the weld.

stringer, carriage – 1) a horizontal structural member; 2) the support member on which stair treads rest.

stringer, housed – a stairway stringer that has dado grooves into which the treads and risers fit. The treads and risers are glued and nailed into the grooves, making a firm, stable stairway that will not squeak. See also stairway, housed.

stringing mortar – the practice of spreading sufficient mortar at one time to lay several masonry units.

string line – a string stretched between points or stakes to mark a straight line, as where an excavation is to be made.

strip – 1) a narrow piece of material; 2) to completely remove paint or other surface coatings; 3) a board that is less than 6 inches wide.

strip flashing – flashing in a continuous narrow sheet which is used to make custom flashing for roofing applications, tops of windows, and other areas where ready-made forms cannot perform the necessary function.

strip flooring – long narrow pieces of wood flooring laid edge to edge in parallel strips.

strip gutter – a section of wood, usually covered with a corrosion-resistant metal, fastened at a right angle to the roof slope near the bottom of the slope. It is used in place of a hung gutter to direct the flow of water from the roof.

stripper – 1) a tool used to trim gypsum drywall panels down to the desired size for installation. The tool has a handle and two opposing serrated wheels that are set apart just slightly less than the thickness of the gypsum drywall panel to be cut. The two wheels score lines parallel to the edge of the wallboard, which create breaks in the front and back paper covering bonded to the gypsum core. That decreases the structural strength of the panel, so a narrow strip of wallboard will break off along these scored lines. Also called an edge cutter; 2) a tool for stripping insulation from electrical wire.

strip reinforcement – a technique for installing gypsum drywall in which strips of wallboard are fastened across studs as a backing for whole drywall panels to be installed. This provides additional stiffness to the second gypsum drywall panel without having to use a full first panel.

stroke – a single pass of the spray gun or paint brush.

strongback – 1) a stiffener, usually a structural member such as a steel beam, used as part of rigging to hoist an object; 2) a reinforcing structural member.

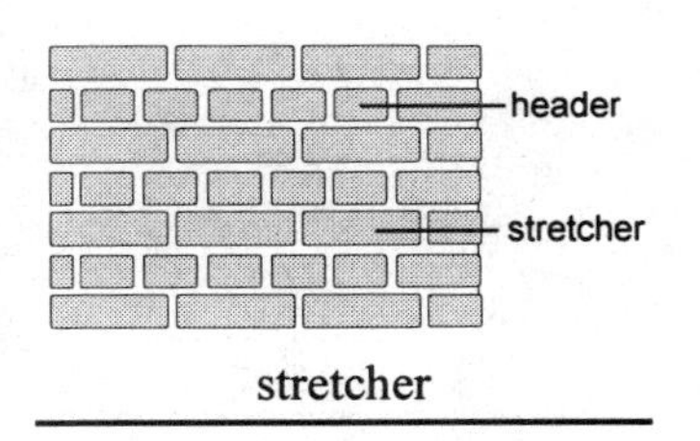

stretcher

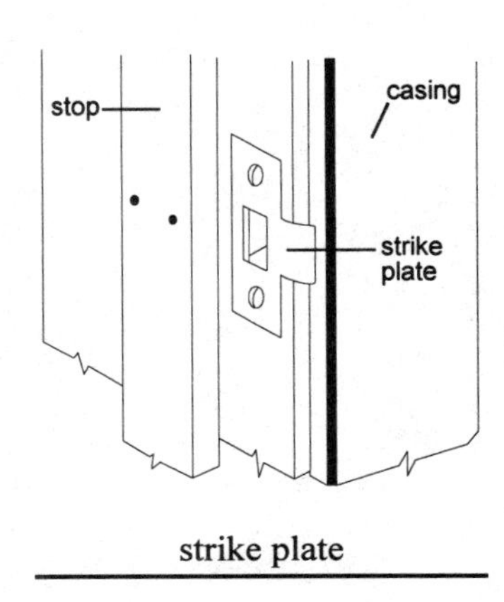

strike plate

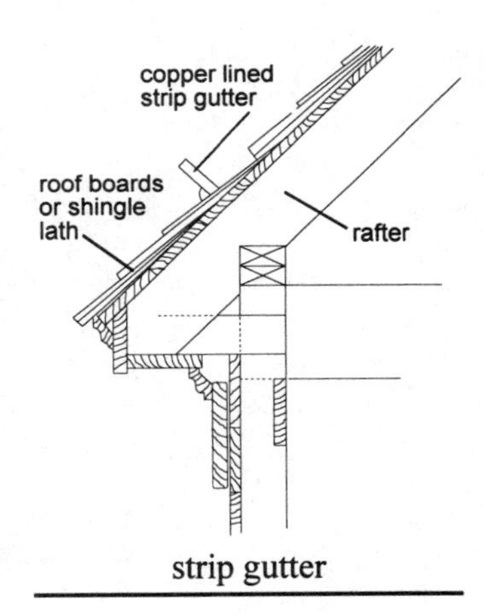

strip gutter

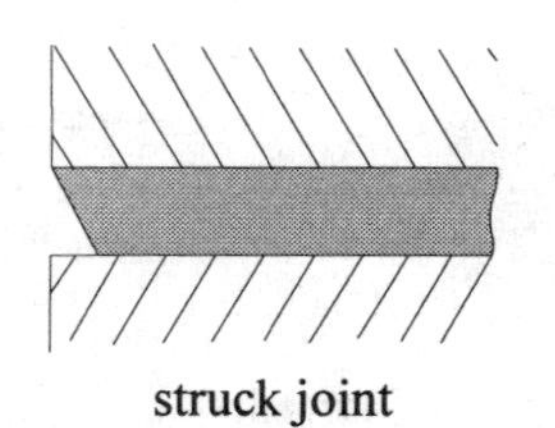
struck joint

strop – 1) a strap around a pulley which transmits the load from the pulley to the hook for lifting; 2) a rope or band used to support a pulley block; 3) a leather strip used to put a fine edge on a blade.

struck capacity – the capacity of an earthmoving bucket when it is full (level with the rim).

struck joint – a mortar joint that is grooved with the edge of a trowel to form an indentation at the bottom of the joint. Also called a troweled joint.

structural – pertaining to load-carrying parts or members; a solid part of a structure or building that is designed and installed to carry a load.

structural attachment – an attachment to an object, such as pipe, that is designed to carry a load. Often, thick sections of metal plate are welded to the outside of a pipe wall to transfer the load from the pipe to these supporting structural attachments. The attachments are particularly useful where the pipe may be subjected to axial movement.

structural block – masonry building block with a compression strength of from 1800 to 3000 psi, sufficient to be used for a wall that will carry a structural load.

structural clay tile – fired clay, in the form of blocks, that are extruded to the required shapes. The tiles are durable, lightweight, fireproof and inexpensive. They have generally been replaced in modern construction by concrete block.

structural composite plywood panel – a plywood panel with a homogeneous, flakeboard core and tongue-and-groove edges, used in applications such as subflooring. It is more economical than regular plywood sheets of the same thickness.

structural drawings – drawings that detail the structure of the building, including connections between members.

structural fiberboard – a fiberboard that is impregnated with asphalt for water resistance. It is available in various thicknesses and several densities. Also called structural insulating board.

structural glazed tile – molded clay block tiles that are glazed on one or more sides and used as finished wall surfaces in buildings. They provide an attractive, easy-care surface and come in a variety of shapes.

structural joist and plank – lumber that is at least 2 inches thick and 6 inches wide or wider.

structural light framing – conventional framing using 2 x 4s and 2 x 6s. This type of framing is used for standard one- and two-story structures. Also called light framing.

structural steel – steel members of various shapes used as load-carrying members in a structure. Steel framing is used for high-rise buildings because it is stronger and more durable than wood. It is also used in situations where the structure is required to be totally fireproof.

structural steel H pile – a steel structural member, whose cross section forms the letter H, that is driven into the ground to a depth determined adequate to support a building or other large structure.

structural steel tubing – hollow lengths of structural steel in the form of round, square or rectangular tubes. Used to fabricate piping system support members and other structures where larger members, such as I-beams, are not required for the loads.

structure – 1) a building or object constructed for a particular use; 2) building materials fastened together to form an integral unit.

structure section – a drawing depicting a cross section or cut through a roadway that shows the layers of material that make up the roadway.

strut – an intermediate structural member used as a brace between other members. The strut adds to the strength of the other members by providing a load path between them.

stucco – a mixture of portland cement, lime, sand and water used as an exterior surface covering for buildings. It is applied in a series of three coats to a total thickness of ¾ inch. The first coat is called the *scratch coat*. It is left with a grooved or cross-hatched surface to provide good adhesion for the thinner (¼ inch) *brown coat*. The last coat is the *color coat*, a thin coat (⅛ inch) that is colored and textured to meet the architectural appearance requirements for the finished siding.

stud – 1) a vertical framing member; 2) a threaded rod fastener, installed with a nut at each end, used to bolt pipe flanges together.

stud arc welding (SW) – an electric arc welding process used to weld a stud (a short length of metal rod) to another piece of metal. It uses the stud as the electrode, so that the electric welding arc jumps from the stud to the metal to which it is being welded.

stud driver – a device that uses an explosive cartridge, similar to a blank cartridge in a small caliber gun, to drive a special fastener into steel, concrete, or other material. Also called a ramset.

stud finder – an ultrasonic or magnetic device used to locate studs behind the wall surface or covering. The ultrasonic device locates studs, the magnet locates nail heads in the studs.

stud guard – a flat metal plate that can be fastened to the edge of a stud to prevent wall covering nails or screws from damaging a pipe or wire passing through the stud.

stud welding – any welding process for the purpose of welding a stud, a short length of metal rod, to another piece of metal.

Sturd-I-Floor – an American Plywood Association trade name for a type of plywood subfloor panel used as a base for interior resilient flooring. The panels are made by several different manufacturers.

S-type fuse – an electrical fuse with a threaded adapter. Each adapter accepts only a fuse of a specific current rating. Once an adapter is installed in a fuse box socket it cannot be removed. As only one specific rating fuse will fit a given adapter, a fuse with the wrong rating cannot be installed in that adapter. This fuse was adopted to prevent the use of a fuse which will pass more current than the circuit can handle. The use of incorrect fuses, which had been fairly common before the S-type fuse, can result in a fuse box fire. Most codes now require this type of fuse for use in buildings with fuse boxes.

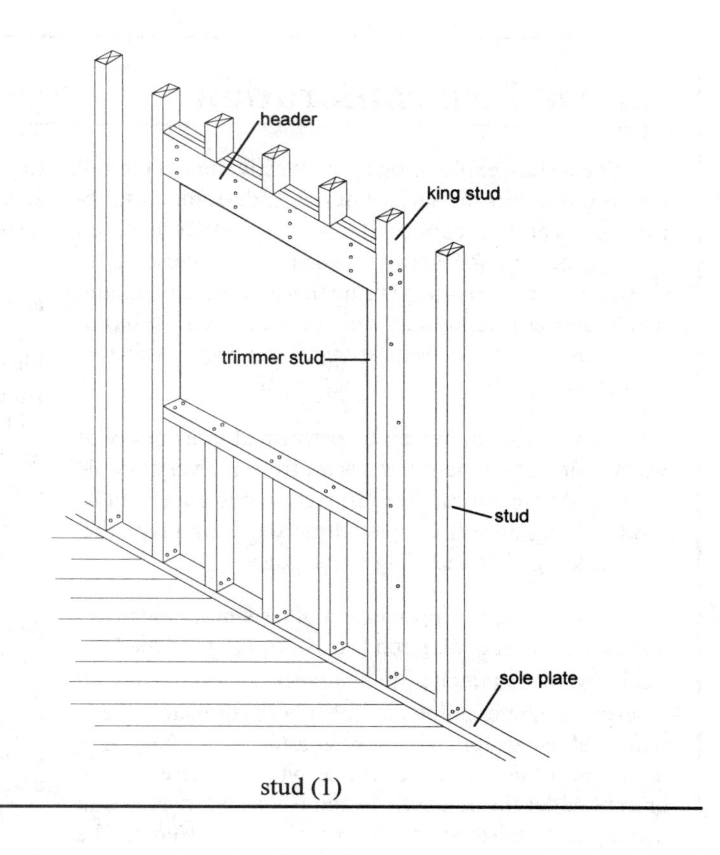

stud (1)

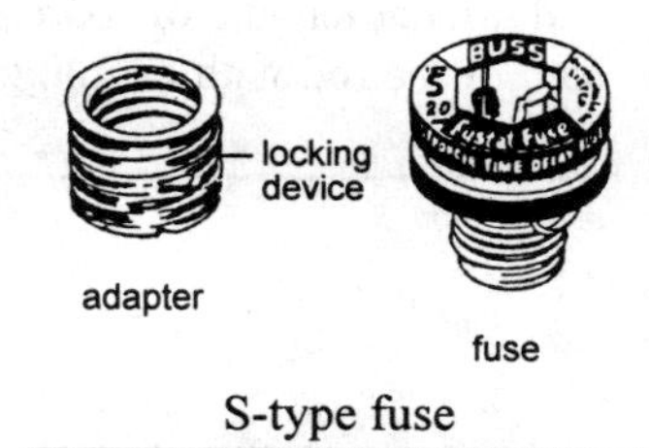

S-type fuse

Styrofoam – a trademark name for a lightweight, rigid polystyrene plastic insulating board.

subbase – 1) leveled and compacted soil below the base for a paved road; 2) the lowest portion of a structural support.

subcontract – work, usually in a particular trade, for which a contractor is hired by another contractor. The primary or general contractor, who works directly for the client, oversees all the work done for the client and is responsible for the work subcontracted out to other trades or contractors.

subcontract bids – bids from contractors, who will perform a defined scope of work, solicited by a contractor who is working as the primary contractor for a client.

Subsurface exploration

Subsurface exploration is an investigation of the soil conditions below the surface of the land. Before a structure is erected, especially a large structure, an investigation of the site is prudent. The investigation should go deep enough to ensure that a foundation failure due to poor soil conditions will not occur after the building is completed. The tests are performed and evaluated by a soils engineer.

This investigation can be performed in a variety of ways, using equipment that varies from the very simple to the highly technical. For instance, a long auger can be used to bring up soil samples if the soil is of a type that will pack together and cling to the auger.

A sounding rod provides another simple means to test the soil. A sounding rod is a 5-foot-long rod made of steel, 5⁄8 to 1½ inches in diameter. One end has an enlarged diameter point, and the other is threaded with a removeable cap for striking with a hammer. The rod is driven into the soil to determine whether there is rock present under the soil surface and if the soil increases in density and resistance as the rod is driven deeper. The greater the density, the harder it will be to drive the rod. Additional lengths of rod can be added to the threaded end of the rod with couplings if the investigation needs to go deeper than 5 feet. If the soil is too rocky to drive a sounding rod into it, another method of investigation will have to be employed.

Wash boring uses water to bore a hole into the soil and bring up samples. The water flows through a hollow boring bit located inside a hollow cylinder. The boring bit is raised and lowered while being turned. The water flows through holes in the sides of the bit and forces the soil up through the hollow cylinder to the top of the excavation. The soil is then examined for type and changes occurring at different depths. The test continues until it is determined that solid, stable soil, such as bedrock, has been reached.

A small-diameter test pit can also be dug to a shallow depth to allow the soils engineer to view the layers of the earth in place, as well as to see the moisture conditions of the soil. This method is usually limited in depth because of the cost involved, unless a large excavation is to be dug as part of the building construction, such as for a deep basement or an underground parking facility.

A more technical subsurface exploration is performed with a portable seismograph. The seismograph generates shock waves, and a device called a geophone detects the refracted and reflected shock wave. The shock

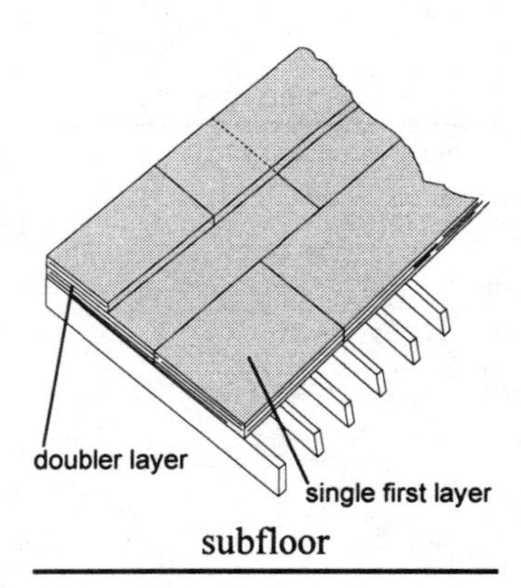

subfloor

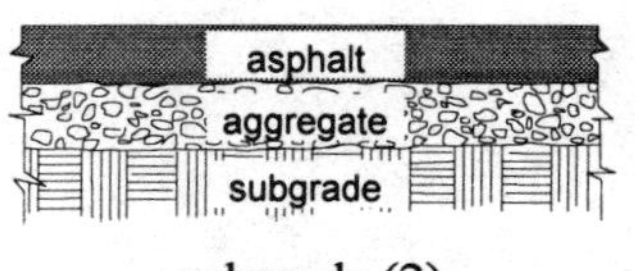

subgrade (2)

subcontractor – a contractor performing work for another contractor.

subdivision – 1) a tract of land that is divided into parcels for individual sale; 2) a tract of homes built by a developer on subdivided land.

subfloor – the layer of flooring that is fastened directly to the floor joists, and to which the finish flooring is applied. It is constructed of plywood sheathing or 2 x 6 planks, and may be a single or double layer.

subgrade – 1) soil structure below grade level, before any surfacing material is placed; 2) compacted soil on which a concrete slab or other structure is placed; 3) the elevation at the bottom of a pipe trench.

sublease – the lease of a property by a lessee to a subsequent lessee for part or all of the lease period.

submerged arc welding (SAW) – an electric arc welding process using an arc from a consumable bare metal electrode to the work, with a layer of fusible material to shield the weld. The welding process is "submerged" under the shielding material.

subrogate – to put in the place of another, as in substituting one party for another in a legal claim or right.

subsidiary – 1) a subordinate part of another; 2) of secondary importance.

subsidy – a grant of money made by the government to assist in the cost of a project it feels has public benefit.

waves are timed by the engineer while viewing the resulting shock wave patterns on an oscilloscope, or on a dial. The pattern and timing will differ with the soil conditions, bedrock giving off the strongest reflection, revealing various subsurface conditions.

Another technical test involves an earth resistivity meter which measures the electrical resistance of the soil at a site. Four electrodes are driven into the soil at carefully measured distances apart. Different materials offer different electrical resistance. The resulting information tells about the soil subsurface structure. Rocks and granular soils have high resistance, and loose materials have low resistance. However, not all soil characteristics can be determined by this method. Test borings are needed to supplement this type of test.

To guarantee structural strength, foundation pilings must be placed deep enough to rest on stable soil. This will ensure that movement occurring in the unstable soil, or soil with a high moisture content, that is closer to the surface will not affect the stability of the building. This is especially important in areas with frequent seismic activity.

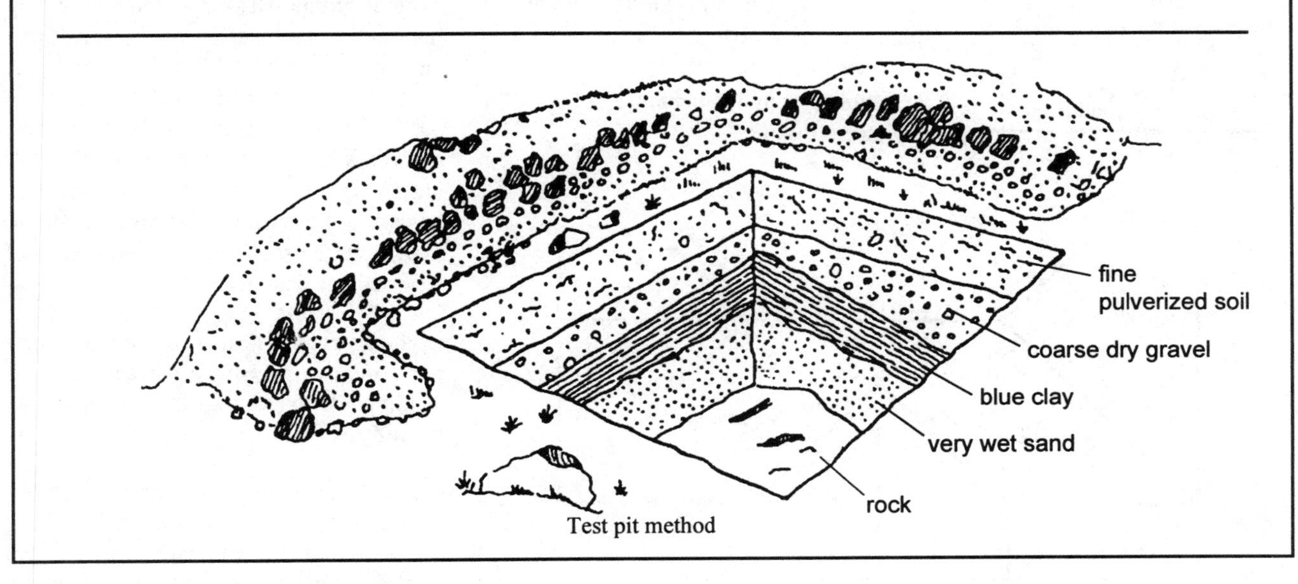

Test pit method

substrate – 1) the underlying surface or material; 2) the supporting material to which a finish is applied.

subsurface drain – a below-grade or underground drain that collects subsurface water and conveys it to a disposal location.

subsurface exploration – tests performed by a soils engineer to examine soil conditions and ground stability at a building site. See Box.

suburban – an area outside the city proper, but within commutable distance to the city, often depending on the city for services.

suction – 1) exerting a force on a solid, liquid or gas by reducing air pressure over its surface; 2) a negative pressure.

summit – 1) the highest point or the peak; 2) the highest elevation of a grade.

summons – a call by an authority to appear at a given place, such as a notice to appear before a court.

sump – 1) a low spot or depression below grade where water collects; 2) a below-grade area used for collecting waste water before it is pumped to another location or drain.

sump pump – a pump used to remove waste water from a collection area.

super – a road with a continuous slope in one direction.

superconductor – a material that conducts electricity with little or no resistance under specific conditions. For example, some materials are being developed that conduct electricity with little or no resistance loss when operating at liquid nitrogen temperature (-320 degrees F) or lower.

superheated steam – steam that is at a higher temperature than its saturation temperature. Boilers, particularly those used for steam-driven turbine-generator units, incorporate superheaters to raise the temperature for increased thermal efficiency.

superimpose – to place over, or on top of another item or surface. Decorators often superimpose a drawing on transparent film showing furniture locations over the plan view of a room to help clients visualize their ideas.

superintendent – a person charged with overseeing work.

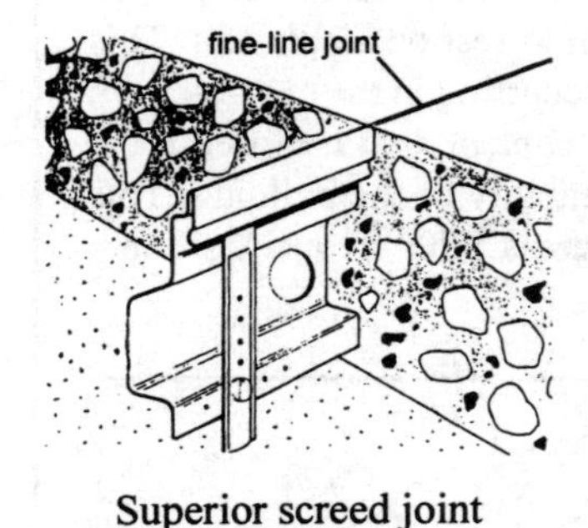

Superior screed joint

supporting valley rafter

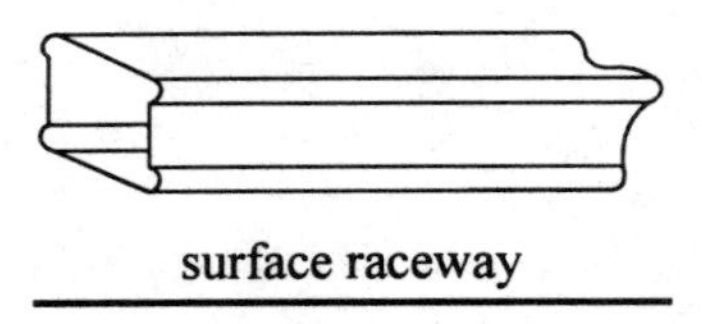
surface raceway

Superior screed joint – a trade name for a product which acts as a combination screed and expansion joint for concrete slabs. It is left in place after the concrete is poured, eliminating the need for form removal, joint repair or filling. It is commonly used for large pours and is manufactured in 10 foot lengths for 4-, 5- and 6-inch thick slabs.

support – a device to hang and restrain piping, cable trays and similar items.

supporting valley rafter – a rafter running at an angle from eave to ridge that strengthens the roof and provides support for valley jack rafters, cripple rafters and the shortened valley rafter adjacent to the same hip.

surety – a written guarantee of performance, such as an agreement to complete a job within a given time period under specifically identified terms.

surface – 1) the top layer or external exposed area of an object or structure; 2) to provide a finish for an exposed area.

surface gauge – a gauge used for scribing lines at a given height, which consists of a base with an adjustable vertical rod projecting from it that holds the scribe. The base of the gauge is placed on a level surface with two pieces of work. By sliding the gauge base on the surface and adjusting the scribe to the correct height, marks can be made on the two pieces of work at identical heights.

surface preparation – any operations performed to prepare a surface, such as sanding a wood surface in preparation for painting.

surface raceway – a preformed rectangular tubing inside which electrical wiring can be routed. The raceway is mounted on wall or ceiling surfaces and may be painted to match the surrounding surfaces. Used when it is impractical to run wiring inside a wall, as can occur when remodeling an existing office or living area. Also called surface metal raceway.

surface testing – tests performed at a site to determine soil characteristics prior to building. *See Box.*

surfacing – 1) adding finish material, such as stucco, to the exterior surface of a wall; 2) depositing weld metal onto a surface to change the properties and/or dimensions of a material.

survey – to take measurements to determine the exact boundaries of a piece of property or an area of land; includes directions, angles, distances, elevations and physical characteristics of the land.

Surface testing

Soil surface testing is required before construction to determine the proper foundation type and size required for the soil conditions at the building site. There are several different tests which can be done to determine various soil characteristics, such as moisture content, density, shear strength and penetration resistance.

A simple test, done with a penetrometer, measures soil penetration resistance. The penetrometer is pushed into the soil up to a pre-marked point. The information on the dial is read and compared with a calibration chart to determine the penetration load, measured in pounds of resistance per square inch. This test determines the hardness of the soil.

Clay soils are often tested for shear strength. A hand vane tester is used to perform this test. It is a hand-held tool made up of a vane with four blades connected to a torsion head via a spring. The vane is pressed into the soil and the torsion head is rotated by the spring at approximately 1 RPM until the clay shears. The maximum spring deflection is read from the torsion head dial, and compared to a calibration curve to determine the soil shear strength.

Moisture tests of soil can be taken using a device that introduces a calcium carbide reagent to a soil sample, causing a gas to form. The amount of gas formed depends on the amount of free moisture in the soil. This gas is confined to a leaktight chamber, and the resultant gas pressure in the chamber is proportional to the amount of moisture. The moisture content is needed to help determine the stability of the soil. The higher the moisture content, the less stable the soil.

A nuclear density meter, which introduces controlled amounts of radiation into the soil, can be used to determine soil density as well as soil moisture content. The meter registers the amount of radiation scattered back from the soil and then gives a reading for whichever function is requested, either a soil density reading or moisture content reading, or both if desired.

Nuclear density tester

Surface testing may have to be combined with deeper subsurface exploration if the soil conditions, building size, or geological location require a more extensive investigation of the building site.

surveying tools – tools used by a surveyor, such as a transit, measuring rod and compass, to take distance measurements and determine angles and elevations for a survey.

surveyor – a trained tradesperson who takes measurements and determines property boundaries and elevations using established procedures.

survey stakes – stakes used to mark boundaries that have been established by survey.

suspend – to hang from supports.

suspended ceiling – lightweight ceiling panels or tiles supported by a metal framework that is suspended by wire hangers from the ceiling joists. Also called a drop ceiling.

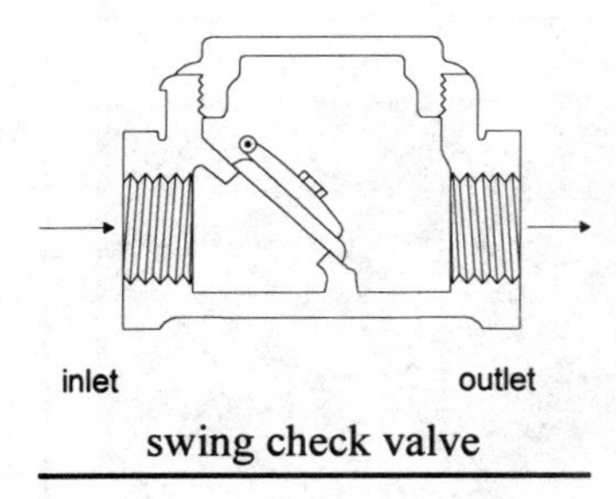

swing check valve

suspended diaphragm – a structural decking system for roofs and floors designed to provide a strong surface.

swag – hanging down or drooping from one point to another, as a hanging ceiling light fixture may be rehung from its installation point to another area of the ceiling by swagging the wiring in an ornamental cording to a hook placed in the desired location.

swage – a hand tool used to shape metal.

swale – a shallow depression shaped to allow runoff of rainwater or other drainage. A swale may be established for a piece of land or a driveway or other passageway.

sweated joint – a soldered joint in copper tubing in which the tubing is fitted into the socket of a fitting or valve. Flux is applied to the inside of the fitting or valve and to the outside of the tubing. When heat is applied to the fitting and solder is touched to the tubing, the solder is drawn by capillary action into the joint.

sweat-equity loan – a loan advanced with the agreement that an owner-builder provides some of the construction labor.

Sweding – a fast method of setting earthmoving grades using three lath stakes, two of which are placed at a correct elevation. The third lath is placed in the center and moved until it is at the same elevation when sighting across the tops of the three laths.

sweep – a long-radius ell plumbing drainage fitting.

sweet gas – natural gas that does not contain hydrogen sulfide.

swell – the volume growth in soil after it is excavated. Compacted soil has less volume than soil that is loose.

swimming pool backwash piping – piping from a swimming pool filter to a sewage drain used to convey filter backwash waste water from a pool.

swimming pool recirculation piping – piping which returns water to the pool after filtering.

swimming pool vacuum fitting – a pump suction fitting located 10 inches or less below the surface of the pool to draw off water for filtering.

swing check valve – a valve which permits flow in only one direction by means of a disc on a hinged swing arm. Fluid can flow in one direction, raising the disc off its seat, but when the flow tries to reverse, gravity and reverse pressure cause the disc to swing back into the seat, blocking the flow.

swinging door – a hinged door that can swing open one or both ways through a doorway.

switch – an electrical device for opening or closing a circuit which allows a light or motor to be turned off or on.

switchboard – a panel with switches and other devices to control electrical circuits.

switchbox – an electrical box mounted in a wall which houses a switch and the conductors connecting the switch to the electrical system.

switch, double pole, double throw – see double pole, double throw switch.

switchgear – a device that controls an electrical circuit by completing or breaking the circuit.

switchgear, metal-clad – see metal-clad switchgear.

switch, knife – see knife switch.

switch, light handle – an electrical switch with a small neon light in the toggle that is on when the switch is off, allowing the toggle to be easily seen in a dark room.

switch loop – electrical wiring run to a light through a switch, in which the light is the last electrical load in the circuit.

switch, single pole, single throw – see single pole, single throw switch.

switch, snap – see snap switch.

switch, weatherproof – see weatherproof switch.

swivel – a connection that permits free rotation between the parts that are connected. Swivel connections are used in a variety of applications, such as in hoist rigging, where the swivel allows the lifting cables to turn, so that they do not become twisted.

symbol – a mark or drawing used to indicate a specific object, material, class or entity, such as the American Plywood Association stamp used to identify grades and types of plywood.

symbol schedule – a legend on a drawing showing the symbols used on the drawing and their meanings.

symmetrical – balanced and evenly proportioned.

synchronous – occurring in phase or at the same time, such as in the connection of a three-phase electrical generator in a three-phase system, where the connection must be timed using a synchronous meter.

system – an assembly of parts and components that work together to perform a function, such as an electrical system or plumbing system.

system curve – a graph showing HVAC (heating, ventilating, and air conditioning) system pressures at various flow rates. The information is used to balance the system in order to attain the temperatures desired for each area on the system.

system ground – the grounding of the neutral (white) wire in an electrical circuit at the service entrance to a building.

T: table saw to Type X gypsum wallboard

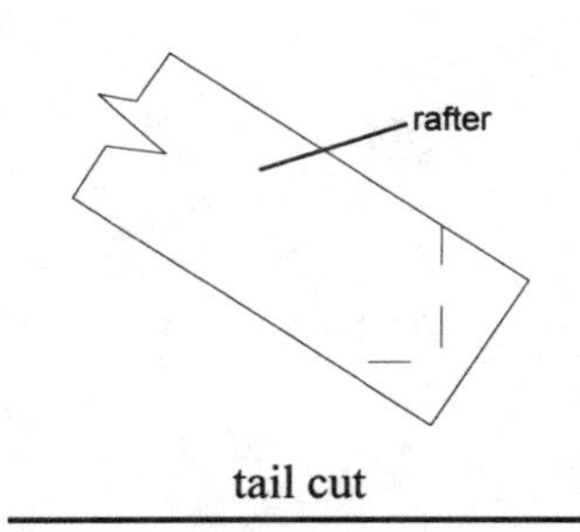

tail cut

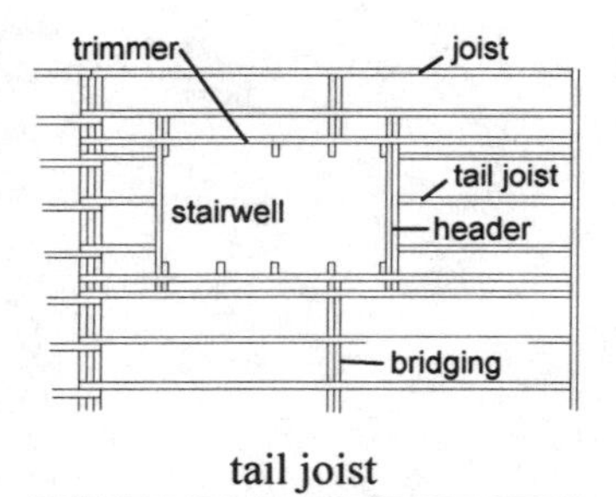

tail joist

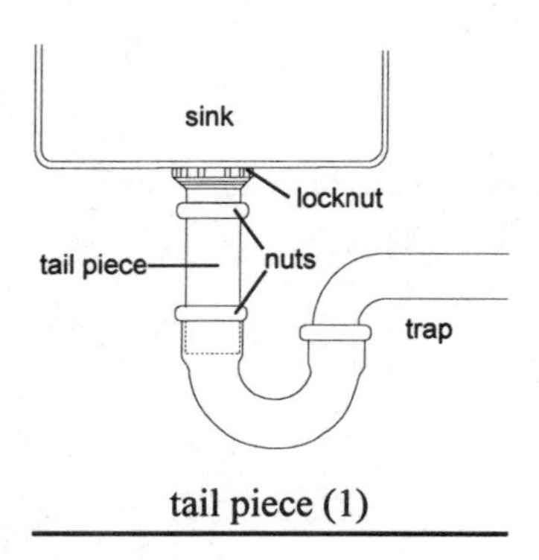

tail piece (1)

table saw – a sturdy work table with the blade of a circular power saw exposed through an opening in the surface. The saw blade can be adjusted to different heights and angles for cutting. Guides are used to aid in manipulating the work past the saw blade.

tachometer – a device which measures the speed, in revolutions per minute, of a machine's rotating parts.

tack cloth – see tack rag.

tack coat – emulsified asphalt sprayed on an existing asphalt surface that is to be covered with a new asphalt coating. The tack coat creates a bond between the old asphalt and the new asphalt.

tack hammer – a small hammer with a magnetic head for holding a tack. The tack, when attached to the head of the hammer, can be driven in with one or two strokes. Used with small fasteners on finish work and upholstery.

tackle – pulleys and ropes required for hoisting heavy materials in construction or loading.

tackless strip – a length of light wood with nail points projecting upward from its surface. The strip is fastened to the floor around the perimeter of a room to hold the edges of wall-to-wall carpeting in place.

tack rag – a lint-free treated cloth used to remove dust from a wood surface prior to painting. To obtain a smooth finish, particularly when varnish is applied, the surface must be dust-free.

tack weld – a short section of weld bead applied to hold parts in the proper position until the permanent weld is made.

tacky – a surface that is slightly sticky, as with fresh paint that has not fully dried or adhesive that is ready to be bonded.

tag line – a guide rope used to position an object laterally that is being moved on a hoist.

tail beam – a short beam or joist that is supported by a wall on one end and a header on the other.

tail cut – a cut made at the overhanging end of a rafter to trim it to the proper length and angle.

tail joist – a short joist that butts against a beam or header, or joist that runs from a header to a support girder or to a side wall.

tail piece – 1) the section of sink drain that fits into the sink outlet and extends below the sink, and to which the drain trap is connected; 2) a short beam or joist. See also tail beam.

tail plumb cut – a vertical cut made on the end of the rafter overhang to trim the rafter to the same length as the other rafters.

tails – an airless spray-painting defect in which thick bands of paint have been applied in areas, rather than a uniform coating thickness. Also called fingering.

tailstock – the part of a lathe that holds and supports one end of the work, but does not provide the motive force to turn the work.

take-off – a list of materials and the quantity required for the assembly of a structure or part. The list is made up from information provided by, or *taken off*, the plans or drawings.

talc – a powder, made from magnesium silicate, used as a preservative coating for rubber and a dry lubricant to coat a variety of surfaces.

tamp – to force the air out and compress a material, such as soil or concrete, into a firm or settled state.

tamper – 1) a manual device with a long handle and a steel grid or mesh base used for compacting wet concrete and forcing large aggregate below the surface. See also jitterbug; 2) a manual or mechanical device used for compacting soil, asphalt or other material by repeatedly striking the surface with a flat plate.

tandem – two or more objects connected in series, such as a truck and trailer.

tang – the projection on a part which serves as the means to connect it to another part, such as the pointed tail on the end of a file onto which the handle is fitted.

tangent – a line or a plane that intersects a curve or a surface at one point.

tangential surface – the surface of wood cut lengthwise from a log and at right angles to the wood rays.

tangent offset method – a surveying method for laying out a land profile in hilly terrain. A vertical curve (also called profile curve) is used to connect two grade lines marking the beginning and end of an upward or downward curve. The vertical curve is tangent to the grade lines. The layout is made by offsets from the tangent. The offset distance from the tangent varies as the square of the distance along the tangent. The tangent-offset method requires only simple arithmetic to construct, and can be used with curves that have both equal and unequal tangents.

tank – a large storage container for liquids, such as water or oil.

tank white – a paint formulated for exterior metal surfaces.

tap – 1) to cut female threads in a predrilled surface that is to be joined to another part with a screw-type fastener; 2) a threading tool used to cut threads on the interior of a part or in a predrilled hole in a part. The tap is made of hardened steel and is threaded and fluted along its length; 3) a connection, or the act of connecting, into a system such as a water supply, or into a circuit or transformer; 4) a water faucet.

tap drill – a drill used to make a hole which will later be threaded with a tap. The drill has a diameter slightly smaller than the tap so that the difference between the two diameters provides just enough material remaining in the opening for the desired depth of the threads.

tape – 1) a strip of flexible material, usually with adhesive on one or both sides, used to join two materials; 2) joint reinforcement used with gypsum wallboard to span the joints between wallboard panels.

tape blister – a defect in gypsum drywall installation in which the bond between the joint tape and joint compound fails.

tape measure – a long, thin strip with dimension markings used for measuring and transferring dimensions, such as room sizes or material lengths or widths. It may be made from a plastic-coated fabric or other flexible, nonstretchable material, or from thin metal that can be retracted into its own storage case.

tape photographing – a defect in gypsum drywall installation in which the joint tape shows through the finished surface.

taper – a gradual, sloped reduction in thickness, width or diameter, as in a wedge or cone.

taper attachment – a lathe attachment that permits feeding the cutting tool into the work at an angle to the work.

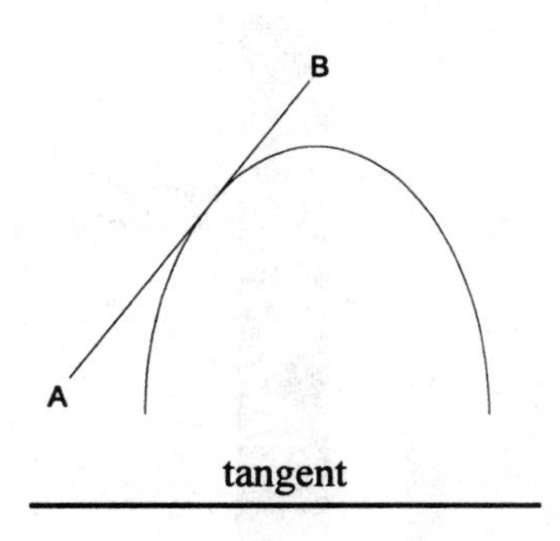

tangent

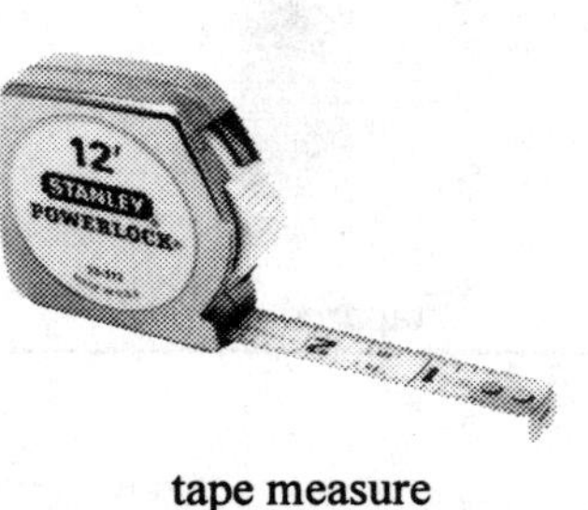

tape measure

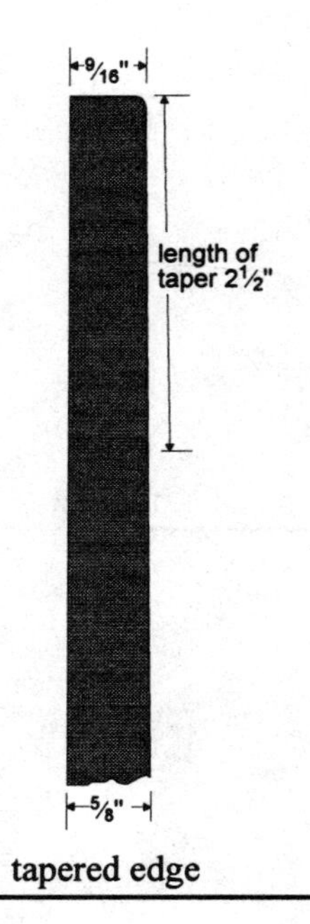

tapered edge

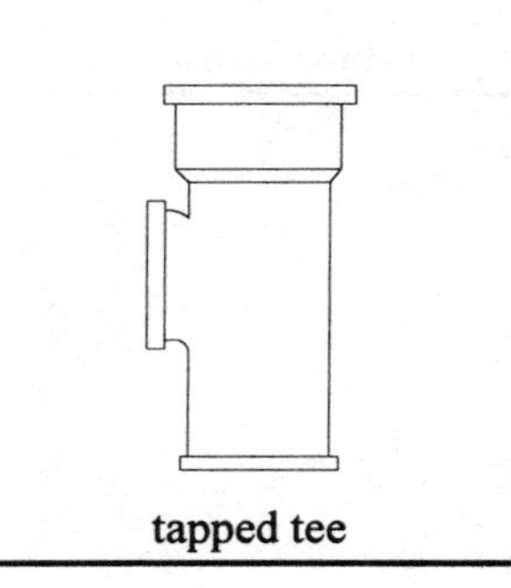
tapped tee

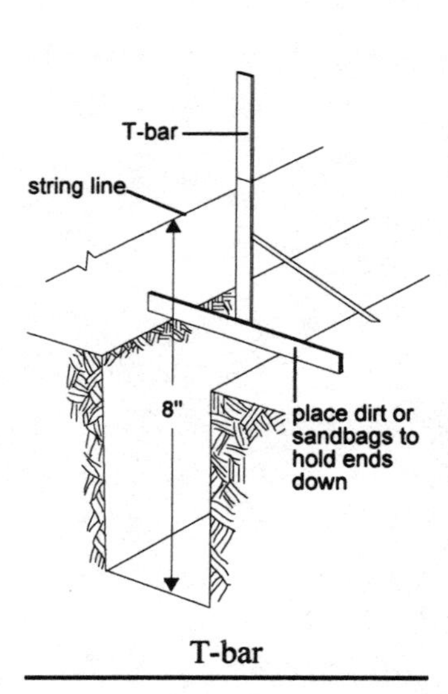

T-bar

tapered edge – the long edge of a gypsum drywall panel which is tapered to form a shallow valley where two panels abut. The valley is filled with compound and taped to make a smooth, level finished surface. See also factory edge.

tapered rim saw blade – a circular power saw blade that has a taper ground into the blade from the outside diameter extending a short distance into the blade. This allows the blade to free-cut thin materials. The remainder of the blade diameter inward to the hub is of a constant thickness to stabilize the blade.

taper ground – see hollow ground.

taper jig – two lengths of metal or wood hinged at one end and held together with a clamping device near the other end. The two pieces, when open, are similar to the letter A, with the angle being infinitely adjustable. The jig is used for cutting material on a taper (usually with the grain) in a table saw. The jig is set at the desired angle and placed along the rip fence of the saw. This holds the work at the desired angle in relation to the saw blade.

taper reamer – a reamer that is tapered from a larger to a smaller diameter along its length. It is used to enlarge holes in thin materials.

tap extractor – a tool with stiff wires that are inserted into the flutes on the outside of a tap that has broken off in a hole. A collar is used to tighten the wires along the flutes so that the tool can be turned, and the tap with it, enabling the tap to be removed.

taping – applying tape, such as drywall tape.

taping knife – see drywall knife.

taping tools – tools designed to apply joint tape and compound and finish the joints during gypsum wallboard installation.

tapped lag bolt – a lag bolt with a tapped hole from the head partway down the length of the bolt. The lag bolt can be screwed into the formwork for a concrete pour and a form tie screwed into the tapped hole on the lag bolt. The form tie then holds the form walls together against the weight of the concrete until the concrete has set.

tapped tee – a tee with one or more short branches having female threads tapped into the branches.

tap wrench – a tool that serves as a handle to hold and turn a tap while the tap cuts threads into metal parts.

tar – a dark, heavy petroleum residue remaining after other products have been distilled off. Tar is used in asphalt paving, and as a sealant for objects that are in contact with high amounts of moisture or water, such as boat hulls.

target – a brightly colored, highly visible marker used in surveying work. The target is placed or held by an assistant at one location and sighted by the surveyor using a transit from another location.

Tarmac – a trade name for an asphaltic binder used in paving.

tarmacadam – a road-paving material consisting of asphalt and crushed rock.

tarpaulin (tarp) – a heavy water-resistant canvas or plastic fabric used as an exterior protective covering.

T-bar – a wood or metal frame in the shape of an upside-down letter T used to suspend a string line along a trench. The crossbar of the tee rests on the ground spanning the trench opening, while the upright of the tee points up from the crossbar. T-bars are placed periodically along the trench opening with the string line fastened to the uprights.

teardrop texture – a sprayed-on texture which forms coarse droplets that are allowed to harden on the surface.

tee – a piping or plumbing fitting resembling the letter T that is used to join three lengths of pipe.

tee plate – a joint reinforcement plate shaped like the letter T that is used to span joints where a column butts against the bottom of a horizontal beam.

telepower pole – a device used in offices or work areas to bring electrical power or other wiring to a work station that is not adjacent to a wall. A telepower pole is installed between the ceiling and the floor of the work area. A junction box at the top of the pole is the connecting point for the wiring which is run through the ceiling and down the pole to the work station.

telescoping gauge – a T-shaped tool used to measure the diameter of a hole. The head of the T is spring-loaded, with spherical ends that can be locked by means of a knurled nut on the end of the stand of the T. The telescoping sections of the head are compressed and inserted into a hole and then allowed to expand across the diameter of the hole. The head is locked into position and withdrawn so it can be measured with a micrometer to determine the hole diameter.

temper – 1) to heat treat a metal in order to strengthen it; 2) to heat or cool make-up air in an HVAC system to adjust the temperature; 3) to add water to mortar or grout to improve its consistency and workability.

tempera – a paint in which egg yolk is used as the vehicle for the pigment rather than oil or water.

tempered glass – glass that is heated to just below melting and then cooled quickly in the air. This causes compression in the outer layers of the glass, which strengthens the glass. See also glass.

template – a pattern or guide used in mechanical drawing or in fabricating standardized parts or designs.

temporary bracing – bracing that is installed during the course of construction to provide temporary support until the structure is completed or able to stand on its own.

tenant – an occupant who dwells in, or conducts business in, a space for which he pays the property owner a fee.

tenon – a projection, cut smaller than the main body of the member, designed to fit into a recess or mortise in another member to form a mortise and tenon joint.

tenon saw – a small, fine-toothed backsaw with a stiffener rib along the back edge of the blade to keep the blade from bending. It is a manual saw used to make the precise cuts needed to shape a tenon.

tensile strength – the ability of a material to withstand stretching or pulling stress without breaking or becoming permanently deformed.

tensile stress – a stress that pulls on a material or structure, such as the stress put on structural members of a building. For example, ceiling joists run between outside walls and serve as plate ties to keep the walls from moving outward at the top.

tensionmeter – a gauge that is used to measure the amount of tension placed on a wire fence during installation.

terminal – 1) the end or termination point of something, such as an electrical wire; 2) a fitting or electrical connector; 3) a building, at the end or intermediate stopping point of a public transportation line, where people can embark, disembark or purchase tickets for travel.

terminal marking – identification on the terminal of an electrical device that indicates the wire material that may be used to connect to it, such as Al-Cu for aluminum or copper.

terminal post – the last fence post in a line, such as at a corner or at the end of the fence.

terminal velocity – a measure of the air circulation capabilities of an HVAC system. The location where the air velocity coming from an outlet decreases to (usually) 50 feet per minute (the terminal velocity) gives an indication of the system's distribution capacity. The information is used to balance the HVAC system air flow.

termite – an insect that burrows into and eats wood, often causing severe structural damage.

termite shield – a metal shield installed on foundation walls and around pipes entering a building to stop termites from reaching the wood parts of the structure.

terne metal – a tin-lead alloy, or alloy coating, applied to steel used in roof covering materials. It protects the steel from corrosion and is cheaper and lighter weight than lead alloy sheeting.

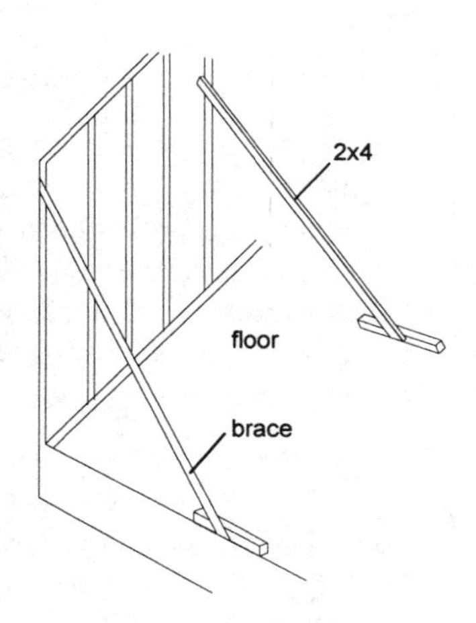

temporary bracing

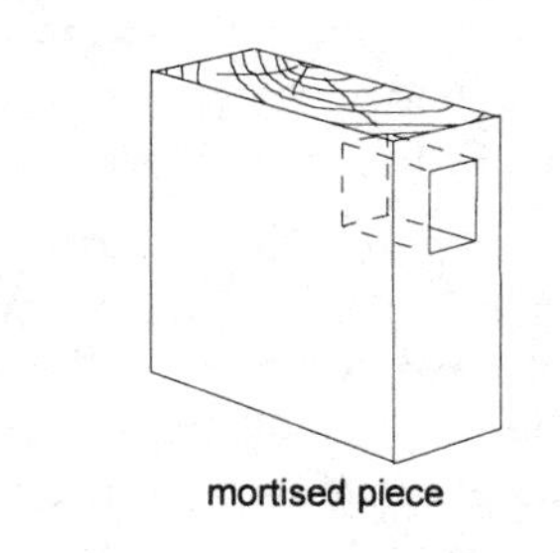

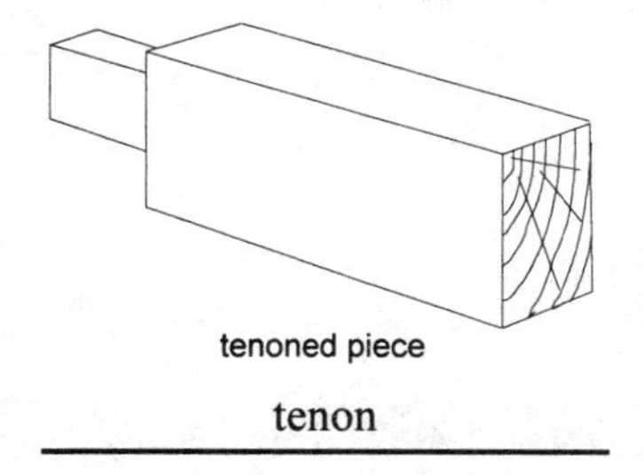

tenon

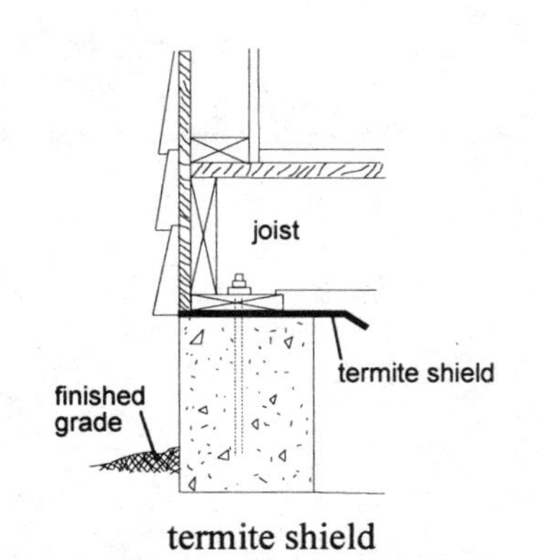

termite shield

Terrazzo

A type of floor surface made by embedding small pieces of marble or other hard stone in a mortar base. The mortar base is poured, and then the stone chips are embedded in the leveled mortar. When hardened, the surface is ground and polished smooth, though sometimes it may be left in a moderate state of roughness for a textured effect. It results in an attractive, durable surface that is produced in various thicknesses, from 3⁄16 inch to 3 inches.

Monolithic terrazzo is a thin terrazzo flooring applied over a concrete slab. The stones are bonded to the slab using a synthetic resin or portland cement base. The smoothed and polished terrazzo is 3⁄16 to ½ inch thick and is often called a one-course or thin-set terrazzo.

Bonded terrazzo is used over steel or wood decking. A concrete slab is laid over the decking (after a waterproof membrane is laid over wood surfaces), then a mortar underbed is poured. Finally the mortar base into which the terrazzo stones are embedded is laid. This surface is ground and polished smooth, resulting in a finished floor that is 1¾ to 3 inches thick. It is also called two-course terrazzo.

Sand cushion terrazzo is a three-course, layered terrazzo. A ¼-inch thick sand bed is laid over a concrete slab. Plastic is laid over the sand, followed by a steel reinforced mortar underbed. A final mortar layer is poured and the terrazzo stones embedded. The surface is then ground and polished smooth. The sand cushions the terrazzo surface making it less likely to crack with settling.

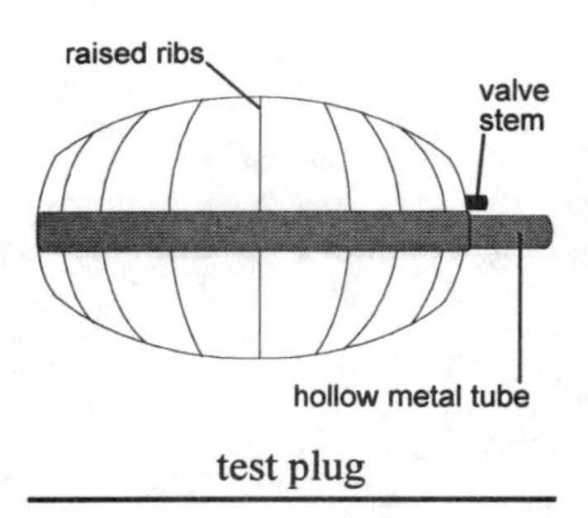

test plug

terrace – 1) flat sections of land ascending in stepped areas on a hillside; 2) a level platform or tiled area adjacent to a house used for outside living, dining or recreation.

terra cotta – a brown fired clay with a red tint used for tiles and building materials.

terrazzo – a type of floor surface made by embedding small pieces of marble or other hard stone in a mortar base. *See Box.*

tessera – glass tiles, approximately ¼ inch thick and ¾ inch square, that appear to be hand crafted because of surface and edge irregularities. They come in a variety of colors and are widely used to make mosaic murals.

test light – a small electrical light, with leads that can be plugged into an electrical outlet, used to determine whether the circuit is live.

test plug – a type of inflatable plug, or a plug that is expanded mechanically by compression, used to seal off a pipeline or pipe-fitting opening so that the pipe system or section may be pressure tested. The plug has a tube through it to permit pressurizing the section of pipeline past the plug.

tetrahedron – a pyramid with three sides and a base.

texture – 1) the visual or tactile surface of a material or a finish coating; 2) a surface decoration that is raised above a flat surface.

theodolite – a surveying instrument used for precision work. It has a compass and an optical micrometer scale that permits horizontal and vertical angle readings to within 10 seconds, and on some models, to within 1 second.

theoretical length – the mathematically calculated length, which may be different from actual length because of the thickness of the material used. For example, a piping layout is done by the centerlines of pipe and fittings. These dimensions do not change with changes in pipe size. However, when a change in direction in the pipeline is made, such as with a 90 degree ell, the overall length from the end of the pipe to the outside edge of the ell will vary with the diameter of the ell used.

therm – the measurement of a unit of heat equal to 100,000 Btu.

thermal – of, or relating to, heat.

thermal coefficient of expansion – a factor designating the amount of contraction or expansion that takes place in a material as the result of temperature changes. It provides a means for the designer to determine the amount of change that will occur in something, such as a piping run, for a given temperature variation. The designer can then decide if there is a need for expansion loops or other devices to be added to the design to ensure that the material or system does not become overly stressed.

thermal conductor – a solid or fluid, such as a metal or water, that allows heat to flow through it.

thermal insulator – a material that resists the transfer or passage of heat.

thermal mass – a material of sufficient density to absorb and retain heat, such as concrete block. Such a material may be used to build a heat-absorbing wall in a passive solar heating system. The wall will absorb heat from the sun through a window during the day, and give up that heat into the building during the night, warming the air around it.

thermal spraying (THSP) – a process of melting, atomizing and spraying molten metal onto a surface to coat the surface. This technique is used to apply a hard surface to another metal to increase its wear resistance, or to build up worn surfaces on metal parts.

Thermax sheathing – a trade name for a rigid insulating board made by bonding foil to both faces of a polisocyanurate foam core. The insulation has an R-value of about 8 per inch of thickness and is used in locations where the maximum amount of insulation must be fit into the smallest space. For example, it might be used to insulate a south-facing building wall that requires an insulating material that will fit between the limits of 2 x 4 studs.

thermit welding (TW) – a welding technique used to weld large cross sections of metal, such as a broken rudder post on a large ship. A form is build around, and spanning, the joint to be welded. It is filled with a mixture of a metal oxide and a powdered metal, such as fine aluminum particles or ferrous oxide. Heat is applied to the mixture to start a reaction which increases the heat, melting both the mixture and the edges of the metal to be welded. The mixture bonds the two edges of the base metal together. When it has cooled, the welded metal can be cut and ground to the desired shape and smoothness. The temperature of the reaction can go as high as 5800 degrees F, but is generally about 4000 degrees F.

thermocouple – a device that permits remote measurement of temperature by generating an electrical current in response to a temperature change. The amount of current can be monitored and converted into a temperature reading once the thermocouple has been calibrated so that the relationship between temperature and electrical current for that device is known. Thermocouples are used to provide remote temperature indications in inconvenient or dangerous locations, such as in the control room of a power plant.

thermoelectricity – electricity that is produced by the direct action of heating two dissimilar metals that are connected.

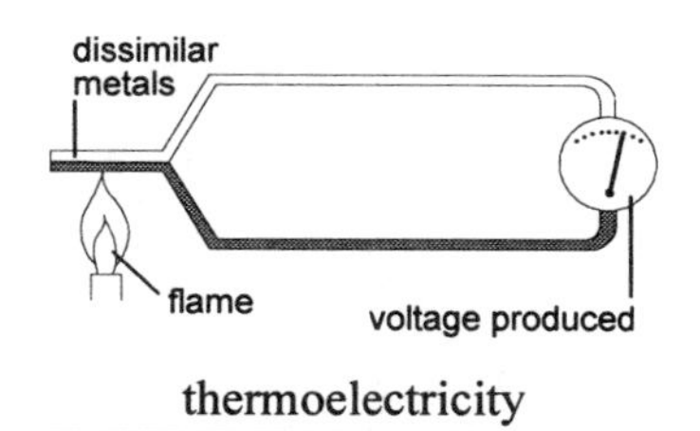

thermoelectricity

thermometer – a device to directly measure the temperature of a gas, liquid or solid.

thermometer, dry-bulb – an instrument to measure the heat in air. This thermometer may function by means of the expansion of a liquid in a graduated glass tube with a reservoir bulb at one end, or by a bimetallic strip that moves an indicator on a scale as the two metals expand at different rates in the presence of heat. This type of thermometer is commonly used to measure outdoor temperatures.

thermometer, wet-bulb – an instrument to measure heat in the water vapor in air. It is similar to a dry-bulb thermometer, but has a water-saturated cloth wick over the bulb that forms a reservoir at the end of the tube containing the liquid. This thermometer is swung through the air to speed the evaporation of the water in the wick. The temperature recorded will vary with the conditions of water vapor content in the air. It is used in conjunction with a dry-bulb thermometer and a calibration chart to determine relative humidity.

thermoplastic – a type of plastic that can be easily softened by heat and that cools to a hard material. The heating/cooling process can be repeated, if necessary. Thermoplastics are used for many products, including pipe, fittings, valves, wire insulation and glue.

thermoplastic-insulated wire – wire, designated Type T, that has a thermoplastic covering and is commonly used for residential electrical systems. It is flame-retardant and suitable for dry locations throughout the house at temperatures up to

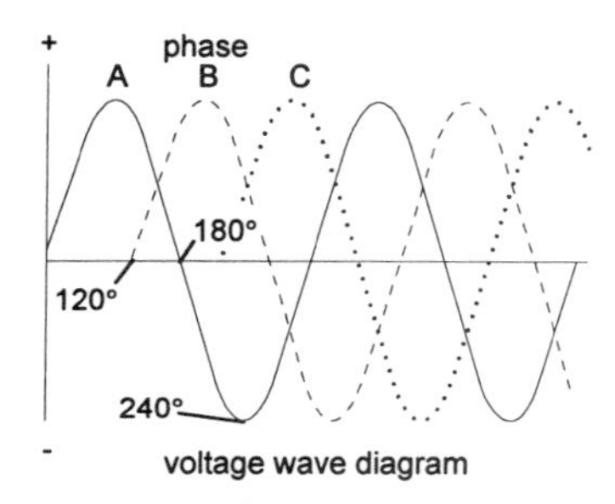

three-phase electrical current

140 degrees F. There are several variations of Type T wire designed for special uses, such as Type TW, which is moisture resistant, Type THW, which is moisture and heat resistant, and Type TF, which is a solid or stranded fixture wire.

thermosetting plastics or resins – materials that can be softened by heat only once, then cool to a temperature-resistant solid. There are several products made with thermosetting plastics or resins, such as reinforced thermosetting resin pipe.

thermosiphon effect – the movement of fluid within a system induced by heating the fluid, causing it to expand and rise as it is heated, or to contract and drop down as it cools and becomes denser.

thermostat – an electromechanical switch that uses a bimetallic strip reacting to changes in temperature to activate or deactivate a heating or cooling system.

thermowelding – a process of joining certain types of plastic fittings, tubing or piping together by melting and fusing them at the juncture.

thickened-edge slab – a combined concrete slab and perimeter foundation used in warm climates for house slab foundations. The thick outer edge adds to the perimeter strength of the foundation.

thickness gauge – a series of metal blades of precision thicknesses which are used for measuring small gaps. Each blade is marked with its thickness in thousandths of an inch. Also called a feeler gauge.

thimble – 1) a stub-in metal sleeve connection to the flue of a chimney that provides the means of connecting the flue to a smoke pipe; 2) a metal sleeve applied to a variety of uses, such as the sleeve placed between the free-spinning wheel and the axle on a wheelbarrow which spreads the load and protects the axle from wear.

T-hinge – a hinge shaped like the letter T used for both decorative and functional purposes, usually on a rustic door.

thinner – a solvent made from petroleum distillates, alcohol or other formulations, depending on its intended function. It is used for thinning oil-base paint and other coatings, or for cleaning paint off brushes and other materials.

thread gauge – a gauge with standard threads used for checking the accuracy of a threaded part.

threads – a series of spiral grooves cut into a fastener, pipe or closure. They are designed to mate with grooves cut into another object in order to join them together. The width and depth of the threads are set by a uniform standard so that objects with external threads can be expected to screw into and fit the equivalent parts having internal threads, such as a bolt in a nut.

three-corner file – a file with a cross section in the shape of a triangle, used for work in tight spots, such as angular notches.

three-four-five method of squaring (3-4-5 method) – a method of squaring a foundation layout. *See Box.*

three-phase electrical current or power – power that is generated with three armature coils (in an alternator), which is the optimum number for generation, balancing output efficiency against mechanical input energy. The three phases follow each other 1/180th of a second apart in a 60 cycle AC circuit. This type of power is most often used in industrial applications to power heavy machinery.

three-prong plug – an electrical plug with two flat prongs for neutral and hot leads and a rounded prong which is used as a ground. This type of plug is used with high current devices that are not double insulated, such as refrigerators, microwave ovens, audio power amplifiers and some types of power tools. Also called a grounded plug.

three-way switch – an electrical switch, used in pairs, which can control a fixture from two different locations, as in operating a stairway light from the top and bottom of the stairs.

three-way valve – a valve with an inlet and two outlets. One or the other outlet may be selected for fluid flow, but not at the same time, or they may both be shut off by the valve. This type of valve is used in chemical plants, refineries and other process systems.

three-wire system – an AC electrical system with two hot leads and one neutral lead. The voltage potential across the two hot leads is twice the voltage potential across one of the hot leads and the neutral lead. Thus, 240-volt power in a residence uses the two hot leads and a neutral lead, while 120-volt power uses one hot lead and a neutral lead. The three-wire, 240-volt power is needed in residential uses for clothes dryers and electric ovens.

threshold – 1) a beveled wood or metal strip over the sill of an exterior door; 2) an entrance to a building; 3) a limit, or the point at which something may begin or end.

threshold limit value (TLV) – the values of airborne concentrations of chemicals, published by the American Conference of Governmental Industrial Hygienists, used to assess the safe or unsafe exposure limits of workers subjected to the chemicals.

throat – 1) a constriction or thinning of a passage, such as in a chimney; 2) a thinning in support or in the concentration of materials at a particular point.

through drying – complete drying and hardening, especially in regard to paint or other coatings.

through single dovetail joint – a dovetail joint in which the dovetail passes completely through the other member being joined.

throw – 1) the distance air travels outward from an outlet in an HVAC system; the distance to the point of terminal velocity; 2) the area that is illuminated by a light fixture.

thrust – an axial force, or a force along the length of an object, such as along a shaft. The main shaft of a turbine or engine has a thrust bearing at the end of the shaft to absorb loads (axial force or thrust) along the length of the shaft.

thrust block – a concrete casting poured at a change of direction in an underground piping system to brace the pipe in position and prevent movement from the impact of the fluid in the pipe against the pipe wall as it changes direction. The thrust of the water moving through the pipeline puts a constant pressure on direction changes and connections along the line. See also kicker block.

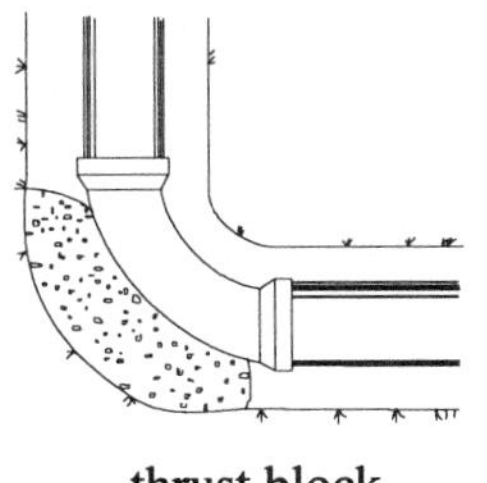

thrust block

thumb screw – a screw, generally machine threaded, that has an expanded and flattened head designed to be turned with a thumb and forefinger rather than a tool. It is used where frequent adjustments are needed, such as on a tubing pinch valve used to control the fluid flow through tubing, or to adjust the height on a tripod.

3-4-5 method

The three-four-five method is a procedure for squaring a foundation layout. Batterboards are installed at the corners of the area where the foundation is to be laid out. String lines are tied between the batterboards at 90 degree angles to establish the foundation outline. Then equal-length diagonal lines are tied between batterboards across the center of the outline. When a corner is square (a true 90 degree angle), a line that is attached to the foundation outline string 3 feet from the corner and attached to the adjacent outline string 4 feet from the same corner will measure 5 feet in length. The 3-4-5 method conforms to the Pythagorean Theorem, in which the sum of the squares of the legs of a right triangle equal the square of its hypotenuse. In this case the right triangle is made up of the lengths of string ($3^2 + 4^2 = 5^2$ or $9 + 16 = 25$). This method is used for both residential and commercial construction.

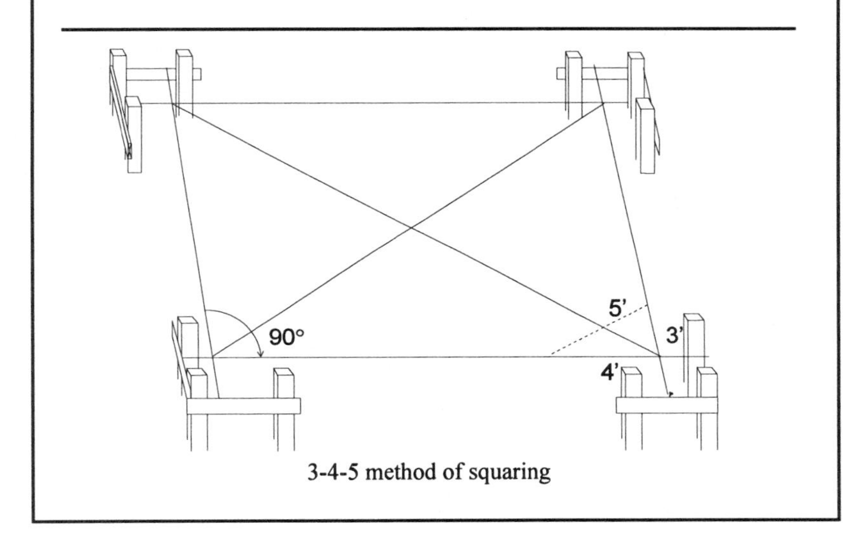

3-4-5 method of squaring

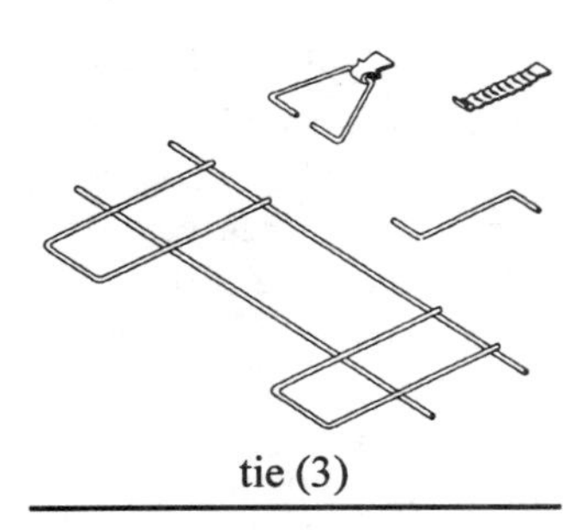
tie (3)

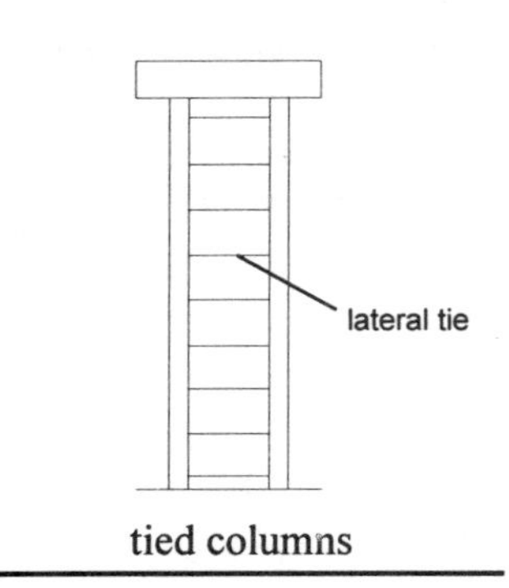

tied columns

tie – 1) a wire wrapped around two objects and twisted to hold them together, such as two pieces of rebar; 2) a wire device used to hold concrete forms together and keep them from spreading when the concrete is placed; 3) a wire or metal strap placed in masonry walls to tie sections of masonry together or to tie veneer to the building structure. See also tie wire.

tie beam – see collar beam.

tie coat – a bonding coat used between two coatings to ensure that they bond together, such as the tack coat applied to old asphalt to create a bond with new asphalt paving.

tied columns – concrete columns that have been poured and have cured, and to which permanent lateral and vertical reinforcement has been added to strengthen the column against shortening and bulging under load.

tie in – to join together, particularly with reference to intersecting planes, such as two roofs or the shingled areas of two roof sections.

tie out – to fix the locations of manholes and other objects in a roadway before paving so that they can be readily located after paving. The "as built" locations are marked on drawings using markers and other locators outside the paved area as reference points.

tie wire – wire used to tie rebar, forms, metal lath or other members together at intersections.

tile – 1) flat square-shaped sheets or slabs of material used as a surface covering. Common tile materials are clay, metal, various types of stone, asphalt and plastic. They are used to cover walls, countertops, roofs and floors; 2) sheets of gypsum, fiberglass or other heat- or fire-resistant material used in between walls or on surfaces for fireproofing or heat protection; 3) curved pieces of material that fit together to form a roof covering or drainage channel; 4) cylindrical pieces of material, usually clay, used for nonpressure drainage applications.

tile, asphalt – a thin flooring tile, usually in 1-foot squares, made of asphalt, often with plastic binders, mineral fillers and pigment. This type of tile is laid in mastic. It is inexpensive and durable, and is easy to install properly. Its durability makes it desirable in high traffic areas of commercial buildings such as supermarkets aisles.

tile, ceramic – flat tile made from clay and other silicon materials, such as sand or quartz.

tile, cork – 1) a resilient floor material composed of ground cork with a resin binder. It can be used in any informal interior floor application, such as in a recreation room; 2) a compressed cork material, available in varying densities, used as a wall covering.

tile cutter – a device for scoring a line in the surface of a ceramic tile and applying a bending force on either side of that line to break the tile. The tile rests on a rubberized pad with a metal spine under the tile at the location of the cutter. After the tile is scored, the worker bears down on the handle of the cutter which has a wing extending out from each side. These wings apply the force evenly to the tile to make a clean break along the scored line.

tile, rubber – a square floor tile made of natural or synthetic rubber with filler material and pigments. This tile is resilient and durable. It is commonly used to cover hard floors, such as concrete, in areas where workers must stand for long periods of time.

tile saw – a metal saw blade with diamond chips bonded to the periphery. The saw blade is water cooled and is used to cut ceramic tile. It is portable, though somewhat cumbersome, and can be rented from most tile stores.

tile, vinyl – a durable floor tile made from vinyl binders and mineral fibers with pigments added for coloring. It is available in 9 x 9-inch or 12 x 12-inch squares, in thicknesses up to ⅛ inch. It is set in mastic and is easy to install.

tilt table – a table for a drill press that can be tilted so that the work to be drilled may be positioned at an angle to the drill. It allows precise and repeatable positioning for drilling holes at other than 90-degree angles.

tilt-up construction – concrete walls or other units that are poured in horizontal forms on or near the site of the structure being built. A 6- to 8-inch form is laid on a concrete slab foundation and lifting eyes are placed in the form. A release agent is applied to the slab to keep the new concrete from sticking. Reinforcement is added and the concrete is poured in the forms. The concrete is screeded, floated, troweled and allowed to set up. When it has cured, it is tilted up on the foundation slab to form a wall.

tilt-up door – a one-piece garage door mounted on spring-loaded pivots that allow the door to swing up and into the garage to open. Also called an overhead door.

timber – 1) cut lumber that is 5 inches or more in the smallest dimension, and is suitable for structural applications; 2) a stand of growing trees.

timber connectors – fastening or connecting devices, often in U or T shapes, cut or bent from ⅛-inch or thicker metal sheets. They are bolted to the intersections of large structural wood members for joint reinforcement.

timber construction – a structure made with timbers as the principal members. Also called wood-frame construction.

time-and-materials contract – a contract in which the contractor is paid for his time, plus the cost of the materials used in completing the job.

time-clock switch – an electrical switch with a built-in clock which can turn an electrical circuit on and off at preset times.

time-lag fuse – an electrical fuse with two elements that will withstand moderate overcurrents, but will blow with large overcurrents. Also called a slow-blow fuse.

time-line – a bar chart or other depiction of time showing changes over a given period. Used in construction to show the time required to complete a job, or a task or set of tasks, that are part of a job.

timer switch – an electrical switch with a built-in clock mechanism that turns the switch off after a period of time. The switch is generally a rotary type with the length of time it will remain on being determined by the amount it is rotated. The switch is manually turned on, and then winds itself off. This type of switch is frequently used for lights in public restrooms or overhead heat lamps in bathing areas. The timer prevents people from leaving lights or heat elements on indefinitely after use.

time-weighted average (TWA) – the exposure of a worker to a potentially harmful atmosphere or substance for what is determined to be a safe amount of time. OSHA and other agencies have set limits on the amounts of time that workers may be exposed to a hazard, and that exposure must be carefully monitored. An example of such a hazard is exposure to radiation in a nuclear power plant. Workers must wear radiation film badges which monitor their exposure. As the radiation slowly exposes the film, the amount of radiation the badge is exposed to can be correlated to the exposure time of the wearer. The badges are monitored carefully to safeguard the workers' health and keep their exposure within the safe range.

time-with-materials-furnished contract – a contract in which the owner supplies the materials and pays the contractor just for his time to perform the work. This type of contract might be used where the client has access to or is in possession of certain materials needed for the work. It can save the client the markup that the contractor may add on to the cost of materials, and may also save the contractor the time and cost of having to get material bids and arrange for deliveries.

tin – a metallic element with the atomic symbol Sn, atomic number of 50 and atomic weight of 118.69. It has a low melting point and is used as a component in various types of solder and alloys.

tinker's dam – a small dam made around a joint in a pipe that is to be flooded with solder.

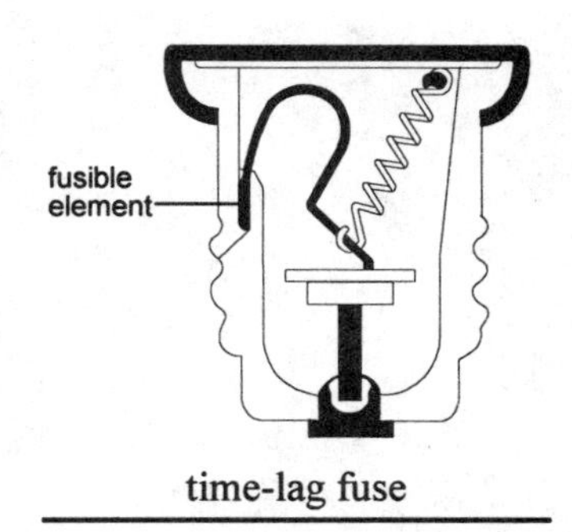

time-lag fuse

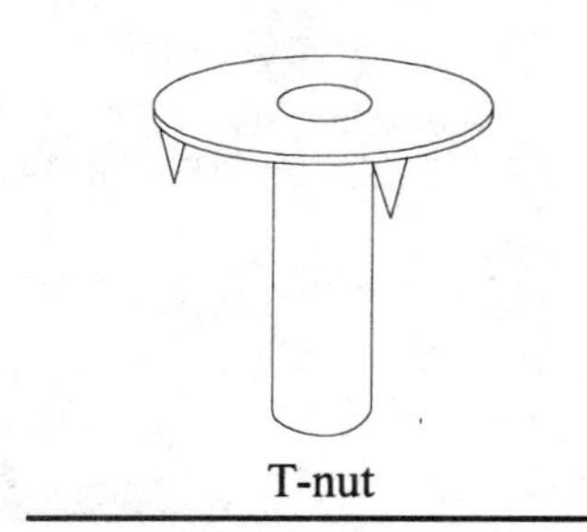
T-nut

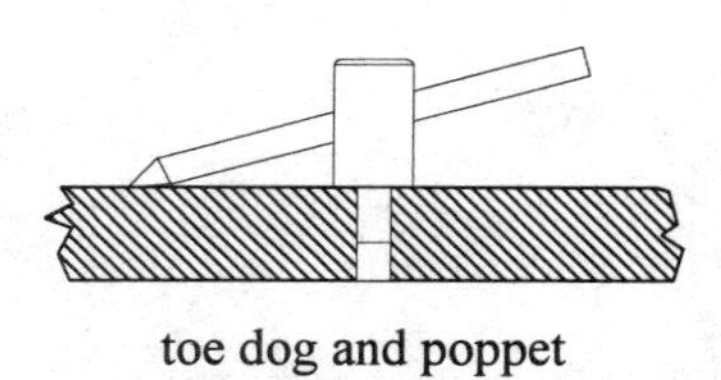
toe dog and poppet

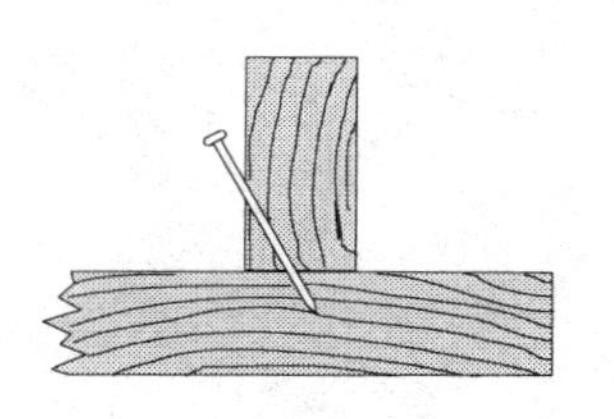
toenailing

tin snips – shears for cutting sheet metal.

tint – 1) a pigment or color additive used with paints; 2) a transparent color, such as one used to color glass; 3) to change color.

tipped-tooth saw blade – a circular saw blade for a power saw with brazed-on carbide tips, slightly wider than the blade thickness. The teeth stay sharper longer than steel teeth. The blades are commonly used for cutting wood.

tipping off – lightly smoothing a fresh coat of paint with the tip of the paint bristles.

titanium – a dark silver metallic element with the atomic symbol Ti, atomic number 22 and the atomic weight of 47.88. It is strong, light and corrosion resistant.

titanium dioxide – a commonly used non-toxic white paint pigment.

title – the elements constituting legal ownership, usually in the form of a legal document such as a deed; a legal right to possession of property.

title block – a block on the face of a drawing or blueprint, located in the lower right corner, that provides identification and revision information about the drawing.

title company – a company that investigates the history of ownership of a property and insures buyers or lenders against disputes regarding title to the property.

title insurance – insurance against problems arising from disputes regarding legal ownership of property or encumbrances against that property that may cloud the title.

title search – an investigation into the history of ownership of a property, including registration records, in order to verify title validity and to check for claims against the property.

title sheet – the first sheet in a set of construction drawings usually providing the name of the architect, the construction firm, an index to the drawings, and a legend for materials.

T-nail – a nail with a rectangular head that is at right angles to the shank, and extends on either side of the shank like the top of the letter T. They are sold in attached strips for use in electric or pneumatic nailers.

T-nut – a short threaded hollow cylinder with a flange on one end that has two or more teeth on its underside. The T-nut is inserted into a hole drilled in wood. The teeth on the flange grip the wood to prevent the T-nut from turning. A machine screw or bolt is inserted in the hole drilled through the wood in which the T-nut is inserted, but from the opposite end of the hole, and is threaded into the nut. The nut acts as a threaded insert and does not have to be accessible from the bolt side of the hole.

toe dog and poppet – a holding device for machine working of metals. It consists of a base (the poppet) that fits into a hole on the working surface, and a threaded insert through a hole in the poppet at a slight downward slant from the horizontal. The poppet is inserted in the vertical hole in the work table. Another piece of metal in the form of a rod with a wedge-shaped point on one end (the toe dog) is placed between the end of the horizontal threaded rod and the work. The threaded rod is adjusted horizontally to cause the toe dog to fit tightly against the work to be held in place.

toenailing – driving nails at an angle near the base of a framing member in order to join it to a second member. The nail should penetrate so that about half the nail is in each member.

toe of slope – the base of a slope.

toe of weld – the point at which the weld metal and the base metal join.

toggle bolt – a type of fastener used with plaster or gypsum wallboard having a hinged, collapsible, threaded nut into which a long machine screw or bolt is threaded. The nut, with the bolt attached, is pressed through a hole drilled into the wall. Once through the wall, the wings on the nut expand to a dimension larger than the hole by means of a spring. The bolt is then tightened, drawing the nut up against the back surface of the wall. Also called a butterfly bolt.

toggle switch – an electrical switch that is actuated by a short lever that moves up and down or back and forth. An example of a commonly-used toggle switch is a wall switch for a light fixture.

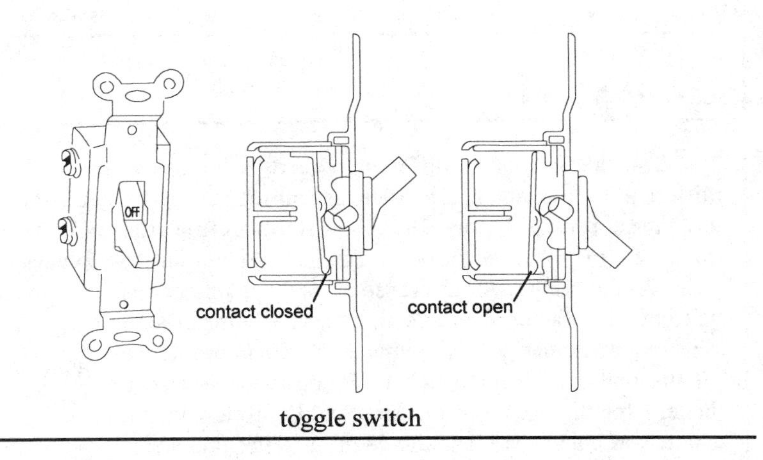

toggle switch

ton – 1) a measure of weight equal to 2000 pounds (U.S. short ton) or 2240 pounds (U.K. long ton); 2) a measure of air conditioning (refrigeration) capacity equal to the removal of heat at a rate of 12,000 Btu per hour.

tongs – a long-handled hinged tool used for grasping and lifting work, especially hot metals.

tongue – 1) the short blade of a framing square; 2) a projection from the edge of an object, such as a wood panel, that is designed to fit into a corresponding groove in a similar object.

tongue and groove – an edge, shaped with a narrow projection along the centerline of its length, designed to form an interlocking joint with a corresponding grooved edge. The tongue fits into the groove to lock the two pieces together. Paneling or flooring members frequently are designed with one tongue edge and one grooved edge so that they can be installed in a continuous interlocking pattern.

tool bit – a cutting bar that has been ground and shaped for use in a metal lathe or other metal-cutting machine.

tooled joint – a mortar joint that is shaped and compressed with a tool other than a trowel. See also masonry joints.

tool holder – 1) a device designed to hold a tool bit in the desired position to cut, such as the device that holds a cutting bit in a milling machine; 2) a rack for storing tools.

tooling – shaping mortar joints with a metal tool, such as a concave jointing tool, to create a particular style of joint.

tool post – the device that positions and holds the tool bit in a metal cutting lathe. Also called a tool holder.

tool steel – a steel, used to make cutting tools, having a high carbon content for added hardness. *See Box.*

tooth – 1) one of the cutting points on a saw blade or other cutting tool; 2) the texture or roughness of a drawing surface, such as drafting film or Mylar; 3) the ability of a paint primer to bond to a surface or to a top coat of paint. Primers are formulated chemically to bond best with certain combinations of paint and surface materials.

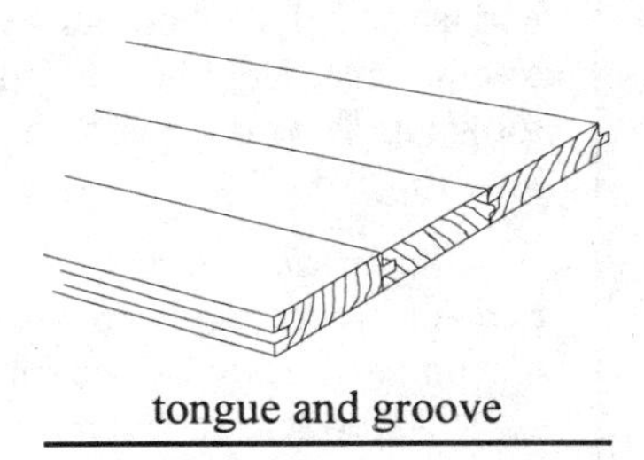
tongue and groove

toothing – a type of brickwork in which alternate courses of brick project from the face of the wall. This is done to improve the bond with brickwork to be installed in the future.

top coat – the final coat of paint, asphalt or other surface treatment that in applied in multiple layers.

topographic map – a map which shows elevations and land contours which include both natural and man-made features.

topographic survey – a survey to create a topographic map in which relative positions and elevations are shown by adding contours.

topographic symbols – symbols used to illustrate various features of the land on a topographic map.

topping compound – a smooth compound that may be used for the final coat or coats over wallboard joints and fastener heads. Topping compound is not suitable for embedding tape because it does not have the adhesive qualities of

Tool steels

Tool steels require different properties for use in different tool applications, such as cutting, shearing, drawing or rolling. The *workability* of a tool steel material to bring it to its desired shape, and the *heat treatability* to achieve the desired characteristics, are considerations used by manufacturers in selecting the proper material to use for a particular tool. Knowing the different properties of the materials from which tools are made is also a benefit for the user. For instance, cold chisels will not shatter on impact; pry bars made of high-strength steels are not brittle and will not break easily under hard use; and hardened-steel blades will maintain a sharp edge for a long time. Having information about the qualities of tool materials may save the user time and money over the long run.

Water-hardening tool steels contain 0.60 to 1.40 percent carbon and are low in cost, tough and easily machinable. They can be surface hardened, but they are subject to warpage and may soften at elevated temperatures. They are commonly used for knives, files, drills, hammers and similar tools.

Shock-resistant tool steels contain chromium-tungsten, silicon-molybdenum, or silicon-manganese. They are hardenable and tough, but they distort easily. They are used for tools such as chisels, where the impact loads are high.

Cold-work steels have high resistance to wear and distortion and average toughness and resistance to softening at high temperatures. They are used for drills, files, planes, putty knives and similar tools. There are three groups of cold-work steels:

1) oil hardening - results in a medium alloy with good machineability and good, but not superior, wear resistance;

2) air hardening - results in an alloy with similar qualities to those of oil hardening;

3) high carbon, high chromium - results in an alloy with poor machineability but superior wear resistance.

Hot-work tool steels have chromium or tungsten as alloying elements. They are air or oil hardenable, resistant to heat softening and deforming and they are the toughest of the tool steels. Hot-work tool steels are moderately machinable and moderately wear-resistant. They are used to make dies for extrusion, forming and casting.

High-speed tool steels contain tungsten or molybdenum, and sometimes have cobalt added to improve their cutting ability. They do not lose much hardness at red heat and have the best characteristics of the tool steels, except for toughness. They are used for drill bits and metal lathe tools.

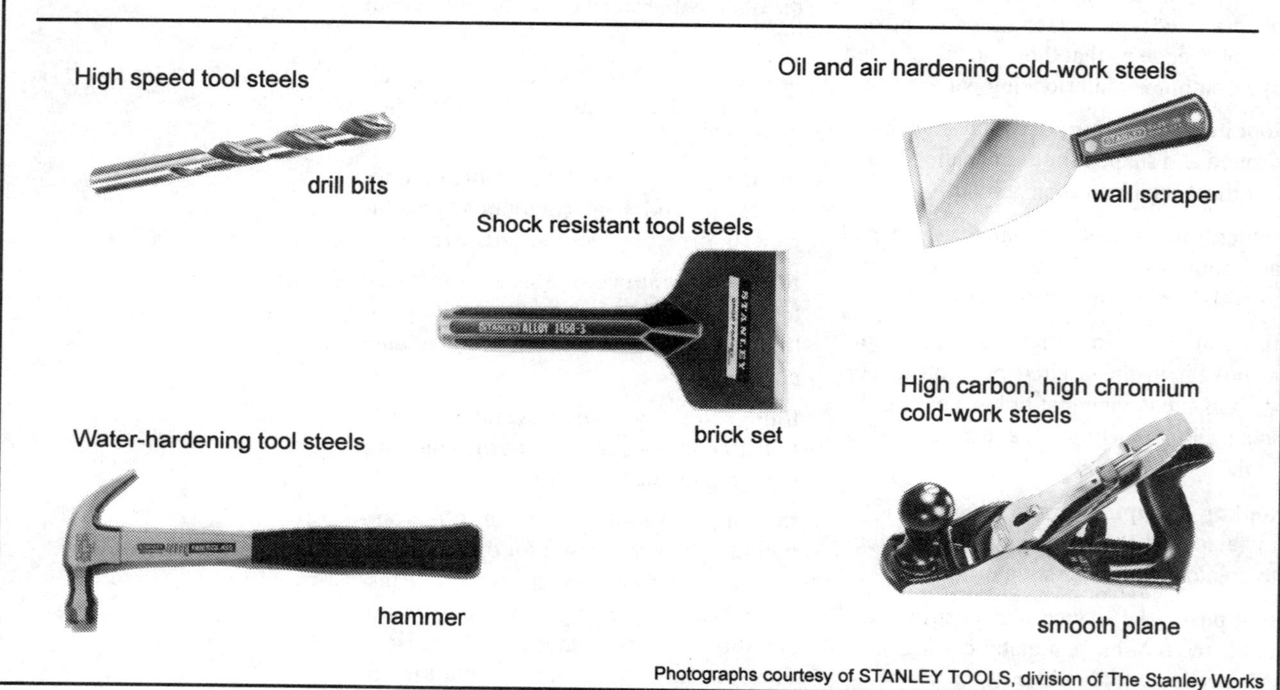

Photographs courtesy of STANLEY TOOLS, division of The Stanley Works

joint compound. However, because it is made of fine granules, it can be spread smooth, and makes an ideal second and third coat of compound over joints and fastener heads. It is also easier to sand than standard joint compound.

top plate – the uppermost horizontal structural member in a framed wall, consisting of 2-by lumber laid in double thickness across the tops of the studs.

top rail – 1) the uppermost horizontal rail of a fence; 2) the top horizontal member of an assembly, such as a panel door or window sash.

torch – a heating device used for welding, soldering and oxygen cutting that is powered by a combustible fuel. It may use either air or pure oxygen as the oxidizer depending on the type of torch. The oxidizer is the source of oxygen that permits combustion of the fuel.

torch brazing (TB) – a heat process, employed to join two or more pieces of metal, using a metal filler with a melting point higher than 840 degrees F but lower than that of the metals being joined. A gas torch is used as the source of heat. Torch brazing may be used to join two pieces of steel with brass as the filler metal.

torch soldering (TS) – a soldering process using a fuel torch for the source of heat.

torpedo level – a small hand-held level used to level and plumb short members. It has rounded, torpedo-shaped ends. Also called a pocket level.

torque – force exerted in a rotating motion.

torque wrench – a type of wrench with a gauge or other means to indicate the amount of rotating force applied to a fastener, such as a nut, as it is turned.

torr – a measurement of vacuum or pressure equal to 1 millimeter of mercury (1 torr = 1 mm Hg).

torsion – the twisting force applied to an object, such as a shaft or beam.

tort – a legal term for a wrong, other than breach of contract, for which a civil action is brought against the perpetrator. For example, a contractor might have a civil action brought against him for damaging adjacent property, such as a fence, while adding on to a house.

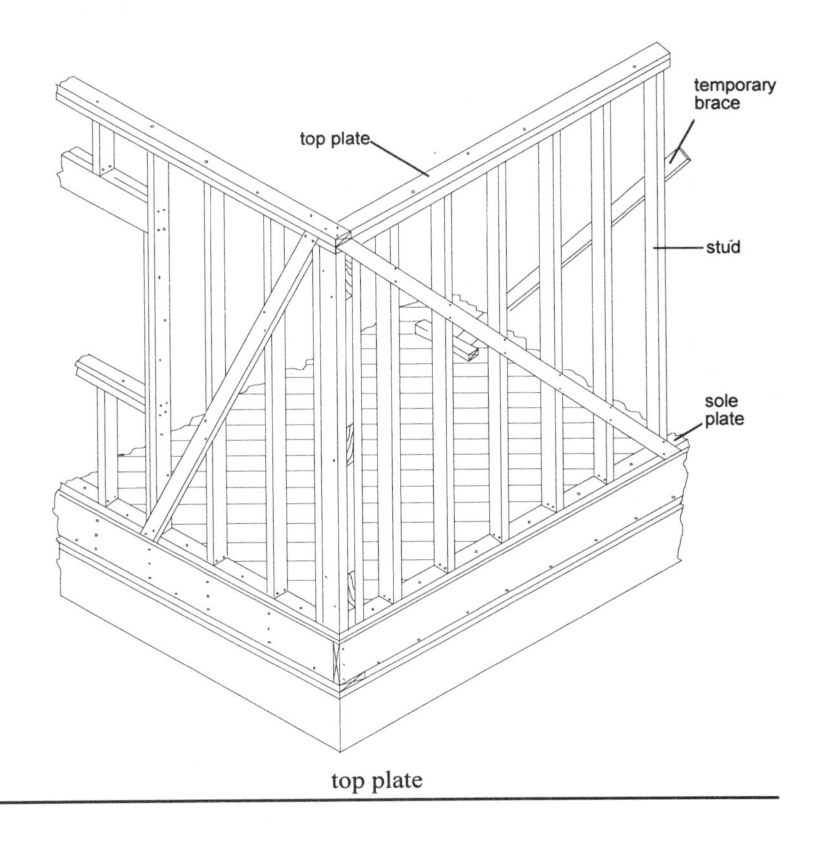

top plate

torus – a semi-circular-shaped design applied to moldings and column bases.

total pressure – the sum of the static and the velocity pressures in an HVAC system.

total rise – 1) the vertical height of a roof measured from the top plate to the top of the roof ridge; 2) the vertical distance from the lower floor to the upper floor used to determine the unit rise in stair building.

total run – 1) the horizontal distance between the outside wall and the intersection of roof rafters at the roof ridge, which is equal to one-half the span of the building for an equal pitch roof; 2) the horizontal distance from the face of the bottom stair riser to the face of the top riser of a staircase.

touch up – to repair small areas, such as with paint, stucco or similar materials, without having to resurface the entire object.

top rail

torpedo level

N

township line

6	5	4	3	2	1
7	8	9	10	11	12
18	17	16	15	14	13
19	20	21	22	23	24
30	29	28	27	26	25
31	32	33	34	35	36

W range line | range line E

township line

S

township

tower – a structure that is higher than it is broad, and usually stands above the surrounding structures. It may be an adjoining section of a larger building or a separate structure set apart.

tower crane – a crane, with a pivoting boom mounted on a tower structure, used for hoisting loads in high-rise building construction.

townhouse – 1) one unit of a multi-unit dwelling structure in which the units share some common walls; 2) an urban dwelling.

township – a 36-square-mile area of public land, based on U.S. geological survey standards, in which the land was laid out in squares, 6 miles on each side, forming 36 sections of 640 acres each.

toxic – poisonous.

T-plate – a metal brace in the shape of a T used to reinforce perpendicular joints of wooden structural members. The plate has predrilled holes for fasteners.

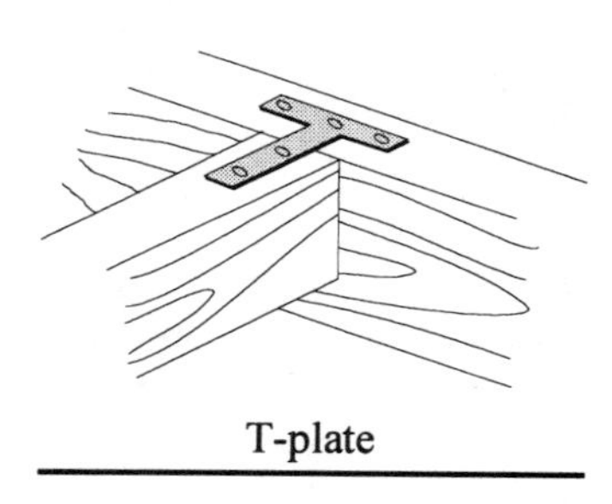

T-plate

tractor-loader-backhoe

trace cloth – a soft, lint-free cloth of closely-woven material, used for applications such as putting a rubbed finish on wood.

tracery – delicate decorative lines.

track dozer – an earthmoving machine, with a scraper blade, that moves on tracks consisting of individual links, or treads, that are driven around a set of wheels. The tracks spread the weight of the machine over their length. This is an advantage when working on very soft, rough or uneven terrain.

tracked backhoe – a backhoe that moves on tracks consisting of individual links, or treads, that are driven around a set of wheels. The tracks spread the weight of the machine over their length, which is an asset when working in very soft, rough or uneven terrain. Also called a track hoe.

tracked excavator – a trench-digging machine that moves on tracks consisting of individual links, or treads, that are driven around a set of wheels. The tracks spread the weight of the machine over their length, which is an asset when working in very soft, rough or uneven terrain.

track lighting – light fixtures that can be positioned at various locations along an overhead track. The track supports the fixtures and also conducts the current to the fixtures.

tract – a defined area of land.

traction – the adhesive friction of a body on a surface over which it moves.

traction engine – an off-track locomotive that can be driven over most surfaces that can support its weight, such as an engine used to pull logs in a logging operation.

tractor – a high-powered track vehicle used for heavy work such as hauling, pulling or grading.

tractor-loader-backhoe (TLB) – an earthmoving machine with a backhoe on one end and a loader bucket on the other. It is a multipurpose piece of equipment, and so is more versatile than a backhoe or a tractor-loader. It is used for both trenching and earthmoving operations.

trade association – an organization of manufacturers, distributors, service contractors or other groups with similar goals or needs. Such associations are usually formed to promote uniform standards within their particular industry and to standardize documents or codes for common benefit.

traffic circle – a roundabout or circular intersection which provides for continuous movement of traffic in one direction, allowing cars to join in from one road and exit to another easily. Also called a rotary.

trailer – a wheeled vehicle without its own motive power designed to be towed behind a powered vehicle. *See Box.*

Trailer

A wheeled vehicle, without its own motive power, designed to be towed behind a powered vehicle. Trailers come in a wide selection of sizes and shapes, small to very large, open or closed, to be used in several capacities from cargo carriers to completely enclosed dwelling units. They are designed to be hauled by cars, trucks or trains. Dwelling units, which are either dependent or independent types, can be used as temporary or permanent office space, storage or living quarters.

A dependent trailer coach is a unit designed for temporary occupancy, with built-in sanitary facilities which are not plumbed for connection to permanent sewage and water supply systems. They are often used as job-site construction offices. An independent trailer coach is a unit designed for permanent occupancy, with kitchen and bathroom facilities and a plumbing system designed for permanent connection to sewage and water supply systems at a trailer park. They can also be hooked up to utilities at a construction site and used for long-term office facilities on a large job, or even home/office facilities for a remote location.

trammel – a straight bar with two adjustable points fastened along its length. It is used for marking large circles by fixing one of the points in place and rotating the other point around the fixed point. The rotating point may hold a pen or pencil to scribe the circle. It is similar in use to a compass, but is used for work on a larger scale.

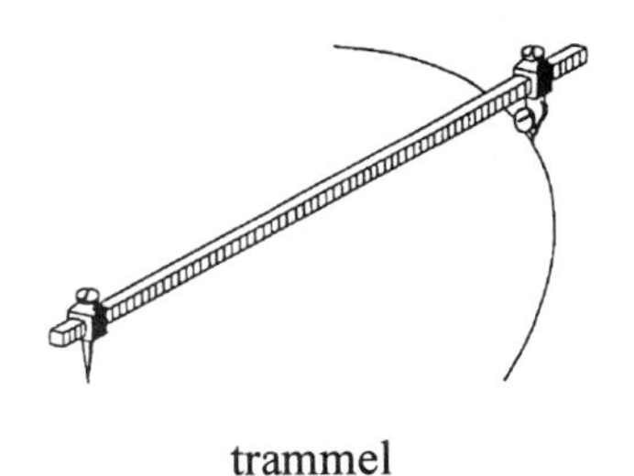

trammel

transducer – a device for converting one type of energy to another, such as a thermocouple.

transfer switch – a switch for connecting an emergency generator to an electrical system normally supplied by a power utility. This switch simultaneously disconnects the utility power supply when connecting the emergency generator to the system. Transfer switches are found in hospitals and other facilities where continuous electrical power is essential.

transformer – an electrical device having a primary and a secondary coil, each with a specific (and often different) number of turns of wire composing the coil. A voltage is generated in the secondary coil by alternating current passing through the primary coil because the magnetic field from the alternating current is moving in relation to the conductor. The voltage generated in the secondary coil is proportional to the voltage in the primary coil with relation to the number of primary and secondary coil windings. Transformers are used to drop the high voltage in overhead power lines to lower voltages that are safe for use in residences and other buildings.

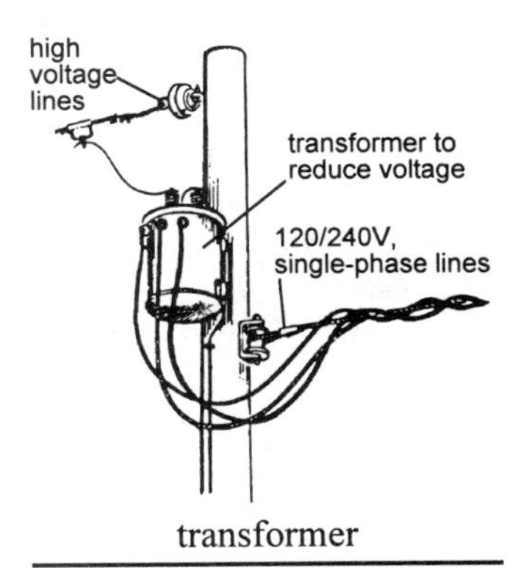

transformer

transient – 1) a temporary oscillation occurring in a circuit due to a change of voltage or load. For example, starting a heavy electrical load, such as a clothes dryer, may cause a momentary dimming of lights in a house; 2) a temporary change in pressure in a piping system.

transit – a telescope with an optical level and an angle-measuring ability that is mounted on a tripod so that it can be rotated over a horizontal east-west axis. It is used in surveying work to establish elevations and measure vertical and horizontal distances.

Transite – a trade-name building material (Johns-Manville Co.) composed of asbestos and cement, used commonly to make pipe.

Transite pipe – asbestos-cement pipe commonly used for water mains. Also called A.C. pipe.

transit level – a type of builder's level including a transit and a spirit level. Also called a builder's level.

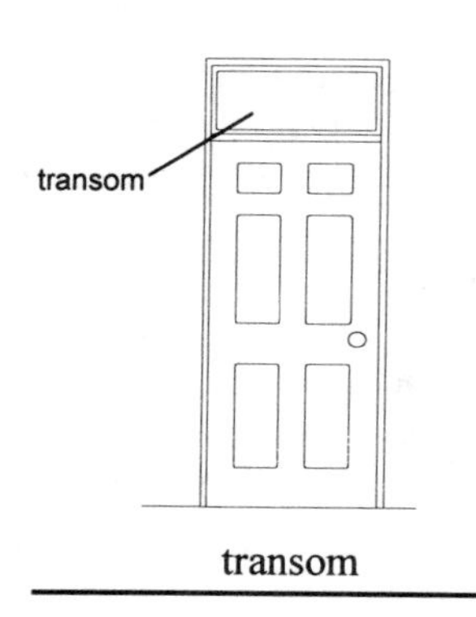

transom

transmission – a gearbox for the purpose of transmitting power at some (usually adjustable) mechanical advantage.

transom – a small window, usually hinged to open and close, located above a door. The window is often hinged at the bottom and opens from the top to allow ventilation to flow into a room.

transparent – clear to the vision; applies to a material that is fine or sheer enough to be seen through.

transport velocity – the air velocity required to keep airborne particles from settling out on surfaces. This information is needed for designing an HVAC system for a *clean room*, one in which particle content must be controlled.

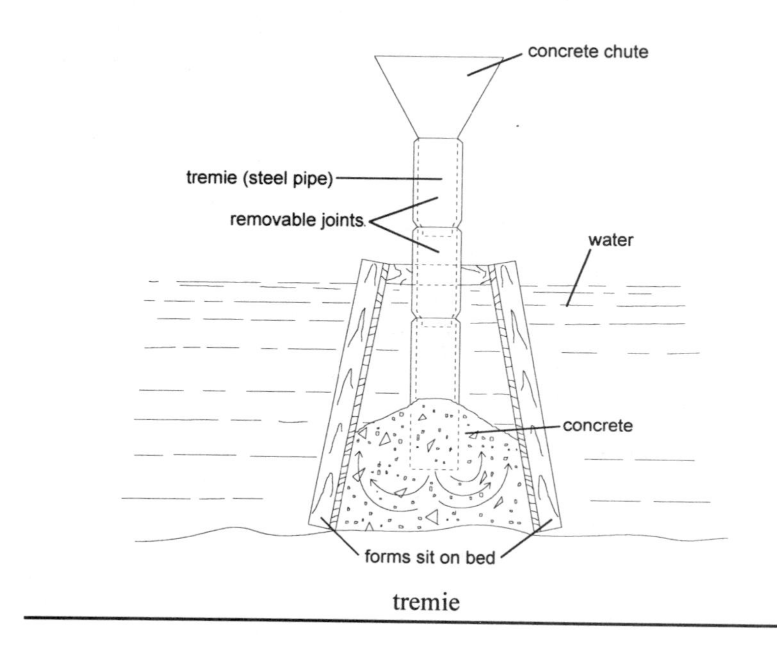

tremie

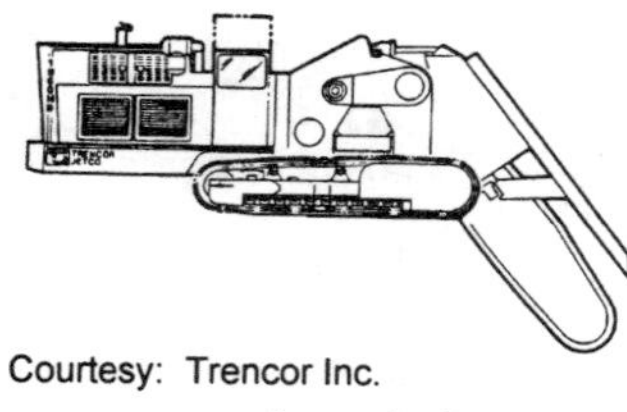
Courtesy: Trencor Inc.
trencher, chain

transverse – crosswise or at right angles to another member; lying across a surface.

transverse surface – the surface of a log that has been cut across its diameter to make round slices, such as those used for decorative garden steps.

trap – a device for catching, holding, or stopping (usually) a gas, liquid or debris. See also particular trap types.

trap door – a hatch or door in a horizontal or near horizontal surface, such as a floor or ceiling.

trap, P – see P-trap.

trap, S – see S-trap.

trap seal – the water seal in a plumbing drainage trap.

trap, steam – see steam trap.

traverse rod – a supporting rod on which draperies are hung. It allows them to be pulled open or closed.

travertine – a form of limestone, deposited by water from springs or streams, that is used for building blocks, interior flooring and trim.

tread – 1) the horizontal boards on stairs which make up the steps; 2) the traction surface of vehicle tires.

trellis – a lattice frame of lath or other lightweight members used as a screen or a support for climbing plants.

tremie – a straight pipe through which concrete can be placed in tall vertical forms, often set under water. The pipe is long enough to reach to the bottom of the concrete form and direct the flow to the desired locations.

trench – a narrow excavation in the ground.

trench box – two plates held apart by spacers used to shore up trench walls and prevent them from collapsing while work is being performed in the trench. Also called trench shield or coffin box.

trencher – a self-powered and self-propelled machine designed to dig trenches for laying pipe or cable. See specific types below.

trencher, canal – a self-powered wheel-type trencher with an arm that shapes the profile of the trench walls to the desired angle. An excavating wheel cuts the trench and a conveyor removes the soil to the side of the excavation.

trencher, chain – a self-powered trench-digging machine that uses an endless loop chain with digging teeth as the means of cutting through the soil and digging the trench. The depth of the trench can be controlled by changing the angle of the arm that carries the cutting chain.

trencher, drainage – a self-powered trench-digging machine that also lays drainage pipe, covers it, and backfills the trench in one pass.

trencher, rock saw – a trenching machine with a self-propelled, large diameter, circular saw blade used to cut a trench through rock.

trencher, wheel – a self-powered trenching machine that has a large diameter wheel with cutting/excavating bucket-type blades around its circumference. The bucket-blades cut into the soil and lift it out of the trench as the wheel rotates. The depth of the trench can be varied by lowering or raising the position of the wheel.

trenching – digging a trench, such as a pipe trench for laying a pipeline below grade.

trepan – to cut a core sample or disc out of a material with a rotating cutter, usually for testing.

trestle – a structural framework of steel or heavy timbers used to support a bridge or roadway from beneath.

triac – an electronic switching device used in dimmer switches to control the current to a light by turning it on for only part of an AC cycle.

triangle – a plane figure having three sides and three angles, with the sum of the angles equaling 180 degrees.

tributyltin – a paint additive used for boat hull paints that controls barnacles and algae growth.

trig – a 4-inch long, thin metal clip with a slot in it used to support the midpoint of a string line set up to align a masonry wall during construction. The trig is placed on a block laid in the center of the wall at the same elevation as the corner blocks. The string line runs from corner to corner with the trig and trig block supporting and holding the center of the line in place.

trim – molding used around an opening or to conceal an edge or joint. Includes door and window trim, baseboards, casings and cornices.

trimmer – a single or double beam or joist rafter to which a header is nailed in framing for a chimney, stairway, or other opening.

trimmer arch – a low-rise brick arch supporting a fireplace hearth.

trippet – a device that is used as a cam to actuate another device on a regular and periodic cycle. For example, a trippet can be a projection on a turning shaft which comes in contact with a lever to turn on an electrical current. As long as the trippet is in contact with the lever, the electrical current stays on. When the slowly turning shaft puts the trippet out of contact with the lever, the current is turned off. This type of device is used as an electrical-mechanical timer for yard lights, turning them on and off automatically.

trolley – the carriage that moves the hoist and hook mechanism along the length of the bridge on a crane bridge. It allows the hoisting device to be moved into position to lift a load, and then carry that load to another location along the length of the bridge.

trolley duct – an electrical bus duct used in areas where drop cord power is needed, such as in a factory where many electrical power tools are in use at various locations. The trolley duct provides overhead power and power drop cords that can be moved or added to locations as needed.

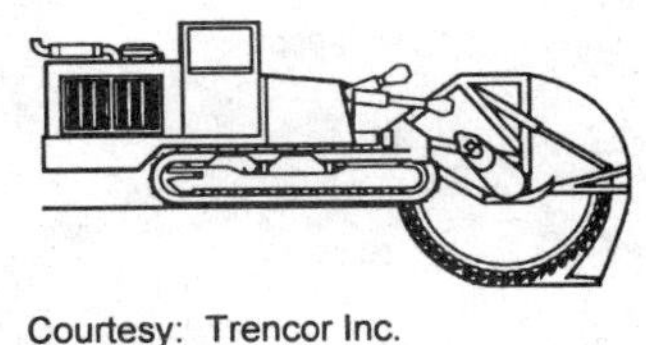

Courtesy: Trencor Inc.

trencher, rock-saw

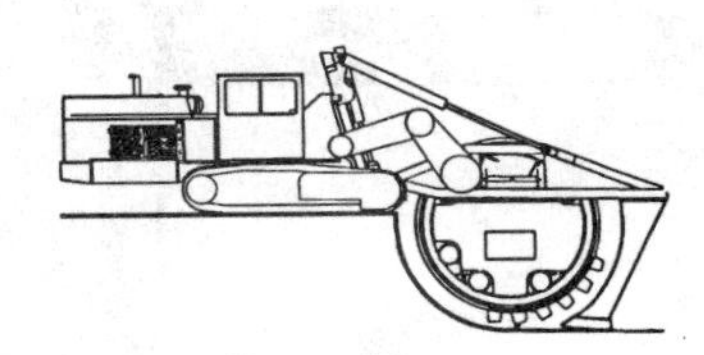

Courtesy: Trencor Inc.

trencher, wheel

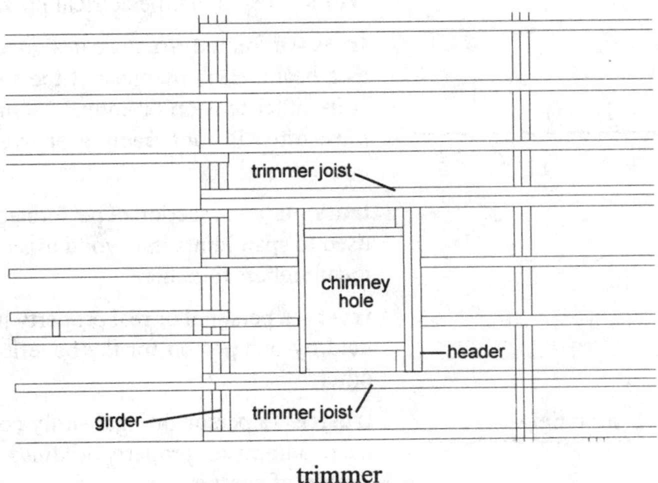

trimmer

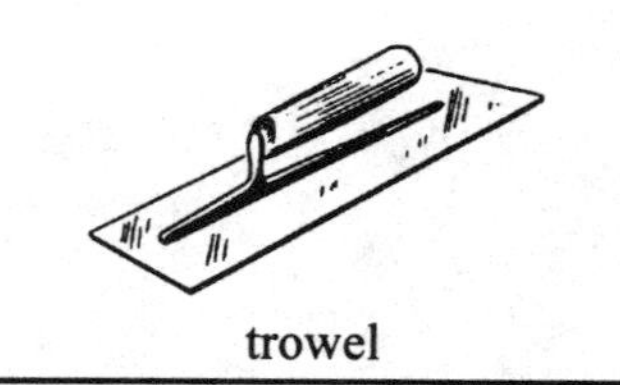
trowel

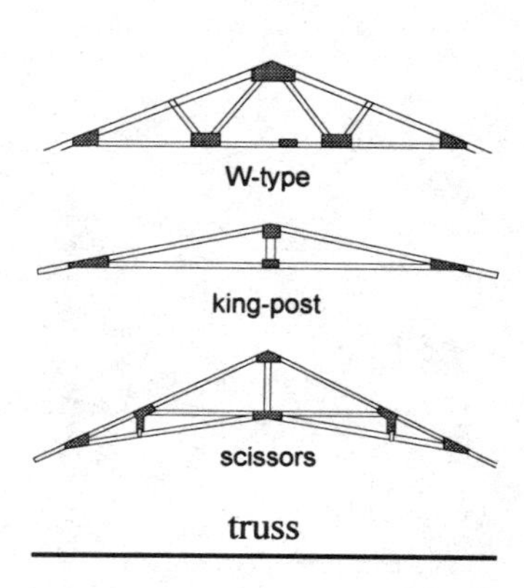

truss

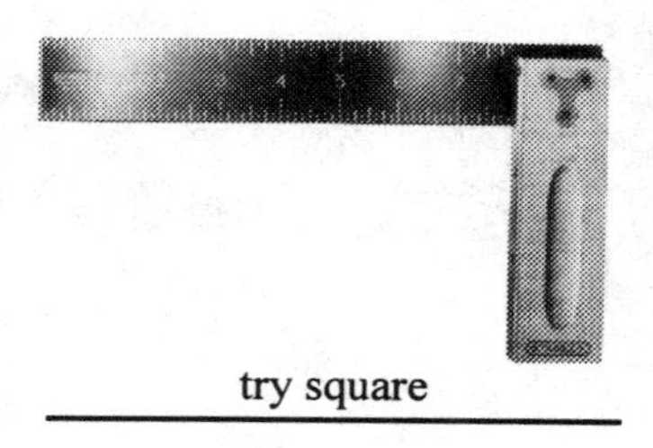
try square

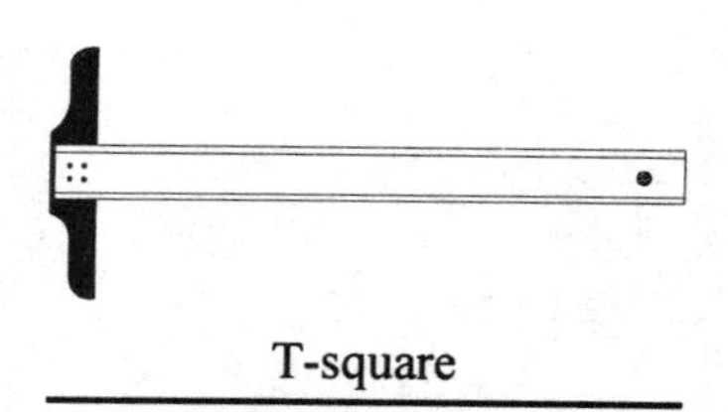
T-square

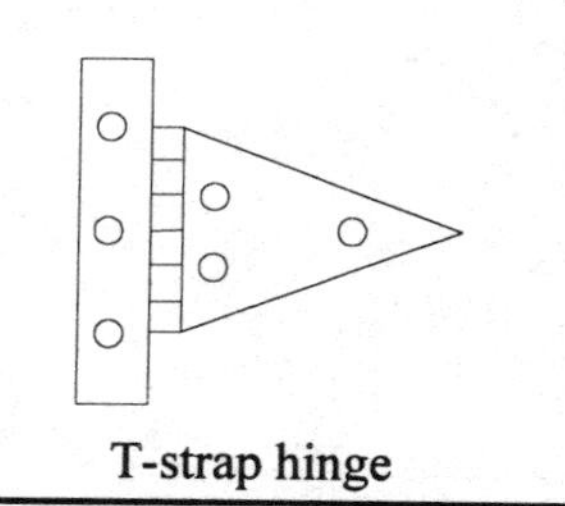
T-strap hinge

trommel – a rotating cylindrical screen used for sorting solid pieces by size, such as sorting pieces of rock for use as aggregate in concrete.

trough – 1) a drainage channel; 2) a framing member set on top of, and at right angles to, the ceiling joists, which brace the bottoms of the structural members that carry the purlins.

trough cable tray – a cable tray that is solid except for a ventilated bottom.

trowel – a hand tool with a flat rectangular or triangular metal surface. The rectangular trowel is used for smoothing and finishing concrete or plaster, the triangular trowel for applying mortar to masonry.

troweled joint – see struck joint.

trowel finish – wall or ceiling texturing created by sweeping a trowel over wet material, such as plaster, stucco or joint compound on a drywall surface.

trowel, power – a gasoline- or electric-powered device for smoothing and finishing wet concrete.

trowel tooth texture – a texture created by brushing the edge of a toothed trowel over a wet surface.

true level – a line perpendicular to a plumb line or other indicator of plumb.

truncate – to shorten by cutting off or not completing, such as a pyramid that does not come to a point at the top.

trunk line – a main electrical power line.

truss – a framed structure designed to act as a beam. Each member of the structure is in either tension or compression. Used most often in roof framing or over wide doorways.

truss plate – a section of perforated metal used to span joints in a wood truss and tie the members together.

trust – a personal or real property interest held by one person for the benefit of another.

trustee – a person or legal entity committed to administer property holdings for the benefit of another.

try square – a metal straightedge with a wooden (or other nonmetal) handle fixed at a right angle to the metal blade. The inside 90-degree angle is used to check and verify square corners in cabinetry and other woodworking applications. Also called a solid square.

T-square – a straightedge with a cross brace forming a letter T at one end. The top of the T is at right angles to the straightedge. It is used as a drafting tool and a guide for laying out and cutting 90-degree angles.

T-strap hinge – a surface mount hinge that is roughly in the shape of the letter T. Often used on fence gates.

tube axial fan – an axial fan mounted in a section of duct or housing. The inside diameter of the tube is just slightly larger than the diameter needed for the fan blades. This arrangement increases the efficiency of the fan.

tube micrometer – a micrometer with a cylindrical anvil designed to be placed inside a tube, sleeve or bushing to permit measuring the wall thickness of the part. The anvil is the fixed part of the micrometer, placed against the object being measured.

tube steel – hollow rectangular-shaped or cylindrical lengths of steel used as structural members.

tube-type finisher – a long cylinder that is dragged behind a slip form paver to trowel the surface of a concrete roadway just after it is poured.

tubing – 1) a hollow cylinder, which may be metallic or nonmetallic, used for conveying fluids. The tubing wall varies in thickness depending on the type of material, such as plastic, steel or copper. Tubing connections also depend on the type of material and can be made by soldering, brazing, compression fittings, flare fittings, etc.; 2) conduit used to protect runs of electrical wire.

tubing cutter – a device with two rollers on one side and a cutting wheel on the other. The tubing is placed between the rollers and the cutting wheel and rotated against the rollers while the cutting wheel is pressed into the tubing wall, eventually cutting through.

tubing fittings – solder joint, compression, or flare ells, tees, and other types of fittings designed to join tubing lengths together.

tubing joint – a connection in a run of tubing, or a connection between tubing and a fitting or piece of equipment. Tubing connections can be made by soldering, brazing or welding some materials, or with compression or flare fittings with other materials. Plastic tubing can be joined by flared or compression fittings and with solvent cement or thermowelding.

tuckpointing – repairing aging and deteriorating unit masonry work by chipping out the broken or weathered mortar, cleaning the joint out, and filling in with fresh mortar.

Tudor arch – see major arch.

Tudor peak – a chamfered section of roof in which a gable end wall terminates before reaching the ridge board and turns into a hip from that point to the ridge.

Tudor plate – the board that spans the common rafters at a gable end, where the gable end terminates and turns into a hip for a Tudor peak roof.

tung oil – a drying oil taken from the seed kernels of the tung tree. Used in making wood finishes. Also called Japanese wood oil or China wood oil.

tungsten – a metallic element with the atomic number 74, atomic symbol W, and the atomic weight of 183.85. Used as an alloying element in various metals.

tungsten carbide – a hard tungsten and carbon compound used as an abrasive on sandpaper and for making cutting blades and saw teeth.

tungsten electrode – a nonconsumable arc welding electrode used in tungsten inert gas welding.

tungsten inert gas welding (TIG) – a type of welding in which a nonconsumable tungsten electrode generates an arc shielded by inert gas, and a separate filler metal rod is melted into the weld. See also gas tungsten arc welding.

tunnel test – a test developed by Underwriter's Laboratory for determining the flame spread of a material. The test utilizes a 25-foot-long tunnel-like test chamber with a nominal inside width of 17½ inches and a nominal inside height of 12 inches. The sides and bottom of the chamber are of insulated masonry and the roof is of another noncombustible material. One side of the chamber has a viewing window. There are two gas burners at one end of a specified design to direct flame at the material being tested so the rate of flame spread on the material can be measured. The chamber is calibrated for flame spread per unit time on red oak flooring. The data gathered on the test material is compared to the data from the test calibration and, using formulas, the flame spread per unit time on the test material is determined.

turbine – 1) a rotary engine driven by an impulse or reaction to a current of liquid or gas, often under pressure; 2) one or more wheels on a shaft which are turned as steam, water or another type of fluid strikes the buckets projecting from the wheel(s). This motion converts kinetic energy to mechanical energy to power equipment such as pumps and generators.

turboelectric – a turbine-driven electrical generator.

turbulence loss – pressure and/or air flow losses caused by air turbulence in an HVAC system. Turbulence in the system may be caused by obstructions to the fluid flow disrupting the smooth flow or it may be a function of the fluid characteristics and the flow rate.

turnbuckle – a rectangular framework with female threads at each end; right-hand threads on one and left-hand on the other so that the turnbuckle can be used to join two threaded metal rods and by turning, loosen or tighten the combined assembly.

turning – the process of shaping materials on a lathe.

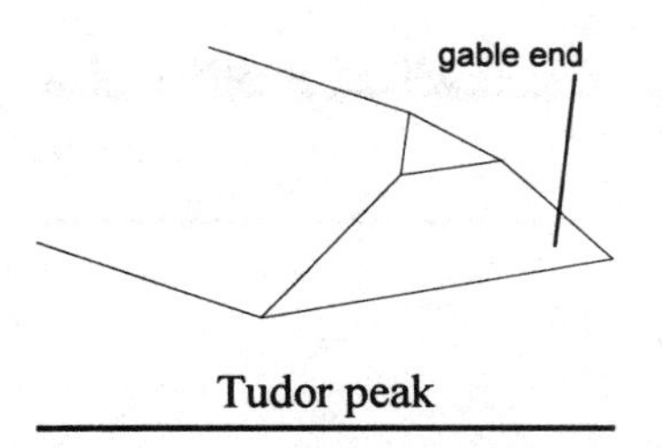

Tudor peak

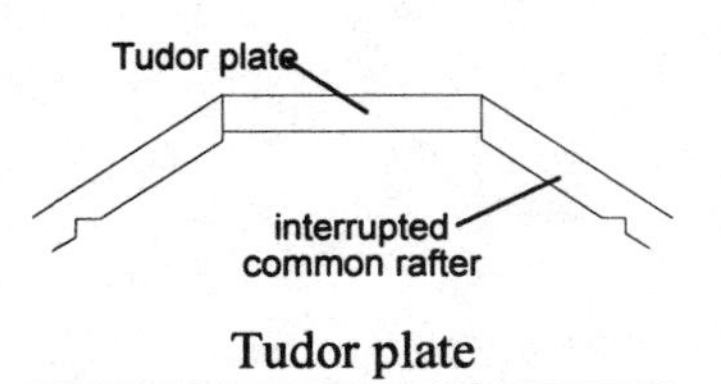

Tudor plate

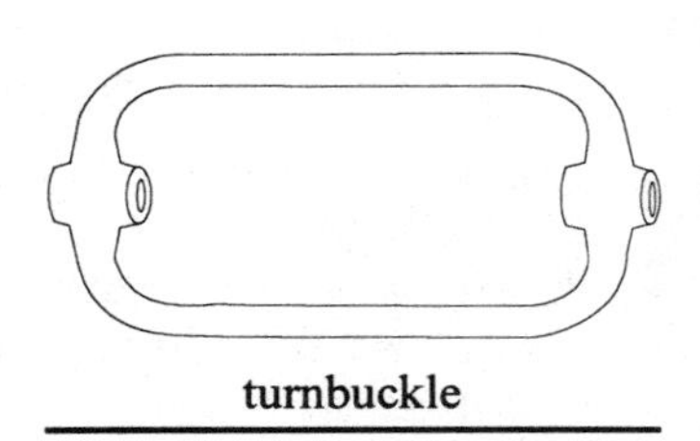
turnbuckle

twist drill

turning point (TP) – a new location of the surveying instrument during a survey.

turning vanes – fixed, curved vanes that redirect the air flow in an HVAC system.

turnkey contract – a job or contract in which the contractor agrees to complete the project to a specified point, often to the point where it is ready for operation and/or occupancy by the client, and assume all the risk for the project.

turn-of-the-nut tightening – establishing a preload on a threaded fastener by bringing it up wrench tight and then turning it another 180 to 270 degrees. Used on ¾-inch or larger diameter bolts.

turpentine – a type of paint thinner and solvent made from pine oil. Used with oil-base paints.

turret lathe – a lathe for shaping metal. There are horizontal and vertical types, both designed to turn large pieces of metal. The turret holds several different tools so changing the cutting tool is easy and quick. The vertical turret lathe is called a Bullard and can accommodate larger pieces of work than the horizontal version. Both have a large chuck which clamps the work and turns it, while a feed mechanism advances the tool in the turret along the work to shape it or bore holes in the material.

twin carbon arc brazing (TCAB) – a process using an electric arc between two carbon electrodes as the source of heat; used to braze metal.

twirl-type laser level – a laser level, used in surveying, that sends laser beams in four directions at once to permit simultaneous leveling in multiple directions.

twist – a spiral warp in a length of material, such as a board.

twist drill – a type of drill bit with spiral flutes along the cutting portion of the shank, and a cone-shaped point. It is designed for cutting metal and other materials.

two-blocking – physical contact between a crane load block and a part of the trolley when the hook is raised to its highest level. As a rule, a load is not lifted this high since there is no further latitude for adjusting the load.

two-man saw – a saw for large trees and logs that has a handle at each end of a long coarse blade so it can be worked by two men, one at each end.

two-way solid structural slab – a concrete slab with two-direction reinforcement inside the slab.

two-wire system – an AC electrical system having a hot lead and a neutral lead, both wires needed to complete the electrical circuit in all 120-volt AC electrical systems.

T-wrench – a type of socket wrench with a T-shaped handle for use with sockets which fit over the heads of nuts or bolts.

Type X gypsum wallboard – a fire-resistant gypsum wallboard that has fiberglass added to the core to increase the natural fire resistance of the gypsum. This type of wallboard is used between an attached garage and the living area or to line stairwells or in other interior wall locations where added fire resistance is needed.

U: U-bolt to UV inhibitor

U-bolt – a bolt shaped like the letter U with both ends threaded along all or a portion of the straight sections. U-bolts are used to clamp items to cylindrical objects, such as clamping pipe to support steel or springs to vehicle axles.

UF wire – underground feeder wire or cable. It is a type of plastic insulated copper wire that is manufactured without fiber spacers that can absorb moisture. It is used in outdoor and underground applications, such as wiring for outdoor lighting, particularly with 120 volt AC supply power.

U-hanger – a metal pipe strap used to secure water distribution piping beneath a building. The ends of the U are bent and have sharp nail-like tips which can be driven into the wooden joists beneath the floor. The pipe rests in the curve of the U.

ullage – the unused tank capacity above the liquid contained in a tank.

ultimate strength – the maximum stress limit of a material before it fails under load.

ultrasonic – vibrations, of the same physical nature as sound waves, but above the audible range of human hearing. Ultrasonic waves used for various applications may be in the range of 50 kHz or higher. Human hearing is about 20 kHz.

ultrasonic examination (UT) – a type of nondestructive examination (NDE) which uses ultrasonic waves to produce a two-dimensional image which can detect structural flaws in a material. *See Box.*

ultrasonic soldering – soldering with the addition of ultrasonic vibrations to remove surface film. This technique is used on metals that are difficult to solder.

ultrasonic stud finder – a tool that locates studs behind a wall surface by means of ultrasonic waves that detect the difference in density in a wall at a stud. The tool is readily available and commonly used in the construction trades.

ultrasonic welding (USW) – a welding method which uses heat energy from ultrasonic waves to join parts. The parts are held together under pressure while the high frequency energy is applied with a UT (ultrasonic) transducer. USW is used to weld small and thin parts during the manufacture of electronic and electrical parts or components.

ultraviolet – light frequency at the violet end of the spectrum that is above the visible range.

underbead crack – a crack in the heat-affected zone of base metal that does not show through to the surface of a weld.

undercoat – a priming coat or protecting coat, such as might be applied to the undercarriage of a vehicle or as a base coat on a surface before the final paint coat.

undercounter lavatory – see lavatory, undercounter.

undercut – a depression or lack of material below the surrounding surface, such as at the edge of a weld bead. If the undercut

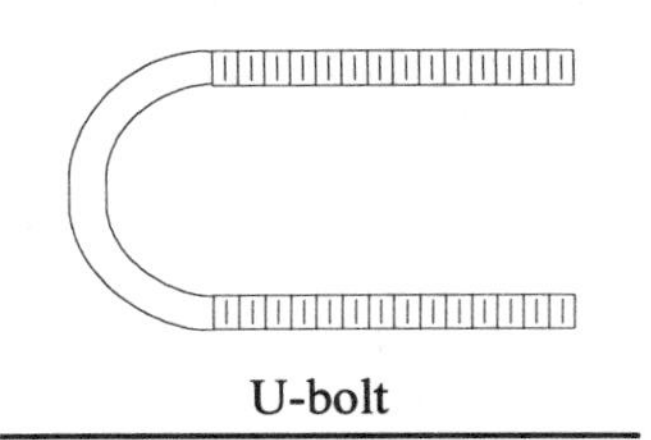

U-bolt

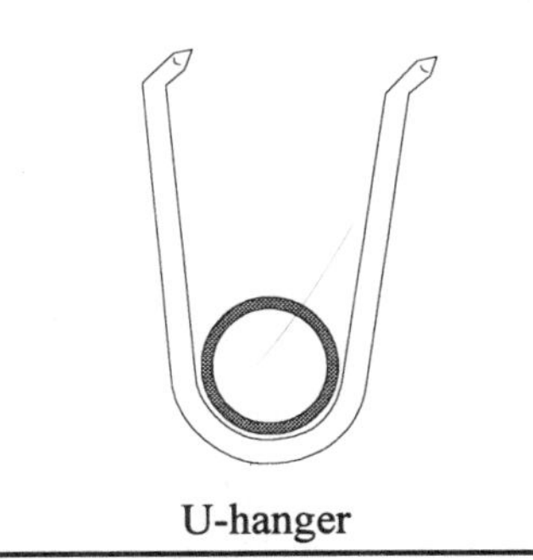

U-hanger

Ultrasonic examination

Ultrasonic or UT examination is a nondestructive volumetric examination using high-frequency sound vibrations. This method is used on objects or materials that will vibrate in order to locate and identify defects or to measure the thickness of the material. The vibrations are initiated by electrical energy, which is converted to mechanical energy by a hand-held transducer that is moved along the surface of the material or object being examined. A liquid is used to couple the transducer to the object's surface for good energy transmission. Because air does not transmit ultrasonic frequencies as efficiently as a liquid, if the object is small enough it may be totally immersed in water. When the energy traveling through the object reaches a defect, or the other side of the object, it will be reflected back to another transducer and converted back to electrical energy.

The electronic signal is displayed on a video tube, similar to a TV screen, and can be interpreted by a trained operator. The frequency of energy is known, and the time for the reflected energy to come back to the transducer is measured. In this manner the defect location and size, or the thickness of the object, can be mapped. Flaws are indicated by a change in the pattern of the reflected wave. As a "volumetric" examination, it can be used to detect flaws throughout the volume of the object being examined. The examination may be performed from many different directions and, where possible, from different sides of the material or object.

Some UT methods use two transducers, others may use one transducer to both send and receive. The former is called through-transmission and the latter, pulse-echo. Piezoelectric transducers are used for UT, operating in a frequency range of 40 kHz to 50 MHz. The higher frequencies are best suited for small defects, and the lower to mid frequencies are usually used for thickness measurement. Calibration is performed on objects of similar materials, called calibration blocks, that have precise dimensions with known defects. UT devices vary in size and portability. Some are hand-held units and others are cart-mounted. An ultrasonic examination can be made on any homogeneous material. It is frequently used to detect flaws and determine thicknesses of ceramics and metals.

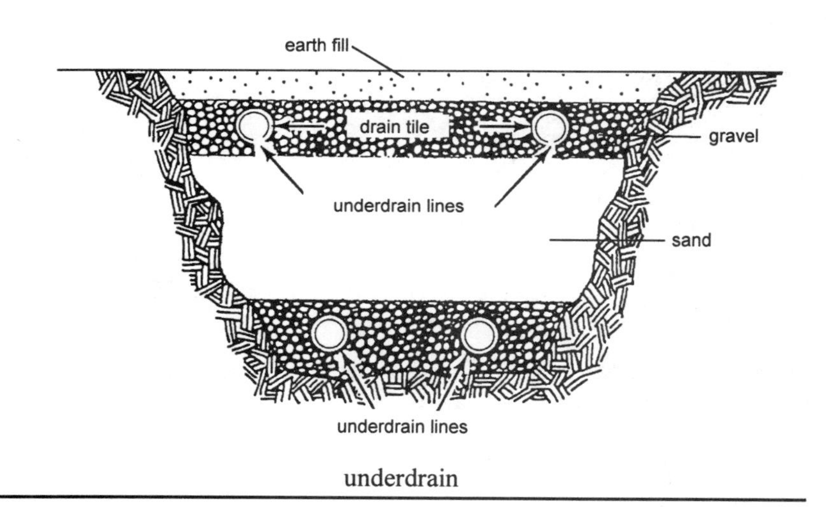

underdrain

underground feeder cable

results in a material thickness below design standards, it will have to be repaired. Stress along an undercut may lead to cracking and subsequent material failure.

undercutting – enlarging a portion of a previously-bored hole to make it a larger diameter at some point away from the end of the hole. For example, undercutting is frequently done to remove excess material between bearings.

underdrain – a drainage system installed under the surface to collect and divert water that accumulates under a slab or road surface, or installed to help prevent saturation in a septic tank drainfield that has poor percolation.

underfloor raceway – hollow metal shapes designed to be mounted under floors for the purpose of routing electrical wiring.

underground feeder cable (Type UF) – electrical cable, with insulated conductors inside an outer sheath, that is suitable for buried installation. The insulation is fire retardant, and resistant to fungus, mildew and moisture.

underground service – electrical power distribution that is run underground rather than from overhead from power lines.

Most new housing tracts use underground service distribution to eliminate unsightly power lines.

underlayment – material laid on top of a subfloor to provide a smooth, even undersurface for the finished floor. This is especially important when the finished floor is of a material that will conform to the surface beneath it, as does vinyl floor covering.

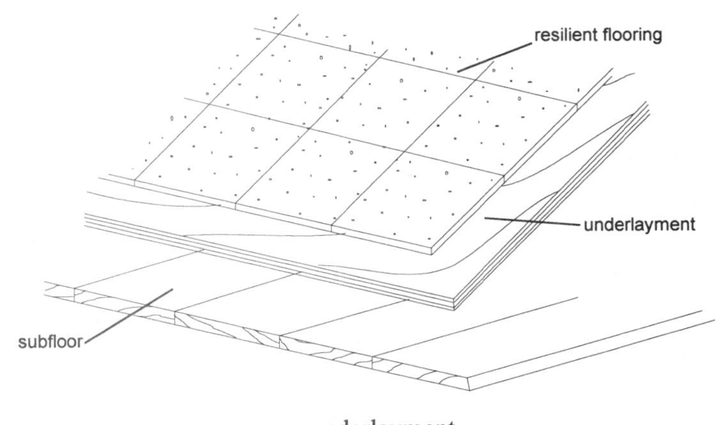

underlayment

undertone – a light or subdued paint color.

Underwriter's knot – a knot tied in the two insulated conductors near the terminals of an electrical plug so that the knot will take any pulling force applied to the wire, protecting the connection at the terminals.

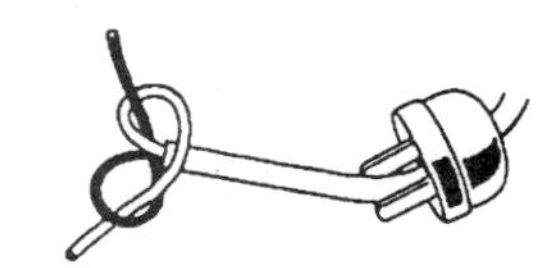

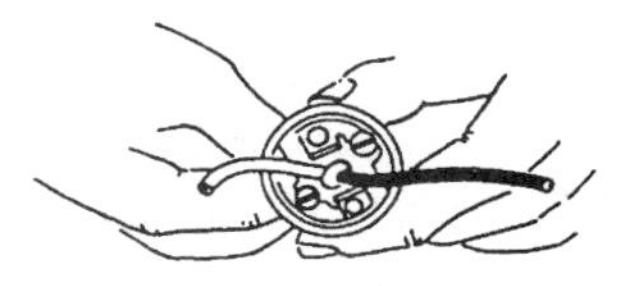

Underwriter's knot

Underwriter's Laboratories (UL) – an independent testing agency which can be hired to test and assess the operation and safety of almost any item or product. Once the item has been tested and approved by Underwriter's Laboratories, the manufacturer can put a UL sticker or tag on the item to show that it has been tested.

unearned income – income that is received for work that has not yet been completed, such as advance payment for a construction job that is in progress.

unemployment insurance – an insurance program, operated by the state and paid for through employer contributions, which is used to provide workers with compensation when they are involuntarily laid off. The purpose is to help workers financially while they look for new employment.

unibit – a drill bit with stepped increases in diameter so that it can drill holes (in thin material) of increasingly large diameters the further it is pushed into the work.

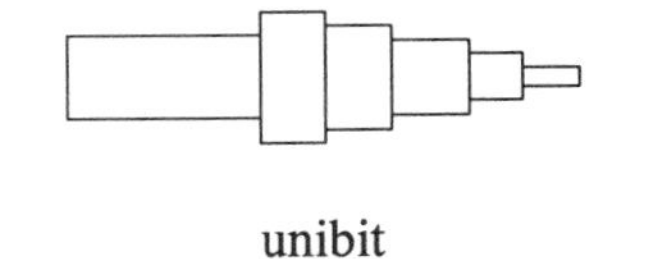

unibit

uniform load – a load that is evenly applied so that its weight is distributed and it is secure in its position.

union – a plumbing fitting consisting of two halves, each having female pipe threads into which lengths of pipe can be screwed. The two halves of the union can be joined by means of a nut on one half which engages threads on the other half, pulling the two seating surfaces into contact as it is tightened. Unions are available in most materials commonly used for plumbing piping as well as in nonconductive (electrically) material for use in situations where galvanic corrosion may take place; 2) a trade organization which workers join to give them political or economic unity.

unit cost – the cost of a single item. The purchase of materials is often based on the cost of one item, such as one pipe fitting, which is multiplied by the number of units (pipe fittings) needed to get an extended cost for the order.

unit heater – a space heater for heating a limited area of a structure, such as one room of a house or a small section of a manufacturing bay.

unit masonry – 1) concrete block; 2) a structure built using concrete or other masonry blocks with or without mortar.

unit rise – 1) the number of inches that a roof slope rises vertically in 12 inches of horizontal distance; for example, a 5⁄12 or 5:12 roof pitch means that the slope rises 5 inches for every 12 inches of horizontal distance; 5 is the unit rise; 2) the vertical distance from the top one stair tread to the top of next on a stairway.

riser
tread
unit rise
stair stringer

unit rise (2)

unit run – the 12 inch level distance over which the roof rise is measured to obtain the roof pitch; for example, in a 5⁄12 or 5:12 pitch, the slope rises vertically 5 inches in each 12 inches of horizontal distance; 12

universal joint

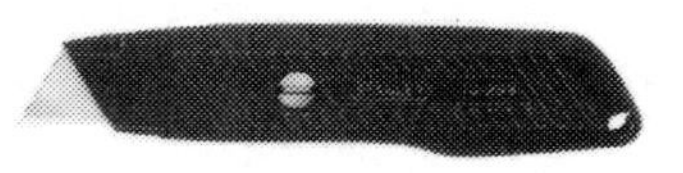

Courtesy: The Stanley Works

utility knife

inches is the unit run. In the English system of measurement, the unit run is always taken as 12 inches.

universal chuck – a lathe chuck with three jaws that move simultaneously, as in the combination chuck, except that the jaws on the universal chuck are shaped to permit gripping the work from the inside or the outside.

universal horizontal milling machine – a milling machine with many combinations of adjustments for different tasks. Complex components such as gear wheels are cut on this type of machine as well as metal parts which are machined, shaped and bored on the milling machine.

universal joint – an adaptor or device that is pinned in two places at 90 degrees to each other so that lateral movement can be taken up in two planes; a universal joint is used in such applications as drive shafts and socket wrench extensions.

universal junior indicator – a lever-type indicator which gives readings on a dial calibrated to gradations as fine as one-thousandth of an inch. It is used by machinists to make precision measurements.

universal vernier bevel protractor – an instrument used in machine shops to measure and lay out precise angles in degrees and minutes. It consists of a graduated dial mounted on a base, and an adjustable blade that can be rotated through various angles that can be accurately read from the dial.

unrippable rock – rock that can only be removed with explosives.

unsuitable material – 1) any material that is inadequate for a specific purpose, such as interior paint on an exterior surface; 2) soil that has absorbed too much water to provide adequate support for a road section. Such soil must be dried before construction can proceed.

unsuitable soil – soil structure that is unstable for the type of building to be done. Creatively designed and executed foundations may be required to offset the inadequacies of the soil.

upright – a vertical structural member.

upset – to deform so as to broaden, as in deforming the head of a nail or a bolt.

upset welding (UW) – an electric resistance welding process with pressure applied to the pieces during the heating process. Used in wire mills to join lengths of wire.

upstream – ahead of something, such as on the suction side of a pump.

urban – relating to or constituting a city.

urban renewal – refurbishment or rebuilding of urban areas.

urea resin glue – a highly water-resistant glue that comes in powder form and is mixed with water; it has high strength when used in tight-fitting joints.

urethane varnish – plastic-based varnishes that are hard and durable. Used on wood in high wear areas, such as floors.

urethane varnish, moisture-curing – an oil-free varnish that cures by reacting with the moisture in the air. Used in damp locations.

useful life – the life expectancy of any system, part, material or structure, such as the life expectancy of a paint or other coating before repainting is necessary for appearance or protection.

U.S. gallon – a liquid measure equal to four quarts, 128 ounces or 3.785 liters.

utilities – the supplies of water, gas and electricity to a building which make it habitable and/or possible to perform necessary household or business activities.

utility knife – a knife with a short, sharp, replaceable blade. The blade may be fixed or retract into the handle.

utility plug – a section of dirt left around a utility line by a trencher to prevent damage to the utility line.

U-value – a unit of measure of heat transfer through a material of known thickness determined by the number of Btu lost per square foot per hour, assuming a 1 degree F difference between the sides of the material. The lower the U-value, the greater the insulating value of the material.

UV inhibitor – a material that limits or blocks ultraviolet light.

V: vacuum to vulcanize

vacuum – 1) negative pressure or air pressure below that of atmospheric pressure; 2) a void.

vacuum brazing – brazing performed in an enclosure in which the atmospheric pressure has been reduced, creating a partial vacuum.

vacuum breaker – a device that permits air to enter a system when a vacuum begins to form in a tank or piping system. *See Box.*

vacuum gauge – a device which measures the vacuum within a container or enclosure.

vacuum pump – a device for pulling air, gases or steam out of a space in order to maintain that space at a negative pressure. Used for vacuum jacket pipe insulation, vacuum brazing or other vacuum processes.

valance – a horizontal panel across the top of a window for the purpose of concealing the top of window curtains, shades, drapery hardware or lighting fixtures.

valance cooler – a convection type cooler that is mounted at a wall-ceiling intersection. It uses chilled water passing through finned tubes to cool the air around the tubes, which is then circulated into the room.

valance lighting – indirect lighting concealed behind a panel valance.

valley – the inside corner junction where two roof slopes meet. *See Box.*

VA loan – a government-backed home loan program, managed by the Veterans Administration, which makes no-money-down loans available to veterans. The program also offers lower interest rates and buying costs than conventional loans, but is limited to an amount set by the VA. The government does not provide the money, but rather guarantees the lender against loss in the case of foreclosure. The purpose of the program is to make buying a home easier for veterans as a reward for their service.

value – the monetary worth of something, or its equivalent worth in goods, services, utility or desirability.

value engineering – a design which evaluates and takes into account all of the functions to be performed by the object being designed, so that it performs in the most cost-effective and efficient manner.

valve – a device for controlling or stopping the flow of a liquid or gas in a piping system. There are different types of valves designed for particular uses and pressure/temperature conditions. See specific valve types.

valve box – a box-like structure that protects and provides access to underground pipe valves. The valve box is fitted around

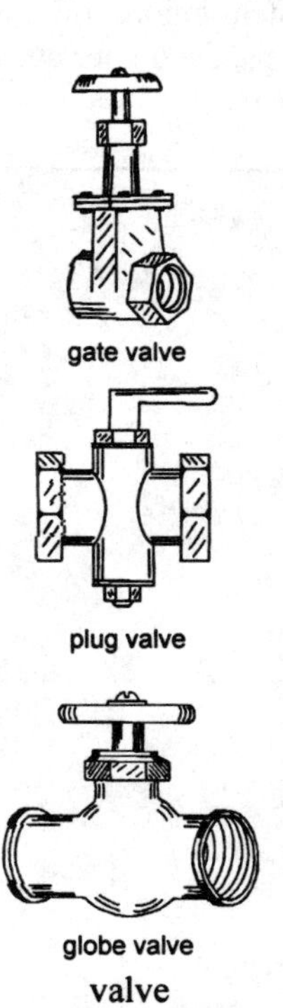

valve

Vacuum breaker

A vacuum breaker is a type of valve that allows air to enter a system, such as a piping system, when the interior pressure drops below atmospheric pressure, preventing a vacuum from forming inside. A vacuum may form when liquid is discharged, drained or otherwise removed from a pipe, piping system or tank without a means for air to enter and replace it. As the liquid is removed, the pressure inside the system drops. If the internal pressure drops below the atmospheric pressure surrounding the system or tank, an external pressure will be exerted on it. If and when that pressure exceeds the capacity of the vessel, the tank or system will collapse inward. The vacuum breaker permits air to enter the system before this can occur. When the inside pressure drops, the higher outside air pressure causes the valve to open.

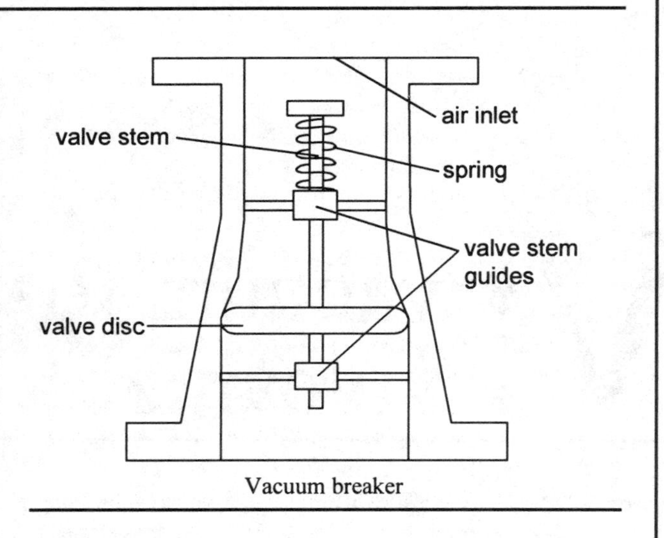

Vacuum breaker

the valve and then the excavation is backfilled around the box. The valve box may have a removable cover, or no cover at all.

vanadium – a metallic element with the atomic number 23, atomic symbol V, and atomic weight of 50.942. It is one of the hardest known metals and is used in steel alloys.

vane axial fan – an air-moving device with fan blades mounted on a shaft that is turned by a motor. The propeller-like blades are designed so that the air flow is along the axis of the shaft. It is mounted in a section of duct that has internal vanes extending from the inside surface of the duct into the airstream. The vanes reduce turbulence in the air stream and direct the air flow along the duct.

vanishing point – 1) the distant point where parallel lines appear to join; 2) the point chosen in a perspective drawing where the horizontal lines converge.

vanity – the floor-mounted bathroom cabinet in which a sink or lavatory is mounted.

vapor – the gaseous state of fluid; small particles of matter suspended in the air.

vapor barrier – a water-resistant material used on a surface or structure to stop or retard the movement of moisture through the surface. Commonly used in walls under drywall, between concrete and wood flooring and under concrete slabs.

vapor barrier, enveloping – a vapor barrier, such as plastic sheeting, that is wrapped around an entire wall.

variable – subject to change or variation.

variable rate mortgage – a loan in which the interest rate varies with a financial index, such as the lender's cost of funds. The loan payments are adjusted at prearranged intervals, such as every six months or once a year. Variable rate loans usually start at a lower interest rate than fixed-rate loans, but cap at a higher rate over time if the interest rates go up.

variable speed drill – an electric drill motor with a variable speed control as part of the trigger; the more pressure that is applied to the trigger, the faster the motor turns.

variance – an approval to deviate from the usual rule, such as in building plans, regulations, specifications, guidelines or zoning requirements. For example, a vari-

Valley

The term valley is used in construction to describe the depressed angle formed by the meeting of two inclined sides of a roof, as well as all the accompanying materials used in building that part of the roof. The *valley rafter*, at the base of the intersection of two sloping roofs, is the main structural support for the valley and the abutting jack rafters or valley jacks. Valley rafter stock is usually one or two sizes bigger than common rafter stock, or it is made up of two common rafters nailed together to obtain the extra needed strength. The *valley jacks* are the rafters that run from the base of the valley up to the hips, or tops of the adjoining slopes.

There are different roofing methods used in covering valley roofs. There are the *closed-cut valley*, the *woven valley* and the *open valley* methods. Both closed-cut roof valleys and woven-roof valleys have shingle courses that meet and overlap to cover the valley. These roofing methods are used with flexible roofing materials, such as asphalt shingles. In the open valley method, the flashing is exposed because the shingle courses stop before they meet at the valley. This roofing method is used with rigid types of roof covering, such as shake shingles.

Valley flashing is installed in the valley of the roof for added protection against leakage and to channel water runoff down the roof. Because of the slant of the roof, drainage is concentrated along the valley, which makes it the most likely place to leak. Valley flashing must be made of a durable material that can withstand exposure to the elements, and it must have a longer life than the roof covering, because it is installed before the roof covering is put on and cannot easily be replaced. Replacing flashing is best done in conjunction with replacing the roof covering. Galvanized steel, laid over 36-inch-wide roofing felt (laid in the same direction as the flashing) is commonly used for valley flashing.

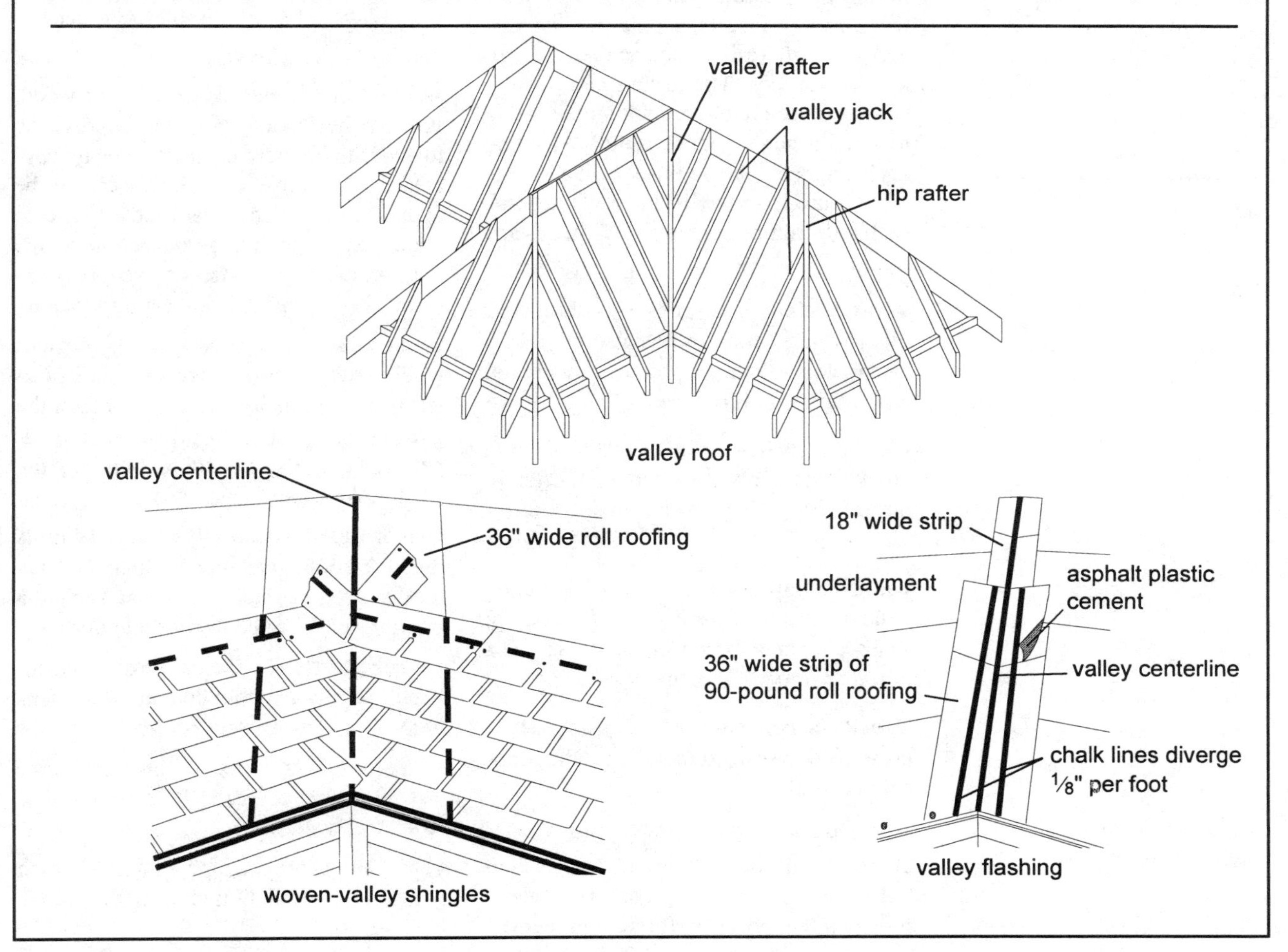

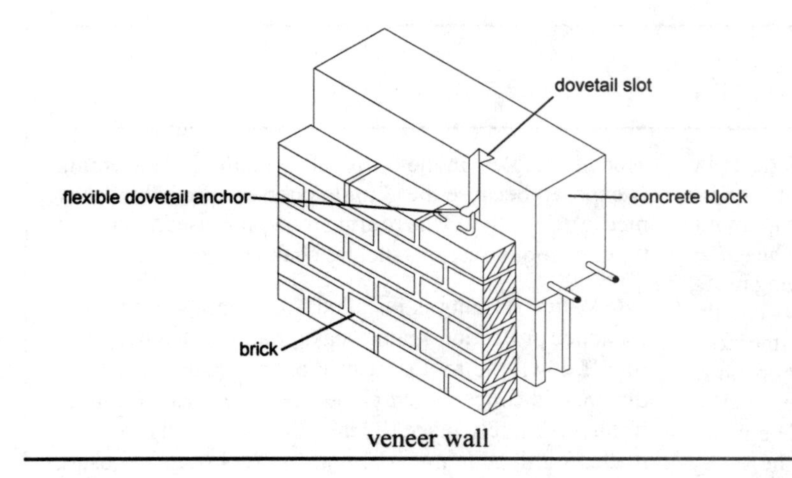

veneer wall

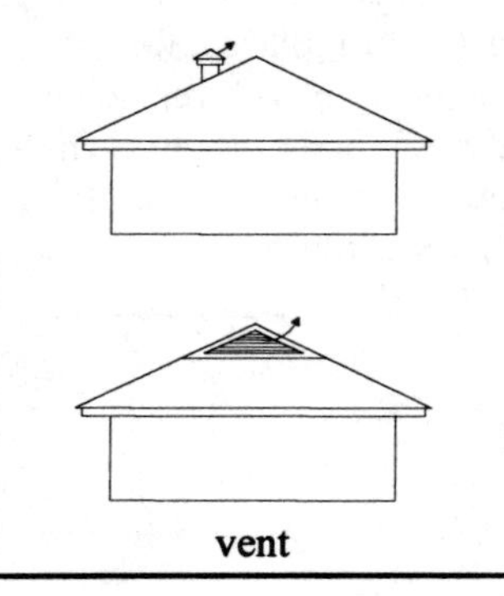
vent

ance would be required to build a structure closer to a property line than is permitted by local building codes.

varnish – a clear, waterproof coating made of a resin or a mixture of resin and a drying oil. There are natural resins made from trees or insects, such as the lac bug, and synthetic resins, such as urethane, vinyl and amides. The drying oils solidify when exposed to air, allowing the coating to dry or harden more rapidly. Natural drying agents include linseed oil and tung oil. It is used to beautify and protect wood, such as furniture, trim, window sashes and floors.

varnish, oleoresin – resins and drying oils in a turpentine or petroleum vehicle. Used most often to protect the natural wood finish on furniture.

varnish, spirit – resin dissolved in a quick-drying solvent. Used most often to protect the natural wood finish on furniture.

vault – 1) an arched or dome-shaped structure usually forming a roof or ceiling; 2) a secure area for storing valuables; 3) a burial chamber, often underground.

V-belt – a belt with a V or truncated V cross section used to transfer rotating motion via pulleys.

V-block – a metal block with a V machined out of the top surface. It is used to hold cylindrical objects, such as dowels or rods, while work is performed on them.

veeing a rafter – cutting the top of a valley rafter in an inverted shallow vee along its length so that the plane of the top of the valley rafter matches flush with the abutting rafters.

vehicle – 1) a conveyance or agent of transmission; 2) the liquid portion of a coating in which other materials are suspended for application.

velocity – speed of motion.

velocity pressure – the force of the air in an HVAC duct, taken in the direction the air is moving. The pressure is measured in inches of water using a manometer, which produces a finer measurement unit than pounds per square inch.

velometer – a device that measures air velocity in an HVAC system.

vena contracta – a restriction in a fluid system.

vendee – the buyer in a sales transaction.

vendor – one who sells.

veneer – a thin surface covering of wood, vinyl, stone, brick, etc. used to improve the look of the furniture or structure being covered. For example, an oak veneer may be applied to a pressed board table top, or a brick veneer applied as an accent to the outside of a house. Masonry veneer is not counted as part of the load-bearing structure.

veneer base – a special type of gypsum wallboard used as the base for veneer plaster. It has a moisture barrier to protect the gypsum core and an absorbant surface paper especially designed to bond to the plaster. Also called blue board.

veneer block – decorative masonry units that are about two inches thick and are used as a veneer facing material, such as a brick veneer applied to a wood building.

veneer plaster – a special plaster formulated to apply in a thin coating. It hardens to an abrasion-resistant surface.

veneer wall – a nonstructural masonry wall providing a facade for a structure or structural wall.

vent – an opening, a pipe, or a duct which permits the circulation of air or the escape of gases.

Vented Suppressive Shielding (VSS) – a proprietary system of blast mitigation and shrapnel containment. VSS can be used to encapsulate an object that is likely to explode so as to contain and lessen the effects of the explosion on the surrounding environment.

vent flashing – see roof jack.

vent header – a vent pipe into which two or more vents connect. The vent header then connects to the vent stack, or vertical pipe vent, which vents to the outside of the building.

ventilation – the supply, circulation and removal of air within an area or enclosure.

vent pipe – any pipe that acts as a vent for an area, system, appliance, etc.

vent stack – the vertical portion of a vent pipe in a building drainage system, tank or other system vented to the atmosphere.

vent system – 1) the system of plumbing drainage vents in a building; 2) a system designed to relieve pressure from a tank, vessel or piping network; 3) a line to route fluid or gases discharged from a relief valve or other device to the desired discharge location or outlet.

venue – the place or county in which an alleged violation of the law has taken place or where a legal cause is brought to trial.

verge – 1) the edge of a roof covering projecting over the gable of a roof; 2) to be on the edge or border.

verify – to gather confirming evidence; to establish proof.

vermicular – designs made of wavy line patterns simulating the tracks of worms.

vermiculate – to decorate with wavy impressed patterns.

vermiculite – a mineral, related to mica, that expands when heated to form a lightweight, highly water-absorbent material that has insulating properties. Used in thermal insulation and as an aggregate in concrete.

vernier – a movable graduated scale with fine gradations, mounted alongside a fixed graduated scale for the purpose of making very fine adjustments. A vernier is based on the principle that the human eye cannot determine the exact distance between two lines, but it can determine when the lines coincide so as to form a continuous line.

vernier caliper – a slider and a fixed projection mounted on a precision ruler. This instrument can be used to take both inside and outside measurements. There are precision gradations marked on the slider which locks into place with a screw adjustment.

vernier depth gauge – an instrument that consists of a precision ruler that can be slid through a base designed to span a hole. There are precision gradations marked on the base and on the ruler, which locks into place with an adjustable screw.

vernier height gauge – a gauge that consists of a slider that can be fixed along the length of a precision ruler. The slider has a projection that has been precision ground and can be adjusted with a fine screw thread to take an exact height reading or to mark an exact vertical location.

vertex – 1) the top or summit; 2) the point of intersection of two or more lines or curves as in the intersection of two tangents to a curve.

vertical – 1) upright; 2) perpendicular to a level surface.

vertical application – the installation of wallboard panels with the length or long edges parallel to the studs.

vertical board siding – see board and batten.

vertical curve – the rate of change in the grade in soil excavation work. Also called a profile curve.

vertical pipe – any pipe that is installed upright, in a position 45 degrees or less from vertical.

vertical position – a welding position where the weld runs up and down in a vertical or near-vertical line.

vertical turret lathe – see turret lathe.

vessel – a hollow or concave container designed to carry or contain liquid or solid matter, such as a cup, jar or tank.

vibration – a rapid, back and forth cyclic movement.

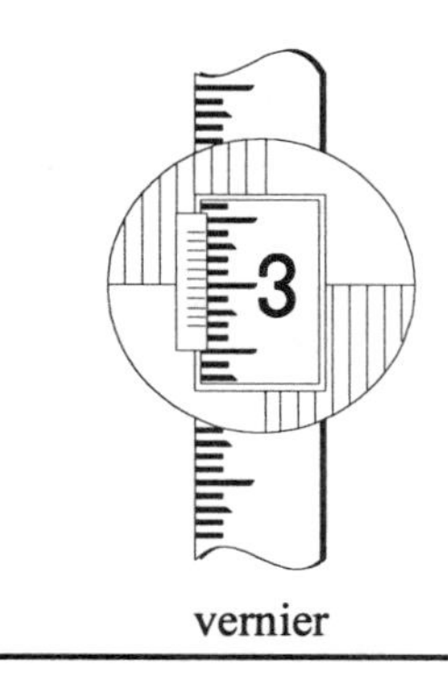

vernier

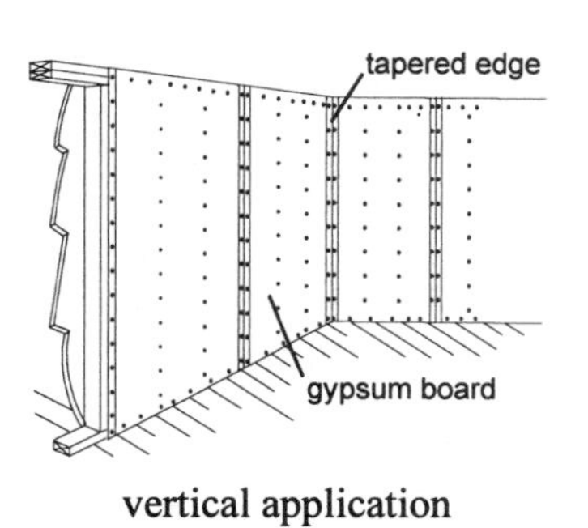

vertical application

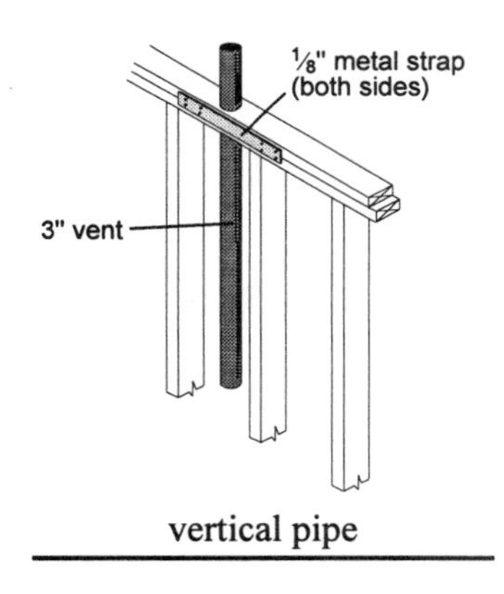

vertical pipe

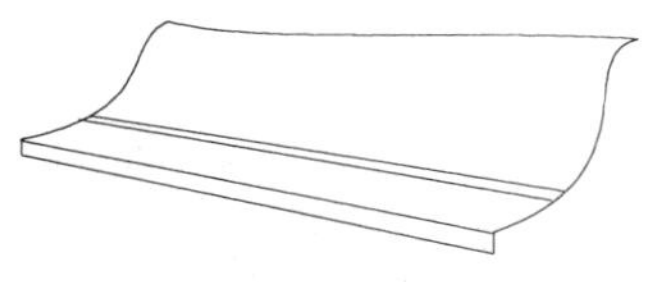
vinyl dual drip cap (VDDC)

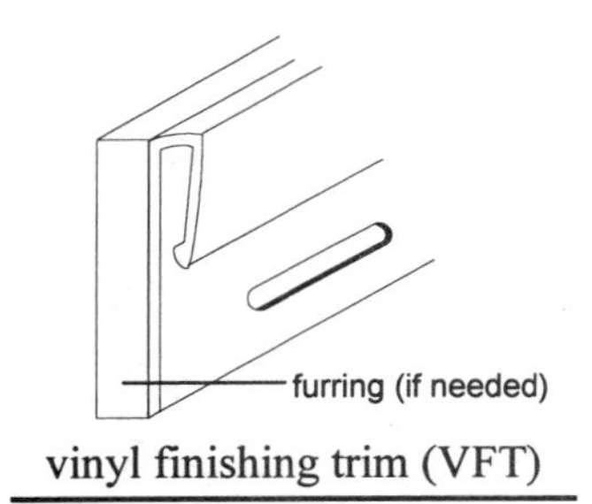

vinyl finishing trim (VFT)

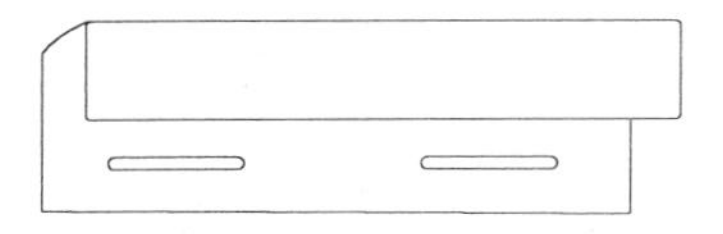
vinyl siding J-channel (VJ)

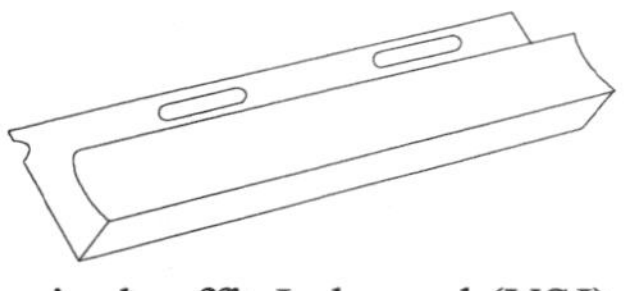
vinyl soffit J-channel (VSJ)

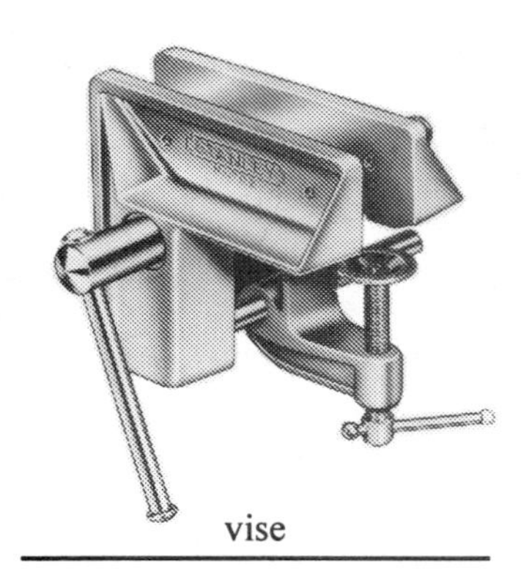
vise

vibrator – 1) a device that creates a continuous shaking motion; 2) a machine used to agitate and consolidate wet concrete.

vibratory plate – a heavy plate with a vibrator used for soil compaction.

vibratory roller – an earth-compacting roller that vibrates as it rolls.

view – 1) a scene or prospect; 2) the subject of a drawing or set of plans portrayed from a particular angle, such as a side view, a front view, or a plan view (the view looking down at the top of the subject).

vinyl – a plastic formed of, or using, univalent compounds of ethylene. Used in making a variety of materials including floor coverings, siding and piping.

vinyl-coated sinker – a construction nail with a vinyl cement coating to lubricate the nail as it is being driven, and to serve as an adhesive to increase its holding power. The holding power of a coated nail when driven into fir is double that of a noncoated nail of the same size.

vinyl dual drip cap (VDDC) – a type of flashing used over window and door headers when installing vinyl siding. The flashing redirects rain water so that it does not run into the door or window.

vinyl finishing trim (VFT) – a type of flashing used under windows to trim the edge and provide a lock for the installation of vinyl siding panels.

vinyl flooring – a floor covering made primarily of polyvinyl chloride (PVC). Vinyl flooring comes in sheets and one-foot-square tiles, and is economical, durable and relatively easy to install.

vinyl siding – building exterior siding boards made from vinyl. Vinyl siding is durable and needs little maintenance. However, colors may fade over time when exposed to direct sunlight.

vinyl siding J-channel (VJ) – a vinyl siding accessory used for door and window trim.

vinyl soffit J-channels (VSJ) – a vinyl siding accessory used for joints on vinyl soffit coverings.

vinyl starter divider (VSD) – vinyl siding flashing used along the edges of the vinyl panels.

vinyl tiling-in bead – a preformed vinyl strip installed with adhesive around the lip of a tub prior to the installation of the tile. It forms a seal against water penetration.

vinyl vertical base drip cap (VDC) – vinyl siding flashing used over doors and windows.

virgin growth – naturally-occurring forest with the original growth of timber. The trees vary in size, age and density.

virgin soil – soil in its natural and undisturbed state.

viscosity – a measure of a liquid's resistance to flow. The higher the viscosity, the thicker and slower the liquid flows.

vise – a mechanism that can be mounted on a bench or other stable surface and used to hold work by clamping the work between adjustable jaws. The jaws are usually opened or closed by a threaded rod that is turned like a screw, bringing the jaws together against the work or releasing them. The work may be further supported by screw type clamps above or below the jaws.

Visegrip – a trademark name for a type of locking pliers. Visegrips are used to clamp parts together, to hold a fastener and keep it from turning or similar operations where a tight grip is needed.

vitreous – of, or relating to, glass; glass-like.

vitrification – the firing of clay to fuse the grains and close the pores of the clay.

vitrified – glazed and fired clay, often in the form of clay pipe used for conduit or underground drainage.

vitrified clay conduit – electrical conduit, similar to vitrified clay pipe, used for underground wiring.

vitrified clay pipe – drainage pipe made from fired clay. It was used for underground piping systems due to its corrosion resistance, but the more durable PVC hub and spigot pipe is now generally preferred over vitrified clay.

VOI – an outside corner post for vinyl siding which accommodates drop-in insulation.

void – 1) a hollow space, such as the hollow space in the core of gypsum wallboard; 2) having no legal standing; invalid.

volatile – 1) evaporating quickly; 2) characterized by rapid unexpected change; explosive.

volt (V) – a measure of the electromotive force that causes current to flow through a circuit, analogous to system pressure in a hydraulic system.

voltage drop – the loss in voltage between the electrical supply and the electrical load because of resistance in the wire.

voltage regulator – a control device used to maintain voltage at a set value.

volt-ampere – a unit of apparent power in an AC electrical circuit, and a unit equal to a watt in a DC circuit.

voltmeter – a meter that indicates the number of volts across electrical terminals. Used to test batteries, electrical circuits and receptacles to determine if they are live.

volume – the cubic capacity of a three-dimensional figure. Volume calculations are essential for such purposes as determining the cubic yards of concrete needed for a pour.

volute – 1) spiral shaped, such as the internal vanes of a centrifugal pump; 2) a spiral or scroll-shaped architectural ornament.

vomitory – the cross-aisle in an auditorium or stadium.

VON – an outside corner post for vinyl siding that is noninsulated.

voussoir brick – wedge-shaped brick formed for use in arches. The shape of brick results in face joints that radiate from a center.

vulcanize – to heat-treat rubber to make it resilient and elastic.

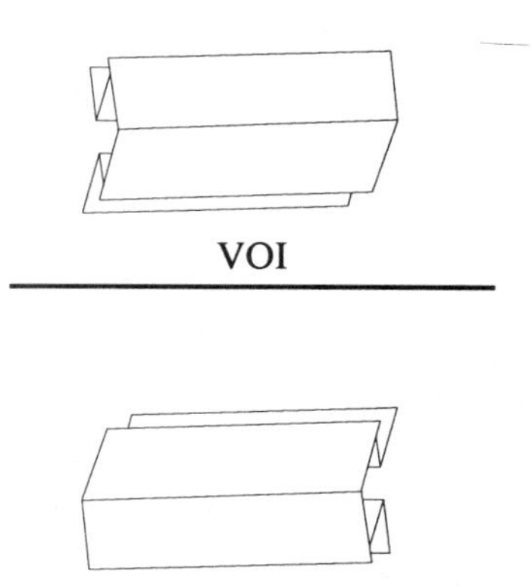

VOI

VON

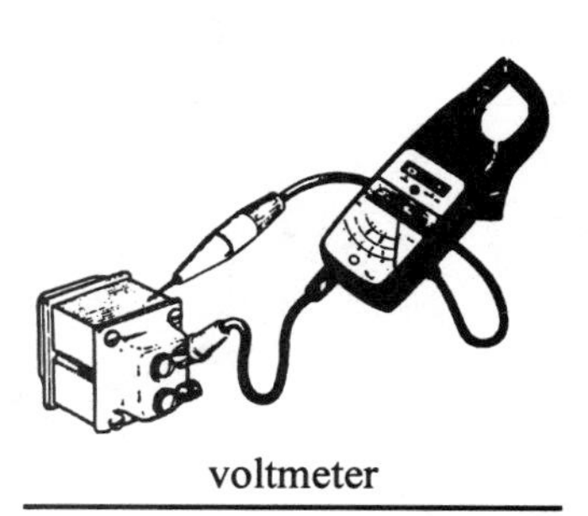

voltmeter

W: waferboard to wythe

waferboard – panels made of wood chips bonded together with resin and compacted under pressure into sheets used for sheathing and paneling.

wainscot – a wall finish for the lower section of an interior wall that is different from the top of the wall section. It is often paneled or tiled, with a finishing molding covering the seam or joint between.

waiver – 1) intentionally giving up a right, claim or privilege; 2) a legal document which is evidence of the giving up of a right, claim or privilege.

waler – reinforcing structural members used to brace concrete forms or piles. They are of 2 x 3 or larger lumber and are installed horizontally on the outsides of the forms, spanning between studs and tied through the forms using wall or form ties. Also called wale, whaler or ranger.

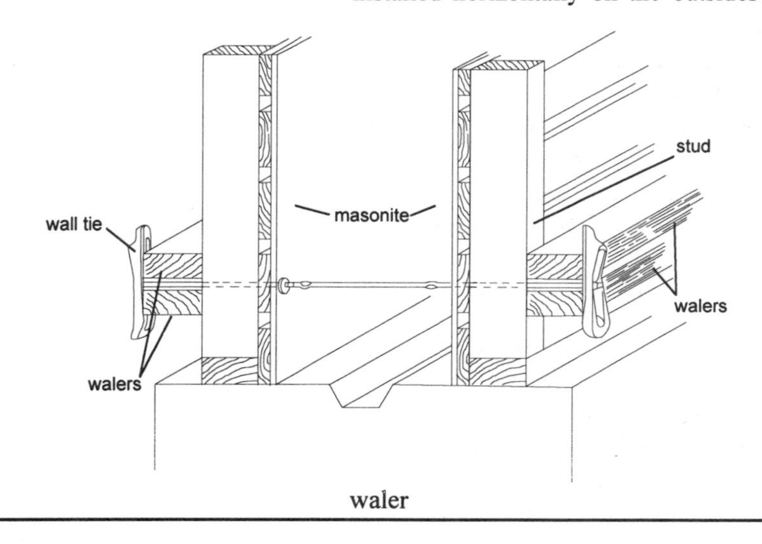

waler

wall – a vertical interior or exterior structure used to enclose an area or to partition off one section from another. It may be an independent structure such as a fence, part of the loadbearing support for a building, or a nonbearing division within a building.

wallboard – a panel designed as wall covering. The panels are made of a variety of materials, such as gypsum drywall, which is the most commonly used interior wall paneling, wood or wood products. See also drywall.

wall box – an electrical box for mounting an outlet or switch in or on a wall, ceiling or other surface.

wall box setback – the distance, measured from the outer surface of a wall, that the front edge of an electrical box is permitted to be set back. If the setback is exceeded, the electrical connections in the box will no longer be totally contained, as other surfaces will be exposed, such as the wall covering edge surrounding the box.

wall cabinet – a cabinet designed to be mounted on a wall and supported by fasteners attached to the structural members of the wall. Wall cabinets are generally elevated off the floor.

wall coping – a preformed cap for masonry walls that has a sloped or peaked top for water runoff.

wall fan – a fan mounted in a wall designed to pull air through the opening in the wall and vent it to the outside. Used in kitchens and baths.

wall footing – a widened section at the base of a wall foundation that bears against the soil and distributes structural loads directly to the soil.

wall furnace – a gas heating system mounted in a wall.

wall heater – a heating system, mounted in a wall, designed to heat the surrounding area.

wall-hung closet tank – a water closet tank supported by a wall bracket and connected to the bowl by a pipe elbow. This type of tank is rarely used today, having been replaced by the combination water closet, in which the tank is mounted on the bowl, and the one-piece water closet. These are easier and faster to install than wall-hung tank units.

wall-hung lavatory – see lavatory, wall-hung.

wall molding – a lightweight section of metal extrusion, shaped like an angle iron, used to fasten the edge tiles of a suspended ceiling to the adjoining walls.

wall plate – 1) a horizontal structural member anchored to a masonry wall. Other structural members may be supported from the wall plate. Also called a head plate; 2) a protective cover over an electrical outlet box which is attached to the box or directly to the wall in which it is mounted.

wall size – a coating used to seal porous walls. Also called sizing.

wall tie – a metal piece used to tie masonry wythes to each other or to another structure. They come in a variety of designs to suit the masonry and type of structure. See also tie.

walnut – a medium to dark brown hardwood used primarily for furniture.

wane – bark on the edge of cut lumber, in many cases counting as defective material.

warp – 1) a variation from straightness; 2) a lumber defect in which the board is twisted due to improper seasoning.

warranty – 1) an assurance of the quality of workmanship in a product and the maker or builder's written guarantee to repair or replace defective parts; 2) a guarantee of good title and the right to undisputed possession of an estate.

wash – a sloped layer of mortar, applied on the top of chimney masonry surrounding the flue, which is designed to route water runoff away from the flue.

wash coat – a very thin coat of varnish, shellac, paint or other covering used as a sealer to prevent color bleeding through the final coating.

washer – a flat circular collar that fits under a bolt head or nut to widen the bearing surface and/or increase the hold of the fastener through friction.

wash primer – a thin coat of paint primer.

waste pipe – see fixture branch.

waste stack – the vertical pipe in a drainage system that receives discharge from fixtures such as sinks, bathtubs and showers, but not from water closets.

waste vent – see stack vent.

water blaster – a device which shoots water out of a nozzle at 2000 to 4000 psig. It is powered by a pump that raises the pressure approximately 50 to 100 times higher than the water main supply pressure. The nozzle is connected to the pump by a high-pressure-rated hose. Water blasters are used to clean cracked and peeling paint and other undesirable coatings from wood, masonry and other surfaces.

water blasting – the cleaning of a surface using a high-pressure water jet.

water chiller – a method of cooling water by passing it over tubes containing a refrigerant or through other types of heat exchangers. This method is used to cool water for drinking fountains, as well as for chilling water for use in cooling system heat exchangers to cool air.

water closet – a fixture designed to receive human waste and flush it into the plumbing waste system to be conveyed away. Also called a toilet. *See Box.*

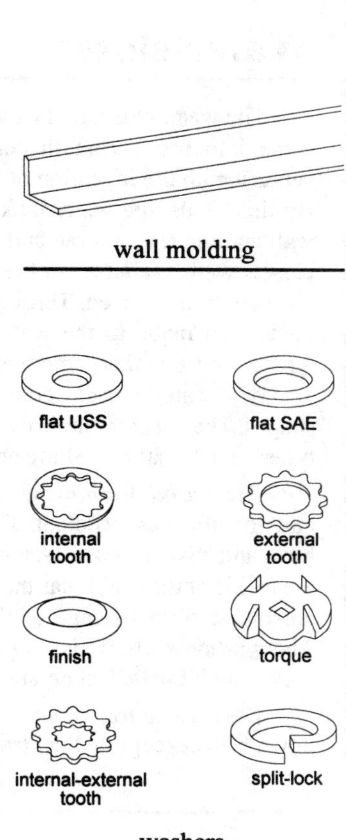

wall molding

washers

Water closet

The water closet that we are familiar with today first came into use during the early part of the twentieth century with the invention of the wall-hung water closet. By the 1920s, the water tank had come down from the wall and the reverse-trap and siphon-jet flush tank water closets were available. In the 1930s the one-piece water closet was introduced. Through the years, improvements have been made to the water closet design to make it quieter, more sanitary, more efficient and easier to repair. Today's water closets come a variety of colors and shapes. They are similar in design, but offer four different types of automatic flushing action.

The *washdown bowl water closet* is the oldest type, and now the least efficient. The trap is in the front of the bowl and it is flushed by water flowing down from around the inside of the rim. It has the smallest water area and the narrowest flush passage of the available water closets. The washdown bowl is a two-piece combination water closet, with the tank separate from the bowl.

The *reverse trap water closet* is similar to the washdown bowl except that the trap is at the rear of the bowl, making the bowl longer and the water area larger. Because the bowl holds more water, it requires less water to flush and operates more efficiently and quietly. The reverse trap water closet is also a combination closet, with separate tank and bowl, and is flushed by water running down from around the inside of the rim.

The *siphon jet water closet* is flushed by a jet of water which is delivered directly to the trap, through a small hole in the bottom of the water closet, while water is simultaneously flowing down from around the rim. The combined water flow starts a siphon to empty the bowl through the trap. This type of water closet has a large water area and a wide flush passage, and operates more quietly than the first two. It is also a two-piece combination water closet.

The *siphon action water closet* is the most efficient and quietest of the various designs. It is similar in action to the siphon jet water closet, but has a higher capacity bowl and larger water surface. It is the only one-piece design, and also the most expensive.

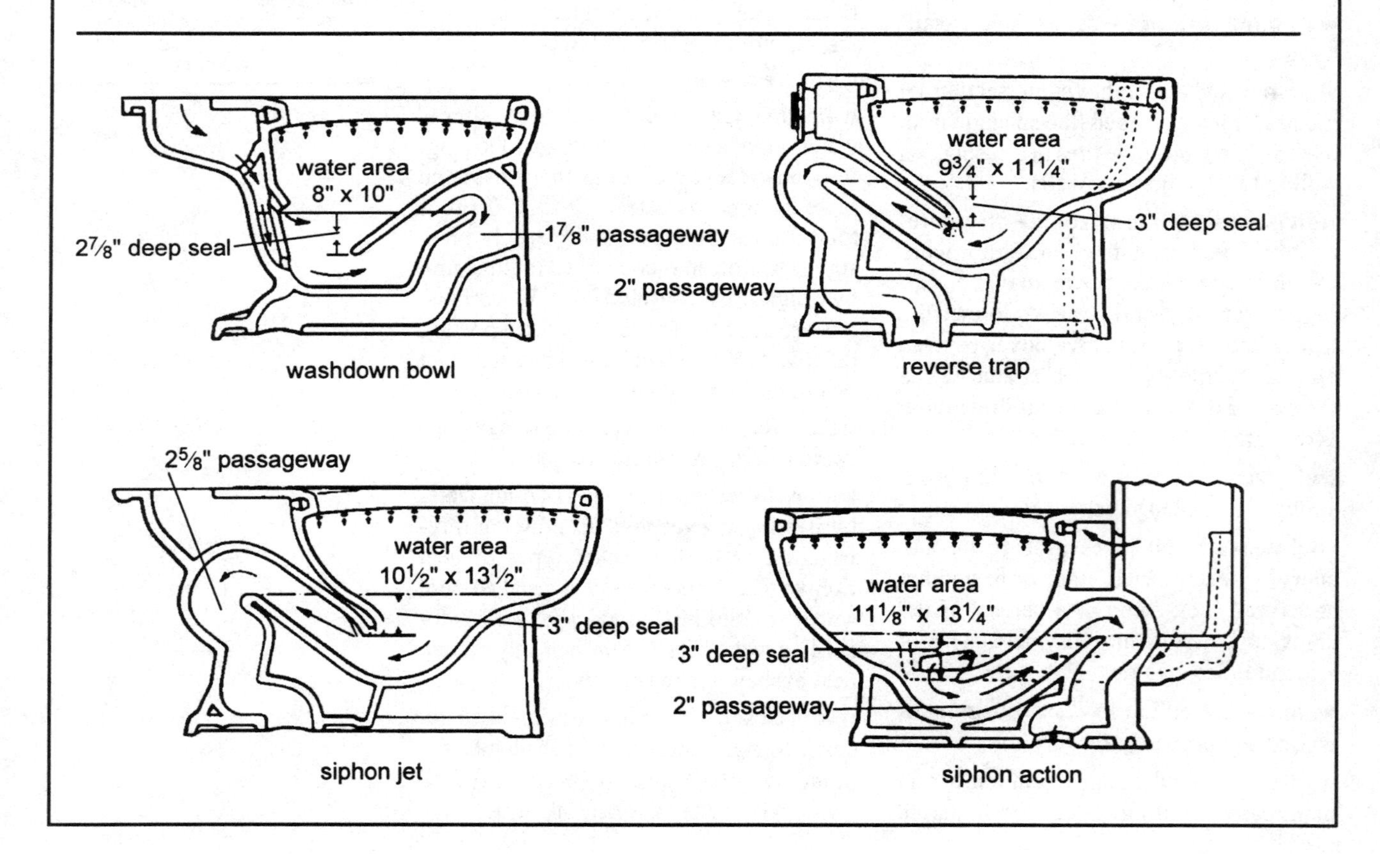

water closet ell – the elbow fitting between a wall-hung water closet tank and the water closet bowl. Also called a flush ell.

water-cooled condenser – a condenser that uses water as a cooling medium to cool refrigerant or steam, depending on the application. They are used in refrigerant plants for cooling compressed refrigerant, in steam power plants, and in steam propulsion plants, such as those on board ships. They condense steam exhausted from turbines and reciprocating engines to increase efficiency.

water gate – a gate used to control water flow in a sluice or dam overflow.

water hammer – a water surge or rapid flow change of water in a steam system which creates noises. When water is carried along in a gas (air), at the same velocity as the gas, and an obstruction such as a change in direction is reached, the mass of the water prevents it from smoothly turning the corner. It strikes the pipe at the turn and creates a hammering sound. Water hammers can physically damage the pipe and cause premature pipe failure.

water heater – an appliance for heating and storing water. It consists of an insulated tank, into which water is fed from the building water system, and a system for heating the water. The heat may be generated by a gas flame, electric heating elements or steam. Some water heaters consist of coils within a wood or coal stove. Water heaters are available in a variety of capacities, with the smallest being 20 gallons. A variable thermostat controls the temperature of the water by cycling the heat source within preset limits.

water main – a large diameter piping system used to transport water from a water district treatment plant to the areas of use. Individual water lines run from the main to homes and buildings.

water meter – a device for measuring the flow of water in a building supply system in order to record consumption.

waterproof – impervious to penetration by water.

waterproofed portland cement – portland cement with waterproof materials, such as methyl metracrylate polymer acrylic, used to coat the cement.

waterproofing – applying a moisture-proof barrier, such as plastic sheeting.

waterproof switch – an electrical switch which is sealed against the elements for use in an outside location. The switch may be made explosion proof as well by the use of gas-tight seals.

water-pump pliers – pliers with jaws that adjust for a range of openings. The jaws are at an angle to the handle to provide better leverage and grip when turning an object. Also called groove-joint pliers.

water resistant – the property of not absorbing water easily.

water-resistant panels – gypsum wallboard panels that have water-resistant asphalt compounds added to the gypsum core. They are also covered with a water-resistant paper so they can be used where moisture is present, such as in a bathroom. The joints are taped with fiberglass mesh tape and sealed with water-resistant joint compound. Also called WR panels, moisture-resistant panels, or greenboard.

water retentivity – the property of mortar to retain water and not lose it to adjacent masonry units.

water service – the potable water supply (drinking water) to a building from the water main.

water service pipe – piping from the water main to the building being supplied.

watershed – 1) a dividing point at which water runoff flows into two different drainage areas; 2) a land area that collects water.

water softener – an appliance installed in a water system which removes minerals from the water.

water spot – a spot caused by minerals remaining on a surface after the water they were dissolved in has evaporated.

water stain – a water-base wood stain.

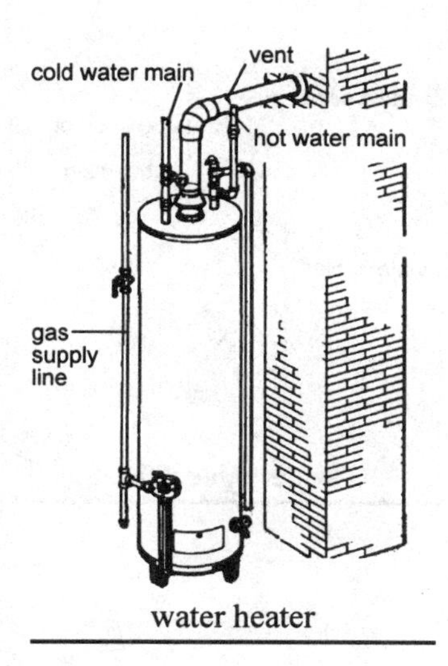

water heater

water-pump pliers

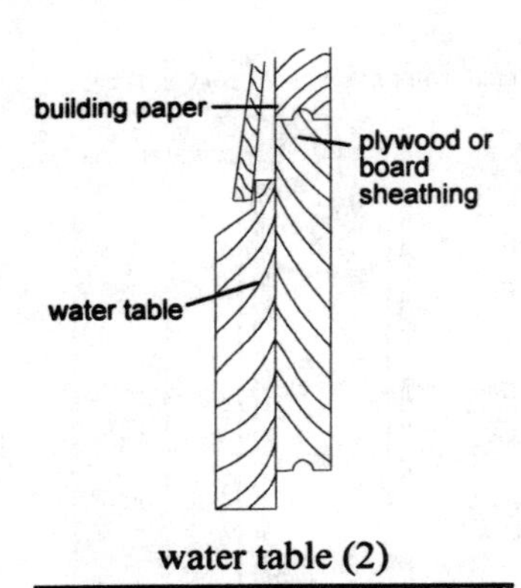

water table (2)

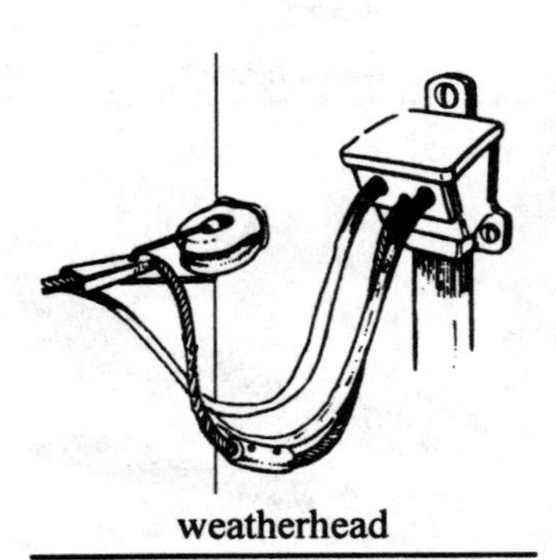
weatherhead

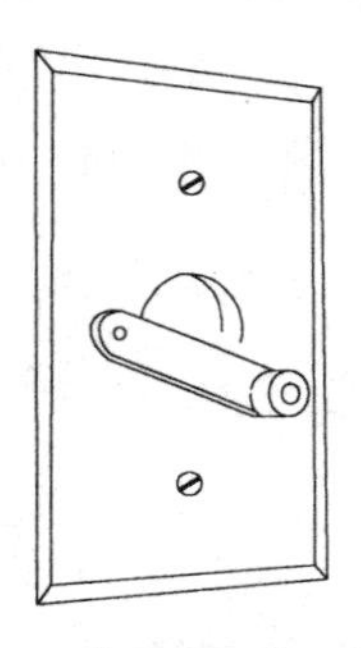
weatherproof switch

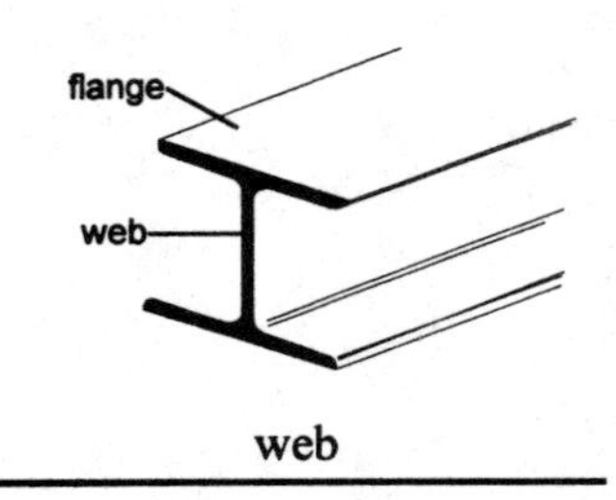

web

waterstop – rubber, plastic or similar material that is inserted in a construction joint to prevent the passage of water through the joint.

water table – 1) the distance down from the soil surface to the soil that retains moisture; 2) a horizontal siding member that projects outward a short distance at the bottom of siding. Its purpose is to direct water that runs down the siding away from the foundation or the wall.

water-thinned paint – a paint that uses water as a vehicle.

watertight – made or constructed to prevent moisture from entering.

watt (W) – the common measurement unit for electric power, named for Sir James Watt (1736-1819) who suggested that the unit of power be the rate at which a horse performs work. One horsepower is defined as 550 foot-pounds. One foot-pound equals 1.356 watts, so one horsepower equals 746 watts. A watt, in terms of electricity, is equal to volts multiplied by amps in a DC circuit and in an AC circuit with a purely resistive load. One watt is the equivalent of 3.41 Btu per hour.

watthour meter – a utility company meter for recording power usage in a building.

wattmeter – a device for measuring the power in an electrical circuit.

wave soldering (WS) – a process where the parts to be soldered are passed through molten solder, soldering many parts at one time. This technique is used to solder components to electronic circuit boards.

wax sticks – sticks of colored wood filler wax used to fill and hide minor blemishes in cabinetry.

W-beam – a steel structural beam with wide flanges; the W stands for wide. Also called wide-flange beam.

weatherhead – a raintight conduit fitting for the service entrance head where overhead electrical power is brought into a building.

weatherproof electrical boxes – boxes that are designed for proper operation when exposed directly to the weather. Entrances into weatherproof boxes are through threaded connectors, and a gasket under the cover plate seals against moisture.

weatherproof switch – an electrical switch designed for outdoor use. It has a cover which seals it against the entry of moisture.

weatherstrip – a seal used around doors and windows which prevents drafts, dust, noise, and moisture from entering the building.

weave bead – a weld bead that is laid in a back and forth pattern along the weld.

web – the central vertical section of an I-beam or similar structural member; the top and bottom sections are called flanges.

web clamp – a woven fabric belt that can be placed around an object to hold it in place for gluing or similar work. It has a tightening mechanism, such as a lever, and a catch to maintain the tension on the belt.

web connection – a connection, usually made of steel, to the web of a beam.

wedge – a tapered or triangular-shaped piece of wood or metal used to brace or hold an object, such as a door, in place.

weep hole – 1) a hole or holes in masonry walls which permit the passage of water and prevent it from building up behind the wall and possibly undermining the foundation; 2) a hole or holes drilled in metal that is to be welded all around. The holes relieve pressure buildup as the welds are made.

weeping joints – mortar joints in which the mortar has been squeezed out slightly and left to set, rather than being troweled out or smoothed.

weir – 1) a small dam; 2) a raised section in a fluid flow path, such as inside a diaphragm valve. The flow in the valve is slowed or sealed off as the diaphragm is pressed against the weir to close the valve.

weld – to fuse metals together by melting the pieces at the interface where they are to be joined. Pressure and/or a filler material may also be used to aid in the fusion.

weld bead – the cooled weld metal that has been deposited in a weld joint.

weld brazing – a combination of electric resistance welding and brazing.

welded switch boxes – electrical boxes made of metal using welded construction.

welded-wire fabric – heavy wire laid out in a grid pattern and welded at the intersections to form a wire mesh mat that is used to reinforce concrete. It comes in sheets and rolls. Also called welded-wire mesh or wire mesh.

welder – 1) a person trained to do welding work; 2) a welding machine.

welder certification – the documentation which proves that a welder has the ability to make particular welds at a standard level of acceptance.

weld gauge – a gauge that shows the specifications of the upper and lower ends of the acceptable size tolerance band for a weld so that it can be measured and checked quickly. This is helpful when a number of welds of the same size and shape are being made. See also go no-go gauge.

weld groove – a groove, formed by the prepared edges of two pieces of metal that are to be joined, before the weld metal is deposited and the weld is made.

welding – applying heat and/or pressure and weld metal to join pieces of metal together.

welding helmet – a protective covering for welders that completely covers the face and has a deeply-tinted viewing plate over the eye section. It prevents injury to the face from heat and weld splatters as well as protecting the eyes from dangerous ultraviolet or infrared light and radiation. Also called a welding hood.

welding machine – a DC power supply or a transformer which converts AC to DC and provides the correct range of amperage for arc welding.

welding position – the position of the weld joint in relation to the floor on which the welder is working. The four basic positions are: horizontal, vertical, flat and overhead.

welding procedure qualification – documented successful weld application with a welding procedure, based on the qualification rules of the American Society of Mechanical Engineers or the American Welding Society. This certifies a particular range of variables for making acceptable welds. Some of these variables include the amperage range, welding position, weld rod material and base material thickness used in the application of a particular weld.

welding procedures – the documents that establish the parameters required for a specific type of weld in accordance with recognized codes and standards, such as those prescribed by the American Society of Mechanical Engineers or the American Welding Society.

welding procedure specification (WPS) – a document that provides all the essential parameters for a specified weld after it has been certified by the welding procedure qualification process. See also welding procedure qualification.

welding process – the type of welding used to produce a weld, such as gas tungsten arc welding.

welding rod – filler metal for use in nonelectric welding processes. The rod is melted into the weld to add strength to the joint.

welding symbol – an industry-recognized symbol used to represent a particular type of weld, such as a rectangle for a plug weld, a circle for a spot weld or a V for a V-groove weld. These symbols are placed along the reference line of an arrow pointing to the weld location. *See Box.*

welding tip – a torch end for an oxyacetylene (gas) welding torch.

welding torch – 1) the nozzle through which oxygen and acetylene or other fuels flow and are mixed prior to being ignited for oxyfuel or gas welding; 2) the nonconsumable electrode holder which forms the electrical connection to the rod and the nozzle for the shielding gas during some types of arc welding.

Welding symbols

A weld symbol indicates a type of weld. A *welding symbol* is a weld symbol with additional information included to show the weld location, its size, the welding procedure to be used and any other details that the draftsman needs to provide on the blueprint or drawing.

The basic arrow symbol is used to show where a weld is required. Weld symbols are added to the arrow to indicate the type of weld, such as a V for a V-groove continuous weld, a rectangle for a plug or slot weld, and a circle with two lines through it for a seam weld. The location of these symbols tells where the weld is to be made in relation to the arrow. If the symbol, such as a V for a V-groove weld, is below the reference line (the line coming off the shaft of the arrow), the weld is made on the side of the material to which the arrow is pointing. If the symbol is on top of the reference line, the weld is to be made on the other side of the material, opposite the arrow. If there is a weld symbol both above and below the reference line, the weld is to be made on both sides. And if there is a circle at the point where the shaft of the arrow meets the reference line, that indicates a weld all around. This method allows the draftsman to show the weld location even where it is not practical to have the arrow point directly at the weld.

Other elements of the welding symbol are indicated by capital letters in the illustration shown on this page. All of these features are not used with every weld, but are shown here to point out the possibilities. Information, such as the length and pitch (center to center spacing) of the weld (L-P), the groove weld size or effective throat (E), and the depth of preparation or size or strength for certain welds (S), is indicated by placing the dimensions above or below the reference line adjacent to the weld symbol. The dimensions are in the same units of measurement, such as metric or English, used for all the dimensions on the drawing. Other dimensions that may be shown are the root opening (R) and the angle of the groove (A). The number of spot welds or projection welds to be made is shown in parenthesis above or below the weld symbol (N).

The completed weld may have a contour that is flush, concave or convex, indicated by a straight or curved line

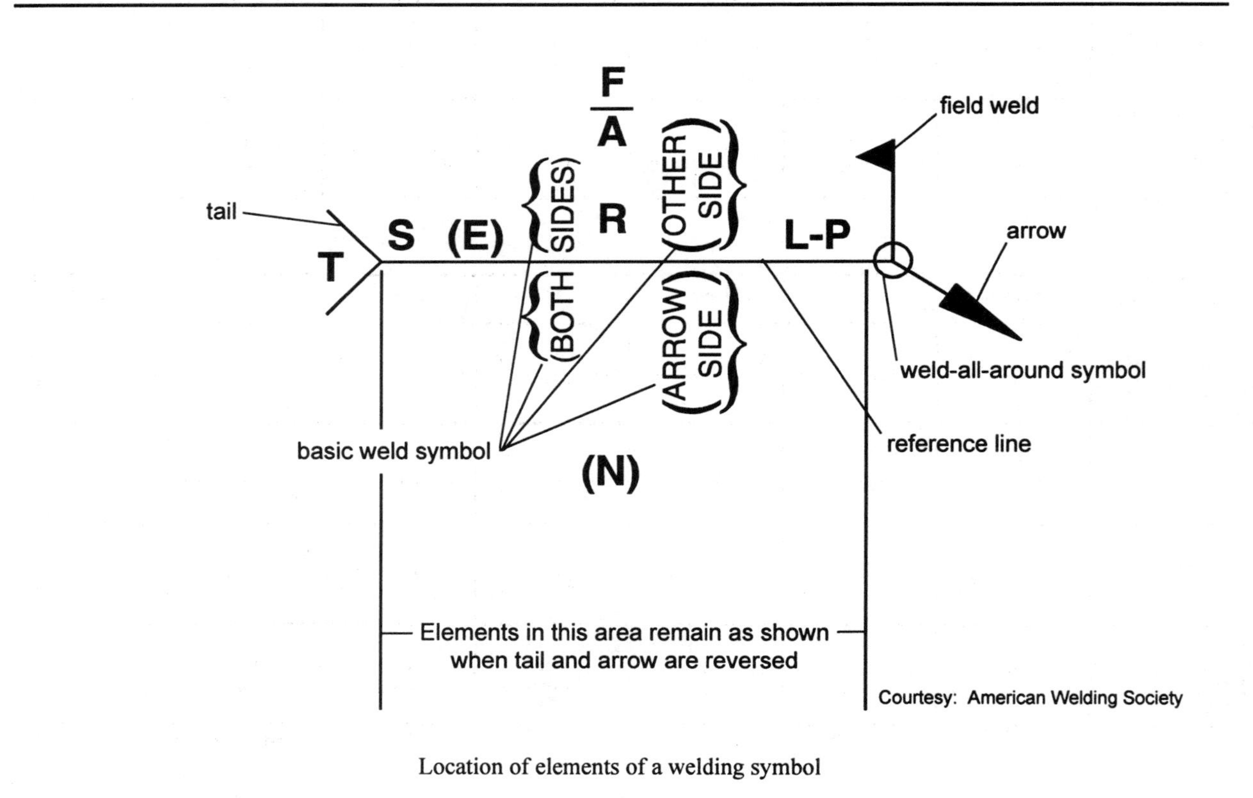

Location of elements of a welding symbol

(–), and may be finished by machining it smooth, grinding the surface with an abrasive wheel, or chipping the surface to remove excess material. These procedures are indicated by an M, G or C placed above, below or next to the contour line (shown at F).

A tail at the end of the reference line (T) provides a space for other necessary information that may not be included in the notes of the drawing, such as the specific welding process to be used. Information does not have to be repeated in the welding symbol when the welds are to be made in accordance with a specification that is already noted. When no additional reference is needed, the tail is omitted from the symbol.

Field welds are welds made at a job site, rather than in a shop. Many times, pieces are fabricated at the shop and shipped to the job site in the largest practical sections. Welds performed in the shop are known as shop welds, and those performed at the job site are called field welds. The field weld symbol, a flag located at the juncture of the arrow and the reference line, is used to show welds which are to be performed at the job site.

Welding symbols, their applications and definitions are updated periodically. Write to the American Welding Society, 550 NW LeJeune Road, Miami, FL 33126 for additional material or changes in this information.

Basic welding symbols and their location significance

Location significance	Fillet	Plug or slot	Spot or projection	Stud	Seam	Back or backing	Surfacing
Arrow side						Groove weld symbol	
Other side				not used		Groove weld symbol	not used
Both sides		not used	not used	not used	not used	not used	not used
No arrow side or other side significance	not used	not used		not used		not used	not used

Location significance	Groove						
	Square	V	Bevel	U	J	Flare-V	Flare-bevel
Arrow side							
Other side							
Both sides							
No arrow side or other side significance		not used	not used	not used	not used	not used	not used

Courtesy: American Welding Society

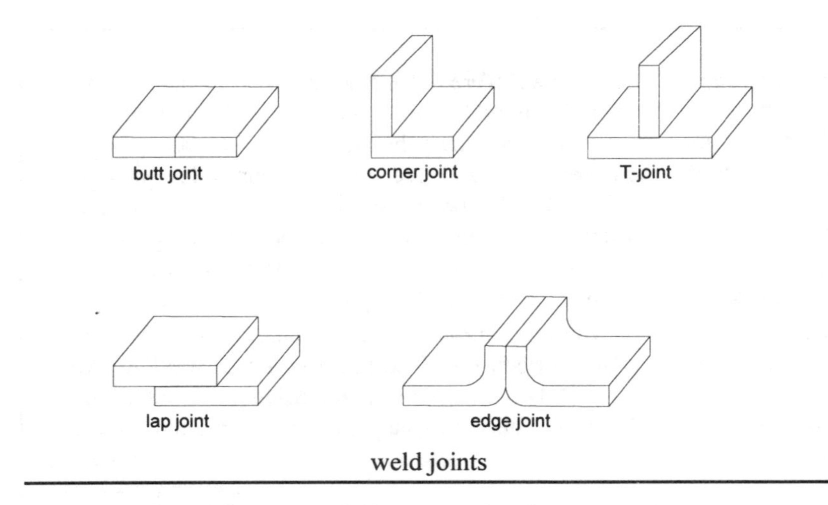

weld joints

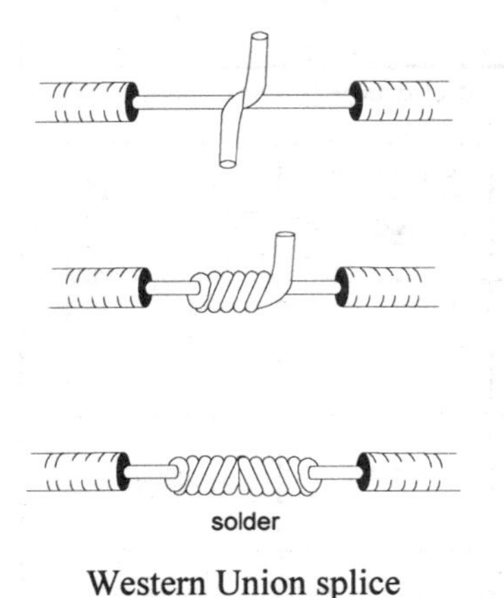

Western Union splice

weld interface – the juncture of the weld filler metal and the base metal or of the two pieces to be joined.

weld joints – the seams or connecting points where pieces of metal have been fused together by welding. There are five basic weld joints: the butt joint, the corner joint, the T-joint, the lap joint and the edge joint.

weld length – the distance that two materials have been fused through welding.

weldment – a completed assembly held together by welds.

weld metal – the total metal melted during the welding process, including filler.

weld pass – one layer of weld metal deposit.

weld penetration – the depth to which the weld metal is deposited.

weld postheat – maintaining the weld and the weld-heat-affected zone at a minimum temperature for a given time in order to allow locked-in stresses to relax.

weld preheat – preheating the base metal (both pieces that are to be joined) to a specified minimum temperature prior to welding to ensure weld quality.

weld preparation – cutting, grinding or machining to shape the edge or end on each of the pieces to be welded, at the location where the weld is to be made. The preparation is designed to ensure that the proper weld penetration and size are achieved. Also called weld prep.

weld rod – weld filler metal in the shape of a rod.

weld size – the dimension of a weld cross section in fractions of an inch.

weld symbol – see welding symbol.

well – 1) a deep hole bored in the ground for the purpose of extracting a fluid, such as water or oil, from the ground; 2) an enclosed area inside a building designed to provide a passage, as for a stairway or air, that is separated from the rest of the building.

wellpoint dewatering – a system used to remove groundwater from an area before beginning excavation or trenching work. *See Box.*

western framing – see platform framing.

Western Union splice – a splice for joining two electrical wires together by overlapping the bared ends and twisting each around the other. Also called a lineman's splice.

wet-aggregate cleaning – a low-pressure water stream carrying friable aggregate (masonry material that crumbles easily). This is a gentle cleaning method suitable for relatively soft surfaces, such as wood.

wet-bulb thermometer – see thermometer, wet-bulb.

wet edge – the point on a newly-painted surface where the paint is still wet enough to blend in paint applied to an adjacent area.

wet-film thickness – the mil thickness of a paint while it is still wet. It is usually an estimated thickness based on material characteristics.

wet-pipe sprinkler system – a fire protection system in which the water supply remains in the pipes under pressure and ready to be released through the sprinkler heads when there is sufficient heat to open the sprinkler heads.

wet-rubbed finish limestone – see limestone, wet-rubbed finish.

wet-sand cleaning – sand blasting with the sand carried in a stream of water rather than air. It has the advantage of being dust-free, but is very abrasive and can be used only for hard surfaces, such as metal.

wet sanding – smoothing with moist sandpaper which is made of a special water-resistant paper and coated with fine abrasive.

wetting characteristics – a description of the ability of a coating to flow over a surface.

wet vent – see fixture branch.

wet wall – a masonry wall that is built using mortar.

Wheatstone bridge – a bridge for measuring electrical resistance. It consists of a conductor joining two branches of a circuit. It can be used to determine an unknown value by comparing it to a known value of electrical resistance.

wheelbarrow – a manual load-moving device that has a triangular frame with a single wheel at the apex in the front and legs and handles at the rear. A square or rectangular container with slanted sides is mounted to the frame. Bulk material is placed in the container for mixing or transporting. When the rear of the device is lifted by the handles, the load can be manually moved about with the weight being carried on the front wheel.

wheelbase – the distance, measured in inches at the centerline, from the front axle to the rear axle of a vehicle.

wheel dresser – a hardened hand tool used for removing particles and smoothing and reshaping the surface of a grinding wheel.

wheeled backhoe – a backhoe that operates on wheels, rather than tracks.

wheel trencher – a trench-digging machine with bucket-shaped blades around the periphery of a large, engine-driven wheel. As the wheel rotates, soil is removed from the trench by the buckets and deposited on the sides of the trench.

whetstone – a fine grindstone for manually putting a sharp edge on a tool.

whip – to tightly wrap the end of a rope with a cord or twine to prevent the rope end from unraveling.

white cedar – a wood from an eastern U.S. swamp conifer used for crates and boxes.

Wellpoint dewatering

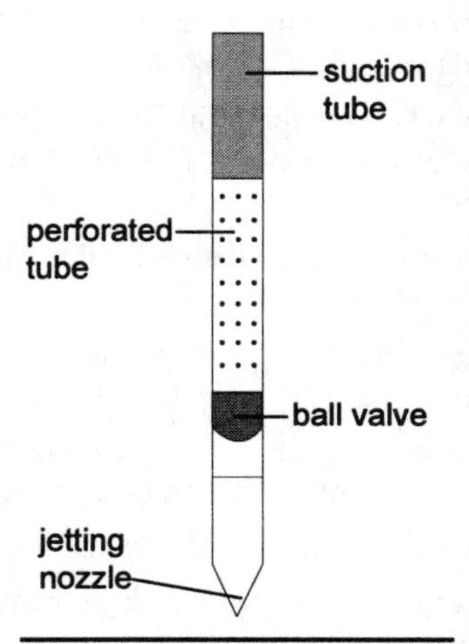

Wellpoint dewatering is a system of removing groundwater from an area before digging a hole or pipe trench. Groundwater causes unstable soil which is dangerous to work in, and impossible to use as backfill. A wellpoint is long tube or pipe which uses water pumped through a jetting device at one end to penetrate the earth to a depth lower than the proposed excavation. Once at the necessary depth, the water jetting is discontinued and the tube closed off at the base by a valve. The middle section of the tube is perforated, allowing ground water to seep in and then be suctioned out through a suction tube at the top. A single wellpoint can be used to dewater a small area, or several may be connected to a pump to lower the water table in a larger area. A wellpoint system works well in sandy soil or marshy swamp areas, but it is difficult to use water jetting to force the pipe through clay or hard soil. In hard, rocky or gravel-type soils other dewatering methods should be employed.

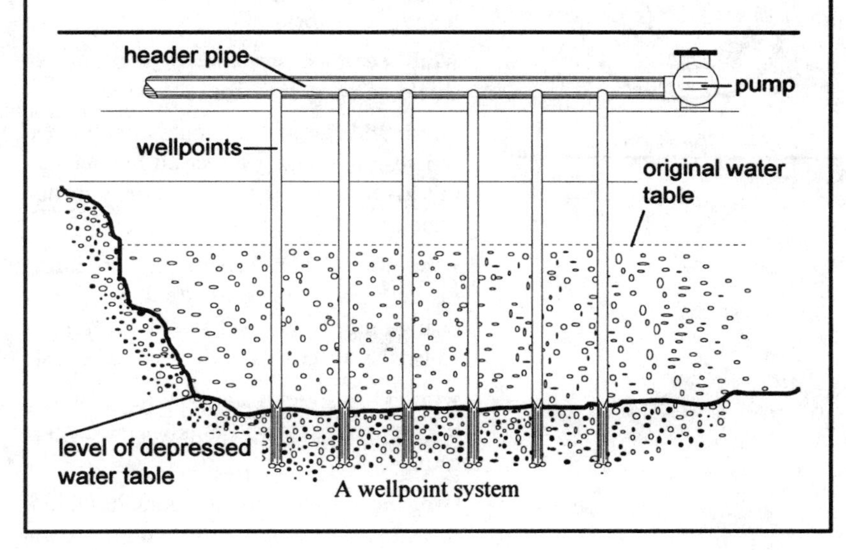

A wellpoint system

white coat – the top or finish plaster coat applied to a surface.

white lead – lead oxide formerly used as a paint pigment.

white metal – 1) a metal surface from which all oxide and scale have been removed; 2) a metal, made of various alloys of tin with antimony and copper, used in the manufacture of bearings. Also called babbitt metal.

white portland cement – cement, ASTM C150 and C175, that is white in color.

white shellac – shellac which is bleached white to provide a clear protective finish for wood.

whitewash – a solution of slaked lime and water that is applied as a paint.

whiting – powdered chalk pigment used in paint and putty.

wicket – a small gate, door, window or other opening, often covered by a grating.

wide-flange beam – see W-beam.

willow oak – an oak tree with hard, heavy wood found in the southern U.S.

winch – a hoist consisting of a cable or rope wound around a drum to which a crank or motor is attached to turn the drum, thus raising the load on the cable. Also called a windlass.

wind – a twisting (or winding) warp in a length of lumber.

windbreak – a structure or plantings to shield an open area from the wind.

wind chill factor – a temperature that expresses the cooling effect on human skin of a given combination of temperature and wind speed.

wind effect – air currents in a building drainage system vent stack that may change the water level in a drain trap, or cause the water seal in the trap to be lost.

winder – a style of stairway run in which the uppermost steps make a 90-degree turn.

winding – 1) warping in doors, or in door or window frames; 2) coils wound around a conductor, which, if moved or turned within a magnetic field, produce electrical energy.

windlass – see winch.

wind load – the force on a structure caused by the wind acting on it. The wind load must be considered in the design of roofs and roof coverings as well as high-rise structures. Also called wind force.

window – a glazed opening in a wall for the purpose of admitting light and/or ventilation.

window frame – the boxed-in framed opening into which a window assembly is installed.

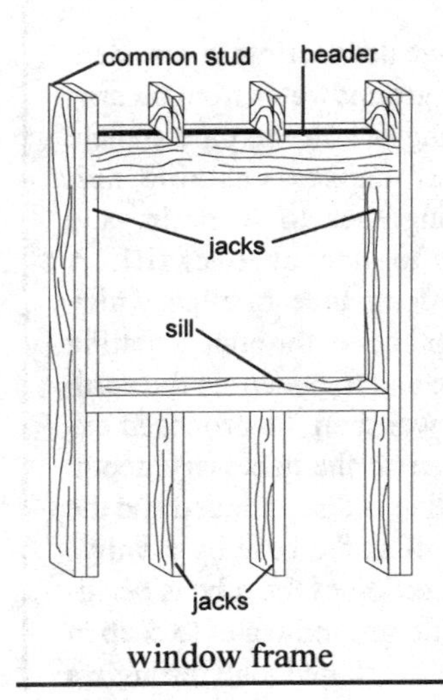

window frame

window shade – a covering for the inside of a window which blocks the entrance of strong light and provides privacy within the building.

window wall – a wall that is largely composed of glass, with the structural members between the glass areas.

windrow – a ridge of loose earth, such as the spill from the edge of a scraper blade.

windshake – separation along the circumference of the growth rings in a log. It is caused by wind stress on a tree during growth.

windshake

winter-degree day – the term which designates days in which the mean temperature is below a set value for purposes of estimating heating fuel requirements for an HVAC system. See also degree day.

winterize – to perform the necessary tasks to adapt something, such as a home, automobile or piece of machinery, for cold weather operation or storage. For example, winterizing a home would involve putting up storm windows and doors, sealing cracks and adding weatherstripping.

wiped joint – a solder joint made by pouring molten solder, which supplies both the heat and the solder, over the joint. A cloth is then used to shape or smooth the solder joint.

wire – metal that has been extruded through a die to form a long solid, flexible strand. The metal used varies according to its intended function. For example, electrical wire is made of copper or aluminum.

wire brads – small diameter finish nails.

wire brush – a hand-held brush with wire bristles used for cleaning surfaces or applying a rough texture to plaster or concrete.

wire cloth – cloth woven from thin wire strands. Used for screening, and as reinforcement in plaster.

wire color coding – standardized colors assigned to electrical wires which designate their functions, such as green for a grounding wire.

wire connections – methods of joining or terminating wire.

wire connectors – devices for interconnecting lengths of wire, or for connecting wire to components.

wire cutter – a plier-type device with cutting edges on the jaws for cutting wire.

wired edge – a sheet metal edge in which the metal has been rolled around a length of wire to make a smooth border.

wiredrawing – the linear erosion across a valve seat caused over time by a small leak in a valve which is under high pressure.

wire fasteners – a fastener which can be secured to a structure and to which wires, such as hanger wires for a suspended ceiling, may be attached or connected.

wire gauge – 1) a standardized numbering system assigned to represent the diameter of a wire. The smaller the wire, the larger the number; 2) a flat, circular piece of metal with standard-size notches around its perimeter which is used to measure the diameter of a piece of wire.

wire glass – two panels of glass with wire mesh embedded between them during manufacturing. The wire serves as reinforcement and prevents the glass from shattering when broken. It is used primarily for security purposes when lights are set into doors.

wire mold – a small metal raceway for electrical wiring that is fastened to the surfaces of walls in remodeling work so that wiring does not have to be pulled through existing walls and is still protected. Also called wire molding.

wire nut – a device that can be used to join two or more stripped ends of electrical wires together with a twisting action. The wire nut has a tapered spring or threads inside to grip the ends of the wires being joined. Also called a solderless connector. See also wire connectors.

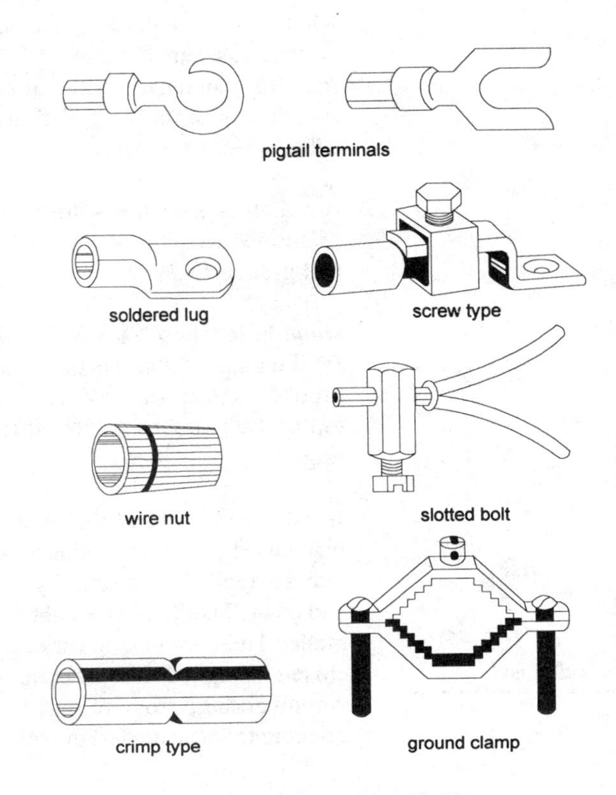

wire connectors and terminals

wire rope – structural cable made from lengths of steel wire twisted together. It is used for the hoist cable on cranes, winches and similar lifting equipment.

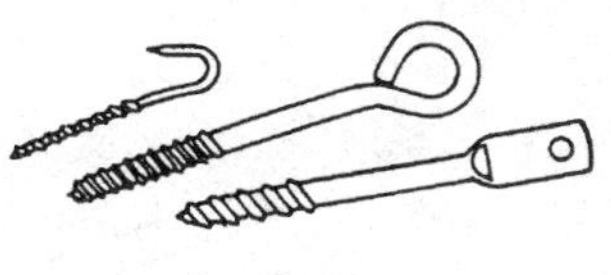
wire fasteners

wire size – see wire gauge.

wire, solid core – wire consisting of a single conductor.

wire, stranded – wire consisting of multiple conductors.

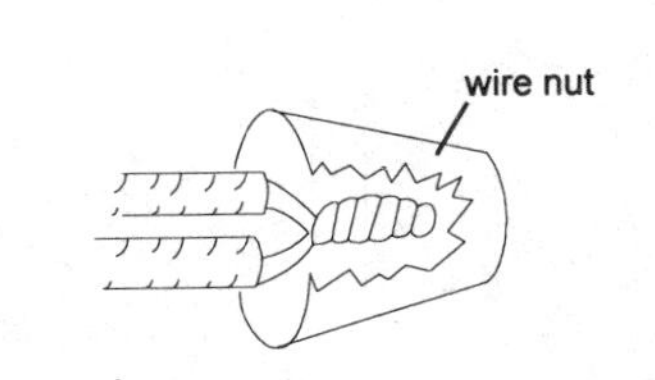

wire nut

wire stripper – a tool for cutting and removing the insulation from the end of electrical wire so that a connection can be made.

wireway – a square or rectangular metal raceway with one hinged or removable side which allows the electrical wire to be laid in place after the wireway is mounted. It is designed to mount to the surface of a wall and is most often used in industrial applications.

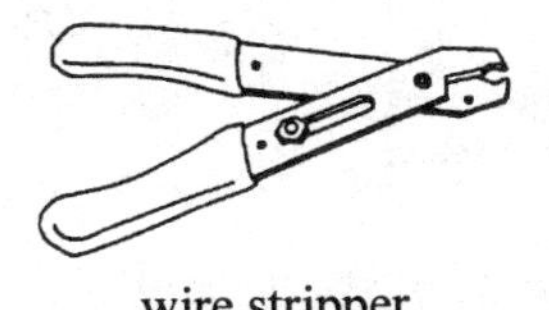
wire stripper

wiring plans – drawings, made by the electrical system designer, which show the installing electrician the locations of switches, receptacles and fixtures. Also called wiring diagrams.

wood block flooring – flooring laid in preformed wood tiles, often with tongue-and-groove edges.

wood butt piling – a wood piling, used for foundations, that can be driven into the ground. A steel band is placed around the top of the piling to prevent it from splitting.

wood chisel

wood chisel – a hand tool with a cutting blade used to cut and shape wood. Force may be applied to the tool by striking the end of the handle with a light hammer or mallet. There are several varieties of wood chisels designed for different facets of woodworking, from rough cutting and gouging to intricate design work.

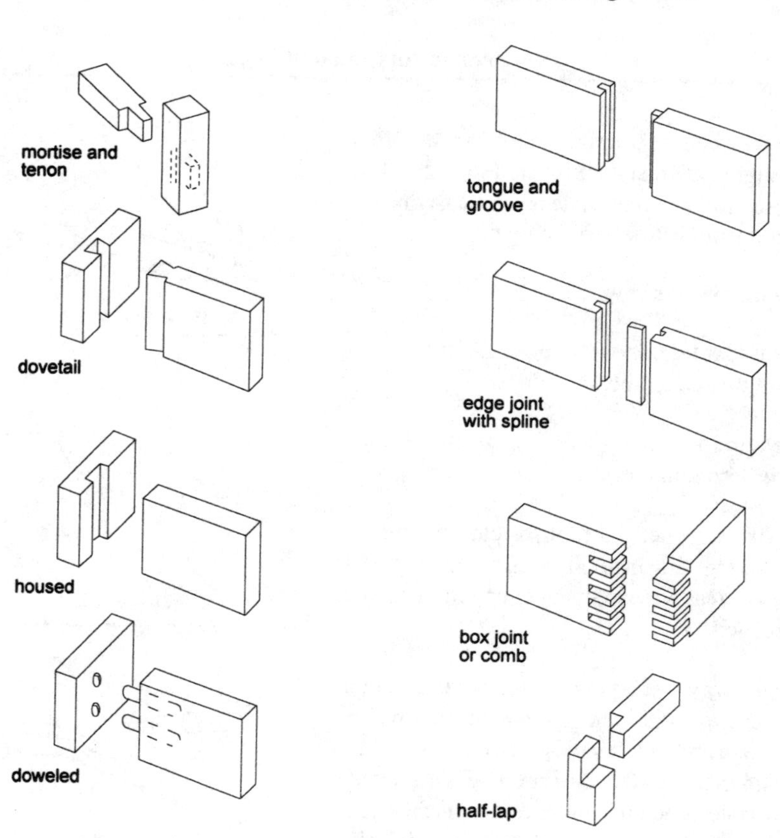

wood joinery

wooden water pipe – a very old type of pipe made from wood staves which were bound together, as in the construction of a wooden barrel, to make hollow wooden cylinders used as water pipe.

wood filler – a putty-like substance used to fill small voids in the surface of wood. It is available in different colors to match a variety of woods. Also called wood putty or Plastic Wood.

wood-frame construction – a type of building construction in which all the loadbearing structural members, such as the studs, plates, joists and rafters, are made of wood.

wood green – a chemical preservative containing copper napthenate, and sometimes arsenic, for use on lumber. It is used to inhibit the growth of fungi and prevent damage by termites and other insects.

wood joinery – the craft of cutting and fitting woodwork using a variety of joint-making techniques. The joints may be fastened together by interlocking design or by cutting to a desired shape and nailing, screwing or gluing the pieces, or a combination of both.

wood lathe – a power machine which holds and turns pieces of wood while they are shaped into the desired form by special wood cutting tools.

wood mallet – a wooden hammer often used with a wood chisel for cutting and shaping wood. The handle of the chisel is lightly struck with the wood mallet to add force to the cut. The mallet may also be used to strike the wood directly, as in forcing a wood joint together.

wood preservatives – chemicals used to protect wood from insects, rot and the growth of destructive organisms. Wood is treated by pressure application, or by soaking, dipping or painting with a preservative solution. There are a variety of preservatives available, each with its own advantages and disadvantages. Creosote, distilled from coal tar, is widely used on fence posts, telephone poles and railroad ties, where its dark color and strong odor are acceptable. Pentachlorophenol in a petroleum distillate is a strong preservative that can applied and then painted over, without leaking through and discoloring the surface. Inorganic salts, such as chromated zinc chloride, have good penetration when dissolved in water. They are not a fire hazard in liquid form, and have minimal odor, but they can cause wood to swell, and some varieties react with metal.

wood putty – see wood filler.

wood sash putty – a glazing material of powdered chalk or white lead in linseed oil, used to hold a pane of glass in a sash.

wood screw – a threaded fastener with a tapered shank ending in a point. The head is larger in diameter than the shank and has a slot for a screwdriver to be inserted to turn the screw. The heads may be flat, oval or rounded, with one slot or a cross slot designed to fit a Phillips head screwdriver. Usually a pilot hole is drilled through the pieces to be joined, and then the screw is inserted in the hole and screwed down tight, fastening the pieces together.

wood sheet piling – sheet piling made from boards that are fastened together with cleats. They are used as shoring for foundation work.

wood shingle – a tapered roofing shingle commonly sawn from cedar.

wood vise – a vise with large flat jaws used for holding wood. The jaws are often faced with wood so that they can hold wood without marring the surface.

woodworking bench – a bench with a vise or vises and other devices designed to hold wood or wood assemblies while they are being worked on.

word processor – a keyboard-operated terminal with a video display that stores material typed into its memory so that the material may be copied, changed, or manipulated into various forms by the operator. Word processing is incorporated into computer operations through software programs or it can be an independent function of the machine, used in a capacity similar to a typewriter, to produce typewritten documents.

workable – 1) something that is reasonable to do; practical; 2) material that is usable or can be manipulated into a usable form.

workers' compensation insurance – a state-required insurance, paid by employers, which covers job-related injuries to employees.

work harden – to increase the surface hardness of a metal through working it by bending or pounding.

working drawings – an accurate drawing used in the making or building of something.

workmanship – the quality of work put into a product or its installation.

workshop – an area within a building dedicated to working material or machinery. Workshops are used for woodworking, metal working, and other types of (usually) light fabrication.

work triangle – a triangle formed by straight lines, either imaginary or drawn on a planned kitchen layout, between the three major work areas of a kitchen: the sink, the stove, and the refrigerator. In a well-designed kitchen, the total distance around this triangle should not be less than 14 or more than 22 feet.

woven hip – a roof covering in which the shingles overlap at the top of the hip.

woven valley – a roof covering in which the shingle courses are overlapped and then interwoven at the roof valley by overlapping alternating courses in opposite directions.

wrecking ball – a large diameter heavy metal ball which hangs from a cable off the end of a crane boom. The operator swings the ball into the walls of a structure to demolish the structure.

wrecking bar – a steel bar with one flat split end and one end that is often curved or hooked to provide mechanical leverage. Used to pry and to pull nails. See also crowbar or pinch bar.

wrench – a hand tool with jaws designed to grip for tightening or loosening a threaded fastener such as a bolt or nut. See also specific wrench types.

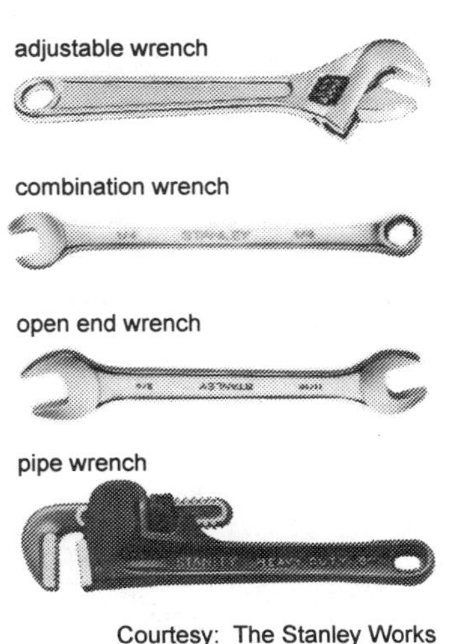

Courtesy: The Stanley Works

wrench

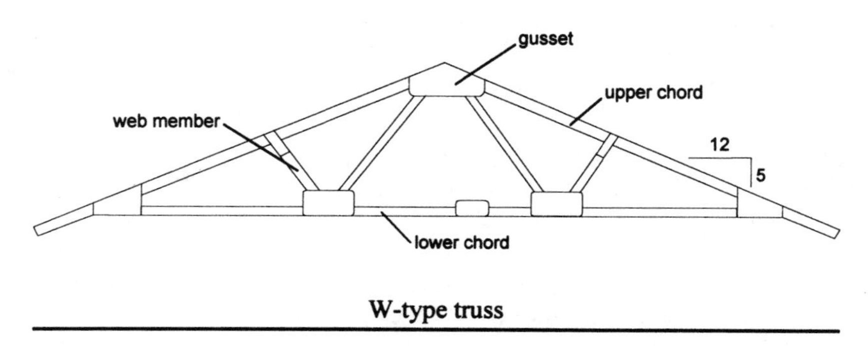

W-type truss

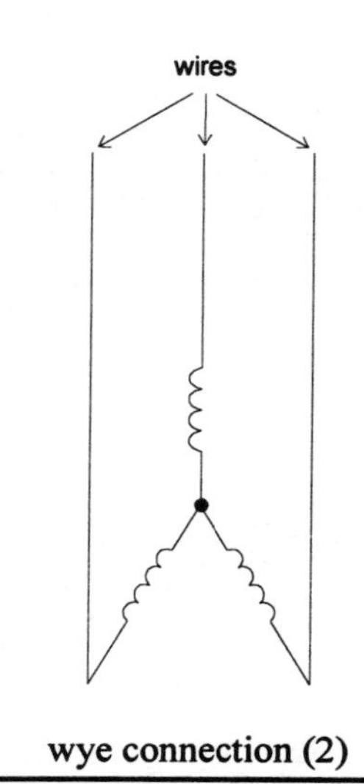

wye connection (2)

wrinkle finish paint – a painting technique or system, created by the inter-coat reaction of a combination of coating media, that forms a wrinkled surface as it dries. The wrinkles give the surface a textured look.

wrinkling – a paint defect in which the surface becomes wrinkled. There are several factors which may cause this problem: applying paint to a surface that is too glossy; using an incompatible primer; applying paint in direct sunlight or during a drastic change of temperature; applying paint over an incompatible coating, such as lacquer over enamel; using too much or the incorrect thinner in the paint; or an improperly-prepared surface. Once wrinkling occurs, the paint must be removed and the surface repainted using the proper materials under the proper conditions.

writ – a legal instrument or formal written order issued by a figure of authority, such as a judge.

wrought iron – a malleable form of iron worked into decorative and functional shapes. Used for such items as ornamental railings, panels and gates.

WR panels – see water-resistant panels.

W-type roof truss – a type of roof truss commonly used in residential construction in which the web members form the letter W. Also called a Fink truss.

wye – 1) a Y-shaped plumbing fitting used where a branch line joins the main line at a 45-degree angle; 2) an electrical schematic of load connections (in the shape of the letter Y) to a three-phase system.

wythe – 1) a vertical masonry section that is one masonry unit thick; 2) a dividing partition between two flues in a single chimney. Also spelled withe.

X: X-brace to xyst

X-brace – a pair of diagonal cross braces used to provide stability against racking or sideward forces.

Xeriscape – a method of landscaping, often focusing on water conservation, which utilizes plants that thrive in arid or semiarid climates.

xerography – a process for copying graphic or printed material by the action of light on an electrically-charged photoconductive insulating surface in which the latent image is developed with a resinous powder.

Xerox – a trademark name for a xerographic copy made on a Xerox brand copier.

x-ray examination – an examination technique which uses radiation passing through an object to produce a photographic image which exposes flaws or defects in the object or material. *See Box.*

xyst – a covered or shaded walkway.

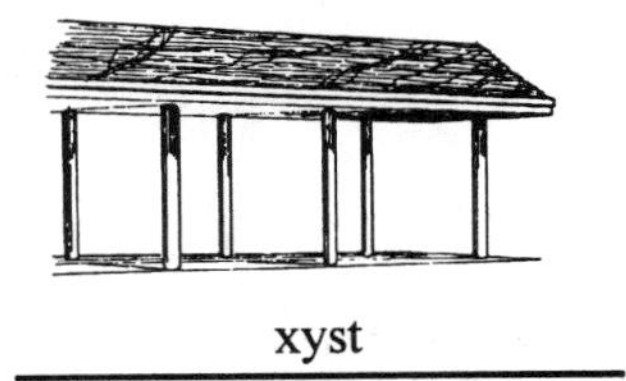

xyst

X-ray examination

Radiography, commonly called x-ray, is a volumetric nondestructive examination technique that places a source of radiation on one side of the object to be examined and a photographic film on the other. Radiation passing through the object makes an image on the photographic film which can be read and interpreted by a trained technician. Flaws in the material show up as an image on the film when the flaw, radiation source and film are in the proper relationship. For this reason, x-rays may be taken from different directions to ensure a complete reading. Lead plates of a specific thickness in relation to the material being examined, and with holes of varying depths drilled through them, are exposed along with the material to provide a known reference. Radiography has a variety of construction uses. One of its most frequent applications is in the examination of welds joining lengths of high pressure steel pipe together. Most governing codes require such examinations to ensure the integrity of joint welds in piping systems installed in power plants, refineries and other industries where joint failure could be hazardous.

Y: yankee screwdriver to yoke vise

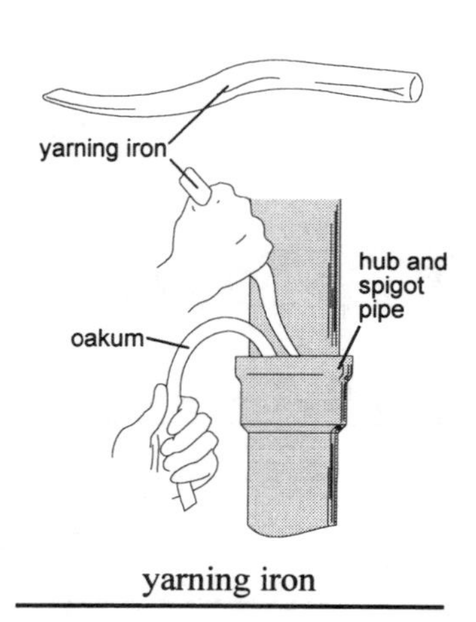

yarning iron

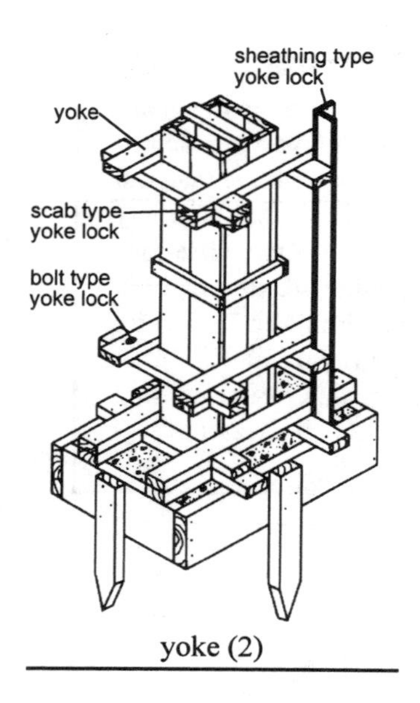

yoke (2)

yankee screwdriver – a ratchet screwdriver with steeply-pitched spiral grooves in the shank. The shank is free to rotate within the handle. A pawl in the handle rides in the spiral groove so that when the handle is pushed along the shank toward the work, the shank rotates, turning the screw.

yard – 1) a lineal measurement equal to 3 feet; 2) a defined outdoor area, with or without a purpose, such as the yard of a house or a lumber yard.

yard lumber – standard lumber less than 5 inches thick used in general frame construction.

yarning iron – a curved iron tool, flattened on one end, used for packing oakum into a cast-iron pipe joint that is to be sealed with molten lead.

Y-connection – an electrical connection in which all three coils in a three-phase generator meet at one point. Also called a wye connection.

Y-fitting – a plumbing drainage fitting shaped like the letter Y, used where a branch enters the run at an angle less than 90 degrees. Also called a wye fitting.

yield – 1) a volume or amount produced or grown, such as the volume of concrete produced per sack of cement or the amount of fruit produced in a crop; 2) a permanent deformation in a material caused by bending or stretching.

yoke – 1) the metal mounting strap on an electrical receptacle; 2) a collar or clamp placed around the form to brace it while a concrete column is poured.

yoke lock – a brace used to hold a yoke on a concrete column form.

yoke vent – a plumbing pipe connecting a waste stack and a vent stack designed to prevent pressure changes in the stacks. The yoke vent slants upward at an angle from the waste stack to the vent stack.

yoke vise – a clamping device that has a fixed lower jaw and an upper jaw that is moveable by means of a screw thread. The vise is hinged on one side so that it can be opened, allowing a length of pipe to be inserted. Also called a hinged pipe vise.

Z: Z-channel to Z-tie

Z-channel – Z-shaped flashing placed along a horizontal edge between abutting vertical exterior wall panels to ensure that any water that infiltrates the joint is directed outward by gravity to the outside of the wall.

zero clearance – well-insulated heating units or metal wood burning fireplace units that do not require a specified distance between the unit and combustible surfaces.

zero line – an imaginary line in earthmoving at which point neither cutting nor filling is required.

Z-frame gate – a sturdy gate design utilizing a diagonal cross support in a Z shape for the main structural members.

zigzag fence – a spit rail fence made of rails that intersect and rest upon one another. It is laid out in a zigzag pattern, with the rails anchored at the joint with a spike or threaded rod. See also snake fence.

zinc – a white metallic element with the atomic symbol Zn, atomic number 30 and atomic weight of 65.37. It is commonly used for galvanizing or coating steel to inhibit oxidation and corrosion.

zinc chromate – a yellow primer and corrosion inhibitor used as a protective undercoat for metals.

zinc oxide – a white compound used as a paint pigment and an ingredient in cosmetics and pharmaceutical preparations.

zinc sulfide – a fluorescent white-to-yellowish compound used as a white pigment and a phosphor.

zinc yellow – see zinc chromate.

zip cord – see SP wire.

zip strip – a vinyl strip used to form expansion joints in concrete slabs. The T-shaped strips are inserted into the concrete immediately after finishing. The flat top strip is peeled away once the concrete has set, exposing the concealed joint.

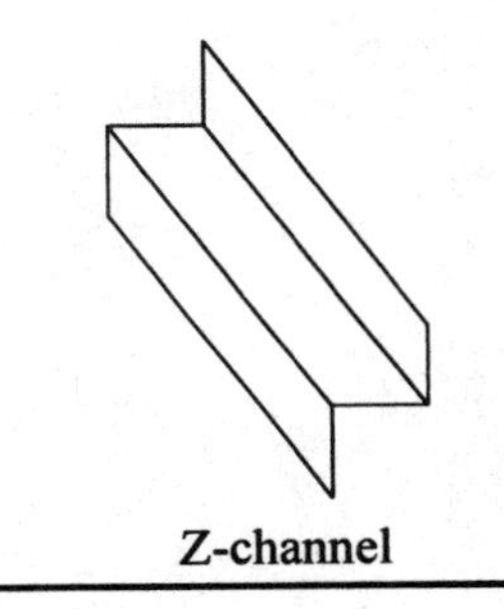

Z-channel

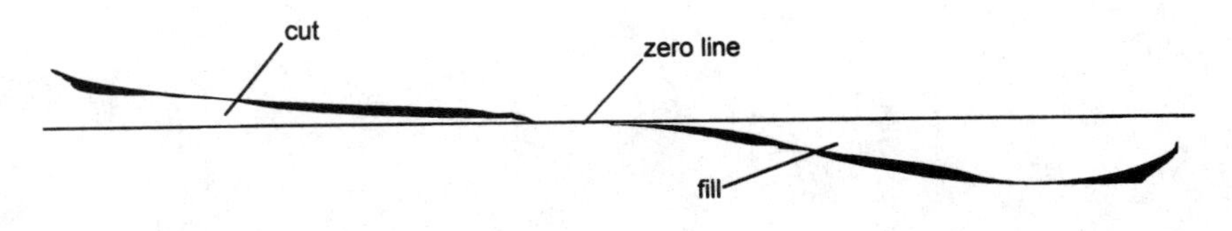

zero line

zonal cavity lighting – a method of calculating the amount of light needed in a room by dividing the room into zones and estimating the needs of each area. The zones are the ceiling cavity, room cavity and floor cavity. Lighting requirements for each zone have been prescribed by the Illumination Engineering Society.

zone, HVAC – a heating and air conditioning system with separate adjustable controls for various parts of a building.

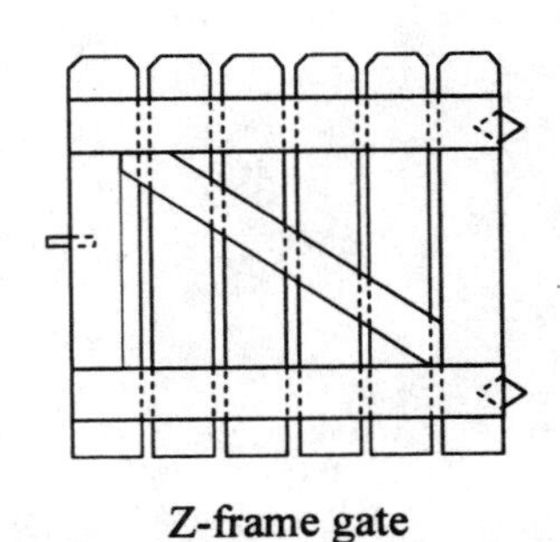

Z-frame gate

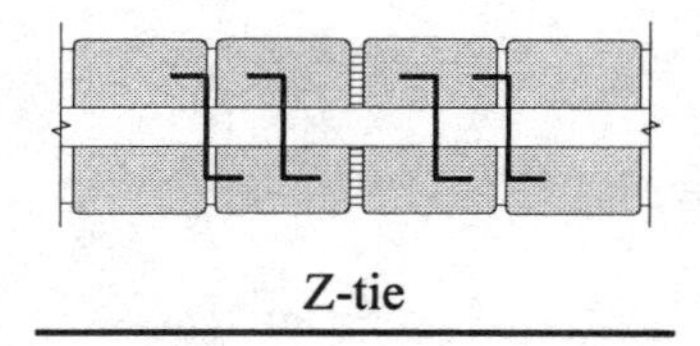

Z-tie

zone switching – electrical switches that operate lights in a section or zone of a room or space.

zoning – local regulations which limit land use and activities within sections of a city. There are usually residential, commercial, manufacturing, recreational and mixed use areas, with those areas also broken down into types of businesses or types of residences.

zonolite – a lightweight insulating material used as loose-fill insulation and as an aggregate in making insulating concrete.

Z-tie – a metal tie used to bond masonry walls together.

Metric Measurement

All of the major industrialized nations except the United States use the International System of Units (SI), or metric system of measurements, as their standard. The U.S. has been slow to adopt the SI system in daily practice, even though use of the metric system was legalized in 1866.

In 1971, a congressional study recommended that the U.S. begin a 10-year conversion to metrics. This was made official by the Metric Conversion Act of 1975, which established a policy of voluntary conversion to the SI system. In an effort to hasten the conversion, the Omnibus Trade and Competitiveness Act of 1988 amended the Metric Conversion Act, and requested that all federal agencies use the SI system for business transactions beginning in 1992. An Executive Order was signed by President Bush in 1991 requiring small businesses to use the metric system. As ever-increasing amounts of work are performed internationally by U.S. firms, including import/export businesses, knowledge of the SI system of measurements has become essential. The following is designed to provide information about SI units, their use, and conversion from customary U.S. units of measure.

The Unit of Measurement

Only one SI unit of measure is used to quantify each physical commodity, greatly simplifying mathematical manipulation by eliminating conversion factors.

The basic SI units are as follows:

Quantity	Unit	Symbol
length	meter	m
mass (or weight)	kilogram	kg
temperature	kelvin* [Celsius or centigrade]	K [C]
electric current	ampere (amp)	A
time	second	s
light intensity	candela	cd

*Degrees Celsius, or centigrade, is more commonly used outside the scientific community. The difference in temperature in degrees K (expressed as K, with no degree (°) sign preceding the K) and degrees Celsius, or centigrade, (expressed as °C), is 373.15. Otherwise, one degree C = one degree K. Kelvin temperature measurement begins at absolute zero, or 0 K, while 0 °C is the freezing point of water. To convert K to °C, subtract 373.15, or to convert °C to K, add 373.15.

Other SI units that are used in the various construction trades are:

Quantity	Unit	Symbol
frequency	hertz	Hz
power, radiant flux	watt	W
energy, work, heat quantity	joule	J
electric potential	volt	V
electric resistance	ohm	Ω
luminous flux	lumen	lm
illuminance	lux	lx
pressure, stress	pascal	Pa
liquid volume	liter	L

SI units change by factors that are multiples of 10. A prefix in front of the unit indicates the number of units:

Units	Symbol	Equivalent
kilo	k	1000
hecto	h	100
deka	da	10
unit (ie. meter, liter, gram)	(m, l, g)	1
deci	d	0.1
centi	c	0.01
milli	m	0.001

Writing with Metric Symbols

SI unit symbols are generally written out in lower case, except for liter (L) and those taken from proper names, as shown in the tables. All SI unit names are lower case. Prefix symbols, such as M for mega (million) and larger units are upper case; for example, MW = megawatt. Prefix symbols for smaller units (of a magnitude 10^3 and lower), are lower case; for example, cm = centimeter.

When writing a combination of a number and a symbol, leave a space between them, such as 33 cm. Symbols do not need an "s" to denote plural, for example 14 meters is written 14 m. Symbols are only used with numbers denoting a particular value or amount, such as 25.4 cm. If numbers are not used, the symbols should not be used; rather the words, such as cubic meters, should be written out. Names and symbols should not be mixed. Symbols do not have periods after them except when they end a sentence.

Decimal numbers, rather than fractions are used with metric values, with a zero preceding a decimal point where the value is less than one. Except in the case of money, spaces rather than commas should be used to separate numbers in groups of threes; for example, 14 630 775 (rather than 14,630,775). This is to avoid confusion internationally, as a comma is used in place of the decimal point in much of the world.

The products of multiple units are shown as the applicable symbols separated by dots above the lower plane of the letters, such as g˙cm.

Meters and millimeters are used in the building industry. A micrometer is used for precision, and a kilometer for long distances. The hectare (equal to 10 000 square meters) is used for land and water area measurement. Square meters are used for other areas, unless the area is very small, in which case square millimeters are used. Cubic meters are used for volumes, except liters are used for liquid volumes (one liter = 1/1000 of a cubic meter).

Plane angles in surveying using SI units are measured in decimal degrees or in degrees, minutes, and seconds. Slope is shown as a ratio, with the vertical first, followed by the horizontal; for example, a rise of 1 meter in 3 meters is shown 1:3.

Basic Conversion Tables

Several standards organizations, such as American Society of Mechanical Engineers (ASME), American Society of Heating, Refrigerating, and Air-Conditioning Engineers Inc. (ASHRAE), American National Standards Institute (ANSI), and American Society for Testing and Materials (ASTM) publish standards with both customary U.S. units and SI units. A good reference is ASTM E621, which is the Standard Practice for the Use of Metric (SI) Units in Building Design and Construction. Sections 3 and 4 of ASTM E390, Standard Practice for Use of the International System of Units (SI), provide rules for usage, conversion, and rounding off of U.S. to SI units. This information can also be found under Section 268: American National Standard Metric Practice in the standards published by the American National Standards Institue or Institute of Electrical and Electronics Engineers (IEEE). There are also calculators and computer programs available to convert units from one form to another.

The following conversion charts show the relationships of commonly used U.S. units to SI units:

Length Measure

U.S.	SI
one mile = 1.609 3 kilometers	one kilometer = 0.62 mile
one yard = 0.914 4 meter	one hectometer = 328.08 feet
one foot = 0.304 8 meter or 30.48 centimeters	one dekameter = 32.81 feet
one inch = 2.54 centimeters or 25.4 millimeters	one meter = 39.37 inches
	one decimeter = 3.94 inches
	one centimeter = 0.39 inch
	one millimeter = 0.039 inch

Area Measure

U.S.	SI
one square mile = 2.590 square kilometers	one square kilometer = 0.3861 square mile
one acre = 4 046.856 square meters or 0.405 hectare	one hectare = 2.47 acres
one square yard = 0.836 square meter	one acre = 119.60 square yards
one square foot = 0.093 square meter	one square centimeter = 0.155 square inch
one square inch = 6.452 square centimeters	
or 645.16 square millimeters	

Volume Measure

U.S.	SI
one acre foot = 1 233.49 cubic meters	one cubic meter = 1.307 cubic yards
one cubic yard = 0.7646 cubic meter	one cubic decimeter = 61.023 cubic inches
one cubic foot = 0.028 317 cubic meter or 28.317 liters	one cubic centimeter = 0.061 cubic inch
one cubic inch = 16.387 cubic centimeters	

Capacity

U.S. dry measure	U.S. liquid measure
one bushel = 35.239 liters	one gallon = 3.785 liters
one peck = 8.810 liters	one quart = 0.946 liter
one quart = 1.101 liters	one pint = 0.473 liter
one pint = 0.551 liter	one gill = 118.294 milliliter
	one fluid ounce = 29.573 milliliter
	one fluid dram = 3.697 milliliter

Capacity (continued)

SI	(U.S. cubic)	(U.S. dry)	(U.S. liquid)
one kiloliter =	1.31 cubic yards		
one hectoliter =	3.53 cubic feet	2.84 bushels	
one dekaliter =	0.35 cubic foot	1.14 pecks	2.64 gallons
one liter =	61.02 cubic inches	0.908 quart	1.057 quarts
one cubic decimeter =	61.02 cubic inches	0.0908 quart	1.057 quarts
one deciliter =	6.1 cubic inches	0.18 pint	0.21 pint
one centiliter =	0.61 cubic inch		0.338 fluid ounce
one milliliter =	0.061 cubic inch		0.27 fluid dram

Mass and Weight

U.S. (avoirdupois)	SI
one short ton = 0.907 metric ton	one metric ton = 1,102 short tons
one long ton = 1.016 metric tons	one kilogram = 2.204 6 pounds
one hundred pounds = 45.359 kilograms	one hectogram = 3.527 ounces
one pound = 0.453 592 kilogram	one dekagram = 0.353 ounce
one ounce = 28.350 grams	one gram = 0.0353 ounce
one dram = 1.772 grams	one centigram = 0.154 grain
one grain = 0.0648 gram	one milligram = 0.015 grain
one pound per square foot = 4.882 428 kilograms per square meter	one kilogram per square centimeter = 14.22 pounds per square inch
one pound per cubic foot = 16.018 5 kilograms per cubic meter	one kilogram per cubic meter = 2.048 pounds per cubic yard
one pound per square foot = 47.880 3 pascal (Pa)	

Temperature

U.S.	SI
one degree Fahrenheit (°F) = (9/5 x °C) + 32	one degree Celsius (°C) = 5/9(°F – 32) or K – 373.15
	one degree Kelvin (K) = 5/9(°F – 32) + 373.15

Energy

U.S.	SI
one British thermal unit (Btu) = 1 055.056 joule (J)	one joule = 0.7376 foot-pounds
one ton of refrigeration = 3.517 kilowatt	
one Btu/($ft^2 \cdot hr \cdot$ °F) [U-value – thermal conductance] = 5.678 263 W/($m^2 \cdot K$)	
one $ft^2 \cdot hr \cdot$ °F/Btu [R-value – thermal resistance] = 0.176 110 $m^2 \cdot K/W$	

Abbreviations

A

A – 1) area; 2) angle; 3) ampere
AAC – air carbon arc cutting
AAW – air acetylene welding
AB – 1) aggregate base; 2) arc brazing
ABS – acrylonitrile-butadiene-styrene
AC or ac – 1) alternating current; 2) air conditioner; 3) arc cutting; 4) armored (electrical) cable
A.C. or ACP – asbestos-cement pipe
ACRS – accelerated cost recovery system
A/cs Pay – accounts payable
A/cs Rec – accounts receivable
A-E – architect-engineer
ALS cable – aluminum-sheathed electrical cable
alt – altitude
alum – aluminum
alw – allowance
amp – ampere
ANSI – American National Standards Institute
AOC – oxygen arc cutting
APA – American Plywood Association
API – American Petroleum Institute
APR – annual percentage rate
apt. – apartment
ARM – adjustable rate mortgage
ASCII – American Standard Code for Information Interchange
ASME – American Society of Mechanical Engineers
ASTM – American Society for Testing and Materials
avdp – avoirdupois
avg – average
AW – arc welding
AWG – American Wire Gauge
AWS – American Welding Society
AWWA – American Water Works Association

B

BB – block brazing
BBC – Basic Building Code
bbl – barrel
BC – begin curve
bd ft – board foot
bdl – bundle
bldg – building
BM – bench mark
B/M – bill of material
BMAW – bare metal arc welding
Btu – British thermal unit
bv – balanced voltage
b/v – brick veneer

C

c – circumference
C – 1) Celsius or centigrade; 2) Roman numeral one hundred; 3) capacitance; 4) hundredweight
CAC – carbon arc cutting
CAD – computer aided design
CADD – computer aided design and drafting
CAM – computer aided manufacturing
CAW – carbon arc welding
CAW-S – shielded carbon arc welding

CAW-T – twin carbon arc welding
cb – circuit breaker
CB – 1) concrete block; 2) catch basin
cc – cubic centimeter
C-C – center-to-center
cd – candela
CD – certificate of deposit
CDA – Copper Development Association
CE – civil engineer
Cem. – cement
cem. fl – cement floor
cem. p – cement paint
Cer. – ceramic
CEW – coextrusion welding
cf – cubic feet
cfm – cubic feet per minute
C.I. or CI – cast iron
CIP – cast iron pipe
cir – 1) circle; 2) circumference; 3) circuit
cir bkr – circuit breaker
circ – circular
CISP – cast iron soil pipe
CL – centerline
cm – 1) centimeter; 2) circular mil
cmp – corrugated metal pipe
CMU – concrete masonry unit
com – common
conc – concrete
constr – construction
contr – contractor
cp – candlepower
CPM – Critical Path Method
cpu – central processing unit
CPVC – chlorinated polyvinyl chloride
CRT – cathode ray tube
CSPE – chlorosulfonated polyethylene (trade name Hypalon)
ctb – ceramic tile base
c to c – center to center
ctr – center
cu – cubic
cu ft – cubic feet
cu in – cubic inch
cu yd – cubic yard
cv – check valve
cw – cubic weight
CWO – cash with order
cwp – circulating water pump
cwt – hundredweight (100 pounds)
cy – capacity
CY – 1) calendar year; 2) cubic yard
cyl – cylinder

D

d – 1) diameter; 2) penny
D – deep
dB – decibel
DB – dip brazing
dbh – diameter, breast height
dbl – double
DC, dc, d-c – direct current
deg – degree
DERM – Department of Environmental Resource Management
dev – development
df – drinking fountain
DFW – diffusion welding
dia – diameter
distr – distribution
dk – deck
do – ditto
doc – document
DPDT – double pole, double throw switch
DPST – double pole, single throw switch
dr – door
dwg – drawing
DWV – drain, waste and vent

E

ea – each
EASP – electric arc spraying
EBC – electron beam cutting
EBW – electron beam welding
EBW-HV – electron beam welding - high vacuum
EBW-MV – electron beam welding - medium vacuum
EBW-NV – electron beam welding (nonvacuum)
EC – end of curve
EEO – equal employment opportunity
eff. – efficiency
E.G. – existing grade
EGW – electrogas welding
EIS – environmental impact statement
el – elevation
emb – embankment
EMT – electrical metallic tubing (conduit)

engr – engineer
EP – edge of pavement
EPA – Environmental Protection Agency
EPDM – ethylene propylene diene monomer
eq – equal
equip – equipment
ERISA – Employee Retirement Income Security Act
ESW – electroslag welding
et – edge thickness
ewc – electric water cooler
exc – excavation
exp – expense
exp. jt. – expansion joint
ext – 1) external; 2) extinguisher; 3) exterior
EXW – explosion welding

F

F – 1) Fahrenheit; 2) farad
FAS – free alongside (ship)
FB – 1) furnace brazing; 2) freight bill
FC – footcandle
FCAW – flux cored arc welding
fco – floor cleanout
fl dr – floor drain
FG – finish grade
FH – fire hose
FHA – Federal Housing Authority
fhc – fire hose cabinet
FICA – Federal Insurance Contributions Act (Social Security withholding tax)
fl – 1) fluid; 2) floor; 3) flush; 4) flashing
FLB – flow brazing
FLOW – flow welding
fl oz – fluid ounce
flr – floor
FLSP – flame spraying
FM – frequency modulation
FNMA – Federal National Mortgage Association (Fanny Mae)
FOB – 1) free on board; 2) freight on board
FOW – forge welding
F1S – faced one side
fp – freezing point
fpm – feet per minute
fps – 1) feet per second; 2) foot-pound-second
freq – frequency
FRM – fixed rate mortgage
FRP – fiberglass-reinforced plastic
frt – freight
ftg – footing
ft lb – foot pound
F2S – faced two sides
FUTA – Federal Unemployment Tax Act
FY – fiscal year

G

g – 1) gram; 2) gravity
ga – gauge
gal – gallon
galv – galvanized
GFI – ground fault interrupter
GFCI – ground fault circuit interrupter
GI – galvanized iron
gm – gram
GMAC – gas metal arc cutting
GMAW – gas metal arc welding
GMAW-P – gas metal arc welding - pulsed arc
GMAW-S – gas metal arc welding - short circuit arc
GMT – Greenwich mean time
GPD – gallons per day
GPH – gallons per hour
GPM – gallons per minute
GPS – gallons per second
gr – 1) grade; 2) gram; 3) gravity; 4) gross
GSA – General Service Administration
GT – gross ton
GTAC – gas tungsten arc cutting
GTAW – gas tungsten arc welding
GTAW-P – gas tungsten arc welding - pulsed arc
gyp – gypsum

H

h – 1) height, high; 2) half; 3) hour
H – henry
ha – hectare
HAP – height above plate
hb – hose bibb
hd – head
HD – heavy-duty
H.D.G. – hot dip galvanized
hex – hexagon or hexagonal
hf – half
HF – high frequency
HFRW – high frequency resistance welding

hgt – height
hp – horsepower
HPW – hot pressure welding
hr – hour
ht – 1) height; 2) heat
htg – heating
htr – heater
HVAC – heating, ventilation and air conditioning
hvy – heavy
HW – hot water
HWCMU – heavyweight concrete masonry unit
Hz – Hertz

I

I – I-beam
IB – 1) iron body; 2) induction brazing
IBBM – iron body, bronze mountings
ICC – Interstate Commerce Commission
ID – inside diameter
IGSCC – intergranular stress corrosion cracking
in. – inch
ins – insurance
INS – iron soldering
IP – initial point
ipm – inches per minute
ips – 1) inches per second; 2) iron pipe size
IRA – individual retirement account
IRB – infrared brazing
IRS – infrared soldering
IS – induced soldering
ISRS – inside screw, rising stem
IW – induction welding

J

J – joule
jb – junction box
J-box – junction box
jnt – joint
jour – journeyman
jt – joint

K

K – 1) kelvin; 2) thousand; 3) kilometer
kc – kilocycles
kg – kilogram
kHz – kilohertz
kip – one thousand pounds
km – kilometer
kV – 1) kilovolt; one thousand volts
kVA – 1) kilovolt amperes; one thousand volt-amperes
kW – 1) kilowatt; one thousand watts

L

l – liter
L – 1) length; 2) liter
lam – laminated
laser – light amplification (by) stimulated emission (of) radiation
lat – latitude
lav – lavatory
lb – pound
LBC – laser beam cutting
LBW – laser beam welding
lel – lower explosive limit
lf – linear feet
LF – low frequency
lg – 1) long; 2) large
lgth – length
lg tn – long ton
LH – 1) left hand; 2) lower half
lic – license
lin – lineal; linear
lin ft – linear feet
liq – liquid
lm – lumen
LNG – liquified natural gas
LOA – length overall
LOC – oxygen lance cutting
long – longitude
LP – low pressure
LPG – liquified petroleum gas
LP gas – liquified petroleum gas
lpW – lumen per watt
ls – limestone
LSI – large scale integrated circuit
LT – long ton
LWCMU – lightweight concrete masonry units
lx – lux

M

m – 1) meter; 2) mass; 3) mile
M – 1) mass coefficient; 2) Roman numeral for one thousand; 3) mega (million)

/M – per thousand
mA – milliampere
MAC – metal arc cutting
mach – machine or machinery
man – manual
matl – material
MATV – master antenna television system
max – maximum
mc – megacycles
MC – metal clad (cable)
MCC – motor control center
mdse – merchandise
meas – measure; measurement
mech – mechanical; mechanics
MEE – machined each end
metal – metallurgy
mfd – 1) manufactured; 2) microfarad
mfg – manufacturing
mfr – manufacturer
mg – milligram
mgr – manager
mgmt – management
mh – manhole
MHz – megahertz
mi – 1) mile; 2) malleable iron
MI – mineral insulated metal sheathed electrical cable
MIG – metal inert gas welding
mil – 1) 1/1000 (0.001) of an inch; 2) million
min – 1) minimum; 2) minute
mktg – marketing
ml – milliliter
mm – millimeter
MOA – machined overall
mpg – miles per gallon
mph – miles per hour
msec – millisecond
MSS – Manufacturers Standardization Society of the Valve and Fitting Industry
mt – mounted
MT – 1) magnetic particle examination; 2) metric ton
mtg – 1) meeting; 2) mortgage
mtn – mountain
mV – millivolt
MV – megavolt
MVA – megavolt-ampere
mW – milliwatt
MW – megawatt
MWh – megawatt-hour

N

n – 1) net; 2) nano-; 3) noon; 4) number
NBFU – National Board of Fire Underwriters
NBS – National Bureau of Standards
NDE – nondestructive examination
NDT – nil-ductility transition temperature
NEC – National Electrical Code
neg – negative
nF – nanofarad
NIC – not in contract
NM – nonmetallic sheathed electrical cable
NMC – nonmetallic sheathed electrical cable
No – number
nps – nominal pipe size
nt wt – net weight

O

O – oxygen
O/A – overall
OAC – oxygen-arc cutting
OAW – oxyacetylene welding
OC – 1) on center; center-to-center dimensions; 2) oxygen cutting
OD – outside diameter
OFC – oxyfuel gas cutting
OFC-A – oxyacetylene cutting
OFC-H – oxyhydrogen cutting
OFC-N – oxynatural gas cutting
OFC-P – oxypropane cutting
oh – overhead
OHW – oxyhydrogen welding
O & P – overhead and profit
OSB – oriented strand board
OSHA – Occupational Safety and Health Administration
OS&Y – outside stem and yoke
oxy – oxygen
oz – ounce

P

PA – 1) public address (system); 2) per annum
Pa – pascal
PAC – plasma arc cutting
PAW – plasma arc welding
pb – pull box
PB – polybutylene plastic
pc – piece

PC – 1) personal computer; 2) point of curve
PCC – point of curve change
pcf – pounds per cubic foot
pci – pounds per cubic inch
pct – percent
pd – paid
PE – polyethylene
perf – perforated
PERT – project evaluation and review technique
PEW – percussion welding
PI – point of intersection
PG – projected centerline grade or projected grade
PGW – pressure gas welding
pl – 1) plate; 2) pipeline
P & L – profit and loss
plbg – plumbing
PMP – perforated metal pipe
PMRA – protected membrane roof assembly
POC – metal powder cutting
PP – polypropylene
PPI – Plastics Pipe Institute
ppm – parts per million
PQR – procedure qualification record
pr – 1) pair; 2) pressure rating
PR – public relations
PSB – polystyrene bead foam board
PSE – polystyrene extruded foam board
psf – pounds per square foot
psi – pounds per square inch
PSI – polystyrene foam bead molded inserts
psia – pounds per square inch absolute
psig – pounds per square inch gauge
PSMI – polystyrene foam bead molded inserts
PSP – plasma spraying
pt – 1) pint; 2) point; 3) part
PT – 1) penetrant examination (liquid); 2) point of tangent
PTFE – polytetrafluoroethylene
PU – polyurethane foam board
PV – photovoltaic
PVA – polyvinyl acetate
PVC – polyvinyl chloride
PVDC – polyvinylidene chloride
PVDF – polyvinylidene fluoride
pwr – power
pwt – pennyweight

Q

q – quarter
QA – quality assurance
QC – quality control
qr – quarter
qt – quart
qty – quantity

R

r – radius
rad – radius
RAM – random access memory
R and R – remove and replace
RB – resistance brazing
RBM – reinforced brick masonry
RCP – reinforced concrete pipe
rd – 1) round; 2) road
red – 1) reducer; 2) reduce
ref – reference
reg – 1) regular; 2) region
req – require
rev – revised, revision
RH – right hand
rm – room
rnd – round
ROM – read only memory
ROW – roll welding
RP – reference point
rpm – revolutions per minute
rps – revolutions per second
RPW – projection welding
RS – 1) reference stake; 2) resistance soldering
RSEW – resistance seam welding
RSW – resistance spot welding
rt – right
RT – 1) radiography; 2) radiographic examination
RW – resistance welding
r/w – right of way

S

s – second
SAW – submerged arc welding
SAW-S – series submerged arc welding
SBA – Small Business Administration
scr – silicon controlled rectifier
SDR – standard dimension ratio

sec – 1) second; 2) secant
SF – square foot (feet)
SG – subgrade
sg – specific gravity
SGT – structural glazed tile
sm – small
SMAC – shielded metal arc cutting
SMAW – shielded metal arc welding
spec – 1) specifications; 2) speculation
sp gr – specific gravity
sq – square
sq ft – square feet
sq in – square inch
sq yd – square yard
SS – 1) Social Security; 2) stainless steel
SSBC – Southern Standard Building Code
SSW – solid state welding
ST – short ton
sta – station
STC – sound transmission class
std – standard
st pr – static pressure
supvr – supervisor
sur – surface
surv – survey
SW – stud arc welding
sy – square yard
sys – system

T

t – 1) ton; 2) time; 3) township
TB – torch brazing
TCAB – twin carbon arc brazing
T&E – tools and equipment
temp – 1) temperature; 2) temporary
T&G – tongue and groove
THSP – thermal spraying
TIG – tungsten inert gas welding
TLB – tractor-loader-backhoe
TLV – threshold limit value
TM – trademark
tol – tolerance
topo – topographical
tot – total
TP – turning point
tpf – taper per foot
ts – tensile strength
TS – torch soldering
TV – terminal velocity
TW – Thermit welding
TWA – time-weighted average
twp – township

U

u – unit
UBC – Uniform Building Code
UCI – Uniform Construction Index
UF – underground feeder wire
UL – Underwriters' Laboratories
UMC – Uniform Mechanical Code
UNC – Unified National Coarse (thread)
UNEF – Unified National Extra Fine (thread)
UNF – Unified National Fine (thread)
UPC – Uniform Plumbing Code
USW – ultrasonic welding
UT – ultrasonic examination
util – utilities
UV – ultraviolet light
UW – upset welding

V

v – 1) volume; 2) velocity
V – volt
VA – 1) Veterans Administration; 2) volt-ampere
vac – vacuum
VAC – volts, alternating current
var – variable
vb – valve box
VCP – vitrified clay pipe
VDC – vinyl vertical base drip cap
VDDC – vinyl dual drip cap
vel pr – velocity pressure
VF – very fine
VFT – vinyl finishing trim
VJ – vinyl siding J-channels
vol – volume
VOM – volt-ohmmeter
VSD – vinyl starter divider
VSJ – vinyl soffitt J-channels
VSS – Vented Suppressive Shielding
VTVM – vacuum tube voltmeter

W

w – 1) with; 2) weight; 3) wide, width

W – 1) watt; 2) weight per square foot of masonry wall surface

WC – water closet

wd – wood

wh – water heater

WH – watt-hour

whr – watt-hour

wk – 1) work; 2) week

w/o – without

wp – working pressure

WPS – welding procedure specification

WS – wave soldering

wt – weight

WWF – welded wire fabric

XYZ

x – 1) cross; 2) multiplication symbol; 3) by

yd – yard

American Society for Testing and Materials Cement Types and Masonry Units

ASTM Type I cement – general purpose portland cement.

ASTM Type II cement – modified cement resistant to moderate sulfate attack.

ASTM Type III cement – cement that achieves high strength in a short time (approximately 7 days).

ASTM Type IV cement – cement that generates low heat during curing.

ASTM Type V cement – cement resistant to high sulfate concentrations.

ASTM Types IA, IIA and IIIA cements – these types correspond to Types I, II, and III, respectively, with the addition of small quantities of air-entraining materials for improved resistance to freeze-thaw cycles and to scaling from chemicals.

ASTM C33 Concrete Aggregates – normal weight (heavy) aggregates; aggregate weighing 125 lbs per cubic foot or more.

ASTM C55 Concrete Building Brick – a concrete brick, either Type N rated at 3500 psi or Type S rated at 2500 psi.

ASTM C 90 Hollow Load-Bearing Concrete Masonry Units – concrete building blocks, which may be Type N, which is rated at 1000 psi or Type S, rated at 700 psi.

ASTM C129 Hollow Non-Load-Bearing Concrete Masonry Units – concrete blocks rated at 350 psi.

ASTM C145 Solid Load-Bearing Concrete Masonry Units – concrete blocks, either Type N rated at 1800 psi, or Type S rated at 1200 psi.

ASTM C331 Lightweight Aggregates For Concrete Masonry Units – aggregate weighing less than 105 lbs per cubic foot.

Practical References for Builders

Basic Engineering for Builders

If you've ever been stumped by an engineering problem on the job, yet wanted to avoid the expense of hiring a qualified engineer, you should have this book. Here you'll find engineering principles explained in non-technical language and practical methods for applying them on the job. With the help of this book you'll be able to understand engineering functions in the plans and how to meet the requirements, how to get permits issued without the help of an engineer, and anticipate requirements for concrete, steel, wood and masonry. See why you sometimes have to hire an engineer and what you can undertake yourself: surveying, concrete, lumber loads and stresses, steel, masonry, plumbing, and HVAC systems. This book is designed to help the builder save money by understanding engineering principles that you can incorporate into the jobs you bid. **400 pages, 8½ x 11, $34.00**

Carpentry Estimating

Simple, clear instructions to help you take off quantities and figure costs for all rough and finish carpentry. Provides checklists, manhour tables, and worksheets with calculation factors built in. Take the guesswork out of estimating carpentry so you get accurate labor and material costs every time. Includes FREE *National Estimator* software, with all the manhours in the book, plus the carpentry estimating worksheets in formats that work with *Word, Excel, WordPerfect*, and *MS Works for Windows*™.
336 pages, 8½ x 11, $35.50

Construction Forms & Contracts

125 forms you can copy and use — or load into your computer (from the FREE disk enclosed). Then you can customize the forms to fit your company, fill them out, and print. Loads into *Word for Windows, Lotus 1-2-3, WordPerfect,* or *Excel* programs. You'll find forms covering accounting, estimating, fieldwork, contracts, and general office. Each form comes with complete instructions on when to use it and how to fill it out. These forms were designed, tested and used by contractors, and will help keep your business organized, profitable and out of legal, accounting and collection troubles. Includes a 3½" disk for your PC. For Macintosh disks, add $15. **432 pages, 8½ x 11, $39.75**

Carpentry for Residential Construction

How to do professional quality carpentry work in residences. Illustrated instructions on everything from setting batterboards to framing floors and walls, installing floor, wall, and roof sheathing, and applying roofing. Covers finish carpentry: installing each type of cornice, frieze, lookout, ledger, fascia, and soffit; hanging windows and doors; installing siding, drywall, and trim. Each job description includes tools and materials needed, estimated manhours required, and a step-by-step guide to each part of the task. **400 pages, 5½ x 8½, $19.75**

Basic Lumber Engineering for Builders

Beam and lumber requirements for many jobs aren't always clear, especially with changing building codes and lumber products. Most of the time you rely on your own "rules of thumb" when figuring spans or lumber engineering. This book can help you fill the gap between what you can find in the building code span tables and what you need to pay a certified engineer to do. With its large, clear illustrations and examples, this book shows you how to figure stresses for pre-engineered wood or wood structural members, how to calculate loads, and how to design your own girders, joists and beams. Included FREE with the book — an easy-to-use version of NorthBridge Software's *Wood Beam Sizing*™ program. **272 pages, 8½ x 11, $38.00**

Drafting House Plans

Here you'll find step-by-step instructions for drawing a complete set of home plans for a one-story house, an addition to an existing house, or a remodeling project. This book shows how to visualize spatial relationships, use architectural scales and symbols, sketch preliminary drawings, develop detailed floor plans and exterior elevations, and prepare a final plot plan. It even includes code-approved joist and rafter spans and how to make sure that drawings meet code requirements.
192 pages, 8½ x 11, $27.50

Illustrated Guide to the 1996 *National Electrical Code*®

This fully-illustrated guide offers a quick and easy visual reference for installing electrical systems. Whether you're installing a new system or repairing an old one, you'll appreciate the simple explanations written by a code expert, and the detailed, intricately-drawn and labeled diagrams. A real time-saver when it comes to deciphering the current NEC.
384 pages, 8½ x 11, $34.75

National Building Cost Manual

Square foot costs for residential, commercial, industrial, and farm buildings. Quickly work up a reliable budget estimate based on actual materials and design features, area, shape, wall height, number of floors, and support requirements. Includes all the important variables that can make any building unique from a cost standpoint.
240 pages, 8½ x 11, $23.00. Revised annually

Planning Drain, Waste & Vent Systems

How to design plumbing systems in residential, commercial, and industrial buildings. Covers designing systems that meet code requirements for homes, commercial buildings, private sewage disposal systems, and even mobile home parks. Includes relevant code sections and many illustrations to guide you though what the code requires in designing drainage, waste, and vent systems. **192 pages, 8½ x 11, $19.25**

Roof Framing

Shows how to frame any type of roof in common use today, even if you've never framed a roof before. Includes using a pocket calculator to figure any common, hip, valley, or jack rafter length in seconds. Over 400 illustrations cover every measurement and every cut on each type of roof: gable, hip, Dutch, Tudor, gambrel, shed, gazebo, and more.
480 pages, 5½ x 8½, $22.00

Residential Wiring to the 1996 NEC

Shows how to install rough and finish wiring in new construction, alterations, and additions. Complete instructions on troubleshooting and repairs. Every subject is referenced to the most recent *National Electrical Code*, and there's 22 pages of the most-needed NEC tables to help make your wiring pass inspection — the first time.
352 pages, 5½ x 8½, $24.50

Contractor's Survival Manual

How to survive hard times and succeed during the up cycles. Shows what to do when the bills can't be paid, finding money and buying time, transferring debt, and all the alternatives to bankruptcy. Explains how to build profits, avoid problems in zoning and permits, taxes, time-keeping, and payroll. Unconventional advice on how to invest in inflation, get high appraisals, trade and postpone income, and stay hip-deep in profitable work. **160 pages, 8½ x 11, $22.25**

Construction Estimating Reference Data

Provides the 300 most useful manhour tables for practically every item of construction. Labor requirements are listed for sitework, concrete work, masonry, steel, carpentry, thermal and moisture protection, door and windows, finishes, mechanical and electrical. Each section details the work being estimated and gives appropriate crew size and equipment needed. Includes an electronic version of the book on computer disk with a stand-alone *Windows* estimating program FREE on a 3½" disk. **432 pages, 8½ x 11, $39.50**

Contractor's Guide to the Building Code Revised

This completely revised edition explains in plain English exactly what the *Uniform Building Code* requires. Based on the newly-expanded 1994 code, it explains many of the changes made. Also covers the *Uniform Mechanical Code* and the *Uniform Plumbing Code*. Shows how to design and construct residential and light commercial buildings that'll pass inspection the first time. Suggests how to work with an inspector to minimize construction costs, what common building shortcuts are likely to be cited, and where exceptions are granted.
384 pages, 8½ x 11, $39.00

Drywall Contracting

How to start and keep your drywall business thriving and do professional quality drywall work. Covers the eight essential steps in making any drywall estimate. Shows how to achieve the six most commonly-used surface treatments, how to work with metal studs, and how to solve and prevent most common drywall problems.
288 pages, 5½ x 8½, $18.25

Blueprint Reading for the Building Trades

How to read and understand construction documents, blueprints, and schedules. Includes layouts of structural, mechanical, HVAC and electrical drawings. Shows how to interpret sectional views, follow diagrams and schematics, and covers common problems with construction specifications. **192 pages, 5½ x 8½, $14.75**

Estimating Home Building Costs

Estimate every phase of residential construction from site costs to the profit margin you include in your bid. Shows how to keep track of manhours and make accurate labor cost estimates for footings, foundations, framing and sheathing finishes, electrical, plumbing, and more. Provides and explains sample cost estimate worksheets with complete instructions for each job phase. **320 pages, 5½ x 8½, $17.00**

Handbook of Construction Contracting

Volume 1: Everything you need to know to start and run your construction business; the pros and cons of each type of contracting, the records you'll need to keep, and how to read and understand house plans and specs so you find any problems before the actual work begins. All aspects of construction are covered in detail, including all-weather wood foundations, practical math for the job site, and elementary surveying. **416 pages, 8½ x 11, $28.75**

Volume 2: Everything you need to know to keep your construction business profitable; different methods of estimating, keeping and controlling costs, estimating excavation, concrete, masonry, rough carpentry, roof covering, insulation, doors and windows, exterior finishes, specialty finishes, scheduling work flow, managing workers, advertising and sales, spec building and land development, and selecting the best legal structure for your business. **320 pages, 8½ x 11, $30.75**

Finish Carpenter's Manual

Everything you need to know to be a finish carpenter: assessing a job before you begin, and tricks of the trade from a master finish carpenter. Easy-to-follow instructions for installing doors and windows, ceiling treatments (including fancy beams, corbels, cornices and moldings), wall treatments (including wainscoting and sheet paneling), and the finishing touches of chair, picture, and plate rails. Specialized interior work includes cabinetry and built-ins, stair finish work, and closets. Also covers exterior trims and porches. Includes manhour tables for finish work, and hundreds of illustrations and photos. **208 pages, 8½ x 11, $22.50**

Estimating Tables for Home Building

Produce accurate estimates for nearly any residence in just minutes. This handy manual has tables you need to find the quantity of materials and labor for most residential construction. Includes overhead and profit, how to develop unit costs for labor and materials, and how to be sure you've considered every cost in the job. **336 pages, 8½ x 11, $21.50**

Land Development

The industry's bible. Nine chapters cover everything you need to know about land development from initial market studies to site selection and analysis. New and innovative design ideas for streets, houses, and neighborhoods are included. Whether you're developing a whole neighborhood or just one site, you shouldn't be without this essential reference.
360 pages, 5½ x 8½, $37.00

National Construction Estimator

Current building costs for residential, commercial, and industrial construction. Estimated prices for every common building material. Manhours, recommended crew, and labor cost for installation. Includes an electronic version of the book on computer disk with a stand-alone *Windows* estimating program FREE on a 3½" disk.
560 pages, 8½ x 11, $37.50. Revised annually

Profits in Buying & Renovating Homes

Step-by-step instructions for selecting, repairing, improving, and selling highly profitable "fixer-uppers." Shows which price ranges offer the highest profit-to-investment ratios, which neighborhoods offer the best return, practical directions for repairs, and tips on dealing with buyers, sellers, and real estate agents. Shows you how to determine your profit before you buy, what "bargains" to avoid, and how to make simple, profitable, inexpensive upgrades. **304 pages, 8½ x 11, $19.75**

Rough Framing Carpentry

If you'd like to make good money working outdoors as a framer, this is the book for you. Here you'll find shortcuts to laying out studs; speed cutting blocks, trimmers and plates by eye; quickly building and blocking rake walls; installing ceiling backing, ceiling joists, and truss joists; cutting and assembling hip trusses and California fills; arches and drop ceilings — all with production line procedures that save you time and help you make more money. Over 100 on-the-job photos of how to do it right and what can go wrong. **304 pages, 8½ x 11, $26.50**

Basic Plumbing with Illustrations, Revised

This completely-revised edition brings this comprehensive manual fully up-to-date with all the latest plumbing codes. It is the journeyman's and apprentice's guide to installing plumbing, piping, and fixtures in residential and light commercial buildings: how to select the right materials, lay out the job and do professional-quality plumbing work, use essential tools and materials, make repairs, maintain plumbing systems, install fixtures, and add to existing systems. Includes extensive study questions at the end of each chapter, and a section with all the correct answers.
384 pages, 8½ x 11, $33.00